BIOCHEMISTRY

BIOCHEMISTRY

SECOND EDITION

FRANK B. ARMSTRONG

North Carolina State University

New York Oxford

OXFORD UNIVERSITY PRESS 1983

Library of Congress Cataloging in Publication Data

Armstrong, Frank Bradley.
Biochemistry.

Bibliography: p.
Includes index.
1. Biological chemistry. I. Title.
QP514.2.A73 1983 574.19'2 81-18922
ISBN 0-19-503109-1 AACR2

Outside front cover: Top view of B DNA depicting one complete turn of the helix (ten base pairs). See P. 233.
Illustration © by Irving Geis.
Inside front cover: The von Liebig Laboratory in Giessen, Germany.
(The Bettmann Archive, Inc.)
Cover design by Charles and Ruth Kaplan.

Printing (last digit): 9 8 7 6 5 4 3 2 1

Printed in the United States of America

PREFACE

The writing of the second edition of *Biochemistry* was influenced significantly by the constructive critiques and suggestions provided by colleagues, students, reviewers, and users of the first edition. With regard to the changes incorporated into the second edition, two new chapters, Lipid Biosynthesis (Chapter 19) and Recombinant DNA Research (Chapter 26), were added, and presentations in many of the other chapters were revised and/or expanded. Some of these changes include a discussion of glycogen synthesis (Chapter 11), the catalytic mechanism of ribonuclease A (Chapter 13), the glyoxylate cycle (Chapter 16), restriction mapping and the newer systems for DNA sequencing (Chapter 21), the mitochondrial genetic code and attenuation (Chapter 22), intracellular transport of proteins (Chapter 23), recent research findings on prostaglandins (Chapter 24), and the mode of action of vitamins A, D, and K (Chapter 25). Particular attention was also given to an expansion of the presentation on nitrogen metabolism (Chapter 20) by including nitrogen fixation, nitrate reduction, and the synthesis and catabolism of all 11 nonessential amino acids. Also, the discussion on purine metabolism is now complemented by one on pyrimidine metabolism.

The expanded second edition may now contain more material than can be conveniently presented in a one-semester, introductory course in biochemistry. If such is the case, it is hoped that individual instructors will select those topics from the book that best suit their treatment of the discipline in meeting the educational needs of their classes.

In preparing the new edition, I was again fortunate to have the assistance of colleagues who provided critical reviews of individual chapters. For their valuable contributions, my sincere appreciation is extended to Wendy F. Boss, William E. Donaldson, H. Robert Horton, Daniel W. Israel, William A. Jackson, Bryan H. Johnson, William L. Miller, David W. Niesel, Joanne M. Ravel, and Samuel B. Tove. The useful critiques furnished by the reviewers selected by Oxford University Press are also appreciated. I recognize with gratitude the help provided by Jeanne B. Koger in proofreading the galleys and by Irene Nunes, who carefully and thoroughly copy-edited the manuscript. The impeccable typing of Anne Richardson, who prepared the manuscript, is also gratefully acknowledged. I express a special thanks to my editor, Michael R. Cook, whose dedicated efforts assured a smooth transition from the first to the second edition. Finally, I extend my appreciation and love to my family; their contributions of patience, support, and encouragement made the task of revising the text much easier.

Raleigh, North Carolina F.B.A.

PREFACE TO FIRST EDITION

Biochemistry is for an introductory course designed to accommodate undergraduates from widely varying curricula. For an instructor, the current broad interest in such offerings in biochemistry poses the challenge of teaching a course that meets the educational needs of majors from preprofessional programs, biological sciences, agricultural sciences, chemisty, engineering, and science education curricula. Such a generalized course has been taught by Frank B. Armstrong at North Carolina State University since 1968 and by Thomas Peter Bennett at the Rockefeller University. It was this mutual interest in teaching introductory biochemistry that prompted us to engage in the collaborative effort that made *Biochemistry* a reality.

The topics covered in the text were selected after seeking the advice of a number of instructors from various campuses who teach courses in biochemistry. A majority of the material, therefore, represents a basic core of instruction that all those consulted considered necessary for an introductory course and that we have found desirable in our own teaching. In the interest of brevity, the text is written in a simple, direct style. Where appropriate, the text is accented with historical notes, definitions, or information pertaining to the relevant aspects of the material being presented; medical anomalies, for example, are tied to inborn errors of metabolism.

Chapters 1 through 5 are introductory in nature and recognize the various backgrounds of students taking their first course in biochemistry. This section includes a brief history of biochemistry, emphasizing the development of molecular biology; a review of the elements needed for life and of biogeochemical cycles; a discussion of water and buffers; chemical topics that are important to the subsequent text; and an overview of cell biology.

Amino acids and proteins are two of the classes of biomolecules discussed in the book's second section, Chapters 6-13. A presentation of the molecular aspects of the structure and biological function of proteins, with emphasis on myoglobin, hemoglobin, and enzymes, precedes a description of other principal types of biomolecules, i.e., carbohydrates, lipids, nucleic acids, and B-vitamins.

The molecular basis of bioenergetics serves as a prelude to the third section (Chapters 14-20), which includes the production of chemical energy by living systems and interrelated topics on intermediary metabolism. The synthesis of ATP is the underlying theme of the chapters on glycolysis, tricarboxylic acid cycle, electron transport, oxidative phosphorylation, and photosynthesis. Carbon metabolism, emphasized in the presentation of the recovery of energy from sugars, is complemented by a discussion of the metabolism of nitrogen-containing biomolecules. This section concludes with a presentation of the molecular aspects of human nutrition.

The last section, Chapters 21-24, returns to some of the key accomplishments attained by molecular studies of biological phenomena. Chapter 21 on hormones includes the impressive strides that have been made toward understanding the molecular basis of hormone action. Chapters 22-24 are devoted to the classes of macromolecules of the Central Dogma, stressing molecular studies on DNA, RNA, and protein synthesis, respectively. A discussion of the coding capacity of the bacteriophage ϕX174 was selected for the book's closing, since the results of these studies succinctly exemplify the revolutionary impact that molecular biology continues to have on the study of biology.

The authors gratefully acknowledge those who aided in the preparation of *Biochemistry.* The following provided constructive critiques of many of the chapters and contributions of information and materials: Drs. Mary Bonneville, Janice Glenn, Thomas Conway, H. Robert Horton, Evan E. Jones, Joseph S. Kahn, Ian S. Longmuir, A. Russell Main, John F. Roberts, and Harold F. Swaisgood. The reviews and information provided by Mr. Hugh Craft are also acknowledged, and a special acknowledgment goes to Mr. Richard R. Randall, who served as an in-house editor for the manuscript. Thanks are given to the graduate students of the Department of Biochemistry at North Carolina State University for their careful help in proofreading the galleys. Appreciation is also expressed for the critical persistence and guidance of Mr. Robert

J. Tilley and his associates at Oxford University Press and for the thoughtful critiques provided by the reviewers selected by the publisher. The excellent typing of Mrs. Anne Richardson, who prepared the copies of the manuscript, is sincerely appreciated. We save for last our thanks to our families, who endured many months with authors-in-residence and who, in spite of this imposition, provided us with continuous support and encouragement.

FRANK BRADLEY ARMSTRONG
Raleigh, North Carolina

THOMAS PETER BENNETT
Philadelphia, Pennsylvania

December, 1978

LIST OF CHAPTERS

CONTENTS

9 Proteins IV: Enzymes 129

10 B-Vitamins as Coenzymes 163

19 Lipid Biosynthesis 367

20 Nitrogen Metabolism 395

But the task of the biochemist is, after all, the study of the physicochemical processes associated with the manifestations of what we call life—not the life of some particular animal or group of animals, but life in its most general sense.

ERNEST BALDWIN, 1937
Preface to *An Introduction to Comparative Biochemistry*

BIOCHEMISTRY

AN HISTORICAL REVIEW

1

INTRODUCTION

The history of biochemistry is intimately associated with the development of the sciences of chemistry in the late eighteenth century and of biology in the nineteenth century. It was not, however, until the early years of this century that biochemistry began to emerge as an independent scientific discipline—an emergence that was completed only several decades ago. Biochemistry, often conveniently described as the study of the chemistry of life, is a multifaceted science that includes the study of all forms of life and that utilizes basic concepts derived from biology, chemistry, physics, and mathematics to achieve its goals. Biochemical research, which arose in the last century with the isolation and chemical characterization of naturally occurring organic compounds, is today an integral component of most, if not all, modern biological research. That a relatively new discipline has had such a profound influence on so many areas of research is a noteworthy accomplishment. To understand this influence, it is necessary to appreciate the historical evolvement of biochemistry and its role in the recent development of molecular biology. The story is one of a science that arose from the study of the basic chemistry of living organisms and developed into a leading participant in the challenging scientific quest to understand the molecular basis of life.

NINETEENTH CENTURY

The development of chemistry as a science in the latter part of the eighteenth century raised certain intriguing questions about what

Friedrick Wöhler.
Wöhler earned worldwide acclaim for his teaching during his 46 years at the University of Göttingen. Between 1845 and 1866, he lectured to 8,250 students. In his lifetime, 13 and 15 editions of his organic and inorganic texts, respectively, were published.

gave life its characteristic features and, in particular, what distinguished organic (compounds of living systems) from inorganic compounds. Publications by German physiologists characterized the difference as *Lebenkraft*—the vital force. Proponents of the *vital force theory* believed that "in living as compared with inanimate nature, elements appear to obey quite different laws." The concept of a vital force was invoked to explain or to relate phenomena encountered only in living plants and animals. For many years, textbooks, journal articles, and most scientists agreed that the chemistry of living organisms was distinct from that of the inanimate world.

It was during the first half of the nineteenth century that organic chemistry became a distinct chemistry of life separate from other branches of chemistry. Probably the separation began with the notable discovery by Friedrick Wöhler in 1828 that urea, a compound found only in living things, could be formed by heating an inorganic compound, ammonium cyanate.

$$\underset{\text{Ammonium cyanate}}{NCONH_4} \xrightarrow{\text{heat}} \underset{\text{Urea}}{H_2N\overset{\overset{\displaystyle O}{\|}}{C}NH_2}$$

As he noted in writing to the great chemist Jöns Berzelius in 1828:

> I can no longer, as it were, hold back my chemical urine; and I have to let out that I can make urea without needing a kidney, whether of man or dog; the ammonium salt of cyanic acid is urea.

The laboratory synthesis of urea, an organic compound, was a serious empirical challenge to the vital force theory. Although there remains historical controversy about the philosophical impact of Wöhler's discovery, in retrospect, it nevertheless represents a landmark in the demarcation of organic chemistry from inorganic chemistry. Following the work of Wöhler and that of other early organic chemists, the notion that life properties were explainable in chemical terms came to be believed by a number of leading scientists.

Justus von Liebig.
By age 36, Liebig headed the world's largest laboratory and school for training chemists; he was one of the early instigators of large-scale research. Today, his laboratory with its furnishings is preserved as a museum in Giessen, Germany.

One eminent inorganic chemist to turn to the new chemistry in the 1820s was Justus von Liebig, who studied the chemistry of animal and plant physiology. He was able to demonstrate that the heat of animal bodies is due to combustion of the food they eat and, therefore, is not a result of a vital force. It was he who introduced artificial fertilizers into agriculture (although, in that era, the importance of nitrogen to plant nutrition was not recognized). His laboratory was the first to categorize foods by the present classification system of carbohydrates, fats, and proteins. He also introduced the concept of *metabolism*—chemical processes that build up or break down substances within an organism. Liebig's work was summarized in his book *Organic Chemistry in Its Application to Agriculture and Physiology,* published in 1840. His views were particularly upsetting to many natural scientists who were not prepared to accept the suggestion of a chemical basis for life processes.

In the 1840s, acceptance of the *cell theory* of Matthais Schleiden and Theodor Schwann, which recognized the cell as the basic structural unit of all organisms, resulted in new studies in the science of physiology which, in turn, succeeded in bringing organic chemistry into an even closer relationship with biology. Collaborative research efforts involving chemistry and physiology did much to lay the foundation for the newly developing science of biochemistry.

Many chemists, with an interest in living matter, began to pursue intensified studies on the isolation and characterization of various molecules found in different cells. One of the most outstanding chemists was Ernst Hoppe-Seyler, who explored the chemistry of blood, cartilage, pus, and various other body materials. Hemoglobin, the red pigment of blood, was one of many substances first isolated and studied by Hoppe-Seyler; it was also the first protein to be crystallized (1864). It was Hoppe-Seyler who established the journal *Zeitschrift für physiologische Chemie* (1877) and who wrote in the first issue that "biochemistry . . . from its first natural and necessary analytical beginnings has grown into a science. . . ." The new science, as viewed by Hoppe-Seyler, was distinct from physiology.

As descriptive knowledge of molecular structures accumulated, an appreciation and understanding of metabolism also continued to develop. This interest in chemical reactions that occur in a cell and in how these reactions are brought about began to provide an understanding of the dynamic aspects of life processes. Such studies yielded a progressive elucidation of *catabolic pathways*—those series of chemical reactions by which compounds are broken down. The processes by which foodstuffs are utilized and how cellular energy is produced were also of prime interest. In fact, it was the continuing controversy during most of the nineteenth century about alcoholic fermentation by yeast and its solution that did much to aid the development of the fledgling discipline of biochemistry.

Matthais Schleiden. (The Bettman Archive) Schleiden was a discontented lawyer who, after an unsuccessful suicide attempt, decided to become a scientist. His textbook, *Grundzüge der wissenschaftlichen Botanik* (1842), revolutionized the teaching of botany by emphasizing plant cytology.

Theodor Schwann. (The Bettman Archive) Schwann's brilliant scientific career lasted only from 1834 to 1839. Severely criticized by German scientific leaders for believing that alcoholic fermentation is caused by a live organism (yeast), he found it impossible to pursue a scientific career in Germany. He moved to Belgium, where he became a professor, inventor, and religious mystic.

Louis Pasteur.
After Pasteur saved the boy Joseph Miester from rabies in 1885, the Pasteur Institute was established by the French government (1888) to treat cases of rabies. In adult life, Miester was a gatekeeper at the Pasteur Institute and, in 1940, killed himself rather than obey the demands of some occupying Nazis to open Pasteur's crypt.

Although yeast was identified (but not readily accepted) as the agent of alcoholic fermentation in the 1830s, a debate about how the organism ferments glucose to alcohol and carbon dioxide continued until the end of the century. Louis Pasteur proposed the existence of "organized ferments" or "formed ferments" which would function only in living yeast cells (a vitalist's view). On the other hand, Liebig believed that "ferment" was a soluble material ("unorganized ferment") and that a particular structural or functional aspect of life was not essential to demonstrate its effects and properties. Liebig's view was that protoplasmic "albuminoid matter" (protein) is decomposed by oxygen and that, in the process, molecular vibrations are communicated to sugar molecules, causing their breakdown to alcohol and carbon dioxide. His thoughts were aligned with a prevalent school of mechanistic thought that albuminoids, because of their energy-rich aldehyde and cyano groups, furnish the energy for intracellular processes. Pasteur proved Liebig's hypothesis wrong when he showed that yeast ferments best in the absence of oxygen.

The idea that soluble catalytic agents (analogous to the extracellular pepsin discovered by Schwann in the 1830s [p. 129]) are responsible for the fermentation was also proposed by some eminent chemists, including Moritz Traube, who, by 1861, was convinced that a chemistry of life was impossible until the resolution of the fermentation theory conflict. Catalytic "unorganized" ferments, however, were never seriously considered since no one had been able to prepare a cell-free extract of yeast that could ferment sugar. It was during this period of controversy that, in 1878, Wilhelm Kühne proposed the term *enzyme* for ferments and, in doing so, introduced a term that subsequently became the general name for organic substances that catalyze biological reactions. It was not until 1897, though, that the experiments of Eduard Buchner (Nobel Prize, 1907) and his brother Hans resolved the fermentation controversy by demonstrating that yeast extracts, prepared free of whole cells, were able to carry out alcoholic fermentation. Because of these results, the idea that vital properties inseparable from the intact cell are responsible for functions of organisms was soon abandoned. The conviction that living organisms are composed of defined chemical substances that interact in understandable chemical processes was strengthened.

Kühne intended for the name *enzyme* (derived from the Greek words meaning *in yeast*) to apply to both yeast ferment and the extracellular catalysts of more complex animals, e.g., pepsin and trypsin. Prophetically, his definition specifically implied that lower and higher forms of life are not so fundamentally different.

The Buchners discovered cell-free fermentation when they attempted to preserve their yeast extracts with sugar, the preservative of jellies and jams.

Eduard Buchner, a major in the German army, died in action on the Romanian front (1917) during World War I.

No recollection of early biochemical research can ignore the definitive contributions made by Emil Fischer. Frequently referred to as the father of biochemistry, Fischer received a Nobel prize in 1902, the second year the prestigious awards were granted.

Emil Fischer and colleagues in his laboratory.

Beginning in the latter part of the century, he followed the organic chemist's approach in his studies on various classes of biological compounds. His demonstration of the specificity of enzymes and his pronouncement of the lock-and-key relationship between an enzyme and its substrate, in 1894, were among the significant contributions he made to scientific thought of that era. His experimentation, in the early years of this century, on proteins revealed that they are composed of a number of different, small building blocks called *amino acids,* which link together to form long chains. He also chemically synthesized a protein containing 18 amino acids and then demonstrated its breakdown by digestive enzymes. The techniques that Fischer developed for determining the chemical structures of substances continued to be useful to biochemical research in the twentieth century.

Emil Fischer.
Fischer's collection of 9,000 reference compounds is housed in the Department of Biochemistry, the University of California at Berkeley. The prized collection was a gift of H. O. L. Fischer, a biochemist and the only one of Fischer's three sons to survive World War I.

FIRST FIFTY YEARS OF THIS CENTURY

After the turn of the century, experimental research on biochemical problems continued to expand and, as time progressed, the true scope of the research potential of the new discipline began to be realized. The first half of the century was required for the science of biochemistry to become an accepted reality—a recognition that was earned by the scientific successes of many eminent biochem-

James B. Sumner.
After Sumner lost his left arm as a result of a hunting accident at age 17, he was discouraged by his teachers from pursuing a career in chemistry. They felt that Sumner was too handicapped for the profession.

Sir Hans A. Krebs.
"Official retirement" for Sir Hans in 1967 merely involved the moving of his laboratory from Oxford University to Radcliffe Infirmary in Oxford. He continued to be actively engaged in research until his death, at age 81, in 1981.

ists. Because it would be difficult to catalogue the numerous accomplishments made by individuals during that period, a brief summary of some of the salient scientific achievements is substituted.

Physiological and chemical emphases continued to dominate biochemical research, especially during the first three decades. The isolation and chemical study of compounds proved to be fruitful and, from these studies, humankind accrued many benefits. Hormones such as epinephrine, thyroxine, and insulin were among the medically important compounds isolated in that era. The identification and subsequent characterization of fat- and water-soluble vitamins, as well as the elucidation of the amino acids required by humans, added greatly to the basic understanding of nutrition. It was truly a golden age for the science of nutrition.

One of the outstanding accomplishments was the first crystallization of an enzyme, urease, by James B. Sumner (Nobel Prize, 1946) in 1926. His success was followed by that of John H. Northrup (Nobel Prize, 1946), who crystallized the digestive enzymes pepsin and trypsin in the early 1930s. Although the importance of enzymes had long been recognized and many had been studied, the feasibility of obtaining them in a pure state was a welcomed breakthrough for enzymology. In time, *in vitro* studies of individual enzymes and their reactions (biological functions) became an increasingly important area of biochemical research.

The analysis of many different types of biological molecules and the determination of their roles in metabolism by *in vitro* methods furnished the numerous pieces of information needed to provide detailed chemical interpretations of biological processes. A stellar example of the success enjoyed by this experimental approach was the clarification of the organized sequence of enzymatic reactions that constitute glycolysis and alcoholic fermentation. The phenomenon of fermentation, which had been intensively studied since the early years of the last century, was explainable in molecular terms by the end of the 1930s. Other famous achievements that occurred in the 1930s were postulations of the urea cycle by Sir Hans A. Krebs and Kurt Henseleit and, later, of the tricarboxylic acid cycle by Krebs (Nobel Prize, 1953). The knowledge gained concerning metabolism did much to support the scientific conviction that biochemistry offered a promising and productive approach to the study of cellular functions.

This knowledge also afforded biochemists insights into the functioning of life processes that hitherto were unavailable to the science of biology. These insights endowed biochemistry with a set of distinctive concepts that, in subsequent years, exerted a profound influence on many research areas, both old and new. Concepts about the energetics of living systems are a key example of the influences that biochemical research has had on scientific thought. By 1950, years of dedicated work by a number of biochemists on the generation and utilization of metabolic energy had produced a basic understanding of the processes involved. What was learned about a cell's ability to transform the chemical energy of nutrient organic compounds into a utilizable form, adenosine triphosphate (ATP), served as the basis of what is now referred to as the *principles of bioenergetics.* These principles form the conceptual cornerstone for present studies on biological energy. By midcentury, the scientific strides made by biochemists toward understanding various aspects of the chemistry of life, e.g., metabolic processes, biological energetics, and enzymatic reactions, earned biochemistry its identity as an independent and mature scientific discipline.

EMERGENCE OF MOLECULAR BIOLOGY

In the 1950s, impressive gains in research continued to be made. Among the most notable achievements was the elucidation of *biosynthetic (anabolic)* pathways which lead to the production of biological compounds. Although catabolic pathways, which account for the breakdown of molecules, had figured prominently in research since the nineteenth century, it was not until after the middle of the twentieth century that a comprehensive understanding of biosynthetic processes was realized. A large measure of the success attained has been credited to the effective use of radioisotopes, introduced into biological research in the mid-1930s. Application of bacterial studies to probe biochemical pathways, particularly synthesis of amino acids, also played a prominent role. By the end of the decade, *intermediary metabolism,* the network of enzymatic reactions that make up the synthetic and degradative machinery of a cell, was well defined.

Linus Pauling.
Pauling also received the 1962 Nobel Peace Prize for his leadership in warning the world about the health hazards of nuclear fallouts and for advocating nuclear disarmament.

The 1950s were an especially productive period for the study of proteins. The decade began with the proposal by Linus Pauling (Nobel Prize, 1954) and Robert B. Corey of the *α-helical conformation,* a secondary protein structure. An equally influential achievement was made by Frederick Sanger (Nobel Prize, 1958) in 1953 with his publication of the first complete amino acid sequence of a protein. In this elegant research, Sanger and his colleagues succeeded in determining the 51 amino acid sequence of the two polypeptide chains of the hormone insulin. The latter part of the 1950s also began to yield the long-awaited answer to the question of how amino acids are incorporated into protein structure.

The realization that a protein (previously thought of as a long chain of amino acids) could have a well-defined structure and that the chemical composition of individual proteins could now be determined opened a new and exciting frontier for research. Biochemical insights and expertise had arrived at the stage where it was now possible to undertake detailed studies on the large-molecular-weight compounds of a cell called *macromolecules.* Two classes of macromolecules (proteins and nucleic acids) were the focal points of these challenging research endeavors. The discovery of the α-helical conformation of proteins prompted James D. Watson and Francis H. C. Crick (joint Nobel Prize, 1962) to attempt to elucidate the structure of deoxyribonucleic acid (DNA). In 1953, their successful efforts resulted in the famous enuncia-

James D. Watson and Francis H. C. Crick. The two-man project to determine the molecular structure of DNA was undertaken by a young postdoctoral (Watson) who, as an undergraduate, was interested in ornithology and a physicist (Crick). (From J. D. Watson, *The Double Helix.* New York, Atheneum; copyright 1968 by J. D. Watson)

tion of the double-helical model. That classic paper (*Nature,* 171:737) is often referred to as the beginning of molecular biology.

To appreciate that last statement, it is necessary to backtrack in time to consider two outstanding contributions made to the science of genetics in the early 1940s. In 1940, George Beadle and Edward Tatum (joint Nobel Prize, 1958) published the *one gene-one enzyme hypothesis,* which states that the function of a gene is to specify the structure of an enzyme. "One gene-one enzyme" refers to their conclusion that each gene is responsible for the structure of one, and only one, enzyme. Explaining genic function in biochemical terms brought the sciences of genetics and biochemistry into a special relationship, called *biochemical genetics,* and added new dimensions to the research in both disciplines. In 1944, a classic scientific paper of the century was published by Oswald T. Avery, Colin M. MacLeod, and Maclyn McCarty. These investigators, studying the phenomenon of transformation in bacteria, identified DNA as the *transforming principle.* That identification also revealed the chemical nature of the gene. The fact that genic material is DNA, however, was not universally accepted until the early 1950s. Consequently, with the announcement by Watson and Crick of their model for DNA, both the chemical nature and the molecular structure of the unit of heredity were hypothesized. The gene (DNA) and its function (protein synthesis) were now amenable to analysis at a molecular level.

Crick named his house at Cambridge the Golden Helix.

Research on macromolecules in the 1950s revolutionized the study of biology. It was during that period that the term *molecular biology* was introduced to describe the driving force of that revolution. Molecular biology is often described as the study of life at the molecular level. That simple definition embodies an experimental approach to the study of life processes that is responsible for a new era in science. In the last few decades, the phrase *explosion of knowledge* has often been used to characterize advances in molecular research. Two distinguishing articles of scientific faith helped to account for that success. One is the belief that knowledge of the three-dimensional structure of a macromolecule provides valuable insights into the mechanism of that molecule's biological function. An excellent example of the validity of this concept is found in the announcement by Watson and Crick on the structure of DNA. In what is probably the most clever understatement in scientific literature, the authors casually stated that the base pairing proposed for the model suggested how DNA replication occurs. The mode of replication (synthesis), postulated by

Watson and Crick in a subsequent publication, proved to be correct. Thus, a molecular representation of a gene furnished a prophetic insight into how genetic material is replicated within a cell.

The second firm belief of molecular biologists is that for all forms of life, the same biochemical processes account for the basic functioning of life. Years of research have proved this belief to be true. For example, except for certain viruses, DNA is the universal genetic material and is replicated by a similar biochemical mechanism in all cell types. Protein synthesis, in whatever form of life it occurs, requires the same 20 amino acids and the same biochemical processes. It is because of this second tenet that molecular biology encompasses all biological sciences. Recognition that basic biological phenomena are universal explains why studies on bacteria and viruses have contributed greatly to fundamental knowledge in biology. These studies using bacteria and viruses, initiated in the 1940s, were originally viewed by many as being on the fringe of biological research. Nevertheless, the early proponents of molecular biology, who initiated these studies on their convictions, correctly anticipated the important generalizations that could be made about molecular aspects of life.

It was a search for new laws of physics in biological phenomena that prompted a school of young physicists and physical chemists, headed by Max Delbrück (Nobel Prize, 1969), to pioneer molecular and genetic studies on bacteria and viruses in the early 1940s.

What specifically occurred in the 1950s to account for the current faith in the molecular approach to the study of life? Two brilliant contributions by Crick in the latter half of the 1950s aided in earning molecular biology its scientific laurels. One contribution was the *Central Dogma* of molecular biology which, in part, is an updated version of the one gene-one enzyme hypothesis since it expresses, in biochemical language, the flow of biological information from gene to protein structure. Briefly stated, the genetic information contained in a gene (DNA) is *transcribed* into a molecule of ribonucleic acid (RNA), another type of nucleic acid, from which it is then *translated* into protein structure. Crick's other contribution was his views on how genetic information is stored in a gene, i.e., chemical identity of the genetic code. These two pronouncements were guiding lights for many molecular studies. By 1965, the genetic code and flow of biological information *(gene expression)* were definable in molecular terms.

Jacob, whose medical studies were interrupted by World War II, was seriously wounded while serving in the Free French forces and is a highly decorated veteran. After the war, he finished his medical training, but his physical disabilities prevented him from fulfilling his original desire to practice surgery.

Another exciting accomplishment occurred in 1961, when François Jacob and Jacques Monod (both awarded a Nobel Prize, 1965) published a molecular explanation for the regulation of procaryotic gene expression, i.e., how genes are turned on and off. The impact the announcement had on biological research cannot be overestimated. In 1963, Jacob and Monod, with Jean-Pierre Changeaux, also proposed a theory to explain the molecular aspects of regulation of the catalytic activity of enzymes. These two reports introduced scientific concepts about the control of genetic and metabolic functions of organisms which rapidly established a new field of scientific inquiry - the study of biological regulation. In enzymology, the regulatory features of enzymes are now examined as intensively as were enzymatic reactions in previous years. In biology, molecular concepts about regulation are the bases for new experimental approaches in studies on processes involved in differentiation and development. At the core of all these endeavors are the predictive insights and experimental techniques furnished by biochemistry.

In the 1970s, the influence of molecular biological research reached a new zenith with the introduction of recombinant DNA research. By the end of the decade, the numerous and varied benefits foreseen by its application to science and industry were beginning to be realized, e.g., production of hormones. As the widespread use of recombinant DNA research continues to develop, it appears that the 1980s will prove to be a decade whose scientific achievements will be as remarkable as those characterizing the preceding ones.

Thus, today, biochemistry continues to occupy a position of ever-increasing importance in biological research. As a basic science, it is in the foreranks of many scientific endeavors that may make a reality of the idea, expressed by some, that "the twentieth century belongs to the biologists." It was only a little over a century ago that Hoppe-Seyler established a journal for those studies in physiological chemistry that he believed should be recognized as the new discipline of biochemistry. If alive today, he would be very proud of the science whose niche he helped to define.

SUGGESTED READINGS

Asimov, I., *A Short History of Biology*. Garden City, N.Y.: Natural History Press, 1964. A well-written introduction to the historical aspects of several biochemical topics.

Cairns, J., G. S. Stent, and J. D. Watson, eds., *Phage and the Origins of Molecular Biology*. Max Delbrück dedicatory volume. Cold Spring Harbor, N.Y.: Cold Spring Harbor Laboratory of Quantitative Biology, 1966. A series of first-rate primary-author summaries of the recent history of molecular biology.

Fruton, J. S., *Molecules and Life*. New York: Wiley-Interscience, 1973. Four innovative and well-connected historical essays about biochemistry, covering the period from 1800 to 1950.

Jacob, F., *The Logic of Life*. New York: Vintage Books, 1976. An excellent history of genetics that encompasses a lucid account of the evolvement of scientific ideas in biology.

Judson, H. F., *The Eighth Day of Creation: Makers of the Revolution in Biology*. New York: Simon and Schuster,

1979. An interesting and informative recounting of the development of molecular biology, as told by the pioneers of the field.

Lipmann, F., *Wanderings of a Biochemist.* New York: Wiley-Interscience, 1971. A unique autobiography which intersperses nostalgic, historical, and biographical information with reprints of many of the author's outstanding contributions to biochemistry.

Needham, J., ed., *The Chemistry of Life: Lectures on the History of Biochemistry.* Cambridge: Cambridge University Press, 1970. A series of essays about different areas of biochemistry and their development.

Srinivasan, P. R., J. S. Fruton, and J. T. Edsall, eds., *The Origins of Modern Biochemistry: A Retrospect on Proteins.* New York: The New York Academy of Sciences, 1979. Proceedings of a conference on the history of protein biochemistry, which include first-hand descriptions of basic scientific discoveries by many of the elder statesmen of the discipline.

Watson, J. D., *The Double Helix.* New York: Atheneum, 1968. A ``best-seller'' personal account of the discovery of the helical structure of DNA, introducing a new form of scientific autobiographical writing.

CHEMICAL ELEMENTS AND BIOGEOCHEMICAL CYCLES

2

INTRODUCTION

For the study of biochemistry, it is appropriate to begin with consideration of the chemical elements upon which life relies for its continued expression. The chemical elements found in living systems are a subset of those present in the earth's crust. However, the proportion of the various elements that constitute the animate and inanimate worlds differs markedly (Table 2-1). The earth's crust contains oxygen (O), silicon (Si), aluminum (Al), and iron (Fe) as the four most abundant elements; however, in life forms, about 99 per cent of the mass of most cells is provided by hydrogen (H), oxygen, carbon (C), and nitrogen (N). The chemical composition of *biomolecules,* organic compounds of living systems, accounts for this difference. Because of the predominance of these four ele-

Table 2-1
Major elements in the earth's crust and the human body (per cent of total atoms)

Earth's crust			Human body	
Oxygen	47	mostly as H_2O	Hydrogen	60
Silicon	28	mostly as H_2O	Oxygen	25.5
Aluminum	7.9		Carbon	10.5
Iron	4.5		Nitrogen	2.4
Calcium	3.5		Calcium	0.22
Sodium	2.5		Phosphorus	0.13
Potassium	2.5		Sulfur	0.13
Magnesium	2.2		Potassium	0.04
			Chlorine	0.03
			Sodium	0.03
			Magnesium	0.01

Table 2-2
Trace elements required by living systems

Aluminum (Al)	Manganese (Mn)
Boron (B)	Molybdenum (Mo)
Chromium (Cr)	Nickel (Ni)
Cobalt (Co)	Selenium (Se)
Copper (Cu)	Silicon (Si)
Fluorine (F)	Tin (Sn)
Iodine (I)	Vanadium (V)
Iron (Fe)	Zinc (Zn)

ments, it is apparent that they were compatible with the evolutionary processes that adapted them for their roles in the chemistry of life, a topic considered in Chapter 4. Two other elements often found in organic matter are phosphorus (P) and sulfur (S). Besides these six elements, five monoatomic ions (Na^+, K^+, Mg^{2+}, Ca^{2+}, and Cl^-) and 16 trace elements (Table 2-2) are now known to be required for life processes. Thus, of the 90 elements contained in the earth's crust, 27 have been identified as essential for living organisms; not all, however, are required by every species. Those needed by humans are discussed in Chapter 25.

All forms of life must obtain elements from an environment that cannot generate new supplies. Continued propagation of the species, therefore, relies on mechanisms that assure the continuous reutilization of chemical elements. These mechanisms, called *biogeochemical cycles,* occur between the inorganic and organic components of the earth and between various forms of life. The study of these cycles is becoming an increasingly important area of biological research. Environmental studies reflect the serious attempts currently being made to understand the balance of natural processes that are essential for the continuation of life.

THE CARBON CYCLE

Carbon occurs in the earth's atmosphere and crust and is dissolved in the earth's water. As shown in Figure 2-1, plankton of the oceans and plants convert carbon dioxide (CO_2) into carbon-containing compounds, such as sugars and amino acids, by the process of photosynthesis (Chapter 18). Carbon is transferred to herbivores when they eat plants. Herbivores, in turn, are consumed by carnivores or by omnivores, such as humans, who may obtain additional carbon by eating other carnivores. At each step, carbon is said to have moved farther along the *food chain.* The chemical energy needed to sustain and propagate life is extracted from these carbon compounds by the process of respiration (Chapters 16 and 17), which occurs in plants, animals, and many microorganisms. A product of respiration is CO_2, which is returned to the atmosphere. Carbon is also recycled by the oxidative action of microorganisms that cause decay of organic matter. These microorganisms occupy a strategic position by serving as the ultimate link in food chains. Some carbon does not immediately return to the carbon dioxide pool but leaves the carbon cycle for millions of years. Such organic material often accumulates in a partially

Herbivore is taken from the Latin words *herba* and *vorus,* meaning *grass* and *eating,* respectively.
Carni is taken from the Latin word *caro,* meaning *flesh.*
Omni is taken from the Latin word *omnis,* meaning *all.*

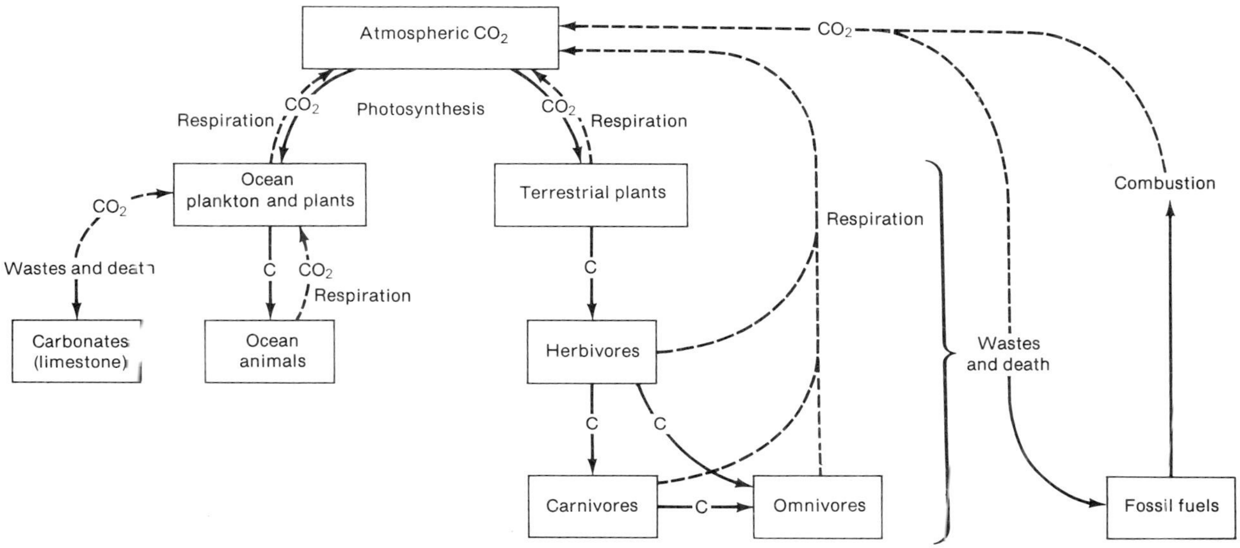

Figure 2-1
The carbon cycle.

degraded state. It is these sidetracked materials that furnish the fossil fuels (peat, coal, oil, and natural gas) which, being hydrocarbons, release CO_2 when they are combusted. Carbon also combines chemically to form minerals such as limestone, resulting in a form of carbon that is less available to living systems.

A key consideration in the carbon cycle is the flow of energy (chemical) in the food chain that proceeds from plants *(producers)* to herbivores *(primary consumers)* and then to carnivores *(secondary consumers)*. At each level (Figure 2-2), only about 10 per cent of the energy obtained is stored in an organism's tissues, i.e., available for transfer to the next level, since the remaining 90 per cent (used for metabolism) is lost as heat to the atmosphere. These substantial losses of energy are receiving serious attention as a result of the world's food and population problems. Some humans, as carnivores, utilize much more of the chemical energy originally produced by plants than do others who, as herbivores, eat vegetable and grain diets. To feed the earth's increasing population, shortening the food chain to minimize energy losses is of practical significance. One complaint of the third-world nations is that affluent countries, with diets high in meat, are utilizing grains edible by humans to feed and fatten cattle and therefore are unnecessarily using food sources needed by other societies.

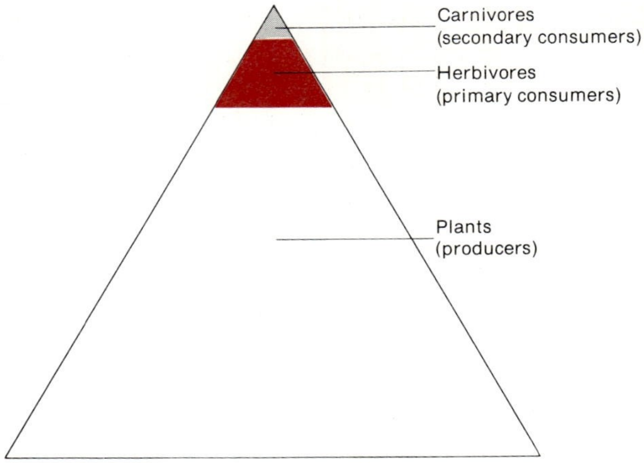

Figure 2-2
The pyramid of energy.

THE NITROGEN CYCLE

Nitrogen is another element that is cycled throughout living materials, atmosphere, and earth. Air is about 79 per cent nitrogen (as nitrogen gas, N_2) and constitutes the major functional reservoir for this element. The sediment and crust of the earth contain compounds of nitrogen, but the nitrogen is less utilizable to organisms in this form. Nitrogen moves through complex pathways, some of which are shown in Figure 2-3. Unlike other atmospheric gases, e.g., O_2 and CO_2, nitrogen cannot be used directly by most organ-

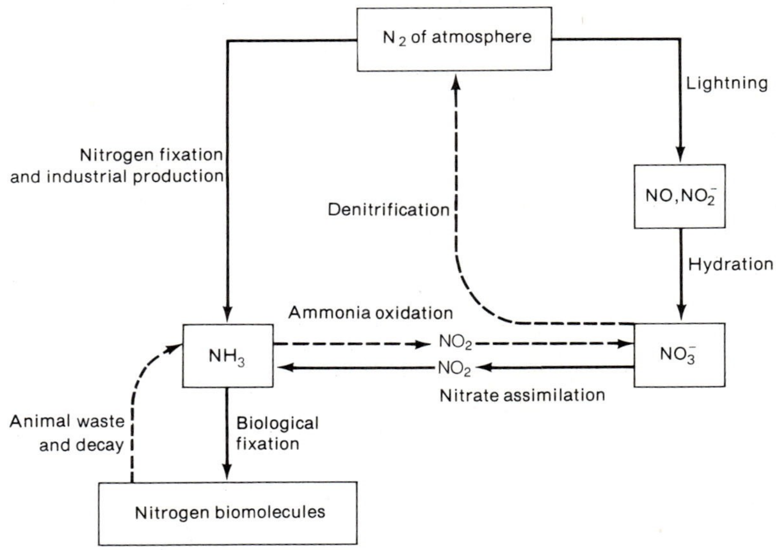

Figure 2-3
The nitrogen cycle.

isms, but must be *fixed* or converted into a useable form by special organisms. Microorganisms that convert nitrogen gas into ammonia (NH_3) are called *nitrogen fixers.* The best known are those which give rise to nodules on the roots of legumes, such as peas and soybeans. An important biological symbiotic relationship exists between these microbes and nodulating plants. Other nitrogen fixers include cyanobacteria (blue-green algae) and some free living bacteria.

The most prevalent form of nitrogen in soil, however, is the nitrate ion (NO_3^-), which is utilized by nonfixing organisms. Nitrates are produced through the oxidation of ammonia by microorganisms and from nitrogen oxides produced by lightning. Soil microbes and plants convert nitrate ions into ammonia, which they then use to synthesize nitrogenous biomolecules. This reduced form of organic nitrogen ($-NH_2$) is transferred from plants to herbivores and then to carnivores and omnivores. The nitrogen found in soil can also be utilized by certain bacteria to produce nitrogen gas *(denitrification).* Today, another important component of the nitrogen cycle is the industrial fixation of nitrogen (p. 396) to produce commercial fertilizers.

THE PHOSPHORUS CYCLE

Figure 2-4 summarizes the phosphorus cycle and flow of this important element through biogeochemical materials. Phosphorus, however, does not cycle as readily as carbon and nitrogen. The major natural reservoirs of phosphorus are phosphate rocks, wastes and fossils of animals, and deposits of excrement from fish-eating birds. Phosphorus is often dissolved from these sources by weathering phenomena. Eventually a large portion of phosphate, because of leaching, finds its way to fresh water and then to the seas, where it is utilized by organisms or deposited in marine sediments. Unlike carbon and nitrogen, which cycle in different chemical forms, phosphorus is almost always found as phosphate (PO_4^{3-}).

THE OXYGEN CYCLE

The major oxygen reservoir is the atmosphere, of which about 21 per cent is oxygen gas (O_2). Oxygen, required by all respiring organisms, occurs in many chemical forms in its cycle. It appears as

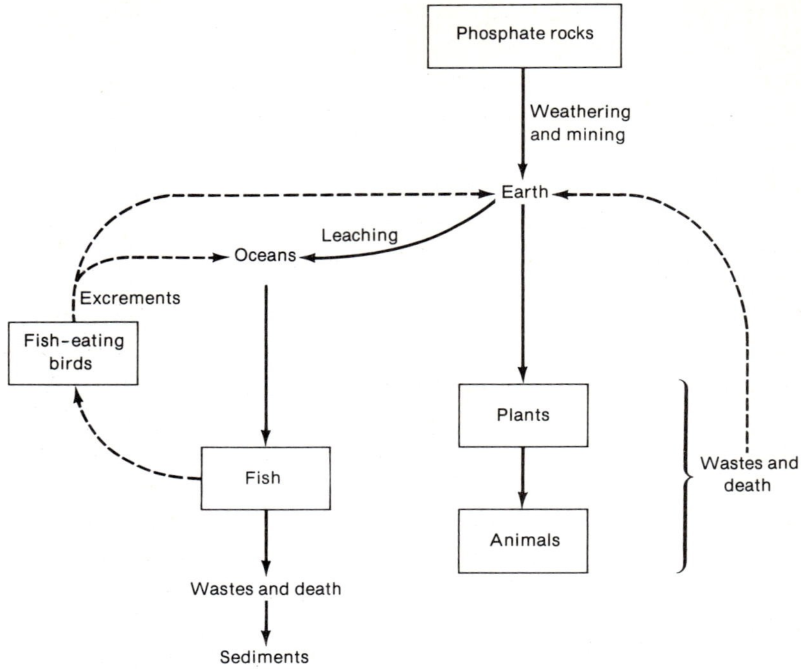

Figure 2-4
The phosphorus cycle.

molecular oxygen in water and is present in such compounds as carbon dioxide and carbon monoxide (CO) and in many biomolecules. Oxygen atoms cycling in these molecules constitute the primary common pool. Oxides of other elements, such as sulfate (SO_4^{2-}) and nitrate ions, are also a source of oxygen for those organisms capable of reducing them to hydrogen sulfide (H_2S) and ammonia. When used in respiration, oxygen reappears in the end products, water and carbon dioxide. Plants and other photosynthetic organisms that utilize carbon dioxide restore oxygen to the atmosphere and complete the cycle.

Figure 2-5
The water cycle in living organisms.

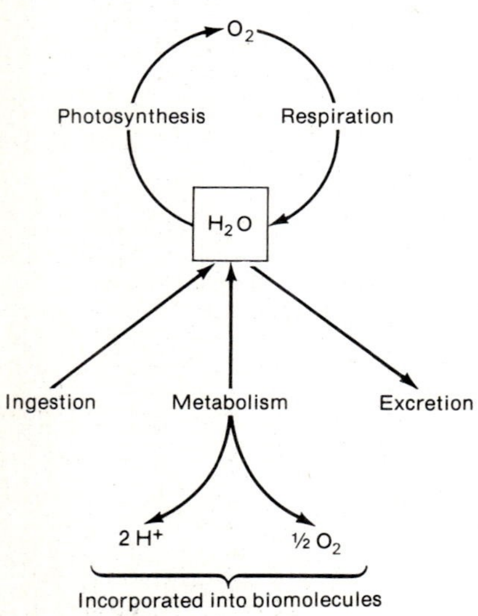

THE WATER CYCLE

Water is the most abundant molecule in living systems (60–95 per cent by weight); its cycle involves both hydrogen and oxygen, elements of the water molecule. The global water cycle involves its evaporation from the seas, its transpiration by plants and animals, and its return to the earth as rain or snow. Precipitation dissolves inorganic soil nutrients, making them available for underground transport and recycling into plants.

The water cycle in living organisms (Figure 2-5) includes both its utilization and generation in life processes. For example, its use in photosynthesis yields oxygen which, in turn, is converted back into water during respiration by plants and animals. A requirement for water in many metabolic processes often results in the incorporation of its hydrogen and oxygen atoms into biomolecules; amino acids produced by degradation (hydrolysis) of proteins are such biomolecules (p. 151). In animals, ingested water is excreted in urine, feces, and perspiration.

MODIFICATION OF BIOGEOCHEMICAL CYCLES

The modification or disruption of these cycles has led to serious ecological problems, destruction of plant life, and the onset of health problems for animals, including humans. Some of the resulting biochemical effects have also endangered species living in particular ecosystems and eventually may affect many more organisms.

Ecology is taken from the Greek words *oikos*, meaning *house* or *habitat*, and *logos*, a science or branch of *knowledge*.

In some cases, synthetic compounds have adversely affected biogeochemical cycles. Such compounds include certain types of insecticides, herbicides, and chemical additives to commercial products. Some of these substances are used intentionally for various reasons, e.g., fertilizers; others are chemicals added to the environment unintentionally, e.g., release of Kepone, an insecticide, from its site of manufacture into the atmosphere and waterways. In either case, the biochemical result proves to be a threat to the delicate balance of ecosystems and to human health. Lake Erie is a dramatic example of the problems that can be encountered.

Lake Erie drains 40,000 square miles of Michigan, Indiana, Ohio, Pennsylvania, New York, and Ontario. In this century, the lake water became polluted as agriculture, industry, and large cities grew. By the late 1960s, Lake Erie represented the first large-scale warning that humans and their activities were endangering the habitability of earth. Algae, industrial wastes, and sewage had turned the lake into brackish soup at some places. One major pollutant was the phosphorus derived from detergents, human wastes, and fertilizers. This element is responsible for *algae blooms,* which occur in spring and summer. When the algae die and decay, so much oxygen is consumed that large parts of the lake become too oxygen deficient to support many fish species. Since 1972, serious efforts have been made to restore the lake.

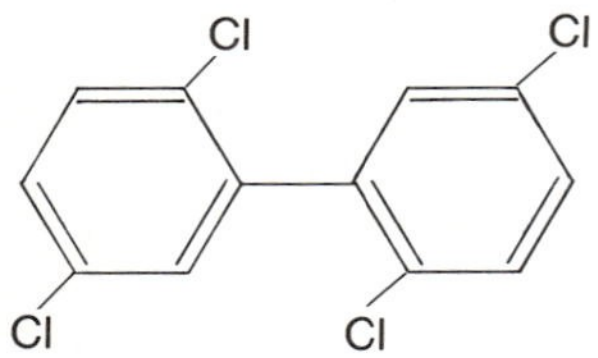

The PCB 2,5,2′,5′-tetrachlorobiphenyl

Over one billion dollars have already been spent to improve sewage treatment facilities; it is estimated that an additional 17 billion dollars will be needed to complete the task. Industrial efforts to abate the pollution problem have also been extensive. The control of municipal sewage and industrial discharges, however, does not represent the most difficult challenge. The runoffs from farms (chemical fertilizers and manure) and cities (street dirt) pose the most serious problems. Smaller discharges of highly toxic substances, such as polychlorinated biphenyls (PCBs), are now recognized as another serious threat to the wildlife in and along the lake. Seepage of these chemicals from landfills (junked electrical equipment and plastics) accounts for the danger. Because Lake Erie's problems are so vast and exceedingly expensive to remedy, its future is still uncertain.

The attempts to understand the role these cycles have in ecosystems might help to prevent future destructive interference with nature. To deal with this complex interdependence, biochemists and other scientists have escalated efforts to study biomolecules in their biogeochemical functions. Such studies will undoubtedly bring about a better understanding and appreciation of the delicate interrelationship that exists between humans and their natural environment.

SUGGESTED READINGS

Barbier, M., translated by M. A. Ferenczi, *Introduction to Chemical Ecology.* New York: Longmans, 1979. A text that examines the chemical relationships existing between living organisms and between the organic and inorganic worlds.

Carson, R., *Silent Spring.* Boston: Houghton Mifflin, 1962. Originally this book was highly controversial because of its dire warnings to the nations of the world about the dangers of pesticides. Today it is considered a modern ecology classic.

Commoner, B., *The Closing Circle.* New York: Alfred A. Knopf, 1971. Written by a leading environmentalist, this book examines the direct relationship between the technological and sociological aspects of society and the current ecological crisis.

Ehrlich, P. R., A. H. Ehrlich, and J. P. Holdren, *Ecoscience.* San Francisco: W. H. Freeman, 1977. A recent book by the authors of *Population/Resources/Environment.*

Henderson, L. J., *The Fitness of the Environment.* New York: Macmilan, 1913. Reprinted in 1958 by Beacon Press (Boston), this classic of biological science foreshadowed the present understanding and appreciation of ecological biochemistry.

Odum, E. P., *Fundamentals of Ecology,* 3rd ed. Philadelphia: W. B. Saunders, 1971. A highly respected textbook on ecology written by an eminent ecologist.

Storer, J. H., *The Web of Life.* Old Greenwich, Conn.: Devin-Adair, 1953. This interesting and easy-to-read book about an ecologist's observations is another highly recommended classic (see p. 364).

Articles

Ambroggi, R. P. "Underground Reservoirs To Control the Water Cycle." *Scientific American,* 236(5):21, 1977.

Brill, W. J. "Biological Nitrogen Fixation." *Scientific American,* 236(3):68, 1977.

Ecology, Evolution, and Population Biology. San Francisco: W. H Freeman, 1974. Readings from *Scientific American.*

Frieden, E., "The Chemical Elements of Life." *Scientific American,* 227(1):52, 1972.

Scheller, S. W., "Nitrogen Fixation." *J. Chem. Ed.* 49:786, 1972.

"The Biosphere." *Scientific American,* 223(3)—September, 1970 issue:

1. A. H. Oort, "The Energy Cycle of the Earth" (offprint #1189), p. 54.
2. G. M. Woodwell, "The Energy Cycle of the Biosphere" (offprint #1190), p. 64.
3. H. L. Penman, "The Water Cycle" (offprint #1191), p. 98.
4. P. Cloud and A. Gibor, "The Oxygen Cycle" (offprint #1192), p. 110.
5. B. Bolin, "The Carbon Cycle" (offprint #1193), p. 124.
6. C. C. Delwicke, "The Nitrogen Cycle" (offprint # 1194), p. 136.
7. E. S. Deevey Jr., "Mineral Cycles" (offprint #1195), p. 148.
8. L. R. Brown, "Human Food Production as a Process in the Biosphere" (offprint #1196), p. 160.
9. S. F. Singer, "Human Energy Production as a Process in the Biosphere" (offprint #1197), p. 174.
10. H. Brown, "Human Materials Production as a Process in the Biosphere" (offprint #1198), p. 194.

WATER AS THE SOLVENT OF LIFE

3

INTRODUCTION

Without water there is no life as we know it. The essentiality of water is a continuous reminder of the aqueous origin of life. It was in the solvent water that the chemical reactions of biological processes evolved. Although not an organic biomolecule, water is the most ubiquitous component in living cells and accounts for 60-95 per cent of their weight. In humans, water is about evenly distributed between the two major compartments (intra- and extracellular) of the body (Table 3-1). Water is required not only for biochemical reactions but also for transporting substances across membranes, maintaining body temperature, producing digestive fluids, and dissolving waste products for excretion.

The maintenance of *water balance,* the equilibrium between water intake and output, is a critical aspect of metabolism. An adult in water balance generally takes in and loses about 2,000 ml of water per day (Table 3-2). Besides the water obtained from foods and liquids, *metabolic water* is also made available through the oxidation of food in the body (p. 337). Oxidation of 100 g each

Table 3-1
Water distribution in the body

Compartment	Total body water (%)	
Intracellular fluids		55.0
Extracellular fluids		
Plasma	7.5	45.0
Interstitial and lymph	22.5	
Dense connective tissue, cartilage, and bone	15.0	

2,000 ml = 2.11 quarts

Table 3-2
Daily water balance in humans

Intake (ml)		Loss (ml)	
As liquids	900	Urine	1,050
In foods	800	Feces	100
Oxidation of foods	300	Evaporation (skin and lungs)	850
	2,000		2,000

of fat, carbohydrate, and protein yields 107, 55, and 41 g of water, respectively. Water losses occur by evaporation (water vapor in expired air and perspiration) and by excretion of urine and feces. If the output of water significantly exceeds its intake, *dehydration* occurs; this condition can result from severe diarrhea, vomiting, fever, or unusually high environmental temperatures. Dehydration can be very serious in young children because their body pool of water is small and therefore can be readily depleted. If water accumulates in tissues, then *edema* occurs. Edema is often observed in children suffering from *kwashiorkor,* a nutritional disease caused by an *inadequate intake of protein.* In the Caribbeans, children with kwashiorkor are called *sugar babies* because of their plump appearance (a deceiving indication of good health). Why edema is a manifestation of kwashiorkor (which is also characterized by diarrhea, i.e., fluid loss) is uncertain. It may be due to a loss of fluid from the blood into tissues because of low concentrations of blood proteins or because of a disturbance in the electrolyte and water balance induced by the malnutrition.

Kwashiorkor is the Bantu word meaning *displaced child* and refers to the fact that kwashiorkor appears in infants when they are no longer nursed by their mothers, i.e., are displaced by the birth of a new baby.

To appreciate the crucial role of water in metabolism, it is necessary to understand those properties of the molecule which are compatible with life processes as they have evolved. Equally as important is an awareness of the ionic environment with respect to the hydrogen ion, which is also essential for the aqueous systems that support life.

PHYSICAL PROPERTIES OF WATER

The physical properties of water differ distinctly from those of other solvents. For example, water as a hydride of oxygen (H_2O) has a higher melting point, boiling point, heat of vaporization, and surface tension than do the comparable hydrides of sulfur (H_2S) and nitrogen (NH_3) (Table 3-3). Such properties are indicative of

Table 3-3
Some physical properties of the hydrides of oxygen, sulfur, and nitrogen

Hydride	Melting point (°C)	Boiling point (°C)	Heat of vaporization (cal/g)
H_2O	0	100	540
H_2S	−85	−60	132
NH_3	−78	−33	327

strong intermolecular forces in liquid water. It is the electrical dipolar nature of the water molecule that accounts for these forces. In a water molecule (Figure 3-1), the highly electronegative oxygen atom attracts the bonding electrons from each of the two hydrogen atoms, polarizing the bonds. A partial positive (δ^+) region is created around each of the two resultant hydrogen nuclei. These partial positive regions and the partial negative (δ^-) region around the oxygen atom make water a dipolar molecule. It is this polar nature of water that allows for electrostatic attraction between its molecules. The result of such an attraction is called a *hydrogen bond,* which occurs between the oxygen atom of one water molecule and a hydrogen atom of another. (As discussed in Chapter 4, the bonding between H and O is just one type of hydrogen bonding that can occur.) Because of the almost tetrahedral arrangement of the oxygen electrons (bond angle 104.5°), each water molecule can potentially hydrogen bond with four other molecules (Figure 3-2). Studies on liquid water reveal an average of 3.4 hydrogen bonds per molecule. The bonding properties which allow water to bind to itself make water a relatively structured solvent (Figure 3-3) and account for its strong internal cohesion as a liquid.

Figure 3-1
The structure of a water molecule.

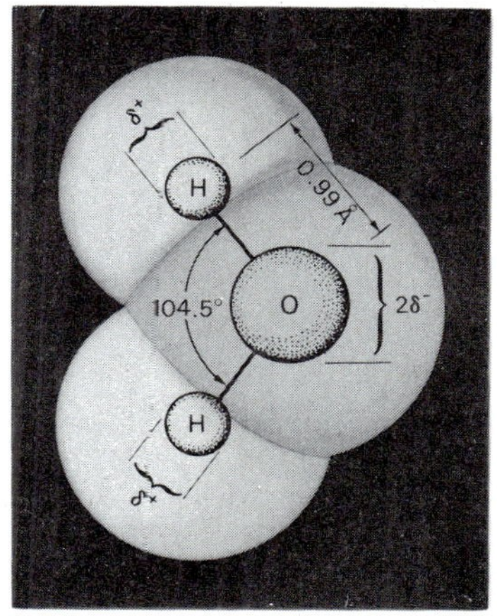

Figure 3-2
Tetrahedral hydrogen bonding of a water molecule, in ice.

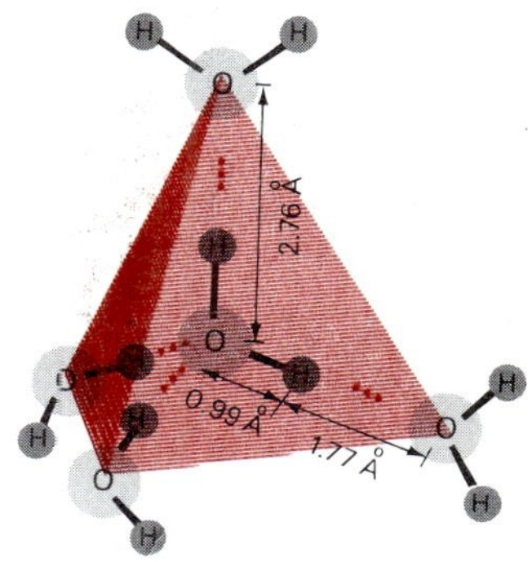

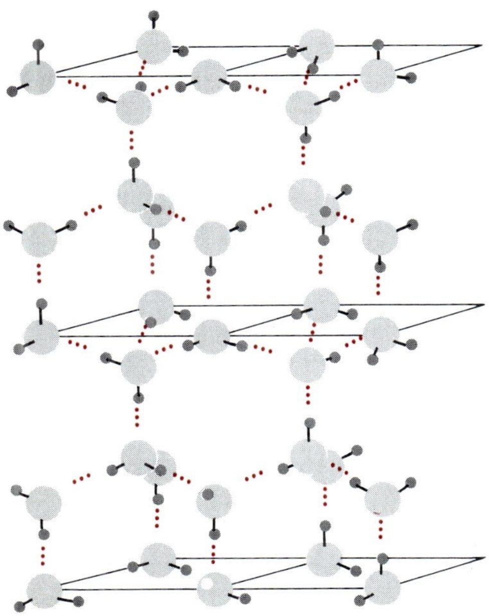

Figure 3-3
The arrangement of water molecules in ice. (After L. Pauling, *The Nature of the Chemical Bond,* 3rd ed. Ithaca, N.Y.: Cornell University Press, 1960)

SOME BIOLOGICAL IMPORTANCES OF WATER

Several rare physical properties of water offer certain biological advantages, two of which are important for the maintenance of the constant internal temperature required by many organisms. One property is water's high *heat of vaporization* (the number of calories absorbed when a gram of liquid is vaporized). The calories removed by evaporation of water provide an effective physiological mechanism by which organisms can dissipate heat. The other property is the high *specific heat capacity* of water (the number of calories required to raise the temperature of one gram of substance 1 C°). As a result, water absorbs heat well and therefore is of great value in helping to keep an organism's temperature constant.

A calorie is the amount of heat necessary to raise 1 gram of water 1 C°.

The fact that water achieves its maximum density at 4°C is also important to many biological systems. The solidification of water at 0°C to form ice produces a less dense phase which floats. Because ice does not sink, bodies of water do not freeze from the bottom upward; this permits aquatic organisms to remain in their normal environment during winter. Also, once frozen, bodies of water thaw more readily because the coldest water is near the surface, exposed to sun and atmosphere.

WATER AS A SOLVENT

Water is an excellent solvent for ionic compounds, such as salts, because the attraction between the ionic components of the molecules and the water dipoles is sufficient to overcome the attraction between the ions themselves. Nonionic polar compounds, such as sugars and simple alcohols, are also very soluble in water. Polar functional groups, such as the hydroxyl group, of nonionic compounds readily hydrogen bond with water molecules, dispersing the compounds among the water molecules.

An interesting phenomenon occurs when *amphipathic molecules,* possessing both polar (hydrophilic) and nonpolar (hydrophobic) groups are dispersed in water. Salts of fatty acids (Figure 3-4) are examples of amphipathic molecules because of their polar head (carboxylate group) and nonpolar tail (hydrocarbon chain). In a dilute aqueous solution, such salts form *micelles* (Figure 3-5). These are aggregations of molecules with polar carboxylate ions on the exterior and nonpolar hydrocarbon chains in the interior, which create an internal hydrophobic environment. Cleansing by soap (alkali salts of fatty acids) is accomplished by the entrapment

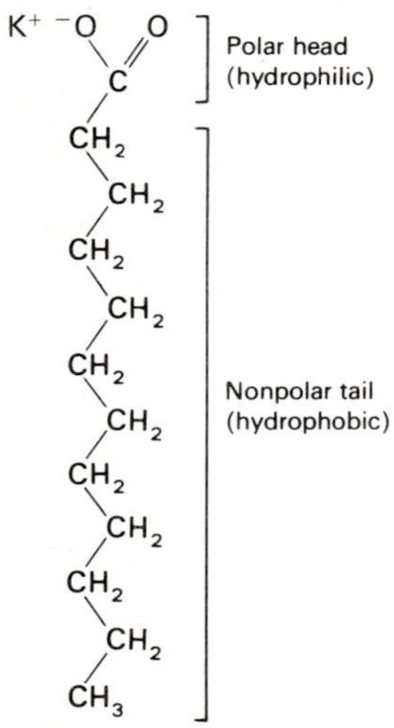

Figure 3-4
Potassium laurate, a 12-carbon fatty acid.

of water-insoluble dirt and grease in the hydrophobic interior of soap micelles. Aside from this practical consideration, micelle formation is important for an understanding of organized biological systems because amphipathic molecules are involved in the formation and structure of membranes.

Figure 3-5
Schematic of a micelle composed of salts of fatty acids. The polar heads (carboxylate ions) are symbolized by ● and the nonpolar tails (hydrocarbon chains) by ∿∿∿∿∿.

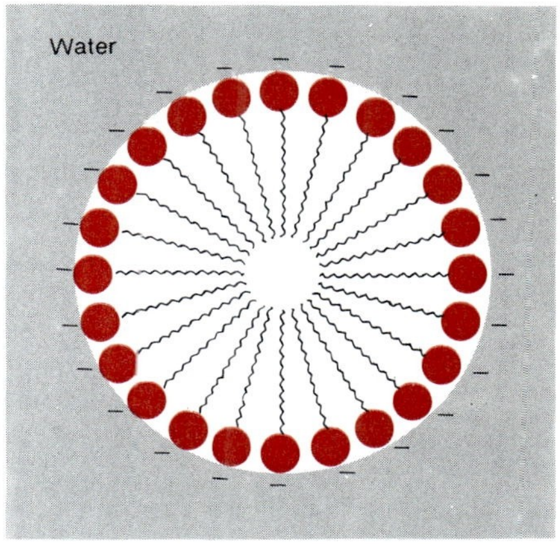

HYDROGEN ION CONCENTRATION OF BIOLOGICAL SYSTEMS

The aqueous environments of biological systems have hydrogen ion (H^+) concentrations that remain remarkably constant. Maintenance of appropriate H^+ concentration is vitally important for the life of any organism because, as will be presented later (p. 36), biochemical reactions are highly sensitive to fluctuations in the concentration of this ion. In the dynamics of metabolism, the presence and production of many biomolecules continuously affect the amount of H^+ present. If there were no mechanisms to control alterations in H^+ concentration, the effective coordination of the many reactions that constitute metabolism would be rapidly lost. Life processes would then cease. Therefore, an understanding of the mechanisms that strictly control H^+ environment is essential for an appreciation of the firmly regulated aqueous systems that life requires.

DISSOCIATION OF WATER

Water itself contributes hydrogen ions to biological systems because it ionizes to a very slight extent to yield a hydrogen ion and a hydroxyl ion (OH^-). The H^+ does not exist as such in aqueous solution but, rather, in association with a water molecule (proton transfer) to form a hydronium ion (H_3O^+).

$$H_2O + H_2O \rightleftharpoons H_3O^+ + OH^-$$

However, for convenience and by convention, the dissociation of H_2O is usually written

$$H_2O \rightleftharpoons H^+ + OH^-$$

The dissociation can be expressed according to the concepts of the *Law of Mass Action,* as follows:

$$K_{eq} = [H^+]\,[OH^-]/[H_2O]$$

This expression states that, at equilibrium, the mathematical product of the concentrations (signified by [ ]) of the products of dissociation (H^+ and OH^-) divided by the concentration of undissociated water (H_2O) is equal to a value known as the *equilibrium constant* (K_{eq}). From this expression, the concentration of H^+ in pure water can be calculated if the K_{eq} is known.

The equilibrium constant for the dissociation of H_2O at 25° C has been accurately determined to be 1.8×10^{-16} *M* (moles/liter), a value that reflects very slight dissociation. The concentration of H_2O in pure, undissociated water is 55.5 *M*, obtained by dividing 1,000 g (weight of 1 liter of H_2O) by 18 g (gram molecular weight of H_2O). This value of 55.5 *M* is generally accepted as the concentration of undissociated water ([H_2O]) in the above equilibrium equation, because dissociation is so slight. It is also usually regarded as the concentration of H_2O in dilute aqueous solutions, since the actual concentration does not differ significantly from that of pure H_2O.

When the values for K_{eq} and [H_2O] are substituted into the above equation, the following ion product is obtained:

$$[H^+][OH^-] = (1.8 \times 10^{-16})(55.5) = 1.0 \times 10^{-14}\ M$$

Since $[H^+] = [OH^-]$ when water dissociates, the following expression is derived:

$$[H^+]^2 = 1.0 \times 10^{-14}\ M$$

$$[H^+] = 1.0 \times 10^{-7}\ M$$

Thus, the $[H^+]$ of pure water is 1.0×10^{-7} *M*. Although the value is expressed only to two significant figures, it is accurate enough for most purposes.

Table 3-4
The pH scale

H^+ (*M*)	pH	
1.0	0	↑
0.1	1	
0.01	2	
0.001	3	Acidic
1×10^{-4}	4	
1×10^{-5}	5	
1×10^{-6}	6	↓
1×10^{-7}	7	Neutral
1×10^{-8}	8	↑
1×10^{-9}	9	
1×10^{-10}	10	Basic
1×10^{-11}	11	
1×10^{-12}	12	
1×10^{-13}	13	
1×10^{-14}	14	↓

pH SCALE

Hydrogen ion concentrations are routinely expressed as pH values. The pH expresses $[H^+]$ as a logarithmic function and is defined as the negative logarithm of the hydrogen ion concentration.

$$pH = \log 1/[H^+] = -\log [H^+]$$

Although pH is an expression of hydrogen ion *activity* (as measured by pH meters), no distinction is usually made between ion activity and concentration. As seen in Table 3-4, pH values offer a convenient means of stating widely varying $[H^+]$ in small positive numbers. The pH scale of 0 to 14 accommodates H^+ concentrations of 1 *M* to 1×10^{-14} *M*. A pH value of 7 ($[H^+]$ of pure water) is considered the *neutral pH.* Increases in $[H^+]$ (pH values smaller than 7) produce *acidic* conditions, and decreases (pH values larger than 7) result in *basic,* or *alkaline,* conditions. In the use of pH values, it is important to remember that the numbers represent a logarithmic function and that therefore a decrease or increase of one pH unit, e.g., pH 7 → pH 6, represents a tenfold difference in $[H^+]$.

BRÖNSTED–LOWRY ACIDS AND BASES

In aqueous systems, the addition or removal of hydrogen ions is best understood in terms of the Brönsted-Lowry concept of acids and bases. A *Brönsted-Lowry acid* is defined as a substance that can *donate* a proton (H^+); conversely, a *Brönsted-Lowry base* is a substance that can *accept* a proton. This broad definition of acids and bases includes many substances that are not usually considered acidic or basic.

In the general expressions:

$$HA \rightleftharpoons H^+ + A^-$$

$$HB^+ \rightleftharpoons H^+ + B:$$

$$HB^- \rightleftharpoons H^+ + B^{2-}$$

HA, HB^+, and HB^- represent Brönsted-Lowry acids, and the anion (A^-), neutral species (B:), and the negatively charged species (B^{2-}), produced by the loss of a proton, are their respective *conjugate bases.* Note that each conjugate base can accept a proton to restore the corresponding acid. Table 3-5 lists some Brönsted-Lowry acids and their conjugate bases. The classification is both simple and useful. Because many biomolecules are capable of donating or accepting protons, they influence the $[H^+]$ in biological systems.

STRONG AND WEAK ACIDS

With respect to acids, there are two general classes—strong and weak. A *strong acid* is defined as a substance that has little affinity for its proton and therefore completely dissociates (ionizes) in

Table 3-5
Some Brönsted–Lowry acids and their conjugate bases

Acid	Undissociated acid (HA)	Conjugate base (A^-)
Acetic acid	H_3CCOOH	H_3CCOO^-
Propionic acid	H_3CCH_2COOH	$H_3CCH_2COO^-$
Pyruvic acid	$H_3CCOCOOH$	$H_3CCOCOO^-$
Phosphoric acid	H_3PO_4	$H_2PO_4^-$
	$H_2PO_4^-$	HPO_4^{2-}
	HPO_4^{2-}	PO_4^{3-}
Ammonium ion	NH_4^+	NH_3
Water	HOH	OH^-

water. Examples of strong acids include mineral acids such as HCl and H_2SO_4. A *weak acid,* on the other hand, is a molecule that displays a high affinity for its proton and does not readily dissociate in water. Organic acids, such as acetic acid (H_3CCOOH), are weak acids. The selective dissociation of weak acids in water is a characteristic which is of great importance to biochemistry because of its role in influencing the $[H^+]$ of an aqueous environment.

IONIZATION OF WEAK ACIDS

The dissociation of the organic acid, acetic acid, is written:

$$H_3CCOOH \rightleftharpoons H^+ + H_3CCOO^-$$

At a given temperature, the following equation can be used to calculate the extent of ionization at equilibrium:

$$K'_a = [H^+]\,[H_3CCOO^-]/[H_3CCOOH]$$

The expression is comparable to that used to describe the dissociation of water, except that the symbol K'_a (apparent ionization constant) is substituted for K'_{eq} (equilibrium constant). The change recognizes the reaction as an ionization. The prime symbol on the K'_a is used to identify the ionization constant as *apparent* (value based on the concentrations of reactant and products) rather than as a *true* value (K'_a corrected for deviation of the system from ideal behavior).

The slight amount of ionization that a 1 *M* solution of acetic acid undergoes ($K'_a = 1.8 \times 10^{-5}$ *M* at 25° C) can be readily calculated by letting $x = [H^+]$ and $[H_3CCOO^-]$ and by not correcting

$[H_3CCOOH]$, which is a relatively large concentration, for the small amount of x produced.

$$x^2 = 1.8 \times 10^{-5}\ M$$

$$x = 4.2 \times 10^{-3}\ M, \text{ or } 0.0042\ M$$

Thus, only slightly over 0.4 per cent of a 1 M solution of acetic acid is ionized at 25° C. The pH of the solution is 2.38, i.e., (− log 0.0042).

HENDERSON–HASSELBALCH EQUATION

In biochemistry, the equilibrium expression is most often written in another form, called the Henderson-Hasselbalch equation, which is obtained as follows:

1. Rearrange the K'_a equation to solve for $[H^+]$:

$$[H^+] = K'_a\,[HA]/[A^-]$$

2. Convert to logarithmic functions:

$$\log\,[H^+] = \log K'_a + \log\,([HA]/[A^-])$$

3. Make the expression negative:

$$-\log\,[H^+] = -\log K'_a - \log\,([HA]/[A^-])$$

4. Define $-\log\,[H^+]$ as pH and $-\log K'_a$ as pK'_a and convert

$$-\log\,([HA]/[A^-]) \text{ to } \log\,([A^-]/[HA])$$

5. Henderson-Hasselbalch equation:

$$pH = pK'_a + \log\,[A^-]/[HA]$$

The equation is very useful for calculating the pH of known solutions and for determining the amounts of an acid and its salt needed to prepare a solution at a given pH.

TITRATION OF A WEAK ACID BY A STRONG BASE

The titration of a weak acid, such as acetic acid, with a strong base results in the acid's complete ionization. A plot of the course of the titration reveals the property of Brönsted-Lowry acids and

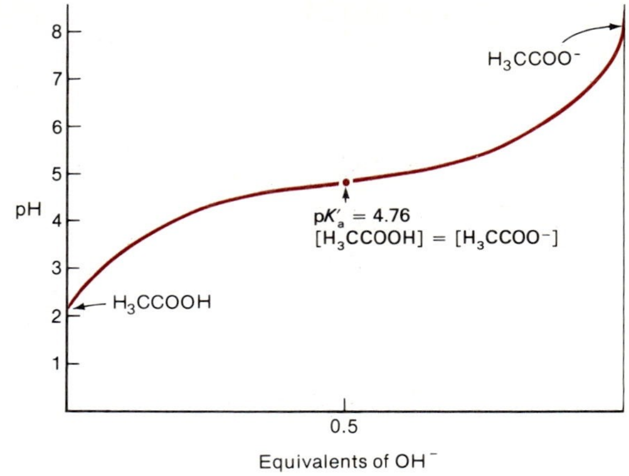

Figure 3-6
Titration of acetic acid, H_3CCOOH, with the base NaOH.

their conjugate bases that makes them useful in biology. When acetic acid (H_3CCOOH) is titrated with NaOH (Figure 3-6), the greatest changes in pH take place at the beginning and end of the titration. The region of least change occurs at the midpoint of titration, when exactly 0.5 equivalents of base have been added. At this point, the concentration of undissociated acid (H_3CCOOH, or HA) is equal to that of its anion (H_3CCOO^-, or A^-). At these particular concentrations of HA and A^-, the pH (4.76) is equal to the pK'_a value. If stated in terms of the Henderson-Hasselbalch equation, when $[A^-] = [HA]$, the term $\log [A^-] / [HA]$ becomes log 1, which is zero. Hence, pK'_a equals pH.

The ability of an acetic acid solution to resist a change in pH is referred to as its *buffering capacity* which, as implied, is maximum when equimolar concentrations of the undissociated acid and its anion are present. These particular concentrations are needed to provide the maximum capability of the solution to neutralize either added acid:

$$H_3CCOO^- + H^+ \rightleftharpoons H_3CCOOH$$

or added base:

$$H_3CCOOH \rightleftharpoons H_3CCOO^- + H^+$$

$$H^+ \xrightarrow{OH^-} HOH$$

A system that can resist a change in pH upon the addition of either acid or base is called a *buffer.* Thus, many substances classified as weak acids can be either acids or bases, depending on pH.

DISSOCIABLE BIOLOGICAL COMPOUNDS

Acetic acid is one of many biological Brönsted-Lowry acids which display a pH-dependent dissociation in aqueous environments. Such ionization properties are employed extensively in living systems to help buffer biological fluids; they are also an important feature of biochemical mechanisms. Table 3-6 lists some dissociable biological compounds and their pK'_a values. It is apparent that ionization characteristics (represented by pK'_a values) are a distinctive property of each compound. Some of the compounds can undergo more than one dissociation. Citric acid has three carboxyl groups which selectively ionize and therefore possesses three pK'_a values.

Table 3-6
Some biological acids and their pK'_a values

Undissociated acid	Formula	pK'_{a1}	pK'_{a2}	pK'_{a3}
Formic acid	HCOOH	3.75		
Acetic acid	H_3CCOOH	4.76		
Pyruvic acid	$H_3CCOCOOH$	2.50		
Lactic acid	$H_3CCHOHCOOH$	3.86		
Succinic acid	$HOOCCH_2CH_2COOH$	4.21	5.63	
Malic acid	$HOOCCH_2CHOHCOOH$	3.40	5.26	
Carbonic acid	H_2CO_3	6.35	10.3	
Citric acid	$HOOCCH_2C(OH)(COOH)CH_2COOH$	3.09	4.75	5.41
Phosphoric acid	H_3PO_4	2.12	7.21	12.66

Phosphoric acid is an example of another compound which has three dissociable protons:

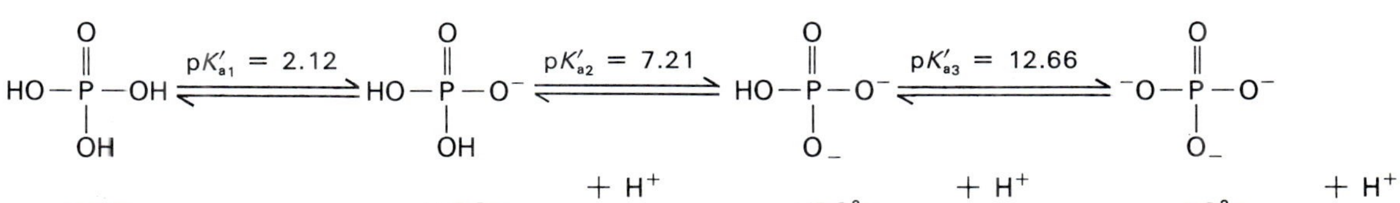

H_3FO_4 $H_2PO_4^-$ HPO_4^{2-} PO_4^{3-}

Note that H_3PO_4 is the conjugate acid of $H_2PO_4^-$, which is the conjugate base of H_3PO_4 and also the conjugate acid of HPO_4^{2-}. In turn, HPO_4^{2-} is a conjugate base (of $H_2PO_4^-$) and a conjugate acid (of PO_4^{3-}). Thus, $H_2PO_4^-$ and HPO_4^{2-} exhibit the characteristics of both an acid and a base—important properties in the biological functioning of inorganic phosphate.

PHYSIOLOGICAL BUFFER SYSTEMS

The pH of different body fluids in humans varies greatly, as exemplified by gastric juices (pH 1-2) and intestinal juices (pH 8-9). In a healthy individual, blood plasma has a pH of 7.40. It is critical that the pH of blood remain constant because relatively small changes of ± 0.2-0.4 of a pH unit result in serious medical conditions which, if not corrected, can lead to death. Blood, because of its heterogeneous population of molecules, possesses an array of biomolecules that behave as Brönsted-Lowry acids and bases. Of importance in the maintenance of well-buffered blood are the numerous plasma proteins and, particularly, the hemoglobin of red blood cells because they serve effectively as proton acceptors and donors. The main *extracellular* buffer system, however, is the *bicarbonate system,* which also provides a means for the necessary removal of the carbon dioxide produced by tissue metabolism.

Carbon dioxide dissolved in an aqueous environment forms carbonic acid (H_2CO_3), which dissociates to produce a H^+ and a bicarbonate ion (HCO_3^-).

$$CO_2 + H_2O \rightleftharpoons H_2CO_3 \rightleftharpoons H^+ + HCO_3^-$$

Carbon dioxide	Carbonic acid	Bicarbonate ion

An acidic condition is normally avoided as a result of an important equilibrium between the H^+ produced and hemoglobin, which removes H^+ from plasma. These equilibria are coupled to one which exists between dissolved CO_2 in the blood and gaseous CO_2 in the lungs.

$$CO_2\ (\text{dissolved}) \rightleftharpoons CO_2\ (\text{gas})$$

This last equilibrium offers an effective mechanism for the removal of CO_2 from the interior environment of the body and also for continuous control of H^+ in the blood. It is estimated that the equivalent of 20-40 liters of 1 *N* carbonic acid is removed by the lungs of an adult every 24 hours.

Thus, the three interrelated equilibria associated with the bicarbonate system can be expressed as follows:

$$\underset{\text{(gas)}}{CO_2} \rightleftharpoons \underset{\text{(dissolved)}}{CO_2} + H_2O \rightleftharpoons H_2CO_3 \rightleftharpoons H^+ + HCO_3^-$$

With respect to the bicarbonate system, a [HCO_3^-] to [H_2CO_3] ratio of 20 to 1 is required for the pH of blood plasma to remain 7.40. The concentration of dissolved CO_2 is included in the [H_2CO_3] value, i.e., [H_2CO_3] = [H_2CO_3] + [CO_2 (dissolved)]. If there is a change in the ratio in favor of H_2CO_3, *acidosis* results. This change can result from a decrease in [HCO_3^-] or from an increase in [H_2CO_3].

Most common forms of acidosis are metabolic or respiratory in origin. *Metabolic* acidosis is caused by a decrease in [HCO_3^-] and occurs, for example, in uncontrolled diabetes with ketosis (excess production of ketone bodies, such as β-hydroxybutyric acid and acetoacetic acid).

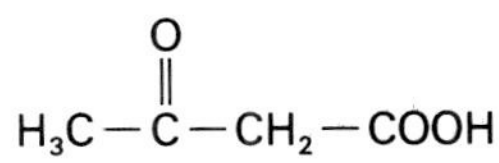

Acetoacetic acid, a ketone body

$$\underset{\beta\text{-Hydroxybutyric acid}}{H_3C-CHOH-CH_2-COOH} + HCO_3^- \rightleftharpoons \underset{\beta\text{-Hydroxybutyrate}}{H_3C-CHOH-CH_2-COO^-} + H_2CO_3$$

Metabolic acidosis can also occur as a result of starvation or high-lipid diets and in certain kidney diseases, poisoning by an acid salt, and cases of vomiting when nonacid fluids are lost. *Respiratory* acidosis is brought about when there is an obstruction to respiration (emphysema, asthma, or pneumonia) or depression of respiration (toxic doses of morphine or other respiratory depressants). If acidosis is not treated promptly, the patient may go into a coma.

Alkalosis results when [HCO_3^-] becomes favored in the bicarbonate/carbonic acid ratio. *Metabolic* alkalosis occurs when the HCO_3^- fraction increases with little or no concomitant change in H_2CO_3. Severe vomiting (loss of H^+ as HCl) or ingestion of excessive amounts of sodium bicarbonate (bicarbonate of soda) can produce this condition. *Respiratory* alkalosis is induced by hyperventilation because an excessive removal of CO_2 from the blood results in a decrease in [H_2CO_3]. Hyperventilation can result from anxiety, hysteria, prolonged hot baths, or the lack of oxygen at high altitudes. Alkalosis can produce convulsive seizures in children and tetany in adults (characterized by sharp flexion of the wrist and ankle joints, muscle twitchings, and cramps).

Intracellular fluids of the body utilize the second ionization of *phosphoric acid* (Table 3-6) as the major buffer system:

$$H_2PO_4^- \rightleftharpoons H^+ + HPO_4^{2-}$$

The pK'_a for dissociation is 7.21, which is close to the pH required by many intracellular environments. Hence, under physiological conditions, $[H_2PO_4^-]$ and $[HPO_4^{2-}]$ are nearly the same and, thus, the buffering capacity of the system is very effective.

SUMMARY

Water, as an essential component of life, has unusual physical properties as a solvent, e.g., high boiling and freezing temperatures, because of the strong intermolecular hydrogen bonding that occurs between water molecules. Because of its polarity and hydrogen-bonding properties, water is an excellent solvent for many ionic and neutral molecules. Amphipathic molecules, e.g., salts of fatty acids, form micelles in aqueous solutions, with the hydrophilic groups of the molecules located on the exterior of the aggregate and the hydrophobic groups in the interior.

There is a slight ionization of water to form hydronium (H_3O^+) and OH^- ions. A hydronium ion results from the transfer of a dissociated H^+ to a water molecule. The concentration of H^+ (H_3O^+) in pure water is 1×10^{-7} *M* which, expressed as the $-\log [H^+]$, is pH 7. The $[H^+]$ of a biological system is critically important since metabolic activities are very sensitive to changes in pH.

Brönsted–Lowry acids and bases are proton donors and acceptors, respectively. A conjugate-base pair refers to a proton donor (HA) and its corresponding proton acceptor (A^-). The dissociation constant K'_a expresses the tendency of an acid, in aqueous solution, to donate protons; the $-\log K'_a$ is called the pK'_a. Weak acids, which display a high affinity for their protons, are of great biological significance since they act as buffers to resist changes in pH and thus help to maintain the appropriate internal pH of an organism. The buffering capacity of a weak acid relies on the relationships (expressed in the Henderson–Hasselbalch equation) between the pH of a solution, the pK'_a of the acid, and the ratio of the concentration of the proton-acceptor and proton-donor species of the acid. The maximum buffering capacity of a weak acid is at the pH that is numerically the same as its pK'_a ($[HA] = [A^-]$). In humans, the main extracellular buffering system is the bicarbonate system (H_2CO_3/HCO_3^- buffer pair), and the principal intracellular system involves the second dissociation of phosphoric acid ($H_2PO_4^-/HPO_4^{2-}$ buffer pair).

QUESTIONS

In the following problems, base the calculations on the concentrations given (assuming an activity coefficient of 1).

1. Calculate the $[H^+]$ of
   (a) blood plasma, pH 7.4
   (b) cow's milk, pH 6.6
   (c) gastric juice, pH 1.5
   (d) tomato juice, pH 4.3
   (e) lemon juice, pH 2.3
   (f) cola drink, pH 2.8
2. Calculate the pH of an aqueous solution of
   (a) 10^{-4} *M* HCl
   (b) 10^{-8} *M* HCl
   (c) 2 m*M* HCl
   (d) 10^{-3} *M* H_2SO_4
   (e) 3 m*M* H_2SO_4
3. Calculate the pH of a solution with a $[H^+]$ of
   (a) 4×10^{-8} *M*
   (b) 3×10^{-3} *M*
   (c) 4×10^{-2} *M*
4. Calculate the $[HPO_4^{2-}]/[H_2PO_4^-]$ ratio in (a) blood plasma, pH 7.4, and (b) muscle cells, pH 6.8.
   In a 0.1 *M* phosphate buffer, what is the concentration (*M*) of $H_2PO_4^-$ and HPO_4^{2-} if the pH is (c) 7.4, (d) 6.8? (e) At which of the two pH values does the 0.1 *M* phosphate buffer have the greater buffering capacity?
5. (a) Calculate the pH of an aqueous solution prepared by combining 50 ml of 0.1 *M* NaOH and 200 ml of 0.1 *M* H_3CCOOH ($pK_a' = 4.76$). (b) Calculate the pH if an additional 25 ml of 0.2 *M* NaOH are added to the solution.
6. (a) If lactic acid ($pK_a' = 3.86$) and its sodium salt are used to prepare a 0.5 *M* buffer, pH 4.0, what are the molar concentrations of the acid and its anion in the buffer? (b) If the same molarity buffer, pH 4.0, is prepared with formic acid ($pK_a' = 3.75$) and its salt, what would their molar concentrations be?

SUGGESTED READINGS

Davenport, H. W., *The ABC of Acid–Base Chemistry,* 6th ed. Chicago: University of Chicago Press, 1974. The contents of the book are aptly described by its title.

Dawson, R. M. C., D. C. Elliott, W. H. Elliott, and K. M. Jones, eds., *Data for Biochemical Research,* 2nd ed. New York: Oxford University Press, 1969. A good reference book containing information useful in biochemical research, this text includes the dissociation constants of many compounds used for buffers.

Dick, D. A. T., *Cell Water.* Washington, D.C.: Butterworths, 1966. A discussion of the properties and functions of water in living systems.

Eisenberg, D. S., and W. Kauzmann, *The Structure and Properties of Water.* New York: Oxford University Press, 1969. A comprehensive monograph about water, for advanced reading.

Fasman, G. D., ed., *CRC Handbook of Biochemistry and Molecular Biology,* 3rd ed. Sec. D, *Physical and Chemical Data,* Vol. 1. Boca Raton, Fla.: CRC Press, 1976. Another good sourcebook for biochemical information pertaining to buffers.

Montgomery, R., and C. A. Swenson, *Quantitative Problems in the Biochemical Sciences,* 2nd ed. San Francisco: W. H. Freeman, 1976. Practice problems concerning acid–base chemistry and buffers.

Segel, I. H., *Biochemical Calculations,* 2nd ed. New York: Wiley, 1976. A good collection of problems concerning acids, bases, and buffers.

SOME BASIC ASPECTS OF THE CHEMISTRY OF LIFE

4

INTRODUCTION

As noted in Chapter 1, the differences between the organic compounds of living systems and the inorganic compounds of the inanimate world were established early in the history of chemistry. *Organic chemistry* was defined as the *chemistry of carbon compounds* because all the molecules which had been isolated from plants and animals were found to contain carbon. Thus, the unique position carbon occupies in the chemistry of life was recognized in the original description of a branch of chemistry which has since expanded to include all carbon compounds, regardless of their origin.

The processes of biological evolution utilized carbon, because of its chemical versatility, as the element for the basic structure of the biomolecules needed for living systems. A fundamental understanding of the chemistry that seeks to explain biological phenomena requires an appreciation of the chemical properties of carbon and of the elements that covalently bond with it. These properties determine the chemical nature of biomolecules. The known reactions and interactions that biomolecules undergo, in turn, furnish the rationale for molecular interpretations of biological processes. Thus, classical concepts of chemistry, especially of carbon chemistry, offer scientific explanations for molecular events in living systems.

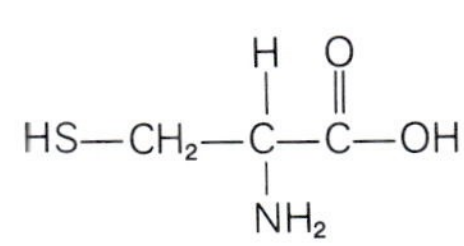

Cysteine (an amino acid). Note the C bonds to C, N, O, H, and S.

BONDING PROPERTIES OF CARBON

Carbon (C) is a small atom (atomic number 6, atomic mass 12) capable of forming stable covalent bonds with a number of ele-

Figure 4-1
Some double and triple bonds formed by carbon.

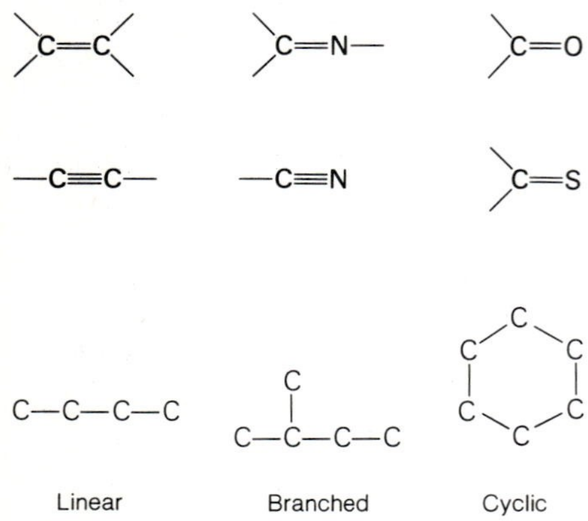

Figure 4-2
Some functional groups found in biomolecules.

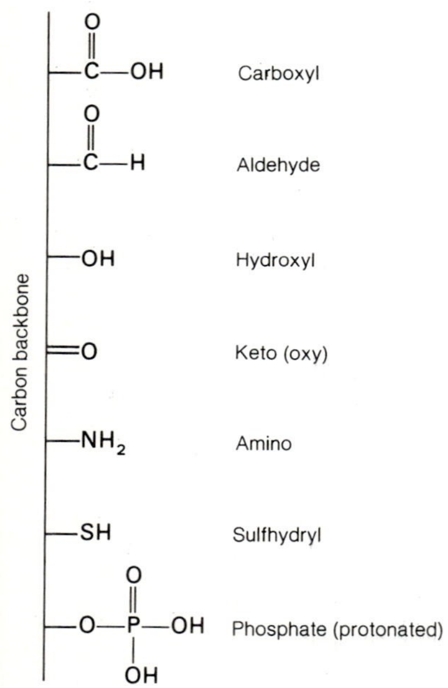

Stereoisomer is taken from the Greek word *stereo*, meaning *solid*, and *isos*, meaning *equal parts* or *sharing equally*. *Enant* is taken from the Greek word *enantios*, meaning *opposite*.

ments. Of major biological importance is the bonding of carbon to other carbon atoms and to nitrogen (N), oxygen (O), hydrogen (H), and sulfur (S). The bondings that can occur among the latter four elements themselves and with phosphorus (P) are also biochemically significant. Carbon, in addition to single-bond formation, forms double bonds with C, N, O, and S, as well as triple bonds with C and N (Figure 4-1). Collectively, these diverse properties of carbon furnish the chemical flexibility to produce a vast variety of different biomolecules.

One distinguishing biochemical feature of carbon is its ability to bond with as many as four other carbon atoms. This property allows different arrays of linked carbon atoms, in linear or branched chain or cyclic arrangements, to serve as backbone structures for biomolecules. The bonding of carbon to N, H, O, and S permits incorporation of functional groups (Figure 4-2) into the structures of biomolecules, which results in even greater chemical variation in biological compounds.

ASYMMETRY OF CARBON COMPOUNDS

Another important characteristic of carbon is its ability to form tetrahedral structures, which introduces the additional parameter of *asymmetry* into biomolecular structures. When a carbon atom is bonded to four different substituent groups (asymmetry), two distinct spatial arrangements of the molecule are possible—one being the mirror image of the other. Figure 4-3 illustrates the *stereoisomerism* of glyceraldehyde, a biomolecule with an asymmetric carbon. The two different mirror-image structures, which are not superimposable, are called *enantiomers.* The *Fischer projection formulas* in Figure 4-3 reflect the difference in asymmetry of the middle carbon of this molecule. A *projection formula* depicts a tetrahedral asymmetric carbon with its four bonds drawn at 90° angles. It is understood that, in the projection, the bond pointing to the top of the page (with CHO) is in the plane of the paper, the bond directly under (CH_2OH) is below the plane, and bonds pointing left and right (H and OH) are above the plane. Another version of the projection is called a *perspective formula* (Figure 4-4).

Many biomolecules possess at least one asymmetric carbon and therefore exist in isomeric forms. This fact is important because most biological reactions require specific three-dimensional structures in the biomolecules they utilize. For example, only one isomer of glyceraldehyde (Figure 4-3) is *biologically active* as an intermediate in alcoholic fermentation.

Because two enantiomers of a compound have identical chemical compositions and many of the same physical properties, e.g., melting points, it is difficult to distinguish one from the other by most analytical methods. One distinctive property, however, is the rotation characteristic that a plane of polarized light exhibits when passed through a solution of an optical isomer. If one isomer rotates plane-polarized light a certain number of degrees to the right (+), as measured in a polarimeter, its mirror-image structure rotates the light the same number of degrees to the left (−). Such *optically active* isomers are referred to as *dextro-* (*d* or +) or *levo-* (*l* or −) rotatory (Latin for right and left, respectively).

Emil Fischer devised a convention whereby a particular asymmetric molecular structure could be designated by placing the prefix D or L before the name of the compound. The convention applies to compounds with the structural form R—CHX—R′, where R—C—R′ is the main chain and X is a functional group. In a proper Fischer projection formula (Figure 4-5), the number 1 carbon of the main chain is written at the top of the formula with the remainder of the chain below it. For the D-isomer, the functional group (X) is written to the right of the asymmetric carbon and H to the left. For the L-isomer the positions of X and H are interchanged. Fischer assigned the D-structure to the dextrorotatory (+ or *d*) isomer of glyceraldehyde and the L-structure to the levorotatory (− or *l*) isomer.

D-(+)-Glyceraldehyde L-(−)-Glyceraldehyde

It is important to realize that this arbitrary assignment of the D-prefix, a specific tetrahedral arrangement of the carbon atom, to the dextrorotatory (+) isomer (the one utilized in alcoholic fermentation) did not establish the *absolute configuration* of the (+)-isomer. Fischer had a 50 per cent probability of being correct. It was not until the 1950s that contemporary physical techniques proved that Fischer's structural assignments, made in the nineteenth century, were fortuitously correct.

The D and L nomenclature is still commonly used in biochemistry (Figure 4-6). Often D and L designations are wrongly interpreted to

Figure 4-3
Stereoisomers of glyceraldehyde.

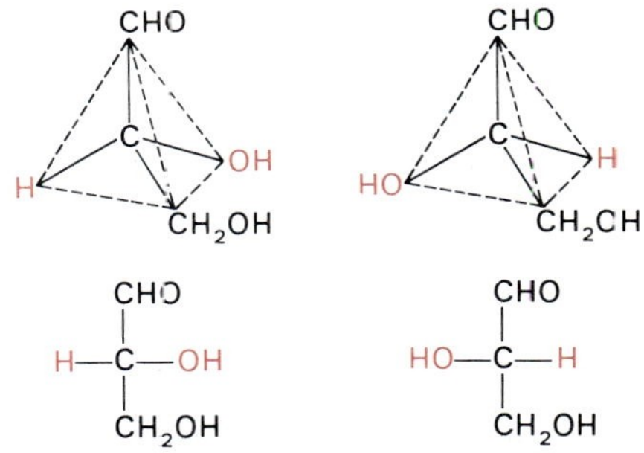

Figure 4-4
Fischer perspective formulas of glyceraldehyde stereoisomers.

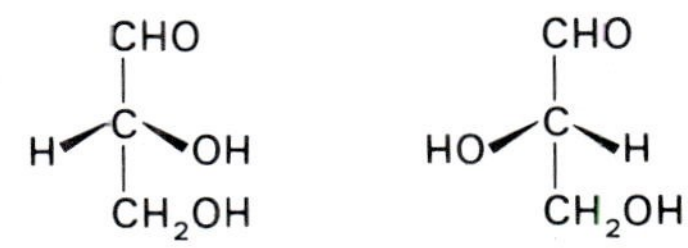

Figure 4-5
Fischer projection formulas of D- and L-stereoisomers.

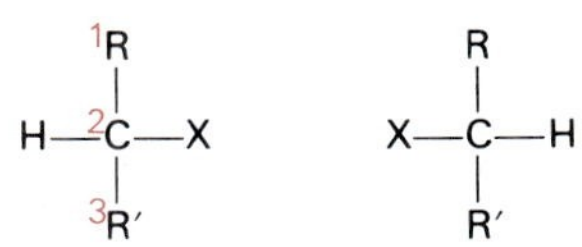

D-Structure L-Structure

Figure 4-6
Fischer projection formulas of L-alanine and D-lactic acid.

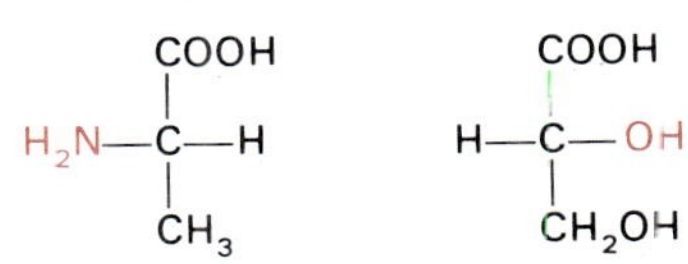

L-Alanine (amino acid) D-Lactic acid

In the R and S classification system, R refers to *rectus* (Latin for *right*) and S to *sinister* (Latin for *left*).

imply that the isomers rotate light to the right (+) or the left (−), respectively. It is the spatial relations of the atoms about the asymmetric carbon in isomers that determine whether they are to be designated D or L. Thus, these symbols do not identify the property of light rotation, i.e., D-isomers can be (−) or (*l*), as well as (+) or (*d*); L-isomers can be (+) or (*d*), as well as (−) or (*l*). To minimize confusion, the symbols *d* and *l* are usually not used today. The nomenclature has limitations because it describes the asymmetry of only one carbon atom in a compound and many biomolecules contain two or more asymmetric carbons. The R and S classification of isomers currently used in chemistry is more useful for defining the asymmetry of biomolecules because it accounts for all asymmetric carbons in an isomer. For example, isoleucine (an amino acid) has two asymmetric carbon atoms (thus four stereoisomers), and the configuration of the biological isomer, L-isoleucine, is more completely described as (2S), (3S)-isoleucine (Figure 4-7).

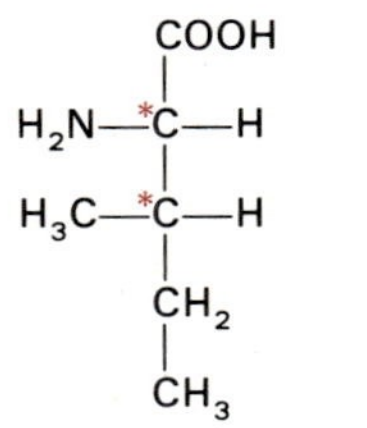

*Asymmetric carbon atom

Figure 4-7
Fischer projection formula of L-isoleucine or (2S), (3S)-isoleucine.

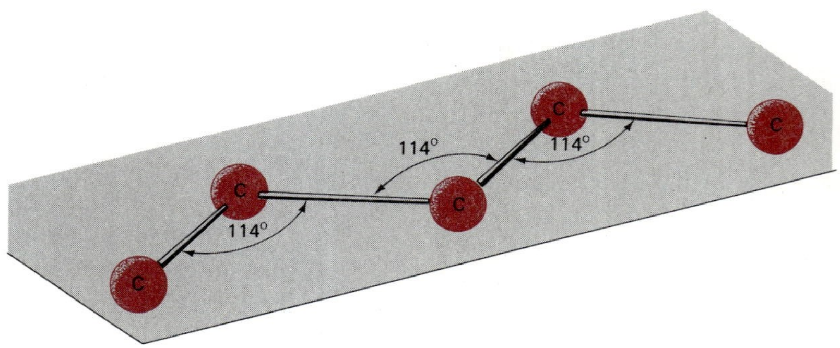

Figure 4-8
Planar zigzag conformation of a linear chain of carbon atoms.

CIS-TRANS ISOMERISM

In molecules consisting of a chain of three or more singly bonded carbon atoms, rotation in space is possible around each bond and many different atomic arrangements or conformations are theoretically possible. Straight-chain carbon compounds, without substituted bulky charged groups, usually assume a *planar zigzag conformation* (Figure 4-8) because of weak forces of interaction in the molecule. In this conformation, carbon atoms are in a plane with C−C bond angles of 114°. Whereas C−C single bonds allow rotation, carbon-to-carbon double bonds (C=C) restrict rotation of the atoms bonded to the two carbons. As a result of this fixed arrangement, molecules with a C=C can display isomerism if two

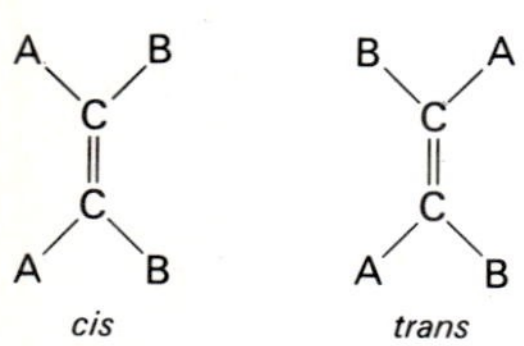

Figure 4-9
Cis and *trans* isomers.

different atoms (substituent groups) are attached to double-bonded carbon atoms; they are called *cis* and *trans* isomers (Figure 4-9). *Cis-trans* isomerism is important because of the isomeric specificity of biological reactions. For example, the biomolecules maleic and fumaric acids are *cis-trans* isomers (Figure 4-10), and each is recognized by biological systems as a different compound.

Figure 4-10
Maleic and fumaric acids.

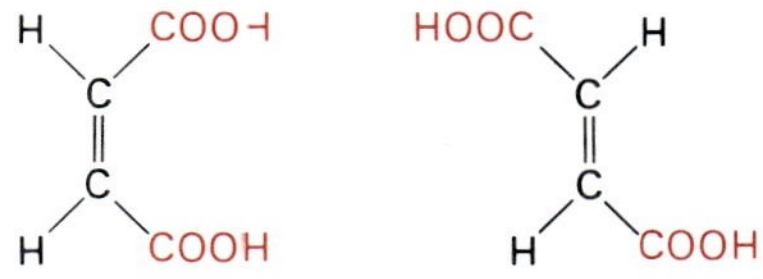

Maleic acid (*cis*) Fumaric acid (*trans*)

NONCOVALENT INTERACTIONS

In addition to covalent bonding, there are weaker forces of interaction that profoundly influence conformation of biomolecules and their function. Van der Waals forces, hydrogen bonds, ionic bonds, and hydrophobic interactions are among the main types of weak, noncovalent forces important in living systems. With respect to bonding, *weak* and *strong* are used to indicate the amount of energy in a bond. Strong bonds, such as covalent bonds found in biomolecules (Figure 4-11), require an average of 100 kcal (kilocalories)/mole, or kcal mol^{-1}, to be cleaved and hence are stable and seldom break under physiological conditions. In contrast, weak bonds, such as hydrogen bonds (Figure 4-12), have energies of 1 to 7 kcal mol^{-1} and are easily broken. Weak bonds are very transitory when they exist singly. However, the overall effect of many weak bonds and other weak forces in a macromolecular structure, such as a protein, greatly influences its native three-dimensional conformation. Noncovalent interactions are also critical to biological function and regulation of biomolecules. Molecular explanations of biological phenomena rely heavily on the understanding of these subtle but significant forces.

Figure 4-11
Energies of some covalent bonds.

Bond	Kcal mol^{-1}
>C—N<	65
>C—O—	82
>C—C<	82
>C—H	95
—O—H	110
>C=C<	145
>C=O	170

VAN DER WAALS FORCES

All types of molecules exhibit van der Waals forces, which arise from the attraction of the bound electrons of one atom for the nucleus of another. When two atoms are far apart, there is a very weak attraction which becomes stronger as the atoms move closer together (Figure 4-13). However, if the atoms move close enough for their outer electron shells to overlap, then a force of repulsion occurs. At a certain distance, defined as the van der Waals radius, there is a balance between the forces of attraction and those of repulsion. Each type of atom has a specific van der Waals radius (Figure 4-14). The van der Waals bonding energy between two

Figure 4-12
Energies of some hydrogen bonds.

Hydrogen bond	Kcal mol^{-1}
—O—H ··· N<	7
—O—H ··· O=	6
—O—H ··· O<	6
>N—H ··· N<	2–4
>N—H ··· O=	2–3

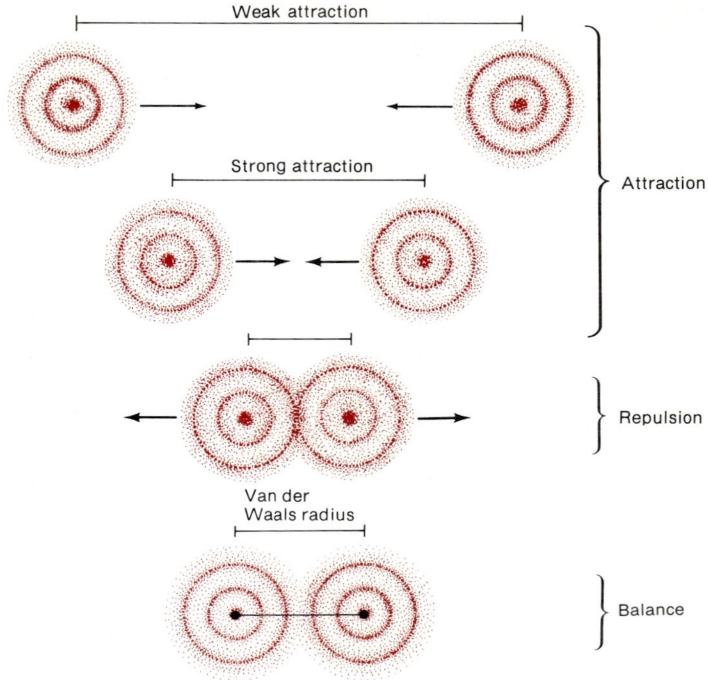

Figure 4-13
Schematic of van der Waals forces of attraction, repulsion, and balance between two atoms.

Figure 4-14
Some van der Waals radii.

Atom or group	Van der Waals radius (angstroms)
H	1.2
CH_3	2.0
N	1.5
O	1.4
S	1.85

average atoms is very small, 1 kcal mol^{-1}, which is only slightly greater than the average thermal energy (0.6 kcal mol^{-1}) of molecules at room temperature.

HYDROGEN BONDS

Hydrogen bonding occurs between a covalently bound hydrogen atom with a partial positive charge and a covalently bound electronegative acceptor atom, such as N, O, or F. Biologically, the most important hydrogen bond structures involve hydrogen atoms covalently bound to oxygen or nitrogen (Figure 4-12). Hydrogen bonds, unlike van der Waals bonds, are highly directional. As discussed on p. 27, water exists in an ordered structure because of the directional orientation imposed on the molecules by hydrogen bonding. In strong hydrogen bonds, the covalent bond of hydrogen points opposite, i.e., 180°, the acceptor atom (Figure 4-12); without this vectorial quality, bond energy is much less. Hydrogen bonds in biomolecules are also more specific than other weak bonds because they require particular complementary groups that donate or accept hydrogen. Hydrogen bonding, both *intramolecular* (within a molecule) and *intermolecular* (between molecules), is common to many macromolecules. The double-helical structure of deoxyribonucleic acid (DNA) is a notable example of a macromolecule with intermolecular hydrogen bonding (p. 232).

IONIC BONDS

Ionic bonds, defined as electrostatic forces acting between oppositely charged groups, frequently occur in and between biomolecules, since positively and negatively charged groups are common. Many functional groups of biomolecules are Brönsted-Lowry acids or bases and are ionized at physiological pH. Molecules with charged groups are usually neutralized by salt formation, i.e., ionic bonding of the biomolecule with an appropriate inorganic cation, Na^+, K^+, or Mg^{2+}, or anion, Cl^- or SO_4^{2-} (Figure 4-15). The average bond energy of ionic bonds in aqueous solution is about 5 kcal mol^{-1}.

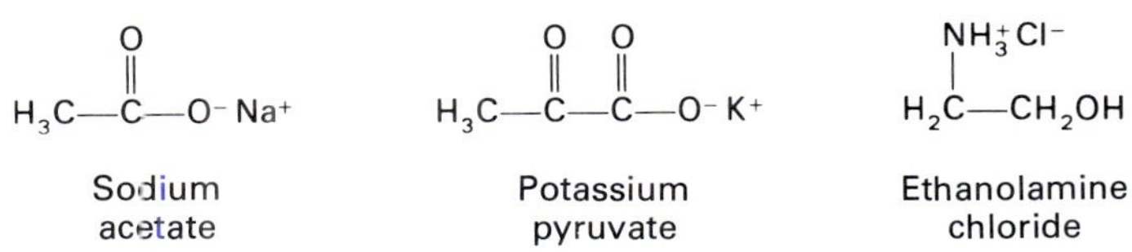

Figure 4-15
Some salts of biomolecules.

Ionic bonding *(electrostatic attraction)* between positively and negatively charged groups in a protein molecule can be influential in determining its shape and function. Biologically important interactions between macromolecules or between many enzymes (proteins) and their substrates are also often brought about by ionic bondings. Groups that bear the same charge (either positive or negative) and that are in close spatial proximity to one another in a macromolecule participate in *electrostatic repulsion.* Such a repulsion affects the conformation of the region of the molecule in which it occurs. Thus, in addition to electrostatic attraction, weak forces of electrostatic repulsion are also involved in determining macromolecular structure.

HYDROPHOBIC INTERACTIONS

The essentiality of hydrophilic (water-loving) properties of biomolecules is obvious. That hydrophobic (water-fearing) characteristics can be valuable may not be as readily apparent. In processes that determine the three-dimensional structure of many macromolecules, especially proteins, the acceptance or rejection by the aqueous environment of the hydrophilic and hydrophobic moieties of large molecules, respectively, exerts a dominant influence on final conformation. As in the formation of micelles (Chapter 3), hydrophobic groups of macromolecules, if in proper spatial relation, will interact (not bond) to the exclusion of solvent molecules (water) and thereby reside in a hydrophobic environment. Hydrophilic groups usually remain exposed to the aqueous environment, where they interact with water molecules. Hydrophobic interactions are considered by many biochemists to be the most influential force in determining the structure of a protein. In the aqueous environment demanded by life, the water-insoluble properties of biomolecules are effectively utilized to the advantage of living systems.

SUMMARY

The element carbon occupies a strategic position in the chemistry of living systems because of its chemical versatility, which includes its capability to bond covalently with a number of other elements and to form single, double, and triple bonds. Carbon, which can link to other carbon atoms, is also ideal for the construction of the backbone structures of a large variety of linear, branched-chain, and cyclic biomolecules.

The asymmetry that a tetrahedral carbon attains when covalently bonded to four different functional groups is of considerable biological importance. In most reactions, which are highly specific, only one stereoisomer of a compound containing one or more asymmetric carbon atoms can be utilized. The same degree of reaction specificity applies to the *cis-trans* isomers that can exist for biomolecules having carbon-to-carbon double bonds.

Among the noncovalent interactions of importance in biological chemistry are van der Waals forces, hydrogen bonding, hydrophobic interactions, electrostatic attraction (ionic bonding), and electrostatic repulsion. These weak forces of interactions are influential determinants of the conformation of biomolecules, e.g., proteins and nucleic acids.

QUESTIONS

1. Which of the following covalent bondings are found in carbon compounds?
   (a) C—C (b) C=C (c) C≡C (d) C≡N (e) C=S (f) C—S (g) C—O (h) C=O (i) C=N (j) C—N
2. For the Fischer projection formulas of the three amino acids shown:

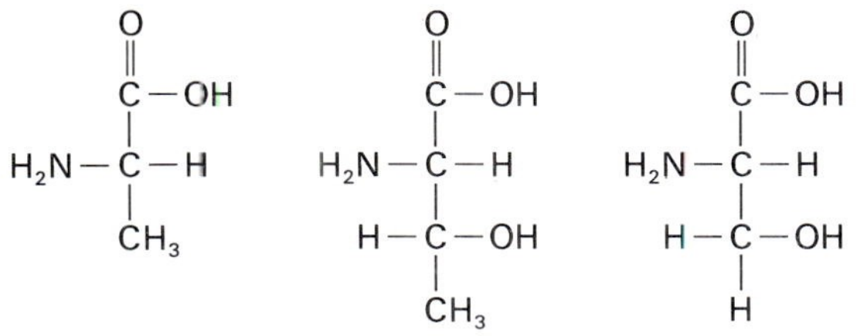

L-Alanine L-Threonine L-Serine

(a) How many asymmetric carbon atoms does each of the biomolecules have?
(b) Draw the Fischer projection formula(s) of the other stereoisomer(s) of each of the amino acids.
(c) Which of the formulas (part b) is the enantiomer of L-threonine?

3. Which of the following carbon compounds have *cis-trans* isomers?
(a) $H_2C{=}CH_2$
(b) $Cl{-}CH{=}CH_2$
(c) $H_2C{=}CH{-}CH_3$
(d) $H_3C{-}CH{=}CH{-}CH_3$
(e) $Cl{-}CH{=}CH{-}CH_3$
(f) $HOOC{-}CH{=}CH{-}COOH$
(g) $(CH_3)_2C{=}CH_2$
(h) $H_3C{-}C{\equiv}C{-}CH_3$
(i) $H_3C{-}CH{=}CH{-}CH{=}CH_2$
(j) $H_2C{=}CH{-}CH{=}CH_2$

4. What type of noncovalent interaction can occur between each of the following pairs of groups found in biomolecules?

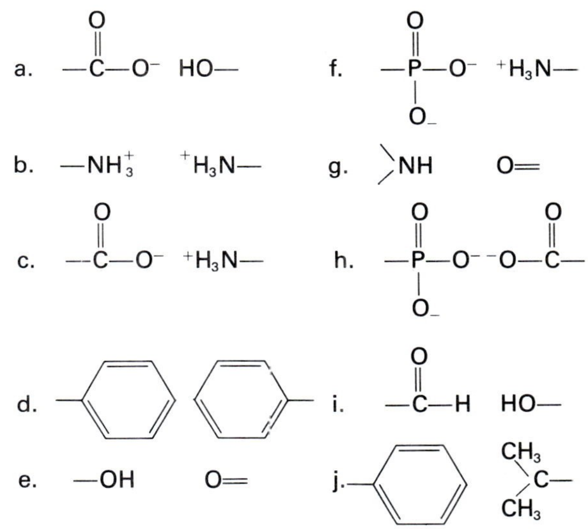

SUGGESTED READINGS

Barry, J. M., and E. M. Barry, *An Introduction to the Structure of Biological Molecules.* Englewood Cliffs, N.J.: Prentice-Hall, 1969. A review of the basic organic chemistry of biomolecules.

Brand, J. C. D., J. C. Speakman, and J. K. Tyler, *Molecular Structure: The Physical Approach,* 2nd ed. New York: Halstead Press, 1975. An advanced undergraduate text that discusses the application of physical methods to analyze molecular structures.

Fasman, G. D., ed., *Handbook of Biochemistry and Molecular Biology,* 3rd ed. Sec. D, *Physical and Chemical Data,* Vol. II. Boca Raton, Fla.: CRC Press, 1976. Good reference source for the van der Waals radii of common elements.

Morrison, R. T., and R. N. Boyd, *Organic Chemistry,* 3rd ed. Boston: Allyn and Bacon, 1973. Contains a well-written chapter on stereochemistry, including a discussion of enantionerism and the tetrahedral carbon and the R and S nomenclature.

Pauling, L., *The Nature of the Chemical Bond,* 3rd ed. Ithaca, N.Y.: Cornell University Press, 1960. A chemistry classic containing a discussion of hydrogen bonding in Chapter 12.

Price, C. C., *Geometry of Molecules.* New York: McGraw-Hill, 1971. A concisely written introduction to elementary and organic chemistry, leading to molecular biology.

Articles

Crick, F. H. C., "Nucleic Acids." *Scientific American,* 197(3):188 (offprint #54), 1957.

Doty, P., "Proteins." *Scientific American,* 197(3):173 (offprint #7), 1957.

5

CELLS: BIOLOGICAL UNITS OF MOLECULAR ORGANIZATION

INTRODUCTION

1 g (gram)	=	0.03527 oz (ounce)
1 cm (centimeter)	=	0.3937 in (inch)
1 cm	=	10^{-2} m (meter)
1 mm (millimeter)	=	10^{-3} m
1 μm (micrometer)	=	10^{-6} m
1 nm (nanometer)	=	10^{-9} m
1 Å (angstrom)	=	10^{-8} cm, or 10^{-10} m

Most biological phenomena of concern to biochemists occur within small, living cells. In the bacterium *Escherichia coli,* for example, all life processes transpire in a cell that weighs 2 × 10^{-12} g and is 2 × 10^{-4} cm in length. The organization of the cellular environment that allows the orderly and continued expression of life is currently studied by biochemists; a past concern by some that biochemistry was becoming "too *in vitro*" for a biological science is no longer valid. In addition to understanding the *chemical* structure and function of biomolecules, it is equally important to comprehend the *organizational* structure and function of the membrane-limited aqueous environments called cells; and attempts to do the latter are now more common than in previous decades. *Where* biochemical processes take place in a cell and *how* these systems function in a coordinated manner are vital aspects of life that cannot be ignored in a meaningful study of biochemistry.

Morph is taken from the Greek word *morphi,* meaning *form.*

Biochemical research in *cell biology,* the study of the morphological and functional organization of cells, has established a special alliance between the two disciplines—analogous to that which arose between biochemistry and genetics in the 1940s. Because of the power of magnification (100,000× or greater) afforded by the electron microscope, cytologists now study detailed structural features of subcellular bodies. Companion studies, using scanning electron microscopy, provide equally pertinent information on cell surfaces and tissue structure. The biochemical approaches complement such studies by determining the molecular composition, structure, and function of cellular components. Collectively, these research endeavors are often referred to as *biochemical cytology.* As molecular studies of life broaden, previously well-defined areas

Cytology is taken from the Greek words *kytos,* meaning *hollow vessel* (cell), and *logos,* a *science* or *branch of knowledge.*

of research continue to blend with one another to produce new, productive scientific relationships.

TWO BASIC TYPES OF CELLS

Pro- and *eucaryote* are taken from the Greek words *karyon,* meaning a *kernel* or *nut* (nucleus), and the prefixes *pro* and *eu,* meaning *before* and *good,* respectively.

Nucleus is Latin for a *little nut* or *kernel. Chromosome* is taken from the Greek words *chroma,* meaning *color,* and *soma,* a *body* (as recognized by staining procedures).

There are two basic classifications of cells: *procaryotic* and *eucaryotic.* Traditionally, the distinguishing feature between the two types is that a eucaryotic cell possesses a membrane-enclosed nucleus and a procaryotic cell does not. This characteristic, however, is only one example of the complexity of cellular organization of eucaryotes that procaryotes do not share.

Procaryotic cells are usually small and relatively simple by cytological standards, and they are considered representative of the first types of cell to arise in biological evolution. Procaryotes include eubacteria, cyanobacteria (blue-green algae), rickettsiae, spirochetes, and mycoplasma organisms. Eucaryotic cells, on the other hand, are generally larger and more complex, reflecting an advanced evolution. Eucaryotes include unicellular protozoa, fungi, most algae, and multicellular plants and animals.

Plasm is taken from the Greek word meaning *form* or *thing molded. Mitochondrion* is taken from the Greek words *mitos,* meaning a *thread,* and *chondros,* a *grain* or *something granular* (first described as a thread-like granule).

Chloroplast is taken from the Greek words *chloros,* meaning *light green,* and *plastos,* an *organized particle* or *granule.*

In eucaryotic cells, different biological functions are segregated in discrete regions within the cell, often in membrane-limited structures. Subcellular structures which have distinct organizational features and functions are called *organelles.* As an organelle, for example, the *nucleus* contains chromosomal DNA and the enzymatic machinery for its expression and replication and the *nuclear envelope* (membrane) separates it from the rest of the cell, which is called *cytoplasm.* There are organelles within the cytoplasm, e.g., *mitochondria,* sites of respiration, and (in some cells) *chloroplasts,* sites of photosynthesis. In contrast, procaryotic cells have only a single cellular membrane and thus no membranous organelles. The lack of many of the distinctive organized structures of eucaryotes does not imply that the internal environment of procaryotes is a homogeneous mixture of biochemical activities. Procaryotic DNA, for example, is contained within a rather autonomous nuclear region, and many enzyme systems are present in ordered arrangements, some bound to the cellular membrane.

One molecular difference between the two types of cells is apparent in their genetic material. Procaryotes have a single chromosome (possibly present in more than one copy), which is simply double-stranded DNA in its composition. Eucaryotes, in contrast, possess more than one chromosome, each consisting of both DNA and protein, e.g., *histones,* in a highly organized structure.

DIVERSITY AMONG CELLS

Although the two basic cell types account for the cellular composition of all living systems, cytological examination readily reveals great variation among different eucaryotic or procaryotic cells. These differences provide an example of the diversity stressed in the *unity and diversity concept* of biology. The classification of cells into many categories illustrates the variety (diversity) of life forms, whereas molecular studies emphasize the sameness (unity) of fundamental life processes. Cellular organization is a distinguishing characteristic of individual organisms and can vary within an organism, i.e., cells of an organism or tissue specialized for different tasks. For this reason, the following presentation of the cellular features of bacterial, animal, and plant cells is only a representative description of the cell biology of these types of organisms and does not include the many variations encountered among and within these organisms.

PROCARYOTIC CELL

Figure 5-1 is an electron micrograph of the bacterium *E. coli,* an aerobic coliform found in the human intestinal tract (p. 598). Within the past 35 years, research on this procaryote and its viruses (bacteriophages) has furnished most of the information on the molecular aspects of genetics. More is known about the

Figure 5-1
Escherichia coli. This classic electron micrograph shows ribosomes (R) and unorganized hereditary material (N) in the nuclear region. The cell wall (CW) is also shown. (Courtesy of Dr. Samuel Conti)

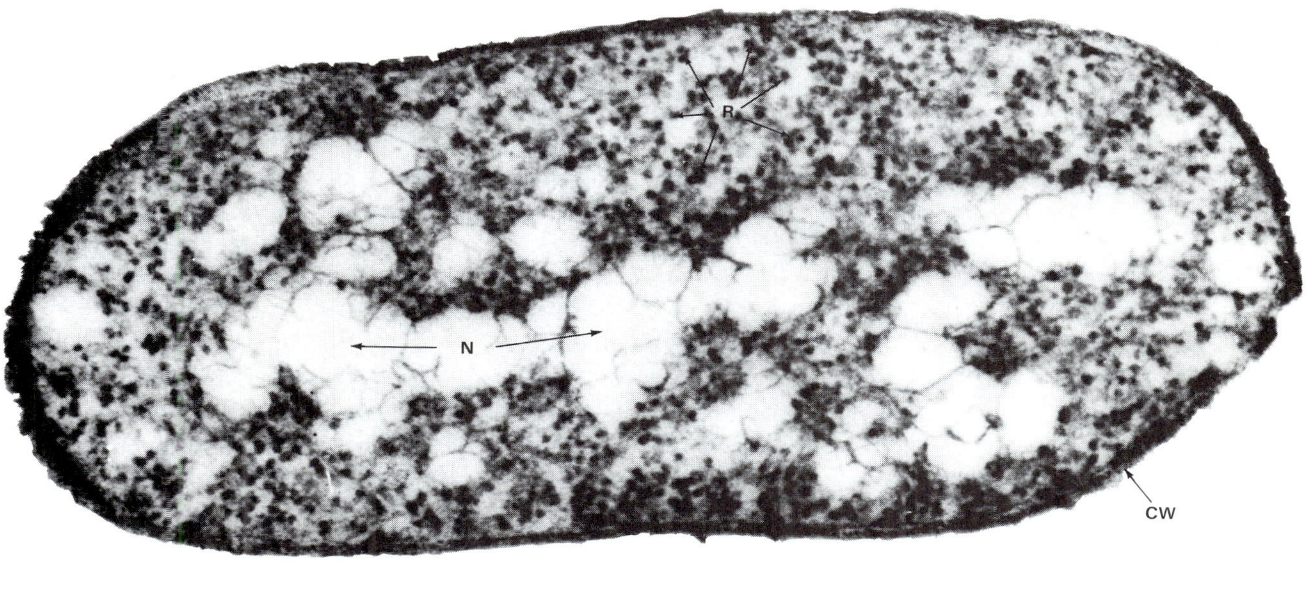

genetics and biochemistry of *E. coli* than about any other organism.

E. coli has a rigid, porous *cell wall* that furnishes form and protection. The outer surface is composed of *lipopolysaccharides* (lipid-carbohydrate molecules). The inner framework is composed of linked arrangements of *polysaccharides* (large molecules composed of carbohydrates) and small peptides (molecules containing amino acids). Some bacteria have *flagella* (for locomotion) and/or *fimbriae* (pili), both of which are protein in composition.

Flagellum is Latin for a *whip*.
Fimbria is Latin for a *border of fibers* or *threads*.
Pilus is Latin for a *hair*.

Closely associated with the cell wall is the *cell membrane*, composed of about equal percentages of lipid and protein. Lipids make up the continuous *nonpolar bilayer phase* that is the water-insoluble boundary of the cell. Nutrients, ions, and water needed by the cell must cross the cell membrane, which exhibits *selective permeability* for many ions and molecules. It is through the membrane that molecules and ions are also excreted. The membrane is therefore a dynamic structure that performs a variety of functions.

Within a cell's interior, genetic material (tightly coiled double-stranded DNA) is located in a *nuclear region*. Genetic information contained in the DNA dictates the biological processes of the bacterium. For cell division, DNA replicates to produce two daughter double-helical molecules.

An *E. coli* cell possesses approximately 15,000 ribosomes—membraneless organelles where protein synthesis occurs. A ribosome is about 20 nm in diameter and can dissociate into a large (50 S) and a small (30 S) subunit, both of which contain about 65 per cent RNA (ribonucleic acid) and 35 per cent protein.

A Svedburg unit of sedimentation (S) is calculated from the sedimentation rate of an organelle or macromolecule in a given centrifugal force. The S value of an organelle or macromolecule is related to its molecular weight and shape.

Many bacteria, like *E. coli*, contain *storage granules*. These granules consist of polymers of biomolecules such as sugars or β-hydroxybutyric acid (Figure 5-2), which serve as reserve fuel supplies for the organism.

Figure 5-2
β-Hydroxybutyric acid.

The soluble portion of a cell's interior is called the *cytosol*, which is very viscous because of its high protein content. Cytosolic proteins are mainly enzymes needed for metabolic reactions. Various inorganic salts and many metabolic biomolecules are also present in cytosol.

The relatively low degree of cellular organization characteristic of a procaryote does not mean that its metabolic activities are simpler than those of eucaryotes. Both types of cells are comparable with regard to the biochemical complexities associated with the basic processes of life, e.g., biosynthetic pathways, production of chemical energy, and DNA replication.

ANIMAL CELL

An electron micrograph of a cross section of a liver cell, a *hepatocyte,* is pictured in Figure 5-3. Because numerous metabolic functions occur in the liver, hepatocytes have been thoroughly studied for many years. The cellular organization of a eucaryotic cell of animal origin is well illustrated by a hepatocyte.

Hepato is taken from the Greek word *hepar,* meaning the *liver.*

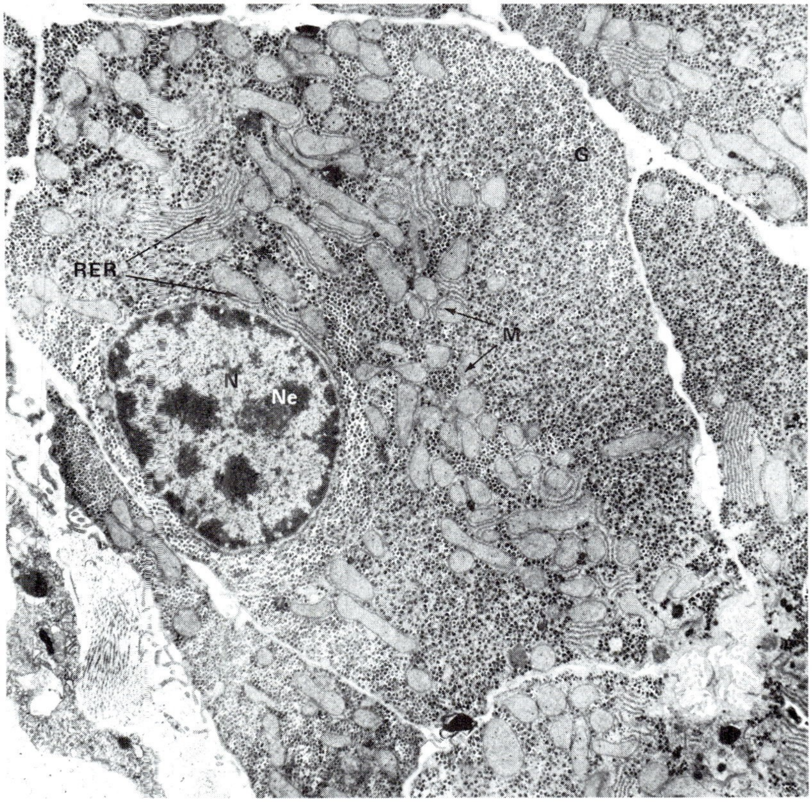

Figure 5-3
Liver cell. This electron micrograph of a "typical" animal cell shows mitochondria (M), nucleus (N), rough endoplasmic reticulum (RER), nucleolus (Ne), and Golgi apparatus (G).

Cell Membrane

The exterior of the cell membrane is referred to as the *cell coat* and is composed of *mucopolysaccharides* (*muco* indicates a protective cellular secretion), *glycolipids* (molecules of carbohydrates and lipids), and *glycoproteins* (molecules of carbohydrates and protein). The sticky nature of cell coats is important for cell-cell recognition and interactions for tissue formation. The *plasma membrane* under the cell coat is about 10 nm thick and is similar to a bacterial membrane in construction, composition, and many of its functions. It is composed of about equal percentages of protein and lipid, with the latter in a bilayer arrangement (p. 215). The membrane is responsible for the selective inflow and outflow of molecules, ions, and water and therefore possesses a number of active-transport systems, e.g., for Na^+ and K^+. Many enzymatic processes are associated with the membrane, as is also true for bacterial cell membranes.

Nucleus

The nucleus (4 to 6 μm in diameter) is the chromosome-bearing organelle of the cell. With the exception of certain specialized cells, such as mature mammalian red blood cells and platelets, the nucleus is present and is an essential component of eucaryotic cells. It is enclosed by a thin *nuclear envelope* (double-membraned) which has openings called *pores* that allow for exchange of substances between nucleus and cytoplasm. Chromosomes, fibers consisting of complex structures of DNA and protein, are in the *nucleoplasm.* An organism's chromosomes preserve the detailed set of genetic information required for the propagation of its species. During the life of an organism, chromosomes continuously furnish genetic instructions for maintenance of the organism's biological existence. Transcription of the genetic code from DNA into RNA, the first cellular transmission of biological information for protein synthesis, occurs in the nucleus (p. 556). The *nucleolus* is associated with a specialized region of a particular chromosome and, in diploid organisms, two are usually observed per nucleus. Nucleoli are characteristically rich in RNA because they are sites where ribosomal RNAs are synthesized. In the early stages of development, when ribosomal RNA synthesis is a dominant cellular activity, the cells of certain organisms possess a great number of nucleoli per nucleus.

Mitochondria

Mitochondria are the energy-producing factories of the cell and constitute about 20 per cent of a hepatocyte's cytoplasmic volume. A liver cell contains about 800 mitochondria, each about 1 μm in diameter. Mitochondria are very dynamic, plastic organelles that continuously move and change shape. A mitochondrion (Figure 5-4) is a *double-membraned organelle,* having *outer* and *inner membranes* which differ in composition and function. The environ-

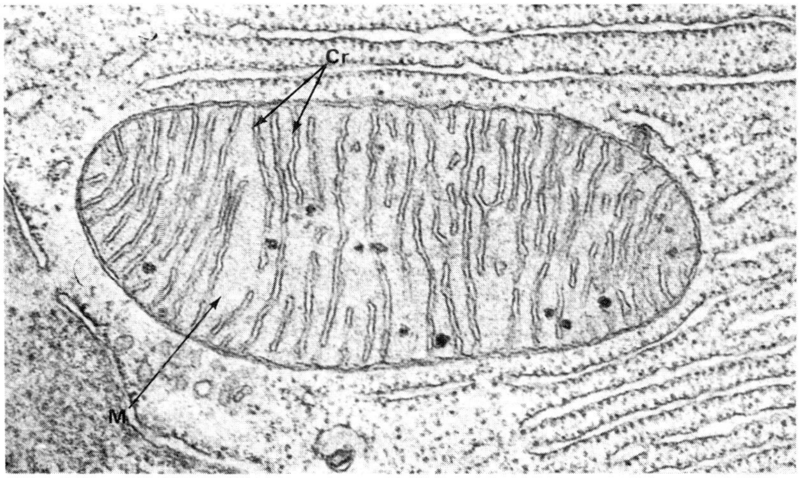

Figure 5-4
Electron micrograph of a mitochondrion. Cristae (Cr) and matrix (M) are identified. (Courtesy of K. R. Porter)

ment within the inner membrane is called the *matrix,* and projections of the inner membrane into the matrix are called *cristae.* The matrix is rich in enzymes. Oxidation of derivatives of carbohydrates, lipids, and amino acids to CO_2 and H_2O *(aerobic respiration)* takes place in mitochondria (p. 316). Some of the chemical energy released by oxidation is converted into *adenosine triphosphate* (ATP) which, in turn, provides the cell with its energy needs. This production of ATP is associated with the inner membrane. Much of the chemical energy is also released as heat. Mitochondria possess DNA *(mitochondrial DNA),* which is required for the synthesis of various macromolecular components, e.g., some inner membrane proteins and the ribonucleic acids (RNAs) needed for mitochondrial protein synthesis.

Matrix is Latin for the *womb* or *uterus,* often used to mean a place where something originates or develops.
Cristae is taken from the Latin word *crista,* meaning a *crest* or *tuft.*

Rough and Smooth Endoplasmic Reticula

The cytoplasm of most eucaryotic cells contains a specialized system of flattened, interconnected single-membraned vesicles and tubules that channel throughout the cytoplasm. This system of *cisternae* and tubules, called the *endoplasmic reticulum* (ER), transports proteins, hormones, and other compounds produced by the cell. Some portions of the endoplasmic reticulum have ribosomes attached to the cytoplasmic outer surface. Such membranes have a granular appearance in electron micrographs (Figure 5-5) and hence are called *rough-surfaced endoplasmic reticulum* (RER). The attached ribosomes are sites for synthesis of proteins, many of which are sequestered in the cisternae for transport to the *Golgi*

Cisternae is taken from the Greek word *kiste,* meaning a *box* or *chest.*
Endo is a Greek prefix for *within* or *inner.*
Reticulum is Latin for *net.*

A
B

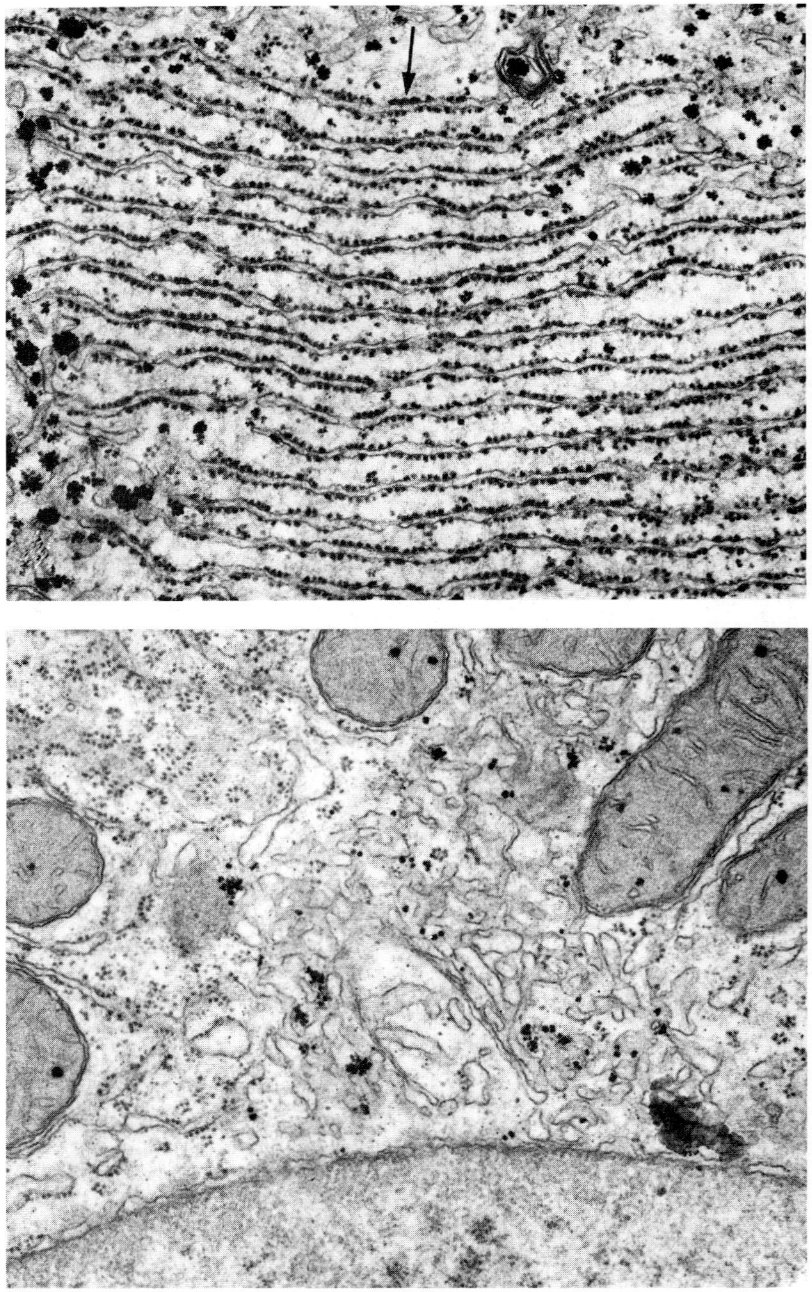

Figure 5-5
(A) Electron micrograph of rough endoplasmic reticulum (RER) with ribosomes (arrow) attached to membrane. (B) Smooth endoplasmic reticulum (SER).

apparatus for export to various cellular locations or from the cell (p. 526). The portions of endoplasmic reticulum devoid of ribosomes are called *smooth-surfaced endoplasmic reticulum* (SER) and are involved in steroid biosynthesis and detoxification mechanisms. In research, the endoplasmic reticulum, when isolated from cells, is referred to as the *microsomal fraction* or *microsomes.*

Golgi Apparatus

The Golgi apparatus (Figure 5-6) consists of flattened, single-membraned vesicles *(smooth cisternae)* which are often stacked. The variety of small vesicles, or vacuoles, associated with the Golgi apparatus is the result of a budding or pinching-off process. The functions of the Golgi apparatus are numerous and all have not been completely defined. In cells that secrete proteins, e.g., pancreatic cells, Golgi vacuoles serve to segregate, concentrate, and store newly synthesized proteins for subsequent secretion. The Golgi apparatus also seems to be involved in the formation of lysosomal and plasma membranes.

The Golgi apparatus is named for Camillo Golgi (Nobel Prize, 1906), who first detected the vesicles in 1898 by staining brain cells of a barn owl with silver salts.

Vacuole is taken from the Latin word *vacuus,* meaning *empty.*

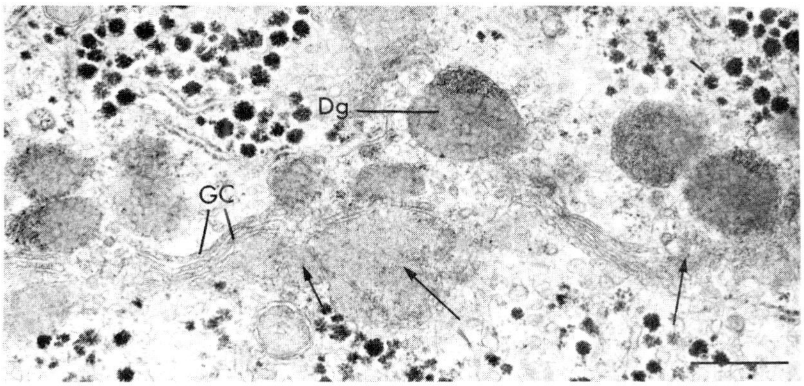

Figure 5-6
Golgi apparatus. This electron micrograph shows the Golgi cisternae (GC) and product being formed (arrows), the so-called dense granules (Dg).

Lysosomes

Lysosomes are small, single-membraned vesicles that are 0.25 to 0.5 μm in diameter. They contain hydrolytic (degradation) enzymes, e.g., ribonuclease, which degrades RNA. The lysosomal membrane is very sensitive to stress; rupture leads to the release of enzymes that degrade cellular components. After the death of a cell, lysosomal enzymes digest cellular contents. In normal intracellular processes, lysosomes digest cellular components needing replacement (p. 522) and foreign matter brought into the cell by

Lyso is taken from the Greek word *lysis,* meaning *loosening* or *dissolution.*

Phagocytosis is taken from the Greek word *phagein,* meaning to *eat* (ingestion by a cell).
Pinocytosis is taken from the Greek word *pinein,* meaning to *drink* (imbibing by a cell).

phagocytosis and *pinocytosis.* Because various subcellular bodies have hydrolytic enzymes and are very heterogeneous in morphology, the identification of lysosomes cannot be made on the basis of the morphology seen in electron micrographs alone, but requires their isolation and subsequent biochemical analysis.

Microbodies or Peroxisomes

Microbodies comprise a group of organelles about the size of lysosomes. They are single-membraned structures about 0.5 μm in diameter. In electron micrographs, the matrix of a microbody often has a dense center as a result of crystalline arrays of enzymes contained by the organelles. These enzymes, e.g., urate oxidase and catalase, catalyze reactions involving hydrogen peroxide (H_2O_2). For this reason, microbodies have been renamed *peroxisomes.* The enzymatic makeup of peroxisomes varies in different cell types. *Catalase* is perhaps the only enzyme found in all peroxisomes. The physiological functions of the variety of known peroxisomes are still not clear. However, catalase activity is critical for the disposal of intracellular hydrogen peroxide (which is toxic) generated by peroxisomal enzymes and other metabolic processes. The peroxisomes are thought to have a role in energy and oxidative metabolism and in other specific metabolic pathways.

PLANT CELL

Figure 5-7 is an electron micrograph of a photosynthetic leaf cell of a higher plant. To synthesize glucose from fixed CO_2 and H_2O, such a parenchymal leaf cell derives its chemical energy from ATP, which is generated by photosynthesis. Some of the glucose is stored as the polymer starch. In the dark, when ATP cannot be produced by photosynthesis, cellular energy is obtained by the oxidation of glucose and other compounds and requires the uptake of molecular oxygen.

The plant cell illustrated contains many of the structures and organelles observed in animal cells, e.g., cell membrane, nucleus, and mitochondria. These cellular components perform the same basic functions in both cell types. There are, however, certain structures of a plant cell that are not, or are rarely, found in animal cells.

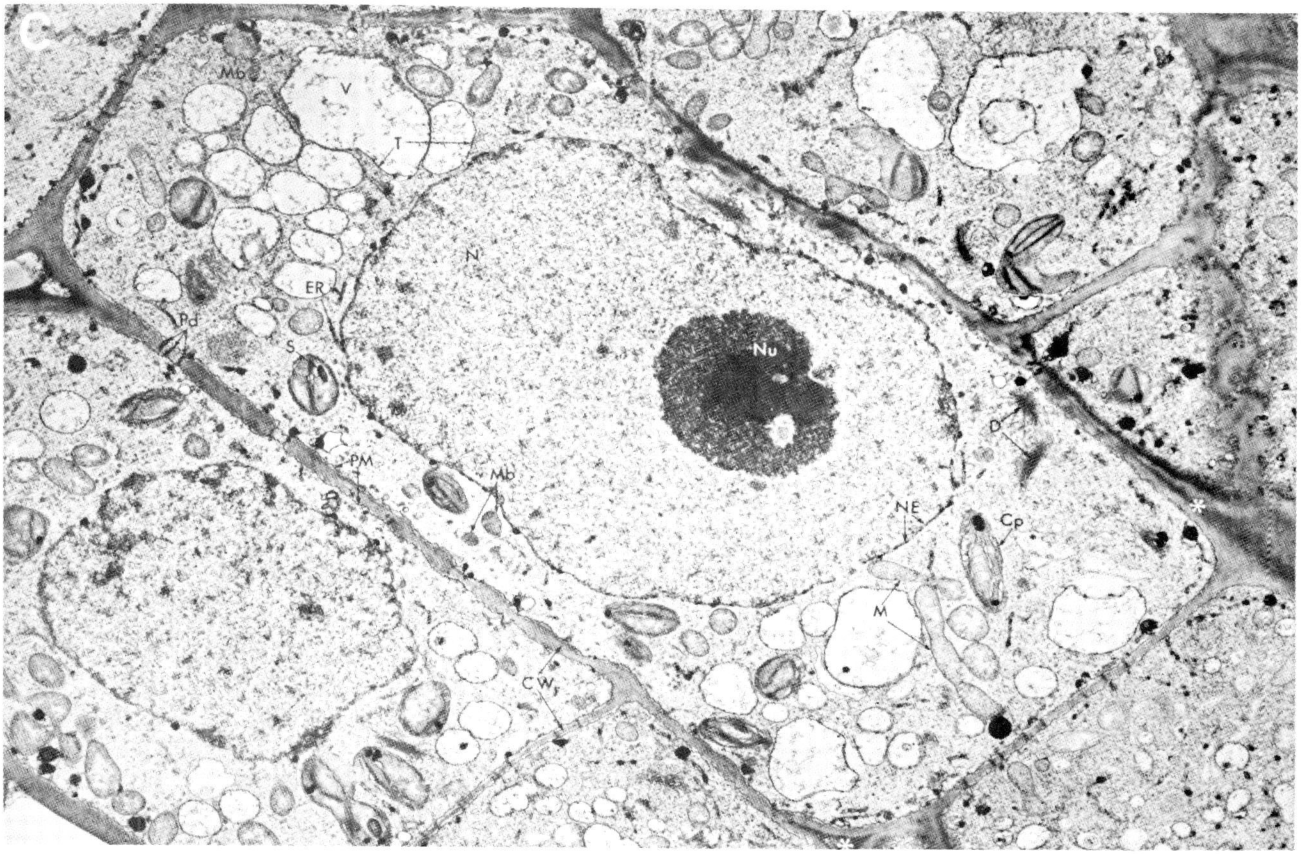

Figure 5-7
Photosynthetic plant cells. This electron micrograph shows the separation of several plant cells by their respective cell walls (CW). Some cytological features include those common to animal cells: plasma membrane (PM), Golgi apparatus (D), mitochondria (M), endoplasmic reticulum (ER), nucleus (N), nucleolus (Nu), nuclear envelope (NE), microbodies (Mb), plasmodesma (Pd), and tonoplast (T). Unique features include chloroplasts (Cp), vacuoles (V), and starch granules (S).

Cell Wall

Plant cells, like bacteria, possess cell walls. The plant cell wall is thicker (≈ 20 nm) and consists of *cellulose* (a polymer of glucose molecules) made rigid by interactions with proteins and other polysaccharides. Cell walls provide protection, strength, and shape to plant tissue.

Vacuoles

Single-membraned vacuoles are conspicuous in the cytoplasm of plant cells. Immature or rapidly dividing plant cells have relatively small vacuoles, often several per cell. Once a cell matures, vacuoles become enlarged by coalescing. Vacuoles appear to be reservoirs for essential metabolites and also for waste products that accumulate during the life of a cell. Among the compounds dissolved in vacuoles are flower, leaf, and fruit pigments; organic acids; sugars; proteins; mineral salts; oxygen; and carbon dioxide. In the animal world, many protozoa contain vacuoles which serve as centers for digestion of foreign substances taken in by the organisms.

Figure 5-8
Glyoxysomic oxidation of glycolic acid.

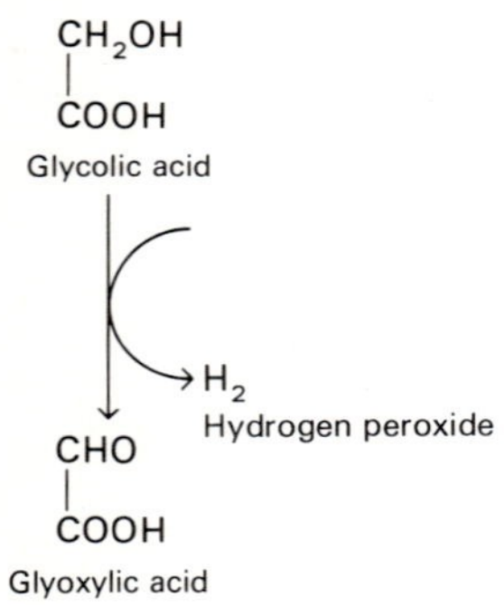

Glyoxysomes

There is a particular type of peroxisome in higher plants and microorganisms, called a *glyoxysome.* In many types of germinating seedlings, the glyoxylate cycle of glyoxysomes (p. 307) performs a critical role in the conversion of seed oil into glucose, which the plant needs for growth. One enzyme of a glyoxysome also oxidizes glycolic acid to produce glyoxylic acid (Figure 5-8). In the reaction, molecular oxygen is converted into hydrogen peroxide, which is then destroyed by peroxisomal catalase. The reaction explains how plants, previously believed to take up oxygen only in the dark, do so during active periods of photosynthesis. As a major by-product of photosynthesis, glycolic acid is readily available for oxidation.

Chloroplasts

Chloroplasts (Figure 5-9) are double-membraned organelles found in green plants and in most other photosynthesizing organisms, e.g., various types of algae. The size, shape, and number of chloroplasts per cell vary considerably between plant species.

The interior of a chloroplast is highly organized. Within the enclosure of the two membranes of a chloroplast, the *stroma* (matrix) contains additional membranes. Chlorophyll, the green pigment of plant cells, is often concentrated in membrane-enclosed structures called *lamellae,* which form *thylakoids,* which appear as flattened sacs. The thylakoids are arranged in stacks called *grana,* which are connected one to another by stroma membranes. Stroma may also contain granules of starch. Like mitochondria, chloroplasts possess DNA *(chloroplast DNA),* which pro-

Lamellae is taken from the Latin word *lamella,* meaning a *thin plate* or *layer.* *Thylakoid* is taken from the Greek words *thylakos,* meaning a *sack* or *pouch,* and *oids,* meaning *in the form.* *Grana* is taken from the Latin word *granum,* meaning *seed* or *grain.*

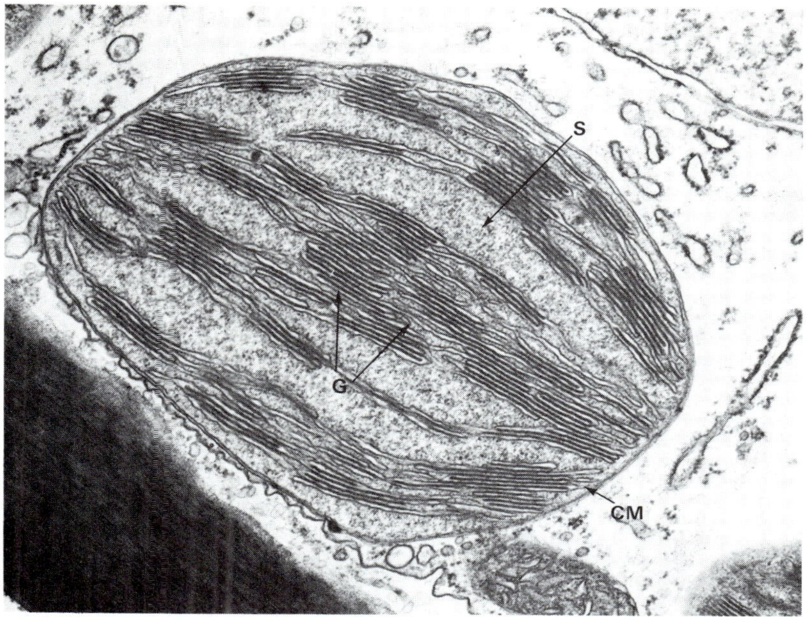

Figure 5-9
Chloroplast of a higher plant cell, shown in cross section, contains grana (G) and matrix, or so-called stroma (S), with a chloroplast membrane (CM) surrounding the whole. The grana consist of stacks of internal membrane systems in the form of flattened sacs called *lamellae* or *thylakoids.* Grana are the site of photosynthetic phosphorylation. (Courtesy of Dr. M. C. Ledbetter)

vides genetic information for some of the biochemical processes of these organelles.

The chloroplasts are sites of the complex processes of photosynthesis which lead to the capture of solar energy and its transduction and eventual storage in various chemical forms. As will be discussed in Chapter 18, there are two basic processes associated with photosynthesis. One depends on light and results in the production of ATP; the other is not light dependent but uses ATP to synthesize carbohydrates from CO_2. The transformation of light energy into chemical energy is the unique characteristic of photosynthesizing organisms, on which all other living systems rely for their existence.

SUMMARY

The two basic classes of cells are called procaryotic and eucaryotic. The characteristic distinguishing the two cell types is the complexity of their cellular organization, with a procaryotic cell having relatively simple cytological features. The metabolic activities of both types of cells, however, are biochemically complex. Bacteria and cyanobacteria are examples of procaryotes; higher animals and plants are representative of eucaryotes. The great variety of life forms is well illustrated by the many cytological differences observed within each of the two basic cell types.

Although both procaryotic and eucaryotic cells have a cellular membrane, the latter also possess intracellular membrane-enclosed structures. These subcellular structures, called organelles, have distinct organizational features and functions. Some of the eucaryotic membrane-limited organelles are the nucleus (containing a cell's genome), mitochondria (the sites of energy production), endoplasmic reticulum (the site of protein synthesis and transport), lysosomes (containing degradative enzymes), and chloroplasts of photosynthetic organisms. The one organelle common to procaryotes and eucaryotes is the membraneless ribosome, the site of protein synthesis. In eucaryotes, most ribosomes are associated with the endoplasmic reticulum.

QUESTIONS

1. What is the difference between cytoplasm and cytosol?
2. Which of the following are single-membraned organelles?
   (a) Chloroplast
   (b) Endoplasmic reticulum
   (c) Golgi apparatus
   (d) Lysosome
   (e) Mitochondrion
   (f) Nucleus
   (g) Peroxisome
   (h) Vacuole

Which of these cellular structures:
   (a) Cell membrane
   (b) Cell wall
   (c) Chloroplast
   (d) Golgi apparatus
   (e) Mitochondrion
   (f) Glyoxysome
   (g) Nucleus
   (h) Peroxisome
   (i) Ribosome
   (j) Rough endoplasmic reticulum
   (k) Smooth endoplasmic reticulum
   (l) Vacuole

3. are characteristic of a photosynthesizing green leaf but not of a hepatocyte?
4. are observed in both a nonphotosynthetic bacterium and a photosynthesizing green leaf?
5. are found in a photosynthetic green leaf, a hepatocyte, and a nonphotosynthesizing bacterium?
6. are organelles where ATP synthesis occurs?
7. contain the enzyme catalase?

SUGGESTED READINGS

Avers, C. J., *Cell Biology.* New York: Van Nostrand, 1981. A highly recommended textbook for the study of cell biology.

DeRobertis, E. D. P., and E. M. F. DeRobertis, *Cell and Molecular Biology,* 7th ed. Philadelphia: W. B. Saunders, 1980. A venerable textbook emphasizing the cell. Each edition becomes more biochemical.

Fawcett, D. W., *An Atlas of Fine Structure—The Cell, Its Organelles and Inclusions.* Philadelphia: W. B. Saunders, 1966. A summary of excellent electron micrographs and interpretive text.

Karp, G., *Cell Biology.* New York: McGraw-Hill, 1979. Another recommended textbook on cell biology.

Ledbetter, M. C., and K. R. Porter, *Introduction to the Fine Structure of Plant Cells.* New York: Springer-Verlag, 1970. This text parallels the Porter and Bonneville book on animal cells.

Novikoff, A. B., and E. Holtzman, *Cells and Organelles,* 2nd ed. New York: Holt, Rinehart and Winston, 1976. A compact but well-written and illustrated presentation of cell biology.

Porter, K. R., and M. A. Bonneville, *Fine Structure of Cells and Tissues,* 4th ed. Philadelphia: Lea and Febiger, 1973. A classic in format and content, this book is excellent for the study and review of animal cells and tissues.

Articles

Albersheim, P.. ''The Walls of Growing Plant Cells.'' *Scientific American,* 232(4):80 (offprint #1320), 1975.

Allison, A., ''Lysosomes.'' New York: Oxford University Press, 1974. A 16-page Oxford Reader.

Brachet, J., ''The Living Cell.'' *Scientific American,* 205(3):50 (offpring #90), 1961.

Chappell, J. B., and S. C. Rees, ''Mitochondria.'' New York: Oxford University Press, 1972. A 16-page Oxford Reader.

DeDuve, C., ''The Lysosome.'' *Scientific American,* 208(5):64 (offprint #156), 1963.

Goodenough, U. W., and R. P. Levine, ''The Genetic Activity of Mitochondria and Chloroplasts.'' *Scientific American,* 223(5):22 (offprint #1203), 1970.

Jordan, E. G., ''The Nucleolus.'' New York: Oxford University Press, 1971. A 16-page Oxford Reader.

Masters, C. J., and Holmes, R. S., ''Peroxisomes—Their Metabolic Roles in Mammalian Tissues.'' *TIBS,** 4:233, 1979.

Neutra, M., and C. P. Leblond, ''The Golgi Apparatus.'' *Scientific American,* 220(2):100 (offprint #1134), 1969.

Nomura, M., ''Ribosomes.'' *Scientific American,* 221(4):28 (offprint #1157), 1969.

Sharon, N., ''The Bacterial Cell Wall.'' *Scientific American,* 220(5):92, 1969.

* *Trends in Biochemical Sciences*

PROTEINS I: AMINO ACID COMPONENTS AND STRUCTURAL FEATURES

6

INTRODUCTION

A eucaryotic or procaryotic cell contains thousands of different proteins, the most abundant class of biomolecules in cells. Because each species of life possesses a chemically distinct group of proteins, millions of different proteins exist in the biological world. The genetic information contained in chromosomes determines the protein composition of an organism and, in this manner, endows members of a species with their macromolecular uniqueness.

As is true of many biomolecules, proteins exhibit functional versatility and are therefore utilized in a variety of biological roles. A number of important functions performed by proteins are listed in Table 6-1. The prominent position occupied by proteins in biological systems was correctly surmised by the Dutch chemist Gerardus J. Mulder, who in 1838 introduced the term *protein,* derived from the Greek word *proteis,* meaning first rank of importance. Mulder commented:

> There is present in plants and animals a substance which . . . is without doubt the most important of the known substances in living matter, and without it, life would be impossible on our planet. This material has been named Protein.

Although biologically active proteins are macromolecules that range in molecular weight from about 6,000 (single protein chain) to several millions (protein complexes), all are polymers composed of covalently linked *amino acids.* The number, chemical nature, and sequential order of amino acids in a protein chain determine the distinctive structure and characteristic chemical behavior of

Table 6-1
Some biological functions of proteins

Function	Examples
Enzymatic activity	Glycolate oxidase of glyoxysomes
	Alcohol dehydrogenase in alcoholic fermentation
Transport	Hemoglobin—oxygen transport in blood (vertebrates)
	Ceruloplasmin—copper transport in blood
Storage	Ferritin—iron storage (spleen)
	Casein—amino acid storage (milk)
Structure	Collagen—fibrous connective tissue (cartilage, bones, tendons)
Contraction	Myosin—thick filament in skeletal muscle
	Actin—thin filament in skeletal muscle
Protection	Antibodies—interact with foreign protein
	Fibrinogen—protein required for blood clotting
Hormonal activity	Insulin—regulator of glucose metabolism
	Growth hormone—required for bone growth
Toxins	Snake venom—hydrolytic (degradation) enzymes
	Clostridium botulinum toxins—lethal bacterial food toxins

each protein. For this reason, an appreciation of the chemical properties of amino acids is a prerequisite for the understanding of biomolecular explanations of the ways proteins function in their biological roles.

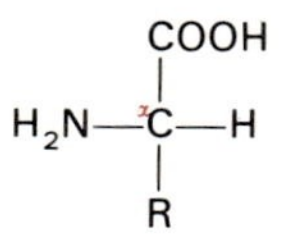

Figure 6-1
General formula of an α-amino acid. R represents a side chain.

TWO GENERAL PROPERTIES OF AMINO ACIDS

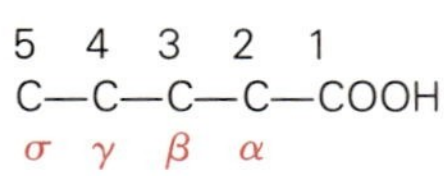

Amino acids are small biomolecules (average MW of about 135) having the general structure depicted in Figure 6-1. All *α-amino acids* are organic acids (α-COOH) containing both an amino (NH_2) group and a hydrogen atom bonded to the α-carbon. They differ from one another by the chemical composition of their R groups (side chains). Their α-COOH and α-NH_2 groups are ionized in solution at physiological pH, with the deprotonated carboxyl group bearing a negative, resonance-stabilized charge and the protonated amino group a positive charge (Figure 6-2). An amino acid in its dipolar state is called a *zwitterion.* The dissociable α-COOH and α -NH_3^+ groups are responsible for the two characteristic pK_a' values of α-amino acids. The ionization of the amino acid alanine is illustrated in Figure 6-3. At low pH values, molecules of alanine bear a net charge of $+1$ because both functional groups are protonated, e.g., at pH 0.35, 99 per cent are positively charged. As the pH is increased ($[H^+]$ lowered), ionization of the carboxyl group occurs ($pK_{a_1}' = 2.35$), creating a net molecular charge of zero. At

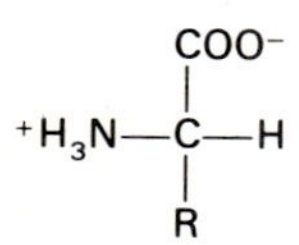

Figure 6-2
Dipole amino acid (zwitterion).

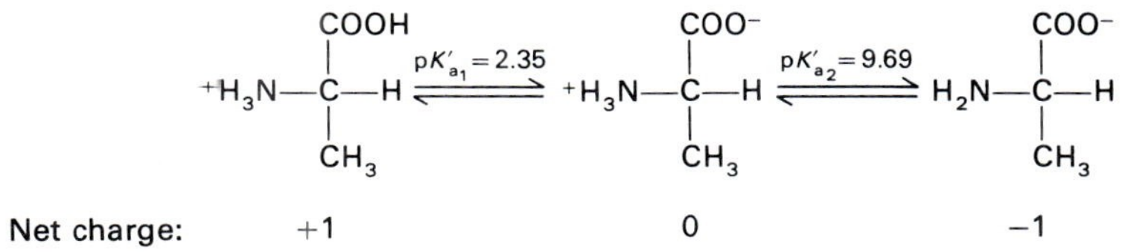

Figure 6-3
The two dissociations of L-alanine.

alkaline pH values, the NH_3^+ group donates its proton (pK'_{a2} = 9.69) and the net charge becomes −1. At pH 11.69, 99 per cent of the alanine molecules are negatively charged. The pH at which the average net charge is zero (called the *isoionic point* or *pI*) is 6.02, which is midway between the two pK'_a values ([2.35 + 9.69]/2). A titration curve reflecting these dissociations is shown in Figure 6-4. Similar pK'_a values are observed for the α-COOH and α-NH_3^+ groups of most amino acids (Table 6-2). An amino acid with a third dissociable group in its side chain exhibits an additional pK'_a value (Table 6-2). The ionization proclivities of individual amino acids confer important biological properties on these biomolecules and also provide chemical properties which are useful in biochemical research, e.g., separation of proteins or amino acids and utilization of amino acids in buffer systems.

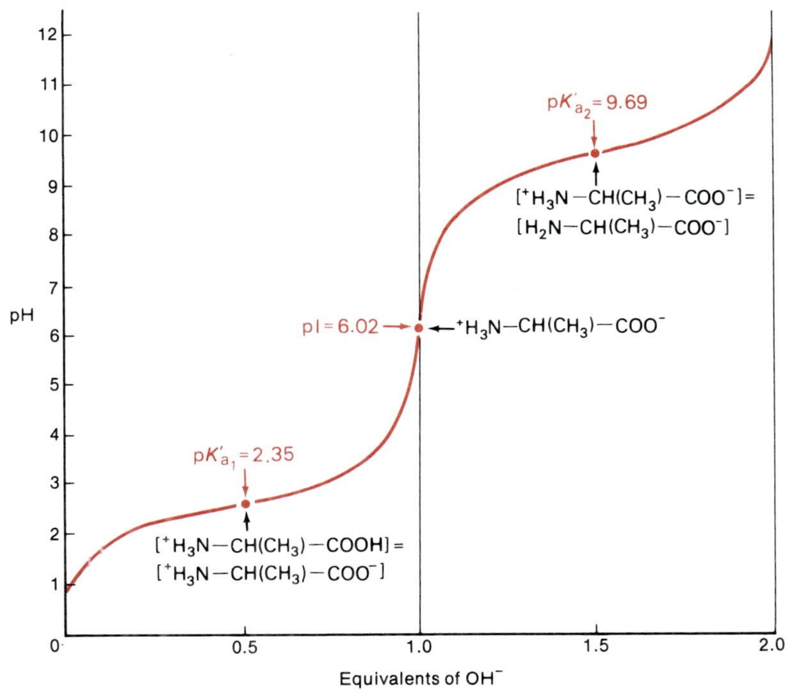

Figure 6-4
Titration curve of alanine.

Table 6-2
pK_a' values, at 25 °C, for the 20 amino acids used for polypeptide synthesis

Amino acid	pK_a' α-COOH	pK_a' α-NH_3^+	pK_a' R group
Alanine	2.35	9.69	
Arginine	2.17	9.04	12.48
Asparagine	2.02	8.8	
Aspartic acid	2.09	9.82	3.86
Cysteine	1.71	10.78	8.33
Glutamic acid	2.19	9.67	4.25
Glutamine	2.17	9.13	
Glycine	2.34	9.6	
Histidine	1.82	9.17	6.0
Isoleucine	2.36	9.68	
Leucine	2.36	9.60	
Lysine	2.18	8.95	10.53
Methionine	2.28	9.21	
Phenylalanine	1.83	9.13	
Proline	1.99	10.60	
Serine	2.21	9.15	
Threonine	2.63	10.43	
Tryptophan	2.38	9.39	
Tyrosine	2.20	9.11	10.07
Valine	2.32	9.62	

Figure 6-5
Stereoisomers of alanine.

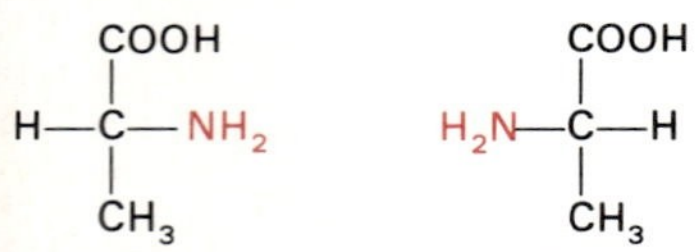

D-Alanine L-Alanine

Figure 6-6
Several structural representations for amino acids.

D-Amino acid L-Amino acid

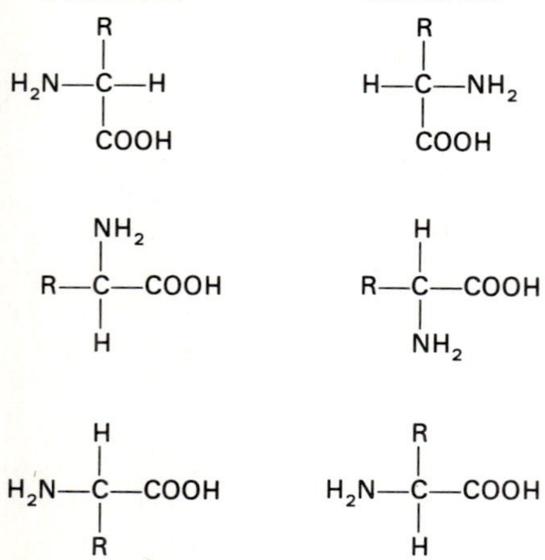

Another distinct property of α-amino acids is the asymmetry of their α-carbon atom. Because four different substituent groups are bonded to this atom, all amino acids except glycine (Figure 6-13) exhibit stereoisomerism. The two optical isomers of alanine are shown in Figure 6-5. The D and L nomenclature and structural formulas were adopted by biochemists from those proposed for glyceraldehyde (p. 43), and the Fischer projection of an α-amino acid is written with the NH_2 (functional) group to the right of the asymmetric carbon for the D-isomer and to the left for the L-isomer. In biochemistry, structural formulas of amino acids are often written not as Fischer projections but in a manner best suited for the particular topic being presented. Figure 6-6 depicts several structural formulas commonly used. Almost all biological functions involving amino acids have strict requirements for L-isomers; however, there is limited biological use of D-amino acids, since they are present in some biological materials, e.g., certain bacterial cell walls and some antibiotics. A few amino acids possess two asymmetric carbons and, of the four optical isomers, only one is usually biologically active.

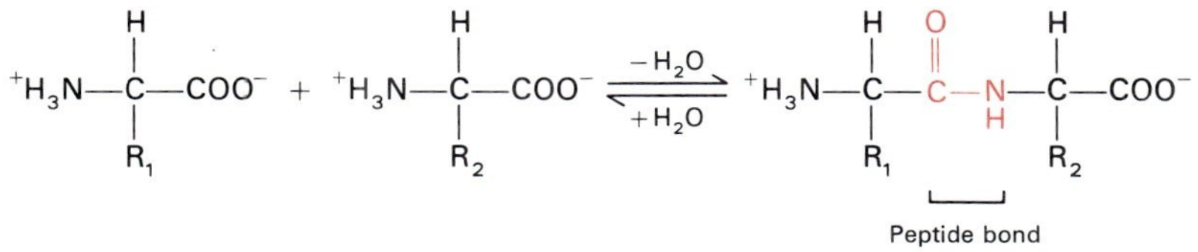

Figure 6-7
Formation of a dipeptide at physiological pH.

GENERAL BONDING FEATURES OF POLYPEPTIDES

The covalent bond that links amino acids together in proteins is called a *peptide bond,* and it is formed by reaction between the α -NH_3^+ group of one amino acid and the α-COO^- group of another (Figure 6-7). In the *dehydration* reaction, a water molecule is also produced; hydrolysis of the peptide bond is required to regenerate the amino acids. Molecules composed of many sequential peptide bonds are called *polypeptides,* and one or more polypeptide chains may be contained in the molecular structure of a biologically active protein. An amino acid unit of a polypeptide is called an *amino acid residue,* e.g., alanyl (alanine) or glycyl (glycine) residue. In a polypeptide structure, the two terminal amino acid residues are the only ones that possess a free α-NH_3^+ or α-COO^- group (Figure 6-8). The terminal residue with a free α-NH_3^+ group is called the *amino-terminal* or *N-terminal residue.* The *carboxyl-terminal* or *C-terminal residue* has a free α-COO^- group and is at the other terminus of the polypeptide. In biological systems, polypeptides are synthesized from the amino terminus to the carboxyl terminus, and the generally accepted convention is to write the amino acid sequence of a polypeptide from left to right, starting with the N-terminal residue (Figure 6-8).

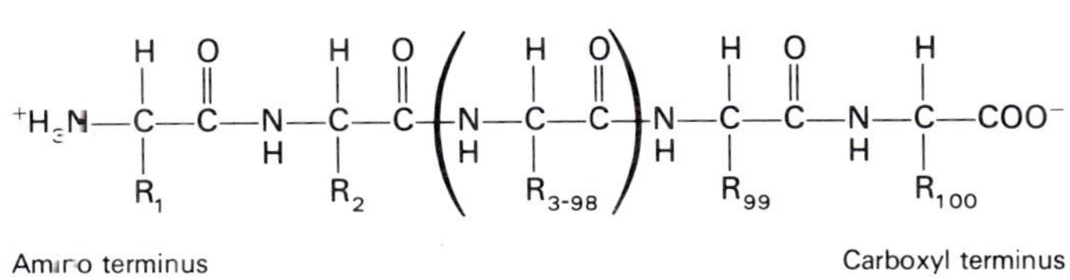

Figure 6-8
Basic structural features of a polypeptide containing 100 amino acid residues.

AMINO ACID RESIDUES OF POLYPEPTIDES

Of over 100 naturally occurring amino acids, only 20 are utilized in polypeptide biosynthesis. Francis Crick coined the term *magic 20* to distinguish this set of amino acids required by all living species for protein synthesis. More than 20 amino acids have been identified in protein structures; however, in all cases, chemical modification of certain of the "magic 20" amino acids (occurring after incorporation into a polypeptide) accounts for these additional residues. For example, certain incorporated prolyl residues (Figure 6-10) can be converted into hydroxyprolyl residues (Figure 6-21) and some seryl residues (Figure 6-14) into phosphoseryl residues (Figure 6-22).

The solubility and ionization properties of R groups are influential traits of amino acids, and collectively they contribute greatly to the native three-dimensional structures of individual polypeptides. It is on the basis of these two characteristics of their R groups that the 20 amino acids are classified into four categories, as follows.

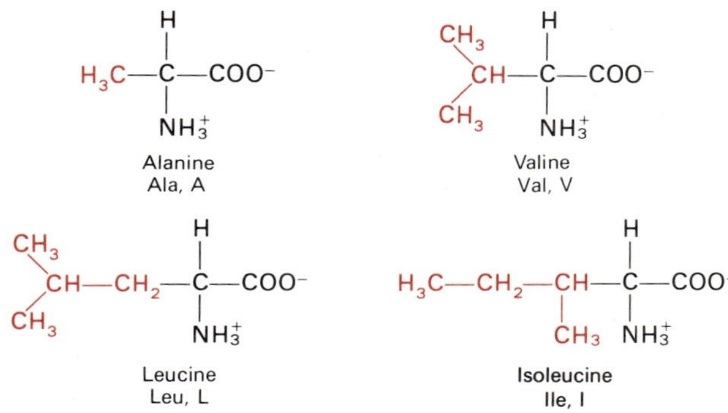

Figure 6-9
Alanine and three branched-chain amino acids.

Nonpolar R Groups

Eight of the amino acids have nonpolar side chains and, as a group, display varying degrees of hydrophobicity. Four (*alanine, valine, leucine,* and *isoleucine*) have aliphatic noncyclic R groups (Figure 6-9). (Structural formulas of the amino acids are shown as the ionized forms that predominate at physiological pH. Conventional three- and one-letter abbreviations are also listed.) Alanine is the least hydrophobic of the eight because of its small methyl side chain. Valine, leucine, and isoleucine are called the *branched-chain amino acids* because of the branching in their aliphatic R groups. The fifth "amino" acid of this group is *proline* (Figure 6-10), which has an aliphatic heterocyclic structure that includes both the R group and the α-nitrogen atom. Thus, proline differs from the other 19 amino acids since it is an *imino acid,* having an *imino* (= NH) rather than an *amino* ($-NH_2$) group. The sixth amino acid, *methionine* (Figure 6-11), has a sulfur atom in its nonpolar side

Figure 6-10
An imino acid.

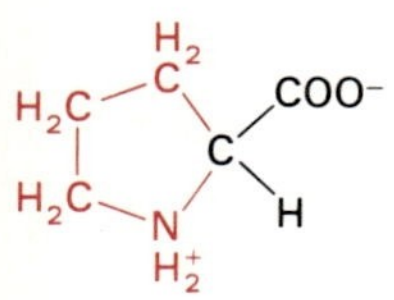

Proline
Pro, P

Figure 6-11
A sulfur-containing amino acid.

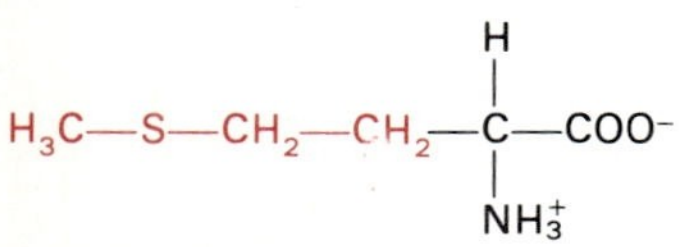

Methionine
Met, M

chain and is one of two sulfur-containing amino acids incorporated into proteins. The remaining two amino acids, *phenylalanine* and *tryptophan* (Figure 6-12), have water-insoluble aromatic rings in their structures. Phenylalanine has a phenyl group in its side chain and tryptophan an indole group (a condensed ring composed of benzene and pyrrole); both are considered *aromatic amino acids.*

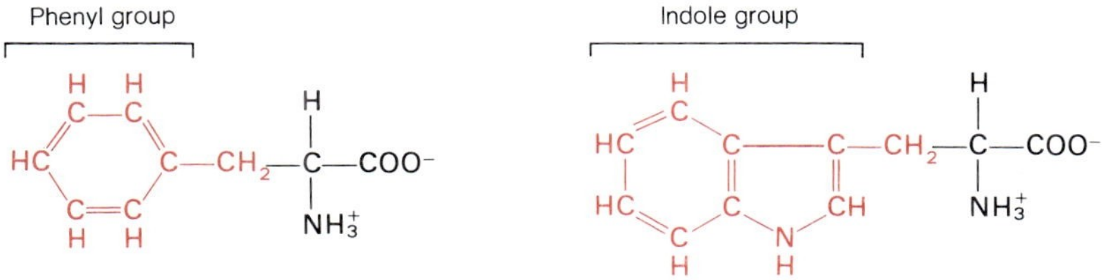

Phenylalanine
Phe, F

Tryptophan
Trp, W

Figure 6-12
Two aromatic amino acids.

Uncharged Polar R Groups

This class contains seven amino acids which are relatively hydrophilic because of the polar functional groups in their side chains. *Glycine,* the nonasymmetric amino acid (Figure 6-13), is sometimes considered nonpolar. However, glycine's small R group (a hydrogen atom) exerts essentially no effect on the hydrophilicity of the molecule. Three amino acids (*serine, threonine,* and *tyrosine*) are hydroxylated, and the OH groups contribute to their polarity (Figure 6-14). Tyrosine, like phenylalanine and tryptophan, is also an aromatic amino acid. *Cysteine,* the other sulfur-containing amino acid (Figure 6-15), is polar because of its sulfhydryl (—SH) group. Often in protein structures, two cysteinyl residues are covalently linked to each other through oxidation of their sulfhydryl groups, which produces a disulfide bond (—S—S—). This oxidized form of cysteine is called *cystine* (Figure 6-15). *Asparagine* and *glutamine* (Figure 6-16) are derived from aspartic acid and glutamic acid (next paragraph), and each has a polar amide group in its side chain.

Figure 6-13
Amino acid with no asymmetric carbon.

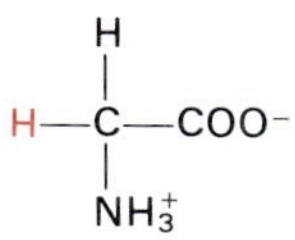

Glycine
Gly, G

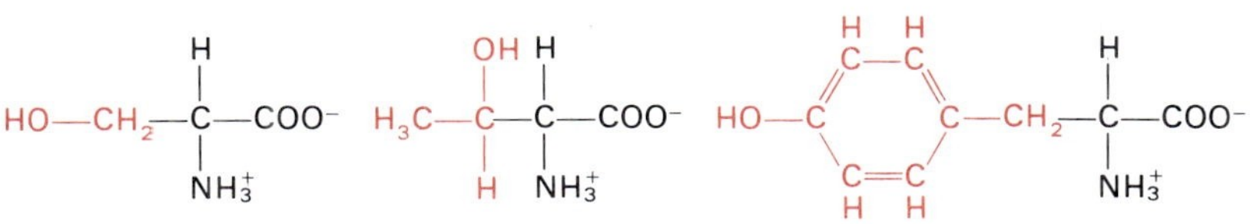

Serine
Ser, S

Threonine
Thr, T

Tyrosine
Tyr, Y

Figure 6-14
Three hydroxylated amino acids.

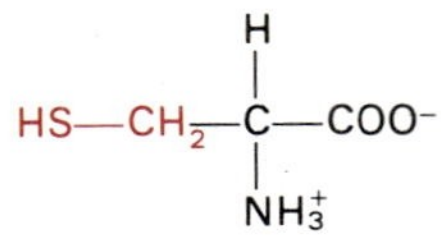

Cysteine
Cys, C

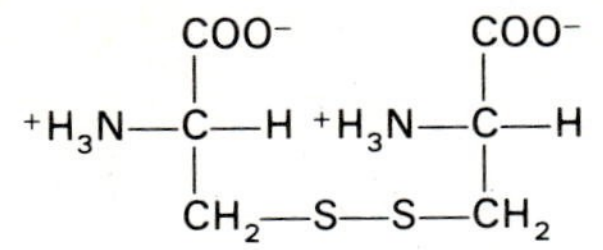

Cystine
Cys Cys

Figure 6-15
Cysteine and its oxidized form.

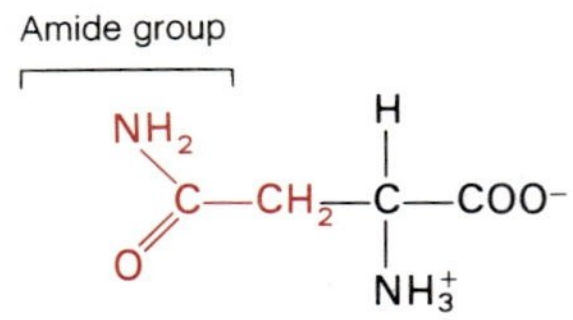

Asparagine
Asn, N

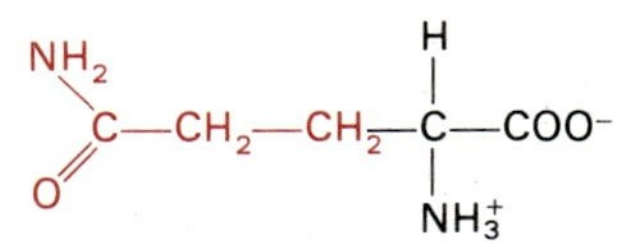

Glutamine
Gln, Q

Figure 6-16
Two amino acids with amide groups.

Figure 6-17
Acidic amino acids.

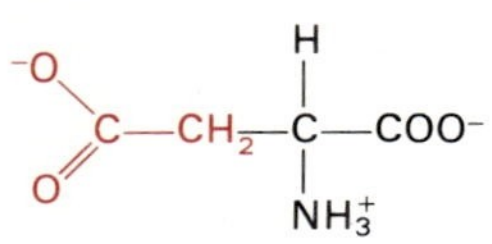

Aspartate
Asp, D

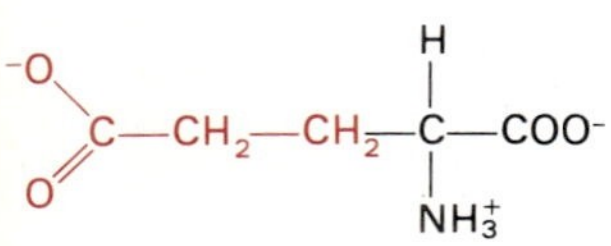

Glutamate
Glu, E

Negatively Charged (Acidic) Polar R Groups

Both *aspartic acid* and *glutamic acid* have a second carboxyl group, which is fully ionized (negatively charged) at physiological pH (Figure 6-17). This ionization contributes significantly to the polarities of their side chains. Aspartic acid (HA), when its α-COOH group dissociates to COO^-, is called *aspartate* (A^-) and, similarly, glutamic acid becomes *glutamate.* Both are referred to as *acidic amino acids* because they donate H^+ when placed in solution.

Figure 6-18
Two basic amino acids.

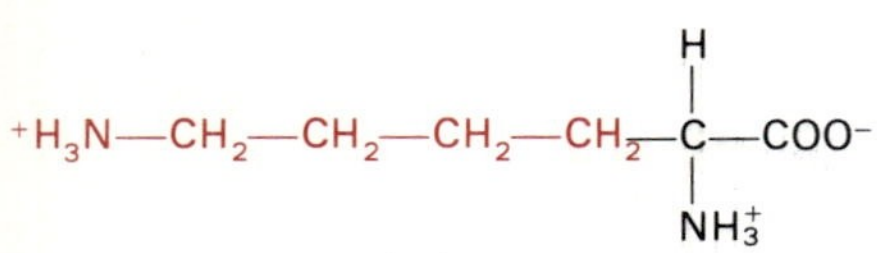

Lysine
Lys, K

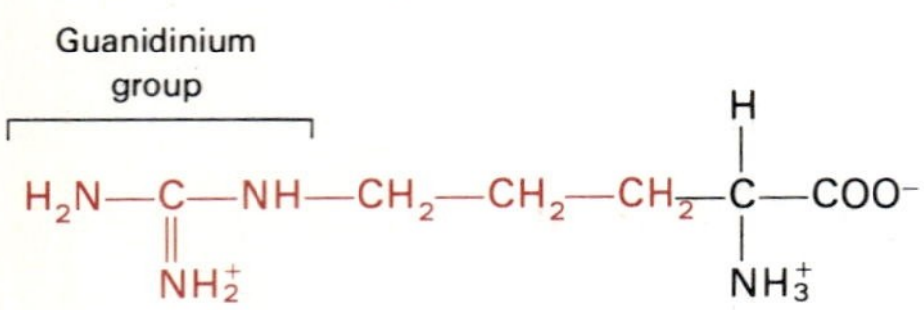

Arginine
Arg, R

Positively Charged (Basic) Polar R Groups

Lysine and *arginine,* two of the three basic amino acids, have R groups that are positively charged at physiological pH (Figure 6-18). Ionic charges are provided by protonation of the amino group of the ϵ-carbon of lysine and of the guanidinium group of arginine. The third amino acid, *histidine* (Figure 6-19), has an imidazolium R group with a pK_a' value of 6.0 (Table 6-2) and, therefore, is less than 10 per cent protonated at pH 7. Histidine's basic properties are clearly marginal. Of the 20 amino acids, histidine is the only one whose isoionic point of about 7.6 is near physiological pH.

Allo Forms of Amino Acids

Isoleucine and threonine are examples of amino acids having two asymmetric carbons. Among the four possible stereoisomers of each, only the L-isomer is used for protein synthesis. The structural

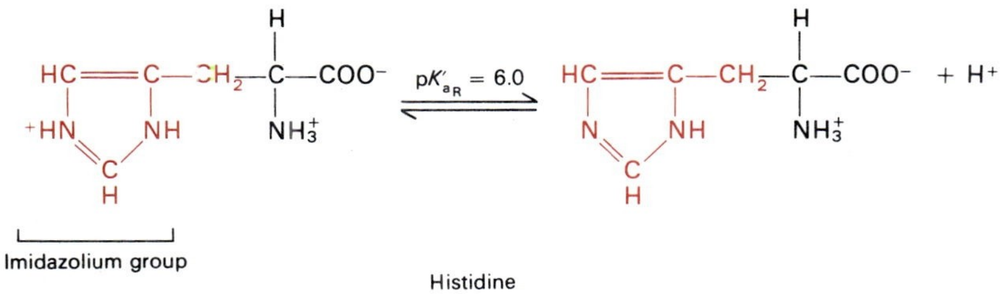

Histidine
His, H

Figure 6-19
Dissociation of R group of histidine.

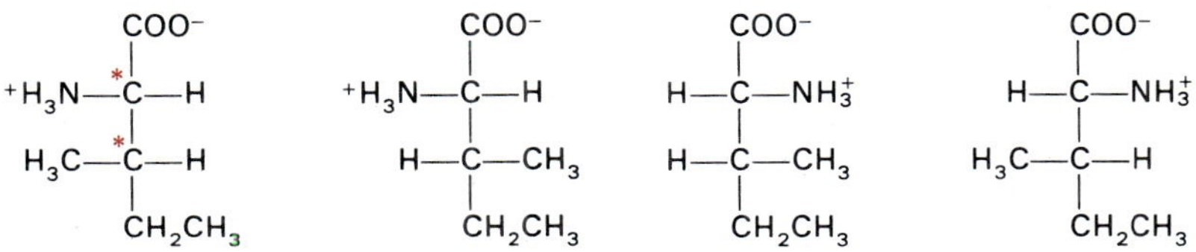

L-Isoleucine (2S), (3S) | L-*allo*-Isoleucine (2S), (3R) | D-Isoleucine (2R), (3R) | D-*allo*-Isoleucine (2R), (3S)

* Asymmetric carbon atom
R & S configurations are also noted

Figure 6-20
Fischer projection formulas of the four stereoisomers of isoleucine.

formulas for the isoleucine isomers are illustrated in Figure 6-20. The *allo* nomenclature identifies the other two isomers (in addition to the D- and L-forms), which are also enantiomers (mirror images). L-Isoleucine and L-*allo*-isoleucine have the same configuration at the α-carbon but opposite configurations at the second asymmetric carbon (β C). The same is true for D-isoleucine and D-*allo*-isoleucine. If a compound has more than one asymmetric carbon, the stereoisomers that are not mirror images of each other are called *diastereomers.* In the case of isoleucine, for example, L-isoleucine is an enantiomer of D-isoleucine and a diastereomer of L- and D-*allo*-isoleucine.

Figure 6-21
Hydroxylated forms of proline and lysine.

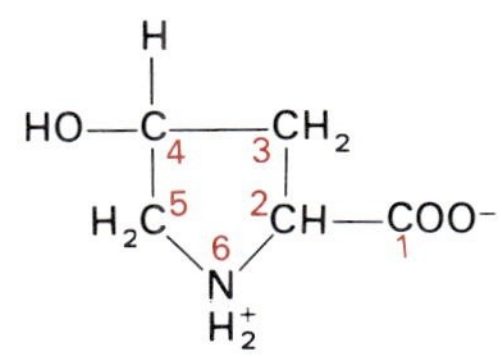

4-Hydroxyproline

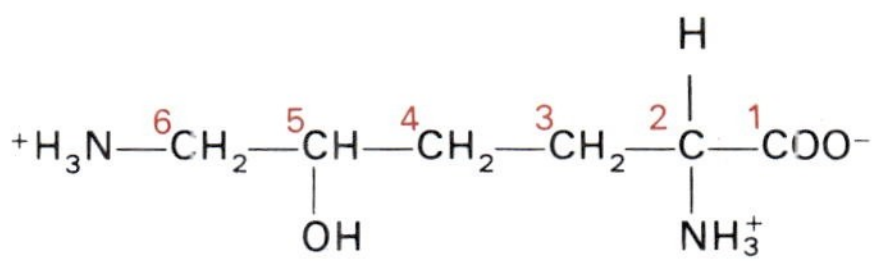

5-Hydroxylysine

SOME OTHER NATURALLY OCCURRING AMINO ACIDS

As stated previously, modified forms of some amino acids are found in certain proteins. Collagen, the major structural protein of higher vertebrates, contains *4-hydroxyproline* and *5-hydroxylysine* residues (Figure 6-21). Some proteins, e.g., casein of milk and certain enzymes, have a phosphorylated form of serine, *phosphoserine* (Figure 6-22). In *Escherichia coli,* protein synthesis is initiated with a derivative of methionine, *N*-formylmethionine (Figure 6-23). Many naturally occurring amino acids are not found in proteins;

Figure 6-22
A phosphorylated amino acid.

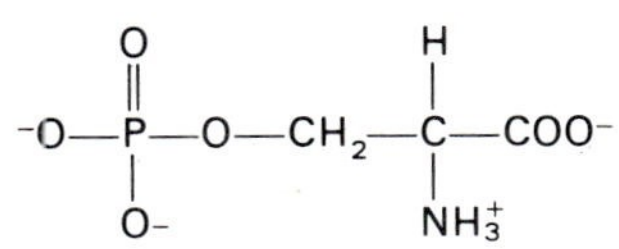

Phosphoserine

Figure 6-23
A formylated amino acid.

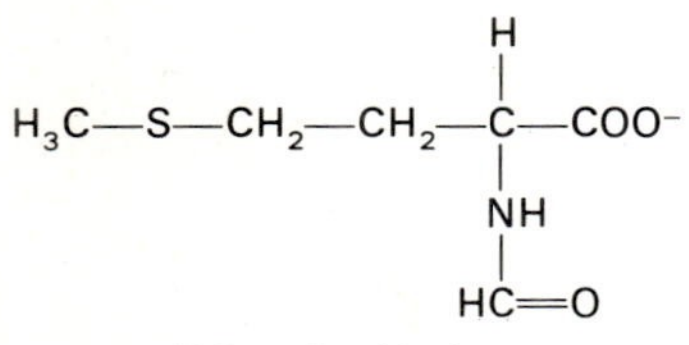

Figure 6-24
Some naturally occurring amino acids not used for protein synthesis.

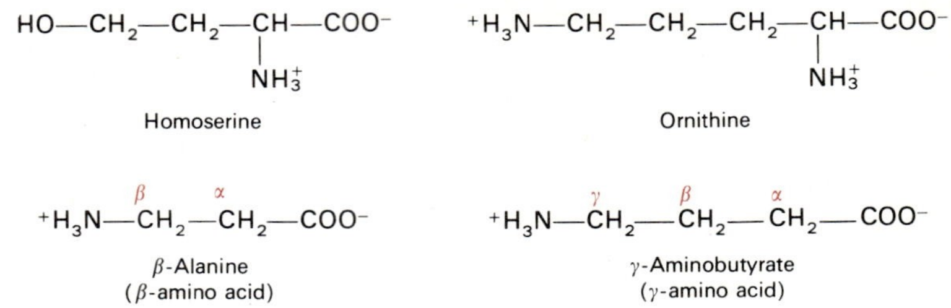

among these are *homoserine, ornithine, β-alanine,* and *γ-aminobutyric acid* (Figure 6-24).

STRUCTURAL LEVELS OF PROTEINS

The native conformation of a protein is determined by interactions between a polypeptide and its aqueous environment, in which the polypeptide attains an energetically stable three-dimensional structure, most often the conformation requiring the least amount of energy to maintain. Such a macromolecule has a highly complex three-dimensional structure. There are four levels of structure that interdependently influence a protein's biologically functional conformation. Three of these structural levels (*primary, secondary,* and *tertiary*) can exist in molecules composed of a single polypeptide chain, whereas the fourth *(quaternary)* involves interactions of polypeptides within a multi-chained protein molecule.

Figure 6-25
The partial double-bond character of a peptide bond that makes it planar.

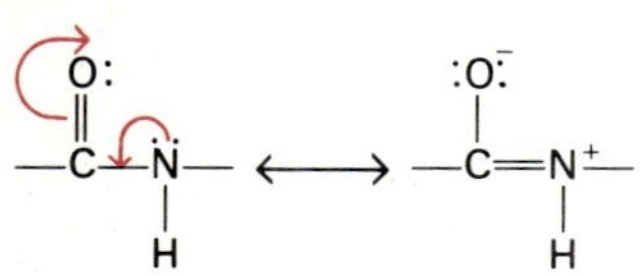

Figure 6-26
Schematic of the rigid peptide-bond unit of a polypeptide and of the bonds linking the peptide groups that have a large degree of rotational freedom.

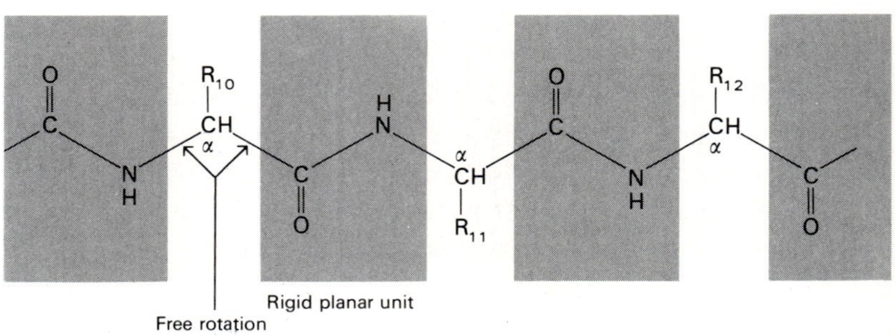

Primary Structure

This level of structure refers to the number and sequence of amino acids in a protein. The covalent peptide linkage is the only type of bonding involved at this level of protein structure. Because each peptide bond has partial double-bond character (Figure 6-25), it is a relatively rigid plane with the hydrogen atom of the amide group *trans* to the oxygen of the carbonyl group. As a result of the double-bond characteristics of the C—N bond, there is no free rotation about these axes (Figure 6-26). However, there is rotational freedom about the single bonds (ϕ and ψ rotational angles) that link each α-C atom to the N and C atoms of peptide bonds (Figure 6-27). This freedom of rotation allows the R groups attached to the α-carbons to have an influential role in establishing the three-dimensional structure of polypeptide chains.

Figure 6-27
Rotational angles ϕ and ψ, which are the rotational angles about the C_α-N bond and the C_α-C bond, respectively. (After R. E. Dickerson and I. Geis, *The Structure and Actions of Proteins.* New York: Harper & Row, 1969)

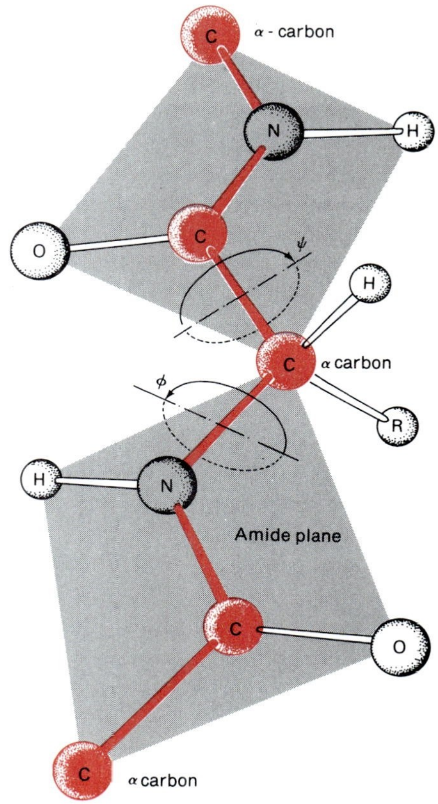

Secondary Structure

This next level generally refers to the amount of structural regularity contained in a polypeptide as a result of hydrogen bonding between peptide bonds. The *α-helical* structure proposed by Linus Pauling and Robert Corey in 1951 for polypeptides depends on *intramolecular* (or *intrachain*) hydrogen bonding between the NH and CO groups of peptide bonds. The hydrogen bonding occurs spontaneously and, as a result, a polypeptide can assume a rod-like structure with well-defined dimensions (Figure 6-28). Along the axis of an α-helix, the rise per amino acid residue is 1.5 Å, and there are 3.6 residues per turn, which makes the pitch of each turn 5.4 Å (Figure 6-29). Residues spaced four apart (on neighboring turns) hydrogen bond, and every NH and CO group of the polypeptide chain segment participates. A helix can be right- or left-handed; α-helices of known polypeptides (L-amino acids) are right-handed. Biologically functional proteins do not usually exhibit 100 per cent α-helical structure. Some have a high percentage of their residues in α-helical structures, e.g., myoglobin and hemoglobin; others have a low percentage, e.g., chymotrypsin and cytochrome *c*. The R groups of the amino acid sequence influence the degree of helical structure a given polypeptide possesses.

Hydrogen bonding

Destabilization of helical conformation by certain residues (Table 6-3) can occur in various ways. A prolyl residue has its α-N atom in a rigid ring system and cannot participate in α-helical structure; instead, it creates a sharp bend in the helix. A sequence of aspartyl

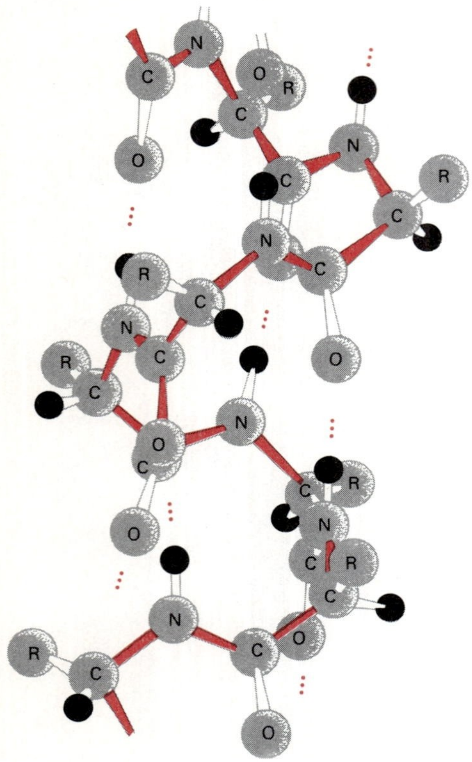

Figure 6-28
Drawing of a right-handed α-helix. (After L. Pauling, *The Nature of the Chemical Bond,* 3rd ed. Ithaca, N.Y.: Cornell University Press, 1960)

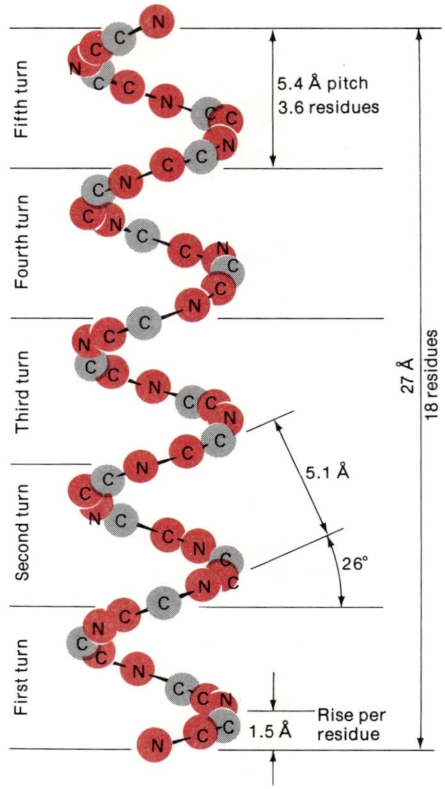

Figure 6-29
Average dimensions of an α-helix: C and N represent the C and N atoms in the peptide bonds; C represents an α-carbon with a side chain attached. (After L. Pauling and R. B. Corey, *Proc. Intern. Wool Textile Res. Conf.,* B. 249, 1955)

Table 6-3
Amino acids affecting α-helical structure

Destabilize α-helix:
Glutamate
Aspartate
Lysine
Arginine
Glycine
Serine
Isoleucine
Threonine
Create bends in α-helix:
Proline
Hydroxyproline

and/or glutamyl residues can destabilize α-helical structure because the negatively charged side chains repel one another (electrostatic repulsion), and the forces of repulsion are greater than those of hydrogen bonding. A cluster of isoleucyl residues, because of steric hindrance imposed by their bulky R groups, also disrupts helical conformation. On the other hand, glycine, with a small hydrogen atom as an R group, is another destabilizer. The lack of a side chain on glycine allows for a great degree of rotation about the amino acid's α-carbon; hence, conformations other than α-helical bond angles are possible. In other words, for a sequence of glycyl residues, the α-helical conformation is not the one of maximal stability.

Pauling and Corey identified a second type of secondary struc-

Figure 6-30
The two β-pleated sheet structures.

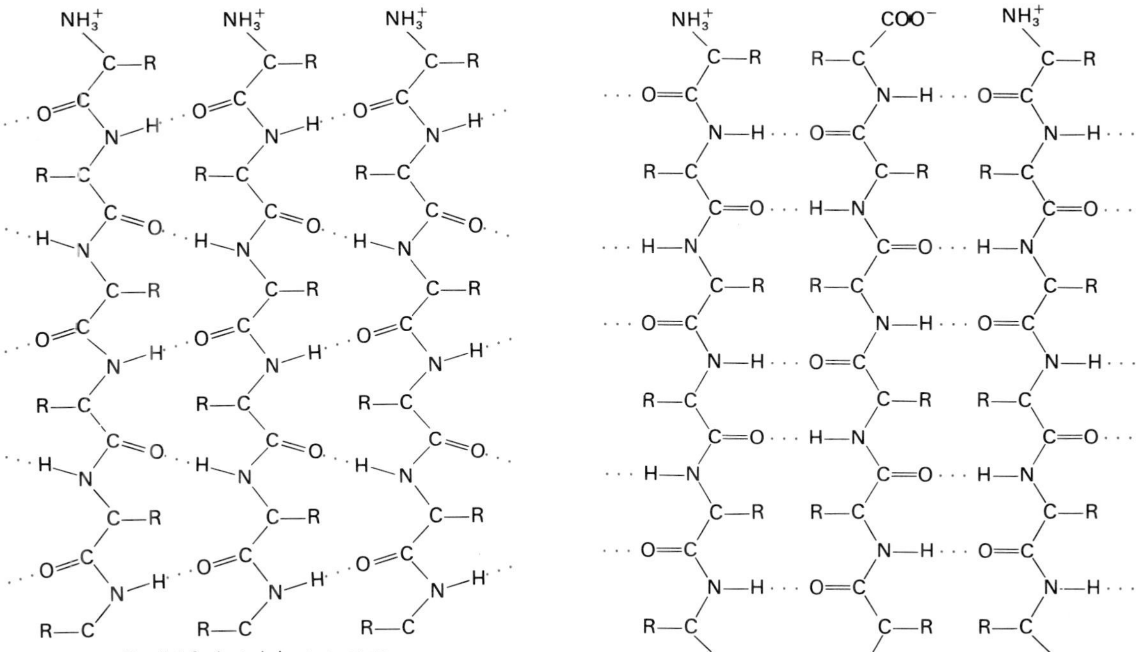

Parallel β-pleated sheet structure

Antiparallel β-pleated sheet structure

ture, called *β-pleated sheet structures,* which depends on *intermolecular (interchain)* hydrogen bonding. One of these structures is the *antiparallel* β-pleated sheet (Figure 6-30), in which the individual polypeptide chain is extended and is maximally bonded to two neighboring polypeptides. In an antiparallel structure, the neighboring polypeptides are aligned in an opposite N-to-C terminus direction. If, in a β-pleated structure, the hydrogen-bonded neighboring polypeptides are aligned in the same N-to-C direction (Figure 6-30), the structure is *parallel β-pleated.* Although a β-pleated sheet structure is usually associated with structural proteins (p. 82), it is also known to occur in the three-dimensional structures of certain globular proteins (p. 83), e.g., the enzymes lysozyme and carboxypeptidase A.

Tertiary Structure

This next level of structure refers to the folding and coiling of a polypeptide to produce a complex, globular molecular shape (Figure 6-31). The convolutions a polypeptide undergoes are specified by the particular set and sequence of R groups in the molecule.

Figure 6-31
Schematic of the tertiary structure of a polypeptide.

Tertiary structure represents the overall effect of many intramolecular forces, including those of primary and secondary structures. In globular proteins whose structures are known, polar R groups, because of their hydrophilicity, are most often located on the molecule's exterior and nonpolar R groups in the interior, where their interactions create a hydrophobic environment (Figure 6-32). The only covalent linkage involved in tertiary structure is the *disulfide bond,* formed by oxidation of the sulfhydryl groups of two cysteinyl residues (Figure 6-15). Noncovalent interactions (Chapter 4) such as electrostatic attractions (salt bridges), hydrogen bonds, and dipole-dipole interactions also contribute significantly to the final structure (Figure 6-32).

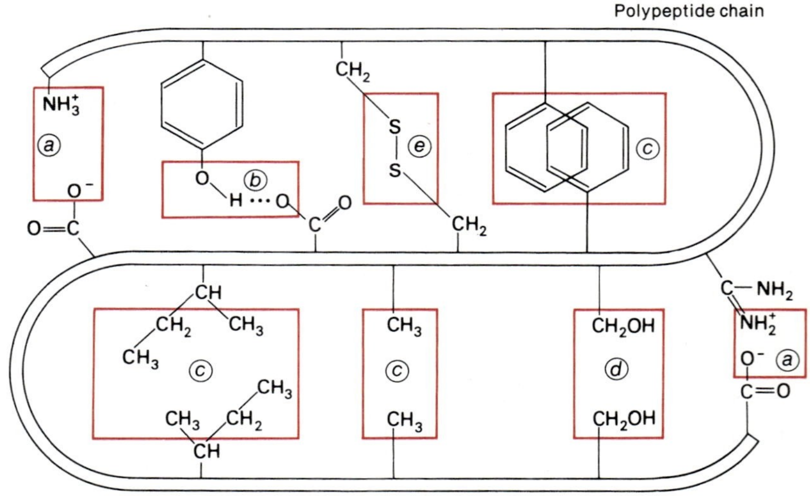

Figure 6-32
Various types of bonds and interactions which stabilize protein structure: (a) electrostatic interaction; (b) hydrogen bonding; (c) hydrophobic interaction of nonpolar side chains; (d) dipole-dipole interaction; (e) disulfide linkage. (After C. B. Anfinsen, *The Molecular Basis of Evolution.* New York: Wiley, 1959)

Quaternary Structure

This fourth structural level concerns interactions by which two or more polypeptide chains associate in a specific manner to form biologically active proteins. Many enzymes are *oligomeric,* consisting, for example, of two or four polypeptides (*protomers* or *subunits*), and their quaternary structures are maintained by specific interactions involving their member protomers. The responsible cohesive forces are usually the same as those involved in the stabilizing of tertiary structure of a polypeptide. For example, some oligomers may rely mainly on hydrophobic interactions of their protomers and others on electrostatic attractions. Quaternary structures are identified as *homogeneous* (containing identical protomers) or *heterogeneous* (dissimilar protomers). Hemoglobin, the oxygen-carrying protein of red blood cells, is a tetramer (four protomers) with a heterogeneous quaternary structure composed of two identical α-chains and two identical β-chains (p. 109). (The designation α- and β-, which are common labels in biochemistry, are used simply to identify two different polypeptide chains and are not to be confused with α- and β- secondary structures or α- and β- amino acids.)

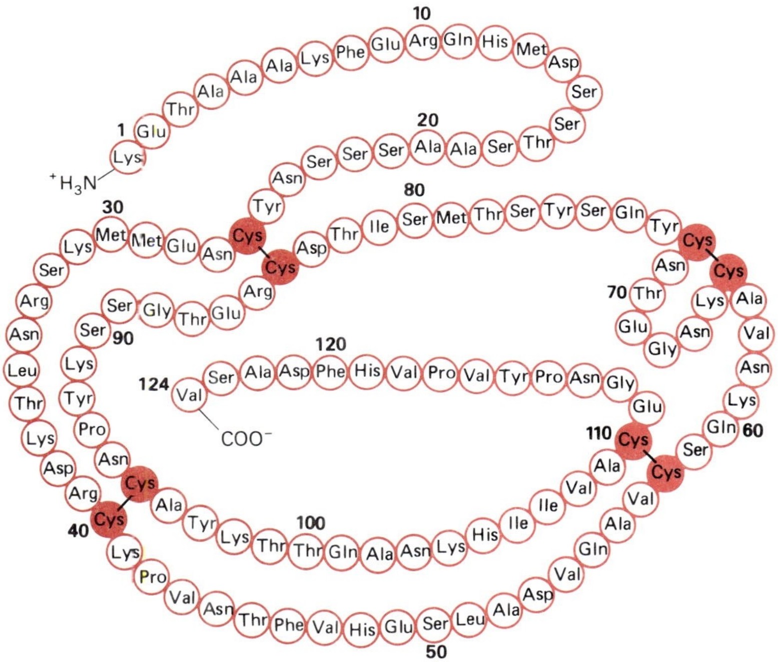

Figure 6-33
Amino acid sequence of ribonuclease A (bovine), with the cysteinyl residues that form the four disulfide bonds identified in red. (After C. W. Hirs, S. Moore, and W. H. Stein, *J. Biol. Chem.*, 235:633, 1960)

SPECIFICATION OF CONFORMATION BY PRIMARY STRUCTURE

Studies by Christian B. Anfinsen (Nobel Prize, 1972) on the pancreatic enzyme ribonuclease A provide a pertinent illustration of the influence of primary structure on three-dimensional structure. Biologically active ribonuclease, a single-chained polypeptide of 124 amino acid residues, has four disulfide bonds (Figure 6-33). Treatment of ribonuclease with β-mercaptoethanol (which reduces disulfide bonds to yield cysteinyl residues) in the presence of urea or guanidine·HCl (denaturing agents which disrupt noncovalent

$HS—CH_2—CH_2—OH$
β-Mercaptoethanol

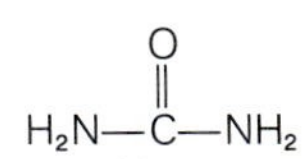
Urea

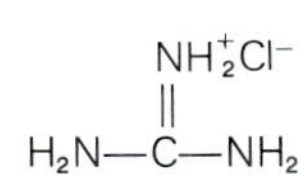
Guanidinium chloride

interactions) causes the enzyme to lose its native conformation and to become a *random coil* (Figure 6-34). This reduced, *denatured* form of the polypeptide has no enzymatic activity. If the reducing and denaturing agents are subsequently removed and the reduced, inactive ribonuclease exposed to air under appropriate conditions, the sulfhydryl groups are reoxidized and the protein slowly regains essentially all its enzymatic activity. The random coil, freed from conditions that prevent conformational interactions, spontaneously re-establishes the biologically active form of the enzyme (Figure 6-34). Of the 105 different intramolecular disulfide bonding arrangements possible among eight cysteinyl residues to produce four disulfide bonds, the specific four required for enzymatic activity are regenerated spontaneously by oxidation in the *renaturation* processes. Thus, it is evident that the amino acid sequence of ribonuclease, specified by the genetic code, furnishes the information needed to establish the enzyme's native conformation, a thermodynamically favored structure. The distinctive conformations associated with proteins, therefore, are determined by the specific arrangement of information contained in the primary sequences of amino acids that were developed during the process of biological evolution.

Figure 6-34
Denaturation and renaturation of ribonuclease A.

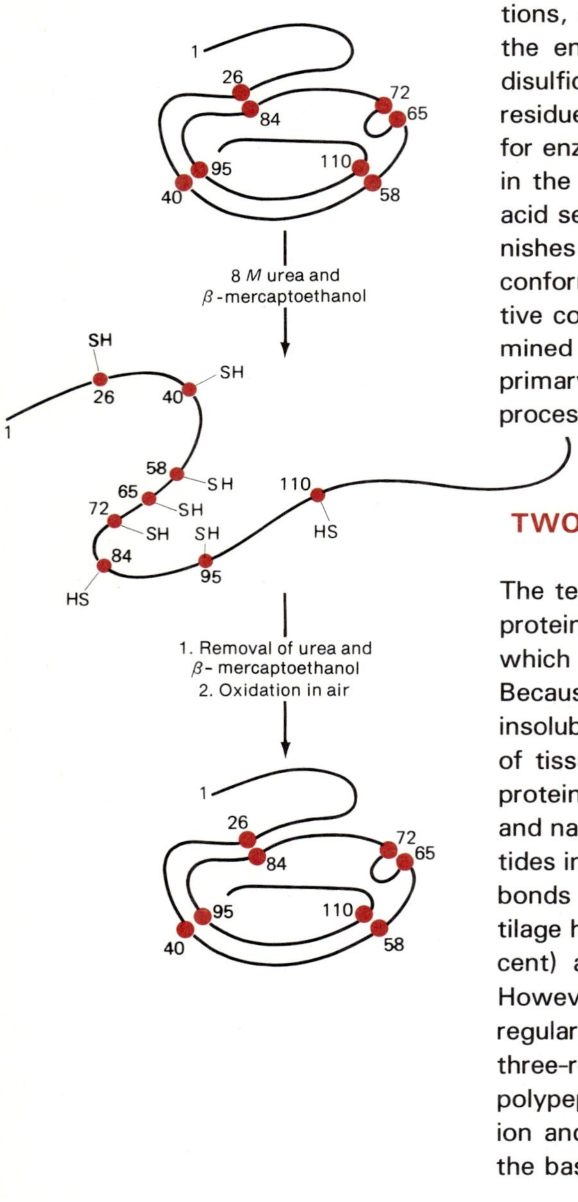

TWO MAJOR TYPES OF PROTEIN CONFORMATION

The terms *fibrous* and *globular* describe the two main classes of proteins. Fibrous proteins are composed of individual polypeptides which are often laterally crosslinked by several types of bonding. Because of their structure, they are physically tough and water insoluble—valuable properties for their role as structural elements of tissue. α-Keratins, collagen, and silk are examples of fibrous proteins. The basic structural unit of α-keratin (protein of hair, fur, and nails) usually consists of three right-handed α-helical polypeptides in a left-handed coil that is stabilized by crosslinking disulfide bonds (Figure 6-35). In contrast, collagen of skin, bone, and cartilage has no α-helix because of its high content of glycine (33 per cent) and proline and hydroxyproline (25 per cent combined). However, since these particular amino acids occur with repeated regularity in the sequence, the polypeptide has a unique, tight, three-residue helical structure that is left-handed. Three of these polypeptide chains are wrapped around one another in cable fashion and crosslinked by hydrogen bonding to form *tropocollagen,* the basic structural unit of collagen (Figure 6-36). Silk fiber, a β-

keratin, is an example of a fibrous structure composed of polypeptides arranged in *antiparallel* β-pleated sheets (Figure 6-30). An interesting phenomenon occurs when hair or wool (α-keratins) is treated with moist heat and stretched because β-keratin structures, with parallel β-pleated sheets (Figure 6-30), are produced.

Globular proteins, unlike fibrous proteins, are water soluble and have spheroidal, compact structures; tertiary and quaternary structural features are usually associated with this class of proteins. Globular proteins are biologically active and constitute the great majority of molecular species of proteins in living systems, e.g., enzymes and antibodies. Studies on the three-dimensional structures of globular proteins are a major aspect of molecular biological research because of the intrinsic relationship between their structure and function.

Figure 6-35
Schematic of the three α-helical coils in keratin.

Figure 6-36
Schematic of the triple-stranded helix of tropocollagen.

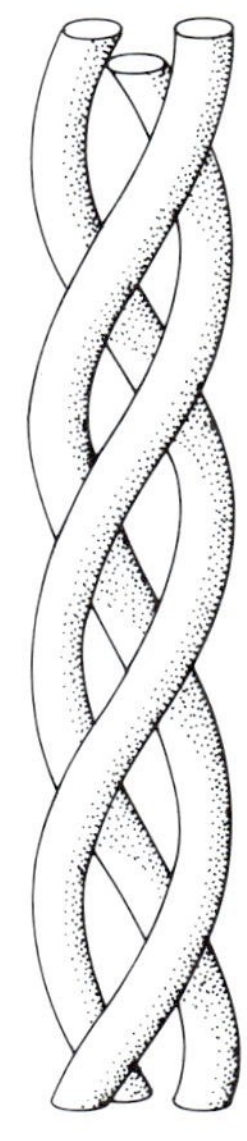

SUMMARY

Proteins are composed of amino acids linked to one another by peptide bonds. An amino acid is generally classified as a molecule having both an α -carboxyl and α-amino group and also a chemically distinct side chain (R group) attached to its α-carbon. Because carboxyl and amino groups are proton donors and acceptors, amino acids characteristically have two pK'_a values; an amino acid has a third pK'_a if its side chain has an additional ionizable group. Another physical property of amino acids is stereoisomerism (D- and L- forms), a result of the asymmetry associated with the α-carbon of all amino acids except glycine. Some amino acids, e.g., isoleucine and threonine, have two asymmetric carbons and therefore four stereoisomers. Only the L-forms of amino acids, with certain exceptions, are used by biological species.

The same 20 amino acids are required by all organisms for protein synthesis. Although modified, e.g., hydroxylated or phosphorylated, forms of certain of these amino acids are found in some proteins, the modifications occur after the amino acids are incorporated into polypeptidyl structure. The "magic 20" amino acids can be classified in the following four groups on the basis of the hydrophilicity or hydrophobicity of their side chains. The eight with nonpolar R groups are alanine, valine, leucine, isoleucine, proline, methionine, phenylalanine, and tryptophan. Proline is unique among the 20 amino acids because it is an imino acid. The seven with uncharged polar R groups are glycine, serine, threonine, tyrosine, cysteine, asparagine, and glutamine. Aspartic acid and glutamic acid are the two with negatively charged (acidic) polar R groups; lysine, arginine, and histidine are the three with positively charged (basic) polar R groups.

There are four levels of protein structure—primary, secondary, tertiary, and quaternary. The primary structure is the number and sequence of amino acids in a polypeptide. Secondary structure refers to the amount of structural regularity in a protein. The α-helical conformation of a polypeptide, produced by intramolecular (intrachain) hydrogen bonding, is one type of secondary structure. Another type is the β-pleated sheet structure, which can exist among neighboring polypeptide chains (by intermolecular hydrogen bonding) or within the three-dimensional structure of a globular protein, e.g., an enzyme. Tertiary structure is the result of the coiling and folding that a polypeptide undergoes in attaining a globular molecular shape. Noncovalent interactions between the R groups of the amino acids in a polypeptide chain greatly influence tertiary structure, e.g., hydrophobic interactions and electrostatic attractions. The formation of disulfide (—S—S—) bonds can also contribute to tertiary structure. The fourth level of structure, quaternary, involves interactions between two or more polypeptides to produce oligomeric forms of biologically active proteins. The same types of noncovalent interactions that stabilize tertiary structure are also involved in the production of quaternary structure.

Proteins are classified as fibrous and globular. The fibrous proteins are composed of polypeptides which are often laterally crosslinked by various types of bonding to yield tough and water-insoluble structural proteins, e.g., α-keratin of hair and skin, collagen of tendons, and silk. Globular proteins are water soluble and have spheroidal, compact structures. They are the biologically active proteins of living systems, e.g., enzymes and antibodies.

QUESTIONS

1. In a repeat of part of Question 2 of Chapter 4, (a) draw the four stereoisomers of threonine; this time identify the D-*allo* and L-*allo*- forms, as well as the D- and L-isomers, of the amino acid.

   Which of the stereoisomers are diastereomers (b) of D-*allo*-threonine and (c) of D-threonine?

2. Draw the formulas to show the dissociations (Table 6-2) of (a) glycine, (b) aspartic acid, and (c) lysine with changes in pH.

   What net charge would each of the three amino acids predominantly have at (d) pH 1 and (e) pH 7? (f) Calculate the isoionic points of the three amino acids.

3. In Table 6-2, which amino acid has a pK_a' value inappropriately listed as that of an α-NH_3^+ group?

4. Which of the amino acids, if any, in Table 6-2 may have an effective buffering capacity at (a) pH 4, (b) pH 6, and (c) pH 7?

5. Why would a sequence of lysyl residues in a polypeptide chain destabilize an α-helical conformation?

6. Given the following peptide:

   Gly—Ala—Asp—Pro—Lys—Met—Cys—Phe—
   Lys—Arg—Glu—Asp—Ser

   what would be its probable net charge (a) at pH 1 and (b) at pH 7?

   How many (c) branched-chain, (d) sulfur-containing, and (e) hydroxylated amino acid residues does it have?

   How many (f) nonpolar R groups and (g) uncharged polar R groups does the peptide have at physiological pH?

   (h) Could the peptide have an intramolecular disulfide bond?

7. Given the following amino acid sequence of a section of a large polypeptide:

—Pro	—Cys	—Thr	—Val	—Tyr	—Arg	—Asn	—Met
1	2	3	4	5	6	7	8
—Glu	—Phe	—Ala	—Asp	—Gly	—Gly	—Gly	—Gly
9	10	11	12	13	14	15	16
—Val	—Met	—Cys	—Ile	—Trp	—His	—Lys	—Asn
17	18	19	20	21	22	23	24

   (a) at pH 7, which residues would likely be in α-helical conformation?

   (b) which residue would produce a bend in an α-helical conformation?

(c) which series of residues would destabilize an α-helical conformation?

If the R groups of the following pairs of residues in the polypeptidyl section could interact, what type of interaction would probably occur?

(d) 12 and 23 (g) 9 and 12 (i) 4 and 17
(e) 10 and 20 (h) 3 and 12 (j) 5 and 9
(f) 2 and 19

8. With regard to the various types of bonds and interactions depicted in Figure 6-32, which amino acid residues are involved in each of the eight examples illustrated?

SUGGESTED READINGS

Dickerson, R. E., and I. Geis, *The Structure and Action of Proteins.* New York: Harper and Row, 1969. An introductory, yet sophisticated, treatment of proteins. The illustrations of the molecular details of protein structure are excellent.

Edsall, J. T., and J. Wyman, *Biophysical Chemistry,* Vol. 1. New York: Academic Press, 1958. An old book that contains a thorough presentation of amino acids as electrolytes.

Fasman, G. D., ed., *Handbook of Biochemistry and Molecular Biology,* 3rd ed. Sec. A, *Proteins,* Vol. 1. Boca Raton, Fla.: CRC Press, 1976. Good reference source for data on amino acids.

Meister, A. *Biochemistry of Amino Acids,* 2nd ed., 2 vols. New York: Academic Press, 1965. A comprehensive treatise on amino acids.

Neurath, H., and R. L. Hill, *The Proteins,* 3rd ed., 3 vols. New York: Academic Press, 1975–1977. The newest edition of a biochemistry classic, emphasizing the structure of proteins and the molecular basis of their biological functionings.

Schulz, G. E., and R. H. Schirmer, *Principles of Protein Structure.* New York: Springer-Verlag, 1979. A good book on the subject.

Wood, W. B., J. H. Wilson, R. M. Benbow, and L. E. Hood, *Biochemistry—A Problems Approach,* 2nd ed. Menlo Park, Calif.: Benjamin/Cummings, 1981. Good problems in Chapters 2 to 5 dealing with amino acids and protein structure.

Articles

Anfinsen, C. B., "Principles That Govern the Folding of Polypeptide Chains." *Science,* 181:223, 1973.

Baldwin, R. L., "The Pathway of Protein Folding." *TIBS,* 3:66, 1978.

Bornstein, P., and H. Sage, "Structurally Distinct Collagen Types." *Ann. Rev. Biochem.,* 49:957, 1980.

Fraser, R. D. B., "Keratins." *Scientific American,* 221(2):86 (offprint #1155), 1969.

Gross, J., "Collagen." *Scientific American,* 204(5):120, 1961.

Pauling, L., R. B. Corey, and R. Hayward, "The Structure of Protein Molecules." *Scientific American,* 191(1):51 (offprint #31), 1954.

Robson, B., "Protein Folding." *TIBS,* 1:49, 1976.

Stein, W. H., and S. Moore, "The Chemical Structure of Proteins." *Scientific American,* 204(2):81 (offprint # 80), 1961.

Ziegler, D. M., and L. L. Poulsen, "Protein Disulfide Bond Synthesis: A Possible Intracellular Mechanism." *TIBS,* 2:79, 1977.

PROTEINS II: DETERMINATION OF AMINO ACID SEQUENCES

7

INTRODUCTION

The successful determination of the amino acid sequence of insulin (Figure 7-1) by Frederick Sanger in 1953 made the goal of complete primary structural analysis of proteins a scientific reality. Because the amino acid sequence of a polypeptide greatly influences its native conformation, knowledge of the sequence provides pertinent chemical insights which aid in the understanding of a polypeptide's macromolecular structure. When sequence determination data are supplemented with those of x-ray crystallographic studies (Chapter 8), the combined results furnish details of

Figure 7-1
Amino acid sequence of bovine insulin.

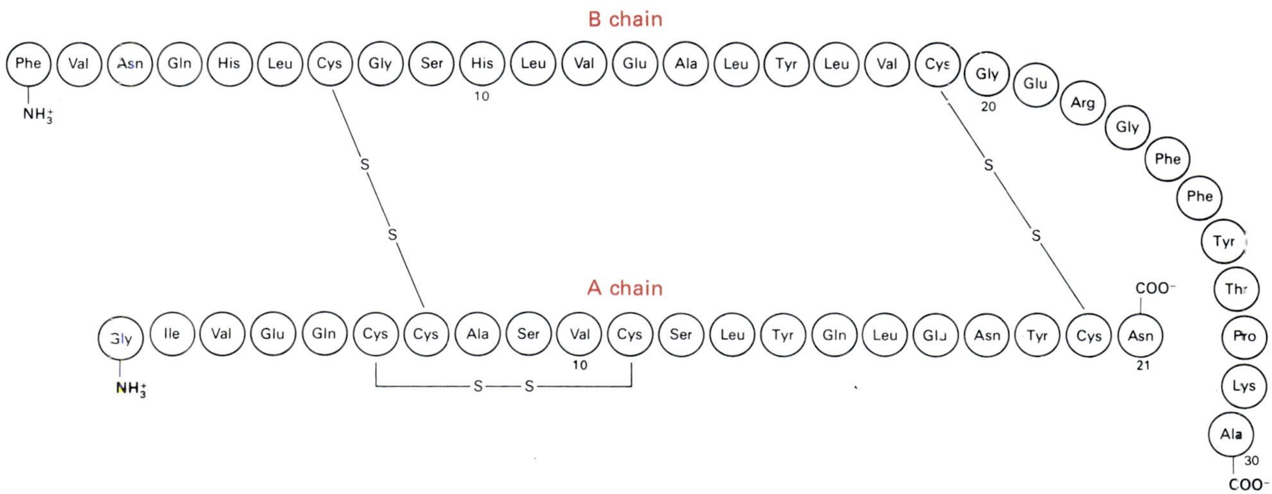

a protein's three-dimensional structure and reveal seemingly minute yet critical chemical properties which help to explain the molecule's biological function. Molecular biological research relies heavily on the analytical powers of these two experimental approaches.

Although Sanger's determination of the order of the 51 amino acid residues in insulin ushered in the era of protein sequencing, it was the subsequent advances in technology and analytical techniques that made sequencing of large polypeptides (e.g., 100 or more residues) a feasible research task. Without the development of the amino acid analyzer and the sequential Edman degradation procedure (p. 98), protein sequencing would not be in its current advanced state. However, the determination of an amino acid sequence is still a time-consuming research task, since it requires several different types of analyses whose results must be collated to obtain the sequence. Sequence determination includes both a number of specific chemical reactions and techniques for separating peptides and identifying amino acids. The experimental procedures utilized are illustrated by the determination of the sequence of the following hypothetical pentadecapeptide, a peptide containing 15 amino acid residues:

$^{+}H_3N$—Ala—Gln—Glu—Lys—Tyr—Met—Ser—Met—Ile—Arg—Val—Cys—Lys—Trp—Gly—COO^{-}

DETERMINATION OF AMINO ACID COMPOSITION

The determination of amino acid composition begins with *acid hydrolysis,* which usually involves treatment of a polypeptide with 6 *N* HCl at 110°C for 24 hours in an evacuated, sealed tube. The amino acids in the resultant mixture are then separated by *ion exchange chromatography,* a procedure that capitalizes on the differential ionization characteristics of the individual amino acids, as well as on the hydrophobicity of their side chains. The classic separation protocol, developed by Stanford Moore and William Stein (joint 1972 Nobel Prize recipients), has been incorporated into an automated system *(amino acid analyzer)* which both separates and quantifies each amino acid in a mixture.

The Moore and Stein separation uses two chromatographic columns; one is called the *short column,* used to resolve the basic amino acids and NH_4^+, and the other the *long column,* which separates the other amino acids. Both columns contain negatively charged sulfonated polystyrene resin in the Na^+ form (Figure 7-2).

Figure 7-2
Sulfonated polystyrene in the Na^+ form. Note the hydrophobic backbone of the resin.

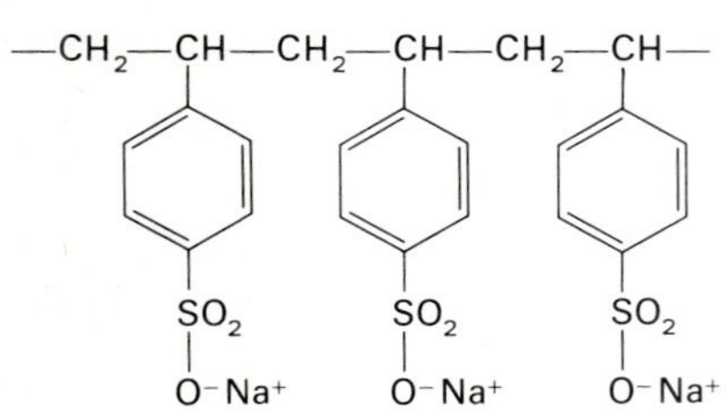

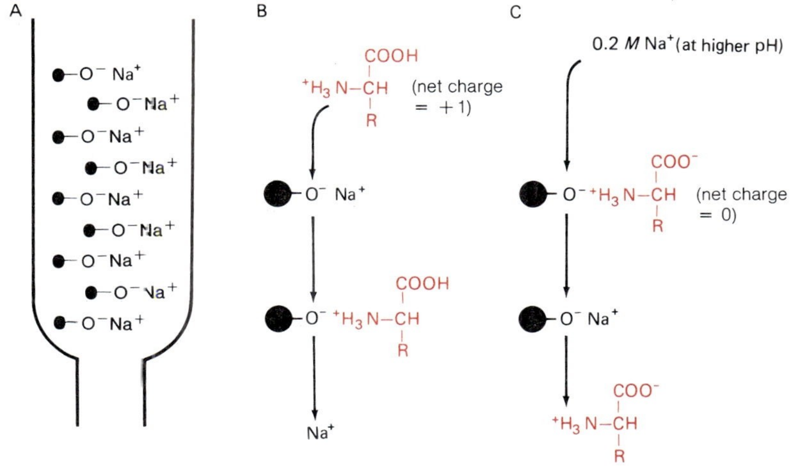

Figure 7-3
Cation-exchange chromatography: (A) column packed with negatively charged, sulfonated polystyrene resin in the Na^+ form (●—O^-Na^+); (B) protonated amino acid exchanging with Na^+; (C) Na^+ replacing an amino acid, which is then eluted.

The usual procedure is to adjust the amino acid mixture to pH 2 (protonating the amino acids) before applying a sample of less than 1 mg of the mixture to a column. *Cation exchange* occurs when the protonated amino acids (and NH_4^+) bind to a column by exchanging with the Na^+ of the sulfonated resin (B of Figure 7-3). Each column is then eluted at a higher pH (which affects the ionization of the amino acids) with sodium citrate buffer. Sodium ions now exchange with the amino acids (C of Figure 7-3), which then selectively elute from the column (Figure 7-4). The short column is eluted with a single buffer, and the long column is a two-step elution process. As stated previously, the hydrophobicity of the resin backbone also contributes to the separation process. Note, for example, that phenylalanine and tyrosine are the last amino acids to elute from the long column (Figure 7-4), with the more hydrophobic phenylalanine eluting after tyrosine.

In the automated system, each eluted amino acid reacts with

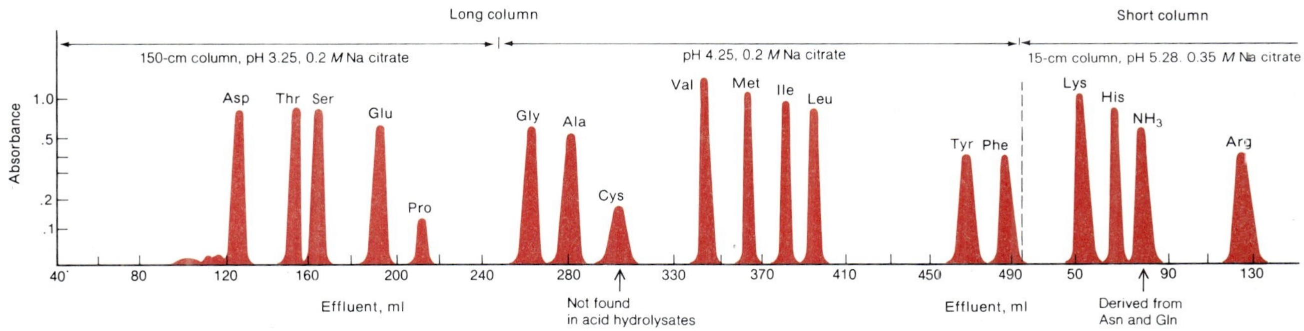

Figure 7-4
Recorded elution profile of amino acids chromatographed by the Moore and Stein procedure, using an amino acid analyzer. (After D. H. Spackman, W. H. Stein, and S. Moore, *Anal. Chem.*, 30:1190, 1958)

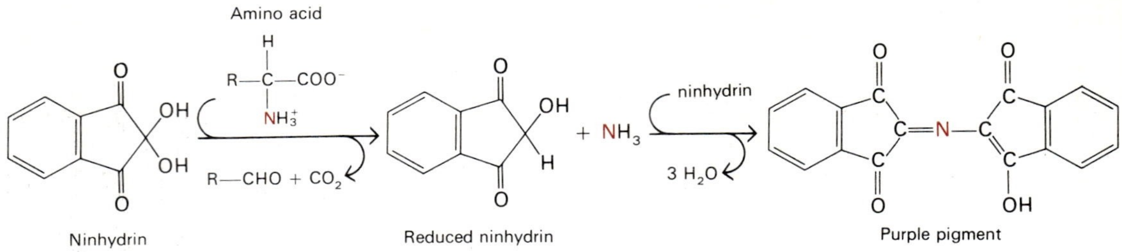

Figure 7-5
The ninhydrin reaction.

the reagent ninhydrin at 100°C (Figure 7-5), which produces a deep blue or purple color (except with proline and hydroxyproline, in which case a yellow color is produced). The intensities of the individual peaks of color that develop are measured by a flow-through colorimeter, and the absorbance data are graphically recorded as a series of peaks on a chart to furnish an analytical pattern of the elutions (Figure 7-4). The area under a given peak is proportional to the amount of the respective amino acid in the applied sample.

The traditional Moore and Stein method just described has been replaced in many laboratories by newer separation techniques and instrumentation. Although some of the newer innovations are based on the same separation principles described for the Moore and Stein procedure, they allow for a more rapid and accurate analysis of very small amounts of amino acids. It is now possible to analyze a mixture containing 1-nanomole quantities (or smaller) of amino acids in about 55 minutes (as opposed to 20-nanomole quantities in about 5 hours by the Moore and Stein method). Also, new analyzers often incorporate a computer-linked integrator to calculate the areas under the peaks and to provide a printout of amino acid concentrations. This inclusion of computer science into amino acid analysis does away with the tedious and time-consuming task of calculating the areas by hand (a particularly welcomed relief to graduate students).

Acid hydrolysis does not yield quantitative recoveries of all amino acids. During acid hydrolysis, the amide bonds of glutamine and asparagine are cleaved to yield glutamic acid and aspartic acid, respectively. Thus, Asp and Asn residues in a protein appear in a single Asp peak following hydrolysis; Glu and Gln residues appear in the Glu peak. Cysteine and tryptophan are largely destroyed by oxidation during hydrolysis, and they must be identified and quantified by additional techniques. Performic acid oxidation, which

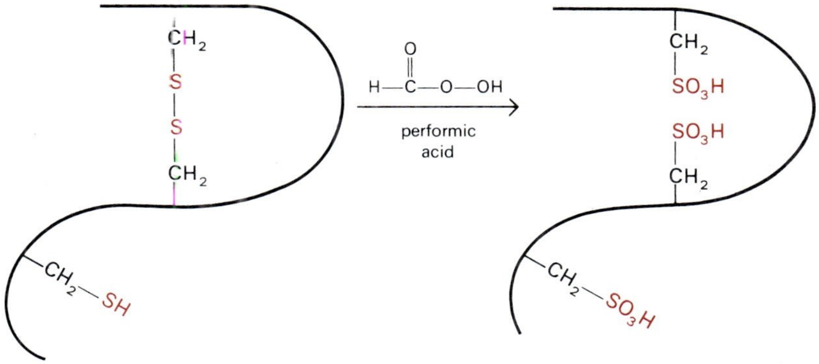

Figure 7-6
Conversion of cysteine and cystine residues of a polypeptide into cysteic acid residues by performic acid oxidation.

converts both cysteine and cystine into stable cysteic acid, is routinely used to quantify these two amino acids (Figure 7-6). Acid hydrolysis in the presence of the reducing agent thioglycolic acid is one method used to avoid oxidative degradation of tryptophan. The complete amino acid composition of the pentadecapeptide listed above would be written as follows:

$HS—CH_2—COOH$
Thioglycolic acid

(Ala, Arg, Cys, 2 Glx, Gly, Ile, 2 Lys, 2 Met, Ser, Trp, Tyr, Val)

The parentheses indicate an amino acid composition of unknown sequence. The three-letter abbreviation *Glx* recognizes that the two glutamic acids (Glu) detected following acid hydrolysis include the glutamine (Gln) content of the sample. Similar uncertainties involving aspartic acid (Asp) and asparagine (Asn) are designated *Asx.*

DETERMINATION OF AMINO-TERMINAL AMINO ACID

The specific labeling of the N-terminal amino acid residue by compounds that covalently bond with NH_2 groups is required for this determination. Sanger introduced the use of *2,4-dinitrofluorobenzene* (*DNFB,* or *Sanger's reagent*), which selectively reacts with NH_2 groups in a polypeptide (Figure 7-7). After treatment with DNFB, subsequent acid hydrolysis yields the N-terminal residue as a dinitrophenyl derivative *(DNP-amino acid),* which is yellow in color. The ϵ-NH_2 groups of lysyl residues in the polypeptide also react with DNFB, and therefore ϵ-DNP-lysine is often present in hydrolysates. An N-terminal lysyl residue can be distinguished from an internal Lys residue because the former possesses two dinitrophenyl moieties, *viz.,* α, ϵ-$(DNP)_2$-lysine, or *bis*-DNP-lysine.

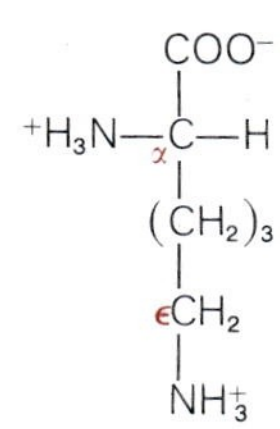

Lysine

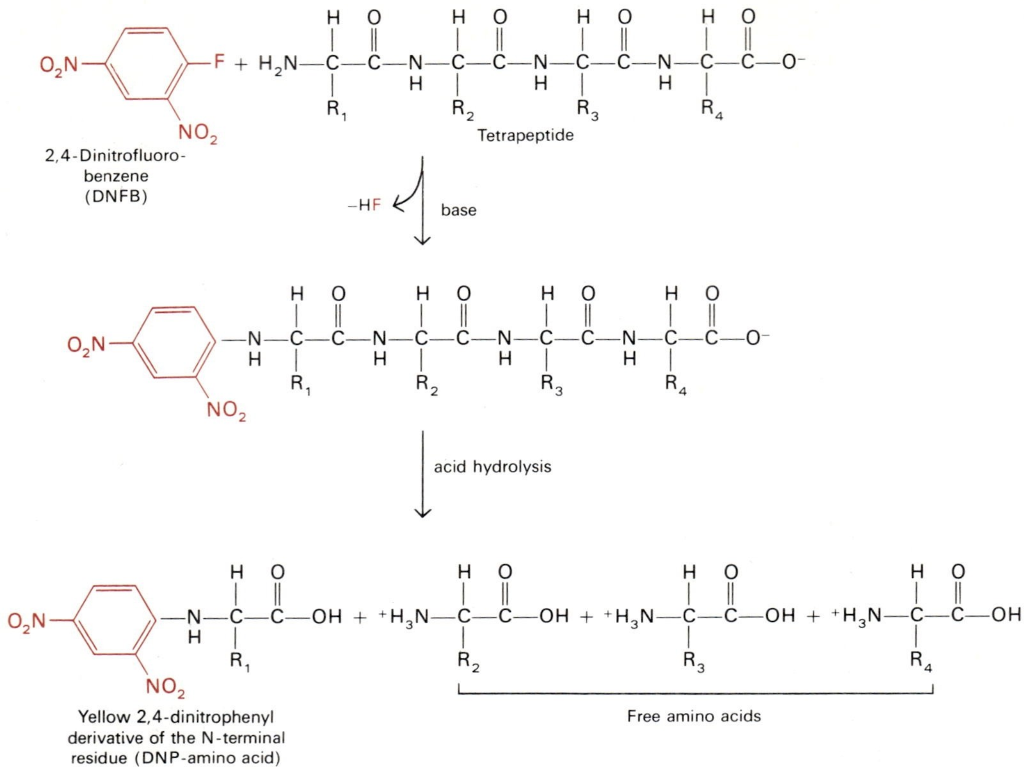

Figure 7-7
Determination of an amino-terminal residue using 2,4-dinitrofluorobenzene (Sanger's reagent).

The DNP derivative of the N-terminal amino acid is identified by separation procedures, usually *thin layer chromatography.* As illustrated in Figure 7-8, this type of separation relies principally on the solubility properties of the various DNP-amino acid derivatives in a given solvent system. Identification of the yellow, unknown DNP-amino acid derivatives is based on the comparison of their motilities in various chromatographic systems with those of known DNP-amino acids in the same systems.

Dansyl chloride also reacts with polypeptides and, after acid hydrolysis, the N-terminal amino acid is present in the hydrolysates as a *dansyl amino acid* (Figure 7-9). Dansyl chloride is advantageously used when only a small quantity of a polypeptide is available for analysis, because dansyl amino acids are highly fluorescent and therefore can be detected in minute quantities by fluorometric methods. Edman degradation (p. 98) is another procedure frequently used for N-terminal determination.

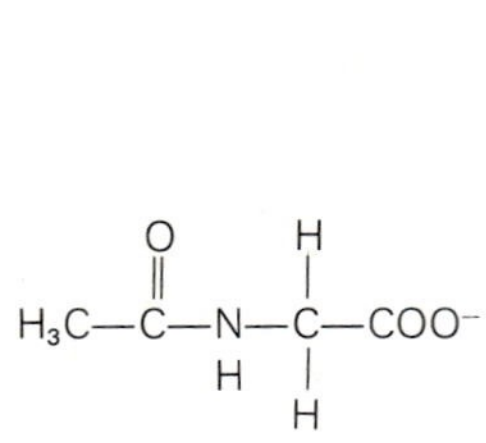

N-Acetylglycine

Many polypeptides have an N-terminal residue with a "blocked" or modified α-NH_2 group, e.g., *N*-acetylglycine of human fetal

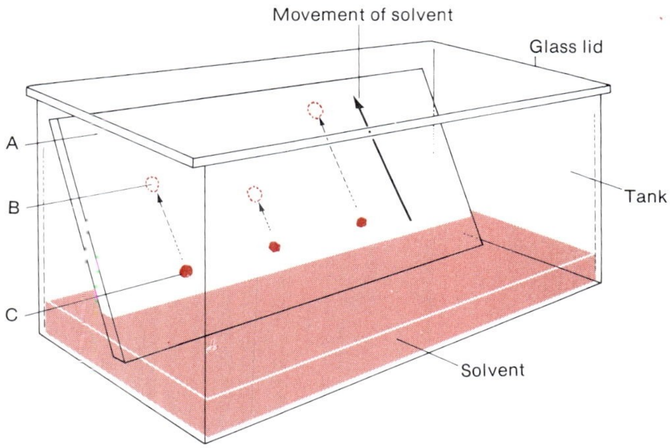

Figure 7-8
Thin layer chromatography: (A) glass plate or plastic sheet coated with thin layer of silica gel or other type of layering material; (B) point to which DNP-amino acid migrated; (C) point where DNP-amino acid sample was applied.

hemoglobin. Neither DNFB nor dansyl chloride reacts with modified NH_2 groups, and other methods must be used for the identification of such N-terminals.

DETERMINATION OF CARBOXYL-TERMINAL AMINO ACID

The usual method for determining a C-terminal amino acid residue is to treat a polypeptide with a carboxypeptidase, an enzyme that hydrolyzes the peptide bond between the penultimate and C-terminal residues (Figure 7-10) and thus releases the latter residue as a free amino acid. Carboxypeptidases are examples of *exopeptidases,* enzymes that degrade proteins either from the N- or the C-terminus, i.e., inwardly from only one end of a polypeptide. Carboxypeptidase A (which readily cleaves a C-terminal residue with an aromatic or a bulky aliphatic R group) is often the enzyme used in C-terminal residue determinations.

The major limitation of the enzymatic method is the substrate specificity of the enzyme; e.g., carboxypeptidase A does not cleave a C-terminal Lys, Arg, or Pro residue. However, comparable enzymes with different specificities can also be used; for example, carboxypeptidase B cleaves C-terminal Lys or Arg residues, and carboxypeptidase C a C-terminal Pro residue. The free C-terminal amino acid is usually identified by thin layer chromatography or with an amino acid analyzer. A sequence of amino acids at the C-terminus can sometimes be determined by analysis of the rates of their release, provided the residues are subject to carboxypeptidase hydrolysis.

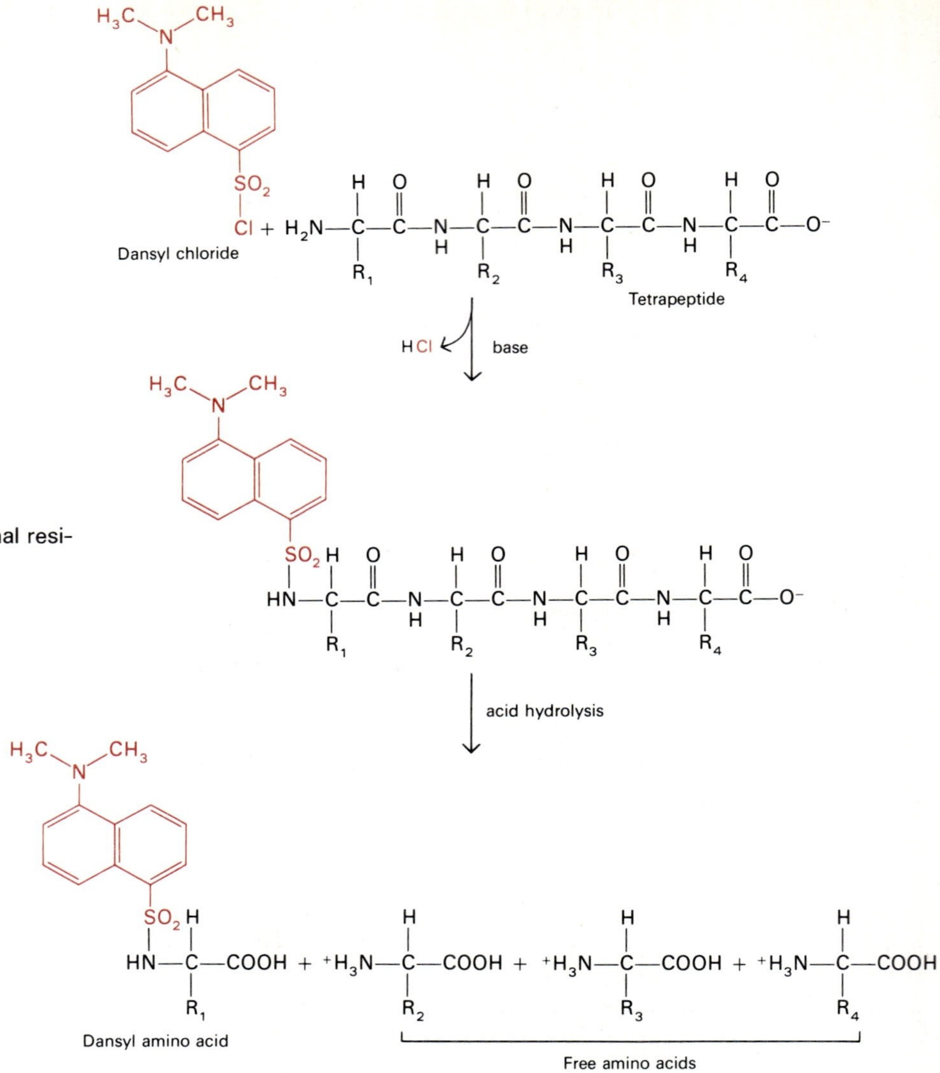

Figure 7-9
Determination of an amino-terminal residue using dansyl chloride.

Figure 7-10
Determination of a carboxyl-terminal residue using carboxypeptidase A.

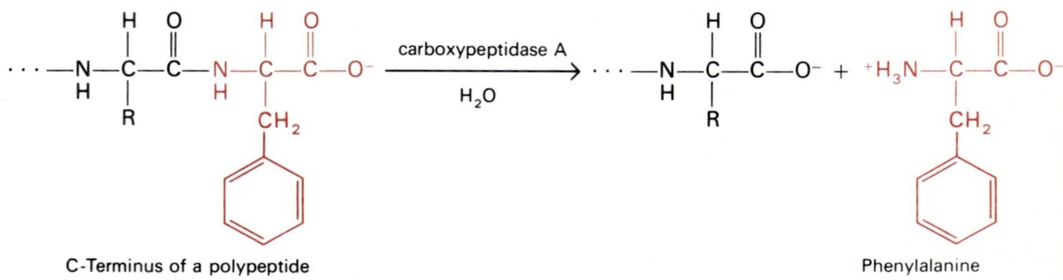

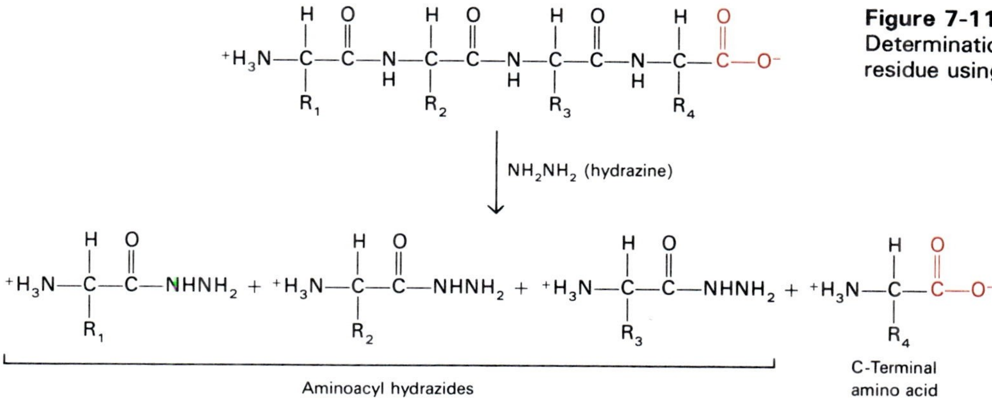

Figure 7-11
Determination of a carboxyl-terminal residue using hydrazine.

Another method for determinig the C-terminal residue involves the treatment of the polypeptide with hydrazine (Figure 7-11), which, by reacting with carbonyl ($>C{=}O$) groups, cleaves every peptide bond in the chain. The products of the reaction are a mixture of aminoacyl hydrazides and the C-terminal amino acid, which is then identified by chromatographic procedures.

After the determination of the terminal residues of the pentadecapeptide under consideration, the information is indicated as follows:

$$^{+}H_3N{-}Ala{-}(Arg, Cys, 2\ Glx, Ile, 2\ Lys, 2\ Met, Ser, Trp, Tyr, Val){-}Gly{-}COO^{-}$$

SELECTIVE FRAGMENTATION OF A POLYPEPTIDE

The next step in determining a sequence is to degrade the polypeptide to produce fragments whose sequences are then independently elucidated. Fragmentation is accomplished with the use of an enzyme or chemical that cleaves specific peptide bonds within the polypeptide chain (Figure 7-12). Enzymes which can be used

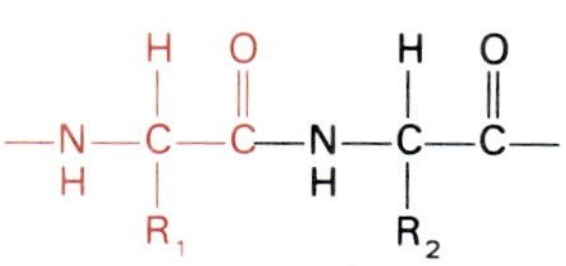

Cleaving agent	Carbonyl residue
1. Trypsin	Lysyl or arginyl
2. Chymotrypsin	Phenylalanyl, tyrosyl, tryptophyl
3. *S. aureus* V8	Glutamyl
4. Cyanogen bromide	Methionyl

Figure 7-12
Four types of specific polypeptide cleavages.

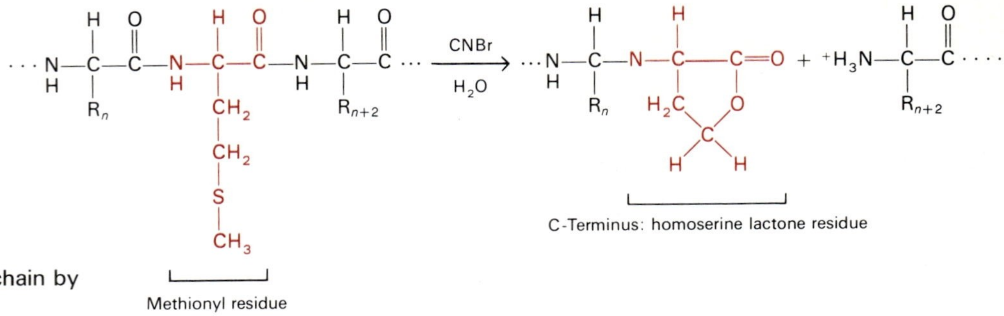

Figure 7-13
Cleavage of a polypeptide chain by cyanogen bromide.

for internal peptide bond cleavage are called *endopeptidases.* Trypsin hydrolyzes a peptide bond when the residue on the carboxyl side bears a positively charged side chain (either lysine or arginine). Chymotrypsin has an analogous specificity for peptide bonds containing aromatic (phenylalanine, tryptophan, or tyrosine) or bulky aliphatic residues, and *Staphylococcus aureus* V8 protease cleaves at glutamyl residues. The chemical cyanogen bromide cleaves when a methionyl residue is on the carboxyl side of the peptide bond, converting methionine into homoserine lactone (Figure 7-13). Treatment of the pentadecapeptide, containing one arginyl and two lysyl residues, with trypsin produces four fragments (a fragment may be a peptide or an amino acid), which then need to be separated.

Molecular sieve column chromatography (Figure 7-14) is frequently used to separate fragments; this method relies on the differences in the molecular weights and shapes of fragments. As indicated in Figure 7-14, separation of the small and large fragments occurs on the column because the small fragments can move *in and out* of the porous particles of the matrix (which retards their elution), whereas the large fragments must travel *between* the matrix particles (which hastens their elution).

Peptide fragments can also be separated by a method that combines *paper chromatography* and *paper electrophoresis* (Figure 7-15). In this technique, the first separation is based on differences in size and solubility (paper chromatography) and the second on differences in size and net electrical charge (paper electrophoresis).

Once the fragments are separated and obtained in pure form, their amino acid compositions and the two terminal residues of each peptide are determined. For the pentadecapeptide, the combination of fragmentation data with those previously obtained furnishes the following sequencing information:

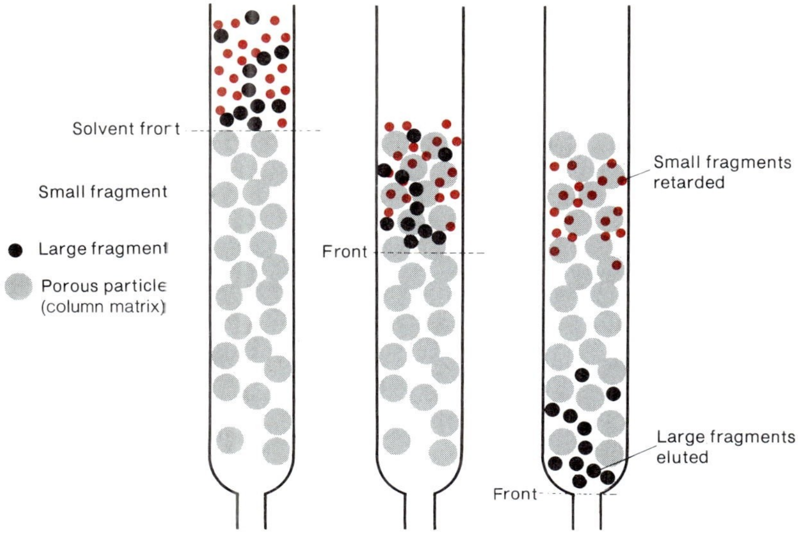

Figure 7-14
Molecular sieve (gel filtration) chromatography. From left to right: the separation of a mixture of small and large fragments on a molecular sieve column.

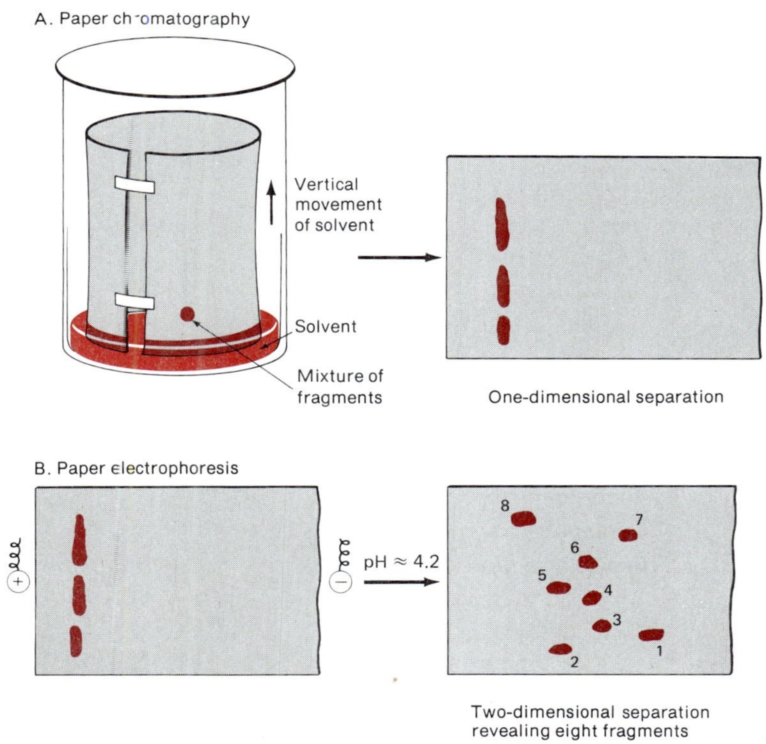

Figure 7-15
Combined paper chromatography (A) and paper electrophoresis (B) separation of a mixture of eight polypeptide fragments.

Fragment T-1: H_2N—Ala—(2 Glx)—Lys—COOH N-terminal fragment
Fragment T-2: H_2N—Tyr—(Ile, 2 Met, Ser)—Arg—COOH
Fragment T-3: H_2N—Val—Cys—Lys—COOH
Fragment T-4: H_2N—Trp—Gly—COOH C-terminal fragment

Peptide T-1 (trypsin fragment 1) is deduced to be the N-terminal fragment because it is the only one of the four having the same N-terminal residue (alanine) as the original pentadecapeptide. The sequences of the two internal residues of peptide T-1 and the four of T-2 remain unknown. Sequences of the dipeptide T-4 and tripeptide T-3 are furnished by amino acid composition and N- and C-termini data. Fragment T-4 is recognized as the C-terminal peptide because unlike the other three, which have the expected Lys or Arg C-terminus (based on trypsin's specificity), it has Gly, the C-terminal residue of the pentadecapeptide.

EDMAN DEGRADATION

To determine the sequence of peptides having four or more amino acids, the *sequential Edman degradation* procedure is used. In this method, *phenylisothiocyanate* is the reagent employed to elucidate

Figure 7-16
The Edman degradation procedure.

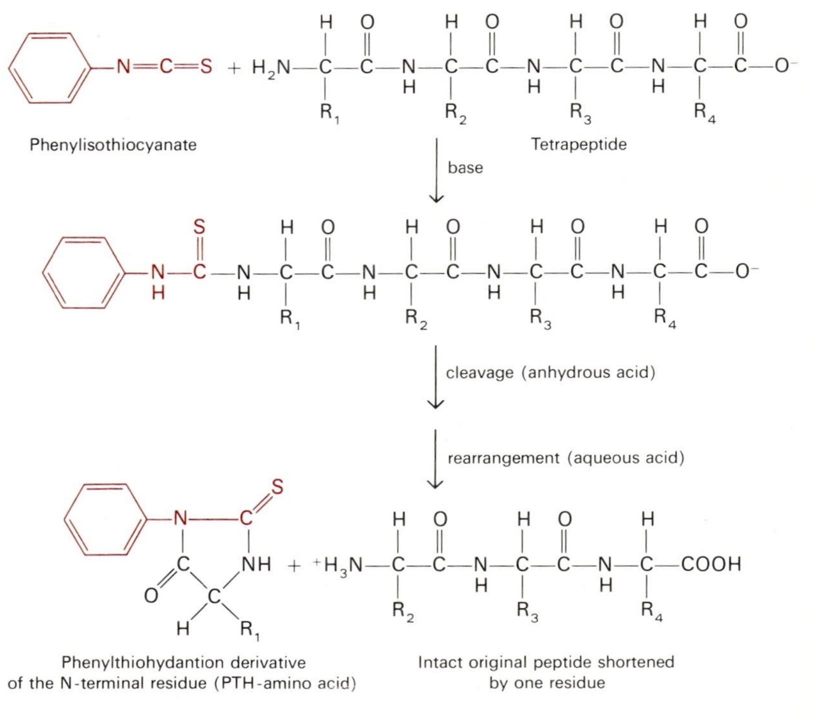

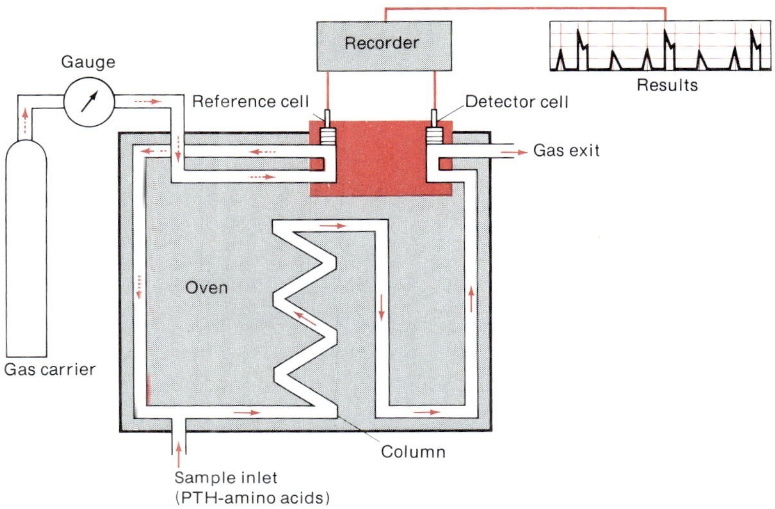

Figure 7-17
Gas chromatography apparatus.

the sequence of a peptide. As illustrated in Figure 7-16, after treatment of the polypeptide with phenylisothiocyanate, the N-terminal amino acid residue can be released as a *phenylthiohydantoin* (PTH) *derivative.* The significant feature of this procedure is that the release of the N-terminal residue does not destroy the remaining portion of the peptide. Thus, the method is of critical importance to sequencing projects because it allows for the repeated determination of newly created N-terminal residues of the peptide. Like the Moore and Stein method for the separation of amino acids, the Edman degradation procedure has been incorporated into an automated system called a *sequenator.* Cycles of repeated Edman degradations can be performed on a polypeptide or peptide with the instrument, with each cycle providing a PTH-amino acid (N-terminal residue). Gas chromatography is frequently used for the primary identification of PTH-amino acids (Figure 7-17). This high-temperature chromatographic procedure relies on the volatility (and some affinity for the column material) that the individual PTH-amino acids display on passage through a column. In the case of the pentadecapeptide, the use of sequential Edman degradation procedures on fragments T-1 and T-2 completes the following sequence identification for all four tryptic digest peptides:

T-1: H_2N—Ala—Gln—Glu—Lys—COOH
T-2: H_2N—Tyr—Met—Ser—Met—Ile—Arg—COOH
T-3: H_2N—Val—Cys—Lys—COOH
T-4: H_2N—Trp—Gly—COOH

Note that the two Glx residues of the amino acid composition data can now be identified as a glutamyl and a glutaminyl residue (T-1), since Edman degradation does not quantitatively hydrolyze amide groups.

REQUIRED SECOND FRAGMENTATION

At this point in the sequencing project, much information has been obtained; however, the sequential arrangement of fragments T-2 and T-3 with respect to T-1 (N-terminal fragment) and T-4 (C-terminal) remains to be established. To accomplish these assignments, fragmentation of the pentadecapeptide by a second method and subsequent analysis of the resultant fragments are necessary. Cyanogen bromide treatment of the pentadecapeptide, which contains two methionyl residues, yields a set of three fragments that are different from those obtained by treatment with trypsin. Repeating all the procedures utilized on the trypsin-digest fragments, the following sequences are obtained:

CB-1: H_2N—Ala—Gln—Glu—Lys—Tyr—Homoserine lactone
CB-2: H_2N—Ser—Homoserine lactone
CB-3: H_2N—Ile—Arg—Val—Cys—Lys—Trp—Gly—COOH

The homoserine lactone residues of the derived peptides are derivatives of the parental methionyl residues of the original structure.

THE COMPLETE SEQUENCE

The sequence of the pentadecapeptide can now be established by comparison of the overlaps and inclusions of the peptide sequences obtained by the trypsin and cyanogen bromide cleavages, as follows:

T-1:	H_2N—Ala	—Gln	—Glu	—Lys											
CB-1:	H_2N—Ala	—Gln	—Glu	—Lys	—Tyr	—Met									
T-2:					Tyr	—Met	—Ser	—Met	—Ile	—Arg					
CB-2:							Ser	—Met							
CB-3:									Ile	—Arg	—Val	—Cys	—Lys	—Trp	—Gly—COOH
T-3:											Val	—Cys	—Lys		
T-4:														Trp	—Gly—COOH

$^{+}H_3N$—Ala—Gln—Glu—Lys—Tyr—Met—Ser—Met—Ile—Arg—Val—Cys—Lys—Trp—Gly—COO^{-}

SUMMARY

The sequencing of a polypeptide begins with elucidation of its amino acid composition, which involves acid hydrolysis and the subsequent separation and quantification of the mixture of amino acids. An amino acid analyzer is commonly used to determine amino acid compositions. Additional information on the polypeptide is obtained by identifying its N- and C-termini. N-Terminal residues are generally determined with the use of 2,4-dinitrofluorobenzene (Sanger's reagent) or dansyl chloride; the latter reagent is advantageously used when only small amounts of a polypeptide are available. C-Terminal residues are frequently determined enzymatically with the use of a carboxypeptidase (exopeptidase). Hydrazinolysis is another method for identifying C-terminal residues.

Once the amino acid composition is known, the polypeptide is selectively cleaved by an endopeptidase, e.g., trypsin or chymotrypsin, with a substrate specificity for certain amino acid residues. A chemical often used for polypeptidyl fragmentation is cyanogen bromide, which specifically reacts with methionyl residues to bring about cleavage and the concomitant conversion of methionine into homoserine lactone (C-terminus of such fragments). After fragmentation, the mixture of peptides is separated, usually by molecular sieve chromatography or by paper electrophoresis and paper chromatography. The amino acid composition and N- and C-termini of each fragment are determined (information that identifies the sequence of di- and tripeptides). The sequences of larger peptides are determined by the sequential Edman degradation procedure, which selectively removes the N-terminal residue and leaves the remaining portion of the fragment intact for further analysis. Thus, by repeated cycles of Edman degradation, a fragment's sequence is obtained.

Collectively, all the various sets of data from the above analyses furnish the complete sequence of the fragments of a polypeptide; however, the positions of many of these fragments in the polypeptide chain remain unknown. It is necessary, therefore, to repeat the polypeptidyl fragmentation with a second cleaving agent, one which has a different residue specificity from the first, to acquire another mixture of fragments for sequence analysis. The sequence of the polypeptide is established by identifying the overlaps between the sets of sequencing data obtained on the two different mixtures of fragments.

QUESTIONS

1. Draw the structures of (a) cysteic acid, (b), α,ϵ-$(DNP)_2$-lysine, (c) dansyl alanine, and (d) PTH-glycine.
2. In the ninhydrin reaction (Figure 7-5), what other type of measurement (besides the development of color) could be used to determine the concentration of an amino acid?
3. What is the predominant ionic species of the following amino acids when each is eluted from a Moore and Stein column? (Use data in Table 6-2).

(a) Glutamate—eluted at pH 3.25
(b) Methionine—eluted at pH 4.25
(c) Lysine—eluted at pH 5.28

4. What is the chemical difference between serine and homoserine? Draw the structure of homocysteine.

5. If a mixture of the following three pentapeptides is electrophoresed at pH 7.5, what would be their pattern of migration toward the anode (+)?
(a) Thr—Phe—His—Asp—Met
(b) Ile—Glu—Asp—Cys—Asp
(c) Glu—Tyr—Asp—Glu—Lys

6. Given the following polypeptide:

$^{+}H_3N$—Gly—Phe—Asp—Lys—Trp—Pro—
Met—Ala—Ser—Val—Thr—Arg—Gly—
Ala—Lys—Ile—His—Met—Leu—Arg—COO^-

which amino acid(s) can be identified by treating the polypeptide with each of the following:
(a) 2,4-dinitrofluorobenzene
(b) carboxypeptidase A
(c) hydrazine
(d) dansyl chloride

7. If the above polypeptide is cleaved by trypsin, (a) how many fragments would be produced? (b) If the amino-terminal residue is determined for each of the fragments, could the N-terminal fragment of the polypeptide be identified (if the N-terminal residue of the polypeptide is known)? (c) Would determination of the amino acid composition and the N- and C-termini of each fragment elucidate the sequence of any of the fragments?

8. Assume that the above polypeptide is cleaved with cyanogen bromide and answer the three questions posed in Question 7.

9. Given the following information, what is the probable sequence of the peptide?
Amino acid composition: (Ala, Leu, Arg, Met, Phe, Thr, 2 Val).
Treatment of the peptide with
- Sanger's reagent yields DNP-Ala.
- trypsin: two fragments (Ala, Arg, Thr) and (Leu, Met, Phe, 2 Val), which when treated with Sanger's reagent yields DNP-Ala and -Val, respectively.
- cyanogen bromide: two fragments, (Ala, Arg, homoserine lactone, Thr, 2 Val) and (Leu, Phe), which when treated with Sanger's reagent yields DNP-Ala and -Leu, respectively.

10. Given the following information about a decapeptide:
- Amino acid composition determined after acid hydrolysis is (Arg, Asx, Gly, Glx, Leu, Lys, Met, Ser, Pro, Tyr)
- Treatment of the peptide with
  (1) dinitrofluorobenzene yields DNP-Gly
  (2) trypsin yields three fragments having the following amino acid compositions:
    T-1: (Asx, Glx, Leu, Lys, Pro)
    T-2: (Arg, Gly, Ser)
    T-3: (Met, Tyr)
  (3) cyanogen bromide results in no cleavage
- Edman degradation (using a sequenator) on fragment T-1 yields Pro, Leu, and Asn after cycles 1, 2, and 3, respectively. Because one of the pumps breaks down after cycle 3, it is necessary to shut down the sequenator.

Based on the above information, what is the probable sequence of the decapeptide?

SUGGESTED READINGS

Blackburn, S., *Protein Sequence Determination.* New York: Marcel Dekker, 1970. A general overview of protein sequencing techniques.

Croft, L. R., *Handbook of Protein Sequences,* 2nd ed. New York: Wiley, 1980. A compilation of over 800 protein sequences.

Croft, L. R., *Introduction to Protein Sequence Analysis.* New York: Wiley, 1980. An up-to-date review of current methods for determining amino acid sequences (paperback).

Dayhoff, M. O., ed., *Atlas of Protein Sequences and Structure,* Vol. 5, Suppl. 3. Silver Springs, Md.: National Biomedical Research Foundation, 1979. A catalogue of all known amino acid sequences.

Fasman, G. D., ed., *Handbook of Biochemistry and Molecular Biology,* 3rd ed. Sec. A. *Proteins,* Vol. III.

Boca Raton, Fla.: CRC Press, 1976. Contains the amino acid sequences of a number of proteins.

Kirchner, J. G., *Thin-Layer Chromatography,* 2nd ed. New York: Wiley, 1978. A comprehensive treatment of the many applications of this technique.

Needleman, S. B., ed., *Molecular Biology, Biochemistry, and Biophysics,* 2nd ed. Vol. 8, *Protein Sequence Determination.* New York: Springer-Verlag, 1975. A comprehensive reference that provides practical information about sequencing procedures.

Needleman, S. B., ed., *Molecular Biology, Biochemistry, and Biophysics.* Vol. 25, *Advanced Methods in Protein Sequencing Determination.* New York: Springer-Verlag, 1977. An authoritative presentation of the principles of certain specialized techniques used in protein sequencing.

Wood, W. B., J. H. Wilson, R. M. Benbow, and L. E. Hood, *Biochemistry–A Problems Approach,* 2nd ed. Menlo Park, Calif.: Benjamin/Cummings, 1981. A good set of sequencing problems in Chapter 3.

Articles

Chou, P. Y., and G. D. Fasman, "Secondary Structural Prediction of Proteins from Their Amino Acid Sequence." *TIBS,* 2:128, 1977.

Dayhoff, M. O., "Computer Analysis of Protein Evolution." *Scientific American,* 221(1):86 (offprint # 1148), 1969.

Stein, W. H., and S. Moore, "Chromatography." *Scientific American,* 184(3):35 (offprint #81), 1951.

Wu, T. T., W. M. Fitch, and E. Margoliash, "The Information Content of Protein Amino Acid Sequences." *Ann. Rev. Biochem.,* 43:539, 1974.

PROTEINS III: STRUCTURE AND FUNCTION

8

INTRODUCTION

In the 1950s, elucidation of the α-helical and β-pleated sheet structures of proteins and the double helix of the nucleic acid DNA was made possible by x-ray crystallographic studies. The value of x-ray crystallography to the study of the more complicated tertiary and quaternary conformations of proteins was established by the definitive determination of the structures of two oxygen-carrying proteins, myoglobin by John Kendrew and hemoglobin by Max Perutz. Twenty-three years of dedicated effort by Perutz (who began the study when he was a graduate student) were required to obtain the structure of hemoglobin. Both scientists shared the Nobel Prize in 1962 for their monumental contributions to science.

Between 1912 and 1915, William Henry Bragg and his son, William Lawrence Bragg, developed the technique of x-ray diffraction by determining the crystalline structure of NaCl. As the joint 1915 Nobel Prize recipients in physics, the Braggs became the only father-son combination to receive the award and W. Lawrence, who was 25 years old at the time, the youngest scientist to be so honored.

The basic experimental design in x-ray crystallographic studies of proteins involves the passage of an x-ray beam through a protein crystal and the detection of its diffraction (scattering) pattern on a photographic film (Figure 8-1). Electrons of the individual atoms of the macromolecule are responsible for the diffraction,

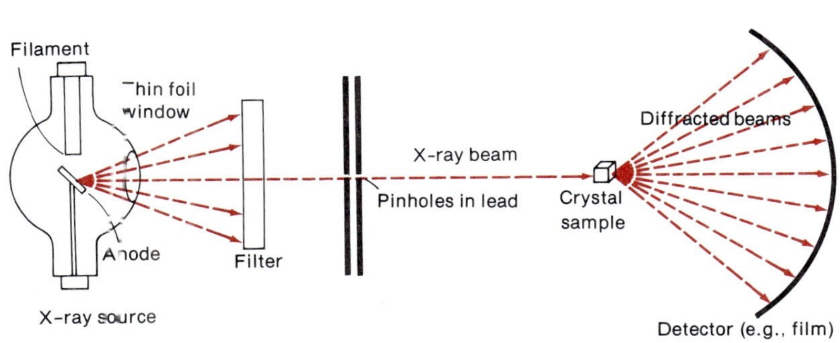

Figure 8-1
Schematic of the diffraction of an x-ray beam by its passage through a crystal.

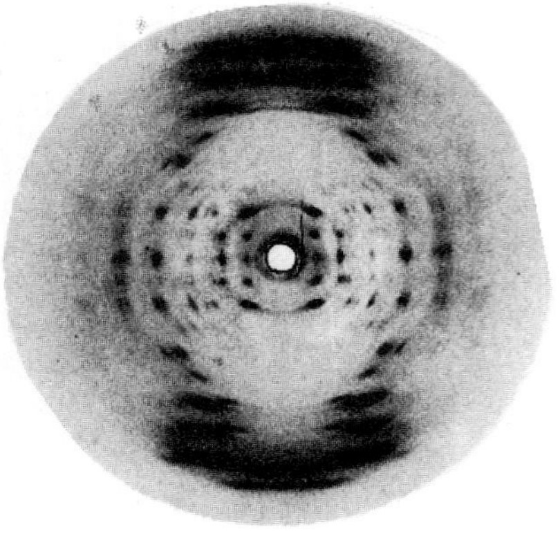

Figure 8-2
An x-ray diffraction pattern of DNA.

with each electron contributing to the overall pattern (Figure 8-2). The intensity of a blackened emulsion spot on the film is proportional to the number of electrons in the atoms responsible for that scatter point, i.e., the heavier the atoms the greater the intensity. These intensities, which collectively indicate the locations of atoms fixed in space in molecular structure, are the basic data used to construct a three-dimensional image of the protein. Separate x-ray diffractions are obtained with the protein containing one or two heavy atoms, e.g., lead or uranium, to provide high-density areas in the diffraction patterns by which the phase of the scattered beams can be determined. Through the use of mathematical analysis (Fourier series) and computers, two-dimensional electron-density patterns of parallel sections of the molecules are constructed (Figure 8-3) which, when stacked one upon another, yield a three-dimensional electron distribution. In x-ray analysis, greater resolution enhances the details of individual atomic arrangements in the protein structure. Resolutions of 6 and 1.4 Å for the myoglobin

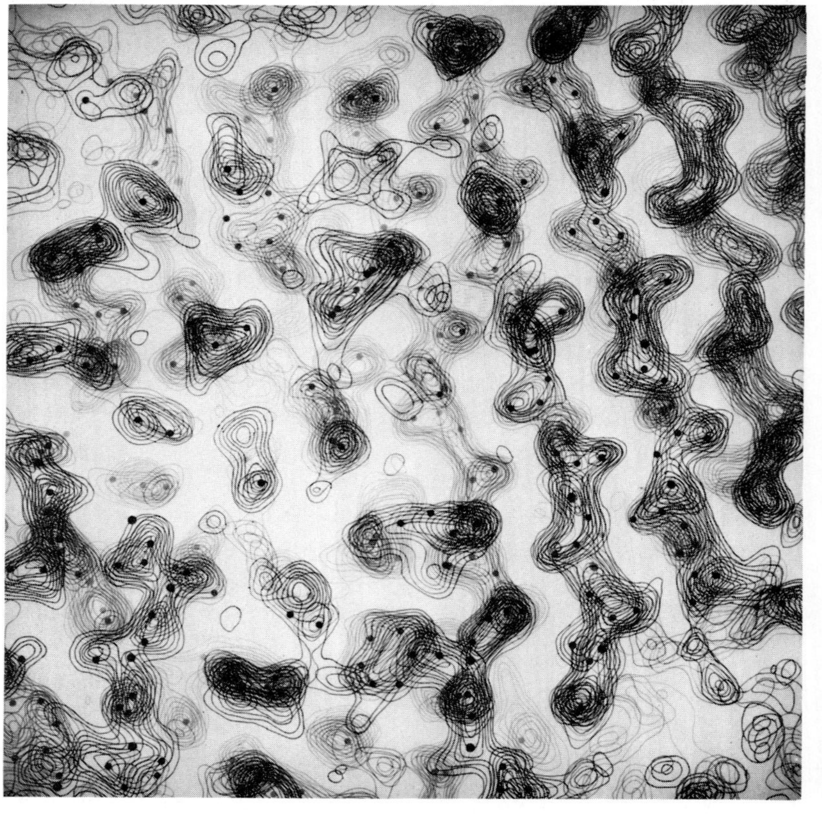

Figure 8-3
A view through several consecutive sections of the three-dimensional electron-density map of the plant protein concanavalin A. Regions of equal electron density are drawn as contours in each section, and consecutive sections, 0.5 Å apart, are stacked to represent the three-dimensional electron density. The dots represent atomic positions. The contours on the sections in the back of the stack appear fainter than those in the front because of the thickness of the plastic sheets on which the map is plotted. (Courtesy, Drs. J. W. Becker, G. N. Reeke Jr., and G. M. Edelman)

structure were provided by measurements of 400 and 25,000 intensities in the diffraction patterns, respectively. The greater the number of intensity data for the latter resolution allowed for an atomically distinct reconstruction of the myoglobin molecule. Knowledge of the amino acid sequence of the protein being analyzed is a valuable asset in x-ray crystallographic studies.

MYOGLOBIN

General Molecular Features

Myoglobin is an oxygen-carrying protein found in muscle, where it serves both as an oxygen reserve and to facilitate oxygen diffusion through the cell. Kendrew elected to study myoglobin of the sperm whale because of the abundance of the protein in the skeletal muscle of diving mammals, which need to have unusually large oxygen reserves. Myoglobin has two molecular components, a single polypeptide chain containing 153 amino acid residues (MW = 17,600) and a *heme group,* an iron-containing biomolecule (Figure 8-4). The organic portion of the heme molecule is a *porphyrin,* containing a tetrapyrrole structure *(porphin)* formed by methene-bridge linkages between four *pyrrole* rings. Various types of porphyrin structures occur in nature. The heme group, or *iron porphyrin,* of myoglobin, hemoglobin, and of most cytochromes (p. 321) is called *iron protoporphyrin IX* and is molecularly characterized by its substituent groups (two proprionate, two vinyl, and four methyl) and their positions in the tetrapyrrole structure. Protoporphyrins *chelate* metal ions, e.g., iron, magnesium, and copper, and are therefore utilized by certain metabolic processes which require a metal ion to form biologically active *chelation complexes* (Figure 8-4). The heme of myoglobin is known as a *prosthetic group* because it is a nonprotein organic molecule which binds tightly to the polypeptide structure. The binding of an iron atom to a heme involves the four nitrogens of the pyrrole rings (Figure 8-4). The bound iron can form two additional bonds, one on each side of the heme plane, called the fifth and sixth coordination positions; it is the sixth coordination position of ferrous iron in myoglobin that binds an oxygen molecule.

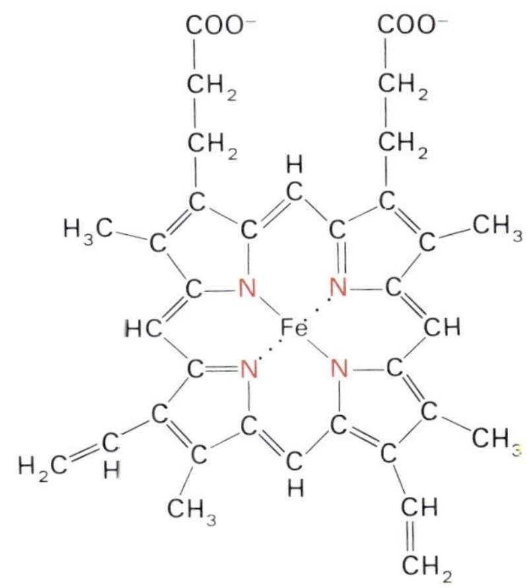

Figure 8-4
A heme group containing protoporphyrin IX and iron (Fe).

Chelate is derived from the Greek word *chele,* meaning the prehensile *claw* of a crab or lobster. A *chelating* biomolecule is one that *binds* metal ions.

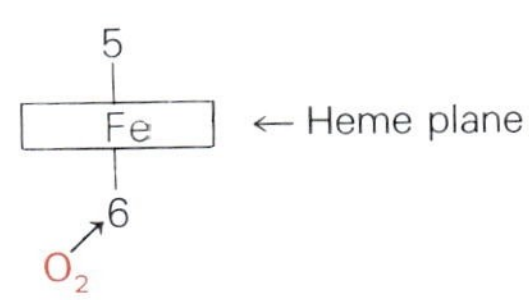

Tertiary Structure

Kendrew provided the scientific world with its first three-dimensional structure of a globular protein, and his model of myoglobin

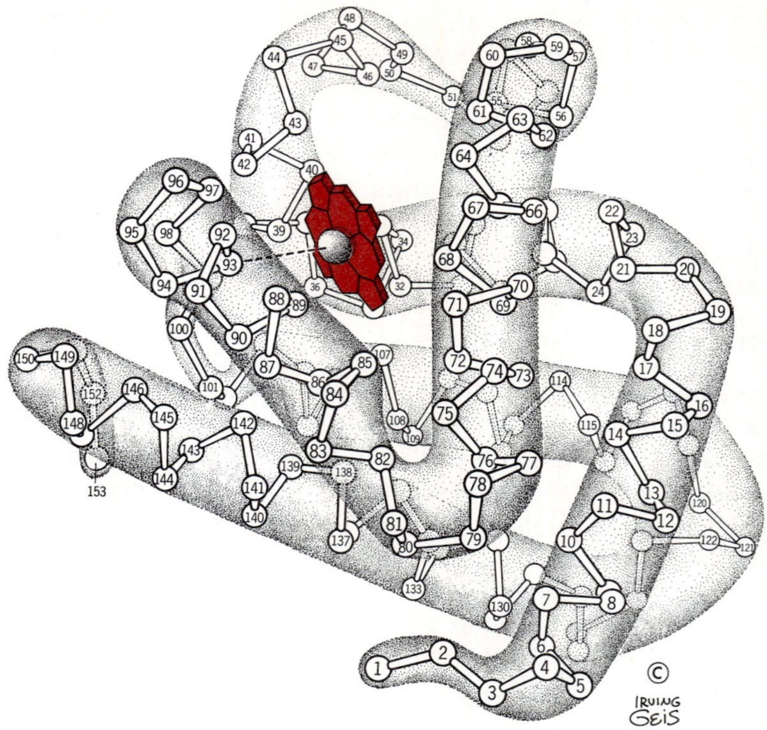

Figure 8-5
Three-dimensional structure of myoglobin (only α-carbon positions are shown). Heme group is in red. (Copyright 1982 by Irving Geis)

(Figure 8-5) offered pertinent insights not only into the complicated tertiary structure of globular proteins but also into the biological functioning of oxygen-carrying macromolecules. Myoglobin, with molecular dimensions of $45 \times 35 \times 25$ Å, is a very compact macromolecule with little empty space in its interior. The secondary structural features previously proposed for other proteins are evident in the globular shape of myoglobin. Eight sections of the molecule exist in right-handed α-helical conformation (designated from the N-terminus as A through H); these helical segments involve approximately 75 per cent of the total amino acid residues. Interposed among the helical segments are five nonhelical regions, each identified by the two segments it joins, e.g., AB is located between helical section A and B. There are also two nonhelical residues at the N-terminus and five at the C-terminus which, as segments, are designated NA and HC, respectively. There are eight terminations of α-helical structure, with the four prolyl residues of the molecule accounting for four of them. All the peptide bonds of the polypeptide chain are planar, with the carbonyl and amide groups being *trans* to each other. The polar residues and some

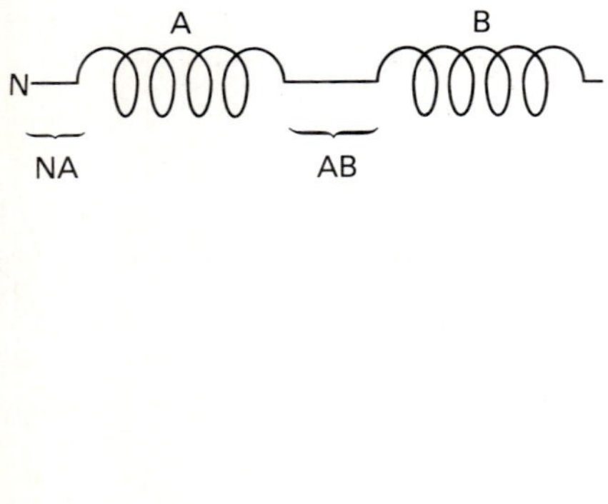

nonpolar residues are located on the exterior of the molecule. The interior of myoglobin contains only nonpolar residues, except for two histidyl residues (out of a total of 12), whose specific interactions with iron are required for biological activity.

The oxygen-binding heme group is located near the surface of the molecule in a crevice which (except for the histidyl residues mentioned above) provides a nonpolar environment (Figure 8-5). The heme plane lies between histidyl residue F8 (eighth residue of helical segment F) and E7, with HisF8 being bound to the iron atom through its fifth coordination position (Figure 8-6). HisE7, on the other side of the plane, is not bound directly to the sixth coordination position of iron, which is the oxygen-binding site. The iron atom is about 0.3 Å out of the heme plane on the HisF8 side. Upon oxygenation, the iron atom descends into the heme plane and the oxygen bound to myoglobin is stabilized by hydrogen bonding to the imidazole ring of HisE7 (Figure 8-6). Reversible oxygenation requires that the iron atom be in the ferrous state (Fe^{2+}), and myoglobin with and without oxygen bound to the Fe^{2+} of heme is called *oxymyoglobin* and *deoxymyoglobin,* respectively. If water occupies the sixth coordination position, the ferrous ion is rapidly oxidized to Fe^{3+}, and the resultant *ferrimyoglobin* or *metmyoglobin* cannot bind oxygen. Hence, the nonpolar character of the heme crevice is an important feature of myoglobin structure, since it helps to prevent water molecules from destroying the oxygen-carrying capabilities of the molecule. The maintenance of the Fe^{2+} state is also aided by enzyme systems which catalyze the reduction of ferric ions ($Fe^{3+} \rightarrow Fe^{2+}$).

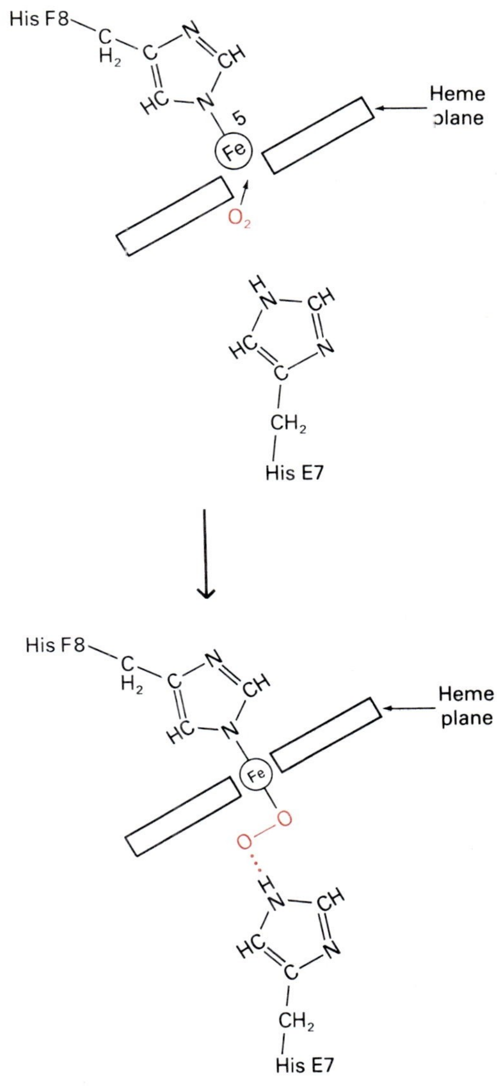

Figure 8-6
Schematic of the binding of oxygen to myoglobin and the bonding relationships of HisF8 with the iron atom and of HisE7 with the bound oxygen.

HEMOGLOBINS

Hemoglobin, which constitutes about 90 per cent of the protein of red blood cells, is a tetrameric protein, i.e., it contains four polypeptide chains. Adult hemoglobin (HbA), the major hemoglobin of adults, has two α and β chains in its structure ($\alpha_2\beta_2$) and thus has a heterogeneous quaternary structure. There is also a minor adult hemoglobin (HbA_2) which normally accounts for about 3 percent of total hemoglobin and has two δ chains instead of β chains ($\alpha_2\delta_2$). HbA and HbA_2 are the postnatal forms of hemoglobin. Prior to birth, embryonic and fetal hemoglobins are used as oxygen carriers. The first hemoglobin to appear in embryonic development is designated $\zeta_2\epsilon_2$; the ζ and ϵ chains are analogous to the α and β chains, respectively. When ζ-chain production ceases (after about

It is estimated that there are about 270 million hemoglobin molecules in a red blood cell.

six weeks), the tetramer $\alpha_2\epsilon_2$ appears, i.e., α replaces ζ. A third embryonic hemoglobin, $\zeta_2\gamma_2$, has also been identified (γ replacing ϵ). These last two hemoglobins represent transition phases leading to the appearance of fetal hemoglobin (HbF), whose tetrameric composition is $\alpha_2\gamma_2$. The two α chains, each with 141 amino acid residues, are a common molecular feature of HbF and HbA. The γ , δ, and β chains, although distinct molecular species, are each composed of 146 residues and have homologous (similar but not identical) amino acid sequences. The γ and δ chains differ from the β chain at 39 and ten amino acids residues, respectively.

α alpha
β beta
γ gamma
δ delta
ϵ epsilon
ζ zeta

Tertiary and Quaternary Structure

In 1959, Perutz succeeded in his ambitious undertaking to determine the three-dimensional structure of hemoglobin (HbA of the horse), which, unlike the myoglobin molecule, has the added

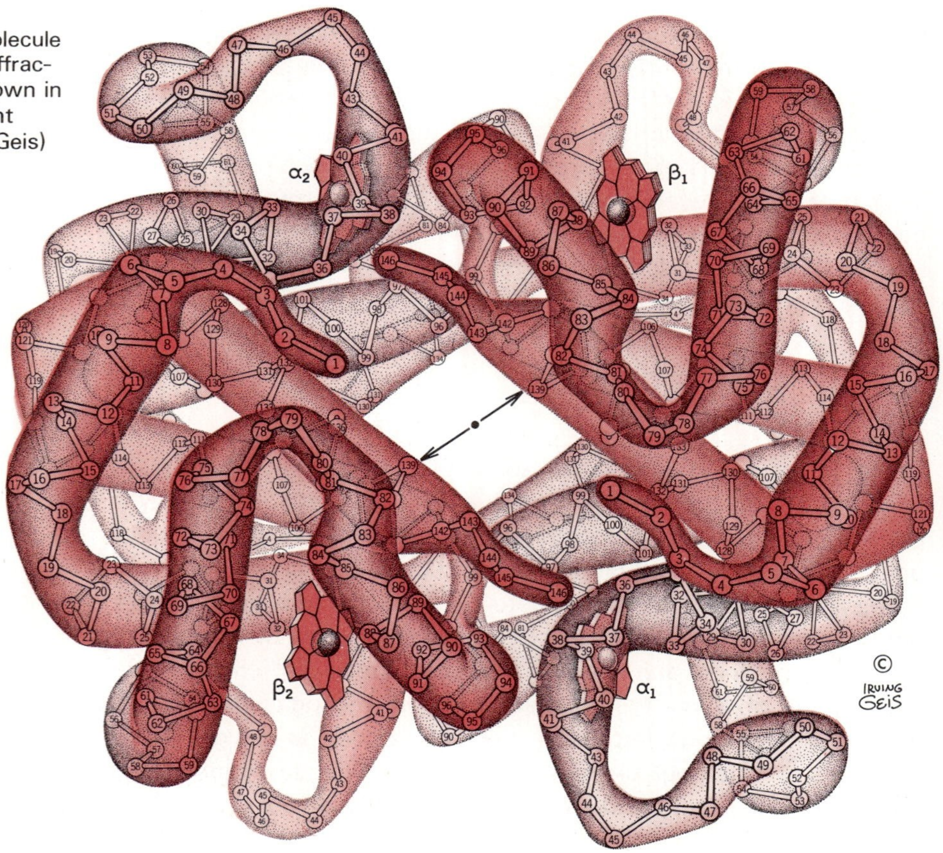

Figure 8-7
Top view of oxyhemoglobin A molecule (horse), as deduced from x-ray diffraction studies. The β chains are shown in dark color and the α chains in light color. (Copyright 1979 by Irving Geis)

dimension of quaternary structure (Figure 8-7). The structure reveals that the two α and two β chains are packed in a molecular form that is almost spherical in shape and about 55 Å in diameter. The majority of the interactions among the four polypeptides is between α and β chains, with only a few between the two α or the two β chains. Each of the chains has a heme group which, like that of myoglobin, is located in a nonpolar crevice near the surface of the molecule. None of the heme groups are close to one another.

The tertiary structures of the α and β chains are very similar to that of myoglobin (Figure 8-8), even though the amino acid sequences of the three polypeptides are distinct, with only 24 of the 141 amino acid positions being identical. By its evolutionary design, the tertiary structure of a molecule functioning as an oxygen carrier must accommodate a heme group in a hydrophobic environment to ensure the reversible oxygen binding that is vital for its biological functioning. A comparison of the α and β chains of hemoglobin and the single chain of myoglobin also illustrates the versatility nature used in employing different amino acid sequences to derive a particular type of tertiary structure.

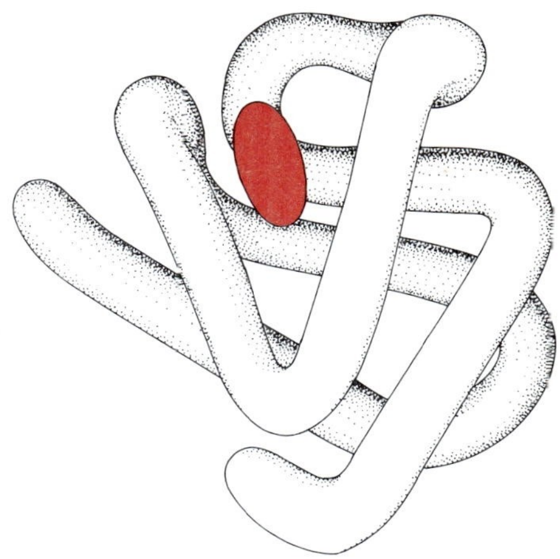

Figure 8-8
Schematic of the three-dimensional structure of the β chain of hemoglobin. Compare with the structure of myoglobin (Figure 8-5).

Biological Significance of Quaternary Structure

Quaternary structure is of critical importance to the proper biological functioning of hemoglobin because it is the tetrameric composition of the oxygen carrier that allows for a variety of functions not observed with single-chained myoglobin. Hemoglobin, in addition to transporting oxygen, also transports CO_2, a waste product of metabolism, to the lungs to be respired. The ability of hemoglobin to bind H^+ (another waste product of metabolism) is also an important physiological function of the macromolecule, since it is essential for the maintenance of physiological pH (Chapter 3). The carrier roles of hemoglobin are closely interrelated, since the binding of oxygen is regulated in part by the amounts of H^+ and CO_2 in the circulatory system. The sensitivity hemoglobin displays toward H^+ and CO_2, as well as toward certain other compounds of metabolic significance, creates a dynamic physiological situation whereby the functions of this carrier molecule are regulated for biological effectiveness and efficiency. The studies on hemoglobin by Perutz and other scientists have yielded molecular insights into nature's use of quaternary structure as a means of providing a protein with the structural information needed to respond to its biological regulators.

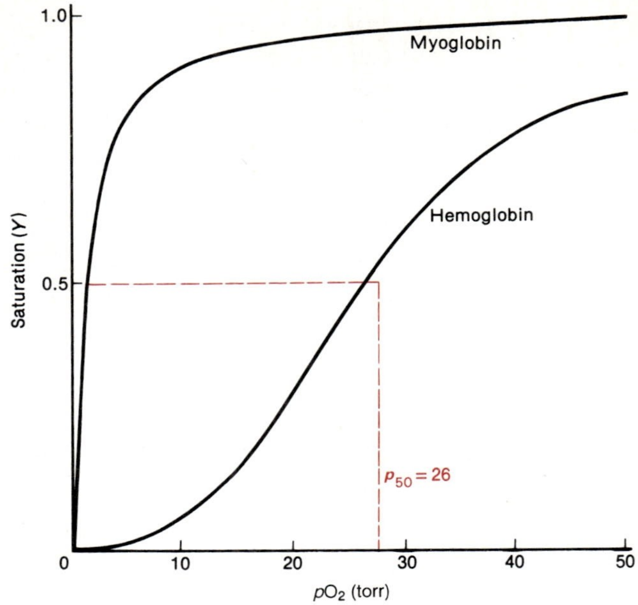

Figure 8-9
Oxygen dissociation curves of myoglobin and hemoglobin.

Oxygen Dissociation Curves

The differences between hemoglobin and myoglobin as oxygen carriers are readily apparent when their oxygen dissociation curves are compared (Figure 8-9). An *oxygen dissociation curve* is a plot of the saturation of the binding sites (heme groups) with oxygen (expressed as *Y*) as a function of the partial pressure of oxygen (pO_2 in torr) to which the protein solution is exposed, i.e., $Y \approx [HbO_2]/[HbO_2] + [Hb]$. Affinity for oxygen is generally expressed as a *p*50 value, which is the partial pressure at which 50 per cent of the sites are oxygenated ($Y = 0.5$). The *p*50s of myoglobin and hemoglobin *in vivo* are approximately 1 and 26 torr, respectively; thus, myoglobin has the greater affinity for oxygen. The shapes of the curves for the two oxygen carriers are also strikingly different, with the curve for myoglobin being *hyperbolic* and that for hemoglobin being *sigmoidal* (S-shaped). The sigmoidal nature of the latter curve reflects the *cooperative binding* of oxygen to hemoglobin, which means that the binding of oxygen to one heme facilitates oxygenation of the other hemes. Because myoglobin has only a single heme group, the phenomenon of cooperative binding is not possible, as evidenced by its hyperbolic curve.

Torr—pressure equal to that exerted by 1 mm Hg (mm column of mercury) at 0°C and standard gravity.

The influence that the binding of an oxygen molecule to hemoglobin has on the subsequent oxygenation of the other heme groups of the molecule is an example of an interaction at one site on a protein affecting another site located in a distinctly different region of the same molecule. Such interactions are generally referred to as *allosteric interactions* or *effects,* and they constitute an important mechanism by which the biological functioning of a protein can be effectively regulated by conformational changes. In addition to hemoglobin, many enzymes are *allosteric proteins* (p. 145). A quaternary structure is usually characteristic of allosteric proteins. The results obtained from the research on hemoglobin have been very fruitful in furnishing molecular details on the types of mechanisms involved in allosterism.

Molecular Aspects of Hemoglobin Oxygenation

As illustrated in Figure 8-10, the three-dimensional structure of deoxygenated hemoglobin *(deoxyhemoglobin)* differs from that of the oxygenated form *(oxyhemoglobin).* In the quaternary structure of deoxyhemoglobin, there are eight additional electrostatic interactions (salt linkages) not found in oxyhemoglobin, making deox-

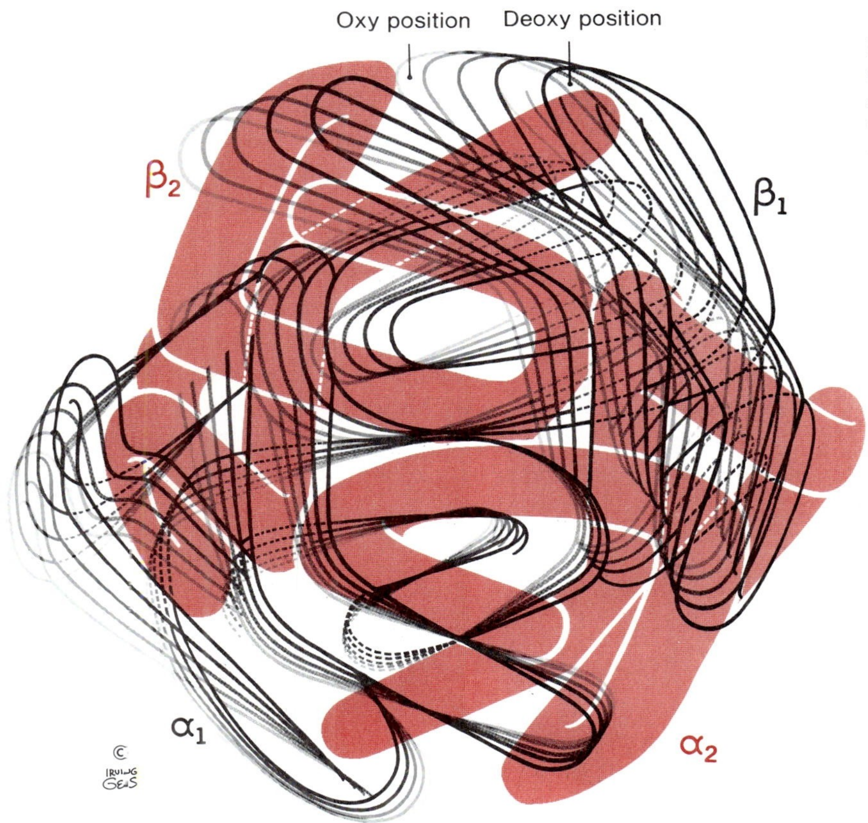

Figure 8-10
Quaternary motion in hemoglobin, showing structural changes that occur when the molecule goes from the oxygenated to the deoxygenated state. Motion of the $\alpha_1\beta_1$ dimer (black and gray lines) relative to that of the $\alpha_2\beta_2$ dimer (red). (Copyright 1982 by Irving Geis)

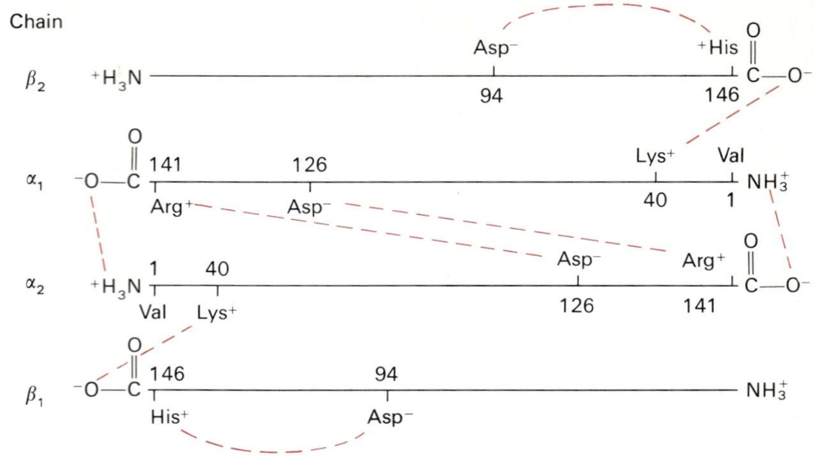

Figure 8-11
Schematic of the eight electrostatic interactions (— —) that occur between the α and β chains of deoxyhemoglobin.

yhemoglobin the more rigid molecule of the two (Figure 8-11). Six of these eight interactions are between chains. The C-terminal residues of the four chains of deoxyhemoglobin are also involved in the electrostatic bondings. As a consequence of oxygenation, a hemoglobin molecule undergoes conformational changes which disrupt the eight electrostatic interactions, and the cooperative binding observed with oxygen is the result of these structural changes. When the first oxygen molecule binds to a heme, the diameter of the Fe^{2+} becomes small enough for the iron atom to fit into the center of the heme group, and the Fe^{2+} is pulled into the plane of the porphyrin ring (Figure 8-12). HisF8, because of its bonding to the iron atom, also moves toward the heme plane, and this slight displacement of the histidine residue initiates a sequential series of conformational changes that ultimately results in the disruption of some of the subunit interactions in the quaternary structure of deoxyhemoglobin. Thus, it is by alteration of molecular shape that the binding of an oxygen molecule facilitates the subsequent binding of additional oxygen to hemoglobin.

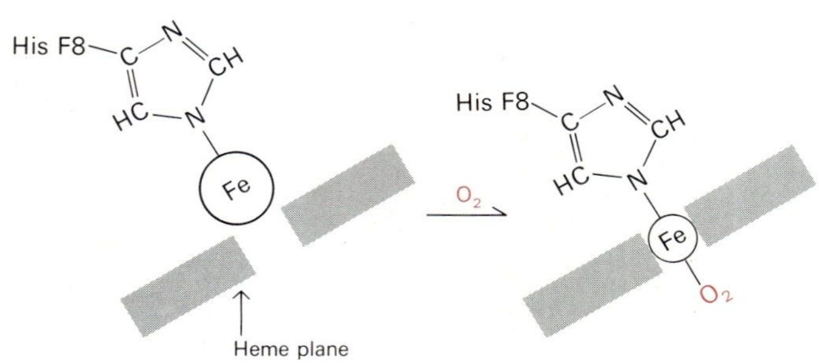

Figure 8-12
Schematic of the movement of the iron atom into the heme plane when oxygenation occurs.

Bisphosphoglycerate as a Hemoglobin Regulator

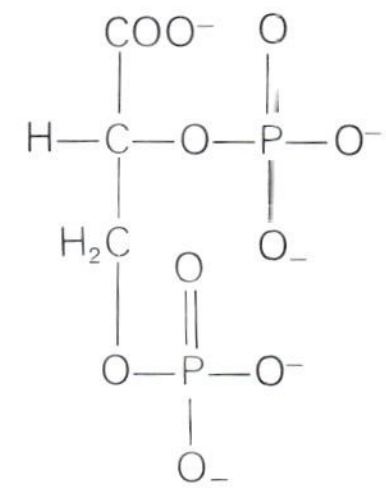

2,3-Bisphospho-D-glycerate

An important physiological biomolecule that regulates the biological functioning of hemoglobin is 2,3-bisphospho-D-glycerate (2,3-P_2-glycerate), which occurs in equimolar ratios with hemoglobin in red blood cells and which markedly reduces the oxygen affinity of hemoglobin. In the presence of physiological levels of 2,3-P_2-glycerate, the $p50$ of hemoglobin is about 26 torr; without 2,3-P_2-glycerate, it is comparable to that of myoglobin, i.e., about 1 torr. 2,3-P_2-Glycerate exerts its regulatory role by binding in the central cavity of deoxyhemoglobin (Figure 8-13). Such binding involves electrostatic interactions between the negatively charged groups of 2,3-P_2-glycerate (net charge of about -4 at physiological pH) and six positively charged residues of the two β chains: the α-amino group of Val NA1, Lys EF6, and His H 21. The interaction of 2,3-P_2-glycerate, which serves to link the two β chains, adds

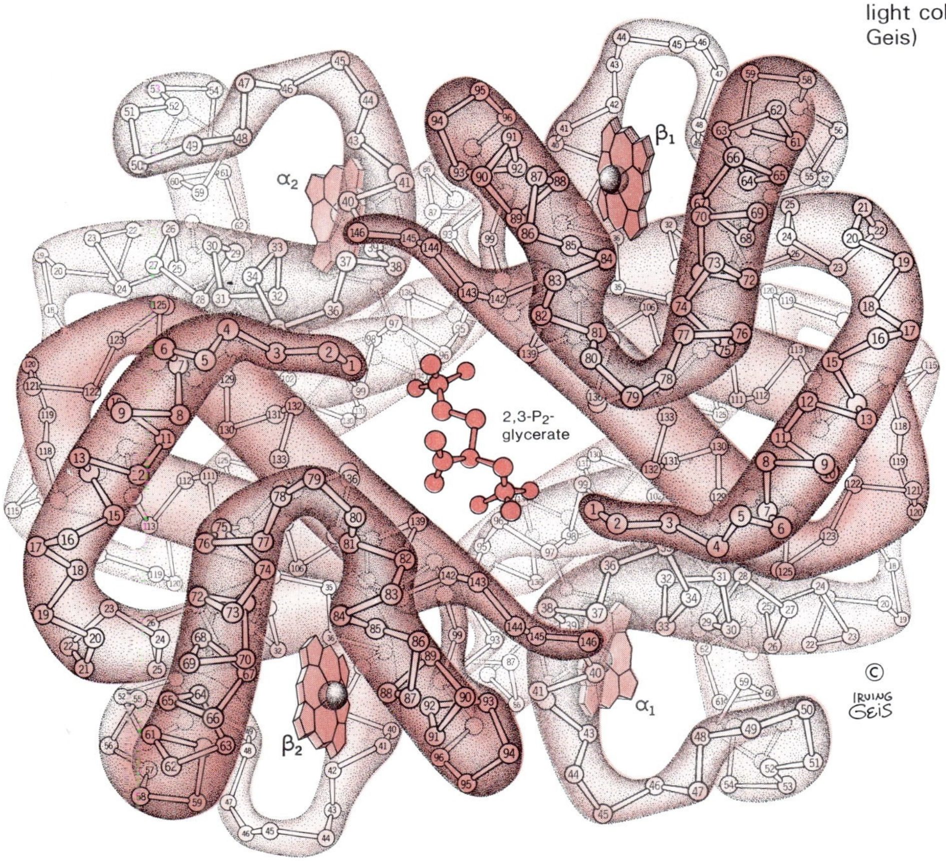

Figure 8-13
Top view of deoxyhemoglobin A (horse), depicting the central cavity with 2,3-P_2-glycerate bound. The β chains are shown in dark color and the α chains in light color. (Copyright 1979 by Irving Geis)

to the number of electrostatic interactions that must be disrupted for oxygenation. As oxygenation occurs, the accompanying conformational changes in hemoglobin make the central cavity too small to accommodate 2,3-P_2-glycerate, which is then expelled. Also, the structural changes that take place during oxygenation destroy the well-defined spatial arrangement of the positively charged residues that allows for the electrostatic bonding of 2,3-P_2-glycerate.

Physiological Importance of Hemoglobin Regulation

Hemoglobin has the dual role of obtaining oxygen from air in the lungs and of delivering it to body tissues. In tissue capillaries, the pO_2 is about 26 torr and, without the lowered oxygen affinity imposed by 2,3-P_2-glycerate regulation, hemoglobin with a $p50$ of 1 torr would effectively oxygenate in the lungs but, on the other hand, would deliver little oxygen to tissues because of insufficient deoxygenation.

The importance of regulation by 2,3-P_2-glycerate in the transfer of oxygen to a fetus is now also recognized. Fetal hemoglobin does not bind 2,3-P_2-glycerate as strongly as does HbA because the γ chain (analogous to the β chain of HbA) has a seryl residue in the H 21 position instead of a positively charged histidine. Because of this particular amino acid difference, HbF has a higher affinity for

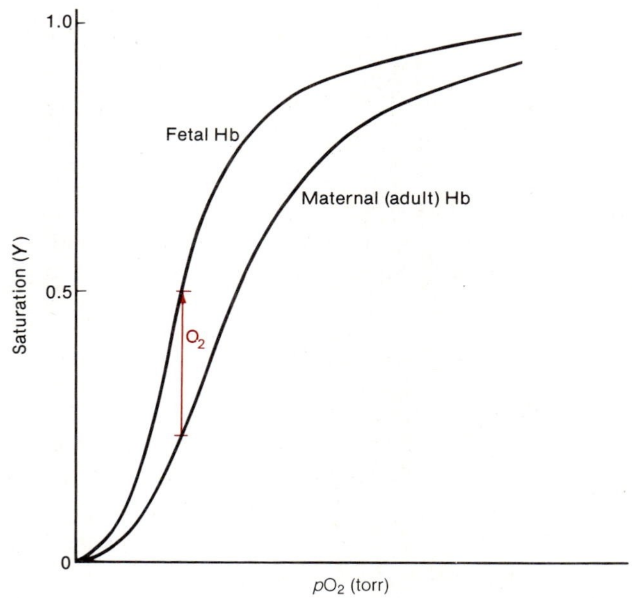

Figure 8-14
Schematic oxygen dissociation curves for HbA (adult) and HbF (fetal), illustrating the higher affinity that HbF has for oxygen. Red arrow represents transfer of oxygen from maternal to fetal hemoglobin.

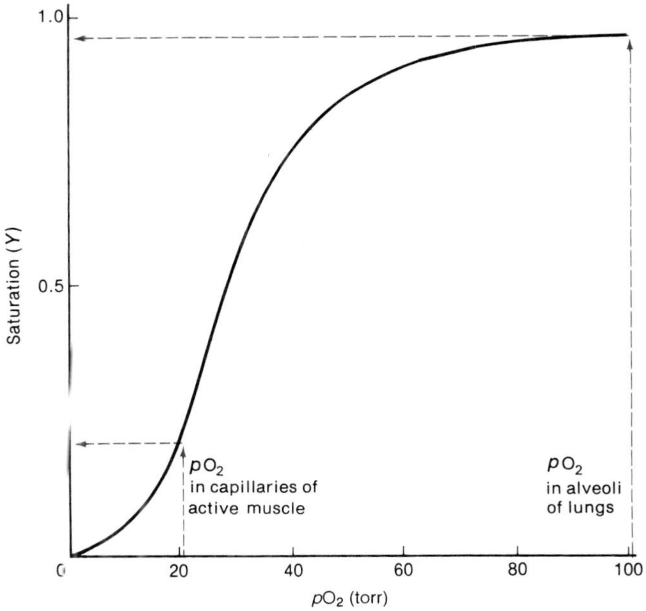

Figure 8-15
Oxygen dissociation curve of hemoglobin, illustrating the difference in oxygen affinity of the molecule in capillaries ($pO_2 \approx$ 20 torr) and in alveoli of the lungs ($pO_2 \approx$ 100 torr).

oxygen than does HbA (Figure 8-14), and this difference is physiologically important because it helps to ensure the transfer of oxygen from HbA (maternal circulation) to HbF (fetal circulation).

The cooperative binding of oxygen to hemoglobin is also biologically important. The sigmoidal characteristics of oxygen dissociation (Figure 8-15) assures effective transport and delivery of oxygen to tissues, since hemoglobin is fully oxygenated in the alveoli of the lungs ($pO_2 \approx$ 100 torr) and significantly deoxygenated in tissue capillaries ($pO_2 \approx$ 20-26 torr). It has been calculated that if oxygen did not bind cooperatively only about half as much oxygen could be delivered to tissues. The cooperative binding of oxygen and 2,3-P_2-glycerate regulation are notable illustrations of the evolutionary development of molecular mechanisms that optimize the biological usefulness of biomolecules.

Hemoglobin as a H^+ and CO_2 Transport Protein

Hemoglobin not only furnishes oxygen to tissues but also transports the waste products of metabolism (H^+ and CO_2) from the tissues. Since the same biomolecule is responsible for both transport systems, it is not surprising that there is a regulatory interplay between the two functions. The metabolic interdependence of the two transport functions of hemoglobin is known as the *Bohr*

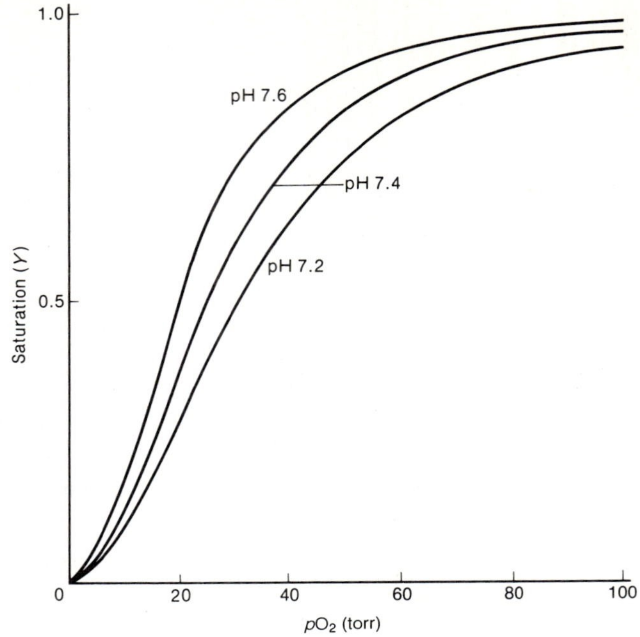

Figure 8-16
Oxygen dissociation curves of hemoglobin at various pH values.

effect. Increased concentrations of CO_2 and H^+ (lowering of pH) decrease the oxygen affinity of hemoglobin (Figure 8-16), and, conversely, increased concentrations of oxygen lower the affinity for CO_2 and H^+. (By contrast, increased concentrations of CO_2 and H^+ have little effect on myoglobin's oxygen affinity.) Under physiological conditions, the interrelationships between the two transport functions are important aspects of each system. Acid (H^+) and CO_2 in the capillaries of actively metabolizing tissue, e.g., muscle, bring about an increased liberation of hemoglobin-bound oxygen, which is needed by the tissues for continued metabolism. In the lungs, the high concentration of oxygen enhances the release of H^+ and CO_2 from hemoglobin for CO_2 exhalation.

Binding of H^+ and CO_2 to Deoxyhemoglobin

In recent years chemical and x-ray studies have been employed to identify the deoxyhemoglobin binding sites for H^+ and CO_2. With respect to H^+, the binding sites were anticipated to be ionizable groups of amino acid residues that, in the deoxyhemoglobin conformation, can accept protons, i.e., have pK_a' values of about 7. The His146 (C-terminal) residues of the two β chains and the α-amino groups (N-terminal residues) of the two α chains (Figure 8-

11) have been identified as four of the H^+ binding sites. The two His122 residues of the α chains have also been implicated as additional binding sites. In the case of CO_2, the evidence is that the α-amino groups (N-termini) of the four polypeptide chains are the binding sites. The binding of CO_2 is accomplished by the formation of a carbamate with unprotonated forms of the α-amino groups.

$$\underset{\text{(N-terminal amino group)}}{R{-}NH_2} + CO_2 \rightleftharpoons R{-}\underset{H}{N}{-}COO^- + H^+$$

Binding of carbon dioxide to hemoglobin

N- and C-Terminal Residues of Hemoglobin

The molecular studies on hemoglobin reveal the substantial contributions that the C- and N-terminal residues make to the conformational stability (ionic interactions), transport function (H^+ and CO_2), and 2,3-P_2-glycerate regulation of deoxyhemoglobin. Of particular interest are the C-terminal His146 of the β chains and the N-terminal Val-1 of the α chains, because their participation in the eight ionic interactions characteristic of deoxyhemoglobin enhances their affinity for H^+ and thereby contributes to their effectiveness as H^+-binding sites. The involvement of the α-amino groups of the α chains of deoxyhemoglobin in ionic bonding and their binding of H^+ and CO_2 are examples of the chemical versatility that a functional group in a macromolecule can exhibit in providing biological activity.

MEDICAL ASPECTS OF HEMOGLOBIN ABNORMALITIES

The contributions that molecular studies on hemoglobin have made to the advancement of medical knowledge are especially noteworthy. Elucidation of the underlying molecular cause of sickle-cell anemia was one of the successful pioneering efforts in science that alerted the medical world to the molecular origin of many diseases.

Sickle-Cell Anemia

Sickle-cell anemia (Hemoglobin S disease) is a severe form of anemia that, if left untreated, often results in death during childhood. The name of the disease is derived from the sickled shape that red blood cells of afflicted individuals assume when exposed to a lowered pO_2 (Figure 8-17). *Sickle-cell crisis* is precipitated when the deformed blood cells clog small blood vessels; the resultant blockage of blood then accentuates the sickling phenomenon because it promotes oxygen deprivation to the afflicted areas of the body.

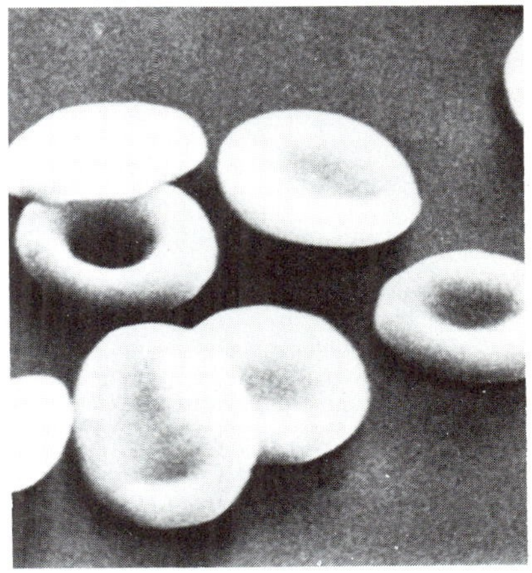

Figure 8-17
Normal (top) and sickled red blood cells (bottom).

Sickle-cell crisis is extremely painful and often results in permanent damage because of resulting strokes.

History of Sickle-Cell Research

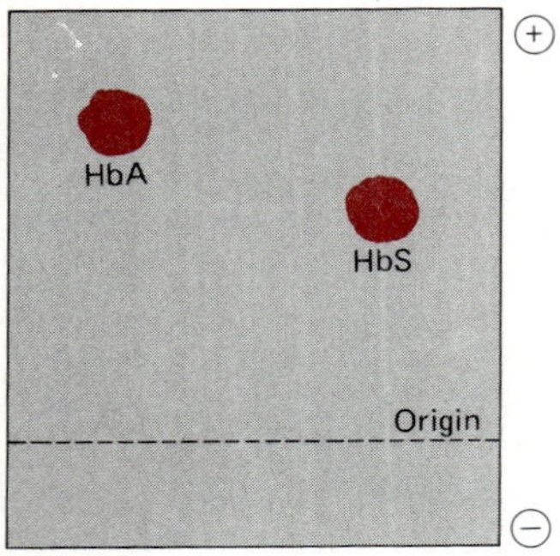

Figure 8-18
Difference between the electrophoretic patterns of hemoglobin A (HbA) and hemoglobin S (HbS) at pH 8.6.

In 1949, Linus Pauling and his colleagues reported a difference between the electrophoretic motilities of normal (HbA) and sickle-cell (HbS) hemoglobins, i.e., HbA migrates at a faster rate toward the anode (+) (Figure 8-18). This observable difference in a physical property of the two hemoglobins prompted the exciting and novel proposal that sickle-cell anemia is a *molecular disease,* one whose cause can be traced to a physiologically abnormal biomolecule. In the wake of Pauling's findings, Vernon Ingram initiated studies to determine the chemical differences between HbA and HbS. The two hemoglobins were fragmented by trypsin digestion, and a *peptide map,* or "fingerprint," of the fragments in each mixture was obtained (Figure 8-19) by paper electrophoresis and paper chromatography (p. 97). The only difference noted between the two peptide maps was the location of peptide 4. The subsequent isolation and determination of the sequences of the two

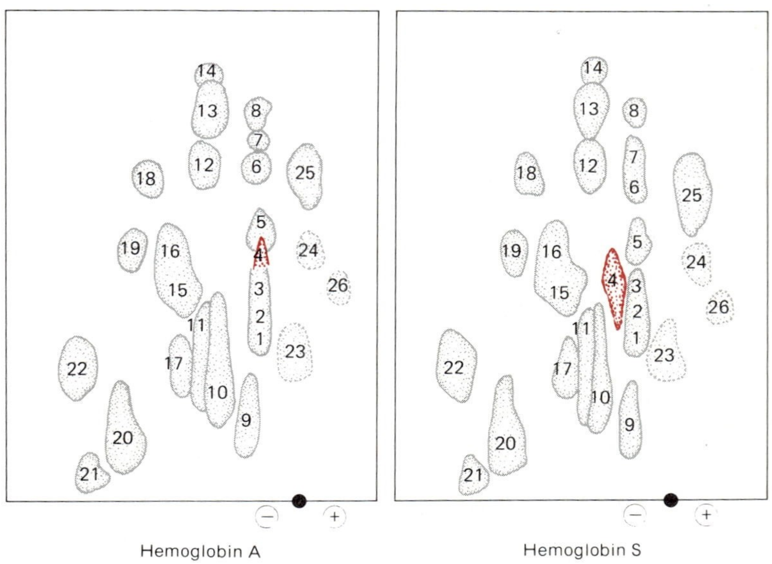

Comparison of the amino acid sequences of peptides 4.

	1	2	3	4	5	6	7	8
Hemoglobin A	H_2N—Val—	His—	Leu—	Thr—	Pro—	Glu—	Glu—	Lys
Hemoglobin S	H_2N—Val—	His—	Leu—	Thr—	Pro—	Val—	Glu—	Lys

Figure 8-19
Peptide maps of tryptic digests of hemoglobin A and hemoglobin S. Electrophoresis of the tryptic digests was followed by paper chromatography. Note the difference in the location of peptide 4 in the two peptide maps. (After Vernon M. Ingram, *Hemoglobin and its Abnormalities.* Springfield, Ill.: Charles C Thomas, 1961)

peptides established that the molecular difference between HbA and HbS was limited to a single amino acid substitution in the β chain. Specifically, HbA has a *glutamyl* residue in position 6 of its β chains, whereas HbS has a *valyl* (Figure 8-19). The revelation that a lethal form of anemia is due to a single amino acid difference in two of the four polypeptides of hemoglobin had a profound influence and brought molecular biology into the realm of medical research.

Molecular Basis of Sickle-Cell Anemia

The hydrophobic group of valine is the key to the sickling phenomenon. The sixth position in the β chain is located on the exterior of the hemoglobin molecule, and therefore HbS has a hydrophobic R group in place of the normal hydrophilic group of glutamate. Because deoxyhemoglobin has a complementary hydrophobic site for the R group of this valine but oxyhemoglobin does not, high concentrations of the deoxygenated form of hemoglobin S induce aggregation. The initial interaction that triggers polymerization involves a *β6 valine* of one deoxyHbS molecule and *β85 phenylalanine* and *β88 leucine* of an adjacent deoxyHbS. Only one of the two β6 valines of a HbS molecule, as indicated in Figure 8-20, participates in this interaction. Once this critical event occurs, other intermolecular interactions take place and, as a result, a double-stranded polymer is formed (Figure 8-20). The double-stranded polymers then form fibrous 14-stranded aggregates which bring about transformation of the shape (sickling) of the red blood cells.

Now that the molecular basis of the anemia is known, intense research efforts are being made to develop medical therapy to prevent the sickling phenomenon or to alleviate sickle-cell crisis. One approach entails the development of therapeutic agents that would inhibit polymerization of deoxyHbS. Among the most promising compounds developed are those that bind to the initial contact regions of HbS molecules which, in turn, prevents the key intermolecular contact needed for polymerization. Included in this list are peptides containing phenylalanine and the benzyl esters of phenylalanine or tryptophan. It is anticipated that, if these agents prove to be therapeutically valuable, they will be less toxic to patients than other compounds, e.g., potassium cyanate, which have been clinically tested.

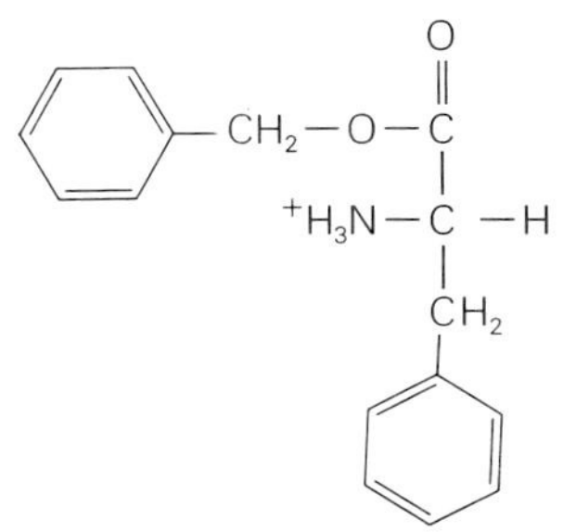

Benzyl ester of phenylalanine

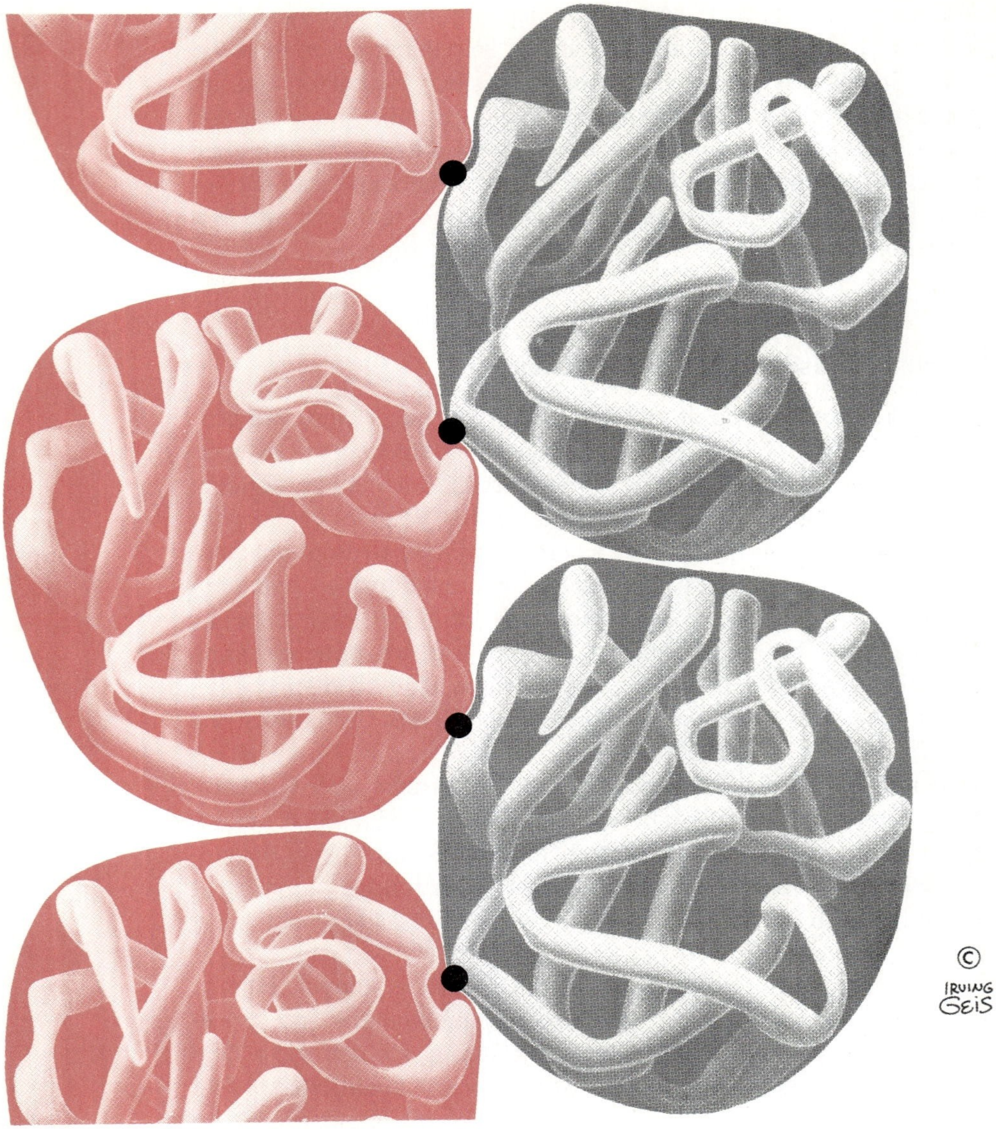

Figure 8-20
The double-stranded polymer formed by deoxygenated HbS molecules when they aggregate, as determined from the crystalline structure of HbS studied by Love and associates (Johns Hopkins University, 1975). The black dots show the position of $\beta 6$ valine, which interacts with $\beta 85$ phenylalanine and $\beta 88$ leucine (hydrophobic pocket) of an adjoining HbS molecule. Other contacts between molecules stabilize the structure of the polymer. (Copyright 1982 by Irving Geis)

	β Chain (HbA)	Altered β Chain (HbS)
β chain (HbA)	HbA (normal)	HbA and HbS (sickle-cell trait) heterozygous
Altered β chain (HbS)	HbA and HbS (sickle-cell trait) heterozygous	HbS (sickle-cell anemia) homozygous

Figure 8-21
Probability of HbS inheritance for each child born to a couple heterozygous for HbA and HbS. Normal: 25%; trait: 50%; anemia: 25%.

Individual Genetics of Sickle-Cell Anemia

Sickle-cell anemia is a genetic disease involving the inheritance of a mutant, physiologically defective hemoglobin. A human, as a diploid organism, receives a set of 23 chromosomes from each parent and, hence, possesses two copies of each gene (except for the XY combination in males). Individuals with the anemia have inherited two copies of the altered gene, one from each parent, for β-chain production and are *homozygous* for the genetic defect (Figure 8-21). Such a person produces only altered β chains and therefore has HbS and no HbA. Individuals inheriting only one mutant β-chain gene are *heterozygous* and produce both HbA and HbS (Figure 8-21); they have *sickle-cell trait.* The trait poses no medical problems because of the production of normal HbA. Only under unusual circumstances, e.g., a depressurized airliner cabin, does the person with the trait face the possibility of sickle-cell crisis.

Population Genetics of Sickle-Cell Anemia

As tragic as the anemia is, the mutation is maintained in certain populations because of the survival advantage it confers. In areas of the world where the malaria-causing protozoan *(Plasmodium vivax)* is endemic, e.g., certain regions of Africa, possession of a copy of the altered gene (heterozygous condition) bestows resistance against the parasitic disease. The protozoan relies on red blood cells for a phase of its life cycle for continued propagation in the body, and HbS red blood cells disrupt the cycle because they lyse when infected by the parasite. Sickle-cell anemia is a vivid illustration of the selective processes employed by nature to adapt the biological fitness of a species to a given environmental niche. What would be considered a detrimental genetic trait in one environment is advantageous in another.

As illustrated in Figure 8-21, there is a 25 per cent probability that a child born to heterozygous parents (HbA/HbS) will be homozygous for HbS. There is a 50 per cent probability that a child of two heterozygous parents or of one heterozygous and one normal homozygous (HbA/HbA) parent will be heterozygous and carry the sickle-cell trait. An environment where malaria and HbS are present selects for heterozygosity and against normal homozygosity.

An Anemia Related to Sickle-Cell Anemia

Among the over 140 known variants of hemoglobin, one called *hemoglobin C* (HbC) is responsible for a moderately severe anemia (hemoglobin C disease) which is clinically very similar to sickle-cell anemia. Interestingly, HbC also has a substituted amino acid at the sixth position of the β chain with *lysine* replacing *glutamate.* Individuals that are heterozygous for HbC and HbS, i.e., possess only these two altered hemoglobins, suffer an anemia that, at times, can be medically as serious as sickle-cell anemia.

Hemoglobin and Molecular Biology

Molecular biology, as a uniting force in science, owes much to the studies on hemoglobin, which serve as an outstanding illustration of research findings that transcend any particular discipline to encompass many. Molecular knowledge of this critical biomolecule and its altered forms has substantiated the concept of the correlation between chemical structure and biological function. It has been possible from the hemoglobin studies to construct an overall scientific evaluation that extends from the level of the chemical consideration of a single amino acid replacement to its impact at the populational level by altering biological fitness.

SUMMARY

The potent analytical powers of x-ray crystallography were effectively used to determine the three-dimensional structures of the oxygen carriers myoglobin and hemoglobin. Myoglobin, a single polypeptide chain with 153 amino acid residues, is a compact, globular molecule of dimensions $45 \times 35 \times 25$ Å. Seventy-five per cent of myoglobin's amino acid residues are located in eight sections that are in right-handed α-helical conformation. The polar and nonpolar residues are primarily associated with the exterior and interior of the molecule, respectively. The heme group, composed of protoporphyrin IX chelated with iron, is situated near the surface of the molecule in a nonpolar crevice. The location of the heme group in a nonpolar environment is important for maintaining the reduced state of iron (Fe^{2+}), required for oxygenation.

Adult hemoglobin (HbA) is a tetramer containing two α and two β chains. The tertiary structure of the α (141 residues) and the β (146 residues) chains is very similar to that of myoglobin; however, the amino acid sequences of the three polypeptide chains do not show comparable similarity. Hemoglobin is almost spherical in shape and about 55 Å in diameter, and the majority of interactions between the four polypeptide chains occurs between the α and β chains. The four heme groups (which are not close to one another) are each located in a nonpolar crevice near the molecule's surface, like that of myoglobin.

The quaternary structure of hemoglobin endows the oxygen carrier with biological properties not observed with myoglobin, which lacks this level of structure. For example, hemoglobin is an allosteric protein whose biological functions are regulated by conformational changes, as evidenced by its cooperative binding of oxygen (the bonding of oxygen to one heme group facilitates the oxygenation of other hemes). The *in vivo* oxygen binding to hemoglobin is also regulated by other physiological factors, one of which is the biomolecule 2,3-bisphosphoglycerate. The electrostatic binding of 2,3-P_2-glycerate within the central cavity of deoxygenated hemoglobin decreases the molecule's affinity for oxygen since 2,3-P_2-glycerate's presence adds to the number of ionic bonds that must be broken for oxygenation.

Hemoglobin (but not myoglobin) also binds hydrogen ions and carbon dioxide, waste products of metabolism. The physiological interplay between the role of hemoglobin as a carrier of O_2 and its role in the transport of H^+ and CO_2, known as the Bohr effect, is of critical importance. Increased $[H^+]$ and/or $[CO_2]$ decrease hemoglobin's affinity for O_2, a mechanism by which actively metabolizing tissues can concurrently have waste products removed and be furnished needed oxygen. Conversely, increased O_2 decreases hemoglobin's affinity for H^+ and CO_2, another valuable mechanism which aids in the release of carbon dioxide in the lungs for exhalation.

The recognition of a variant form of hemoglobin, HbS, introduced the concept of a molecular disease (one whose cause is traceable to the production of an abnormal biomolecule). An individual homozygous for HbS suffers from sickle-cell anemia, a fatal disease if left untreated. Molecular studies have revealed that HbS differs from normal HbA by a single amino acid substitution in the β chain (β6 Glu of HbA $\rightarrow$ β6 Val of HbS). The presence of the nonpolar R group of Val on the exterior of HbS is responsible for hydrophobic interactions between deoxygenated HbS molecules, which form the long filaments that ultimately precipitate sickle-cell crisis. In areas of the world where malaria is endemic, heterozygosity for HbS is selected in populations because it confers resistance to the disease. The studies on hemoglobin and its variant forms provide a classic example of the basic but relevant structure–function concept of molecular biology.

QUESTIONS

1. Which of the following characteristics are associated with myoglobin (M), hemoglobin (H), both of them (B), or neither of them (N)?
   (a) Majority of structure in α-helical conformation
   (b) Oxygen carrier
   (c) Carbon dioxide carrier

(d) Not an allosteric protein
(e) Heme group(s) in a polar crevice(s)
(f) Protoporphyrin IX
(g) Sigmoidal oxygen-dissociation curve
(h) Tertiary structure
(i) Quaternary structure
(j) Blocked N-terminal residue

2. Which of the following statements are true about the three-dimensional structure of fully oxygenated hemoglobin *in vivo?*
   (a) A molecule of 2,3-P_2-glycerate is located in the interior cavity.
   (b) An oxygen atom is bound to each of the four heme groups.
   (c) The ferric ions are located in the planes of the heme groups.
   (d) Much of its α-helical conformation is lost.
   (e) It is a less rigid molecule than deoxyhemoglobin.

3. Which of the following amino acid residues are probably located in the interior of a globular protein under physiological conditions?

(a) Arg	(c) Phe	(e) Asn	(g) Met
(b) Thr	(d) Glu	(f) Val	(h) Asp

4. How would the oxygen affinity of hemoglobin be affected by each of the following *in vitro* treatments?
   (a) Increase in $[H^+]$ from 2 to 4 $\times$ 10^{-8} *M*
   (b) Increase in pH from 7.1 to 7.3
   (c) Increase in 2,3-P_2-glycerate from 1 to 4 $\times$ 10^{-4} *M*
   (d) Increase in pO_2 from 20 to 60 torr
   (e) Increase in pCO_2 from 10 to 30 torr
   (f) Dissociation into 2 α and 2 β chains

5. How would the *in vitro* treatments a through e listed in Question 4 affect the oxygen affinity of myoglobin?

6. Upon electrophoresis at pH 8.6, would HbA or HbC migrate faster toward the anode?

7. Would a fetus homozygous for HbS have normal fetal hemoglobin?

8. How does HbA inhibit the sickling of red blood cells in individuals heterozygous for HbS?

9. On p. 124, a statement was made that a child of a person with sickle-cell trait and a person homozygous for HbA would have a 50 per cent probability of bearing the trait. In a diagram analogous to that in Figure 8-21, illustrate how this probability is derived. What is the probability of a child being homozygous (a) for HbA or (b) for HbS?

SUGGESTED READINGS

Bunn, H. F., B. G. Forget, and H. M. Ranney, *Human Hemoglobins.* Philadelphia: W. B. Saunders, 1977. A book devoted to what is currently known about human hemoglobins.

Cantor, C. R., and P. R. Schimmel, *Biophysical Chemistry.* Part I, *The Conformation of Biological Macromolecules.* San Francisco: W. H. Freeman, 1980. Excellent discussions and illustrations of the structures of macromolecules.

Dickerson, R. R., and I. Geis, *The Structure and Action of Proteins.* New York: Harper and Row, 1969. This text provides an introductory approach to the contemporary treatment of proteins. Excellent illustrations complement the text.

Glusker, J. P., and K. N. Trueblood, *Crystal Structure Analysis: A Primer.* New York: Oxford University Press, 1972. A good introduction to x-ray crystallography.

Holmes, K. C., and D. M. Blow, *The Use of X-ray Diffraction in the Study of Protein and Nucleic Acid Structure.* Huntington, N.Y.: R. E. Krieger, 1979. A good introduction to a valuable technique in molecular biology (a reprinting).

Perutz, M. F., *Proteins and Nucleic Acids: Structure and Function.* Amsterdam and New York: Elsevier, 1962. A very readable introduction to the interrelationship between proteins and nucleic acids. Some sections are outdated, but they give a historical perspective to molecular biology.

Articles

Allison, A. C., ''Sickle Cells and Evolution.'' *Scientific American,* 195(2):87 (offprint #1065), 1956.

Edelstein, S. J., ''Cooperative Interactions of Hemoglobin.'' *Ann. Rev. Biochem.,* 44:209, 1975.

Kendrew, J. C., ''The Three-Dimensional Structure of a Protein Molecule.'' *Scientific American,* 205(6):96 (offprint #121), 1961.

Kendrew, J. C., ''Myoglobin and the Structure of Proteins.'' *Science,* 139:1259, 1963.

Kilmartin, J. V., ''The Bohr Effect of Human Hemoglobin.'' *TIBS,* 2:247, 1977.

Martin, D. L., and J. E. Hichey, ''Sickle-Cell Anemia, Hemoglobin Solubility and Resistance to Malaria.'' *J. Chem. Ed.,* 44:177, 1972.

Perutz, M. F., ''The Hemoglobin Molecule.'' *Scientific American,* 211(5):64 (offprint #196), 1964.

Perutz, M. F., ''Hemoglobin Structure and Respiratory Transport.'' *Scientific American,* 239(6):92 (offprint # 1413), 1978.

Perutz, M. F., ''Regulation of Oxygen Affinity of Hemoglobin: Influence of the Globin on the Heme Iron.'' *Ann. Rev. Biochem.,* 48:327, 1979.

Zuckerlandl, E., ''The Evolution of Hemoglobin.'' *Scientific American,* 212(5):110 (offprint #1012), 1965.

PROTEINS IV: ENZYMES

9

INTRODUCTION

Although the name *enzyme* was proposed for a biological catalyst by Wilhelm Kühne in 1878, the discovery of an enzyme is generally attributed to Anselme Payen and Jean-François Persoz, chemists working in a sugar factory in Paris. In 1833 the two investigators reported obtaining from malt extract a reusable factor they called *diastase* (now known as amylase) which converted starch into sugar. Within the next few years, Theodor Schwann succeeded in extracting pepsin, which digests meat (protein), from gastric juice, and he later identified trypsin, a peptidase in digestive fluids. Hence, the notion of diastases (the early name for all enzymes) was soon extended to animals. By 1837, the famous chemist Jöns Berzelius, recognizing with remarkable foresight the catalytic nature of biological diastases, wrote:

Payen, a chemist-industrialist, was the owner of the sugar factory. When he and Persoz called the first known biological catalyst "diastase," they introduced the now historical use of the *ase* suffix in the naming of most enzymes.

> There are well-justified reasons to suppose that in living animals and plants, thousands of catalytic processes take place between the organic fluids and tissues, whence, we get the formation of countless heterogenous chemical compounds . . . it is likely that some time in the future we shall find out that the cause of it all is the catalytic power of the tissues forming the organs of the living body.

The importance of enzymes was therefore suggested early in the history of biochemistry. Ironically, although enzymes and proteins were both intensively studied by biochemists in the 1800s, Sumner's crystallization of urease in 1926 was required before the protein nature of enzymes was seriously considered. Indeed, the consensus at that time in Europe was that enzymes were *not* proteins, because of the findings of the influential chemist Richard Wilstätter

(Nobel Prize, 1915), who, in the 1920s, reported that he could not detect protein in purified enzyme preparations from yeast. Proteins associated with enzymes were thought of as inert carriers of the true enzymes, which were believed to be small, biologically potent molecules analogous to hormones. With hindsight, it is realized that the protein assays used in that era were not sensitive enough to detect the small amounts present in Wilstätter's purified preparations. The nonprotein nature of enzymes was so entrenched in scientific thinking that decades passed before their polypeptidyl composition was unequivocally accepted.

The history of enzymes and that of DNA differ in that the controversy concerning DNA in the 1940s (p. 232) was not about its chemical structure but about its biological function. Interestingly, the argument used against DNA's being genetic material was reminiscent of the enzyme controversy, *viz,* the observed genetic activity of DNA preparations was allegedly due to the small percentage of protein contamination and not to the DNA, assumed to be genetically inert.

GENERAL CATALYTIC PROPERTIES OF ENZYMES

Enzymes are remarkably effective catalysts, responsible for the thousands of coordinated chemical reactions involved in biological processes of living systems. Like an inorganic catalyst, an enzyme accelerates the rate of a reaction by lowering the energy of activation required for the reaction to occur. However, unlike a simple inorganic catalyst, an enzyme lowers the energy of activation by replacing a large activation barrier with multiple lower barriers. This property is illustrated in Figure 9-1 where, as noted, the difference between the energy coordinates of the enzymatic and nonenzymatic reactions is the amount of energy ($\Delta G^{\ddagger}$) needed for the reactant (substrate) to attain the activated state. Note that for both reactions, the difference between the *free energy* of A (reactant) and B (product), called the ΔG *of the reaction,* is the same.

As a catalyst, an enzyme is not destroyed in the reaction and therefore remains unchanged and is reusable. An outstanding feature of an enzyme as a catalyst is its *substrate specificity,* which determines its biological function. Many enzymes have only one biological substrate *(absolute substrate specificity);* others have a broader specificity and utilize a group of structurally similar biomolecules as substrates *(relative group specificity).* Glucose 6-phosphatase is an example of an enzyme with a specificity for D-

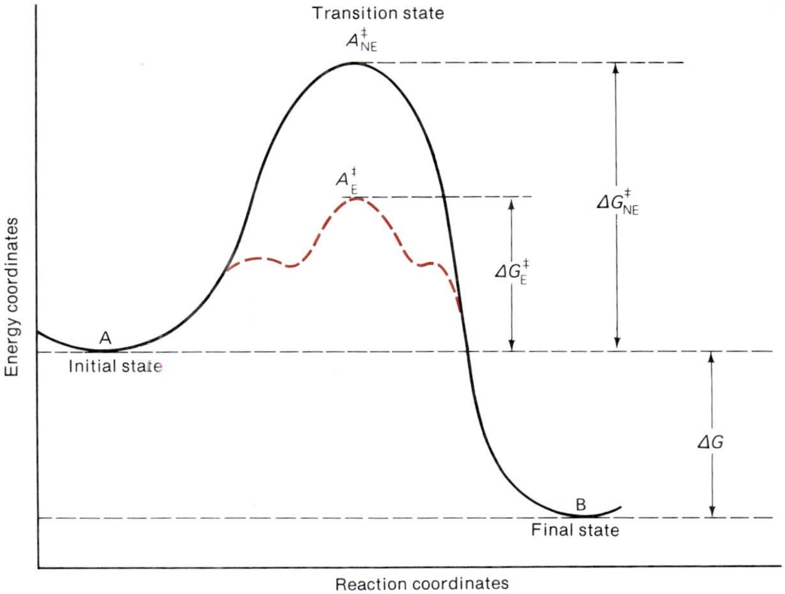

Figure 9-1
Energy diagram for a reaction A → B. A is the reactant and B the product. $A^{\ddagger}_{NE}$ and $A^{\ddagger}_{E}$ are the activated complexes in the nonenzymatic and enzymatic reactions, respectively. $\Delta G^{\ddagger}_{NE}$ and $\Delta G^{\ddagger}_{E}$ are the free energies of activation for the nonenzymatic and enzymatic reactions, respectively. ΔG is the overall free-energy change of the reaction.

glucose 6-phosphate; it catalyzes the hydrolytic removal of the phosphate moiety (Table 9-2). Conversely, acid and alkaline phosphatases can act upon various phosphorylated substrates and therefore exhibit broad specificity. Another critical biological feature of enzymatic reactions is that their substrate and catalytic specificities ensure synthesis of only specific biomolecular products without the concomitant production of by-products, in contrast to many reactions of organic chemistry.

Some enzymes are referred to as *simple proteins* because they require only their protein structure for catalytic activity. Other enzymes are *conjugated proteins* since they each require a nonprotein component, called a *cofactor,* for activity. The enzymes that need metal cofactors e.g., Mg^{2+}, Fe^{2+}, or Zn^{2+}, are called *metalloenzymes.* Other enzymes have organic biomolecules, *coenzymes,* as cofactors. The coenzyme forms of B-vitamins (Chapter 10), for example, are required for many reactions. If a cofactor (metal ion or coenzyme) is firmly bound to an enzyme, it is called a *prosthetic group,* e.g., the B-vitamin biotin in its coenzyme form is covalently bonded to a lysyl residue of the enzyme pyruvate carboxylase (p. 283). The complex composed of an *apoenzyme* (protein portion of an enzyme) and its cofactor(s) is called a *holoenzyme.*

CLASSIFICATION OF ENZYMES BY CATALYTIC FUNCTION

The international symbolic language of chemistry was orginally developed by Berzelius, who proposed that an element be identified by the initial letter or first two letters of its Latin name.

Berzelius was correct in his 1837 prediction that thousands of catalytic processes occur in a biological system. A little over a century later, Beadle and Tatum's one gene-one enzyme hypothesis emphasized the concept that each of these thousands of reactions is catalyzed by a specific enzyme, a proposal that is now accepted as generally true. In the 1960s, the International Union of Biochemistry (IUB) established a Commission on Enzyme Nomenclature to adopt a systematic classification and nomenclature for the ever-increasing number of enzymes being identified and described. The Commission, identifying enzymes by the types of reactions they catalyzed, defined six major classes (Table 9-1); examples of each class are listed in Table 9-2. Oxidation-reduction reactions are catalyzed by *oxidoreductases,* and reactions involving the transfer of a group utilize transferases. *Hydrolases* employ H_2O to cleave covalent bonds, and this class includes enzymes that degrade polysaccharides, proteins, and nucleic acids. *Lyases* cleave or remove groups from compounds by electron rearrangements (eliminations) and therefore create double bonds in one of the products. If a lyase reaction is reversible, a group may be added to a double bond (deaminase reaction in Table 9-2). *Isomerases* catalyze internal rearrangements within a substrate and therefore do not involve the addition or removal of groups. *Ligases* catalyze the formation of various types of covalent bonds to synthesize biomolecules and require an input of chemical energy, provided by hydrolysis of bio-

Table 9-1
Six major classes of enzymes

Class number	Enzymes	Catalytic function
1	Oxidoreductases	Oxidation-reduction reactions
2	Transferases	Group transfer reactions
3	Hydrolases	Hydrolytic reactions
4	Lyases	Reactions involving elimination of a group by the cleaving of a bond (leaving a double bond) or the addition of a group to a double bond
5	Isomerases	Reactions involving isomerizations
6	Ligases	Reactions joining together two molecules coupled with the hydrolysis of high-energy pyrophosphate bonds

Table 9-2
Representative enzymes of the six major classes and their reactions

Enzyme class		General reaction
1. Oxidoreductases	Type of oxidation	
Alcohol dehydrogenase	Alcohol to aldehyde	$H_3CCH_2OH \rightarrow H_3CC(=O)H + 2H\cdot$
Succinate dehydrogenase	Double bond information	${}^-OOCCH_2CH_2COO^- \rightarrow {}^-OOCC(H)=C(H)COO^- + 2H\cdot$
2. Transferases	Group transferred	
Phosphotransferase	Phosphoryl	$RO-P(=O)(O^-)-O^- + HOR' \rightarrow ROH + {}^-O-P(=O)(O^-)-OR'$
Aminotransferase	Amino	$R-CH(NH_3^+)-COO^- + R'-C(=O)-COO^- \rightleftharpoons R-C(=O)-COO^- + R'-CH(NH_3^+)COO^-$
Hydrolases	Bond hydrolyzed	
Peptidase	Peptide	$R-C(=O)-N(H)-R' + HOH \rightarrow R-C(=O)-O^- + {}^+H_3N-R'$
Phosphatase	Monophosphate ester	$R-O-P(=O)(O^-)-O^- + HOH \rightarrow R-OH + HPO_4^{2-}$
4. Lyases	Group removed	
Decarboxylase	Carbon dioxide	$R-C(H)(NH_3^+)-COO^- \rightarrow R-CH_2(NH_3^+) + O=C=O$
Deaminase	Ammonia	$R-CH_2CH(NH_2)R' \rightleftharpoons RCH=CHR' + NH_3$
5. Isomerases	Group isomerized	
Epimerase	C-3 of a five-carbon sugar	D-Ribulose 5-phosphate $\rightleftharpoons$ D-xylulose 5-phosphate
Racemase	α-Carbon substituents	L-Alanine $\rightleftharpoons$ D-alanine
6. Ligases	Covalent bond formed	
Acetyl-CoA synthetase	C-S	Acetate + CoA−SH + ATP $\rightleftharpoons$ acetyl−S−CoA + AMP + PP_i
Pyruvate carboxylase	C-C	Pyruvate + CO_2 + H_2O + ATP $\rightleftharpoons$ oxaloacetate + ADP − P_i

molecules such as nucleoside triphosphates, e.g., ATP. As an energy donor ATP (Table 9-2), is hydrolyzed by some ligases to yield AMP and PP_i (inorganic pyrophosphate) and by other ligases to yield ADP and P_i (inorganic phosphate).

In the official nomenclature, every enzyme is distinctly identifiable by its formal name and by a four-component number. Alcohol dehydrogenase (Table 9-2) is identified in scientific reports as alcohol:NAD^+ oxidoreductase, E.C. 1.1.1.1. (The notation E.C. refers to Enzyme Commission.) The first number 1 refers to Class 1 (oxidoreductase); the second, to the type of group oxidized (1 = alcohol); the third, to the oxidizing agent (1 = the coenzyme NAD^+); and the fourth, to the specific reaction (1 = alcohol dehydrogenase). Because official names are often lengthy, the trivial names of enzymes are generally used after initial identification.

ENZYME ASSAYS

The absorption spectra of NAD^+ and NADH.

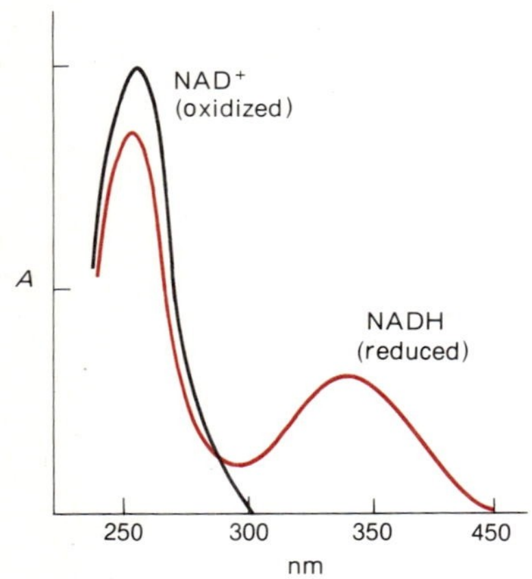

To study an enzyme, it is necessary to have an assay by which to measure its catalytic activity. Assays are designed to measure the rate of product formation or the rate of substrate disappearance. An assay that measures product formation is preferred because it involves a direct measurement, as opposed to the indirect measurement obtained from an assay relying on the determination of substrate disappearance. Frequently, the amount of product produced in a given length of time is measured in a *fixed-time assay.* The manner in which the amount of product is determined depends on its chemical and physical properties. If the product is colored or can undergo a reaction to produce a colored solution, then absorbance of light at an appropriate wavelength can be measured *(colorimetric assay)* and related to the concentration of product present at the time of sampling. *Spectrophotometric assays* are especially useful because the progress of the reaction can be monitored continuously in *kinetic assays.* The activities of enzymes that utilize NAD^+ or NADH as coenzymes are generally assayed spectrophotometrically because NADH, but not NAD^+, has an absorbance peak at a wavelength of 340 nanometers (nm), which is in the ultraviolet region of the spectrum. Hence, increases or decreases in absorbance which measure appearance or disappearance, respectively, of NADH can be conveniently monitored in a spectrophotometer. A *radioactive assay* is another commonly used method for monitoring enzymatic activity. By using a substrate labeled with the appropriate radioisotope, e.g., ^3H, ^{14}C, or ^{32}P, accurate determinations can be made of minute quantities of isolated, radioactive product.

Optimal conditions are also determined for an enzymatic assay, including determination of the optimum pH, temperature, and ionic strength for the reaction. Because enzymes are globular proteins, most are thermolabile and begin to denature (indicated by loss of enzymatic activity) at temperatures between 45° and 50°C (Figure 9-2). Ionic strength and pH (Figure 9-3) are also important parameters because they specify the charges of amino acid residues that may influence the three-dimensional structure of an enzyme and, thus, its catalytic activity. The appropriate concentrations of enzyme and of cofactors (if needed) to use in a particular volume of assay mixture are also established empirically.

Figure 9-2
Hypothetical temperature-activity profile of an enzyme.

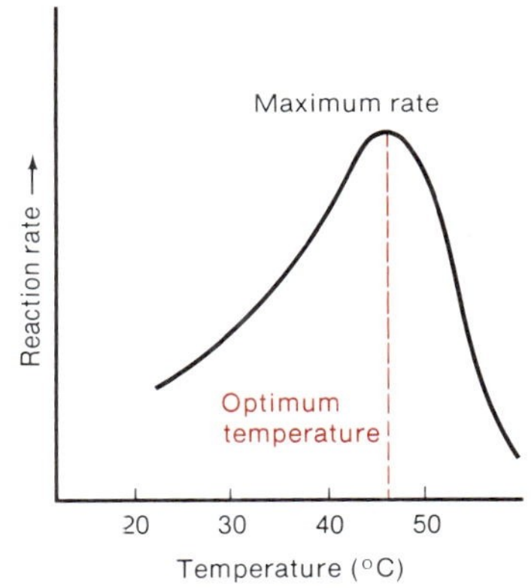

ENZYME REACTION RATES

When increasing concentrations of substrate are used in a series of assays, standardized by the above criteria, a plot of v (velocity, or reaction rate) *vs.* substrate concentration often produces a hyperbolic curve (Figure 9-4). Velocity is usually expressed as units (μmoles of product produced or substrate transformed per minute) or specific activity (units per milligram of protein). An analysis of the plot reveals reaction characteristics of the enzyme. The hyperbolic curve analogous to the oxygen-dissociation curve of myoglobin (p. 112), is representative of the reaction kinetics of a noncooperative enzyme. The plot shows that the velocity increases with substrate concentration until maximum v (V_{max}) is approached asymptotically, after which larger concentrations of substrate do not significantly enhance the reaction rate. In the lower region of the curve, the reaction approaches *first-order kinetics,* meaning that v is a direct function of substrate concentration because the active sites of the enzyme molecules are not saturated. At the plateau at the upper portion of the plot, the reaction approaches *zero-order kinetics* because the active sites of all the enzyme molecules are saturated and the reaction rate is therefore independent of further increases in substrate concentration. For the intermediate portion of the curve, as the enzyme approaches substrate saturation, kinetics are mixed zero and first order in substrate concentration. Routine enzyme assays are generally designed to follow zero-order kinetics to avoid the influence of substrate concentration on reaction velocity. Under such conditions, measured rates are directly proportional to the concentration of the enzyme itself.

Figure 9-3
Hypothetical pH-activity profile of an enzyme.

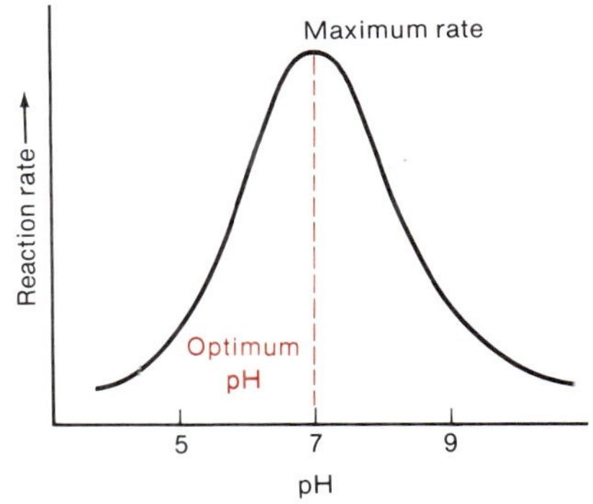

Figure 9-4
Effect of substrate concentration on the reaction rate of an enzymatic reaction.

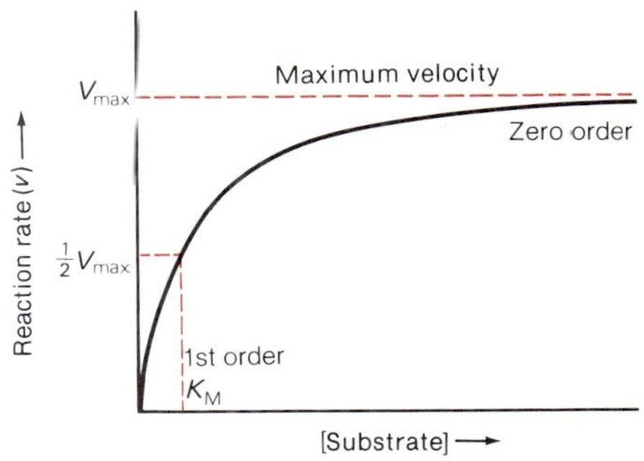

Table 9-3
K_M values for some enzymes

Enzyme	Substrate	K_M (M)
Catalase	H_2O_2	2.5×10^{-2}
β-Galactosidase	Lactose	4×10^{-3}
Penicillinase	Benzylpenicillin	5×10^{-5}
Glutamate dehydrogenase	α-Ketoglutarate	2×10^{-3}
	NH_4^+	5.7×10^{-2}
	NADH	1.8×10^{-5}
Hexokinase	Glucose	1.5×10^{-4}
	Fructose	1.5×10^{-3}

The substrate concentration needed for half-maximum velocity ($\frac{1}{2}V_{max}$) is called the K_M *value (Michealis constant)* and is expressed in units of substrate concentration (moles per liter or M). K_M may be considered an approximate measure of the affinity of an enzyme for its substrate; the lower the K_M, the higher the affinity. If two enzymes have K_M values of 1×10^{-3} M and 1×10^{-5} M, the one with the 10^{-3} M value requires a substrate concentration 100-fold greater to attain its $\frac{1}{2}V_{max}$ than that required by the one with the 10^{-5} M value. The K_M value is a characteristic property of an enzyme and, as seen in Table 9-3, a wide range of values is found among enzymes and their substrates.

MICHAELIS–MENTEN EQUATION

The discovery by Michaelis that keratin, the major ingredient of hair, is soluble in thioglycolic acid made possible the development of the home permanent industry.

In 1913 Leonor Michaelis and Maud L. Menten proposed a general theory of enzyme action and kinetics. The theory explains the course of an enzymatic reaction as follows:

$$E + S \underset{k_{-1}}{\overset{k_1}{\rightleftharpoons}} ES \underset{k_{-2}}{\overset{k_2}{\rightleftharpoons}} E + P$$

Enzyme E first reacts with substrate S to form an enzyme-substrate complex ES, which, in a second step, yields enzyme and product P. Both reactions are considered reversible, and the four elementary rate constants are designated k_1, k_2, k_{-1}, and k_{-2}. Most enzymes, however, show kinetic behaviors that are more complex. For example, a reaction may have two substrates (S_1 and S_2), two products (P_1 and P_2), and two ES complexes, as follows:

$$E + S_1 \rightleftharpoons ES_1 \xrightleftharpoons{+\,S_2} ES_1S_2 \rightleftharpoons E + P_1 + P_2$$

Reactions may also involve more than one intermediate complex:

$$E + AB \rightleftharpoons EAB \rightleftharpoons \underset{+B}{EA} \rightleftharpoons E + A$$

In the above reaction, EAB and EA represent intermediary complexes.

Although the limitations of the Michaelis-Menten expression for an enzymatic reaction are recognized, it is still useful for general analysis of enzyme kinetics. Michaelis and Menten used the expression to develop an equation, later modified by George E. Briggs and John B. S. Haldane, that describes a hyperbolic curve obtained with increasing substrate concentrations, such as that depicted in Figure 9-4. The Michaelis-Menten equation is expressed as follows:

$$v = \frac{V_{max}[S]}{[S] + K_M}$$

The derivation of the equation is summarized in Figure 9-5. The two main assumptions are that, in measurements of initial velocity, rates of formation and decomposition of ES achieve a steady state ($d[ES]/dt = 0$) and that an insignificant amount of product is present ($k_{-2}[P] = 0$). According to the Briggs-Haldane model, K_M is defined as $(k_2 + k_{-1})/k_1$, a ratio of constants for the breakdown and formation of ES. V_{max} is represented as the rate achieved when the total amount of enzyme ($[E]_t$) is saturated with substrate, i.e., $V_{max} = k_2([E]_t)$, and v is the rate attained with the enzyme-substrate complex ([ES]) at a particular [S], i.e., $v = k_2[ES]$.

Figure 9-5
Derivation of the Michaelis–Menten equation under the steady-state approximation. (After Briggs and Haldane)

1. $E + S \underset{k_{-1}}{\overset{k_1}{\rightleftharpoons}} ES \underset{k_{-2}}{\overset{k_2}{\rightleftharpoons}} P + E$

Assume: $k_{-2}[P] = 0$

2. $\dfrac{d[ES]}{dt} = k_1[E][S] - k_{-1}[ES] - k_2[ES]$

Assume: $[E] = [E]_t - [ES]$

3. $\dfrac{d[ES]}{dt} = k_1([E]_t - [ES])[S] - (k_2 + k_{-1})[ES]$

Assume: $\dfrac{d[ES]}{dt} = 0$

4. $k_1([E]_t - [ES])([S]) - (k_2 + k_{-1})[ES]) = 0$

5. $k_1[E]_t[S] = k_1[ES][S] + k_2[ES] + k_{-1}[ES]$

6. $k_1[E]_t[S] = (k_1[S] + k_2 + k_{-1})[ES]$

7. $[ES] = \dfrac{k_1[E]_t[S]}{k_1[S] + k_2 + k_{-1}} = \dfrac{[E]_t[S]}{[S] + \dfrac{k_2 + k_{-1}}{k_1}}$

Define: $v = k_2[ES]$

8. $v = \dfrac{k_2[E]_t[S]}{[S] + \dfrac{k_2 + k_{-1}}{k_1}}$

Define: $K_M = \dfrac{k_2 + k_{-1}}{k_1}$ and $V_{max} = k_2[E]_t$

9. $v = \dfrac{V_{max}[S]}{[S] + K_M}$

Under conditions in which [S] is very large relative to K_M, the equation can be reduced to $v \approx V_{max}$, resulting in a zero-order reaction. At the other extreme, when [S] is very small relative to K_M, its contribution to the denominator becomes negligible so that $v \approx V_{max} \times [S]/K_M$. Thus, at very low [S], the rate of the reaction approaches first order. The fact that K_M is the substrate concentration which yields half maximal velocity can be stated algebraically by substituting $\frac{1}{2}V_{max}$ for v in the Michaelis-Menten equation:

$$\frac{V_{max}}{2} = \frac{V_{max}[S]}{[S] + K_M}$$

Under these conditions, the equation can be rearranged to:

$$[S] + K_M = 2[S] \qquad \text{or} \qquad K_M = [S]$$

LINEWEAVER-BURK PLOT

A convenient means of evaluating K_M and V_{max} is to plot kinetic data as the reciprocals of v and [S]. Such a *double reciprocal plot* (Figure 9-6) was proposed by Hans Lineweaver and Dean Burk in 1934. If the reciprocal of each side of the Michaelis-Menten equation is taken, the following expression can be derived:

$$\frac{1}{v} = \frac{K_M}{V_{max}} \times \frac{1}{[S]} + \frac{1}{V_{max}}$$

The equation is equivalent to the mathematical expression for a straight line: $y = ax + b$. Thus, ideally a Lineweaver-Burk plot of kinetic data yields a straight line with a slope $= K_M/V_{max}$ and the $1/[S]$ and $1/v$ intercepts $= -1/K_M$ and $1/V_{max}$, respectively. The plot has the advantage that there is no need for measurements of v at extremely high concentrations of substrate (often difficult to obtain experimentally) since [S] can be extrapolated to the value of $1/[S] = 0$. Figure 9-6, in addition to illustrating a plot for normal kinetics, also depicts the type of plots obtained when an enzyme exhibits either *substrate inhibition* or *substrate activation.* Inhibition at very high levels of substrate is a common occurrence.

Several other linear plots for kinetic data have been developed. The Woolf and Eadie-Hofstee plots and the equations from which they are derived are presented in Figure 9-7. The Woolf plot facilitates curve fitting because the more reliable data points (at lower [S]) assume the greatest importance in extrapolation. The Eadie-

Figure 9-6
Lineweaver–Burk plots: a graph of $1/v$ versus $1/[S]$, with (I) and (II) depicting substrate activation and substrate inhibition, respectively.

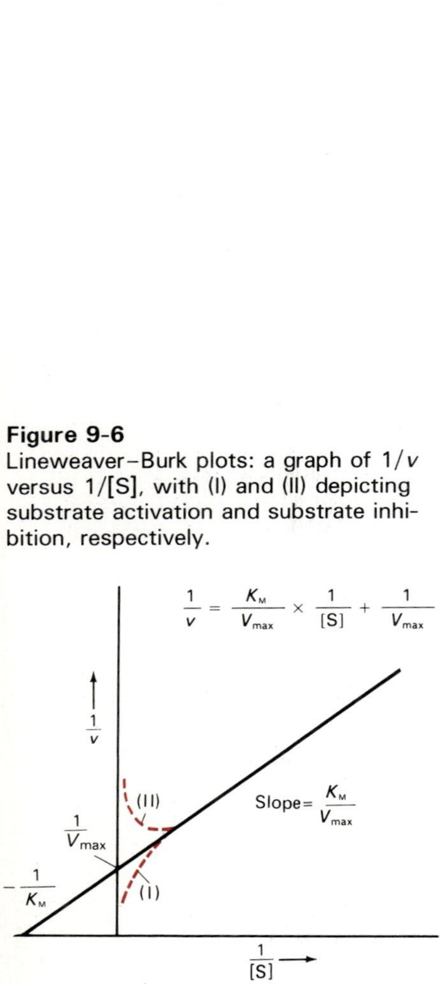

Hofstee plot yields K_M and V_{max} in a direct manner and exposes departures from linearity that may be masked by other graphs.

Although the foregoing discussion of the Michaelis-Menten equation was based on the simplest mechanistic model for enzyme catalysis (transformation of a single substrate into a single product via a single ES complex), hyperbolic initial velocity steady-state kinetics are exhibited by many enzymes mechanistically more complex. Thus, the Michaelis-Menten equation expressed in terms of the two kinetic constants V_{max} and K_M is broadly applicable to a wide variety of enzymes. Of course, the specific rate constants (k_1, k_{-1}, k_2, k_{-2}, k_3, etc.) which make up these kinetic constants (V_{max} and K_M) vary, depending on the assumed mechanism. Whatever the combination of specific rate constants may be, however, V_{max} is *always* the limiting, or maximum, rate (enzyme molecules saturated with substrate) and K_M is the substrate concentration required to achieve ½V_{max} (half saturation of enzyme molecules with substrate) under steady-state conditions.

Figure 9-7
(A) Woolf plot: a graph of [S]/v versus [S].
(B) Eadie–Hofstee plot: a graph of v versus v/[S].

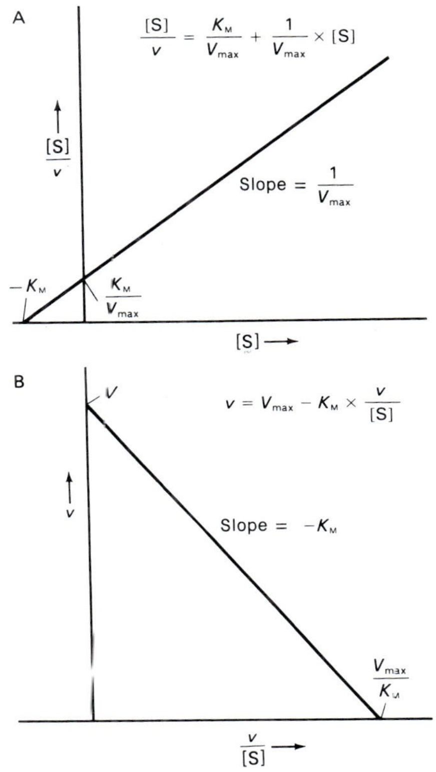

ENZYME INHIBITION

Substances that specifically decrease the rate of enzymatic reactions are called *inhibitors,* and, in enzymology, inhibitory phenomena are studied seriously because of their importance to many different areas of research. For many years inhibitors have been classified as either *competitive* or *noncompetitive,* terms which represent two general types of *reversible* inhibition.

Competitive Inhibition

As implied by its name, a competitive inhibitor has been classically envisioned as a compound that competes with a natural substrate of an enzyme for the active (substrate-binding) site. Such an inhibitor is almost always structurally similar to the natural substrate and, by mimicry, binds to the enzyme and precludes catalytic activity. Competitive inhibition is reversible and can be overcome by increasing substrate concentration. The effectiveness of a competitive inhibitor is determined by the relative affinities the enzyme has for the substrate and inhibitor.

Oxaloacetate and malonate are competitive inhibitors of succinate dehydrogenase, which catalyzes the conversion of succinate into fumarate (p. 304), and, as seen in Figure 9-8, the substrate and these two inhibitors have similar structures. The pharmaceut-

Figure 9-8
Substrate and inhibitors of succinate dehydrogenase.

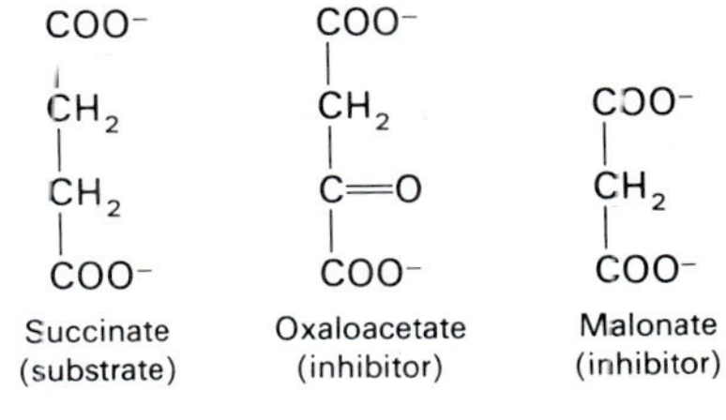

Figure 9-9
p-Aminobenzoic acid and sulfanilamide.

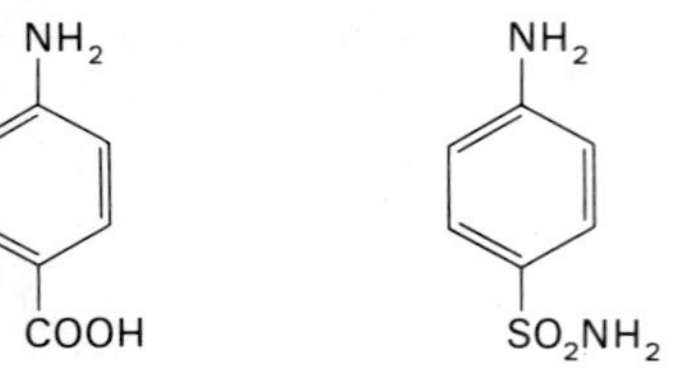

p-Aminobenzoic acid Sulfanilamide (sulfa drug)

ical industry relies heavily on the concept of competitive inhibition in the synthesis of drugs. Sulfanilamide, a sulfa drug (Figure 9-9), is an antibacterial agent because it effectively competes with p-aminobenzoic acid, which is required for synthesis of the essential metabolite folic acid (p. 174) by some pathogens. Sulfanilamide can inhibit the bacterial enzyme selectively because humans require the B-vitamin folic acid (cannot synthesize the vitamin) from p-aminobenzoic acid and, therefore, do not have the comparable enzyme. Much of the chemical therapy used to combat cancer utilizes compounds structurally designed to compete with substrates needed for the enzymatic processes of DNA replication in cancer cells, and, in this manner, cell division can be prevented.

Noncompetitive Inhibition

Noncompetitive inhibition is generally characterized as an inhibition of enzymatic activity by compounds that bear no structural relationship to the substrate and therefore the inhibition cannot be reversed by increased concentration of substrate. Unlike competitive inhibitors, reversible noncompetitive inhibitors cannot interact at the active site but must bind to some other portion of an enzyme or enzyme-substrate complex. This type of inhibition encompasses a variety of different inhibitory mechanisms and is therefore not amenable to a simple description.

Inhibition of enzymatic activity by heavy metal ions, e.g., Ag^+, Hg^{2+}, or Pb^{2+}, is often given as an example of reversible noncompetitive inhibition. Heavy metals form mercaptides with sulfhydryl (—SH) groups of enzymes:

$$E{-}SH + Ag^+ \rightleftharpoons E{-}S{-}Ag + H^+$$

The established equilibrium inactivates enzymes that require a sulfhydryl group for activity. Because of the reversibility of mercaptide formation, the inhibition can be relieved by removal of the heavy metal ion. In the medical treatment of lead poisoning, advantage is taken of the metal's affinity for SH groups. Appropriate sulfhydryl compounds are administered to interact with the metal in the circulatory system and form mercaptides, which are then excreted.

Irreversible Inhibition

Irreversible inhibitors usually inactivate enzymes by covalently bonding to their active sites *(active-site directed-affinity labeling).*

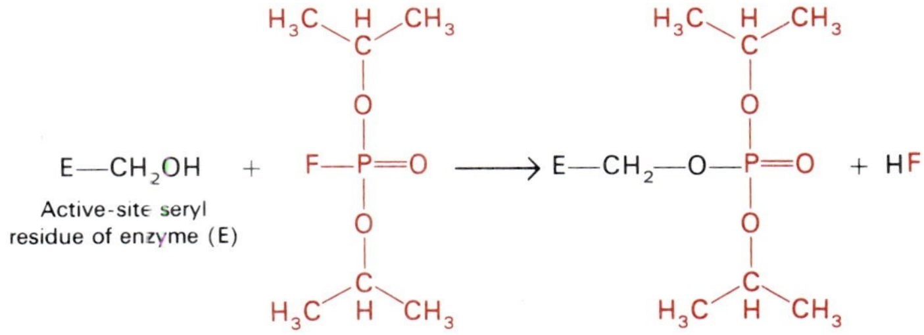

Diisopropylphosphofluoridate (DIFP)

Diisopropylphosphoryl-enzyme (DIP-enzyme)

Figure 9-10
Enzyme inhibition by diisopropylphosphofluoridate.

Although irreversible inhibition was once categorized and tested as noncompetitive inhibition, it is now recognized as a distinct type of inhibition. Organophosphorus compounds, e.g., diisopropylphosphofluoridate (DIFP or DFP), are potent irreversible inhibitors of enzymes that have active seryl residues (p. 150) at their catalytic sites (Figure 9-10). DIFP is closely related chemically to nerve gas, whose lethality is due to the inactivation of acetylcholinesterase, an enzyme critical for the transmission of nerve impulses. Organophosphorus compounds are the toxic ingredients of certain insecticides, e.g., malathion and parathion.

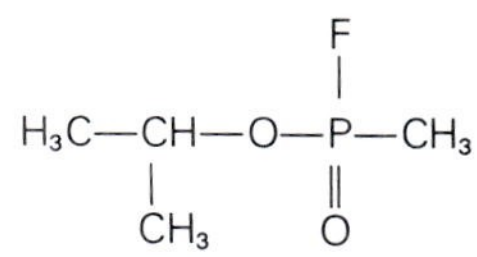

Sarin, a nerve gas

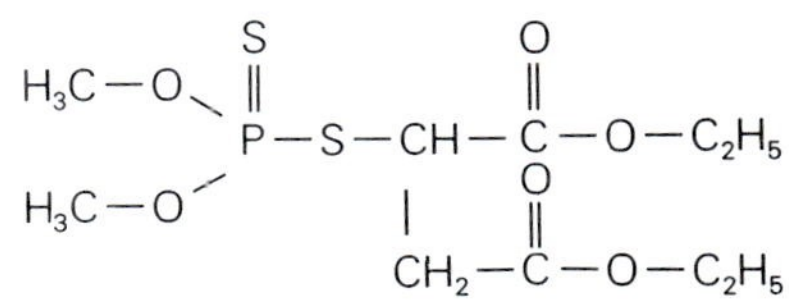

Malathion

LINEWEAVER–BURK PLOTS OF INHIBITION

Figure 9-11 summarizes the types of Lineweaver-Burk plots observed for competitive and noncompetitive inhibition. Analysis of the straight lines obtained in the absence and presence of a truly

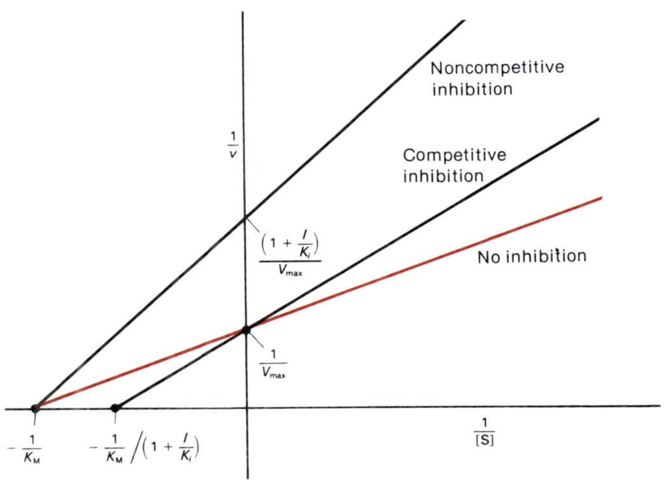

Figure 9-11
Hypothetical Lineweaver–Burk plots of enzymatic reactions, illustrating the effects of a competitive and noncompetitive inhibitor on an enzymatic reaction.

competitive inhibitor reveals that competitive inhibition alters the apparent K_M but not the V_{max} of the reaction. These characteristics reflect the fact that competitive inhibition can be relieved by increased substrate concentrations. Thus, in the presence of a competitive inhibitor, V_{max} is attainable (same V_{max} values); however, a greater substrate concentration is needed to reach $\frac{1}{2}V_{max}$ (different K_M values). With regard to strictly noncompetitive inhibition, V_{max} is decreased but K_M is not changed. Noncompetitive binding of such an inhibitor to a percentage of enzyme molecules or enzyme-substrate complexes in an assay ensures that the V_{max} will be less than that for a comparable number of active enzyme molecules in an uninhibited assay (difference in V_{max} values). On the other hand, an active enzyme molecule, whether in an assay that has or does not have the noncompetitive inhibitor, will have the same K_M for the substrate (same K_M values). In some cases, reversible inhibition of enzymes by an inhibitor leads to inhibition kinetics intermediate between strictly competitive and noncompetitive kinetics; such patterns of inhibition (increased K_M and decreased V_{max}) are referred to as *mixed.* It should also be noted that the Lineweaver-Burk plots for many irreversible inhibitors are comparable to those obtained with reversible noncompetitive inhibitors.

ACTIVE SITE AND SUBSTRATE BINDING

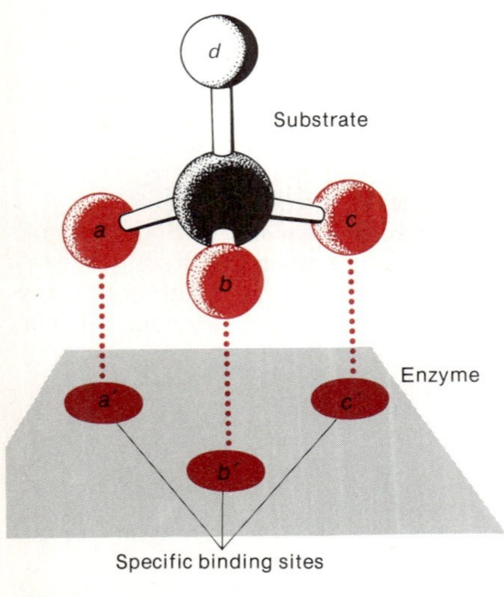

Figure 9-12
Schematic of the proposed three-point interaction between an enzyme and its substrate.

The three-dimensional structures that have been elucidated for some enzymes allow for a detailed description of their active sites. Although an enzyme typically has more than 100 amino acid residues in its structure, only a small number of residues make up the three-dimensional site that comes directly in contact with the substrate. Such an *active site* is often a crevice, or cleft, in the molecule's exterior. In some cases, the crevice, containing the substrate- and, if needed, coenzyme-binding sites, may provide a largely hydrophobic environment, which effectively excludes most water molecules. Some polar residues, essential for substrate binding and catalytic activity, are also contained within the active site; largely hydrophobic environment can be advantageous for the specific catalytic functions of certain polar residues.

To assure the geometrically precise substrate binding and orientation of catalytic groups that are necessary for enzyme action, at least three points of specific interactions are required between the substrate and active site (Figure 9-12). In the 1890s, Fischer

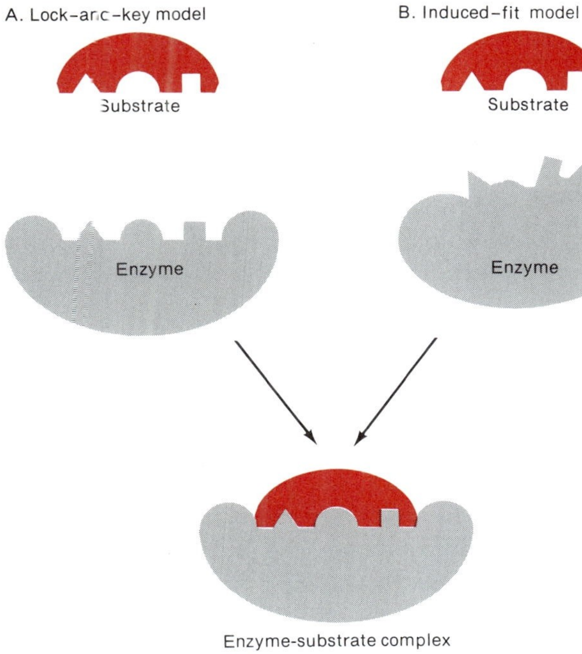

Figure 9-13
Schematic of the two proposed models for the interaction between an enzyme and its substrate.
(A) Lock-and-key model: active site of enzyme and substrate have complementary "shapes."
(B) Induced-fit model: the active site of the enzyme becomes complementary to that of the substrate as the substrate binds.

proposed the *lock-and-key model* (Figure 9-13), which accounts for substrate binding by the matching of complementary substrate and active-site structures. Over the years this model has proved valuable in studies on the stereospecificity of enzymatic reactions. More recently, Daniel Koshland proposed a modification of the lock-and-key model by suggesting a hand-in-glove type of relationship between some enzymes and their substrates, as the result of an induced fit. The *induced-fit model* (Figure 9-13) is an interpretation that considers the binding site of an enzyme not as a rigid structure but rather as one that changes in conformation upon substrate binding to produce a precise enzyme-substrate fit. Thus, Koshland's model incorporates dynamic properties into substrate binding.

QUATERNARY STRUCTURE AND ISOZYMES

The biologically active form of an enzyme having only one polypeptide chain in its molecular structure is referred to as *monomeric;* an enzyme with more than one chain in its structure is called *oligomeric.* Table 9-4 lists examples of enzymes of varying poly-

Table 9-4
Protomeric structure of some enzymes

Enzyme	MW	Number of protomers (MW)
Ribonuclease	13,700	1
Trypsin	23,800	1
Alkaline phosphatase	80,000	2(40,000)
Aspartate aminotransferase	100,000	2(50,000)
Threonine deaminase *(Salmonella)*	194,000	4(48,500)
Aspartate transcarbamoylase *(E. coli)*	310,000	6(34,000)
		6(17,000)
Lactate dehydrogenase	150,000	4(35,000)
Urease	483,000	6(83,000)

peptidyl composition. Alkaline phosphatase, a dimer, and threonine deaminase, a tetramer, have *homogeneous quaternary structures* because each has only one type of protomer (monomer or subunit). Aspartate transcarbamoylase, a dodecamer, has two different types of protomers and hence a *heterogeneous quaternary structure.* Lactate dehydrogenase (LDH) is an example of an enzyme that can have either a homogeneous or heterogeneous quaternary structure. LDH is an *isozyme,* an enzyme that occurs in multiple forms in an organism at different life stages and/or in different organs and tissues. The isozymic nature of LDH was described by Clement Markert, who identified five forms of the enzyme in rats. Genetic analysis established that the observed LDH isozymes represent the following five theoretically possible tetrameric arrangements of two different protomers (M and H):

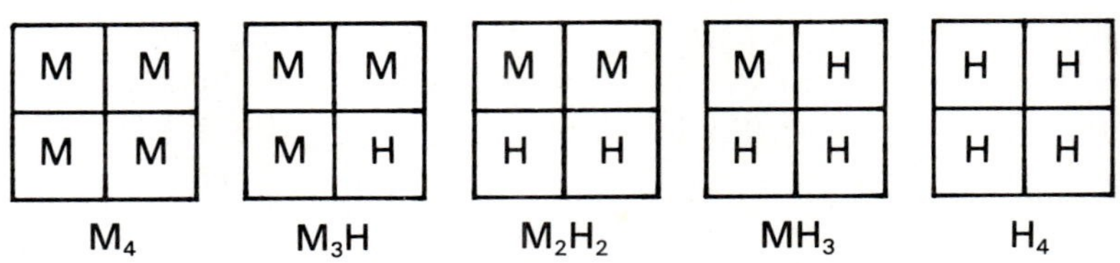

The homogeneous M tetramer of LDH is predominant in skeletal *muscle* tissue and the homogeneous H tetramer in *heart* tissue. Blood analysis of LDH isozymes is of significant clinical importance in diagnosing heart attacks and evaluating their seriousness. As a result of a myocardial infarction, the LDH-isozyme pattern of heart muscle is detectable in blood serum as early as six hours after the initial attack of pain. The amounts of LDH released from the damaged organ correlate more closely with infarct size than do any of the other serum enzymes estimates.

Although varying combinations of different protomers provide one molecular explanation for a particular set of isozymes, their occurrence can also result from other processes, e.g., chemical modifications or conformational changes. Efforts to elucidate the biological significance of the many known isozymes currently constitute an active area of research.

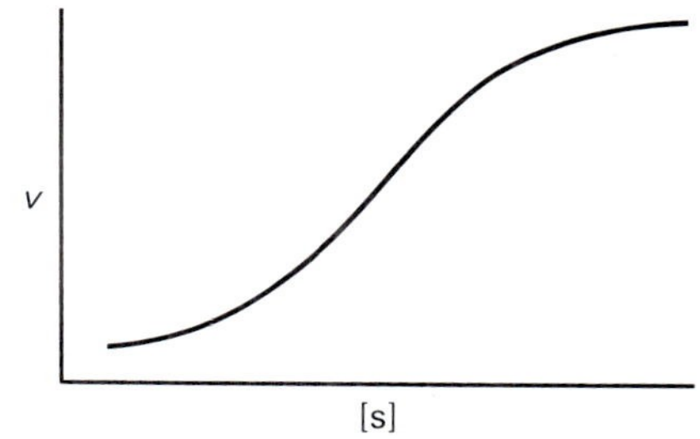

Schematic of an allosteric enzyme's activation by its substrate.

ALLOSTERIC ENZYMES

The general molecular concepts describing allosterism in hemoglobin (Chapter 8) also apply to allosteric effects observed for enzymes. *Allosteric enzymes* are key metabolic enzymes whose catalytic activities are regulated by conformational changes in their three-dimensional structures. Allosteric modulation is accomplished by *allosteric effectors,* molecules which increase (*positive effector,* or *allosteric activator*) or decrease (*negative effector,* or *allosteric inhibitor*) the reaction rate of an enzyme. These effectors, often structurally dissimilar to the substrate, bind reversibly and noncovalently to *allosteric sites,* which are distinct from the substrate-binding site(s). An allosteric enzyme may have one or more positive effectors and/or one or more negative effectors. Allosterism is an effective mechanism by which enzymatic activities can be controlled to ensure that biological processes remain coordinated at all times to meet the immediate metabolic requirements of a cell. Allosteric inhibition is considered by some to be a form of noncompetitive inhibition.

Control of enzymatic activity can also be accomplished in some instances by chemical modification of certain amino acid residues of an enzyme, involving addition or removal of a specific group. Bacterial glutamine synthetase (p. 310) and mammalian glycogen phosphorylase (p. 549) are examples of enzymes whose regulation entails *covalent modification.*

CATALYTIC EFFECTIVENESS OF ENZYMES

Enzymes are usually much more effective as catalysts than are the inorganic or organic catalysts commonly used in chemistry. One of the most catalytically potent enzymes, carbonic anhydrase, has a *turnover number* of 600,000 per second (Table 9-5). This means that when an enzyme molecule is fully saturated with its substrate

Table 9-5
Turnover numbers of some enzymes

Enzyme	Turnover number/sec
Carbonic anhydrase	600,000
Acetylcholinesterase	25,000
Lactate dehydrogenase	1,000
β-Galactosidase	208
Phosphoglucomutase	21
Tryptophan synthetase	2

(CO_2), 600,000 substrate molecules are transformed into product (H_2CO_3) in 1 second. Not all enzymes, however, are that active, and the turnover numbers of most enzymes range from 1 to 10^4 per second for their natural substrates. Note that the turnover number, or *catalytic coefficient* (k_{cat}), of an enzyme is its V_{max} per unit enzyme concentration; thus $k_{cat} = V_{max}/[E]_t$. Although biochemists have learned much about mechanisms of enzyme catalysis by chemical studies on active sites and by sequence and x-ray crystallographic analyses of enzymes, satisfactory explanations to account for the amazingly high rates of catalysis exhibited by most enzymes are still being sought.

ACID-BASE CATALYSIS

The principles of acid-base catalysis, well identified for organic chemical reactions, are useful in explaining fundamental chemical mechanisms of some enzymes. Two types of acid-base catalysis, general and Lewis, are of importance to studies on enzyme mechanisms. *General acid-base catalysis* relies on the Brönsted-Lowry concept of acids (proton donors) and bases (proton acceptors). In this type of catalysis, a specific group on the enzyme is viewed as an acceptor or donor of protons. Amino, carboxyl, sulfhydryl, imidazolium, and phenolic (tyrosine) groups of enzymes can serve as acids and/or bases. *Lewis acid-base catalysis* stresses the Lewis concept of acids and bases to explain certain enzymatic mechanisms. *Lewis acids* are defined as electron-pair acceptors; thus, they constitute *electrophilic (electron-seeking) groups.* Conversely, Lewis bases are electron-pair donors, *nucleophilic (nucleus-seeking) groups.* Certain metal ions, e.g., Mg^{2+}, Mn^{2+}, and Fe^{3+}, are electrophilic. Hydroxyl (e.g., serine), imidazole (histidine), carboxylate (glutamate or aspartate), and sulfhydryl (cysteine) groups are nucleophilic because each has an electron pair which could interact with an electron-deficient center of a substrate molecule. Such reactions of nucleophilic and electrophilic groups are important to certain catalytic mechanisms. Histidine is versatile in acid-base catalysis because its imidazole group can serve as a Lewis base (strong nucleophile) and, in its protonated form, the imidazolium ion can behave as a Brönsted-Lowry acid. An illustrative example of acid-base catalysis is the mechanism proposed for the peptidase chymotrypsin, one of the most thoroughly studied enzymes.

CHYMOTRYPSIN

Activation of Chymotrypsinogen

Chymotrypsin is one of several proteolytic digestive enzymes synthesized in the pancreas as *zymogens,* inactive precursors of enzymes (Table 9-6). The prefix *pro* or suffix *ogen* added to an enzyme's name designates the zymogenic state. Chymotrypsinogen, other zymogens, and digestive enzymes, including ribonucleases and lipases, are stored in lipid-protein membranes (called *zymogen granules*) in a mammalian pancreas and, when needed for digestive purposes, are secreted into a duct that leads into the duodenum, where they are activated. The biosynthesis of proteolytic enzymes as zymogens and their storage in membranes are necessary precautions to avoid enzymatic hydrolysis of pancreatic tissue (acute pancreatitis).

Table 9-6
Some pancreatic zymogens

Zymogen	Active enzyme
Chymotrypsinogen	Chymotrypsin
Trypsinogen	Trypsin
Procarboxy-peptidases	Carboxy-peptidases
Proelastase	Elastase

Chymotrypsinogen is composed of a single polypeptide chain with 245 amino acid residues and five disulfide bridges (Figure 9-14). The activation of the zymogen to produce chymotrypsin first involves the selective cleavage by trypsin of the peptide bond linking Arg 15 and Ile 16, yielding π-chymotrypsin. This single hydrolysis of a peptide bond converts the inactive form of the enzyme into active π-chymotrypsin, which then proceeds to cleave selectively other π-chymotrypsin molecules to produce α-chymotrypsin, the stable form of the active enzyme. Chymotrypsin catalyzes the removal of two dipeptides (residues 14 and 15 and residues 147 and 148) from π-chymotrypsin in the production of α-chymotrypsin. As a result, α-chymotrypsin, which has a molecular weight of 25,000, consists of three chains (designated A, B, and C); the chains remain convalently linked by disulfide bonds (Figure 9-14).

Three-Dimensional Structure of Chymotrypsin

The three-dimensional structure of chymotrypsin (Figure 9-15), which has been elucidated at 2 Å resolution by the x-ray crystallographic studies of David Blow, is a compact ellipsoid having the dimensions 51 × 40 × 40 Å. The structure has complex folding but, unlike myoglobin and hemoglobin, has little α-helical structure. Instead, parts of the secondary structure resemble an antiparallel pleated sheet arrangement (p. 79) because the three polypeptides, tending to be fully extended, are often aligned antiparallel

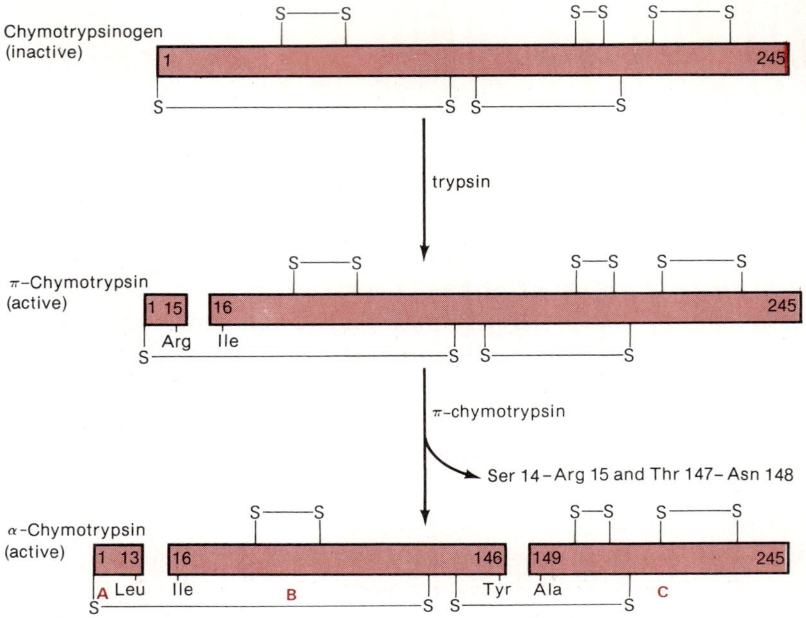

Figure 9-14
Activation of chymotrypsinogen.

to one another, with hydrogen bonding between peptide groups of adjacent strands.

Substrate Specificity of Chymotrypsin

Chymotrypsin catalyzes protein hydrolysis in the small intestine by cleaving peptide bonds on the *carboxyl side* of amino acids with *aromatic* (phenylalanine, tyrosine, and tryptophan) or *bulky hydrophobic* (e.g., methionine) R groups. As revealed by x-ray crystallographic studies using an enzyme-competitive inhibitor complex, the substrate-binding site includes a nonpolar pocket; therefore, substrate specificity is determined by hydrophobic interactions between the substrate's nonpolar R group and this site. Chymotrypsin can also hydrolyze ester linkages (Figure 9-16) by essentially the same catalytic mechanism as that used for protein hydrolysis. Ester hydrolysis, although not of physiological importance, offers certain advantages for *in vitro* kinetic studies and has been used extensively in research on chymotrypsin. Hydrolysis of esters of ρ-nitrophenol (Figure 9-17) provides a convenient assay method for certain studies because an easily quantifiable yellow product (ρ-nitrophenate ion) is produced.

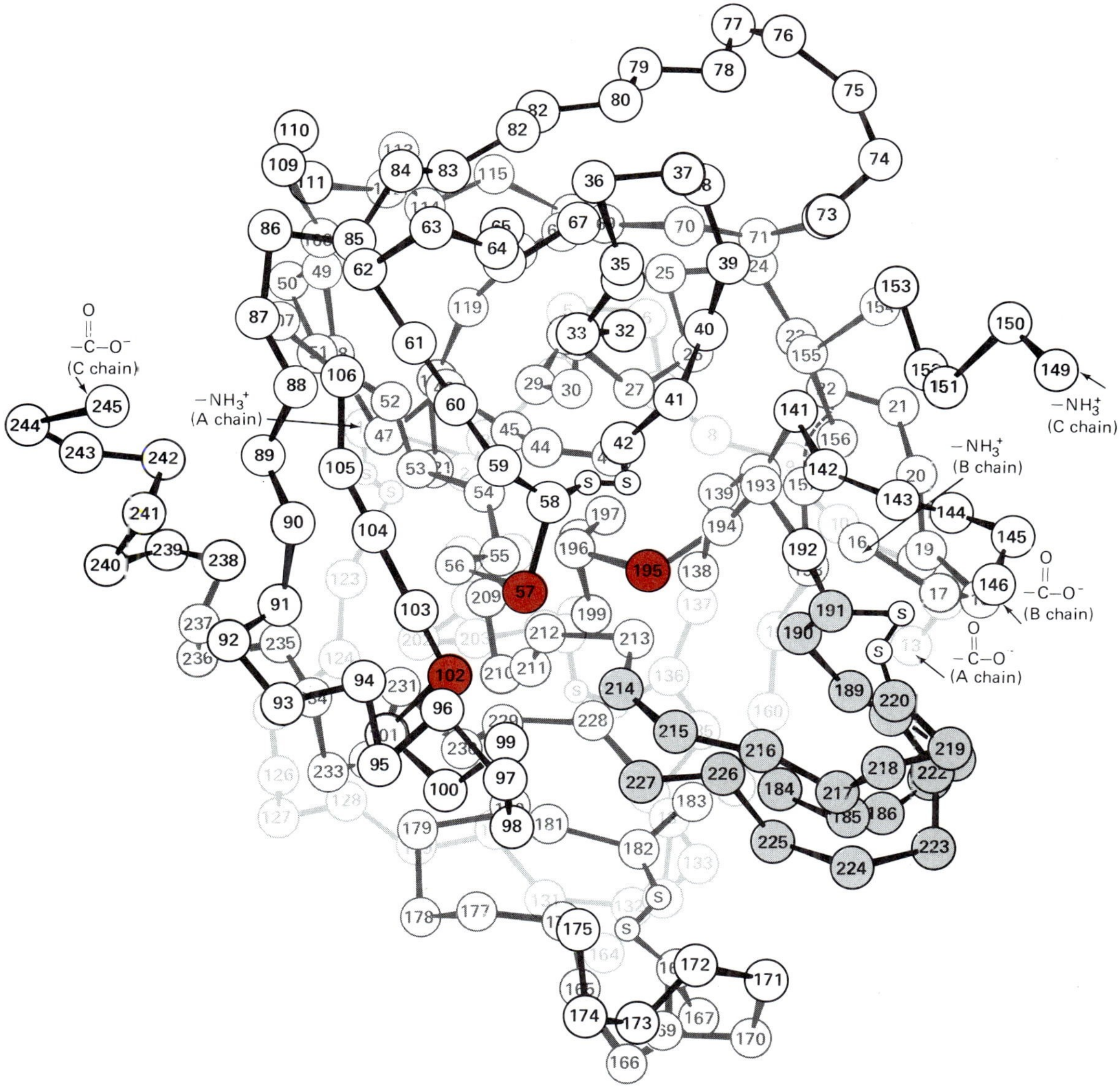

Figure 9-15
Three-dimensional structure (α-C atoms) of α-chymotrypsin. Residues of the catalytic triad (Asp 102, His 57, and Ser 195) are indicated in red. The hydrophobic pocket for the substrate is indicated by the dark residues. (After R. E. Dickerson and I. Geis, *The Structure and Action of Proteins.* New York: Harper and Row, 1969)

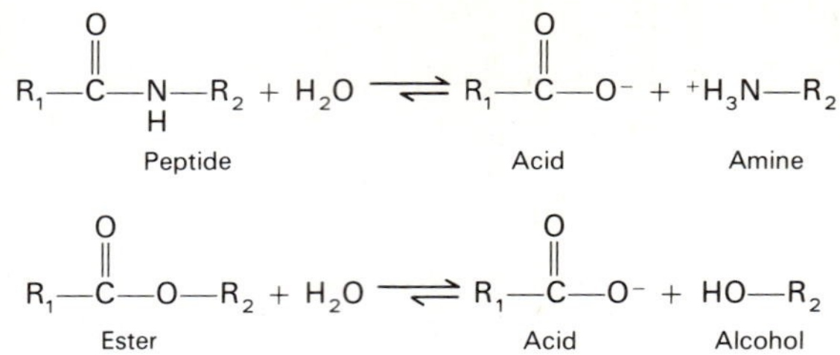

Figure 9-16
Hydrolysis of a peptide bond and of an ester bond.

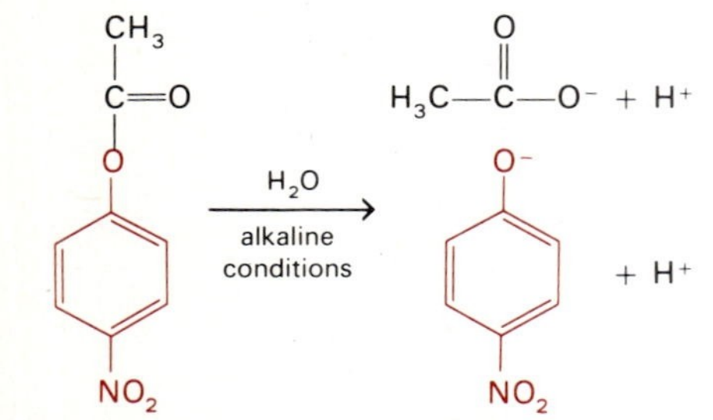

p-Nitrophenylacetate *p*-Nitrophenate (yellow)

Figure 9-17
Hydrolysis of a *p*-nitrophenol ester.

PROPOSED CATALYTIC MECHANISM OF CHYMOTRYPSIN

"Charge Relay Network"

Knowledge of the sequence and three-dimensional structure of chymotrypsin and the results of chemical and kinetic studies have allowed biochemists to propose a chemical explanation for the catalytic mechanism of the enzyme. Chymotrypsin is regarded as an *"active" serine protease,* having an unusually nucleophilic seryl oxygen (Ser 195) at its active site, which is necessary for the enzyme's catalytic action. A number of enzymes require a reactive seryl residue for their activity, e.g., acetylcholinesterase, chymotrypsin, trypsin, thrombin, and elastase; they are readily identifiable by their rapid inactivation by DIFP (p. 141). That only one of the 27 seryl residues of α-chymotrypsin is a strong nucleophile is explainable by a specific spatial relationship between three residues (Asp 102, His 57, and Ser 195) that creates what David Blow termed a *"charge relay network"* (Figure 9-18) based on hydrogen bonding. The hydrogen bonding that occurs between the buried Asp 102 and His 57 and between His 57 and Ser 195 establishes an equilibrium that permits an effective withdrawal of the proton of the hydroxyl group of Ser 195 (at the catalytic site), making the oxygen atom of that residue a strong nucleophile. The hydrogen-bonding resonance of the imidazole ring (conjugate base form) of His 57 is the key to the effectiveness of the proposed system.

Figure 9-18
Proposed "charge relay network" in chymotrypsin, involving the hydrogen bonding of Ser 195 and His 57 and Asp 102.

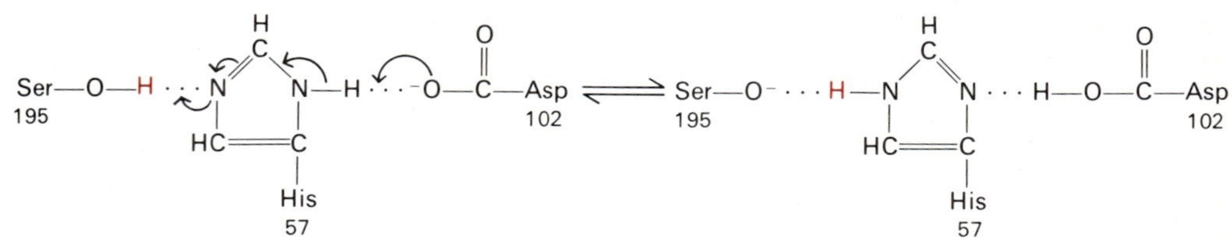

While the positions of C, N, and O atoms within the tertiary structure of chymotrypsin could be accurately assessed by x-ray crystallography, the bonding of the H atoms had to be inferred, since the locations of the hydrogen atoms cannot be determined by this technique. Recent evidence obtained by neutron diffraction studies, however, has established the positions of the hydrogen atoms within the "catalytic triad," i.e., the three invariant catalytic residues Asp 102, His 57, and Ser 195. The carboxylate ion of Asp 102 is *not* protonated, but retains its charge and forms an ion pair with the imidazolium ion of His 57. Thus, strictly speaking, the Asp-His-Ser catalytic triad is not a "charge relay system" in the sense originally postulated by Blow.

Double-Displacement Mechanism for the Hydrolysis of Peptide Bonds

Hydrolysis of a peptide bond by chymotrypsin is proposed to occur by a two-step displacement (Figure 9-19) with an amine being produced first, followed by production of an acid. Detailed kinetic analyses show that the mechanism of hydrolysis involves initial

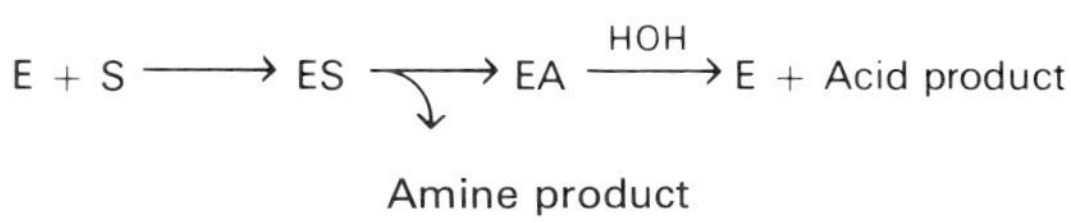

Figure 9-19
Two-step production of the products of peptide bond hydrolysis by chymotrypsin.

acylation of the enzyme, with release of the amine product, followed by *deacylation* of the enzyme to provide the acid product (Figure 9-20). Catalysis commences when the oxygen atom of Ser 195, in its strongly nucleophilic state (Lewis base), initiates a nucleophilic attack on the partially positive carbonyl carbon of the substrate (1 of Figure 9-20); this results in the formation of an unstable tetrahedral transition state (2 of Figure 9-20). The enzyme-substrate transitional state rapidly decomposes, with cleavage of the C—N covalent bond, to yield an acyl-enzyme intermediate (esterified Ser 195) and the first product, an amine (3 of Figure 9-20). Thus, formation of the acyl-enzyme intermediate is the result of a nucleophilic displacement, i.e., one nucleophilic substituent ($-NH-R_2$) of the carbonyl carbon is replaced by another ($-O-CH_2-$Ser 195). For production of the second product (acid), deacylation of the enzyme occurs by a nucleophilic attack by the oxygen of H_2O on the acyl carbon of the acyl-enzyme intermediate (4 of Figure 9-20), which creates a second unstable tetra-

1. Nucleophilic attack on carbonyl C of substrate by O of Ser 195
2. Unstable tetrahedral transition state
3. Production of amine product and acyl-enzyme intermediate (acylation)
4. Entrance of water and nucleophilic attack on carbonyl C by O of water
5. Second unstable tetrahedral transition state
6. Production of acid product (deacylation)

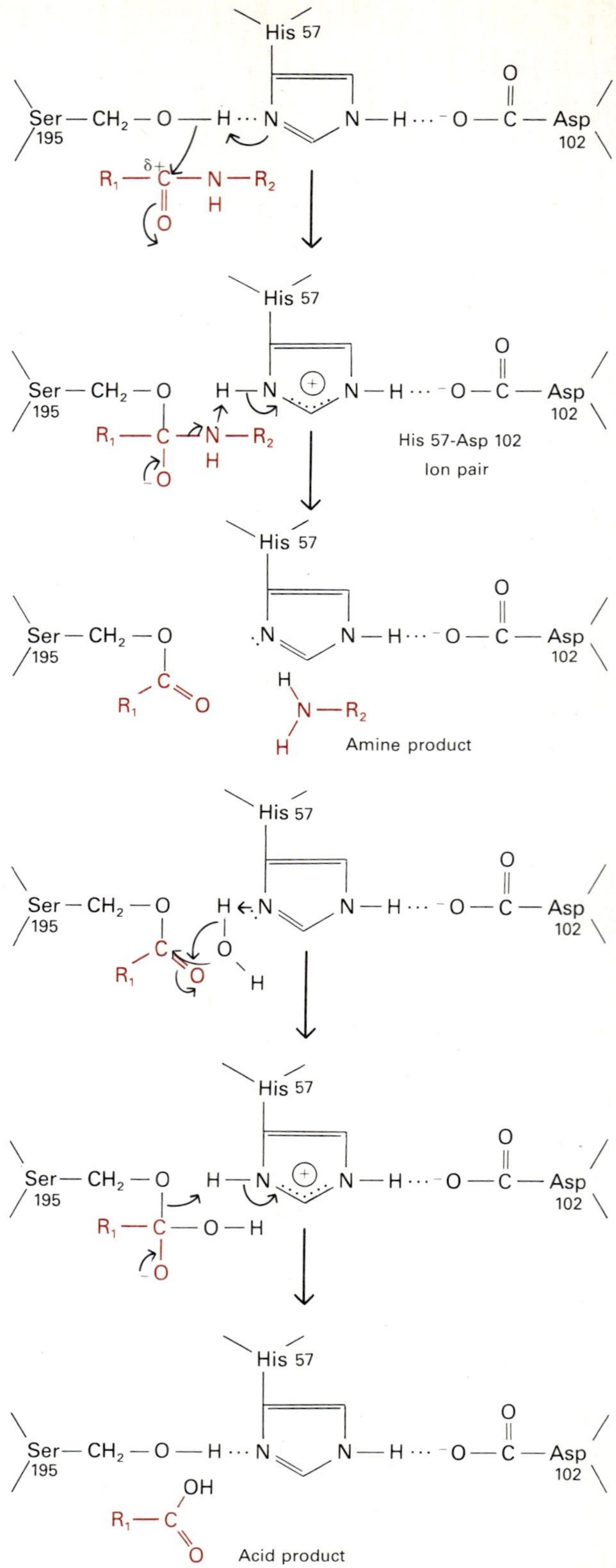

Figure 9-20
Proposed mechanism of hydrolysis by chymotrypsin.

hedral transition state (5 of Figure 9-20). The C—O Ser 195 bond is cleaved to yield an acid and the restored enzyme with its activated Ser 195 (6 of Figure 9-20). Hydrogen bonding (catalytic triad) and Lewis acid-base catalysis (nucleophilic attacks) provide the chemical basis for the double-displacement catalytic mechanism of chymotrypsin. Another enzymatic mechanism involving acid-base catalysis, that of ribonuclease A, is presented on p. 239.

BIOCHEMICAL EVOLUTION

Biochemical evolution is the study of the origin and history of biomolecules which, in the case of proteins, makes use of the comparative analyses of biological properties, amino acid sequences, and three-dimensional structures. Comparisons of the biological and physical properties of four proteases—chymotrypsin, trypsin, elastase, and the microbial enzyme subtilisin—provide the following proposal for their genetic histories. Chymotrypsin, trypsin, and elastase are believed to have a common ancestral gene which, by the process of gene duplication during biological evolution *(divergent evolution),* gave rise to the three genes that now code for the proteases, each with a different substrate specificity. This conclusion is based on the facts that the three proteases have (1) a DIFP-reactive serine at the active site, (2) the same sequence about the active-site serine (Figure 9-21), (3) the same invariant Asp-His-Ser

-Gly-Asp-Ser-Gly-Gly-Pro- (chymotrypsin, trypsin, and elastase)

-Gly-Thr-Ser-Met-Ala-Ser- (subtilisin)

Figure 9-21
Residues around the active-site serine of subtilisin and of several other serine proteases.

triad in their tertiary structures (Figure 9-22), (4) amino acid sequences that are about 40 per cent identical, and (5) very similar three-dimensional structures (Figure 9-23). The differences in substrate specificities of the three enzymes are accounted for by minor modifications of the substrate-binding sites. By several of these criteria, subtilisin differs markedly from the other three proteases.

Asp 102	His 57	Ser 195	(chymotrypsin)
Asp 32	His 64	Ser 221	(subtilisin)

Figure 9-22
Residues making up catalytic triad of subtilisin and of chymotrypsin.

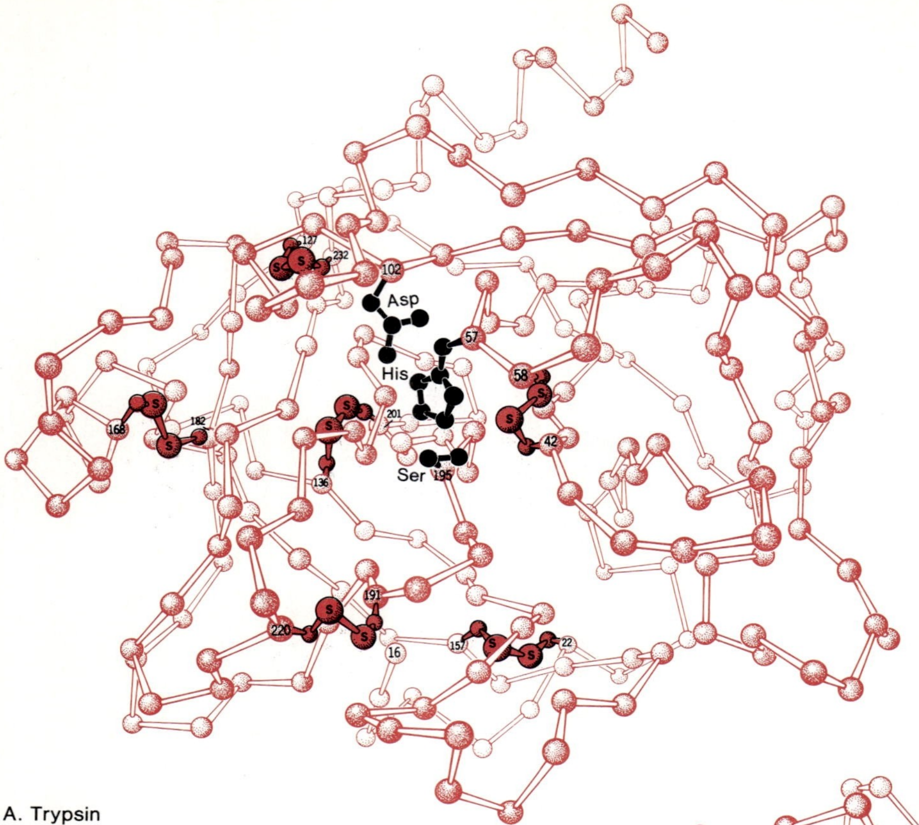

A. Trypsin

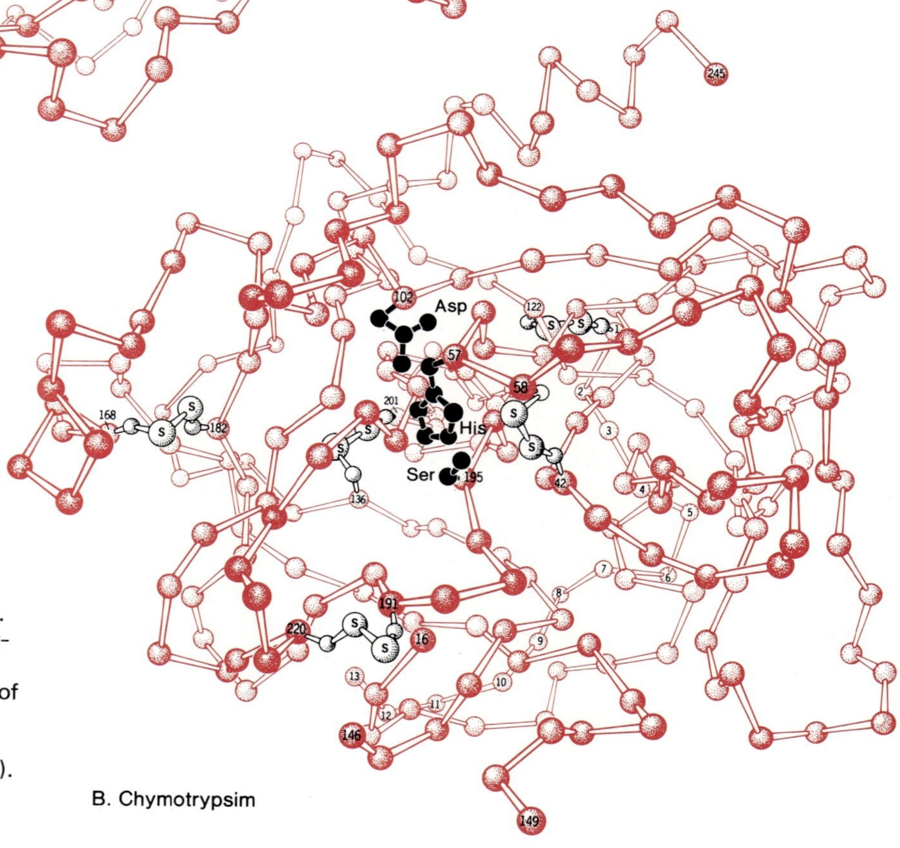

B. Chymotrypsim

Figure 9-23
Tertiary structures of four serine proteases. (A) Trypsin. (B) Chymotrypsin. (C) Elastase. (D) Subtilisin. Only the α-carbon positions are identified in the molecules, except for the side chains of the Asp, His, and Ser residues of the catalytic triad and for the disulfide bonds (indicated by the dark linkages). (Copyright 1982 by Irving Geis)

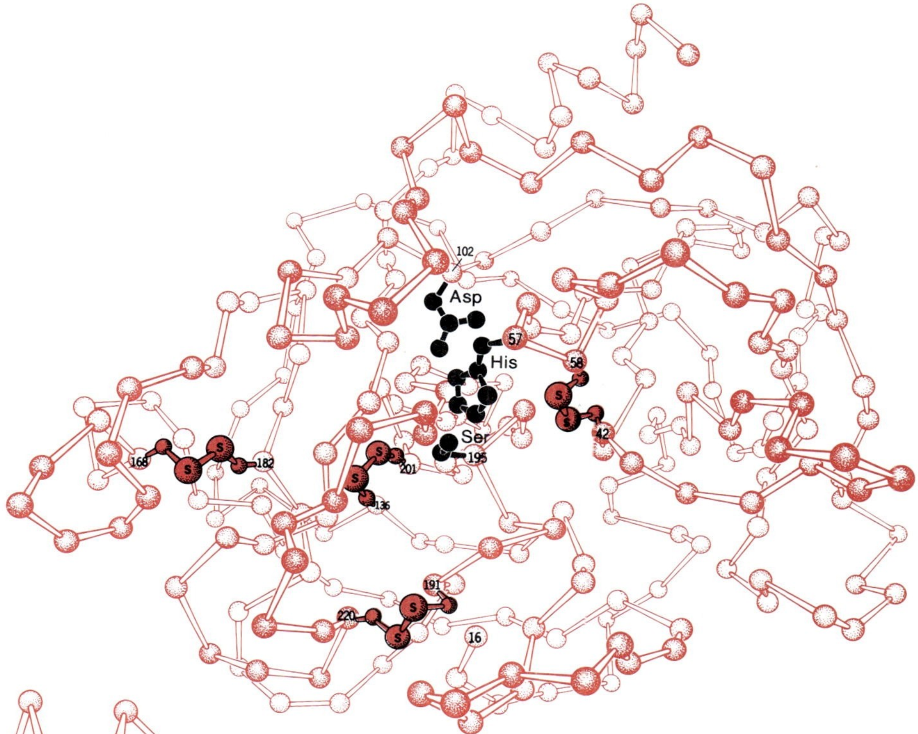

C. Elastase

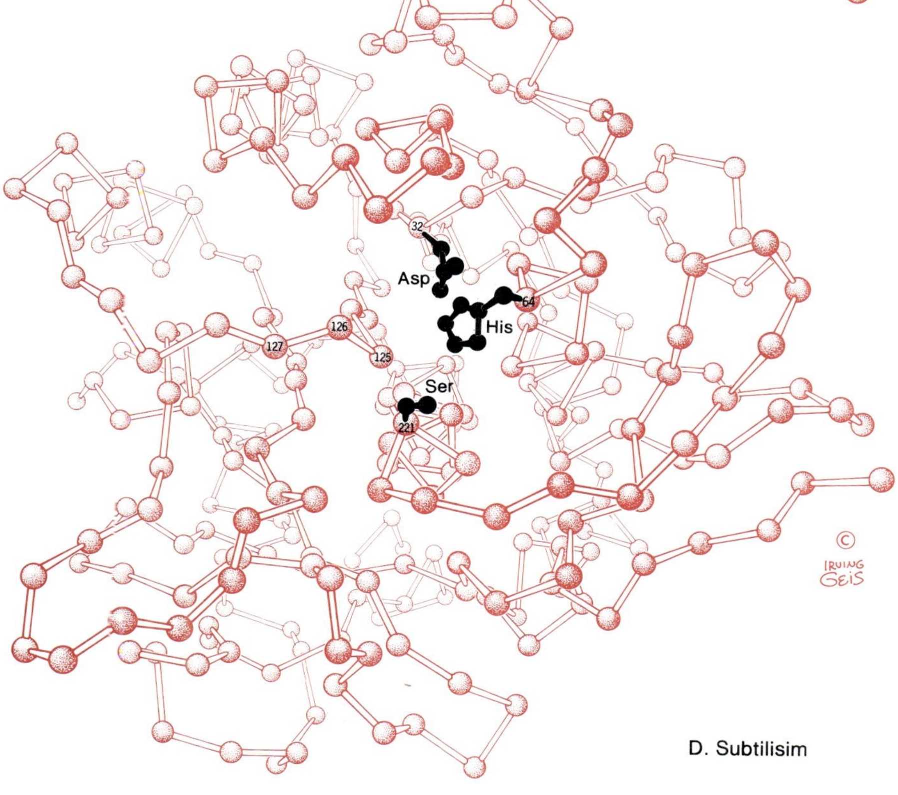

D. Subtilisim

Although subtilisin has an active-site serine (DIFP-reactive) and an Asp-His-Ser network, the sequence around the active-site serine (Figure 9-21) and the primary structural positions of the three residues of the catalytic triad are different (Figure 9-22). Subtilisin's amino acid sequence and overall three-dimensional structure are also strikingly different from those of the trypsin-like proteases (Figure 9-23). For example, unlike the other three proteases, subtilisin has no disulfide bonds. The conclusion is that subtilisin's evolutionary history is distinct from that of the other three proteases. This example of two independent but parallel evolutionary processes is called *independent convergent evolution.* Nature developed two distinct molecular types of hydrolytic proteases that make use of the same catalytic mechanism, i.e., a double-displacement mechanism involving an active-site serine and an Asp-His-Ser network. Similar comparative studies on the oxygen-carrying proteins, myoglobin and the hemoglobins, suggest that the genes that code for myoglobin and the individual polypeptides of the tetrameric hemoglobins, e.g., α, β, and γ chains, are all derived from the same ancestral gene.

ENZYMOLOGY AND INDUSTRY

Much of the early interest in enzymology was developed by scientists associated with the food, wine, and beer industries, e.g., Pasteur, Payen, and Persoz. As a result, scientific and industrial objectives were often inseparable. Pasteur was perhaps best known to the French nation as the ``saviour of the wine industry'' because his pasteurization process salvaged an ailing industry beset with problems of microbial contamination. The close ties between enzymology and industry still continue, and the specific catalytic abilities of many enzymes are effectively used for commercial and medical purposes. Proteases of fungi and bacteria are used in the baking industry to control the breakdown of gluten (protein) of grains, e.g., wheat and rye. As seen in Table 9-7, proteases are also employed in detergents (alcalase), meat tenderizers (bromelain and papain), cheesemaking (rennin), and in the brewing industry (papain) to remove small amounts of protein that cause turbidity in chilled beer. Because hydroxyprolyl residues create bends in collagen helices, which contribute to the ``tough and rubbery'' texture often associated with cooked meat, treating the meat with a protease (bromelain or papain) prior to its cooking hydrolyzes peptide bonds and thus ``tenderizes'' it. Medically, pro-

Table 9-7
Commercial uses of enzymes

Enzymes	Use
Proteases	
Alcalase	Detergent additive to remove protein stains
Bromelain	Meat tenderizer
Papain	Stabilizer for chill-proof beer; meat tenderizer
Pepsin	Digestive aid in precooked foods
Trypsin, ficin, and streptodornase	Wound debridement
Rennin	Cheesemaking aid
Carbohydrases	
Amylase	Digestive aid in precooked foods
Amyloglucosidase	Production of dextrose from starch
Cellulase and hemicellulase	Preparation of liquid coffee concentrates Conversion of sawdust to sugar (World War II)
Glucose isomerase	Production of fructose and high-fructose syrups from hydrolyzed cornstarch
Glucose oxidase	Removal of glucose from egg solids
Invertase	Prevention of granulation of sugars in soft-centered candies
Lactase	Prevention of lactose crystals in ice cream Production of lactose hydrolyzed milk for consumption
Pectinase	Clarification of wine and fruit juices
Catalase	Removal of hydrogen peroxide following "cold pasteurization" of milk for cheesemaking
Lipase	Flavor production in cheese
Lipoxygenase	Bread whitening

teases (streptodornase, ficin, and trypsin) are used to clean wounds. Other classes of enzymes are also extensively used for industrial and medical purposes (Table 9-7).

The use of immobilized enzymes for the production of commercial quantities of compounds is another promising avenue for the use of enzymes in industry. An *immobilized enzyme* is one that is physically entrapped or covalently bonded by chemical means to an insoluble matrix, e.g., glass beads, polyacrylamide or cellulose. Immobilization of an enzyme often greatly enhances its stability, which makes its prolonged catalytic life a valuable industrial trait. The use of column chromatography, which uses enzymes bound to the column material, is also industrially attractive because it offers an effective production process, i.e., substrate can be added to the column and the desired product collected in the eluate. Currently, in the United States, immobilized glucose isomerase is successfully being used in the production of large quantities of high-fructose corn syrup.

SUMMARY

Enzymes are the remarkably efficient biological catalysts responsible for the thousands of reactions needed for the chemical processes of life. All enzymes are proteins and can be classified either as simple proteins, whose catalytic activities rely on the enzyme's polypeptidyl structure, or as conjugated proteins, which require a nonprotein component(s) for activity. The nonprotein components of enzymes, called cofactors, include metal ions, e.g., Mg^{2+}, and organic biomolecules (coenzymes), e.g., B-vitamins. If a cofactor is tightly bound to an enzyme, it is called a prosthetic group. An enzyme is classified by the type of reaction it catalyzes, and the six major classes of enzymes are oxidoreductases, transferases, hydrolases, lyases, isomerases, and ligases.

In an enzyme-catalyzed reaction, increasing the substrate concentration increases the reaction rate (v). However, after the active sites of the enzyme molecules are saturated with substrate, the reaction rate becomes independent of substrate concentration (zero-order reaction); the maximal velocity theoretically attainable under these conditions is called V_{max}. The substrate concentration at which the reaction rate is half that of V_{max} is called the K_M

Excerpt from *The Andromeda Strain* by Michael Crichton

In another room, Leavitt was carefully feeding similar chips into a different machine, an amino-acid analyzer. As he did so, he smiled slightly to himself, for he could remember how it had been in the old days, before AA analysis was automatic.

In the early fifties, the analysis of amino acids in a protein might take weeks, or even months. Sometimes it took years. Now it took hours—or at the very most, a day—and it was fully automatic.

Amino acids were the building blocks of proteins. There were twenty-four (sic) known amino acids, each composed of a half-dozen molecules of carbon, hydrogen, oxygen, and nitrogen. Proteins were made by stringing these amino acids together in a line, like a freight train. The order of stringing determined the nature of the protein—whether it was insulin, hemoglobin, or growth hormone. All proteins were composed of the same freight cars, the same units. Some proteins had more of one kind of car than another, or in a different order. But that was the only difference. The same amino acids, the same freight cars, existed in human proteins and flea proteins.

That fact had taken approximately twenty years to discover.

But what controlled the order of amino acids in the protein? The answer turned out to be DNA, the genetic-coding substance, which acted like a switching manager in a freight-yard.

That particular fact had taken another twenty years to discover.

But then once the amino acids were strung together, they began to twist and coil upon themselves; the analogy became closer to a snake than a train. The manner of coiling was determined by the order of acids, and was quite specific: a protein had to be coiled in a certain way, and no other, or it failed to function.

(Michaelis constant). Each enzyme displays a characteristic K_M for its substrate(s). The relationship between substrate concentration, v, V_{max}, K_M, and the initial reaction rate of an enzyme reaction is mathematically expressed in the Michaelis–Menten equation. The K_M and V_{max} values of an enzymatic reaction are often determined by plotting the reciprocals of [S] and v values (Lineweaver–Burk plot). Ideally, the Lineweaver–Burk plot yields a straight line with a slope equal to K_M/V_{max}, and the $1/v$ and $1/S$ intercepts are equal to $1/V_{max}$ and $-1/K_M$, respectively.

The three general types of inhibition of enzymatic activity are called competitive, noncompetitive, and irreversible. Competitive and noncompetitive inhibition are reversible phenomena. A competitive inhibitor lowers the rate of an enzyme-catalyzed reaction by competing with the substrate for the active site to form an enzyme-inhibitor complex; this type of inhibition can be reversed by increasing the substrate concentration. A noncompetitive inhibitor interacts not with the active site but, instead, with another region of the enzyme or enzyme-substrate complex. An irreversible inhibitor usually inactivates an enzyme by covalently bonding to an essential group in its active site.

An isozyme is an enzyme that occurs in multiple forms within an organ-

Another ten years.

Rather odd, Leavitt thought. Hundreds of laboratories, thousands of workers throughout the world, all bent on discovering such essentially simple facts. It had all taken years and years, decades of patient effort.

And now there was this machine. The machine would not, of course, give the precise order of amino acids. But it would give a rough percentage composition: so much valine, so much arginine, so much cystine and proline and leucine. And that, in turn, would give a great deal of information.

Yet it was a shot in the dark, this machine. Because they had no reason to believe that either the rock or the green organism was composed even partially of proteins. True, every living thing on earth had at least some proteins—but that didn't mean life elsewhere had to have it.

For a moment, he tried to imagine life without proteins. It was almost impossible: on earth, proteins were part of the cell wall, and comprised all the enzymes known to man. And life without enzymes? Was that possible?

He recalled the remark of George Thompson, the British biochemist, who had called enzymes "the matchmakers of life." It was true; enzymes acted as catalysts for all chemical reactions, by providing a surface for two molecules to come together and react upon. There were hundreds of thousands, perhaps millions, of enzymes, each existing solely to aid a single chemical reaction. Without enzymes, there could be no chemical reactions.

Without chemical reactions, there could be no life.

ism. An isozyme pattern (forms) may vary during the development of an organism and/or in different tissues. Like hemoglobin, many enzymes are allosteric, and their activities are regulated by positive and/or negative effectors which increase and/or decrease the reaction rates, respectively. Allosteric enzymes are generally key metabolic enzymes, and control of their activities is critical for the coordination of the metabolic activities of a living cell.

The principles of acid-base catalysis that apply to organic reactions have proved useful in explaining the catalytic mechanisms of a number of enzymes. The proposed catalytic mechanism for chymotrypsin is illustrative of the biochemical importance of general acid-base catalysis (involving Brönsted–Lowry acids and bases) and Lewis acid-base catalysis (involving electrophilic and/or nucleophilic groups). Chymotrypsin is a digestive peptidase which is synthesized in the pancreas as a zymogen (inactive precursor of an enzyme), called chymotrypsinogen. After the zymogen is transported into the duodenum, four selective hydrolytic cleavages convert chymotrypsinogen into α-chymotrypsin, the stable, active form of the enzyme. α-Chymotrypsin is a serine protease, i.e., has a highly nucleophilic seryl oxygen (essential for catalysis) at its active site. Ser 195 is a strong nucleophile (Lewis base) because of the hydrogen bondings that occur between the following three residues: Asp 102, His 57, and Ser 195. The two-step cleavage of a peptide bond by chymotrypsin is initiated by a nucleophilic attack on the carbonyl carbon of the substrate by the oxygen atom of Ser 195, resulting in the formation of an acyl-enzyme intermediate and the first product, an amine. A nucleophilic attack by the oxygen atom of a water molecule on the acyl carbon of the acyl-enzyme intermediate than occurs to yield the second product, an acid, and the restored enzyme.

Comparison of the results of the detailed biological and chemical studies on the four serine proteases chymotrypsin, trypsin, elastase, and subtilisin provides evidence that the first three enzymes share a common ancestral gene—an example of divergent evolution. Although the fourth protease, subtilisin, also possesses an active-site serine and a hydrogen bonded network involving an aspartyl, a histidyl, and a seryl residue, the enzyme appears to have evolved independently of the other three. The evolution of two distinct types of serine proteases with the same catalytic mechanism is representative of independent convergent evolution.

The specialized catalytic capabilities of enzymes have been used for industrial and medical purposes for many years and continue to increase in their commercial importance. The current development of immobilized enzymes is introducing new industrial applications for these biological catalysts.

PROBLEMS

1. Which of the following statements about chymotrypsin are true?
   (a) Has a quaternary structure
   (b) Is a simple protein
   (c) Has a prosthetic group
   (d) Displays relative-group specificity
   (e) Is a lyase
   (f) Is a metalloenzyme

(g) Is a zymogen
(h) Requires a coenzyme
(i) Has a majority of its residues in α-helical conformation

2. Although trypsin was not discussed in detail in the chapter, which of the statements listed in Question 1 are also probably true about this protease?

3. Which of the following statements are true for both (a) chymotrypsin and subtilisin and (b) trypsin and elastase?
   (1) Are serine proteases
   (2) Believed to be derived from a common ancestral gene
   (3) Are hydrolytic enzymes
   (4) Are digestive enzymes of animals
   (5) Have an aspartyl-histidyl-seryl catalytic network
   (6) Have disulfide bonds

4. Assuming an enzyme behaves according to the Michaelis–Menten equation, what percentage of V_{max} would be observed at [S] (a) K_M, (b) $\frac{1}{2}K_M$, (c) $2K_M$, (c) $\frac{1}{10}K_M$, and (e) $10K_M$?

5. Draw Lineweaver–Burk plots using the following experimental information:

[S] (mM)	v–no inhibitor (μmoles min^{-1} mg^{-1})	v–inhibitor present (μmoles min^{-1} mg^{-1})
3.0	2.29×10^3	1.83×10^3
5.0	3.20×10^3	2.56×10^3
7.0	3.86×10^3	3.09×10^3
9.0	4.36×10^3	3.49×10^3
11.0	4.75×10^3	3.80×10^3

What are the K_M and V_{max} values of the (a) uninhibited reaction and of the (b) inhibited reaction? (c) Is the inhibitor competitive or noncompetitive?

6. Which of the reactions listed below are catalyzed by an isomerase, lyase, hydrolase, or transferase?
   (a) Protein → amino acids
   (b) Histidine → histamine + CO_2
   (c) Glucose + ATP → glucose 6-phosphate + ADP
   (d) Glucose 6-phosphate → glucose + H_3PO_4
   (e) $H_3CCOCOOH$ + 2H· → $H_3CCHOHCOOH$
   (f) $H_3CCOCOOH$ → H_3CCHO + CO_2
   (g) 3-Phosphoglycerate → 2-phosphoglycerate
   (h) Tryptophan → tryptamine + CO_2
   (i) L-Lysine → D-lysine
   (j) Acetylcholine → acetic acid + choline
   (k) 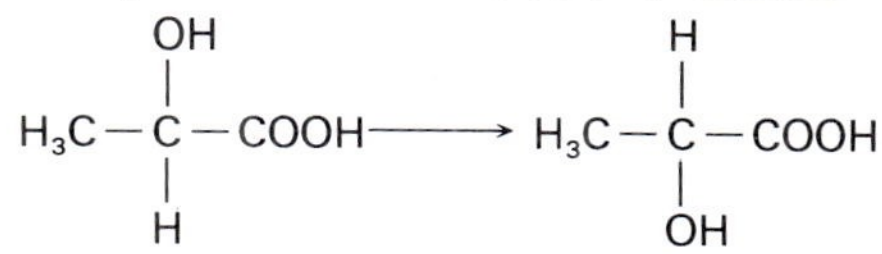

   (l) Glu + oxaloacetate → α-ketoglutarate + Asp

7. Why is pepsin useful as a digestive aid in precooked foods?

SUGGESTED READINGS

Bender, M. L., and L. J. Brubacher, *Catalysis and Enzyme Action.* New York: McGraw-Hill, 1973. A good introductory paperback on the structure and function of enzymes.

Boyer, P. D., ed., *The Enzymes.* New York: Academic Press 1973. Paperback, student edition, containing comprehensive reviews on enzymes and coenzymes. The emphasis is on structure, control, kinetics, and mechanisms.

Dixon, M., and E. C. Webb, *Enzymes,* 3rd ed. New York: Academic Press, 1979. An advanced, comprehensive text on general enzyme properties—a classic among enzyme texts.

Dugas, H., and C. Penney, *Bioorganic Chemistry—A Chemical Approach to Enzyme Action.* New York: Springer-Verlag, 1981. A text that integrates organic chemistry and biochemistry, with special emphasis on enzyme catalysis.

Haldane, J. B. S., *The Enzymes.* Cambridge, Mass.: MIT Press, 1965. A reprint of the 1930 classic text on enzymes.

Pitcher, W. H. Jr., *Immobilized Enzymes for Food Processing.* Boca Raton, Fla.: CRC Press, 1980. A discussion of immobilized enzyme systems and an evaluation of their application to food processing.

Walsh, C., *Enzymatic Reaction Mechanisms.* San Francisco: W. H. Freeman, 1979. A good treatment of enzymes and their mechanisms.

Zeffren, E., and P. L. Hall, *The Study of Enzyme Mechanisms.* New York: Wiley, 1973. The text discusses allosterism, coenzymes, catalysis, and kinetics of enzymes in an advanced, but not difficult, manner.

Articles

Chan, W. W. -C., "The Relationship Between Quaternary Structure and Enzyme Activity." *TIBS,* 1:258, 1976.

Gabriel, O., "Common Mechanistic and Structural Properties of Enzymes." *TIBS,* 3:193, 1978.

Koshland, D. E. Jr., "Protein Shape and Biological Control." *Scientific American,* 229(4):52 (offprint # 1280), 1973.

Kraut, J., "Serine Proteases: Structure and Mechanism of Catalysis." *Ann. Rev. Biochem.,* 46:331, 1977.

Mildavin, A., "Mechanism of Enzyme Action." *Ann. Rev. Biochem.,* 43:357, 1974.

Mosbach, K., "Enzymes Bound to Artificial Matrixes." *Scientific American,* 224(3):26 (offprint #1216), 1971.

Neurath, H., "Protein-Digesting Enzymes." *Scientific American,* 211(6):68, 1964.

Phillips, D. C., "The Three-Dimensional Structure of an Enzyme Molecule." *Scientific American,* 215(5):78 (offprint #1055), 1966.

Stroud, R. M., "A Family of Protein-Cutting Proteins." *Scientific American,* 231(1):74 (offprint #1301), 1974.

Sumner, J. B., "The Story of Urease." *J. Chem. Ed.,* 14:255, 1937.

Tang, J., "Pepsin and Pepsinogen: Model for Carboxyl (Acid) Proteases and Their Zymogens." *TIBS,* 1:205, 1976.

Wróblewski, F., "Enzymes in Medical Diagnosis." *Scientific American,* 205:(2):99, 1961.

B-VITAMINS AS COENZYMES

10

INTRODUCTION

One of the most rewarding scientific accomplishments of biochemistry in the first half of this century was the isolation and chemical characterization of the *B-vitamins* and the identification of their individual roles as *coenzymes* in enzymatic reactions. The landmark advances made during this period toward understanding human nutritional requirements at a molecular level have had worldwide beneficial effects on the health and well-being of people. It was an exciting era of research because much of the knowledge about vitamins and their functions was derived from diverse lines of investigation, with each producing findings that cumulatively furnished a comprehensive biochemical understanding of vitamins. The history of vitamin research includes contributions by biochemists, microbiologists, chemists, nutritionists, and medical researchers; the different organisms studied included humans, chickens, pigeons, rats, yeast, and bacteria. Although not generally viewed as such at the time, the basic unifying concept of molecular biology was well substantiated by these studies, which revealed that many of the B-vitamins are nutritional requirements for a broad spectrum of living organisms.

The name *vitamin* (originally spelled vitamine) was proposed by Casimur Funk in 1912 to describe a *vital amine (thiamin),* found in concentrates of rice husks, which cured a nutritional disease called *beriberi* that had been identified in the Far East. As more vitamins (biomolecules which serve as necessary growth factors) were discovered, two broad classes, called *water soluble* and *fat soluble* were established. Thiamin, the original water-soluble B-vitamin, was renamed vitamin B_1 when new water-soluble growth

Table 10-1
B-Vitamins, their coenzymes, and enzymatic functions

B-Vitamin	Coenzyme form	Enzymatic function
Thiamin (B_1)	Thiamin pyrophosphate (TPP)	Aldehyde-group removal or transfer
Nicotinamide	Nicotinamide adenine dinucleotide (NAD^+)	Hydrogen transfer
	Nicotinamide adenine dinucleotide phosphate ($NADP^+$)	Hydrogen transfer
Riboflavin (B_2)	Flavin adenine dinucleotide (FAD)	Hydrogen transfer
	Flavin mononucleotide (FMN)	Hydrogen transfer
Pyridoxal (B_6)	Pyridoxal phosphate	Amino-group transfer, carboxyl-group transfer, racemization
Biotin	Biocytin	Carboxyl-group removal or transfer
Pantothenic acid	Coenzyme A (CoA)	Acyl-group carrier or transfer
Folic acid	Tetrahydrofolic acid (FH_4)	One-carbon transfer
Vitamin B_{12}	Coenzyme B_{12}	1,2-shift of hydrogen atom, methyl-group carrier
(Lipoic acid	Lipoamide	Acyl-group transfer)

factors were described and assigned the B designation, e.g., vitamins B_2 and B_6. The *B-series,* or *B-complex, of vitamins,* therefore, is a group of *water-soluble nutrients* which, being excretable, are *required daily in small amounts* (milligram quantities or less) for the normal growth and good health of humans and many other organisms. Table 10-1 summarizes the accumulated knowledge about the B-vitamins and their metabolic functions as coenzymes. Additional information on the water- and fat-soluble vitamins is presented in Chapter 25.

THE B-VITAMINS

Thiamin (Vitamin B_1)

Following its discovery as the antiberiberi factor, studies on the isolation and characterization of thiamin culminated in 1935 with the elucidation of its chemical structure (Figure 10-1) by Robert R. Williams. The coenzyme form of the vitamin is *thiamin pyrophosphate* (Figure 10-1), previously called cocarboxylase; it serves as the cofactor in several important reactions in carbohydrate metabolism involving the *removal* or *transfer of an aldehyde group,* e.g., pyruvate decarboxylation. The reactive site of thiamin pyrophosphate (TPP) is the *C-2 of the thiazole ring,* which can readily lose a proton to become a *carbanion* (Figure 10-2). The mechanism for the decarboxylation of pyruvate (Figure 10-3) by pyruvate decarboxylase, an enzyme in the anaerobic pathway for glucose utilization, exemplifies how TPP functions as a coenzyme. By a series of

The monies that Williams received from his patents on the chemical synthesis of thiamin were used to establish the Williams–Waterman Fund for the Combat of Dietary Disease, which supported nutritional research and field programs to abolish malnutrition.

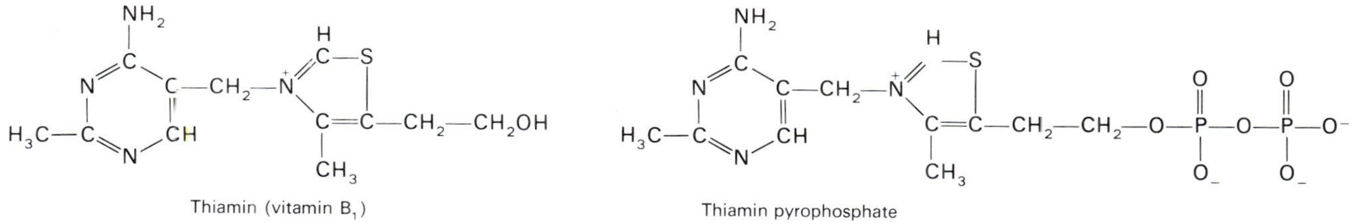

Figure 10-1
Thiamin and its coenzyme form, thiamin pyrophosphate.

electron rearrangements, the adduct formed between pyruvate and TPP leads to the decarboxylation of the three-carbon acid to yield *hydroxyethylthiamin pyrophosphate,* which then cleaves to produce acetaldehyde and to regenerate TPP. Thiamin-catalyzed reactions require Mg^{2+}.

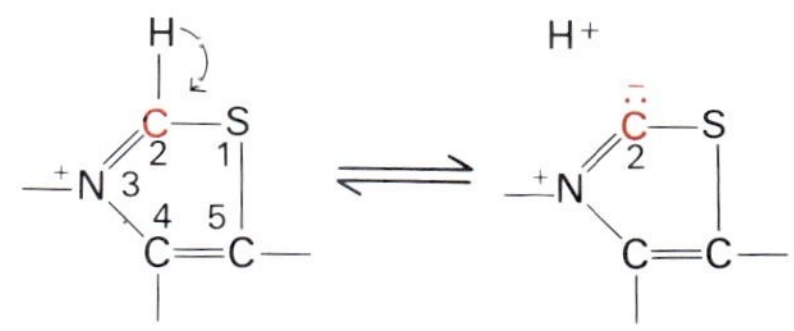

Figure 10-2
Formation of a carbanion in the thiazole moiety of thiamin pyrophosphate.

Nicotinamide (Niacinamide)

The coenzyme role of *nicotinamide* (Figure 10-4), commercially called *niacinamide* (to avoid any misassociation with the alkaloid nicotine of tobacco), was well known by enzymologists before it was recognized as the vitamin responsible for the prevention of *pellagra* (Italian word for *rough skin*) in humans and black tongue in dogs. As early as 1904, it was realized that an unidentified *coenzyme I* was required for the alcoholic fermentation of sugar by

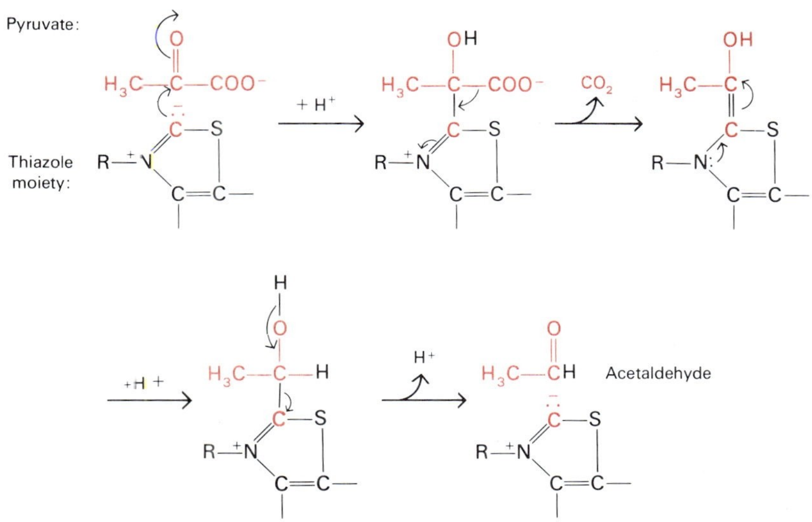

Figure 10-3
Mechanism for the decarboxylation of pyruvate. R is the pyrimidine portion of the thiamin pyrophosphate molecule.

Figure 10-4
Nicotinamide and nicotinic acid.

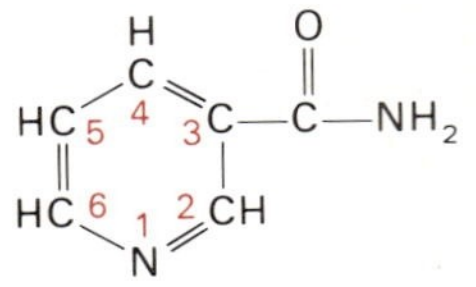

Nicotinamide
(Niacinamide)

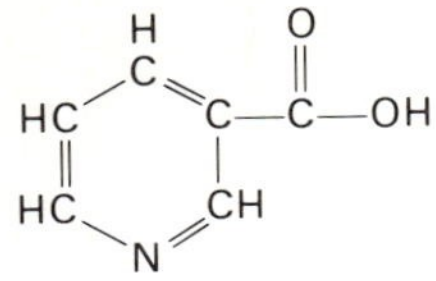

Nicotinic acid
(Niacin)

Joseph Goldberger, an Austrian-American physician of the U.S. Public Health Service, was instrumental in proving that pellagra is not caused by infectious bacteria or protozoa (as commonly believed until after his experiments, which began in 1915) but, instead, by a nutritional deficiency. Pellagra is characterized by the "3d's"—diarrhea, dermititis, and dementia. Goldberger and his coworkers produced pellagra in volunteers from a Mississippi prison farm by feeding them a diet lacking meats and milk. The investigators then attempted unsuccessfully to contract the disease by exposing themselves to the pellagrins, their clothing, bedding, and excreta. Goldberger called the missing nutrient the P-P (pellagra-preventing) factor and, after his death, it was sometimes called vitamin G in his honor.

Although Woolley was blind since young adulthood, undaunted, he pursued a science career that established him as one of the most prominent biochemists in the United States.

In 1912, Funk's concentrate of antiberiberi factor was principally nicotinic acid, contaminated with thiamin. He isolated nicotinic acid but lost interest in the compound when it proved ineffective in treating beriberi.

yeast and, in 1933, the coenzyme was isolated by Hans von Euler-Cheplin (Nobel Prize, 1929). The following year, Otto Warburg (Nobel Prize, 1931) isolated a closely related coenzyme, *coenzyme II,* and then identified nicotinamide as a component of both coenzymes. From nutritional studies, it was established by 1920 that pellagra (a disease that plagued the southern region of the United States) was due to a nutritional deficiency that could be alleviated by adding meat, milk, and eggs to a predominantly starchy diet. Soon after the nicotinamide content of coenzymes I and II was discovered in 1935, Conrad Elvehjem and D. Wayne Woolley identified *nicotinic acid* (Figure 10-4) as the *antipellagra factor* and, by 1937, the combined findings of these independent nutritional and enzymatic studies had identified both a new B-vitamin and its function as a coenzyme. There is an interesting nutritional relationship between nicotinic acid and tryptophan since many organisms can synthesize the vitamin from the amino acid; hence, for these organisms, diets high in tryptophan can spare some of the requirement for nicotinamide.

One coenzyme form of nicotinamide, *nicotinamide adenine dinucleotide* (NAD^+, formerly coenzyme I), differs from a second, *nicotinamide adenine dinucleotide phosphate* ($NADP^+$, formerly coenzyme II), because NAD^+ lacks the phosphate group bonded to the C-2′ of the ribose attached to adenine (Figure 10-5). Although NAD^+ and $NADP^+$ are the abbreviations proposed by the International Commission on Enzymes in 1964, the older abbreviations DPN^+ (diphosphopyridine nucleotide) and TPN^+ (triphosphopyridine nucleotide) are still sometimes used for NAD^+ and $NADP^+$, respectively.

NAD^+ and $NADP^+$ serve as coenzymes for a class of enzymes, called *pyridine-linked,* or *nicotinamide nucleotide-linked, dehydrogenases,* that catalyze oxidation-reduction reactions. In these dehydrogenase reactions, which are sometimes physiologically irreversible, two hydrogen atoms are generally removed from (oxidation) or added to (reduction) a substrate. The *NAD^+-linked dehydrogenases* are usually associated with *aerobic respiratory*

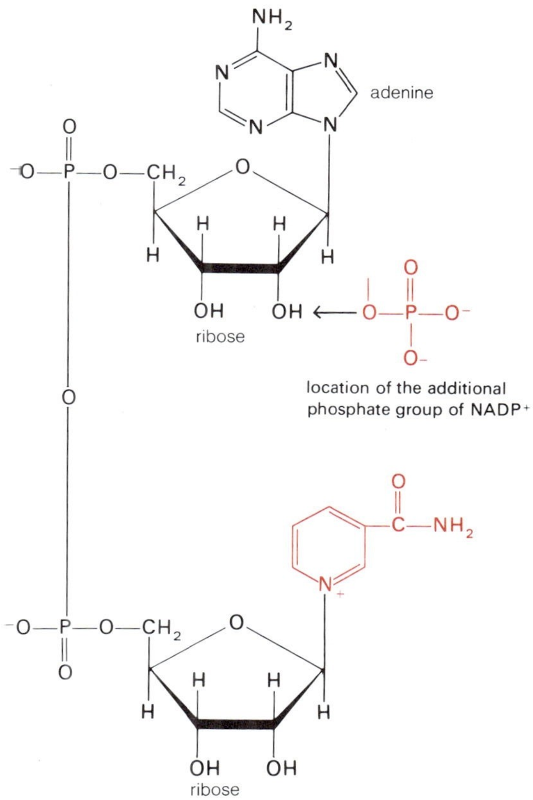

Figure 10-5
Nicotinamide adenine dinucleotide (NAD^+).

processes and the *$NADP^+$-linked dehydrogenases* with *biosynthetic reactions.*

The nicotinamide moiety of NAD^+ and $NADP^+$ serves as the hydrogen atom (electron) carrier by accepting or donating a *hydride ion* ($1H^+$, $2e^-$), as illustrated in Figure 10-6. The general reaction catalyzed by an NAD^+- or $NADP^+$-linked dehydrogenase is written:

$$\text{Substrate}\cdot H_2 + NAD^+\ (NADP^+) \rightleftharpoons \text{substrate} + NADH\ (NADPH) + H^+$$

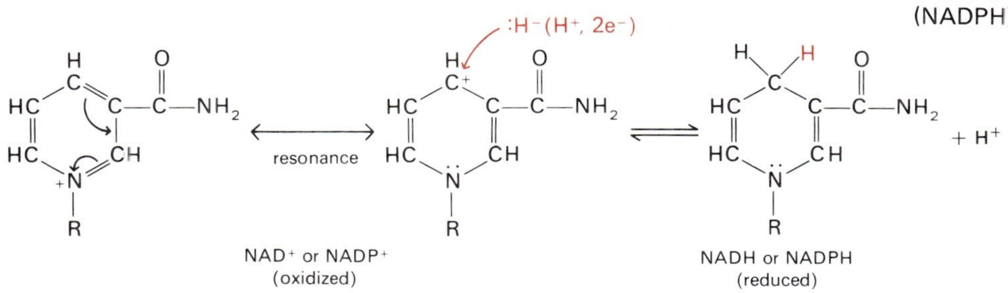

Figure 10-6
Hydride ion acceptance by NAD^+ or $NADP^+$. R is the remaining portion of the NAD^+ ($NADP^+$) or the NADH (NADPH) molecule.

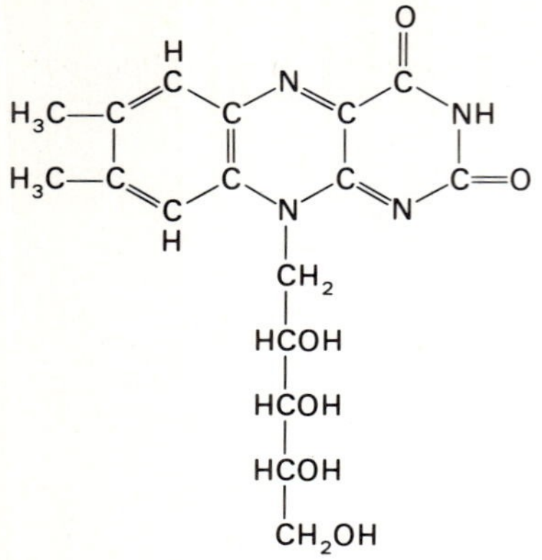

Figure 10-7
Riboflavin.

and accounts for the H^+ that is released when two H atoms ($2H^+$, $2e^-$) of a substrate are accepted as a hydride ion ($1H^+$, $2e^-$) by NAD^+ or $NADP^+$. Conversely, when hydrogen atoms are donated to a substrate by NADH or NADPH, a H^+ is taken up from the environment.

Riboflavin (Vitamin B_2)

The structure of vitamin B_2, *riboflavin* (Figure 10-7), contains *D-ribitol (ribo-)*, a five-carbon sugar alcohol, and an *isoalloxazine (flavin)* derivative. The two coenzyme forms of riboflavin, like those of nicotinamide, participate in oxidation-reduction reactions; *flavin mononucleotide (FMN)*, or *riboflavin monophosphate,* is one of the coenzymes, and *flavin adenine dinucleotide (FAD)* is the other (Figure 10-8). In a strict sense, the mono- and dinucleotide designations given to the riboflavin coenzymes are not chemically correct,

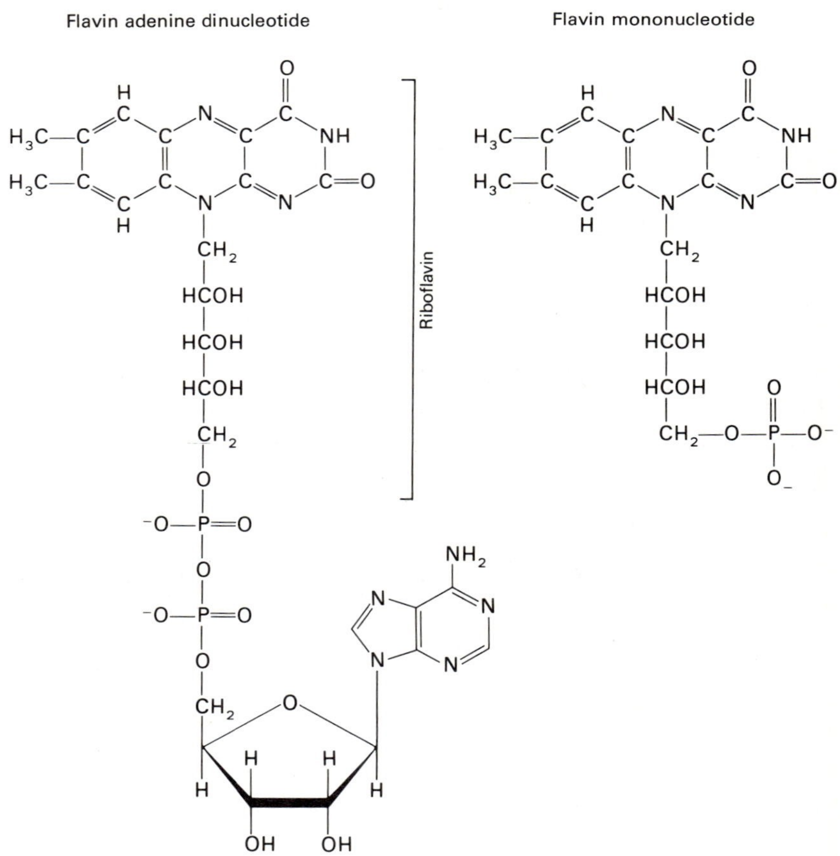

Figure 10-8
Two coenzyme forms of riboflavin.

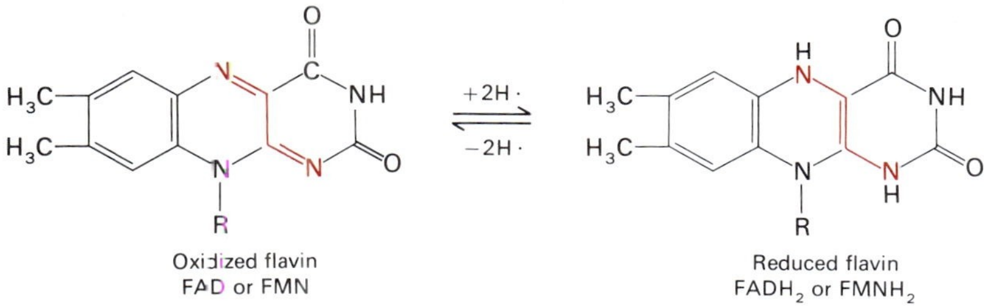

Figure 10-9
Oxidized and reduced forms of the isoalloxazine moiety of FAD and FMN. R is the remaining portion of the FAD (FMN) or $FADH_2$ ($FMNH_2$) molecule.

since the carbohydrate attached to the flavin is a polyhydric alcohol (reduced sugar) and not the aldopentose, D-ribose.

FMN and FAD are coenzymes for a class of enzymes called *flavin-linked dehydrogenases,* or *flavoproteins,* some of which also have metal cofactors, e.g., iron or molybdenum *(metalloflavoproteins).* The flavin-linked dehydrogenases, which catalyze a variety of reversible and irreversible cellular oxidation-reduction reactions, use the isoalloxazine moiety of FMN or FAD as the acceptor or donor of two hydrogen atoms (Figure 10-9).

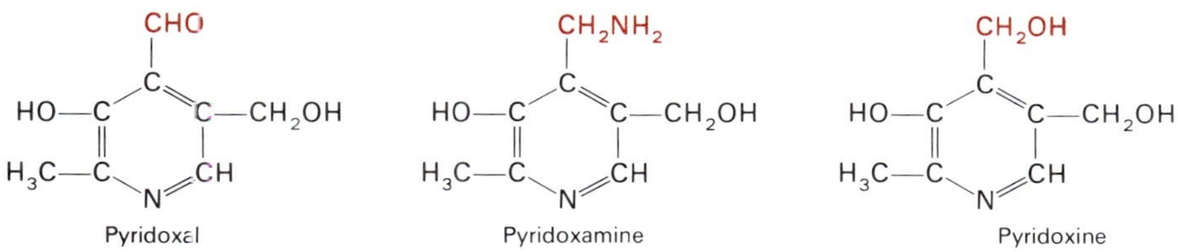

Figure 10-10
Forms of vitamin B_6.

Vitamin B_6

Pyridoxal, pyridoxamine, and *pyridoxine* are referred to as the *B_6-group of vitamins* (Figure 10-10). Vitamin B_6 deficiency in rats, which in severe cases causes convulsions, was shown by early studies to be cured by pyridoxine. Later studies established that pyridoxine is converted biologically into pyridoxal and pyridoxamine, which are then phosphorylated to produce the coenzyme forms of the vitamin, *pyridoxal phosphate* and *pyridoxamine phosphate* (Figure 10-11).

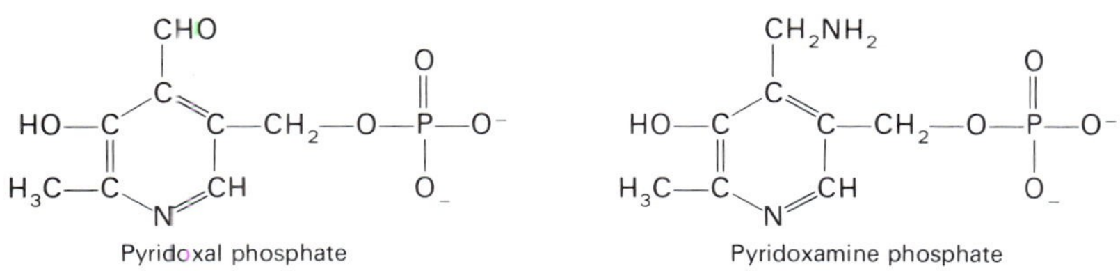

Figure 10-11
Coenzyme forms of vitamin B_6.

Figure 10-12
Schiff-base formation involving pyridoxal phosphate.

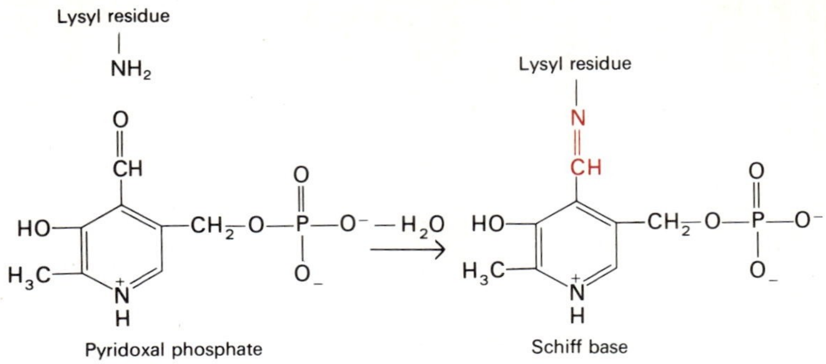

Enzymes that require pyridoxal (pyridoxamine) phosphate as a coenzyme are of particular importance to amino acid metabolism and, unlike other B-vitamin coenzymes, catalyze several distinct types of reactions, e.g., transaminations, decarboxylations, and racemizations. As a coenzyme, pyridoxal phosphate is covalently bonded in a *Schiff-base formation* to the *ϵ-amino group* of a specific *lysyl* residue of an apoenzyme (Figure 10-12). Studies by Esmond Snell and colleagues elucidated the molecular details of the various coenzymatic functions of pyridoxal (pyridoxamine) phosphate. The catalytic mechanism for transamination (aminotransferase) reactions is illustrated in Figure 10-13. The general reversible transaminase reaction is written:

$$\underset{\text{Amino-donor substrate}}{R_1-\overset{\displaystyle H}{\underset{\displaystyle NH_3^+}{C}}-COO^-} + \underset{\text{Amino-acceptor substrate}}{R_2-\underset{\displaystyle O}{\overset{\|}{C}}-COO^-} \rightleftharpoons R_1-\underset{\displaystyle O}{\overset{\|}{C}}-COO^- + R_2-\overset{\displaystyle H}{\underset{\displaystyle NH_3^+}{C}}-COO^-$$

and usually involves the transfer of an α-NH_2 group from an *amino-acid donor* to an *α-keto-acid acceptor.* In the reaction (Figure 10-13), the amino acid displaces the ϵ-amino group contributed by the apoenzyme to form a Schiff base with the bound coenzyme. Subsequent hydrolysis of a Schiff-base intermediate results in the formation of *pyridoxamine phosphate* and the *α-keto-acid analogue* of the amino acid, i.e., the NH_2 group is transferred to pyridoxal phosphate. As indicated in Figure 10-13, reversal of this series of reactions transfers the coenzyme-bound NH_2 group to the second substrate, an α-keto acid, to produce the analogous amino acid. In pyridoxal phosphate-catalyzed reactions, the formation of a Schiff base between the substrate and the coenzyme is a pre-

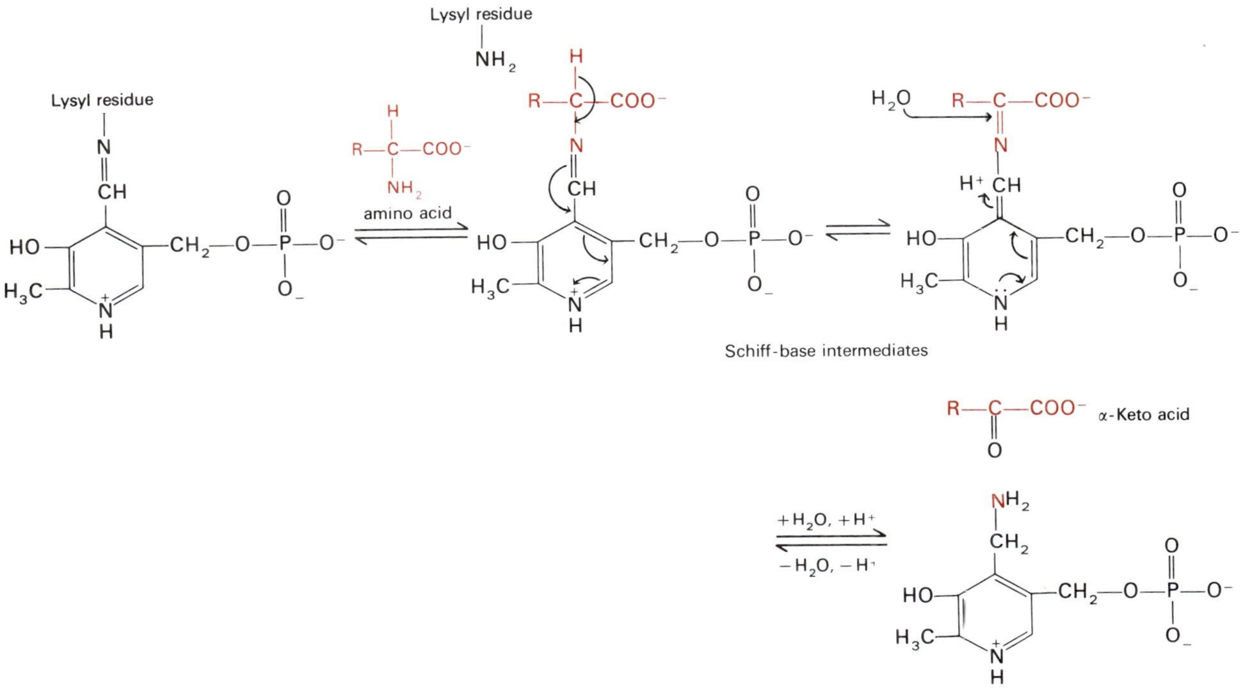

requisite, and it is the subsequent electron rearrangements that the Schiff base undergoes which determine the type of reaction catalyzed, e.g., transfer of an amino group or loss of the carboxyl group of the substrate.

Figure 10-13
General features of the mechanism of an aminotransferase (transamination) reaction.

Biotin

Biotin (Figure 10-14) was first isolated in 1935 by Fritz Kögl and Benno Tönnis from liver concentrates as a growth factor for yeast. In animals, the biotin requirement (μg amounts) is furnished by intestinal bacteria which synthesize the vitamin. To study biotin deficiency in laboratory animals, it is necessary to create a deficiency artificially by feeding them raw egg whites, which contain a protein, *avidin,* that tenaciously binds biotin and prevents its absorption from the intestinal tract.

The coenzyme form of biotin is a prosthetic group because the B-vitamin is covalently bonded by an amide linkage to the *ε-amino group* of a specific *lysyl* residue of the apoenzyme; *ε-N-biotinyllysyl* (Figure 10-15) is called *biocytin.* Biotin-requiring enzymes catalyze the *incorporation (carboxylation)* or *transfer of CO_2 (trans-*

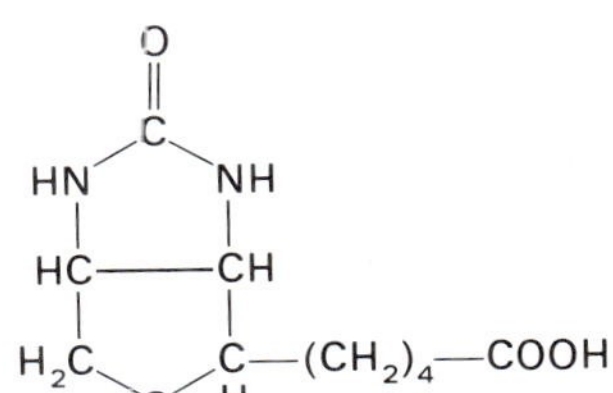

Figure 10-14
Biotin.

Figure 10-15
Coenzyme form of biotin.

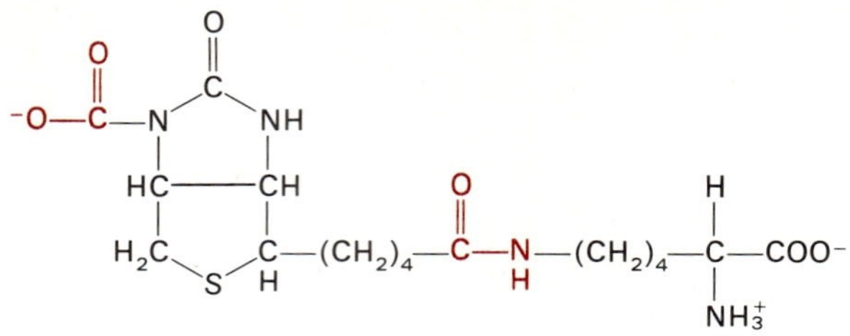

N-Carboxybiotinyllysine
(*N*-Carboxybiocytin)

carboxylation). In the carboxylation reactions, ATP and Mg^{2+} are required and biotin, as *N*-carboxybiotinylsine (Figure 10-15), serves as a carrier of CO_2. The two steps in the carboxylation reactions can be written:

Step 1: Biotin-enzyme + HCO_3^- + ATP → carboxybiotin-enzyme + ADP + P_i
Step 2: Substrate + carboxybiotin-enzyme → carboxylated substrate + biotin-enzyme

Pantothenic Acid

Roger J. Williams is the brother of Robert R. Williams, who elucidated the structure of thiamin. Although both shared similar interests in the B-vitamins and nutrition, their research careers were independent of each other and never involved collaboration.

Another B-vitamin isolated from liver as a growth factor for yeast is *pantothenic acid* (Figure 10-16); it was isolated in 1938 by Roger J. Williams. Because of its ubiquitousness, the vitamin was named for the Greek word meaning *from everywhere.* The coenzyme form of pantothenic acid is *coenzyme A* (Figure 10-17), so called because it was first described as a cofactor for certain enzymatic *acetylation* reactions. Coenzyme A (CoA or CoA-SH) is biologically important as a *carrier* or *donor of acyl groups,* with the sulfhydryl (—SH) group of the β-mercaptoethylamine portion (Figure 10-17) of the coenzyme serving as the attachment site of the acyl group. The two general types of reactions catalyzed by enzymes requiring coenzyme A are:

$$\text{Formation of acyl-CoA: } R-\overset{\displaystyle O}{\overset{\|}{C}}-O^- + HS-CoA \underset{}{\overset{-H_2O}{\rightleftharpoons}} R-\overset{\displaystyle O}{\overset{\|}{C}}-S-CoA$$

$$\text{Transfer of acyl group: } R-\overset{\displaystyle O}{\overset{\|}{C}}-S-CoA + \text{substrate} \rightarrow R-\overset{\displaystyle O}{\overset{\|}{C}}-\text{substrate} + HS-CoA$$

Figure 10-16
Pantothenic acid.

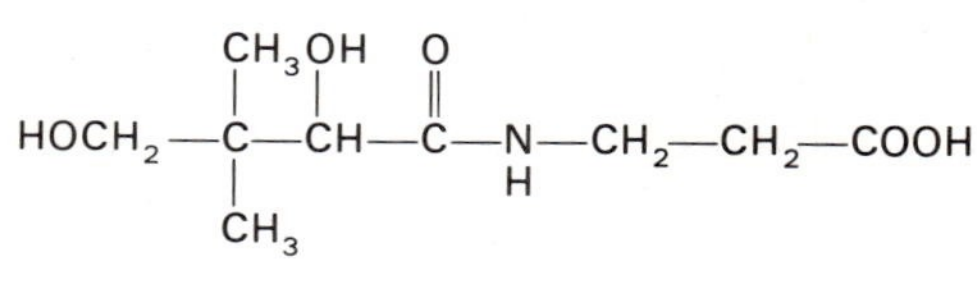

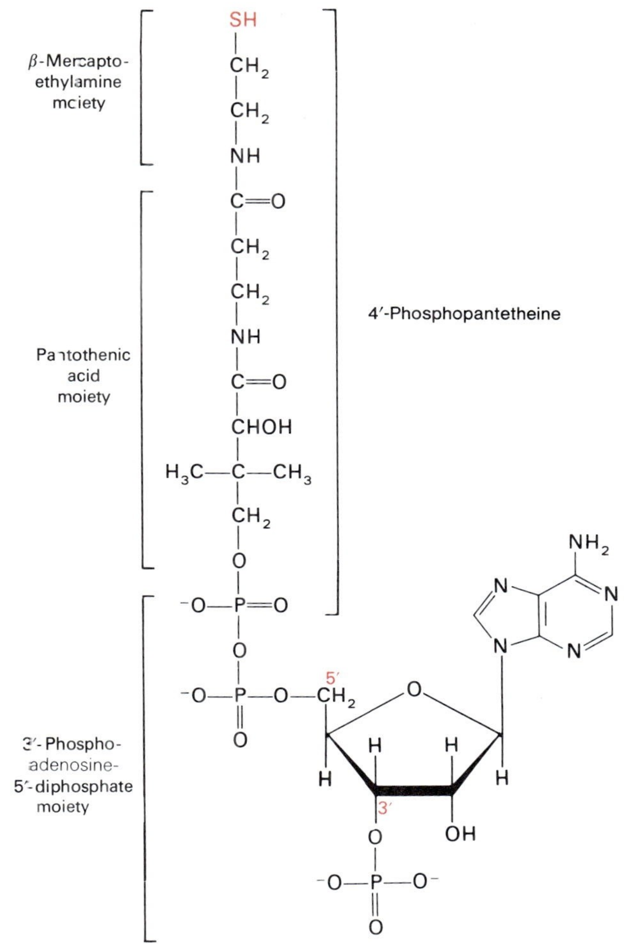

Figure 10-17
Coenzyme A (CoA-SH).

The thioester bond of an acyl-CoA has an energy potential (p. 302) comparable to that of a phosphoanhydride bond of ATP (p. 255) because, unlike an oxygen ester bond, the thioester bond is not stabilized by resonance and is unstable. In reactions involving the transfer of an acyl group from CoA to a substrate, cleavage of the thioester bond provides the energy to drive the reaction.

Folic Acid

Folic acid (Figure 10-18) was first isolated from spinach leaves, and its name was derived from the Latin word for leaf, *folium.* In its structure, the vitamin contains a *pteridine derivative,* p-*aminobenzoic acid,* and *glutamic acid* (Figure 10-18). Another name

Figure 10-18
Folic acid.

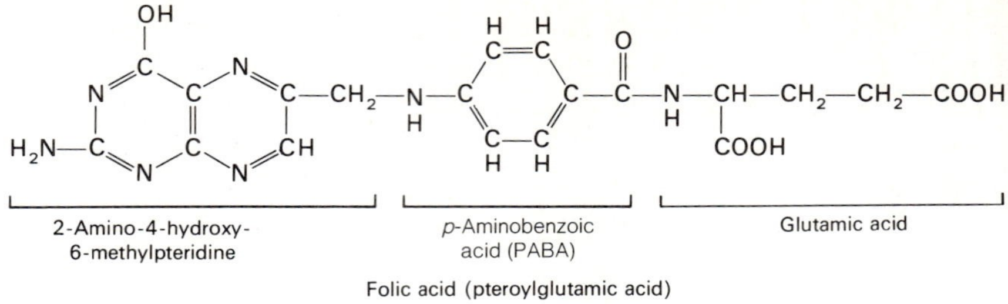

for folic acid is *pteroylglutamic acid.* Sulfa drugs are antibacterial agents because they effectively compete with *p*-aminobenzoic acid as a substrate (competitive inhibition) in the biosynthesis of folic acid by some pathogens (p. 140). The coenzyme form of folic acid is *tetrahydrofolate* (Figure 10-19), which acts as an *intermediate carrier* in the enzymatic transfer of *one-carbon (1-C) groups.* Tetrahydrofolate, abbreviated FH_4 or THF, carries 1-C units in several forms, e.g., as a formyl or methyl group, and there are a variety of 1-C derivatives of FH_4 (Figure 10-20), with a particular derivative being required for each type of reaction catalyzed by an FH_4-requiring enzyme.

Figure 10-19
Tetrahydrofolate (TH_4 or THF).

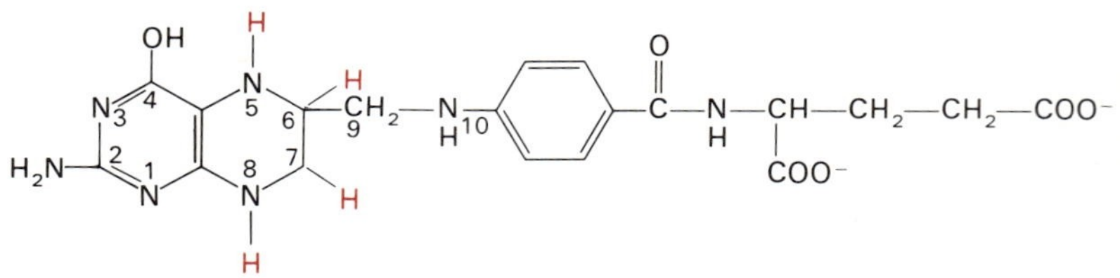

Vitamin B_{12} (Cyanocobalamin)

The isolation and characterization of *vitamin B_{12}* (Figure 10-21), the last B-vitamin isolated, was a very difficult research task. In the 1920s, it was discovered that the addition of liver to a patient's diet cures *pernicious anemia.* Isolation of the anemia-curing factor of liver proceeded slowly because the disease could not be produced in experimental animals and, hence, liver-factor concentrates had to be assayed by using human patients with pernicious anemia. It was not until 1948 that E. Lester Smith of England and Edward L. Rickes and Karl Folkers of the United States isolated the liver factor, vitamin B_{12}, in crystalline form. The determination of the structure of vitamin B_{12} then became the challenging problem and, in 1957, Dorothy C. Hodgkin (Nobel Prize,

1964) established the unusually distinct structure by a combination of chemical and x-ray diffraction studies. Vitamin B_{12} has a *corrin ring system,* structurally resembling the *porphyrin ring system* (p. 107), with a *cobalt atom* coordinated to the inner nitrogen atoms (Figure 10-21). Additionally, vitamin B_{12} contains an odd

1
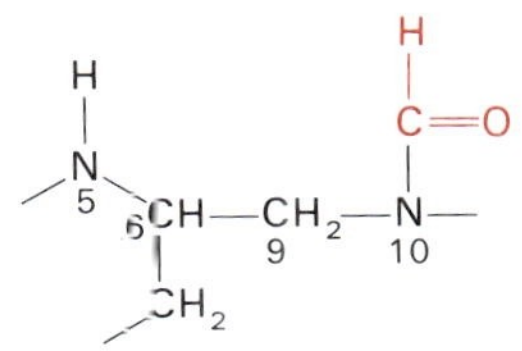
N^{10}-Formyltetrahydrofolate (N^{10}-Formyl-FH_4)

2
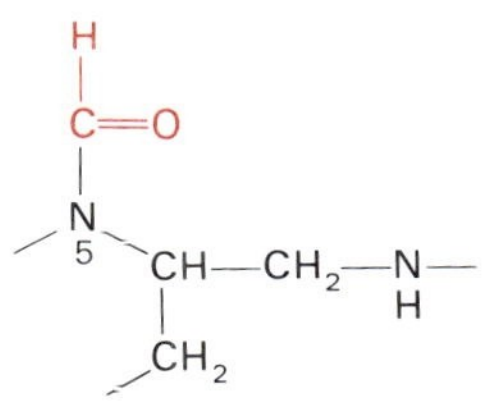
N^5-Formyl-FH_4

3
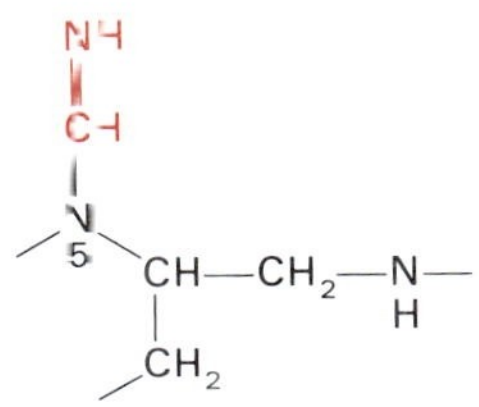
N^5-Formimino-FH_4

4
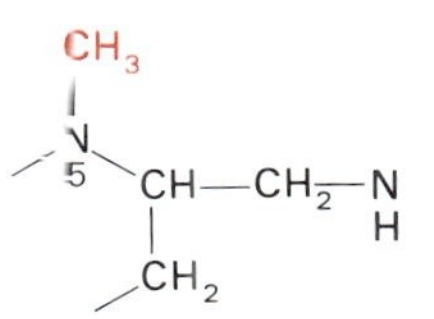
N^5-Methyl-FH_4

5
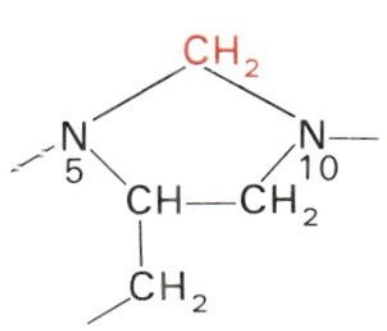
N^5, N^{10}-Methylene-FH_4

6
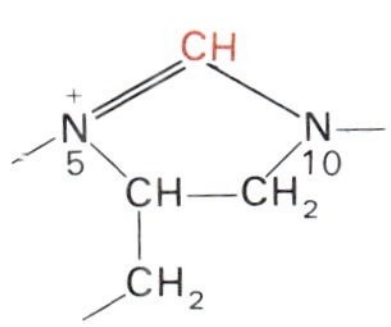
N^5, N^{10}-Methenyl-FH_4

Figure 10-20
Partial structures of six 1-carbon derivatives of tetrahydrofolate.

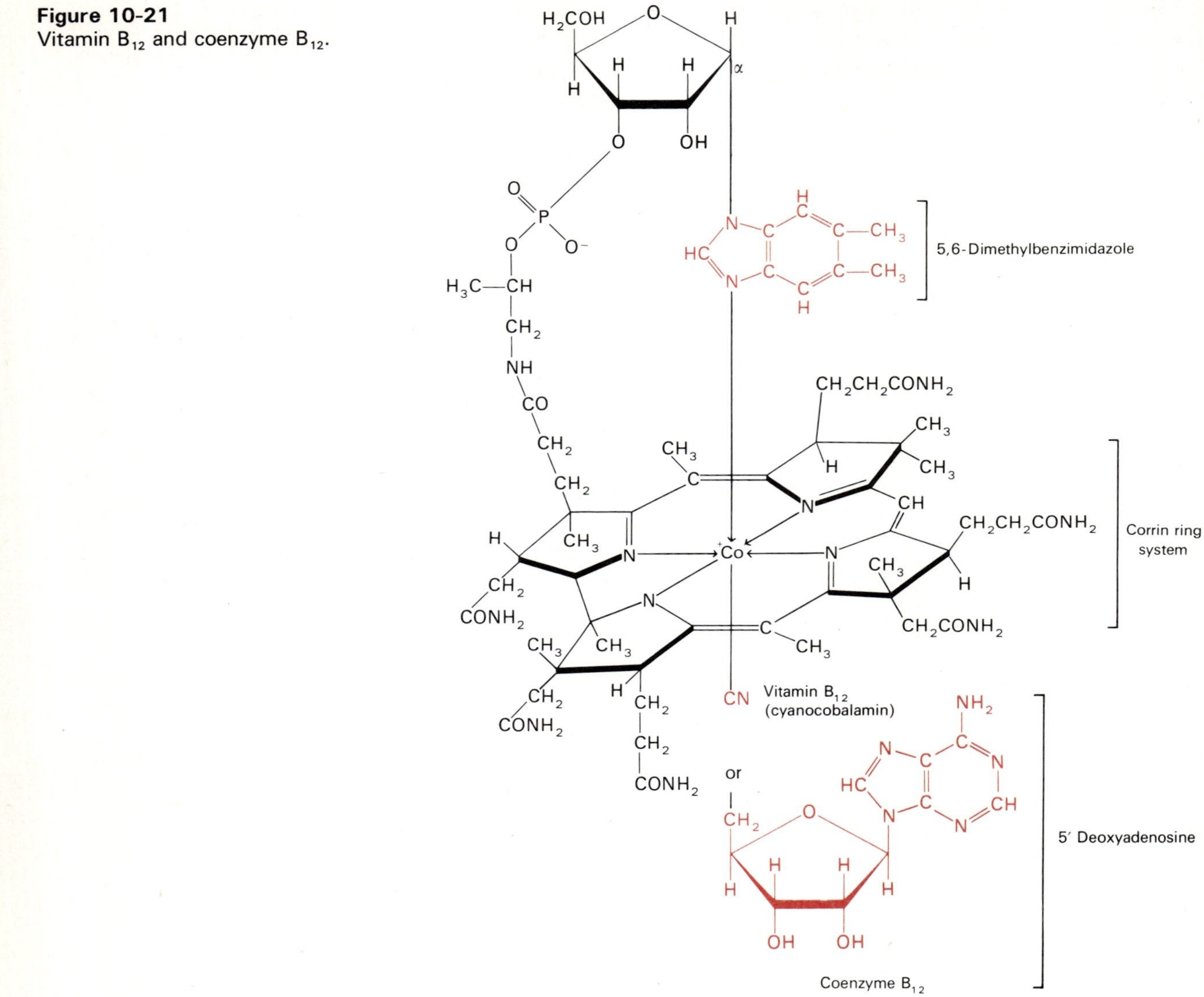

Figure 10-21
Vitamin B_{12} and coenzyme B_{12}.

ribonucleotide that has the base *5,6-dimethylbenzimidazole* linked to ribose in an α-N-*glycosidic linkage,* rather than the β-linkage found in other nucleotides. The ribonucleotide is bonded both covalently by an ester linkage to the corrin ring system and coordinately to the cobalt atom. Because vitamin B_{12} is isolated with cyanide occupying one of the coorindation positions of cobalt, the vitamin is also called *cyanocobalamin.* In the coenzyme form of vitamin B_{12}, called *coenzyme B_{12},* the cyanide ion is replaced by a *5-deoxyadenosyl group* (Figure 10-21).

There are two types of enzymatic reactions that require coenzyme B_{12}. The first type (Figure 10-22) catalyzes a *1,2 shift of a hydrogen atom* from one carbon atom of the substrate to the next with a concomitant 2,1 (reverse) shift of some other group, e.g., hydroxyl or alkyl group. In the second type of reaction, coenzyme B_{12} serves as a *carrier of a methyl group,* obtained from *N^5-methyltetrahydrofolate,* to the appropriate acceptor molecule. In the reaction, a methyl group occupies the 5-deoxyadenosyl coordination position of coenzyme B_{12}. An example of this second type of coenzyme B_{12} reaction is the methylation of homocysteine to produce methionine (Figure 10-23).

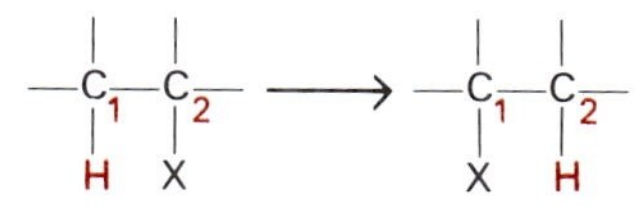

Figure 10-22
1,2-Shift of a hydrogen atom catalyzed by coenzyme B_{12}-requiring reactions.

Vitamin B_{12} is one of the most potent biological compounds known, with only about 0.1 μg needed to be absorbed daily by humans. Although a daily dietary intake of the vitamin is easily obtainable, vitamin B_{12} deficiency is not rare because some individuals are unable to absorb the vitamin from the intestinal tract. Gastric juice contains a mucoprotein, called the *intrinsic factor,* that specifically binds vitamin B_{12} and transports the nutrient into intestinal cells for its subsequent transfer to body tissues. Individuals who lack or have a deficiency of intrinsic factor cannot absorb ingested vitamin B_{12} and, as a consequence, suffer from pernicious anemia because the vitamin is necessary for normal blood formation. As a genetic abnormality, lack or a deficiency of intrinsic factor is an example of an *inborn error affecting transport* of an essential nutrient into the body. Elderly people often have diminished production of intrinsic factor and, to avoid an anemic condition, must receive vitamin B_{12} by injection, because taking the vitamin orally would not alleviate the transport deficiency.

Figure 10-23
Involvement of coenzyme B_{12} in methionine biosynthesis.

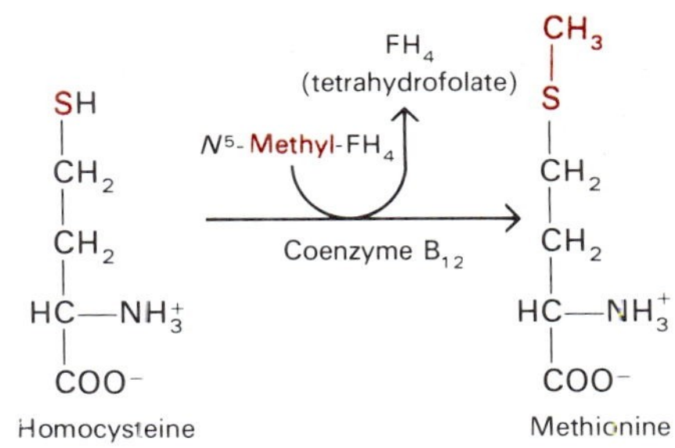

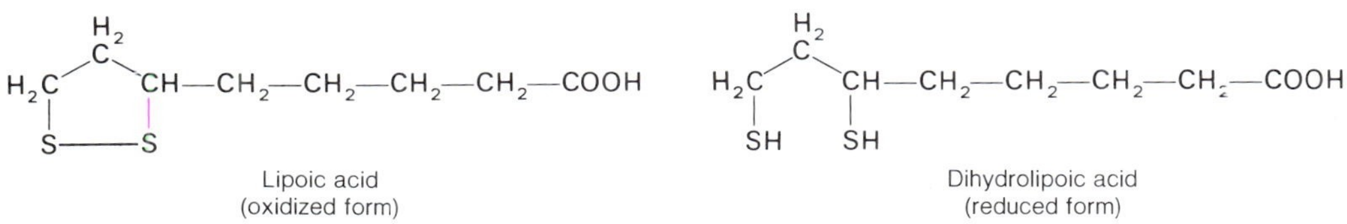

Figure 10-24
Oxidized and reduced forms of lipoic acid.

Lipoic Acid (Thioctic Acid)

Lipoic acid (Figure 10-24) was crystallized in 1951 by Lester J. Reed and Irwin C. Gunsalus and their colleagues. When first isolated, lipoic acid (also called *thioctic acid*) was believed to be a B-vitamin; however, the current evidence is that animals synthesize the small quantities they need and, thus, have no dietary requirement for the biomolecule. Often classed with the B-vitamins

Such small quantities of lipoic acid are present in tissues that ten tons of water-insoluble liver residue were used by Reed and Gunsalus to obtain about 30 mg of the biomolecule.

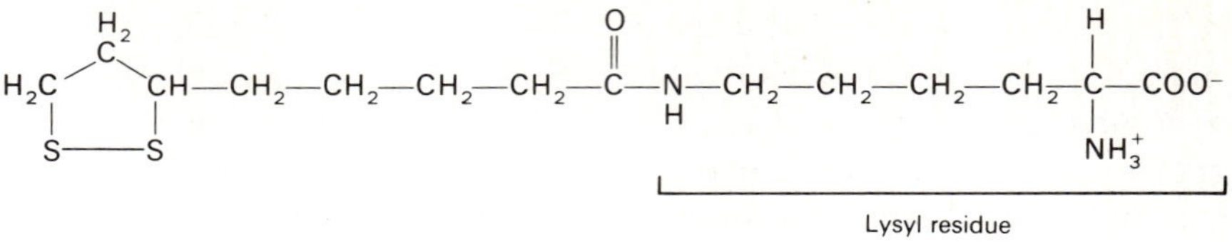

Figure 10-25
Lipoamide (ϵ-*N*-Lipoyllysine).

because of its coenzymatic function, lipoic acid is referred to as a *pseudo-vitamin.* There are two forms of lipoic acid; one is *oxidized lipoic acid,* which is a cyclic disulfide (Figure 10-24), and the other is *dihydrolipoic acid,* the reduced form with two sulfhydryl groups (at C-6 and -8).

The coenzyme form of lipoic acid, like that of biotin, is covalently bonded by an amide linkage to the ϵ-amino group of a specific lysyl residue of the apoenzyme; ϵ-*N-lipoyllysine* is called *lipoamide* (Figure 10-25). As a coenzyme, lipoic acid functions in two key oxidative decarboxylations in the aerobic utilization of carbohydrates for energy (Chapter 16) by transferring an acyl group, donated by thiamin pyrophosphate (TPP), to CoA-SH (Figure 10-26).

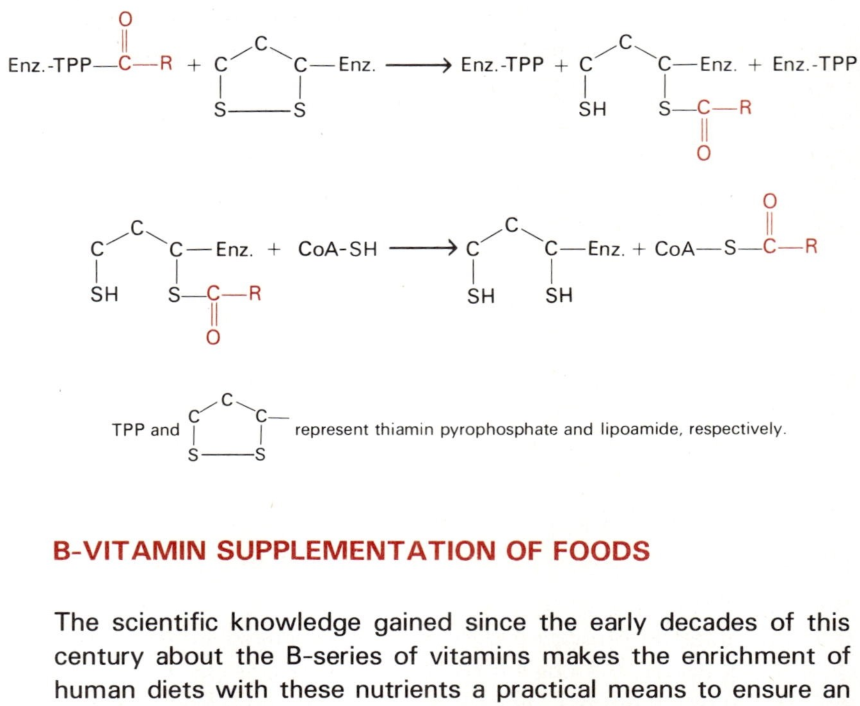

Figure 10-26
Schematic of the function of lipoic acid in acyl-group transfer reactions.

B-VITAMIN SUPPLEMENTATION OF FOODS

The scientific knowledge gained since the early decades of this century about the B-series of vitamins makes the enrichment of human diets with these nutrients a practical means to ensure an adequate daily intake of the vitamins. In the United States, as well as in several other countries, *recommended daily allowances* (RDA) have been established for the B-vitamins (p. 569). The RDA of a vitamin is a recommended quantity that significantly exceeds the daily *minimal* requirements as determined by nutritional studies. The destruction of B-vitamins that occurs in the commercial processing of some foods can now be remedied by supplementation of the product with B-vitamins, sometimes in amounts that exceed the natural vitamin content of the unprocessed food item, e.g., white flour. The four vitamins most often included in food supplementation are thiamin (US RDA = 1.1 to 1.5 mg), riboflavin (1.3 to 1.7 mg), nicotinamide (14 to 19 mg), and vitamin B_6 (2.0 to 2.2 mg). More recently, in dry cereals, folic acid (0.4 mg) and vitamin B_{12} (0.003 mg, or 3 μg) have been added to the vitamin enrichments. The two B-vitamins for which no human deficiency has ever been detected are biotin and pantothenic acid. It appears that all types of diet or the intestinal flora (bacteria) furnish adequate amounts of these two vitamins.

SUMMARY

The B-vitamins are a group of water-soluble biomolecules that are necessary for the normal growth and good health of humans and many other organisms. The eight B-vitamins are thiamin, nicotinamide, riboflavin, pyridoxal, biotin, pantothenic acid, folic acid, and vitamin B_{12}. As required nutrients, they are needed daily in milligram or microgram quantities. Based on the results of nutritional studies, daily recommended allowances (RDA) have been established for the B-vitamins.

Thiamin (B_1), as the coenzyme thiamin pyrophosphate, is required for enzymatic reactions that remove or transfer an aldehyde (—CHO) group. Nicotinamide adenine dinucleotide (NAD^+) and nicotinamide adenine dinucleotide phosphate ($NADP^+$) are the coenzyme forms of nicotinamide; both coenzymes are utilized in oxidation-reduction reactions which are catalyzed by pyridine-linked dehydrogenases. Riboflavin (B_2) also has two coenzyme forms, flavin adenine dinucleotide (FAD) and flavin mononucleotide (FMN), which are required for oxidation-reduction reactions catalyzed by flavin-linked dehydrogenases.

Pyridoxal phosphate, the coenzyme form of pyridoxal (B_6), is needed for enzymatic reactions in amino acid metabolism, e.g., transaminations, decarboxylations, and racemizations. Pyridoxal phosphate's linkage to an enzyme entails a Schiff-base formation with an ϵ-amino group of a specific lysyl residue. The B-vitamin biotin is utilized in carboxylation and transcarboxylation reactions. As a coenzyme, biotin is linked covalently by an amide bond to an ϵ-amino group of a lysyl residue of the enzyme. Pantothenic acid is contained in the structure of coenzyme A (CoA), which serves as a carrier or donor of acyl (R—CO—) groups. Like CoA, lipoic acid (although not classified as a B-vitamin) functions as a coenzyme in reactions that transfer an acyl group. And, like biotin, lipoic acid occurs covalently linked to an ϵ-amino group of a lysyl residue of the enzyme.

Tetrahydrofolate (FH_4), the coenzyme form of folic acid, is used as an intermediate carrier of one-carbon (1-C) groups. A 1-C group is carried by tetrahydrofolate as a methyl, methylene, or formyl derivative of the coenzyme. Vitamin B_{12}, cyanocobalamin, is required for enzymatic reactions that catalyze a 1,2 shift of a hydrogen atom from one carbon atom to an adjacent carbon atom in the substrate. The vitamin also serves as a carrier of a methyl group, which it accepts from a coenzyme form of folic acid and donates to the substrate.

QUESTIONS

1. Which B-vitamin(s), if any, corresponds to each of the following descriptions?
   (a) Has a five-carbon sugar alcohol
   (b) Has two sulfur atoms
   (c) Has a thiazole ring
   (d) Has a pteridine derivative
   (e) Has cobalt
   (f) Has an isoalloxazine derivative
   (g) Is an antipellagra factor
   (h) Is the antiberiberi factor
   (i) Is used to treat pernicious anemia
   (j) No known human deficiencies
   (k) Dietary requirement for many organisms spared by tryptophan
   (l) Binds to avidin
   (m) Biosynthesis inhibited by sulfa drugs
2. Which coenzymes discussed in the chapter
   (a) have an adenyl nucleotide in their structures?
   (b) are involved in acyl-transfer reactions?
   (c) are involved in decarboxylation reactions?
   (d) are involved in oxidation-reduction reactions?
3. Which coenzyme(s), derived from a B-vitamin,
   (a) accepts or donates two hydrogen atoms?
   (b) accepts or donates a hydride ion?
   (c) forms a Schiff base with a lysyl residue?
   (d) is a vitamin in an amide linkage with a lysyl residue?
   (e) is a monophosphorylated form of a vitamin?
   (f) relies on a carbanion for its reactivity?
   (g) carries an acyl group covalently bonded to a sulfur atom?
   (h) carries a group covalently bonded to a nitrogen atom or atoms?
4. In Table 9-2, two oxidoreductase reactions are listed. Given the information that alcohol dehydrogenase requires NAD^+ as a coenzyme and that succinate dehydrogenase requires FAD, rewrite the two reactions to include the coenzymes.
5. (a) Which three reactions listed in Table 9-2 require pyridoxal phosphate as a coenzyme? (b) Which reaction requires biotin?
6. Pyridoxal phosphate is a coenzyme for enzymatic reactions that decarboxylate amino acids (see Table 9-2). Given the following abbreviated structure of a Schiff-base intermediate between pyridoxal phosphate and

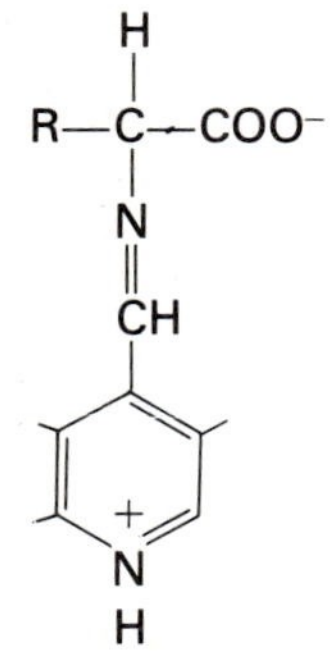

an amino acid, diagram the first electron rearrangement that leads to CO_2 production, then the second rearrangement (including the addition of a water molecule and a proton) that results in the production of an amine and the regeneration of pyridoxal phosphate.

SUGGESTED READINGS

Blakley, R. L., *The Biochemistry of Folic Acid and Related Pteridines.* Amsterdam: Elsevier, North Holland, 1969. A good coverage of the biochemistry of one of the B-vitamins.

Briggs, M. H., ed., *Vitamins in Human Biology and Medicine.* Boca Raton, Fla.: CRC Press, 1981. Comprehensive review of recent development in vitamin research.

Dolphin, D., *Vitamin B_{12},* Vols. 1 and 2. New York: Wiley, 1981. A comprehensive reference source about the B-vitamin and its biological functions.

McCormick, D. B., and L. D. Wright, *Vitamins and Coenzymes.* Part D, *Methods Enzymol.,* Vol. 62. New York: Academic Press, 1979. This volume includes articles on thiamin, lipoic acid, pantothenic acid, biotin, and pyridoxal.

McCormick, D. B., and L. D. Wright, *Vitamins and Coenzymes.* Parts E and F, *Methods Enzymol.,* Vols. 66 and 67. New York: Academic Press, 1980. Part E contains numerous articles on nicotinic acid, riboflavin, and folic acid. Part F has articles on vitamin B_{12}.

Needham, J., ed., *Chemistry of Life: Eight Lectures on the History of Biochemistry.* New York: Cambridge University Press, 1970. This text contains an intriguing chapter by L. J. Harris about the early history (up to 1938) of vitamins.

Sebrell, W. H. Jr., and R. S. Harris, eds., *The Vitamins: Chemistry, Physiology, Pathology, and Methods,* 2nd ed., Vols. 1 to 5. New York: Academic Press, 1967–1972. An excellent, comprehensive reference source.

Smith, E. L., *Vitamin B_{12},* 3rd ed. New York: Wiley, 1965. A good reference book on vitamin B_{12}, written by the English scientist who isolated the vitamin.

Wagner, A. F., and K. Folkers, *Vitamins and Coenzymes.* New York: Wiley-Interscience, 1964. Chapter 1 has an informative discussion about the evolution of the vitamin theory. The technical aspects of the chemistry and modes of action of vitamins and coenzymes are stressed throughout the book.

(Other references pertaining to the nutritional aspects of the B-vitamins are included in the Suggested Readings for Chapter 25.)

Articles

Benkovic, S. J., "On the Mechanism of Action of Folate- and Biopterin-Requiring Enzymes." *Ann. Rev. Biochem.,* 49:227, 1980.

Benkovic, S. J., and C. M. Tatum Jr., "Mechanisms of Folate Cofactors." *TIBS,* 2:161, 1977.

Obermayer, M., and F. Lynen, "Structure of Biotin Enzymes." *TIBS,* 1:169, 1976.

Plaut, G. W. E., C. M. Smith, and W. L. Alworth, "Biosynthesis of Water-Soluble Vitamins." *Ann. Rev. Biochem.,* 43:899, 1974.

Wood, H. G., "The Reactive Group of Biotin in Catalysis by Biotin Enzymes." *TIBS,* 1:4, 1976.

Wood, H. G., and R. E. Barden, "Biotin Enzymes." *Ann. Rev. Biochem.,* 46:385, 1977.

CARBOHYDRATES

11

INTRODUCTION

Carbohydrates are widely distributed in nature and, on the basis of mass, are the most abundant class of biomolecules. Carbohydrates, commonly called sugars, are formed during photosynthesis (Chapter 18) and are thus the first intermediates in the incorporation of inorganic carbon (CO_2), hydrogen, oxygen, and solar energy into living matter. This conversion of solar energy into the chemical energy of biomolecules makes carbohydrates the primary source of metabolic energy for living organisms. Carbohydrates also serve as carbon sources for the synthesis of other biomolecules and as the polymeric storage forms of energy. Additionally, they are components of many structural and cellular secretory materials and are components of nucleotides (Chapter 13), which, in turn, are also utilized in a variety of functions. Thus, like other biomolecules, carbohydrates exhibit functional versatility.

Originally, only "hydrates of carbon" having the empirical formula $C_nH_{2n}O_n$ were classified as carbohydrates. This definition soon proved inadequate, however, because certain substances, e.g., formaldehyde (CH_2O) and lactic acid ($C_3H_6O_3$), conform to the empirical formula yet are not considered carbohydrates. On the other hand, other compounds, like deoxyribose ($C_5H_{10}O_4$) and glucosamine ($C_6H_{13}O_5N$), are classified as carbohydrates even though they have other empirical formulas. *Carbohydrates* are now often defined as *polyhydroxyaldehydes* or *polyhydroxyketones and their derivatives*. A carbohydrate is an aldehyde ($-CHO$) if the carbonyl oxygen is associated with a terminal carbon atom and a ketone ($>C=O$) if the carbonyl oxygen is bonded to an internal carbon. This definition avoids classification by empirical formula and

Number of carbon atoms in a sugar	Name
3	Triose
4	Tetrose
5	Pentose
6	Hexose
7	Heptose

Figure 11-1
Naming of carbohydrates.

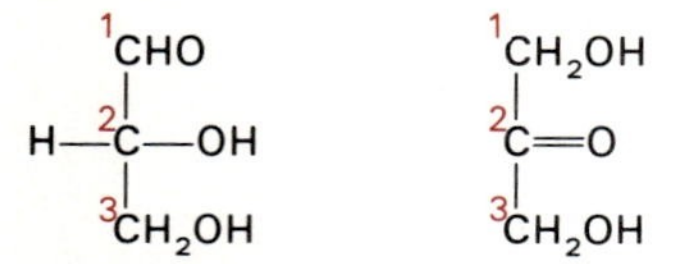

D-Glyceraldehyde an aldotriose

Dihydroxyacetone a ketotriose (no asymmetric carbon)

Figure 11-2
Two 3-carbon sugars.

Figure 11-3
Enantiomers of glucose, an aldohexose.

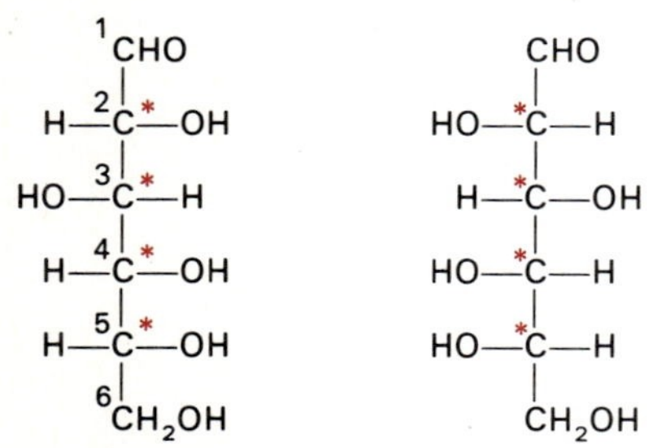

D-Glucose

L-Glucose

*asymmetric carbon atom

encompasses derivatives such as deoxy (Figure 11-3) and amino (Figure 11-4) sugars.

In nature, carbohydrates occur as *monosaccharides* (individual or simple sugars), *oligosaccharides,* and *polysaccharides.* Oligosaccharides are generally defined as molecules containing two to ten monosaccharide units, and polysaccharides are larger polymeric carbohydrates, some of which have molecular weights of several millions. It is within the context of these three classifications that the broad subject of carbohydrates is presented.

MONOSACCHARIDES

Nomenclature and Fischer Projections

A monosaccharide is identified by the number of carbon atoms it contains (Figure 11-1) and by its carbonyl functional group, i.e., *aldose* if an aldehyde and *ketose* if a ketone. The smallest carbohydrates (Figure 11-2) are generally considered the three-carbon sugars, glyceraldehyde (the *aldotriose*) and dihydroxyacetone (the *ketotriose*). In the case of glyceraldehyde, there are two stereoisomers, the D- and L-forms (Chapter 4). Stereochemically, all sugars can be related to one of these two isomers; in the classification of monosaccharides having more than one asymmetric carbon atom, the symbols D and L always refer to the configuration of the asymmetric carbon most distal to the carbonyl carbon (analogous to the one in glyceraldehyde). In biological systems, the D-forms of sugars predominate.

The systematic nomenclature for monosaccharides becomes cumbersome for larger sugars because they have two or more asymmetric centers and, hence, increased numbers of stereoisomers. Thus, trivial, or common, names of these carbohydrates are generally used. Glucose (also called dextrose), the most prevalent organic compound in nature, is an *aldohexose* containing four asymmetric carbons (Figure 11-3) and is therefore one of 16 possible stereoisomers (2^4). In a Fischer projection of a sugar, a D or L assignment is based on the asymmetry at the penultimate carbon atom of the molecule, which is C-5 in an aldohexose. As in the formula for D-glyceraldehyde (Figure 11-2), the hydroxyl group is written to the right of C-5 to designate the D form of glucose. The enantiomer (mirror image isomer) of D-glucose is L-glucose (Figure 11-3), which has the opposite configuration at *each* of the four asymmetric centers.

D-Galactose and D-mannose, two other aldohexoses frequently

found in living organisms, are also stereoisomers of D-glucose (Figure 11-4). As in the case of D-glucose, the placement of the hydroxyl group of C-5 identifies these sugars as D-isomers. D-Glucose and D-galactose are also referred to as *epimers* because the two monosaccharides differ only in the configuration at a single carbon atom (C-4). D-Glucose and D-mannose are also epimers because the difference is again limited by the asymmetry at a single carbon atom (C-2). However, D-galactose and D-mannose are not epimers, since they differ in the asymmetry at both C-2 and C-4. D-Fructose, sometimes called levulose, is another common biological sugar (Figure 11-5), and it is one of eight isomeric *2-ketohexoses,* having three asymmetric carbons.

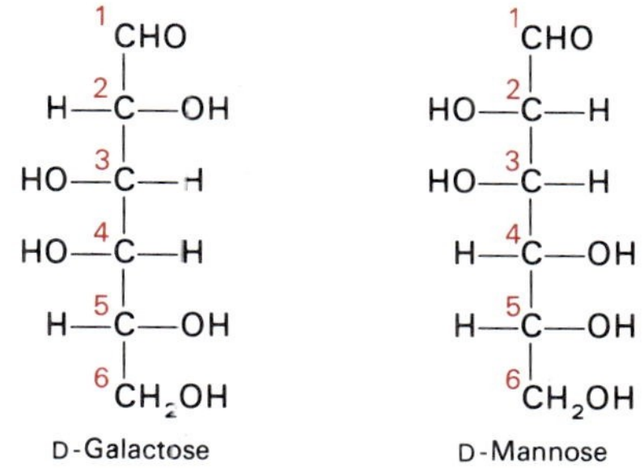

Figure 11-4
Two naturally occurring aldohexoses.

Closed-Ring Structures

In solution, D-glucose is not a linear but a closed-ring molecule because of the internal *hemiacetal* structure resulting from a reaction between the aldehyde at the C-1 position and the hydroxyl (alcohol) group of C-5 (Figure 11-6). Formation of the hemiacetal creates an additional asymmetric carbon (C-1) in the molecule, and hence there are two isomers (α and β) of the ring structure. The new asymmetric center is called the *anomeric carbon* and the α- and β-isomers, *anomers.* Thus, hemiacetal formation doubles the number of stereoisomeric aldohexoses from 16 to 32. In a Fischer projection of D-glucose (Figure 11-6), the anomeric α and β forms are indicated by writing the hydrogen atom of C-1 to the left and right, respectively. The reverse is true for L-glucose. As a general rule, the hydrogen atoms attached to the asymmetric carbon determining D or L configuration and to the anomeric carbon are *erythro* (written on the same side of the structure) in an α-form and *threo* (written on opposite sides) in a β-form.

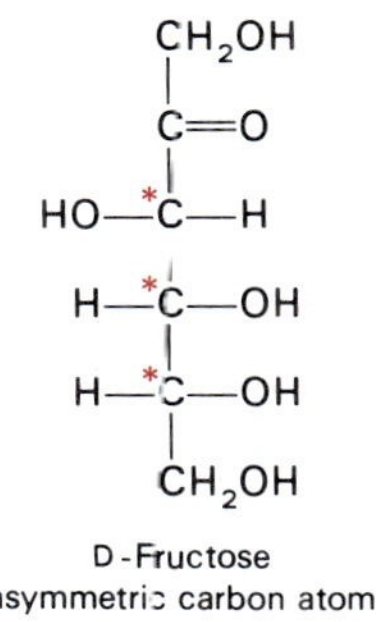

Figure 11-5
A 2-ketohexose.

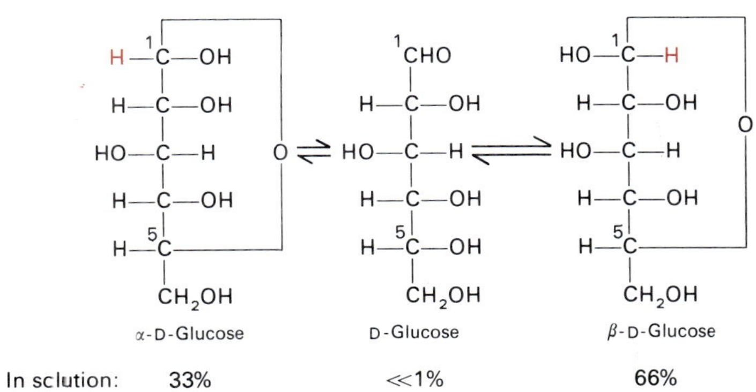

Figure 11-6
Mutarotation of glucose (as Fischer projections).

As indicated in Figure 11-6, α- or β-forms of D-glucose are interconvertible; this spontaneous phenomenon is called *mutarotation,* and it occurs with anomers of many monosaccharides. With respect to D-glucose, either anomer when dissolved in water will slowly undergo mutarotation until an equilibrium mixture consisting of about one third α-D-glucose and two thirds β-D-glucose is attained.

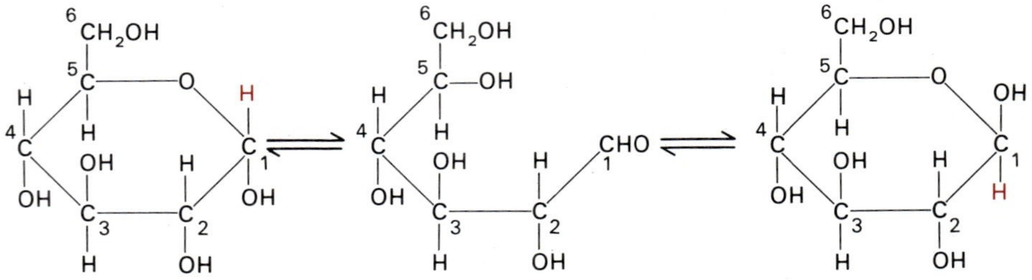

Figure 11-7
Mutarotation of glucose (as Haworth projections).

Haworth Projections

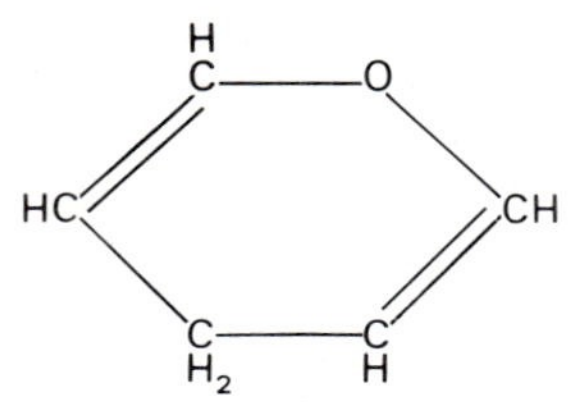

A more representative projection of the closed-ring structures of monosaccharides is that first suggested by Walter H. Haworth (Nobel Prize, 1937) in 1925. Figure 11-7 illustrates the Haworth projections of α- and β-D-glucose. Abbreviated versions of this type of projection for D-glucose either would not show the carbons except C-6 or, additionally, would omit the hydrogens and indicate hydroxyl groups by short lines (Figure 11-8). Haworth projections more closely approximate the predominant "chair" structure of glucose (Figure 11-9) and other hexose sugars that exist in solution. Because the six-membered ring of glucose is similar to that of *pyran,* it is called the *pyranose form* of glucose; the anomers are referred to as either α- or *β-D-glucopyranose.*

In the Haworth structures depicted in Figure 11-7, the *α-form* of the sugar has the hydroxyl group written *below* the anomeric

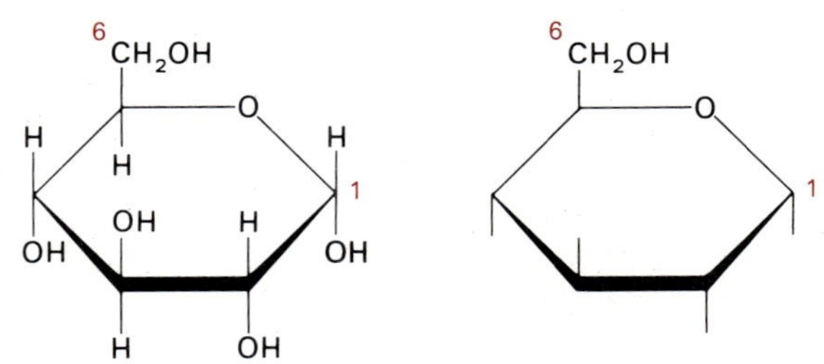

Figure 11-8
Abbreviated Haworth projections of α-D-glucopyranose.

carbon and the *β-form above.* With regard to asymmetric centers 2, 3, and 4 of glucopyranose, a hydroxyl group written *below* the carbon corresponds to one written to the *right* in a Fischer projection and, conversely, one written *above* corresponds to the *left.* The rule for the remaining asymmetric carbon (C-5) of the molecule, however, is different. When the hemiacetal linkage is written to the *right* of C-5 (as in Figure 11-7), the $-CH_2OH$ group written *above* C-5 identifies the monosaccharide as a D-isomer; conversely, if the $-CH_2OH$ is written *below,* the isomer has an L-configuration.

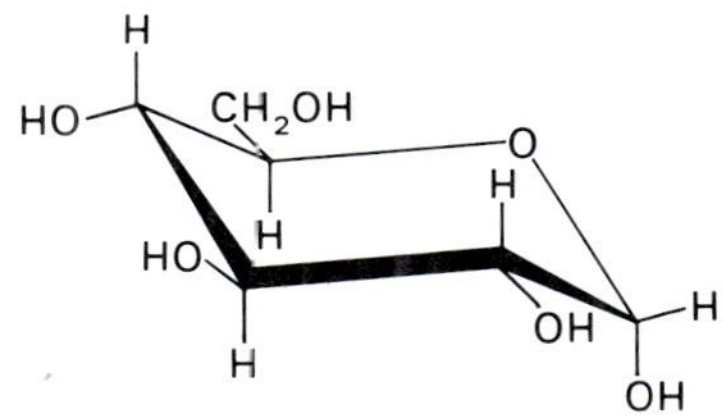

Figure 11-9
"Chair" form of α-D-glucopyranose.

Internal *hemiketal* bonding between the keto group of C-2 and the hydroxyl group of C-5 in D-fructose results in formation of a five-membered ring (Figure 11-10), similar to that of *furan.* This type of ring structure is thus called the *furanose form* of a sugar, and the two anomers of D-fructose are termed *α- and β-- D-fructofuranoses.* Note particularly the respective orientations of the $-CH_2OH$ and $-OH$ groups bound to C-2. The furanose and pyranose forms of monosaccharides are the thermodynamically stable structures, and only very small amounts of the linear structures of such sugars are ever present in solution. It is estimated that 90 per cent of cellular sugars are in closed-ring structures, with hexoses commonly as six-membered rings and pentoses as five-membered rings.

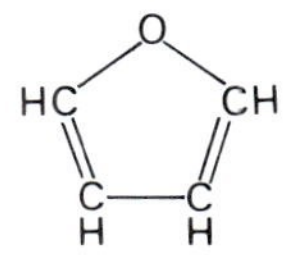

Furan

Figure 11-10
Haworth projection of β-D-fructofuranose.

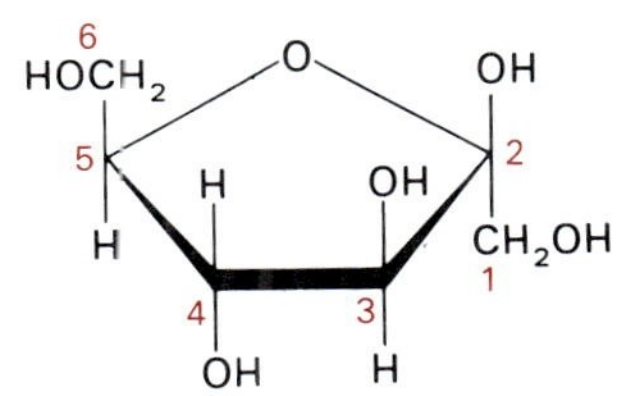

Other Monosaccharides of Importance

Both types of nucleic acid, DNA and RNA, have specific pentoses in their structures. D-Ribose is the sugar component of RNA and 2-deoxy-D-ribose that of DNA (Figure 11-11). Deoxyribose is a *deoxy sugar,* one that has a hydrogen atom substituted for a hydroxyl group at one or more of its carbon atoms. Deoxy sugars provide an example of one of several distinct classes of modified monosaccharides, which have specific structural alterations that

Figure 11-11
D-Ribose and D-deoxyribose.

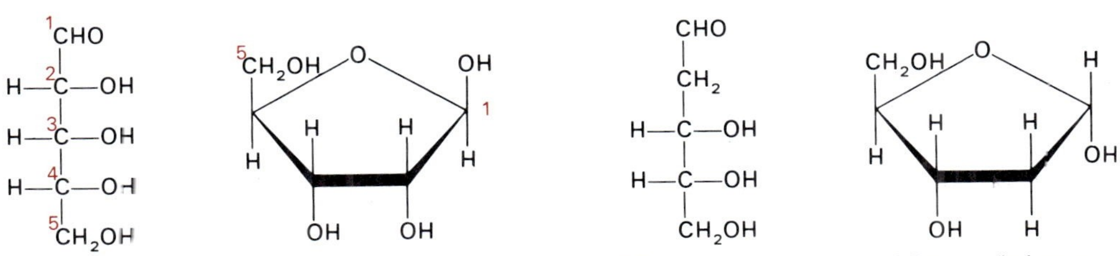

D-Ribose β-D-Ribofuranose 2-Deoxy-D-ribose α-2-Deoxy-D-ribofuranose

Figure 11-12
Two modified monosaccharides.

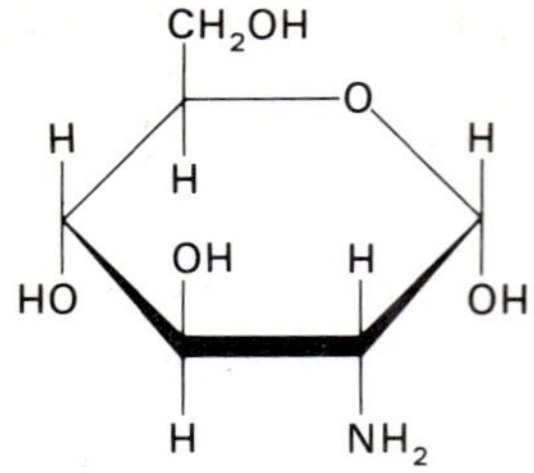

D-Glucosamine

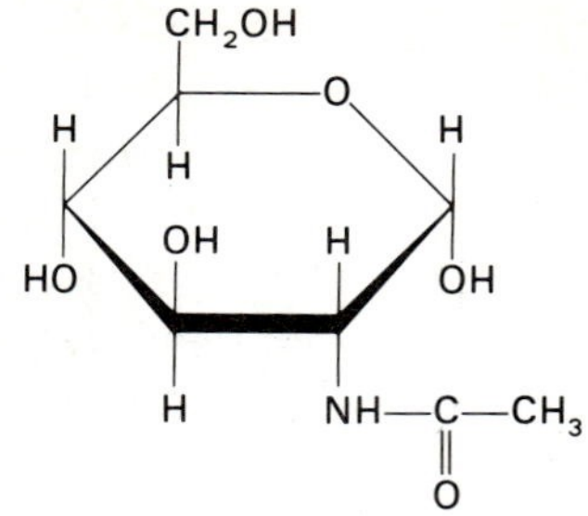

N-Acetyl-D-glucosamine
(α-anomer)

accommodate their specific biological functions. Another class of modified monosaccharides is that of the amino sugars, e.g., D-glucosamine (Figure 11-12), which have an amino group substituted for a hydroxyl group. *N*-Acetyl-D-glucosamine (Figure 11-12) is a common derivative of glucosamine. Amino sugars frequently occur in large quantities in structural materials. Chitin of the exoskeletons of invertebrates, e.g., crabs and lobsters, is a linear polymer of *N*-acetyl-D-glucosamine. Other substituted monosaccharides include phosphorylated forms of sugars, e.g., D-glucose 1-phosphate and D-glucose 6-phosphate (Figure 11-13). Phosphorylated sugars are often utilized for intracellular metabolic processes.

Figure 11-13
Two phosphorylated forms of glucose.

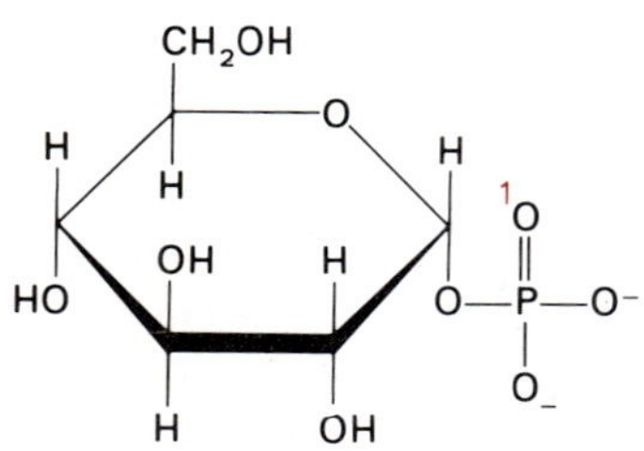

D-Glucose 1-phosphate
(α-anomer)

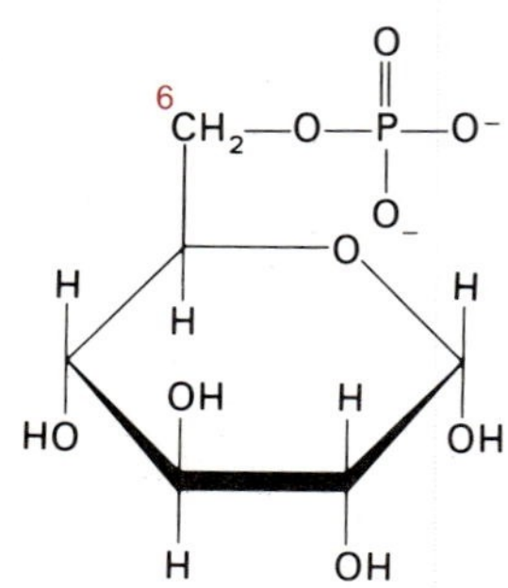

D-Glucose 6-phosphate
(α-anomer)

Figure 11-14
A uronic acid.

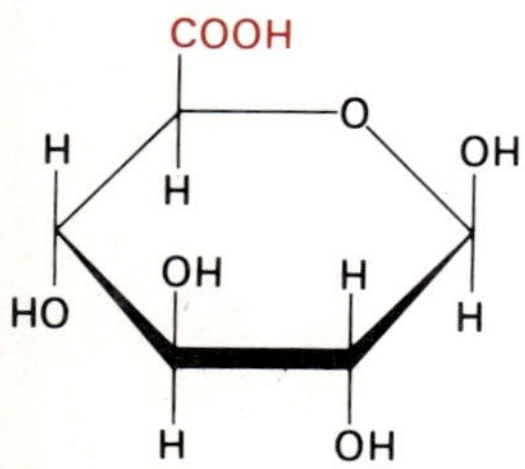

β-D-Glucuronic acid

Oxidized (acidic) and reduced (alcoholic) forms of sugars are also present in living systems. *Uronic acids,* such as glucuronic acid (Figure 11-14), are often constituents of acid mucopolysaccharides, high-molecular-weight compounds frequently found in large quantities in structural tissues such as bone, cartilage, and skin. Another oxidized derivative of glucose is D-gluconic acid (Figure

11-15) which, unlike glucoronic acid, has a C-1 instead of a C-6 carboxyl group. Gluconic acid is an example of an *aldonic* acid. D-Mannitol (Figure 11-16) is a naturally occurring polyhydric alcohol (a reduced sugar); its structure is the same as that of mannose (Figure 11-4), except for the alcohol group at the C-1 position. Mannitol was the first *alditol* crystallized from plants.

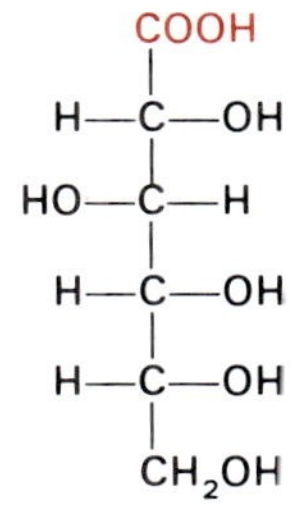

D-Gluconic acid

Figure 11-15
An aldonic acid.

OLIGOSACCHARIDES

In nature, the most abundant oligosaccharides are the disaccharides sucrose and lactose. Sucrose (table sugar) occurs in plants, where it is synthesized from D-glucose and D-fructose (Figure 11-17). A *glycosidic linkage* between the anomeric C-1 of α-D-glucose and the anomeric C-2 of β-D-fructose joins the two monosaccharides by an oxygen bridge, producing an α-(1-2) linkage. In Figure 11-17, the fructose structure is written for convenience with its anomeric carbon (C-2) facing that of glucose and, at first approximation, may appear to be in an α-linkage instead of a β-linkage. Lactose, the carbohydrate of mammalian milk, is composed of D-galactose and D-glucose (Figure 11-18). In this disaccharide, the glycosidic linkage between the anomeric C-1 of β-D-galactose and the nonanomeric C-4 of D-glucose is β-(1-4).

The synthesis of lactose by lactose synthetase, a heterogeneous dimer, is a novel example of the modification of catalytic specificity

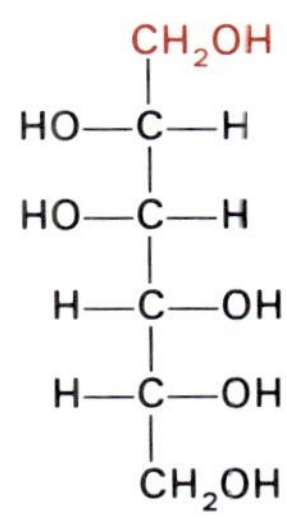

D-Mannitol

Figure 11-16
An alditol.

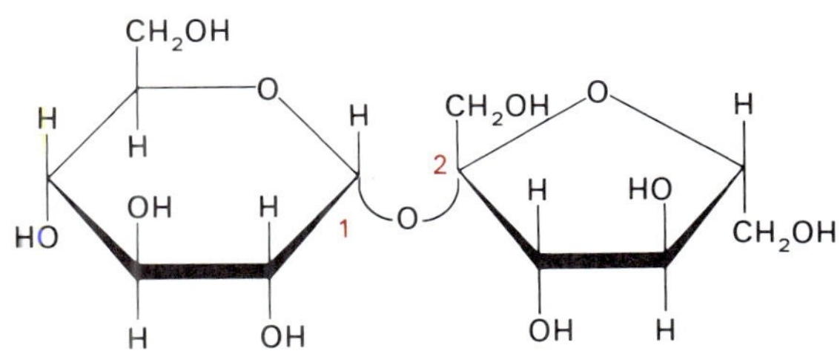

O-α-D-Glucopyranosyl-(1-2)-β-D-fructofuranoside

Figure 11-17
Sucrose.

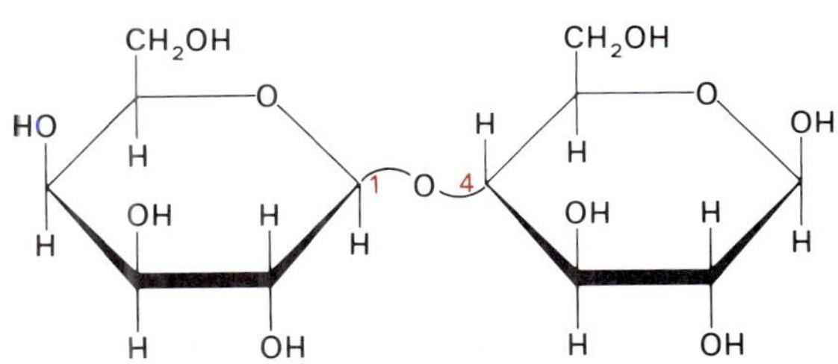

O-β-D-Galactopyranosyl-(1-4)-β-D-glucopyranose

Figure 11-18
Lactose.

by dimer formation, a form of allosteric conformational change that may be more common than has heretofore been realized. One of the two protomers is an enzyme (galactosyl transferase) that occurs widely in animal tissues, including mammary glands during pregnancy; by itself, it catalyzes the following reaction:

$$\text{UDP-galactose} + \text{N-acetylglucosamine} \rightarrow \text{N-acetyllactosamine} + \text{UDP}$$

UDP is uridine diphosphate (p. 195), which serves as a molecular carrier of carbohydrates in certain enzymatic reactions. For milk production, the second protomer of lactose synthetase, α-lactalbumin, is synthesized specifically in mammary tissue, and this protein's interaction with galactosyl transferase alters the substrate specificity so that the dimeric enzyme catalyzes the synthesis of lactose in the presence of glucose:

$$\text{UDP-galactose} + \text{glucose} \rightarrow \text{lactose} + \text{UDP}$$

α-Lactalbumin occurs only in mammary tissue; hence, lactose appears to be unique to mammalian milk.

Maltose (Figure 11-19) and cellobiose (Figure 11-20) are two disaccharides which do not occur naturally but are commercial degraded products of starch and cellulose, respectively. Both saccharides have two D-glucosyl residues linked by a 1-4 glucosidic linkage; the single structural difference between the two disaccharides is that the linkage in maltose is α-(1-4) and that in cellobiose

Figure 11-19
Maltose.

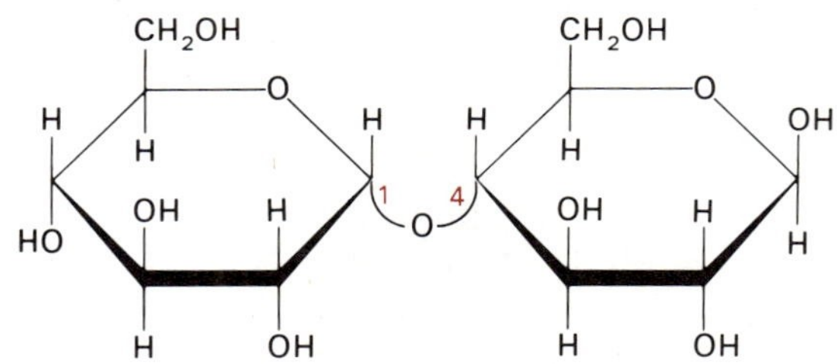

O-α-D-Glucopyranosyl-(1-4)-β-D-glucopyranose

Figure 11-20
Cellobiose.

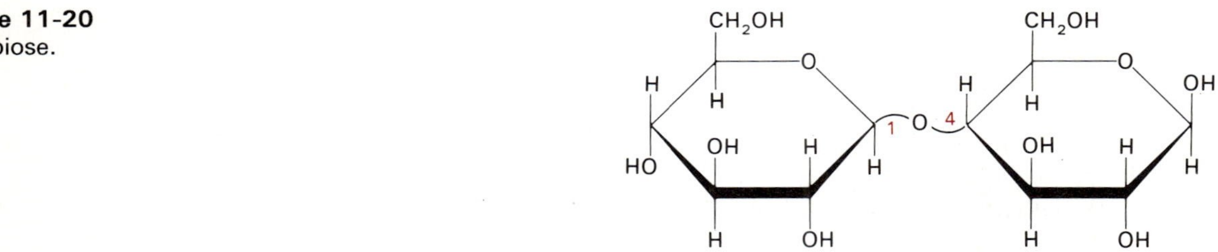

O-β-D-Glucopyranosyl-(1-4)-β-D-glucopyranose

is β-(1-4). This seemingly small difference serves as a pertinent illustration of the high degree of specificity frequently encountered in biological systems. Polymers of D-glucose in α-(1-4) linkage serve as readily available energy supplies for plants and animals, whereas analogous polymers in β-(1-4) linkage are structural components and are not degraded by most living systems, which lack the enzymatic capability to hydrolyze the β-(1-4) glycosidic linkage. Ruminants, e.g., cows, utilize cellulose as a food source only because the bacteria in their stomachs can digest the polysaccharide. Even termites rely on the bacteria in their guts to degrade wood. If it were not for the ability of certain bacteria and fungi to hydrolyze the β-(1-4) linkages found in plant polysaccharides, continued accumulation of dead vegetation would pose serious ecological problems.

Sugars as Reducing Agents

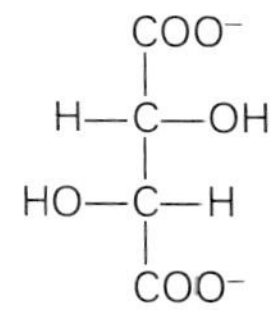

Tartrate

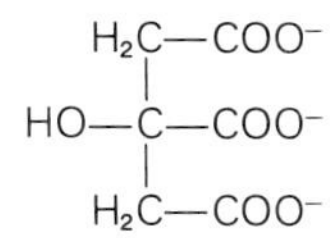

Citrate

As α-hydroxyaldehydes and α-hydroxyketones, most monosaccharides and disaccharides are effective reducing agents. This property of carbohydrates has for many years served as a chemical basis for clinical tests for sugars, e.g., to detect glucose in the urine of suspected diabetics. The Fehling and Benedict tests, for example, capitalize on the ability of reducing sugars to reduce cupric ions (Cu^{2+}). Both the Fehling's solution and the Benedict's solution are alkaline solutions of Cu^{2+}, complexed with tartrate (Fehling) or citrate (Benedict) ions; this complexing of copper is needed to maintain its solubility. If a fluid being tested contains a reducing sugar, the cupric ion is reduced to the cuprous state (Cu^{+}) which, upon heating, precipitates as cuprous oxide (Cu_2O). This reduction of copper is seen as the disappearance of the deep-blue color of the test solution (due to Cu^{2+}) and the appearance of a red precipitate (Cu_2O). In recent years, the traditional methods of testing glucose as a reducing sugar have been replaced by a more definitive assay that uses the enzyme glucose oxidase, which specifically utilizes glucose (but not other reducing sugars) as a substrate.

Most disaccharides are reducing sugars because one of their two anomeric carbons (reducing aldehyde or ketone group) is not in glycosidic linkage. An exception is sucrose (a *nonreducing sugar*), which, as illustrated in Figure 11-17, has both of its anomeric carbons in the bond that links the two monosaccharides. The finding that sucrose is not a reducing sugar was a key observation in earlier studies that elucidated the disaccharide's unusual structure.

Table 11-1
Relative sweetness of some sugars

Sugar	Relative sweetness
Sucrose	100
Glucose	74
Fructose	173
Galactose	32
Maltose	32
Lactose	16

Sugars and Other Molecules as Sweeteners

As sugars, mono- and disaccharides are generally thought of as having a sweet taste. However, as indicated in Table 11-1, the degree of sweetness varies widely among carbohydrates. When the sweetness of the sugars listed is compared with that of sucrose (reference value of 100), the monosaccharide fructose rates as the sweetest sugar. Because of this characteristic, fructose is often the sugar of choice in the manufacture of candy. *High-fructose* food products also capitalize on that sugar's sweetness; for example, as "diet food" items, they provide the desired sweetness but less sugar (fewer calories). The taste of sweetness is not restricted to carbohydrates, as evidenced by the long-established use of saccharin as a non-nutritive sweetener. Although the chemical structure of saccharin (Figure 11-21) is distinct from that of a carbohydrate, the molecule is nevertheless 500 times as sweet as sucrose. There are also proteins that are sweeter tasting than sugar. One of these proteins is *monellin,* which is present in the sap of serendipity berries, fruit of the West African *Dioscoreophyllum comminsi.* This sweet protein is composed of two protomers (one having 42 amino acid residues and the other 50 residues) and is 2,500 times as sweet as sucrose. Monellin's sweetness requires the undissociated protein, since neither of the protomers has a sweet taste; it appears that the quaternary structure of monellin is essential for its sweet taste. In the case of the African plant Katemfe *(Thaumatococcus daniellii),* its intense sweetness is attributed to two proteins, which together are 1,600 times sweeter than sucrose.

Figure 11-21
Saccharin.

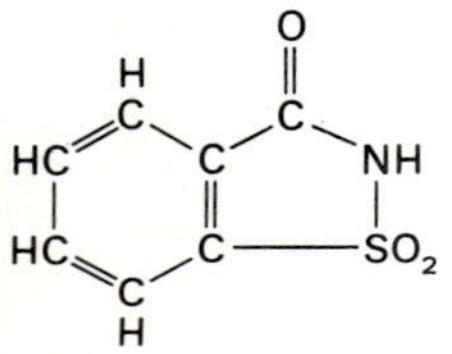

The newest low-calorie sweetner to be approved for commercial use is a dipeptide, called *aspartame,* containing *aspartic acid* and the *methyl ester of phenylalanine.* The peptide is about 200 times-sweeter than sucrose and does not have the bitter aftertaste often associated with saccharin. Sold under the tradenames NutraSweet or Equal, aspartame is expected to be widely used in the food industry, e.g., in cold cereals, gelatins, and puddings.

POLYSACCHARIDES

Starches and Glycogen (Homoglycans)

The best-known polysaccharides are the polymers of D-glucose, which serve as storage forms of energy (plant starches and animal glycogens) or as structural materials (cell wall cellulose of plants).

Table 11-2
Biochemical characteristics of four classes of polysaccharides composed of D-glucose

Characteristic	Amyloses	Amylopectins	Glycogens	Celluloses
Monomeric unit	D-Glucose	D-Glucose	D-Glucose	D-Glucose
Linkage	α-(1-4)	α-(1-4) and α-(1-6)	α-(1-4) and α-(1-6)	β-(1-4)
Branching	None	$\approx$4%	$\approx$9%	None
Molecular weight	4×10^3–1.5×10^5	5×10^4–1×10^6	$\approx 5 \times 10^6$	2×10^5–2×10^6

Because these polysaccharides contain only one type of sugar, they are called *homoglycans.* Starch is a mixture of linear (amylose) and branched (amylopectin) polymers. The characteristics of these two components of starch are summarized in Table 11-2. *Amylose,* which is 20 to 30 per cent of most natural starches, is an unbranched polymer of glucosyl residues in α-(1-4) glycosidic linkage. Maltose (Figure 11-19) is the repeating disaccharide unit of amylose. *Amylopectin* is a branched form of amylose, with every twentieth to twenty-fifth residue serving as a branch point by bonding to an additional glucosyl residue in α-(1-6) linkage (Figures 11-22 and 11-23). The glycogen of animals has the structural characteristics of amylopectin except that it is a larger and more highly branched macromolecule (Table 11-2). In humans, glycogen is most abundant in liver and muscle; the amounts of glycogen present in these tissues depend on an individual's nutritional status and health condition.

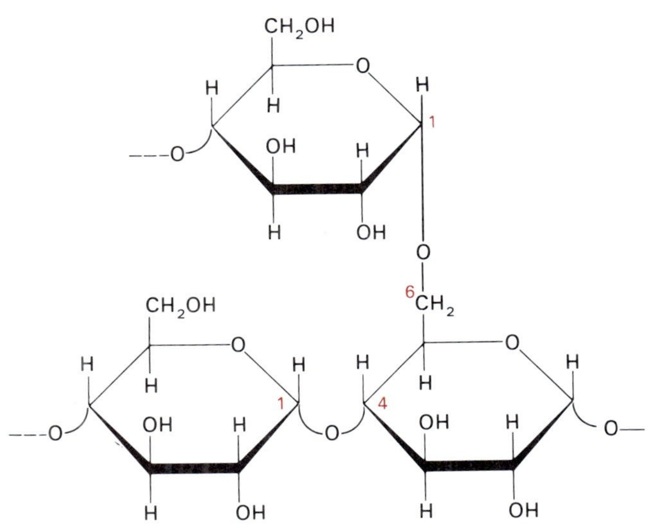

Figure 11-22
Branch point in amylopectin.

Figure 11-23
Schematic of the branched structure of amylopectin.

Glycogen Synthesis

For glycogen synthesis in animals, an activated form of glucose is required for the addition of glucosyl residues to the main polysaccharide chain. The activated form (Figure 11-24) is *uridine diphosphate glucose (UDP-glucose),* a molecule containing glucose esterified to a phosphate group of the diphosphate derivative of uridine (p. 229). The synthesis of UDP-glucose, catalyzed by *glucose 1-phosphate uridyltransferase,* proceeds as follows:

$$\alpha\text{-D-Glucose 1-phosphate} + \text{UTP} \rightleftharpoons \text{UDP-D-glucose} + \text{PP}_i$$

Glucose, phosphorylated at C-1, and uridine triphosphate (UTP) serve as the substrates, and the products are UDP-glucose and inorganic pyrophosphate (PP_i). In glycogen synthesis (Figure 11-25), UDP-glucose donates its glucosyl residue to the nonreducing terminus of the main glycogen chain in a reaction catalyzed by *glycogen synthase.* The reaction produces an α-(1-4) glycosidic linkage between the C-1 of the added glucosyl residue and the hydroxyl group at C-4 of the terminal glucosyl residue. An α-(1-4) polyglucose chain of at least four residues is required by the enzyme as a *primer* for the reaction. In plants, the synthesis of amylose (unbranched starch) occurs by a two-step process similar to that described above. However, in many plants, *adenosine diphosphate glucose (ADP-glucose)* rather than UDP-glucose is the preferred glucose donor. The two enzymes required for amylose

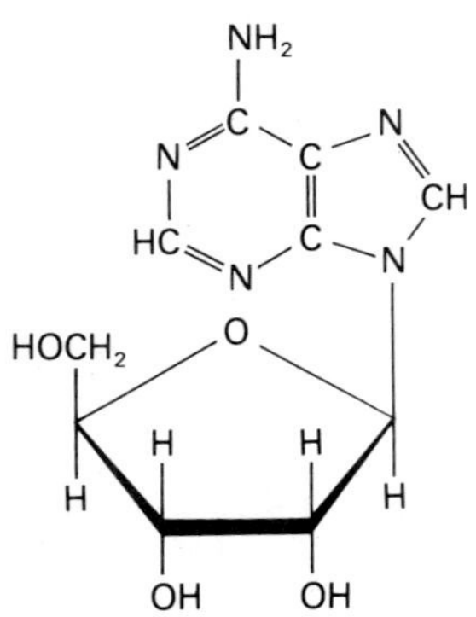

Adenosine

Figure 11-24
Uridine diphosphate glucose.

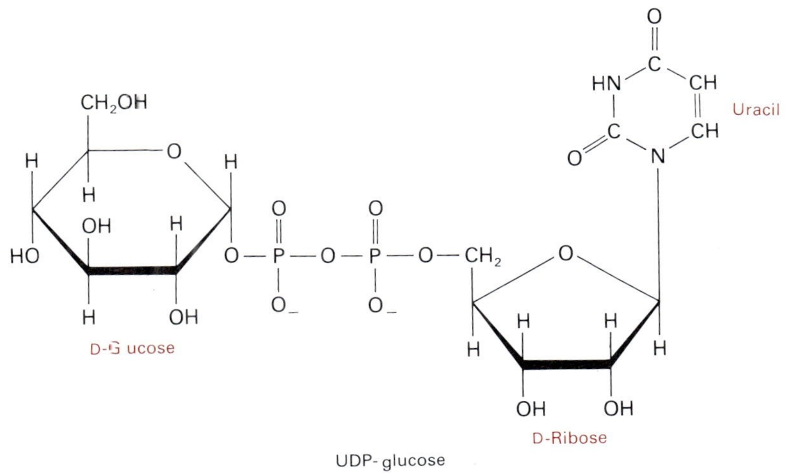

Figure 11-25
Glycogen synthase reaction in animals. R represents the remaining portion of UDP-glucose or of glycogen.

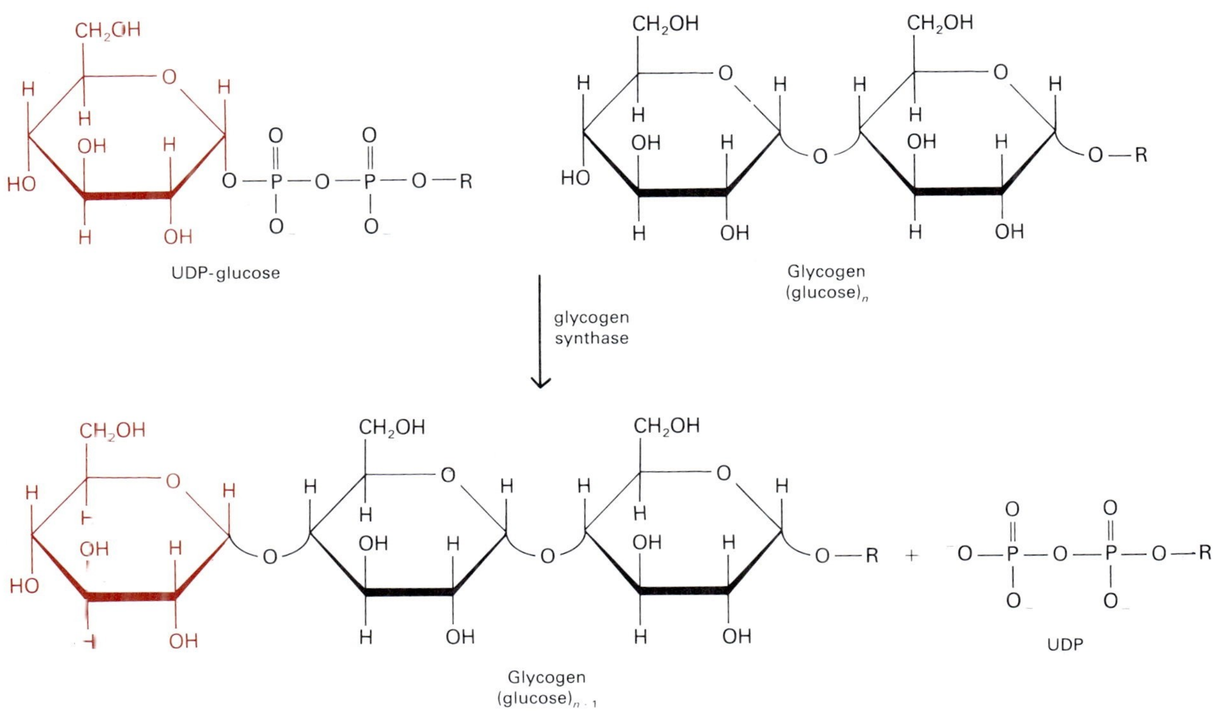

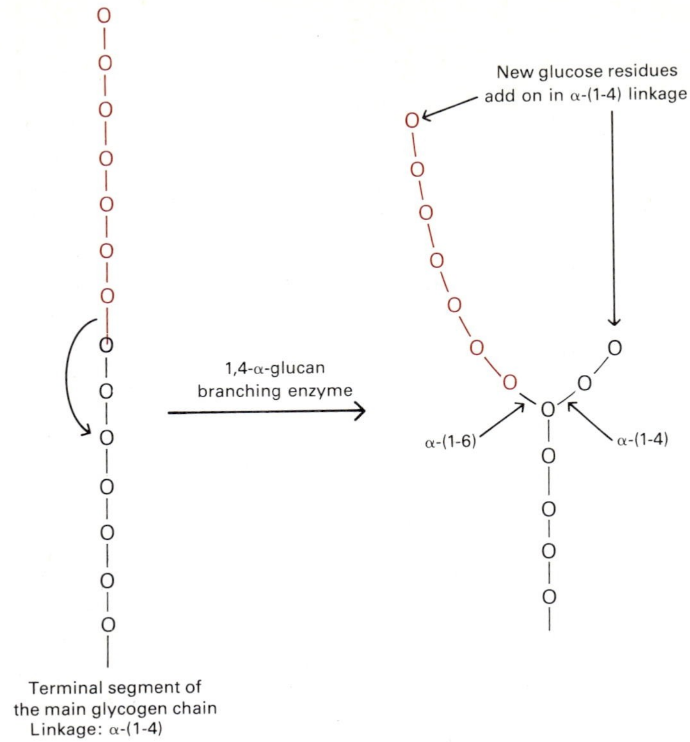

Figure 11-26
Schematic of the action of 1,4-α-glucan branching enzyme. Circles represent glucosyl residues; those in red denote the fragment that is transferred to form a new branch.

production are *glucose 1-phosphate adenyltransferase* and *starch synthase.*

Since glycogen synthase catalyzes only the formation of α-(1-4) glycosidic linkages, the branch points of glycogen [involving α-(1-6) linkages] are produced by another enzyme, called the *1,4-α-glucan branching enzyme.* To produce an α-(1-6) glycosidic bond (Figure 11-26), the enzyme catalyzes the transfer of an oligosaccharide containing six or seven glucosyl residues from the end of the glycogen chain to the hydroxyl group of the C-6 of a glucosyl residue of the same or a different glycogen molecule. The reaction creates both a branch point and a new terminus for adding glucosyl residues.

Payen, codiscoverer of diastase, also isolated a compound common to cell walls of higher plants, which he named "cellulose." His naming of the polysaccharide introduced the *ose* suffix into the nomenclature of carbohydrates.

Cellulose (Homoglycan)

Cellulose, an unbranched glucosyl polymer, contains β-(1-4) linkages, and the repeating disaccharide unit is cellobiose (Figure 11-20). Some characteristics of this plant polysaccharide, by far the most abundant macromolecule in nature, are listed in Table 11-2. In plant cell walls, cellulose occurs in densely packed fibrils cemented together in a matrix of three other polymeric materials: hemicellulose, pectin, and extensin. *Hemicellulose* is a polymer of the aldopentose D-xylose (Figure 11-27) in β-(1-4) linkage, with side chains containing other sugars. *Pectin,* with α-(1-4) linkages, is composed of D-galacturonic acid residues (Figure 11-28) whose C-6 carboxyl groups are methylated to various degrees. Pectins, along with sugar, furnish the gelling characteristics of jellies and preserves made from fruit. *Extensin* is a protein rich in hydroxyproline and is covalently bonded to cellulose fibrils. The rigid wall structures of plant cells both protect the cell and provide form and support for plant tissues.

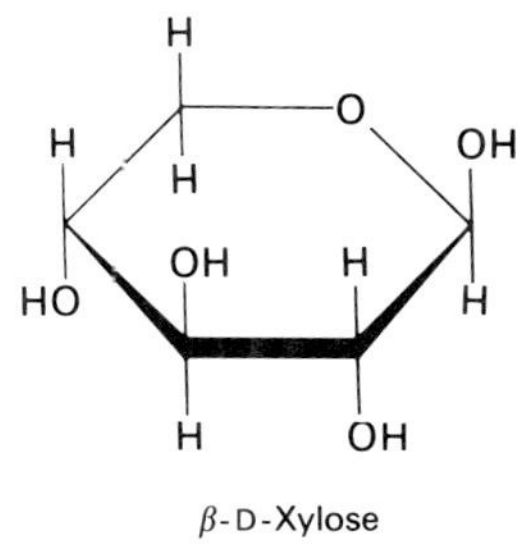

Figure 11-27
Predominant monosaccharide of hemicellulose.

Conformations of Cellulose and Starch

The β-glycosidic linkages of cellulose play a critical role in determining the structural conformation of the polysaccharide which, in turn, accounts for its strength as a fiber. As illustrated in Figure 11-29, because of its β-linkages, cellulose assumes an extended, rigid conformation with each glucosyl residue turned 180° from the two residues it is linked to in the molecule. As a result of this conformation, a network of hydrogen bonds is established between adjacent cellulose molecules. As illustrated, the hydroxyl groups of C-2, C-3, and C-6, as well as the hemiacetal oxygen, participate in this hydrogen-bonding system. Thus, it is this extensive hydrogen bonding coupled with the rigid β-conformation that makes cellulose a tough, structural polysaccharide. On the other hand, amylose (starch), with its α-glycosidic linkages, does not extend itself but, rather, forms a left-handed helix (Figure 11-29).

The structure of starch, postulated by Charles Hanes in 1937, was the first proposal of a helical conformation for a biological macromolecule.

Heteroglycans

Heteroglycans are polysaccharides composed of more than one type of monosaccharide unit and, like homoglycans, are widespread in nature. Examples of heteroglycans in humans are the simple acid mucopolysaccharides *hyaluronic acid* and *chondroitin sulfate A,* abundant in connective tissues. Hyaluronic acid is a repeating disaccharide unit of *N*-acetyl-D-glucosamine and D-glucuronic acid in alternating β-(1-3) and β-(1-4) linkages (Figure 11-30). Chondroitinsulfate A consists of a repeating disaccharide unit

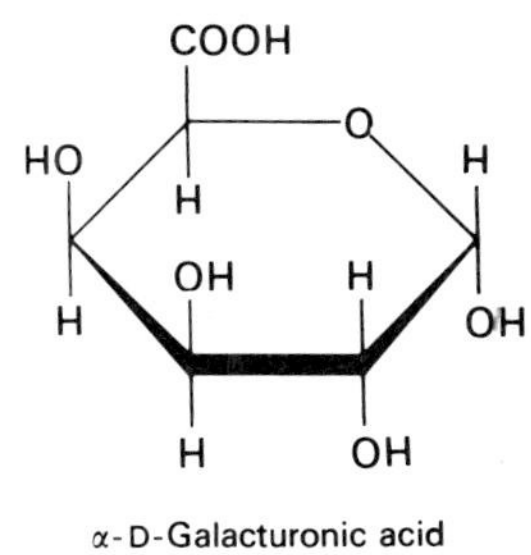

Figure 11-28
Predominant monosaccharide of pectin.

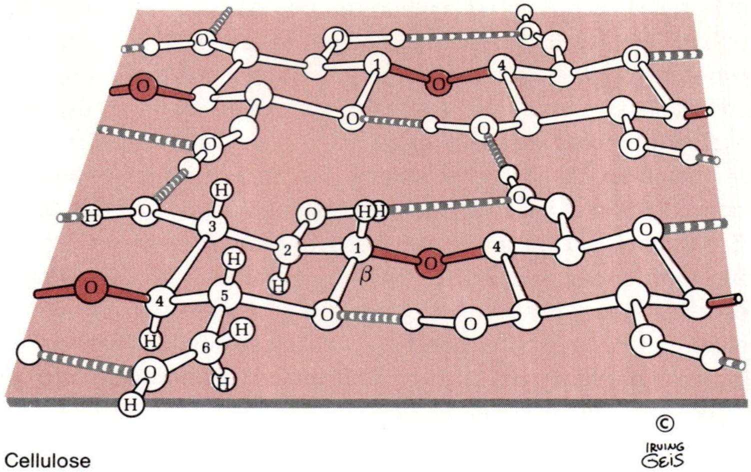

A. Cellulose

Figure 11-29
Structural conformations of cellulose and starch (amylose). A. Cellulose. B. Starch. (Copyright 1982 by Irving Geis)

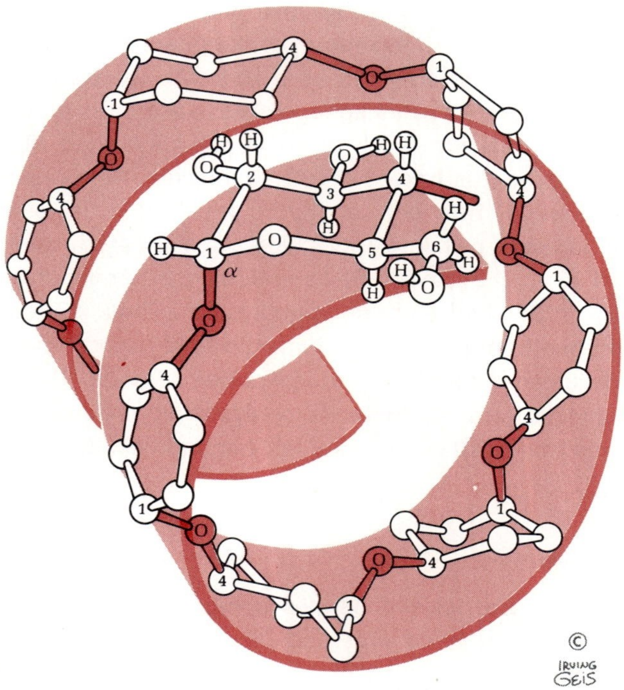

B. Starch

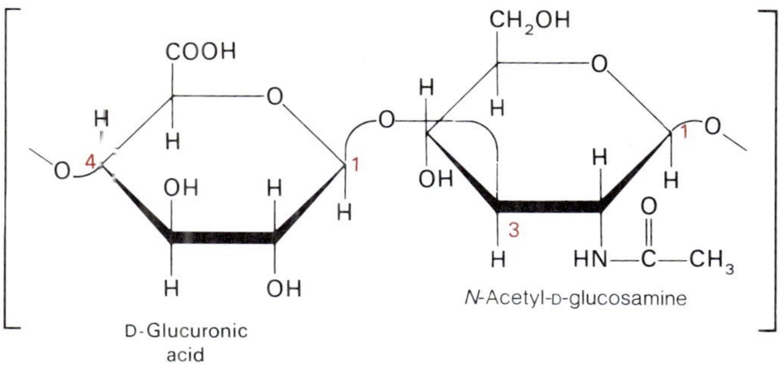

Figure 11-30
Repeating disaccharide unit of hyaluronic acid.

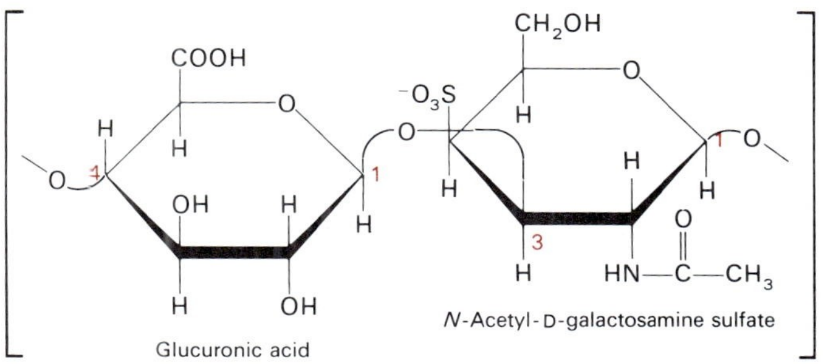

Figure 11-31
Repeating disaccharide unit of chondroitin sulfate A.

of *N*-acetyl-D-galactosamine sulfate and glucuronic acid, also in alternating β-(1-3) and β-(1-4) linkages (Figure 11-31).

Chemical and Functional Versatility

The chemical identification of naturally occurring polysaccharides is a continuous research endeavor because a great variety of different molecular types exists in plant and animal species. The same can be said for the general study of carbohydrates in living systems, since they are ubiquitous and are utilized for many purposes by all organisms. In addition to their occurrence in the three categories based on molecular size, carbohydrates also occur covalently bonded to other types of biomolecules, e.g., lipids (*glycolipids,* p. 210) and proteins (*glycoproteins,* p. 524). The diverse chemical properties of biomolecules are well demonstrated in nature's uses of carbohydrates. As a result of evolutionary processes, the *chemical versatility* of the carbon atoms and functional groups of simple monosaccharides was extensively utilized to generate an array of compounds, including polymers, which greatly enhanced the *functional versatility* of this class of biomolecules.

THE MANY USES OF POLYSACCHARIDES

For many years, polysaccharides have been extensively utilized for numerous commercial and medical purposes. This important aspect of polysaccharides is well illustrated by the applications that have been found for the two classes of D-glucose polymers, dextrins and dextrans.

Dextrins

The name *dextrin* (starch gum) refers to the various products obtained from the partial breakdown of starch, e.g., by heating it in an autoclave (pressure cooker) or by limited hydrolysis (enzymatic or chemical). Different dextrins vary in size and amount of branching. In the textile industry, dextrins are used for thickening dye pastes and mordants (chemicals that fix dyes in materials), printing tapestries, sizing fabrics, and preparing felts. In the publishing industry, they are used as glues and mucilages and in the manufacture of printer's ink and the sizing of paper. For medical purposes, dextrins serve as excipients (inert fillers) of dry extracts and pills and in the preparation of emulsions and dry bandages. They also are added as modifiers to milk and milk products in infant formulas and as a source of carbohydrate in the diets of animals. These polymers of D-glucose are additionally utilized in the production of matches, fireworks, and explosives.

Dextrans

Dextrans are polysaccharides produced by certain bacteria, e.g., *Leuconostoc mesenteroides,* when they are grown on sucrose. All dextrans are chains of D-glucosyl residues in α-(1-6) glycosidic linkages, with α-(1-3) branching. The various types of dextrans differ in chain length and degree of branching. Like dextrins, dextrans also have a variety of important uses. In the candy industry, they are utilized to prepare soft-center candies and, in the paint industry, the mixed ethers and esters of dextrans are used in lacquers. Research laboratories commonly use specialized preparations of dextrans as column materials for chromatographic purposes. Dextrans also have important medical applications. In the treatment of persons in shock, dextran is administered as a plasma volume extender to hold water in the bloodstream, i.e., to avoid a possible fatal drop in blood volume and pressure. Dextran-iron complexes are used medically to stimulate red blood cell formation or to increase the hemoglobin content of the blood. In veterinary medicine, dextran-iron complexes are employed for the treatment of iron deficiency in animals. Additionally, the sodium salt of the sulfuric acid ester of dextran (dextran-sulfate complex) is used therapeutically as an anticoagulant.

Additional Comments

The above brief discussion illustrates how the physical, chemical, and nutritive properties of polysaccharides produced by organisms have been advantageously adapted for a variety of uses by humans. The ability of many polysaccharides to absorb water is a physical property that has found numerous applied uses. Some examples are the use of pectins and agar as gelling agents and the use of dextran as a plasma volume extender and as a matrix for column chromatography. The chelation of dextran with iron atoms (dextran-iron complex) and its reactivity with sulfuric acid (dextran-sulfate complex) are chemical properties that have proved to be of medical importance. The nutritive or non-nutritive properties of polysaccharides are also selectively used. Whereas dextrin (D-glucosyl residues in α-glycosidic linkage) can be used to supplement the carbohydrate content of human and animal foods, indigestible cellulose or pectin can be added to food products to enhance their natural fiber (roughage) content. The development of these various uses of polysaccharides demonstrates how the properties of biomolecules can be astutely exploited for functions that far exceed the specific biological roles for which they were intended.

SUMMARY

Carbohydrates, called sugars, are the primary products of photosynthesis and are also the predominant energy source for living systems. They are generally defined as polyhydroxyaldehydes (aldoses) or polyhydroxyketones (ketoses) and their derivatives, such as amino sugars and sugar alcohols. Monosaccharides are individual sugars, oligosaccharides are carbohydrates containing two to ten monosaccharide units, and polysaccharides are large carbohydrate polymers. Since they have one or more asymmetric carbons, monosaccharides occur as stereoisomers; the D-isomers of sugars (stereochemically related to D-glyceraldehyde) are predominantly used by organisms.

Three commonly occurring six-carbon aldoses (aldohexoses) are D-glucose, D-galactose, and D-mannose; a common six-carbon ketose (ketohexose) is D-fructose. The two five-carbon aldoses (aldopentoses) D-ribose and D-deoxyribose are the sugars found in RNA and DNA, respectively. In solution, aldohexoses spontaneously undergo hemiacetal formation (between C-1 and C-5) to produce a six-membered ring structure, the pyranose form of a sugar. Fructose forms an internal hemiketal structure (C-2 and C-5) to yield a five-membered ring (furanose form). Pentoses, e.g., ribose, generally form furanose structures, too. Most cellular sugars exist in ring structure. Hemiacetal or hemiketal formation creates a new asymmetric center (anomeric carbon) and thus generates two (α and β) additional isomeric forms.

The disaccharides sucrose and lactose are two of the most abundant oligosaccharides. Sucrose (table sugar) is composed of D-glucose and D-fructose joined by an α-(1-2) glycosidic linkage. Lactose (sugar of mammalian milk) contains D-galactose and D-glucose in β-(1-4) linkage. Maltose, the disaccharide unit of starch, has D-glucosyl residues in α-(1-4) linkage. The structure of cellobiose, the comparable disaccharide unit derived from cellulose, is identical to that of maltose, except that the glycosidic linkage is β.

The most common polysaccharides are cellulose, plant starches, and animal glycogens. Cellulose, a structural polysaccharide of plants, is the most abundant macromolecule in nature; it is a linear polymer of D-glucosyl residues in β-(1-4) linkage. Starches are composed of two molecular components, amylose [linear polymer of D-glucose in α-(1-4) linkage] and amylopectin [branched D-glucosyl polymer in α-(1-4) and α-(1-6) linkages]. Glycogens are similar in their chemical structures to amylopectin, except that they are more highly branched and are larger polymers. Starches and glycogens serve as storage forms of energy.

Glycogen synthesis in animals utilizes UDP-glucose (uridine diphosphate glucose), which donates its glucosyl residue to the C-4 hydroxyl group of the terminal glucosyl residue of the main glycogen chain. It is in this manner that the α-(1-4) glycosidic linkages are produced. To create the α-(1-6) branch points of glycogen, oligosaccharides of six or seven glucosyl residues are transferred from the end of a glycogen chain to the C-6 hydroxyl group of a glucosyl residue of the same or of a different glycogen molecule.

Cellulose, starches, and glycogens are homoglycans, since each contains only one type of monosaccharide unit. The acid mucopolysaccharide hyaluronic acid is an example of a heteroglycan, a polysaccharide with more than one type of monosaccharide unit. Hyaluronic acid is composed of repeating disaccharide units of *N*-acetyl-D-glucosamine and D-glucuronic acid in alternating β-(1-3) and β-(1-4) linkages. Polysaccharides are used extensively for industrial, medical, and nutritive purposes. Some of the applications include their use as glues and adhesives, gelling agents, sizing for fabrics and paper, dye and varnish preparations, carbohydrate additives for foods, and anticoagulants and plasma volume extenders.

QUESTIONS

1. Given the following Fischer projections of six monosaccharides:

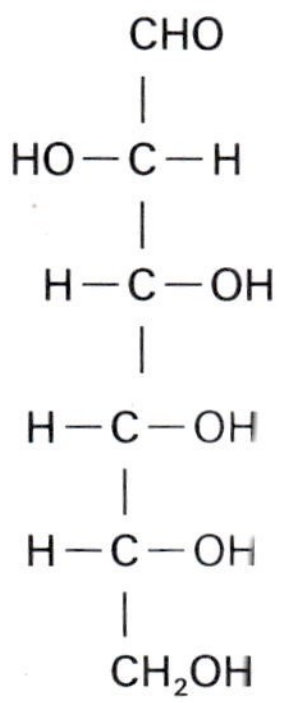
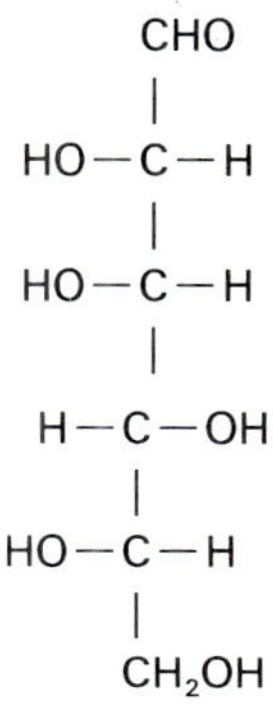
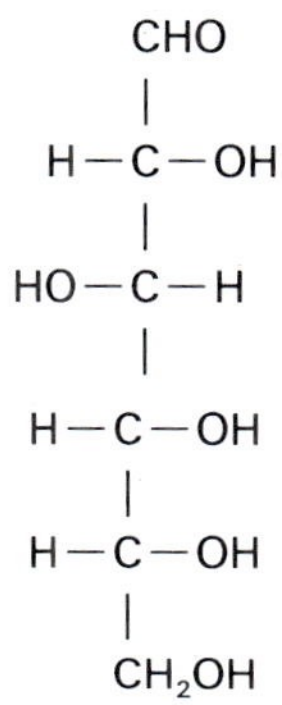

1 2 3

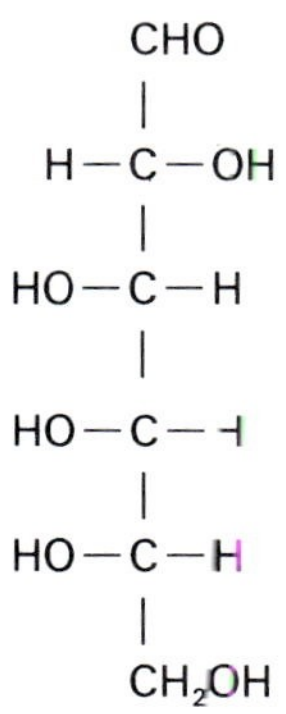
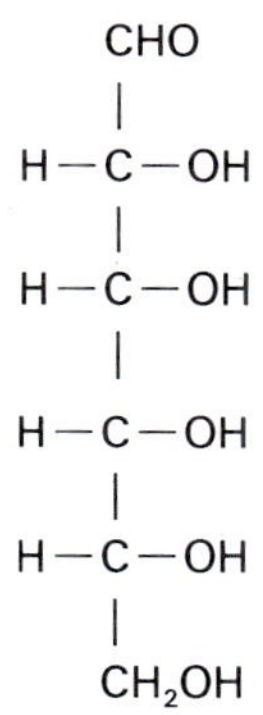
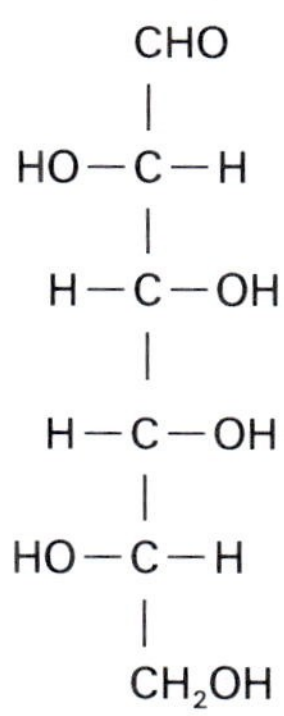

4 5 6

(a) What is the general classification of the six carbohydrates?
(b) How many asymmetric carbons does each have?
(c) What are D-isomers?
(d) Is one of the sugars D-glucose?
(e) Which, if any, are enantiomers of each other?
(f) Which, if any, are epimers of sugar 2?
(g) Draw the Haworth projection of the α-anomer of sugar 1.

2. Draw the Fischer projection of (a) D-ribitol and (b) D-glucitol (alditols of D-ribose and D-glucose, respectively) and of (c) D-glucaric acid (D-glucose with both C-1 and C-6 oxidized).

3. Below is the Haworth projection of the disaccharide gentiobiose:

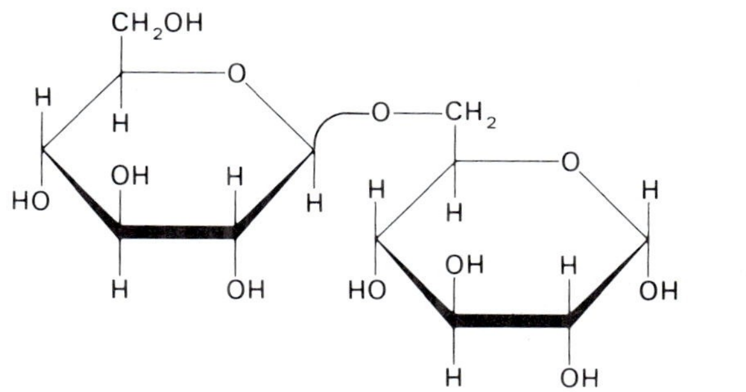

(a) What two carbohydrates are contained in the structure?
(b) What two carbon atoms are in glycosidic linkage?
(c) Is the glycosidic linkage α or β?
(d) Is the free anomeric carbon in an α or β configuration?

4. Draw the Haworth projection for the basic repeating disaccharide unit of a hemicellulose [D-xylose in β-(1-4) linkage].

5. Draw the Haworth projection for the repeating unmethylated disaccharide unit of pectin [D-galacturonic acid in α-(1-4) linkage].

6. Refer to Table 9-7 in answering the following:
(a) Write two general reactions to illustrate the conversion of cornstarch into fructose.
(b) Write the general enzymatic reactions for the conversion of sawdust into sugar.
(c) Why is amylase effective as a digestive aid in precooked foods?
(d) A majority of the adults in the world suffer gastric difficulties when they drink any appreciable quantity of milk. A deficiency of what enzyme is responsible for this milk intolerance?

7. Draw the structure of the sweetener aspartame, which is a dipeptide containing aspartic acid and the methyl ester of phenylalanine.

SUGGESTED READINGS

Aspinall, G. O., ed., *Carbohydrate Chemistry,* Vol. 7 of *IRS Organic Chemistry* series. London: Butterworths, 1976. Good chemistry text on carbohydrates.

Florkin, M., and E. H. Stotz, eds., *Carbohydrates,* Vol. 5 of *Comprehensive Biochemistry.* New York: American Elsevier, 1963. A reference treatise covering all aspects of the chemistry of carbohydrates (mono-, oligo-, and polysaccharides).

McIlroy, R. J., *Introduction to Carbohydrate Chemistry.* Reading, Mass.: Butterworths, 1967. A short, general overview of carbohydrates, covering their classification, structures, synthesis, and biological oxidation.

Morrison, R. T., and R. N. Boyd, *Organic Chemistry,* 3rd ed. Boston: Allyn and Bacon, 1973. This text contains two excellent chapters (34 and 35) on carbohydrates.

Pigman, W., and D. Horton, eds., *The Carbohydrates.* 4 vols., 2nd ed. New York: Academic Press, 1972. A valuable compilation of chemical and biochemical information on carbohydrates.

Articles

Frohwein, Y. Z., "A Simplified Proof of the Constitution and the Configuration of D-Glucose." *J. Chem. Ed.,* 46:55, 1969.

Hudson, C. S., "Emil Fischer's Discovery of the Configuration of Glucose." *J. Chem. Ed.,* 18:353, 1941.

Kirkwood, S., "Unusual Polysaccharides." *Ann. Rev. Biochem.,* 43:401, 1974.

Kretchner, N., "Lactose and Lactase." *Scientific American,* 227(4):70 (offprint #1259), 1972.

Okuda, J., "Activites of the D-Glucose Anomers." *TIBS,* 3:161, 1978.

Robyt, J. F., "Mechanisms Involved in the Biosynthesis of Polysaccharides." *TIBS,* 4:47, 1979.

Sharon, N., "Carbohydrates." *Scientific American,* 243(5):90, 1980.

LIPIDS AND MEMBRANES

12

INTRODUCTION

Lipids are characterized as biomolecules that are insoluble or sparingly soluble in water and that are extractable by solvents. In the past, lipids were not a particularly attractive subject for biochemical research because it was difficult to study these nonpolar compounds using the conventional separation and quantitative techniques developed for the biochemical study of water-soluble components of the aqueous environment of life. The roles of lipids as storage forms of energy and as structural components of membranes were also thought of as uninteresting because they seemed to lack the diverse metabolic roles of other biomolecules, e.g., carbohydrates and amino acids. Today, with the aid of new techniques such as thin layer and gas-liquid chromatography, lipid research has attained a key role in biochemical research, especially in the molecular studies on membranes. Once viewed as inert structures, membranes are now recognized as functionally dynamic; the biochemical understanding of their cellular functions in molecular terms is a key to the explanation of various biological phenomena, e.g., the mechanism of active transport systems and of receptor systems for external stimuli. The structural role of lipids has also been revised to include their specific functions in the biological operation of membranes. Thus, the structure-function theme of molecular biology, initiated with nucleic acids and proteins, has been extended to the protein-lipid structures of membranes.

Classically, lipids are identified as *saponifiable* (meaning those that are hydrolyzable by heat and alkali to yield salts of fatty acids and other molecular components) or *nonsaponifiable*. The former class includes the neutral fats of animals and plants and various types of lipids found in the membranes of biological systems. The latter class includes terpenes, steroids (Chapter 19) and fat-soluble vitamins (Chapter 25).

TRIACYLGLYCEROLS

Structural Components

Figure 12-1
General structure of a triacylglycerol. R_1, R_2, and R_3 are fatty acid residues. Numbering of carbons: 1 or α, 2 or β, 3 or α'.

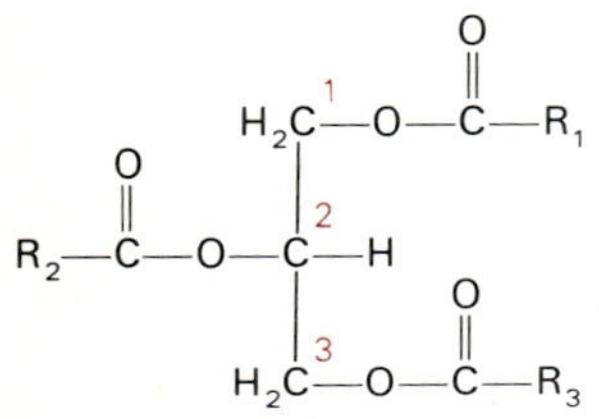

The general structure of a triacylglycerol (neutral fat), shown in Figure 12-1, contains three fatty acids bonded by ester linkages to a glycerol molecule (Figure 12-2). If fatty acids R_1 and R_3 are different, C-2 of the molecule is asymmetric and R_2 is written to the left of C-2 in a Fischer projection to designate the L-configuration of naturally occurring triacylglycerols. Carbon atoms are designated either by *stereospecific numbers (sn)* 1, 2, and 3 or by an older nomenclature, α, β, and α'.

A fatty acid (Table 12-1) is a long hydrocarbon chain with a terminal carboxyl group; over 70 naturally occurring fatty acids have been identified. The predominant triacylglycerols of higher plants and animals have fatty acids with an even number of carbon atoms, 14 to 22 carbons in length. *Saturated fatty acids* contain no carbon-to-carbon double bonds in their structures, whereas *unsaturated fatty acids* have one or more which, with rare exceptions, are in the *cis* geometrical configuration (p. 44). The most abundant unsaturated fatty acids have one or two double bonds (*monoenoic* and *dienoic* fatty acids, respectively); however, *olefinic* fatty acids with three *(trienoic)* and four *(tetraenoic)* also occur naturally (Table 12-1).

Unsaturated fatty acids have substantially lower melting points than saturated fatty acids. For example, the saturated 18-carbon (C_{18}) fatty acid (stearic acid) has a melting point of 70°C; a monoenoic form (oleic acid) melts at 13°C, and a dienoic form (lino-

Figure 12-2
Saponification of a triacylglycerol.

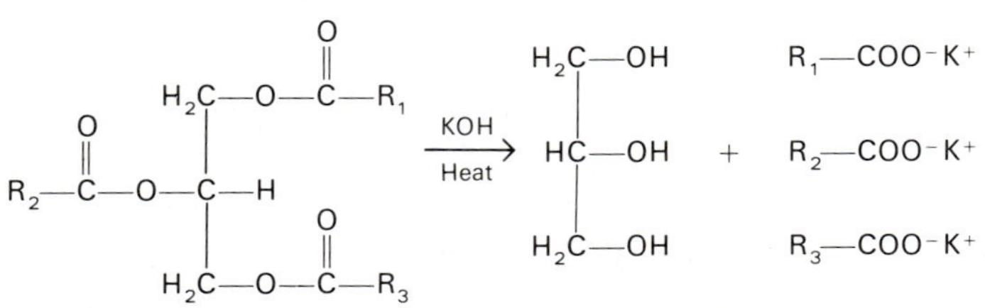

Table 12-1
Some naturally occurring fatty acids

Common name	No. of C atoms	Structure	Melting point (°C)
Saturated fatty acids			
Myristic	14	$H_3C(CH_2)_{12}COOH$	54
Palmitic	16	$H_3C(CH_2)_{14}COOH$	63
Stearic	18	$H_3C(CH_2)_{16}COOH$	70
Unsaturated fatty acids			
Palmitoleic	16	$H_3C(CH_2)_5CH\overset{cis}{=}CH(CH_2)_7COOH$	−0.5
Oleic	18	$H_3C(CH_2)_7CH\overset{cis}{=}CH(CH_2)_7COOH$	13
Linoleic	18	$H_3C(CH_2)_4CH\overset{cis}{=}CHCH_2CH\overset{cis}{=}CH(CH_2)_7COOH$	−5
Linolenic	18	$H_3CCH_2(CH\overset{cis}{=}CHCH_2)_2CH\overset{cis}{=}CH(CH_2)_7COOH$	−10
Arachidonic	20	$H_3C(CH_2)_4CH\overset{cis}{=}CHCH_2)_3CH\overset{cis}{=}CH(CH_2)_3COOH$	−50

lenic acid) at −5 °C (Table 12-1). Plant triacylglycerols (plant oils) are liquid at room temperature because they have a larger proportion of unsaturated fatty acids than animal triacylglycerols (e.g., lard), which are solid or semi-solid at the same temperature. This difference in the content of unsaturated fatty acids has received much attention in recent years because of the implication that excessive daily dietary intake of saturated fatty acids and cholesterol is associated with the occurrence of heart diseases. As a result, medical and nutritional advisors recommended a reduction of animal fats (and cholesterol) in the diet, with a higher proportion of lipid intake being triacylglycerols high in polyunsaturated fatty acids (*polyunsaturates,* i.e., fatty acids containing two or more double bonds). A lower intake of fat is also an effective means of reducing the caloric content of a diet because, on a weight basis, over twice as many calories (energy) are obtained from fats as from carbohydrates and proteins (p. 337).

Saponification

Treatment of a triacylglycerol with heat and alkali, called *saponification,* results in an irreversible hydrolysis of the molecule to produce glycerol and the alkali salts of three fatty acids (Figure 12-2). Triacylglycerols are also hydrolyzable by acid; however, hydrolysis by this procedure is reversible. Saponfication of triacylglycerols is the historical origin of the soap industry. *Soaps* are Na^+ and K^+

salts of fatty acids, e.g., sodium oleate and potassium stearate, and the cleansing action of soaps relies on their formation of micelles in aqueous solutions (p. 28). The other product of triacylglycerol saponfication, glycerol (glycerin), has long been of importance to the cosmetic industry as a major lubricative ingredient of skin creams, hand lotions, and soaps.

LIPIDS OF MEMBRANES

Lipids are major components of the membranes of living systems. Two types of saponifiable lipids in membranes have a phosphate group in their structures and are therefore referred to as *phospholipids.* One of these types has glycerol as the parent compound (the *phosphoglycerides*) and the other has sphingosine (the *sphingolipids*). Two other important lipid components of membranes are the carbohydrate-containing *glycolipids* and the steroid *cholesterol* (Figure 12-3), the latter being a nonsaponifiable lipid of eucaryotic origin (p. 378) found in the cellular membranes of animals.

Figure 12-3
Cholesterol.

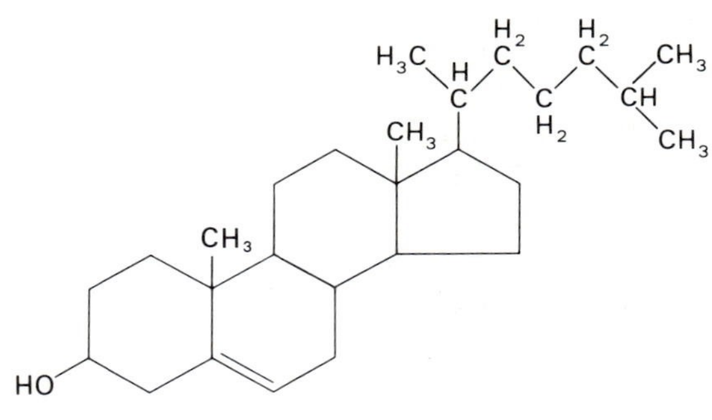

Phosphoglycerides

Phosphoglycerides occur almost exclusively in cellular membranes of plants and animals. A phosphorylated form of glycerol, sn-*glycerol 3-phosphate* (Figure 12-4), provides the carbon backbone, and the immediate precursor of phosphoglycerides is *phosphatidate* (Figure 12-4), which has two fatty acids esterified to the C-1 and C-2 atoms of the glycerol 3-phosphate molecule. Only very small amounts of phosphatidate are found in membranes. A *phosphoglyceride* is a phosphatidate molecule to which a polar alcohol

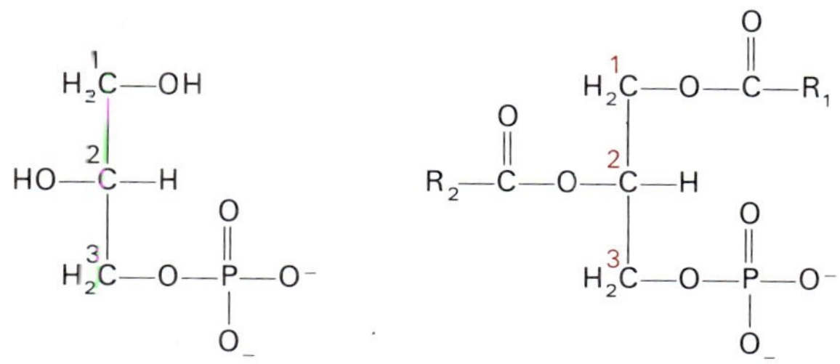

sn-Glycerol 3-phosphate   Phosphatidate

Figure 12-4
Phosphorylated glycerol precursors of phosphoglycerols. R_1 and R_2 are fatty acid residues.

(R—OH) is bonded by an ester linkage to the C-3 phosphate (Figure 12-5). There are many different individual phosphoglycerides, depending on the two fatty acids and alcohol contained in their structures. Table 12-2 lists some of the alcoholic components of phosphoglycerides; *phosphatidyl choline* (Figure 12-5) and *phosphatidyl ethanolamine* are the most abundant phosphoglycerides in higher plants and animals. As indicated in Figure 12-5, phosphoglycerides frequently have a saturated fatty acid at C-1 and an unsaturated fatty acid at C-2. A distinctive feature of the molecular configuration of phosphoglycerides is that all have a *polar head* and *two nonpolar tails;* they are, therefore, *amphipathic* (p. 28)—a critical property for their structural role in membranes. The charged polar heads make phosphoglycerides the most hydrophilic of lipids. *Phosphatidyl sugars,* having a polar carbohydrate instead of an alcohol, occur in microorganisms and plants.

Table 12-2
Alcohol components of some phosphoglycerides

Phosphoglyceride	Alcohol (at physiological pH)
Phosphatidyl choline	$HO-CH_2CH_2N^+(CH_3)_3$
Phosphatidyl ethanolamine	$HO-CH_2CH_2NH_3^+$
Phosphatidyl serine	$HO-CH_2CHCOO^-$ NH_3^+
Phosphatidyl inositol	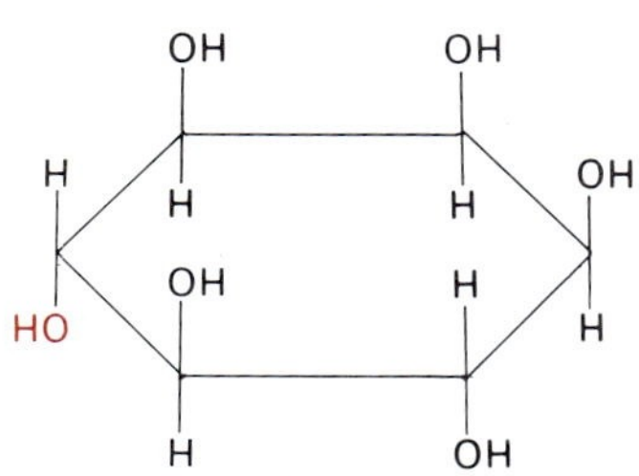

Red hydroxyl groups bond to a phosphatide.

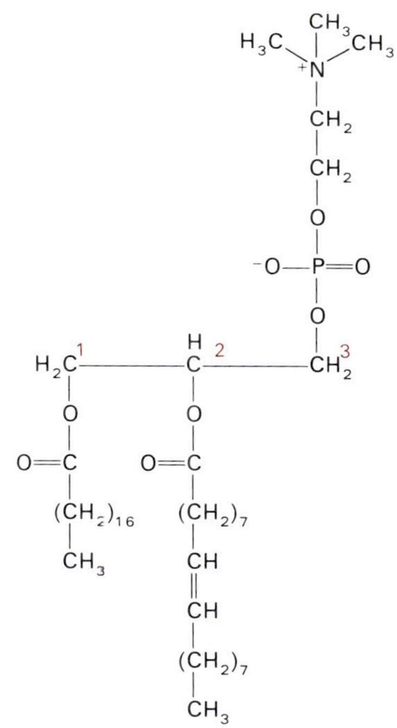

A phosphatidyl choline
(1-stearic-2-oleoyl-phosphatidyl choline)

Figure 12-5
A phosphoglyceride.

Sphingolipids

Sphingolipids are lipids that contain *sphingosine* (Figure 12-6), an amino alcohol with a long, unsaturated hydrocarbon chain instead of glycerol. Sphingolipids are found in plants and animals and are present in large amounts in nerve and brain tissue. *Sphingomyelin* (Figure 12-7) has a fatty acid linked by an amide bond to the amino group of sphingosine and has phosphoryl choline attached to the terminal hydroxyl group by ester linkage. Sphingomyelin is the most abundant sphingolipid and is the only sphingosine-derived phospholipid in membranes. Like phosphatidyl choline (Figure 12-5), sphingomyelin has a polar head and two nonpolar tails, one being the long unsaturated hydrocarbon portion of sphingosine.

Figure 12-6
Sphingosine.

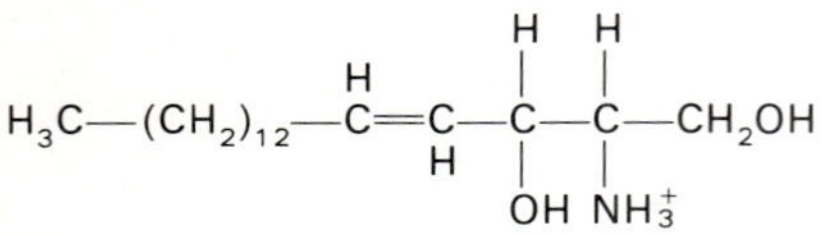

Figure 12-7
Sphingomyelin. The sphingosine moiety is shown in red, and R is a fatty acid residue.

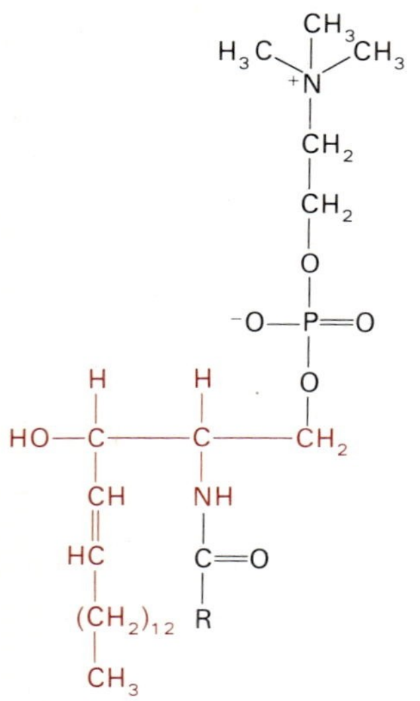

Glycolipids

Glycolipids are generally characterized as lipids that contain a polar carbohydrate head, such as D-glucose or D-galactose, but not a phosphate group. Glycolipids may be derived from glycerol or sphingosine and often are classified as glycerides (e.g., glycosyldiacyl glycerols) or as sphingolipids (e.g., cerebrosides). *Cerebrosides* are *simple glycolipids* containing a sugar, a fatty acid, and sphingosine (Figure 12-8); they are synthesized by the addition of a sugar to a *ceramide* (sphingosine with a fatty acid attached). Cerebrosides are found in the membranes of the brain and nervous system and are especially abundant in the myelin sheath. The 24-carbon fatty acid depicted in Figure 12-8 is *lignoceric acid,* one of the predominant 22-to-26-carbon fatty acids of cerebrosides. Glycolipids, whether derived from glycerol or sphingosine, have the characteristic polar head (carbohydrate) and two nonpolar tail con-

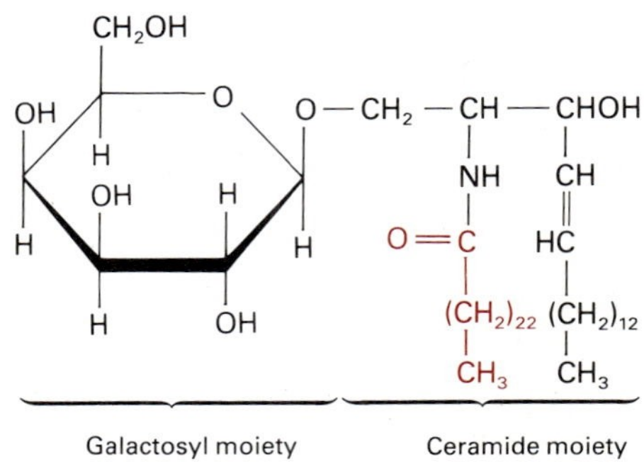

Figure 12-8
A galactocerebroside. The fatty acid residue, in red, is that of lignoceric acid.

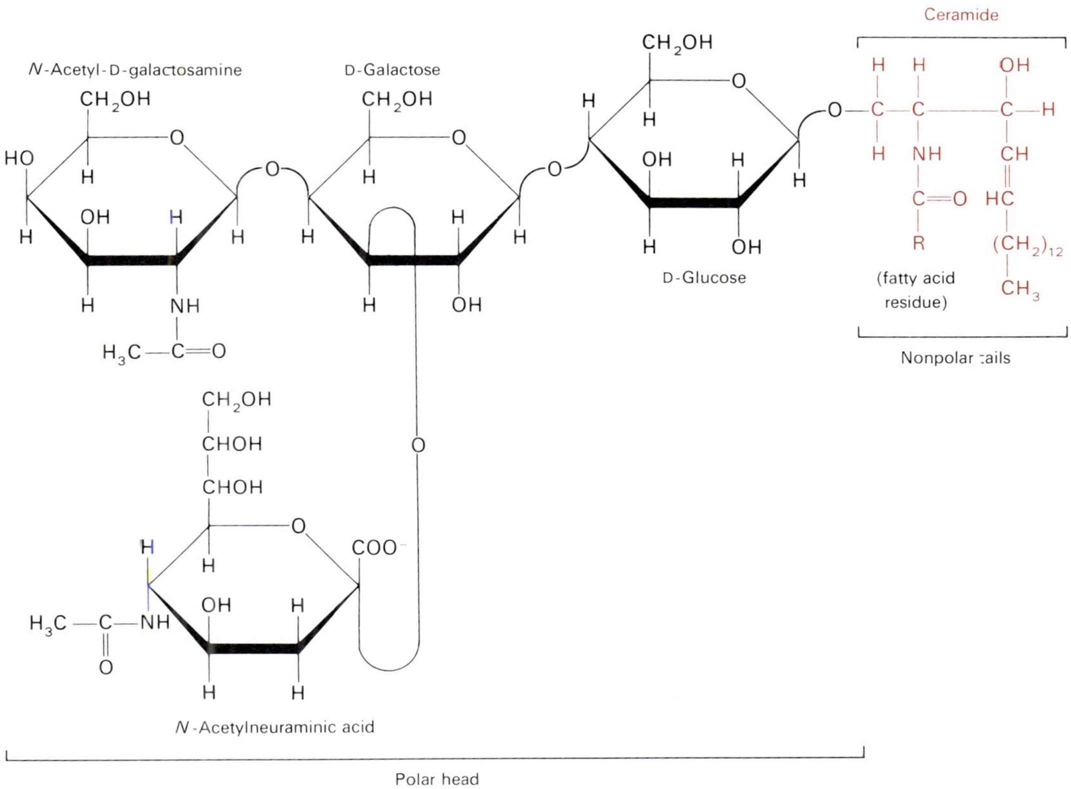

Figure 12-9
The ganglioside G_{M_2}.

stituents of most membrane lipids. *Gangliosides* are a more complex class of sphingosine-derived glycolipids because of the oligosaccharide nature of the polar head. The ganglioside G_{M_2} illustrated in Figure 12-9 has the two nonpolar tails of a ceramide and a polar head containing residues of D-glucose, D-galactose, and two carbohydrate derivatives, *N*-acetyl-D-galactosamine and *N*-acetylneuraminic acid (sialic acid). The outer surfaces of membranes of nerve cells are rich in a number of different molecular species of gangliosides.

ABNORMAL LIPID METABOLISM

A number of abnormalities of lipid metabolism, due to impaired catabolism of body lipids, result in the accumulation of lipids in viscera, brain, and blood vessels. The disorders, as a group, are called *lipid storage diseases* or *lipidoses* and are recognized as *inborn errors of metabolism,* i.e., inherited diseases resulting from defective or mutated genes which produce impaired enzymes. In terms of Mendelian genetics, known inborn errors are classified as *X-linked* (altered gene located on the X chromosome) or *autosomal* (located on a chromosome other than an X or Y chromosome). In the diploid state, the abnormality is considered *dominant* if only one altered copy (allele) of the gene is required *(heterozygous condition)* to manifest the disease syndrome and *recessive* if two copies of the defective gene are required *(homozygous condition).* Sickle-cell anemia (p. 119) is an example of an autosomal recessive trait, since two copies of the altered gene for hemoglobin β-chain production, located on an autosomal chromosome, are needed to produce the classical clinical symptoms of the anemia.

Tay-Sachs Disease (G_{M_2}-Gangliosidase-Deficiency Type I)

The oldest medically described lipid storage disease is Tay-Sachs disease, described in 1881. An infant with Tay-Sachs disease appears to be normal during the first few months after birth; however, failure to develop coordinated motor activity is readily apparent by the seventh or eighth month. After one year of age, mental and motor deterioration progresses rapidly and, by 18 months, the child's failing condition is characterized by dementia, blindness, and often deafness and convulsions. Death eventually comes between ages three and four. Biochemical and genetic research have established that Tay-Sachs disease is an autosomal recessive trait involving an inability to degrade the membrane ganglioside G_{M_2} (Figure 12-9). Afflicted infants have a deficiency in *hexosaminidase A activity,* and the normal catabolic removal of an *N*-acetyl-galactosamine residue from G_{M_2} to produce G_{M_3} does not occur (Figure 12-10). During the neonatal period, there is very active turnover of gangliosides during brain development, and a failure to degrade G_{M_2} results in the ganglioside's continued accumulation in the brain. The brain of a Tay-Sachs disease victim has 100 to 300 times the normal G_{M_2} content.

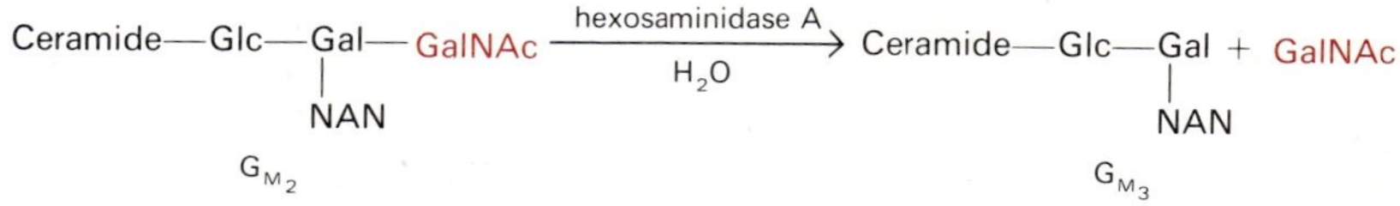

Figure 12-10
Biochemical block in Tay-Sachs disease. Glc = glucose; Gal = galactose; NAN = *N*-acetylneuraminic acid; GalNAc = *N*-acetylgalactosamine.

Fabry Disease (Angiokeratoma)

Fabry disease is an example of an X-linked recessive trait, since the altered gene is located on the X chromosome. A woman having a copy of the defective gene (heterozygous condition) is an unaffected *carrier* of the trait. There is a 50 per cent probability that each son she bears will suffer from the syndrome, since his one X chromosome (of his XY pair of sex chromosomes) is obtained from his mother, i.e., one of her two X chromosomes. Because of a deficiency in *ceramidetrihexosidase* (Figure 12-11), an enzyme normally produced by the kidneys, the disease is characterized by cellular accumulation of ceramidehexosides. Clinical symptoms include purplish skin nodules, attacks of abdominal pain, and burning sensation in hands and feet. Vascular lesions of a lipoidal nature in the kidneys cause progressive renal insufficiency, which is the usual cause of death of affected males between ages 30 and 50. Kidney transplants offer a medical means of providing the normal enzyme to patients suffering from Fabry disease.

Figure 12-11
Biochemical block in Fabry disease.

$$\underset{\text{Ceramide trihexose}}{\text{Ceramide-glucose-galactose-galactose}} \xrightarrow[H_2O]{\text{ceramidetrihexosidase}} \underset{\text{Ceramide lactoside}}{\text{ceramide-glucose-galactose}} + \text{galactose}$$

Gaucher Disease (Glucocerebrosidosis)

Gaucher disease is an autosomal recessive lipidosis brought about by an inability to degrade glucocerebrosides because of a deficiency of the catabolic enzyme *glucocerebrosidase,* normally produced in the spleen (Figure 12-12). Glucocerebroside accumulation results in hepatosplenomegaly (severely enlarged liver and spleen) and neurologic deterioration. The cerebroside may constitute up to 4 per cent of the splenic dry weight of a patient. Gaucher disease type II (infantile cerebral) results in the death of an infant patient before one year of age.

$$\underset{\text{Glucocerebroside}}{\text{Ceramide-glucose}} \xrightarrow[H_2O]{\text{glucocerebrosidase}} \text{ceramide} + \text{glucose}$$

Figure 12-12
Biochemical block in Gaucher disease.

Niemann-Pick Disease (Sphingomyelin Lipidosis)

The classic infantile type of Niemann-Pick disease (Type A) is also an autosomal recessive trait and is caused by the lack of normal degradation of sphingomyelin (Figure 12-7) into phosphocholine and a ceramide because of a lack of *sphingomyelinase* activity. Accumulation of sphingomyelin and cholesterol in tissues leads to hepatosplenomegaly, mental retardation, and neurologic deterioration. Niemann-Pick disease shares the tragic element of Tay-Sachs and Gaucher diseases in that the affected child seldom lives to age three.

Impaired Biosynthesis of Lipids

The above four medical syndromes are examples of inborn errors of metabolism concerning catabolic enzymes in lipid metabolism. Roscoe O. Brady and his colleagues, who have contributed greatly to the biochemical understanding of lipid storage diseases, have also reported on the medical consequences of an impairment of ganglioside biosynthesis *(anabolic gangliosidosis).* As expected, the abnormality resulted in severe neurological dysfunction. The infant patient, with defective ganglioside synthesis, began having convulsions and respiratory difficulties soon after birth and died at three and a half months. Autopsy revealed a severe lack of brain myelin and a spongy deterioration of the central nervous system. The higher gangliosides (G_{M_1}, $G_{D_{1a}}$, and G_{T_1} in Figure 12-13) were absent from the brain, and only the gangliosides G_{M_3} and G_{D_3} were present. Deficiency of the amino-sugar transferase that synthesizes G_{M_2} from G_{M_3} (and presumably G_{D_2} from G_{D_3}) was established. This particular anabolic gangliosidosis, expressed as an inability to

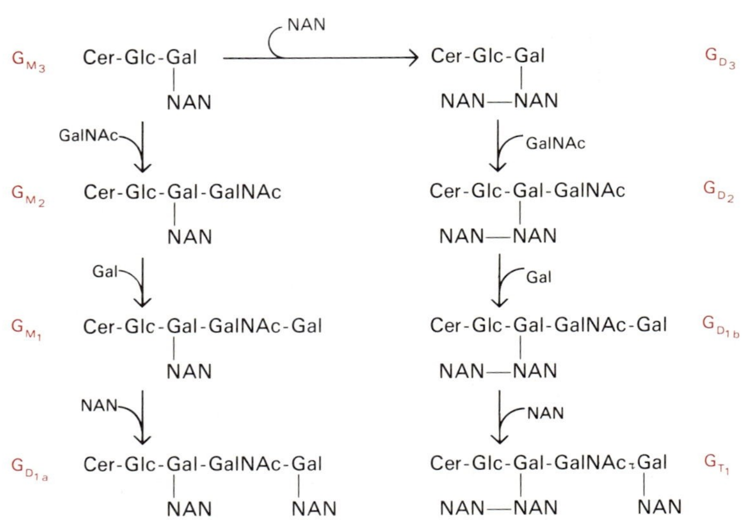

Figure 12-13
Simplifed scheme of the biosynthetic pathway for gangliosides. Cer = ceramide; Glc = glucose; Gal = galactose; NAN = *N*-acetylneuraminic acid; GalNAC = N-acetylgalactosamine. Ganglioside designations are in red.

add an *N*-acetylgalactosamine residue to G_{M3}, is biochemically the reverse of the metabolic block characterizing the catabolic gangliosidosis Tay-Sachs disease.

MEMBRANES

Membranes are highly organized molecular assemblies that constitute a water-insoluble barrier between a cell and its environment and also between the cytosol and the internal compartments of eucaryotes. Membranes are essential components of all living systems, and, as monitors of the external environment of cells, they are required to perform numerous crucial functions. One critical cellular role of membranes is to act as a selective permeability barrier and regulate the inward and outward flow of compounds and ions. Membranes are also responsible for a cell's response to environmental information, e.g., chemotactic movement of bacteria toward a food supply, and for receiving physiological instructions, e.g., hormone receptor sites on membranes of target organs. Although each type of membrane has its distinctive structure and functions, all share a number of general features. Membranes are *sheet-like structures* formed by *cooperative noncovalent interactions* between *lipids* and *proteins,* their major molecular constituents. Only a few molecules thick (usually 60 to 100 Å), membranes rely on their lipid components to be effective barriers to polar molecules and ions. Their structural proteins and many of their lipids participate in specific biological functions, e.g., pumps, receptors, and enzymes. Membranes are also *asymmetrical,* meaning that the molecular composition of the outer and inner surfaces are different.

Bilayer Lipid Membranes

In an aqueous environment, the amphipathic nature of membrane lipids, mainly phospholipids and glycolipids, dictates specific molecular arrangements to accommodate the intramolecular hydrophilic and hydrophobic properties of the biomolecules. One possible arrangement is the micelle (p. 28), characteristically formed in aqueous media by fatty acids; however, lipids with a polar head and two nonpolar tails characteristically assemble in bilayer formations (Figure 12-14). In a bilayer aggregation (membrane), the polar heads of the lipids constitute the two outer surfaces exposed to aqueous environments and the nonpolar tails, by

Figure 12-14
Schematic of a portion of a bilayer membrane formed by phospholipids.

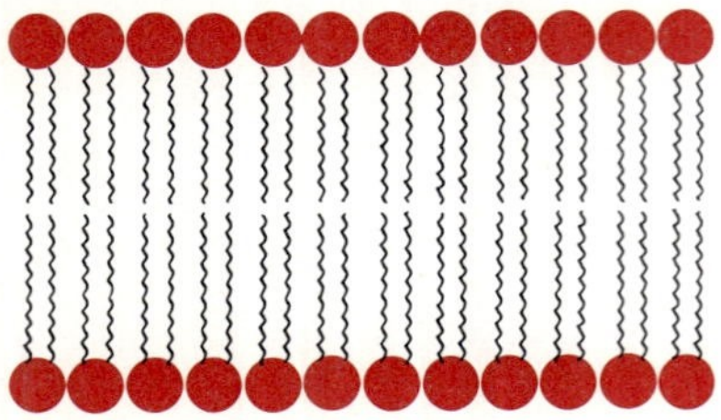

hydrophobic interaction, create an internal nonpolar environment that is impermeable to soluble components of the aqueous environment. The continuous sheets formed by bilayer aggregations (unlike micelles, which have diameters of 200 Å or less) are of sufficient dimensions to serve as membranes to enclose cells or organelles. Two other structural advantages of bilayer arrangements of lipids, based on the maintenance of favorable energy states, are that they form *compartments,* that is, they avoid having open ends with exposed nonpolar tails, and they are *self-sealing,* i.e., they have no holes to expose the hydrophobic interior to the aqueous environment.

The high content of unsaturated fatty acids in membrane lipids is a biologically important structural feature. Saturated fatty acids, which can exist in a great variety of conformations, usually assume the extended form (Figure 12-15), which requires the least amount of energy to maintain. Unsaturated fatty acids, on the other hand, have a bend of about 30° in the hydrocarbon chain (Figure 12-15), imposed on the molecule by the *cis* configuration of the double bond. Two or more double bonds in *cis* configuration create more bends and significantly shorten the length of the molecule (Figure 12-15). These conformational features of unsaturated fatty acids are critical for *membrane fluidity.* In bilayer formations, saturated fatty acids, because of their extended form and greater length, pack tightly to produce highly ordered, rigid structures, whereas the packing of unsaturated fatty acids, with their bends and shorter length, creates a relatively disordered structure, referred to as a *fluid bilayer membrane* (Figure 12-16).

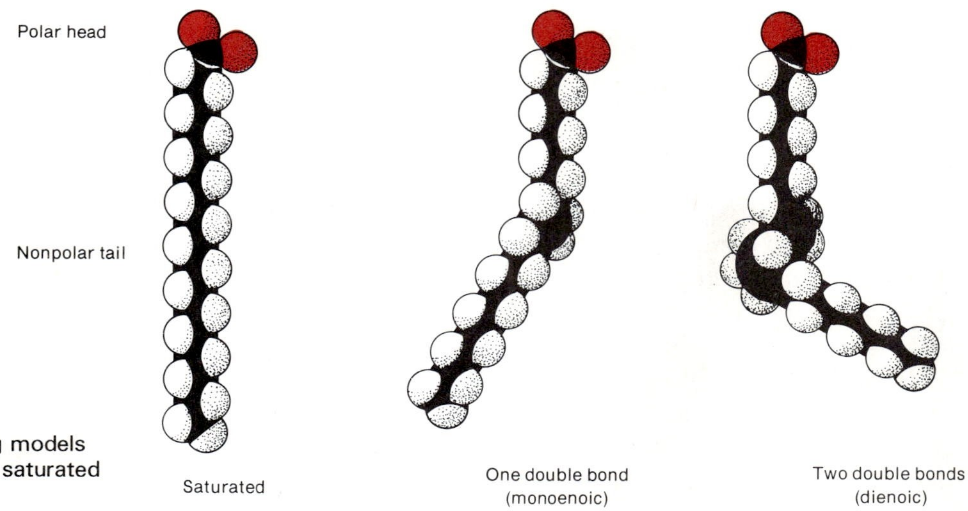

Figure 12-15
Illustration of three space-filling models depicting the conformations of saturated and unsaturated fatty acids.

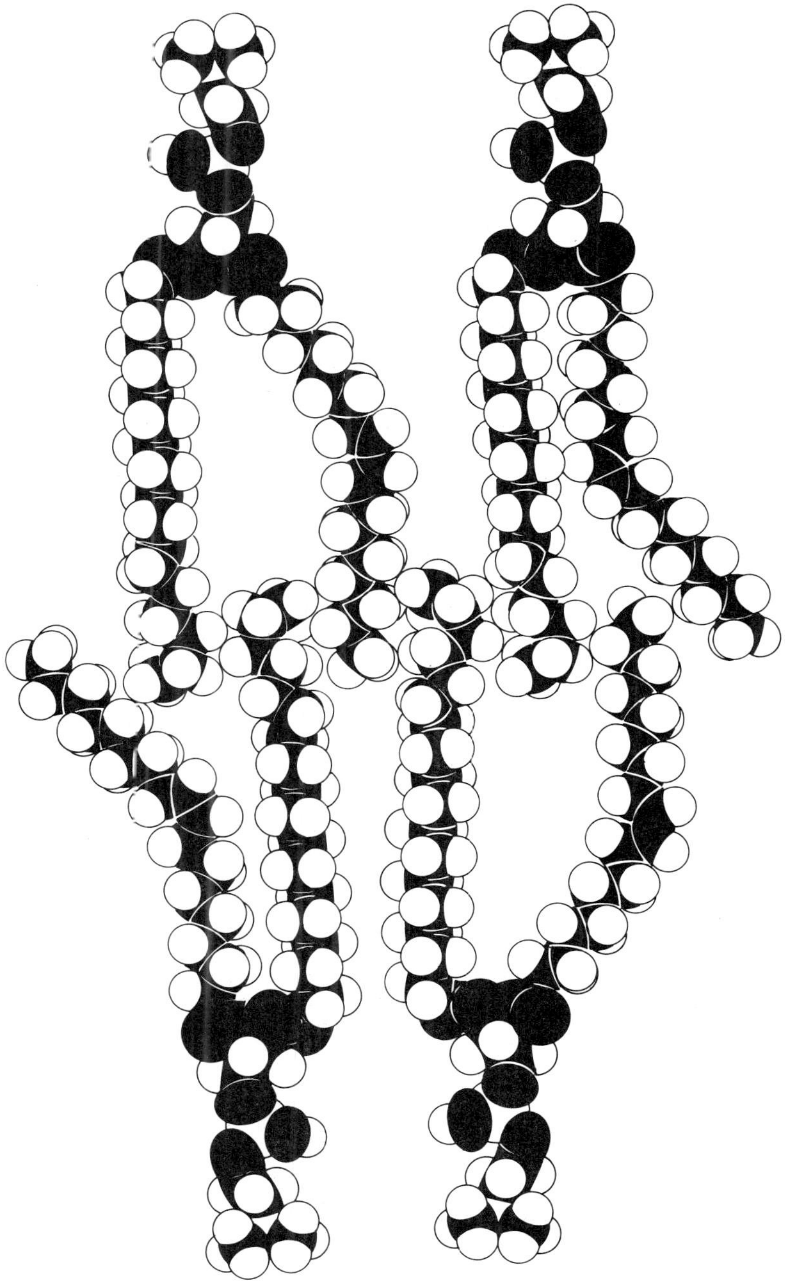

Figure 12-16
Schematic depicting a section of fluid bilayer membrane, created by the packing of membrane lipids containing saturated and unsaturated fatty acids.

Bimolecular lipid membranes are the result of a spontaneous *self-assembly process* that utilizes a number of different noncovalent interactions to attain and retain a stable structure. In addition to the dominant hydrophobic interactions, occurring between the hydrocarbon tails of the lipids, van der Waals attractive forces also favor close packing between hydrocarbon chains. The polar heads of lipids also participate in hydrogen bonding and electrostatic interactions with the aqueous environment. Hence, the same types of molecular interactions that determine protein structure are also critical for the formation and stabilization of bilayer lipid membranes.

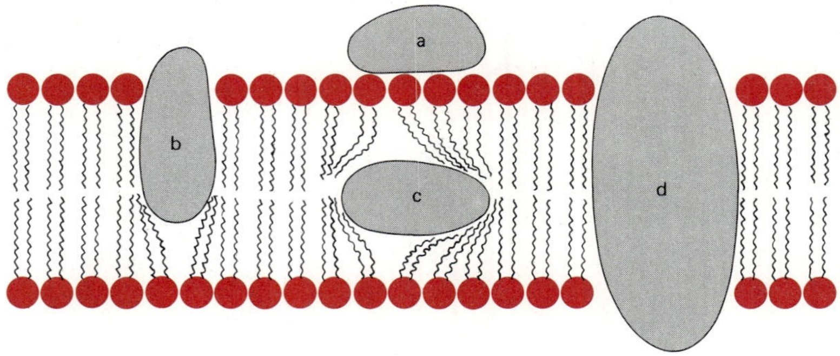

Figure 12-17
Schematic of the various types of proteins associated with membranes. (a) A peripheral membrane protein, (b) an integral membrane protein projecting into the membrane, (c) an internal integral membrane protein, (d) an integral membrane protein transversing the membrane.

Proteins of Membranes

On a weight basis, proteins account for about 20 to 80 per cent of membrane composition. For many years, membrane structure was believed to consist of a bimolecular layer of lipids sandwiched between two layers of surface proteins. It is now known that the protein components of membranes are not limited to this type of static structural arrangement but, rather, occur in diverse arrangements in membranes. Membrane proteins are generally classified as *peripheral* or *integral* (Figure 12-17), as judged by their dissociation properties. Peripheral proteins are those that easily dissociate from membranes, e.g., by salt extraction; they are believed to be surface proteins bound by electrostatic interaction or hydrogen bonding. Conversely, integral proteins are difficult to separate from membranes, often requiring treatment with a detergent to disrupt hydrophobic interactions of the lipid bilayer. Such separation treatment implies that integral proteins participate in nonpolar interactions with membrane lipids. As membrane components, integral proteins are thought to occupy internal positions, project into the bimolecular lipid layer, or transverse the membrane (Figure 12-17).

Some proteins of membranes are *glycoproteins,* which, like glycolipids, contain one or more carbohydrate residues. In the case of glycoproteins, the sugar residues are linked to the proteins by the bonding of a carbohydrate, usually *N*-acetylglucosamine or *N*-acetylgalactosamine, to a seryl, threonyl, or asparaginyl residue (Figure 12-18). (The synthesis of glycoproteins containing oligosaccharide units attached to asparaginyl residues is discussed in Chapter 23.) Research on membranes provides evidence that carbohydrate moieties of both glycoproteins and glycolipids are always located on the external surface of a membrane (Figure 12-19). It is believed that the carbohydrates aid in orienting the glycoprotein into its appropriate position in the membrane. These surface carbohydrates also have critical roles in the intra- and extracellular transport of proteins (p. 524) and in the process of *intercellular* recognition, which, for example, is required for tissue formation.

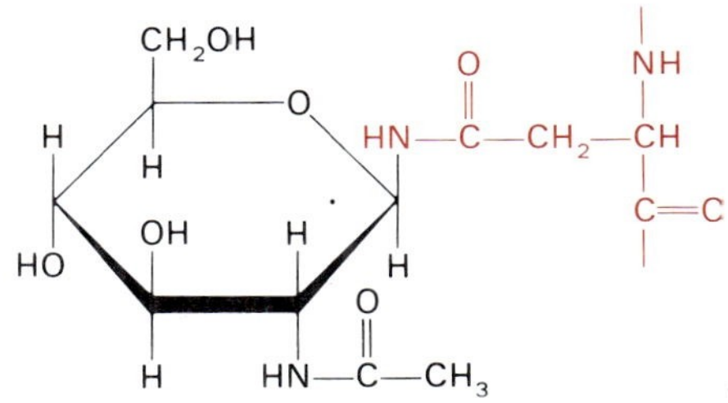

Figure 12-18
One type of carbohydrate-protein binding. *N*-Acetylglucosamine is linked to the amide group of an asparaginyl residue (in red).

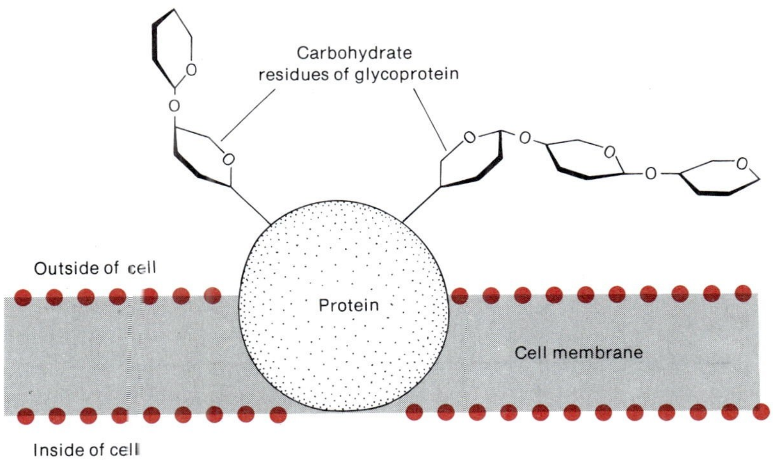

Figure 12-19
Schematic of the location of carbohydrate residues of glycoproteins (and glycolipids) on the outer surface of membranes.

Fluid Mosaic Model of Membranes

The current view of membranes as dynamic structures is embodied in the *fluid mosaic model of membranes,* proposed by S. Jonathan Singer and Garth Nicholson in 1972 (Figure 12-20). In the model, the bimolecular lipid layer is viewed both as a *permeability barrier* and, because of its fluid nature (established by unsaturated fatty acids), as a *solvent* for integral proteins. Specific lipid-protein inter-

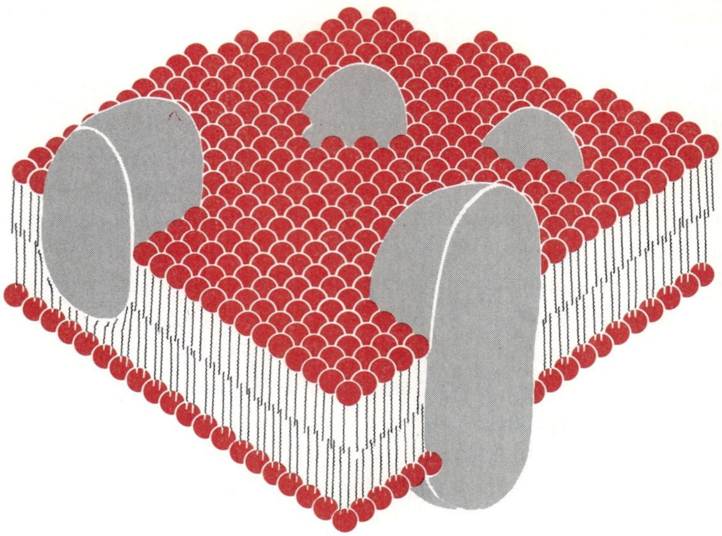

Figure 12-20
Fluid mosaic model of a membrane. (After S. J. Singer and G. L. Nicholson, *Science* 175:723, 1972. Copyright 1972 by the American Association for the Advancement of Science)

actions are also proposed to be necessary for the biological functioning of certain membrane proteins, i.e., lipid functions involve more than being a solvent. The protein components of membranes are depicted as being arranged in various mosaic patterns and, unless restricted by specific interactions, free to move laterally within the lipid solvent. This latter dynamic aspect of membrane proteins also extends to membrane lipids. Such *lateral diffusion,* i.e., movement along the plane of the bilayer with which a protein or lipid is associated, has been calculated to occur at rapid rates for proteins (several microns per minute) and even faster rates for lipids (several microns per second). *Transverse diffusion,* movement from one surface of a membrane to the other, can occur only at a very slow rate, e.g., several hours for a phospholipid to transverse a membrane. It is doubtful that all proteins, e.g., glycoproteins with hydrophilic carbohydrate components, are capable of transverse diffusion. The molecular features of the Singer-Nicholson model emphasize the fact that the old concept of membranes as biologically inert structures has given way to an appreciation of these macromolecular assemblies as complicated, versatile structures, each molecularly programmed to perform many specific biological functions.

SUMMARY

Lipids are biomolecules that are insoluble or sparingly soluble in water. They are generally characterized as saponifiable, e.g., neutral fats, or as nonsaponifiable, e.g., steroids; the former class is hydrolyzable by heat and alkali (saponification) and the latter class is not. Neutral fats of animals and plants contain three fatty acids esterified to glycerol and are called triacylglycerols. Fatty acids are long hydrocarbon chains with a terminal carboxyl group; they are classified as saturated or unsaturated, with the latter having one or more double bonds in their hydrocarbon chains and the former having none. Triacylglycerols of plants have a higher content of fatty acids with two or more double bonds ("polyunsaturates") than do those of animals. Saponification of triacylglycerols produces soap (metal salts of fatty acids) and glycerol, which is used in skin creams and lotions.

Membranes always have a significant percentage of lipid in their structure. Two major types of saponifiable lipids in membranes, phosphoglycerides and sphingolipids, are called phospholipids because both classes contain a phosphate group in their structures. Alkaline hydrolysis of a phosphoglyceride yields glycerol, two fatty acids, an alcohol (e.g., choline), and a phosphate. Saponification of sphingomyelin produces sphingosine, a fatty acid, an alcohol, and a phosphate. Sphingomyelin is the only sphingosine-derived phospholipid found in membranes. Membrane lipids also include glycolipids, which can be derivatives of either glycerol or sphingosine. Glycolipids contain a carbohydrate instead of a phosphate and an alcohol. A sphingosine-derived glycolipid is called a cerebroside. An important biological feature of glycolipids, phosphoglycerides, and sphingomyelin is that each type of lipid has a polar head (an ionized phosphoryl alcohol or a carbohydrate) and two nonpolar tails (two fatty acids or one fatty acid and the long hydrocarbon tail of sphingosine). Cholesterol is a nonsaponifiable lipid found in the membranes of animals.

Gangliosides of the membranes of brain and nervous system are cerebrosides with additional carbohydrate moieties in their polar head. Inborn errors of metabolism involving the catabolism of cerebrosides and gangliosides produce very severe clinical manifestations and frequently result in the early death of an afflicted individual.

Membranes are sheet-like structures which are formed by cooperative noncovalent interactions between lipids and proteins. The lipids of membranes constitute a water-insoluble bimolecular aggregation, with their polar heads exposed to aqueous environments and their nonpolar tails creating an internal nonpolar environment. The proteins associated with a membrane are in a mosaic pattern, with some being located on the outside (peripheral proteins) and others either located within or projecting into the lipid layer or transversing the membrane (integral proteins). The fluid mosaic model of membranes views a membrane as a dynamic structure with its permeability barrier (bimolecular layer of lipid) also serving as a solvent for integral proteins and lipids. Lipids and integral proteins in the bimolecular layer are free to move laterally within their lipid solvent. The carbohydrate moieties of glycolipids and glycoproteins, located on the outside of a membrane, are believed to be important in assuring the proper orientation of proteins in a membrane's structure; they also have a critical role in the phenomena of cell-to-cell recognition and interaction.

QUESTIONS

1. Which of the molecular components in the following list are present in the structures of both a triacylglycerol and a phosphatidyl choline?
   (a) Only one fatty acid
   (b) At least two fatty acids
   (c) Three fatty acids
   (d) Glycerol
   (e) Sphingosine
   (f) Phosphate
   (g) Alcohol
   (h) Sugar

2. Which components from the list in Question 1 do a galactocerebroside and sphingomyelin have in common?

3. Which components from the list in Question 1 do a phosphatidyl ethanolamine and sphingomyelin have in common?

4. Which of the following would *not* be found in a ceramide?
   (a) A fatty acid
   (b) Alcohol
   (c) Carbohydrate
   (d) Glycerol
   (e) Phosphate
   (f) Sphingosine

5. With respect to ganglioside G_{M_2} depicted in Figure 12-9
   (a) what is the glycosidic linkage between *N*-acetylgalactosamine and galactose?
   (b) what is the glycosidic linkage between galactose and glucose?
   (c) which carbon atoms in *N*-acetylneuraminic acid and galactose are linked to each other?
   (d) Given the information that *N*-acetylneuraminic acid (NAN) is *N*-acetylated mannosamine (NH_2 at C-2) linked to pyruvate ($CH_3-\overset{\overset{\displaystyle O}{\|}}{C}-COO^-$), which carbon atoms in NAN are derived from mannose and which from pyruvate?

6. Which of the following lipids are found in animal membranes?
   (a) Cholesterol
   (b) Glycolipids
   (c) Phosphoglycerides
   (d) Sphingolipids
   (e) Triacylglycerols

7. Which of the following descriptions are generally true of membranes?
   (a) Consist of a bimolecular layer of lipid sandwiched between two layers of proteins
   (b) Are asymmetrical
   (c) Are sheet-like structures
   (d) Are formed by covalent bonding between lipids and proteins
   (e) Contain glycolipids and glycoproteins
   (f) Contain a high content of saturated fatty acids in their lipids

8. Which of the following statements are consistent with the fluid mosaic model of a membrane?
   (a) A bimolecular lipid layer is viewed as a solvent for peripheral proteins.
   (b) Integral proteins, but not lipids, exhibit lateral diffusion.
   (c) All membrane proteins can exhibit transverse diffusion.
   (d) Enzymes are associated with it.
   (e) The term *mosaic* specifically refers to the arrangement of the lipids in a membrane.
   (f) Carbohydrate moieties are associated with the outside of a membrane.

SUGGESTED READINGS

Ansell, G. B., J. N. Hawthorne, and R. M. C. Dawson, *Form and Function of Phospholipids,* 2nd ed. New York: American Elsevier, 1973. A thorough treatment of phospholipid biochemistry.

Bittar, E. E., ed., *Membrane Structure and Function,* 3 vols. New York: Wiley-Interscience, 1980. A comprehensive coverage of membranes, including their biogenesis, chemical characteristics, permeability, and transport mechanisms.

Bondy, P. K., and L. E. Rosenberg, eds., *Metabolic Control and Disease,* 8th ed. Philadelphia: W. B. Saunders, 1980. Chapter 9, by Roscoe Brady, on spingolipodystrophies is highly recommended.

Chapman, D., *Introduction to Lipids.* London: McGraw-Hill, 1969. A general introduction to the chemistry of lipids, including nomenclature, structure, and chemistry of various lipid groups.

Fasman, G. D., ed., *Handbook of Biochemistry and Molecular Biology,* 3rd ed. Sec. C, *Lipids, Carbohydrates and Steroids,* Vol. 1. Boca Raton, Fla: CRC Press, 1976. Good reference source for physical and chemical data on fatty acids and triglycerides.

Gunstone, F. D., *An Introduction to the Chemistry and Biochemistry of Fatty Acids and Their Glycerides.* New York: Halstead Press, 1975. A general introduction to fatty acids and their derivatives, including a discussion of their synthesis, metabolism, structure, and basic properties.

Gurr, A. I., and A. T. James, *Lipid Biochemistry: An Introduction,* 2nd ed. New York: Halstead Press, 1976. A good introductory text.

McKusick, V. A., and R. Claiborne, eds., *Medical Genetics.* New York: HP Publishing, 1973. Chapter 25 on Tay-Sachs disease offers a hopeful message about the accurate screening for adult carriers and the prevention of afflicted infant births.

Tanford, C., *The Hydrophobic Effect: Formation of Micelles and Biological Membranes,* 2nd ed. New York: Wiley-Interscience, 1980. Most comprehensive and current account of an important subject.

Articles

Ashwell, G., and A. G. Morell, "Membrane Glycoproteins and Recognition Phenomena." *TIBS,* 2:76, 1977.

Capaldi, R. A., "A Dynamic Model of Cell Membranes." *Scientific American,* 230(3):26 (offprint #1292), 1974.

Fox, C. F., "The Structure of Cell Membranes." *Scientific American* 226(2):30 (offprint #1241), 1972.

Friedmann, T., "Prenatal Diagnosis of Genetic Disease." *Scientific American,* 225(5):34 (offprint #1234), 1971.

Lodish, H. F., and J. E. Rothman, "The Assembly of Cell Membranes." *Scientific American,* 240(1):38 (offprint #1415), 1979.

Op den Kamp, J.A.F., "Lipid Asymmetry in Membranes." *Ann. Rev. Biochem.,* 48:47, 1979.

Sharon, N., "Glycoproteins." *Scientific American,* 230 (5):78, 1974.

Singer, S. J., and G. L. Nicholson, "The Fluid Mosaic Model of the Structure of Membranes." *Science,* 175:720, 1972.

Wickner, W., "The Assembly of Proteins into Biological Membranes: The Membrane Trigger Hypothesis." *Ann. Rev. Biochem.,* 48:23, 1979.

Yamakawa, T., and Y. Nakai, "Glycolipids at the Cell Surface and Their Biological Functions." *TIBS,* 3:128, 1978.

Zwaal, R. F. A., R. A. Demel, B. Roelofsen, and L. L. M. Van Deene, "The Lipid Bilayer Concept of Cell Membranes." *TIBS,* 1:112, 1976.

NUCLEIC ACIDS

13

INTRODUCTION

In 1869 Friedrick Miescher, a young Swiss studying with the eminent Hoppe-Seyler in Germany, isolated nuclei from pus cells and found that they contained a hitherto unknown phosphate-rich substance which he named *nuclein.* He continued his studies on salmon sperm, a prime source of nuclein, and in 1871 prophetically wrote:

> It seems to me that a whole family of such phosphorus-containing substances, differing somewhat from each other, will emerge, as a group of nuclein substances, which perhaps will deserve equal consideration with the proteins.

When nuclein was established to be acidic in nature, its name was changed to *nucleic acid.* Research on these biomolecules in the first decades of this century revealed that they, like proteins, are polymers. The monomeric unit of a nucleic acid is called a *nucleotide;* thus, nucleic acids are also referred to as *polynucleotides.* A nucleotide can be hydrolyzed to a nitrogenous base, a five-carbon sugar, and phosphoric acid. Additionally, it was discovered that there are two types of nucleic acid: *ribonucleic acid* (RNA), for many years called pentose nucleic acid (PNA), and *deoxyribonucleic acid* (DNA). RNA differs chemically from DNA in that it has ribose as its pentose and DNA has a 2-deoxyribose. RNA also has the nitrogenous base uracil, whereas DNA has thymine, a methylated derivative of uracil. This latter distinction is not absolute as once thought, since it is now known that certain RNAs also contain thymine. Continued studies on nucleic acids revealed that the nucleotide units are linked to one another by phosphodiester bonds

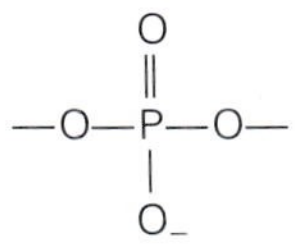

Phosphodiester linkage

forming macromolecular structures which, in the case of DNA, can have molecular weights of billions. Until the 1940s, RNA and DNA were generally believed to be of plant and animal origin, respectively; however, both types of nucleic acids are present in all plants and animals. Viruses also contain nucleic acids; however, unlike a plant or animal, a virus has either RNA or DNA, but not both.

Although the chemistry of nucleic acids was seriously studied after their discovery, 75 years passed before the biological significance of these macromolecules was realized. The suggestion made by Avery and his colleagues, in 1944 (p. 11), that DNA is genetic material was the first specific biological role proposed for a nucleic acid. The announcement, although belatedly appreciated, laid the cornerstone for a new and productive era of biological research on nucleic acids. As for RNA, it was not until 1957 that a specific cellular function for this nucleic acid (RNA involvement in protein synthesis) was established. (It should be noted, however, that RNA had been identified earlier as the genetic material of some viruses.) The emergence of molecular biology (p. 9) emphasized the biological eminence of both DNA (the gene) and RNA, whose different cellular species have prominent roles in protein synthesis (gene expression). Understanding the chemistry and the biochemical functioning of nucleic acids has been one of the most meaningful achievements of twentieth century science.

LEVELS OF NUCLEIC ACID STRUCTURE

Nitrogenous Bases

Nucleic acids contain two classes of nitrogenous bases, *purines* and *pyrimidines.* The two purines of RNA and DNA are *adenine* and *guanine,* which chemically are amino- and oxy-substituted purine molecules (Figure 13-1). With respect to pyrimidines (Figure 13-2), *cytosine* is common to both RNA and DNA, whereas *uracil* occurs in RNA, with *thymine* being its counterpart in DNA. The

Figure 13-1
Purines.

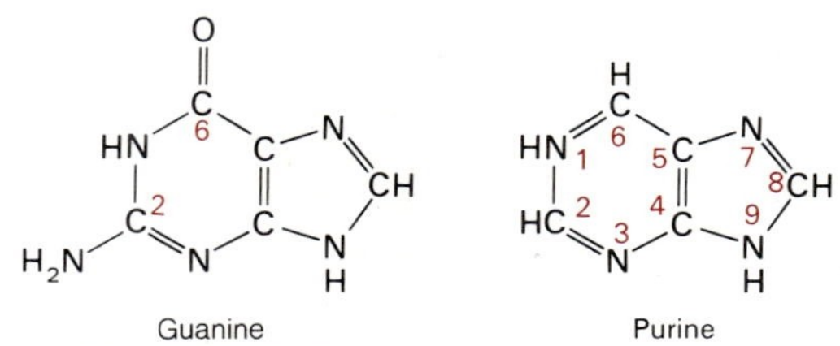

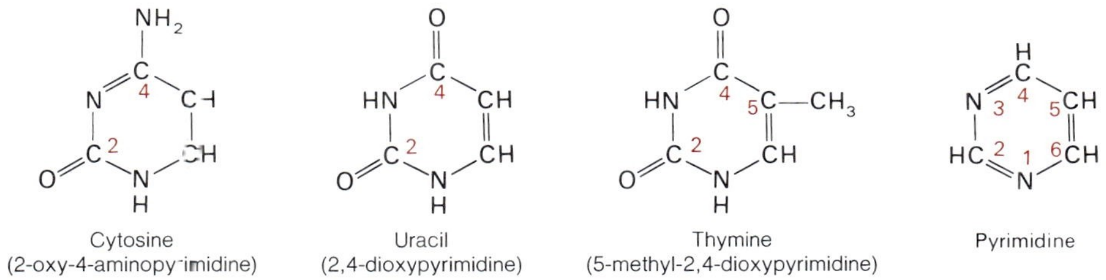

Figure 13-2
Pyrimidines.

three bases are substituted pyrimidine molecules (Figure 13-2). Because of the tautomerism exhibited by purines and pyrimidines, a pH-dependent equilibrium exists between the keto (lactam) and enol (lactim) forms of the bases. The two tautomeric structures for cytosine are depicted in Figure 13-3. At physiological pH, the keto forms of the bases predominate (Figures 13-1 and 13-2).

Until recently the above five nitrogenous bases were believed to account for the total base composition of animal and plant nucleic acids. It is now known that other bases, called *modified* or *minor bases,* also occur in polynucleotide structures. For example, transfer RNAs (tRNAs) are a class of nucleic acids which contain a significant percentage of modified bases in their structure (Figure 13-21). Figure 13-4 depicts some of the modified purines which have been identified in nucleic acids. Methylation is the most common form of purine modification, and methylated purines occur in varying amounts in both RNA and DNA. It is now believed that particular methylation patterns of purines in DNA may be a characteristic

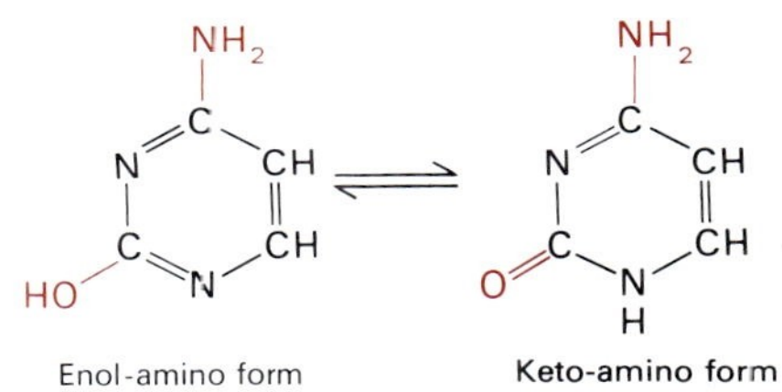

Figure 13-3
Two tautomeric forms of cytosine.

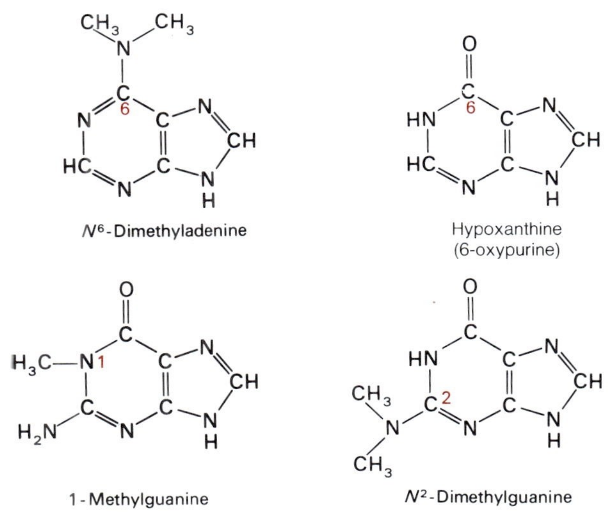

Figure 13-4
Some modified purines.

Figure 13-5
Some modified pyrimidines.

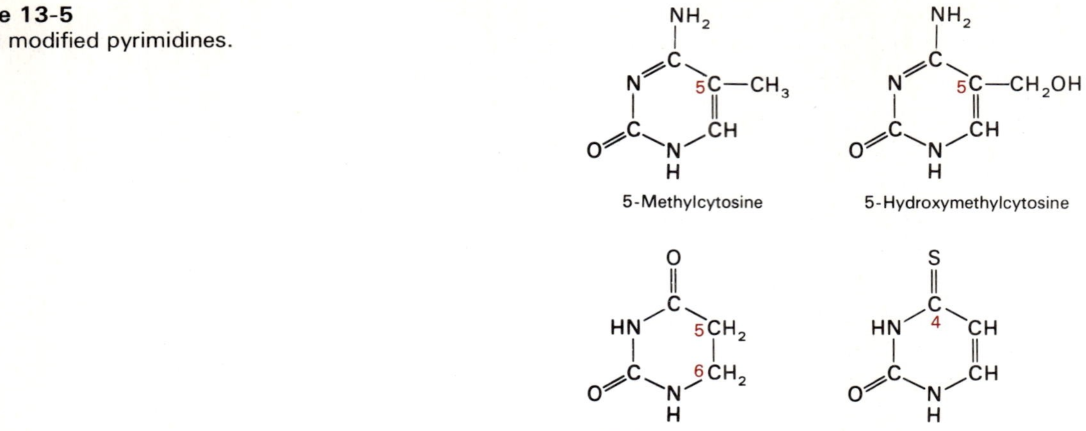

feature of the genetic material of many bacteria (Chapter 21). Some naturally occurring forms of modified pyrimidines are shown in Figure 13-5. 5-Methyl- and 5-hydroxymethylcytosine have been detected in various plant and animal DNAs, respectively. The DNA of T-even bacteriophages of *Escherichia coli* has no cytosine but instead has 5-hydroxymethylcytosine and its glucoside derivatives.

Nucleosides

Nucleosides are compounds that have a purine or pyrimidine covalently bonded to D-ribofuranose *(ribonucleosides)* or to 2-deoxy-D-ribofuranose *(deoxyribonucleosides)* in an *N-β*-glycosidic linkage (Figure 13-6). Bonding involves the hemiacetal group of C-1′ of the pentose and the N-9 nitrogen atom of a purine or N-1 of a

Figure 13-6
A ribo- and a deoxyribonucleoside.

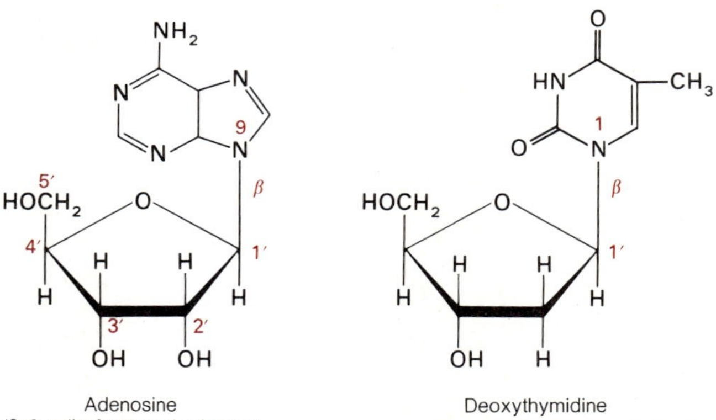

pyrimidine. In chemical nomenclature, the carbon atoms of a sugar in a nucleoside are identified by primed numbers to distinguish them from the atoms of the nitrogenous base. One nucleoside found in tRNAs that has a different linkage is *pseudouridine* (Figure 13-7), which has the C-1′ of ribose attached directly to the C-5 of uracil. The uracil moiety at physiological pH has one oxy group in the keto form and the other in the enol form.

Table 13-1 lists the common names of the major nucleosides of RNA and DNA. Thymidine is the historical name of the deoxyribonucleoside of thymine, long thought to be the only naturally occurring nucleoside of that base. Now that thymine has been identified in tRNAs, ribothymidine and deoxythymidine are the preferred and least confusing names.

Figure 13-7
Pseudouridine. Note the absence of a true nucleosidic bond.

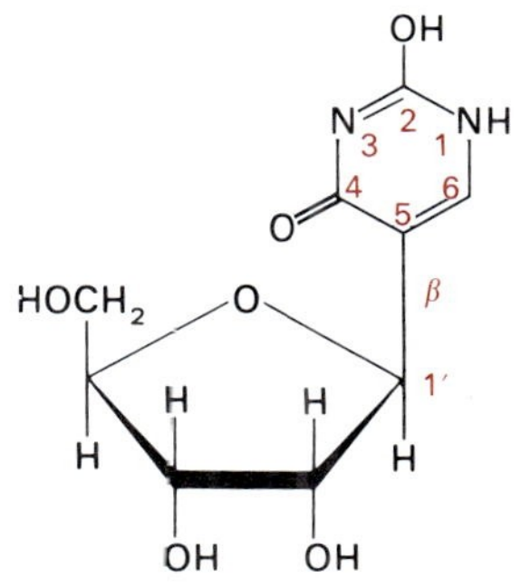

Table 13-1
Trivial names of nucleosides

Base	Ribonucleoside	Deoxyribonucleoside
Adenine	Adenosine	Deoxyadenosine
Guanine	Guanosine	Deoxyguanosine
Uracil	Uridine	Deoxyuridine
Cytosine	Cytidine	Deoxycytidine
Thymine	Ribothymidine or thymine ribonucleoside	Deoxythymidine or thymidine

Nucleotides

Nucleotides are phosphate (phosphoric acid) esters of nucleosides. There are several classes of nucleotides since the phosphate ester can be at the 2′-, 3′-, or 5′-carbon of a ribonucleotide or at the 3′ - or 5′-carbon of a deoxyribonucleotide. Naturally occurring nucleotides are commonly 5′-monophosphates (Figure 13-8). Table 13-2 lists the two trivial names and abbreviations used for each of the 5′-nucleotides of the five major bases of RNA and DNA; their classification as acids is an alternative nomenclature.

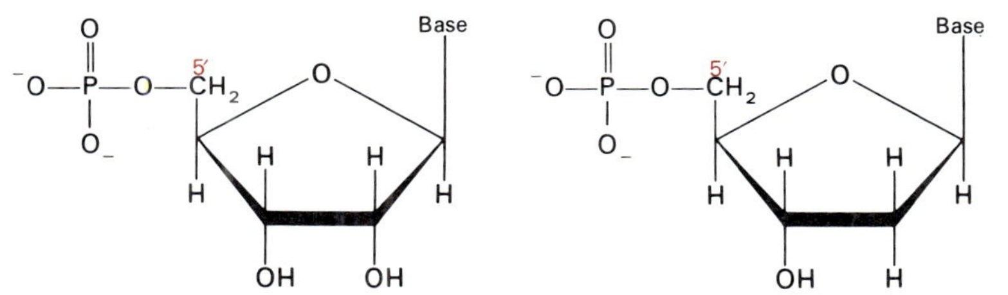

Figure 13-8
General structures of a ribo- and a deoxyribonucleotide.

Table 13-2
Nucleoside 5′-monophosphates of RNA and DNA

Ribonucleoside 5′-phosphates	Deoxynucleoside 5′-phosphates
Adenosine 5′-monophosphate 5′-Adenylic acid, AMP	Deoxyadenosine 5′-monophosphate 5′-Deoxyadenylic acid, dAMP
Guanosine 5′-monophosphate 5′-Guanylic acid, GMP	Deoxyguanosine 5′-monophosphate 5′-Deoxyguanylic acid, dGMP
Cytidine 5′-monophosphate 5′-Cytidylic acid, CMP	Deoxycytidine 5′-monophosphate 5′-Deoxycytidylic acid, dCMP
Uridine 5′-monophosphate 5′-Uridylic acid, UMP	Deoxythymine 5′-monophosphate 5′-Deoxythymidylic acid, dTMP

The di-and triphosphate derivatives of monophosphonucleosides also occur naturally. Nucleoside triphosphates are substrates for the synthesis of nucleic acids (Chapters 21 and 22) and the triphosphates of ribonucleosides (ATP, GTP, CTP, and UTP) furnish the energy needed for many biochemical reactions. Adenosine 5′-mono-, di-, and triphosphates (Figure 13-9) are a critically important group of biomolecules because of their key roles in the conservation and utilization of chemical energy in all biological systems. As shown in Figure 13-9, the phosphorus atoms of ATP are identified as α, β, and γ. Ribonucleotides of adenine, as described in Chapter 10, also occur in the structure of certain coenzymes, such as FAD and NAD^+. Another nucleotide of adenine, cyclic AMP (Figure 13-10), is an intracellular mediator of the physiological action of many hormones (Chapter 24) and a metabolic regulator of some biological processes in *E. coli* (Chapter 22). Ribonucleotides of other nitrogenous bases and their derivatives are

Figure 13-9
Mono-, di-, and triphosphorylated forms of adenosine.

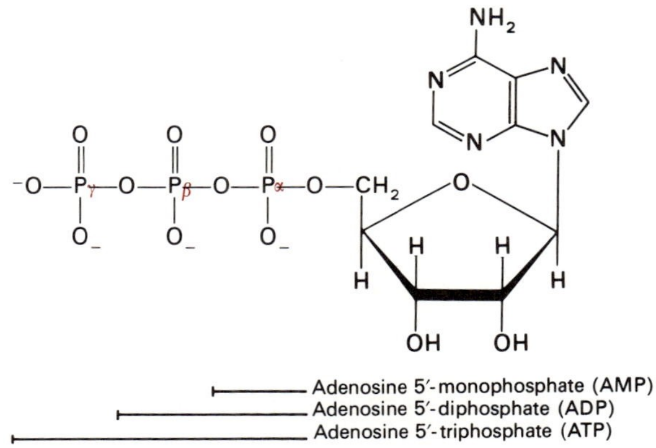

Figure 13-10
Cyclic AMP (or cAMP).

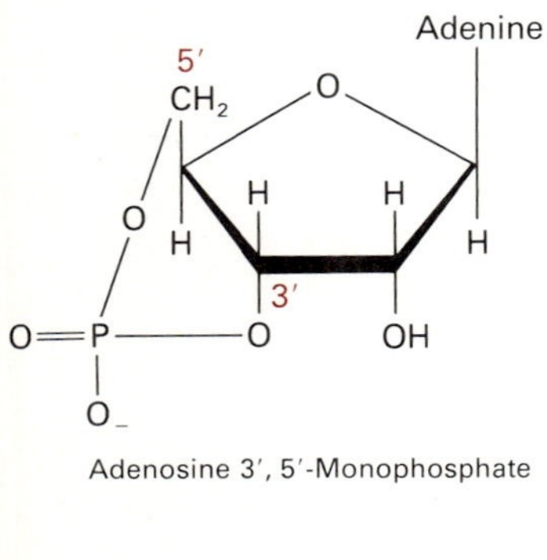

Adenosine 3′, 5′-Monophosphate

also used for intermediary metabolic purposes, e.g., uridine 5′-diphosphate linked to glucose (UDPG) is a glucose donor (p. 195). Thus, as a class of biomolecules, nucleotides are utilized for a variety of biologically distinct functions. This is especially true of adenine-containing molecules because they, like glucose and its derivatives, are extensively involved in many aspects of the chemical process of life.

Polynucleotide Structure

Figure 13-11 shows the structure of a tetranucleotide fragment of RNA containing the four major nitrogenous bases of this nucleic acid. The individual nucleotides are joined by phosphodiester bonds which involve the 3′- and 5′-carbon atoms of the ribosyl

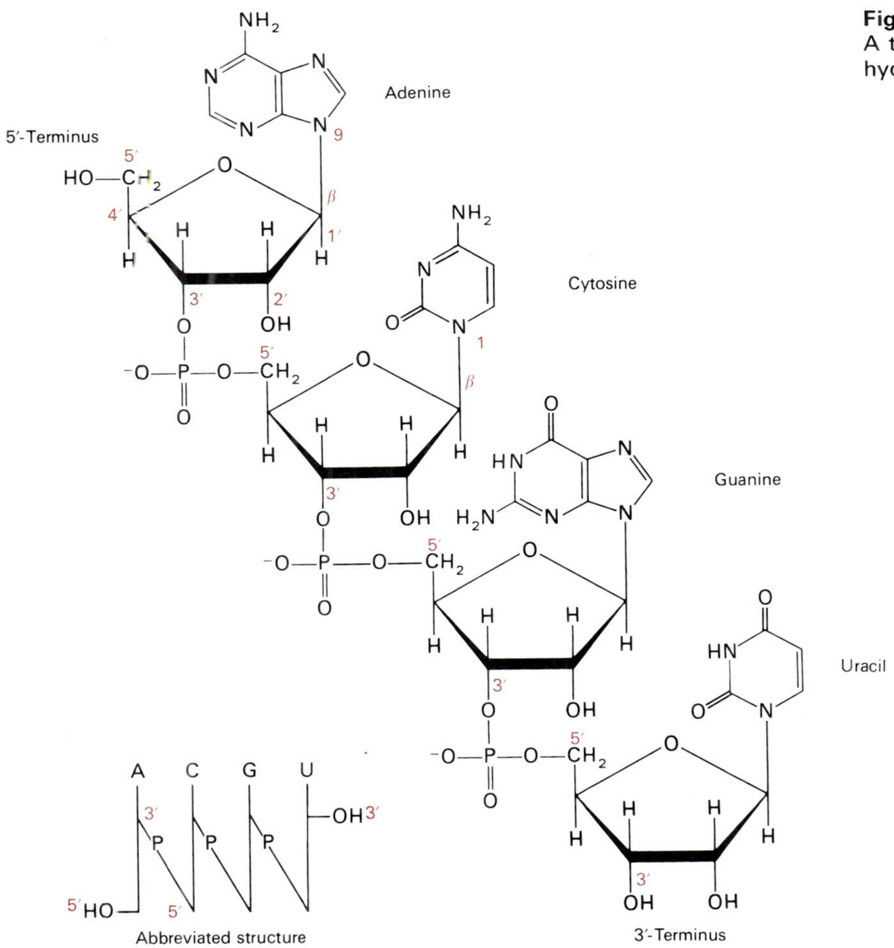

Figure 13-11
A tetranucleotide with 3′- and 5′-hydroxyl termini.

residues and are called *3′-5′-phosphodiester linkages.* The same 3′-5′-diester linkage occurs between the deoxyribosyl residues of DNA. Like proteins, many nucleic acids, being linear polymers, have two termini, one called the 5′-terminus (or 5′-end) and the other the 3′-terminus. As illustrated in Figure 13-11, the ribosyl moiety at the 5′-terminus has a free C-5′ (not in phosphodiester linkage) and the ribosyl residue at the 3′-terminus a free C-3′. It is not unusual, however, for cellular polynucleotides to have the terminal C-5′ in a monophosphate ester linkage.

For convenience, polynucleotides are usually depicted or written in an abbreviated form. The abbreviated structure of the tetranucleotide illustrated in Figure 13-11 is shown in the same figure. The tetranucleotide can also be written 5′-ApCpGpU-3′ or 5′-A-C-G-U-3′, i.e., forms that emphasize the sequence and content of the nitrogenous bases of the molecule. To indicate a 5′-phosphate group, the same tetranucleotide would be written 5′-pApCpGpU-3′ or 5′-pA-C-G-U-3′; a ''p'' following the ''U'' of the tetranucleotide would designate a 3′-phosphate group.

Until the 1950s tetranucleotides with the four different bases in varying arrangements, such as the one shown in Figure 13-11, were believed to be the basic repeating unit of RNA structure and the analogous tetradeoxynucleotides the basic unit of DNA. The concept that DNA of chromosomes possessed the genetic information of the cell was not seriously considered because the macromolecule's assumed monotonous structure of repeating tetranucleotides was thought incapable of having the chemical versatility required to account for the thousands of diverse hereditary traits. Proteins were the candidate of choice because they have a greater number of different monomeric units (amino acids) which can be arranged in an almost endless variety of sequences. Thus, because of the tetranucleotide hypothesis, it was difficult for many scientists to accept the idea of DNA as genetic material when it was first proposed in 1944 by Avery and his colleagues. The subsequent finding that DNA bases are not arranged in repetitive tetranucleotide sequences but occur in many different sequences aided in dispelling doubts about the biological function of the nucleic acid.

STRUCTURE OF DNA

Elucidation of the double-helical structure of DNA by Watson and Crick was made feasible by several pertinent findings of other scientists. That most DNAs possess a common and possibly the same three-dimensional structure was suggested by the similarity in x-ray diffraction patterns obtained by Maurice Wilkins (Nobel Prize, 1962) and Rosalind Franklin on DNA from different sources. Their findings further indicated a helical structure containing two or more polydeoxyribonucleotides. Independently, Erwin Chargaff had provided the crucial observation that, in DNA obtained from a wide variety of organisms, the molar ratio of adenine to thymine and that of guanine to cytosine were close to unity. These results indicated that a specific relationship must exist between the two bases within each of the ratios. The results of titration studies had also suggested that hydrogen bonding between bases of the deoxynucleotides was responsible for the association of the long DNA strands of the molecule. Watson and Crick, using these available data to construct models of DNA, succeeded in suggesting what is now confirmed as the correct structure, the double-helical model (Figure 13-12). In this structure, two strands of DNA are wound together about a common axis in a right-handed double helix. The spirals of the helix are the deoxyribosyl residues linked by phosphodiester bonds, and the bases of the deoxyribonucleotides project perpendicularly into the center of the helix. The double helix is maintained by hydrogen bonding between bases of the two DNA strands. The hydrogen bonding specifically occurs between an adenine of one strand and a thymine of the other and, similarly, between guanine and cytosine (Figure 13-13). The two DNA strands are thus *complementary* because every base of one strand is matched by its complementary hydrogen-bonding base on the other. The two strands are also *antiparallel,* meaning that the 3′-5′-diphosphodiester linkages are in opposite directions in the two strands, as indicated in Figures 13-12 and 13-15.

As illustrated in Figure 13-14, two forms of the double helix, called A DNA and B DNA, have been described. In the A form, the base pairs are about 20° inclined with the axis of the helix and, as a result, there is a central hole in the helix (part B of Figure 13-14). In the B form of DNA, the base pairs lie about normal (perpendicular) to the helix axis and therefore occupy the central portion of the helix. (The drawing on this book's cover depicts a top view of one complete turn of the B DNA helix, which is ten base pairs.) It is believed that most of a cell's DNA is in the B form.

The complementary nature of the two strands suggested a mode of DNA replication (duplication) to Watson and Crick. They proposed that replication occurs by the synthesis of complementary strands on each of the two parent strands (Figure 13-15) to

Figure 13-12
Double-helical model of DNA. The two ribbons represent the phosphate (P) and deoxyribose (D) moieties of the two DNA strands. A ::: T and G ::: C represent adenine-thymine and guanine-cytosine base pairs, respectively.

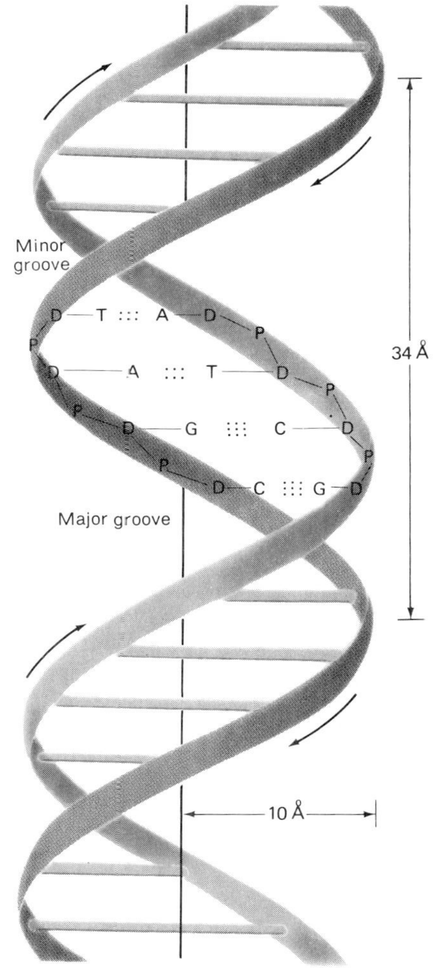

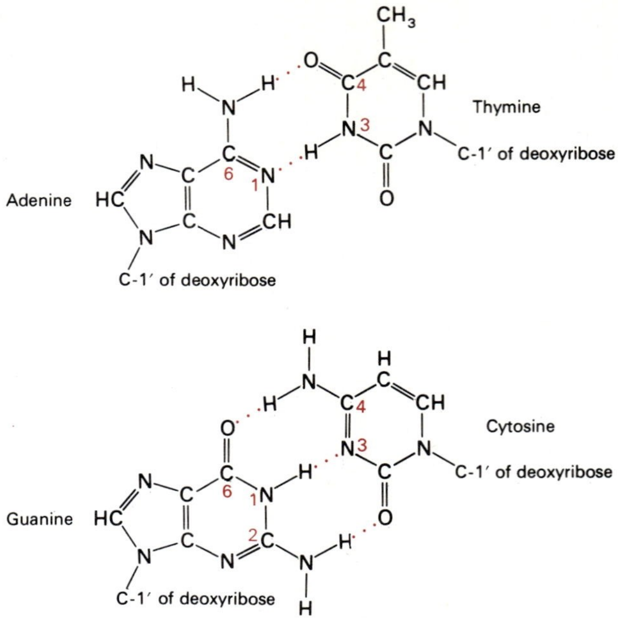

Figure 13-13
Base pairing by hydrogen bonding between adenine and thymine and between guanine and cytosine in a DNA double helix. An adenine-thymine pair has two hydrogen bonds and a guanine-cytosine pair has three.

yield two daughter DNA molecules, each having one newly synthesized and one parental strand in its structure. This proposal for a *semiconservative* mode of replication was subsequently proved to be correct by Matthew Meselson and Franklin Stahl.

MOLECULAR SPECIES OF RNA

RNA, unlike DNA, has several distinctly different cellular molecular species. The following three classes of cellular RNAs have been identified: *messenger RNA (mRNA), ribosomal RNA (rRNA),* and *transfer RNA (tRNA).* As a percentage of the total RNA content of a cell, rRNA constitutes 75 to 80 per cent because of the thousands of ribosomes within a cell, tRNAs between 10 and 15 per

Figure 13-14
A and B forms of DNA. A. Side views of A DNA and B DNA. Purines and pyrimidines (shown in red) are illustrated as rectangular and square slabs, respectively, and hydrogen bonds as dotted lines. In A DNA the base pairs rotate with the helix, and in B DNA the helix passes between the base pairs. B. Top views of A DNA and B DNA, showing that in A DNA the base-pair arrangement creates a central hole through the helix, whereas in B DNA the base pairs are in the central region of the helix. (Copyright 1982 by Irving Geis)

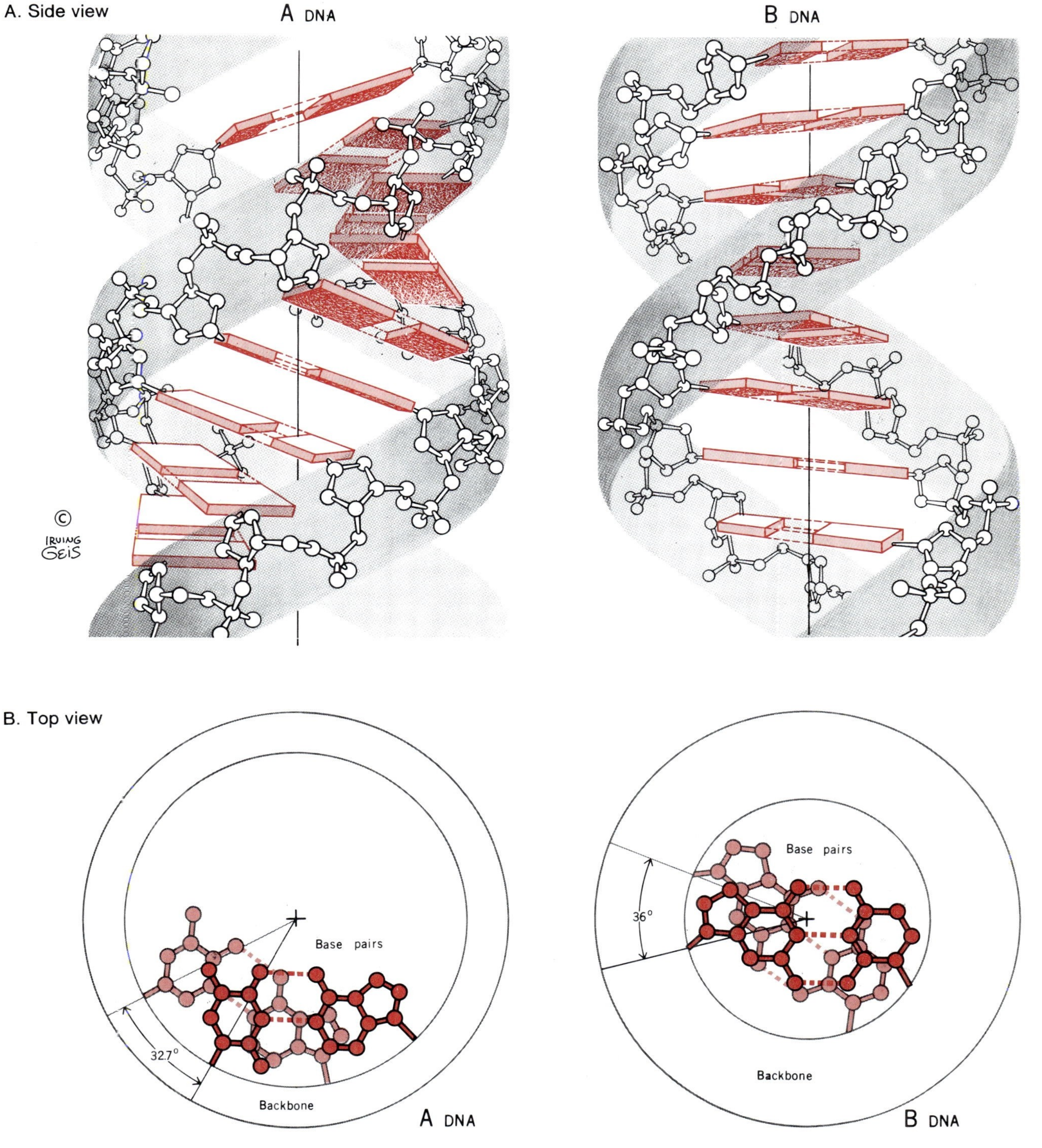
A. Side view
A DNA
B DNA
©
IRVING
GEIS
B. Top view
Base pairs
32.7°
Backbone
A DNA
Base pairs
36°
Backbone
B DNA

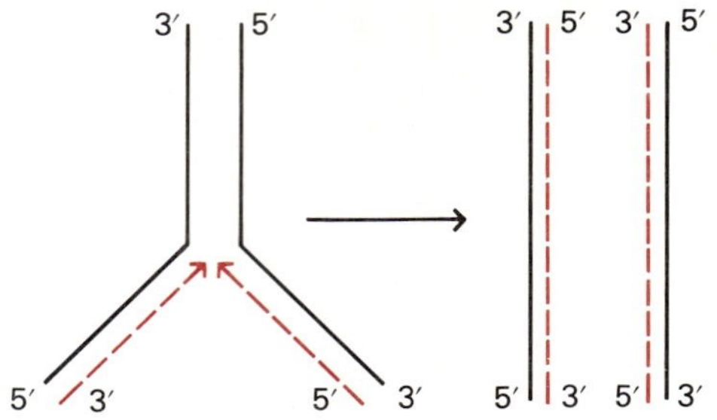

Figure 13-15
Diagram of the semiconservative replication of DNA. Solid lines represent the two parental strands of DNA; the dashed red lines, the newly replicated (daughter) strands.

cent, and mRNA about 5 to 10 per cent. In eucaryotes, all three molecular species are primarily cytoplasmic. Eucaryotic cells, however, also contain nuclear RNAs, many of which are precursor molecules for the cytoplasmic species. There may also be other RNAs whose functions have not yet been elucidated. The three known types of RNA are involved in cytoplasmic protein synthesis. Messenger RNAs, as implied by the name, serve as the cytoplasmic messengers of genes and carry genetic information (genetic code) for the synthesis of polypeptides (p. 12). Ribosomal RNAs are structural components of ribosomes (p. 513), the organelles on which synthesis of polypeptides takes place. Transfer RNAs are responsible for transporting activated amino acids to ribosomes for their subsequent incorporation into polypeptide structure (p. 504). Once the role of RNA in protein synthesis was discovered, the problem of how amino acids became incorporated into protein, previously thought to be a process involving only amino acids, was soon solved.

SEQUENCING OF tRNAs

The tRNAs are the smallest of the three RNA species, containing between 70 and 90 nucleotide residues and having molecular weights of about 25,000. The first nucleotide sequence of a nucleic acid to be established was that of an alanine-specific tRNA from yeast. The task was accomplished by Robert Holley (Nobel Prize, 1968) and his associates and, as a scientific endeavor, ranks with Sanger's landmark determination of the amino acid sequence of insulin. The techniques that Holley developed for tRNA sequencing are analogous in principle to those used for protein sequencing, i.e., selective cleavage of the macromolecule and subsequent analysis of the fragments produced. Like acid hydrolysis, which cleaves all peptide bonds in a protein, *alkaline hydrolysis* cleaves all 3′, 5′ -phosphodiester bonds in *ribonucleic acids.* Thus, Holley's RNA sequencing technique involved (1) alkaline hydrolysis and subsequent identification and quantification of the nucleotides, (2) partial hydrolysis (by specific enzymes) and sequence determination of the oligonucleotide fragments produced and, finally, (3) "overlapping" the oligonucleotide sequences obtained to deduce the complete polynucleotide sequence.

RNA HYDROLYSIS

There is an important difference between the hydrolysis of proteins and nucleic acids. The hydrolysis of peptide bonds in polypeptides always yields free α-carboxyl and α-amino groups and, as previously discussed, proteases are classified as exo- or endonucleases, depending on their peptide bond specificities. In the case of primary bonds of nucleic acids (3′,5′-phosphodiester bonds), *either* of the ester linkages may be cleaved by hydrolysis, depending on the hydrolytic mechanism of the catalyst. Thus, as shown in Figure 13-16, hydrolysis may produce a free 5′-phosphate monoester and a 3′-hydroxyl group (classified as an *a* cleavage) *or* a free 5′-hydroxyl group and a 3′-phosphate monoester (a *b* cleavage). Consequently, a *nuclease* (an enzyme that degrades nucleic acids) may be classified as an *a* or *b* hydrolase, whether an *endonuclease* or *exonuclease.* In a terminology analogous to that of proteases, an endonuclease hydrolyzes internal phosphodiester bonds and an exonuclease initiates hydrolysis from either the 3′- or the 5′-terminus of a polynucleotide.

Figure 13-16
Schematic of *a* and *b* cleavages of nucleic acids.

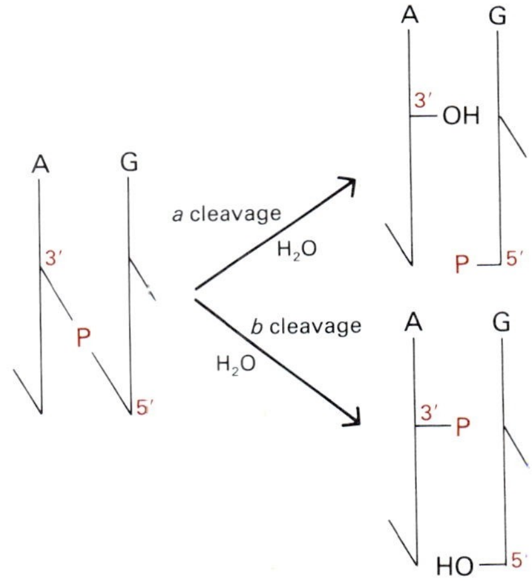

ALKALINE HYDROLYSIS

It was discovered in the 1940s that RNAs (but not DNAs) are rapidly hydrolyzed, when heated in alkaline solutions, to yield a mixture of 2′- and 3′-mononucleotides. This observation provided a key to the understanding of the mechanism of hydroxide-ion catalysis of RNA hydrolysis. As depicted in Figure 13-17, the first step in alkaline hydrolysis is the base-catalyzed abstraction of the proton from the 2′-hydroxyl group of each ribonucleotidyl residue to yield 2′,3′-cyclic monophosphates. Subsequently, *random* hydrolysis of these cyclic phosphodiester intermediates produces a mixture of 2′- and 3′-mononucleotides.

A. Production of 2′, 3′-cyclic monophosphate intermediate

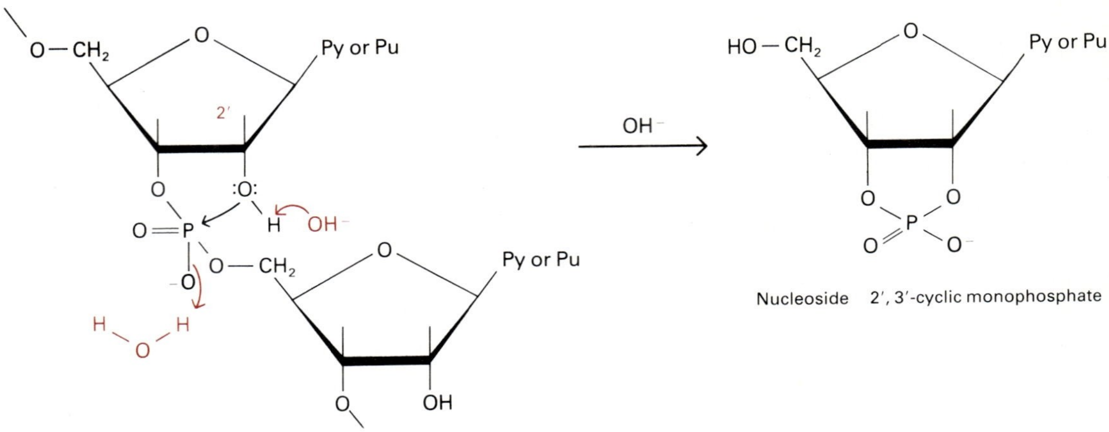

B. Random hydrolysis of 2′, 3′-cyclic monophosphate

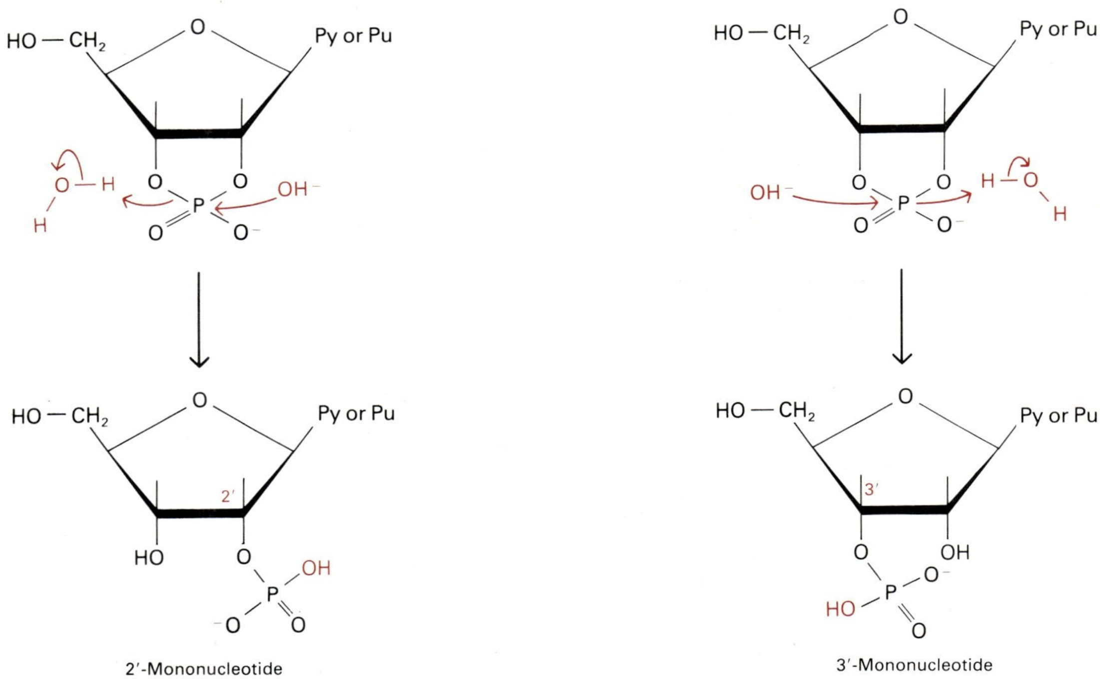

Figure 13-17
Mechanism of alkaline hydrolysis of RNA.

PROPOSED CATALYTIC MECHANISM OF RIBONUCLEASE A

Ribonuclease A (the first enzyme to be sequenced) is a pancreatic endonuclease (p. 81) of the *b* class, specific for the phosphodiester bonds on the *3′ side of pyrimidine ribonucleotide* residues. As in the case of chymotrypsin (Chapter 9), combined data obtained from chemical modification and x-ray diffraction studies led to an understanding of the enzyme's structure and catalytic mechanism. Figure 13-18 depicts the tertiary structure of bovine pancreatic ribonuclease and also shows its binding of substrate.

Three residues in the active site which participate in the hydrolysis mechanism are *His12, Lys41,* and *His119.* Hydrogen bonds link the pyrimidine moiety of a susceptible nucleotide residue to

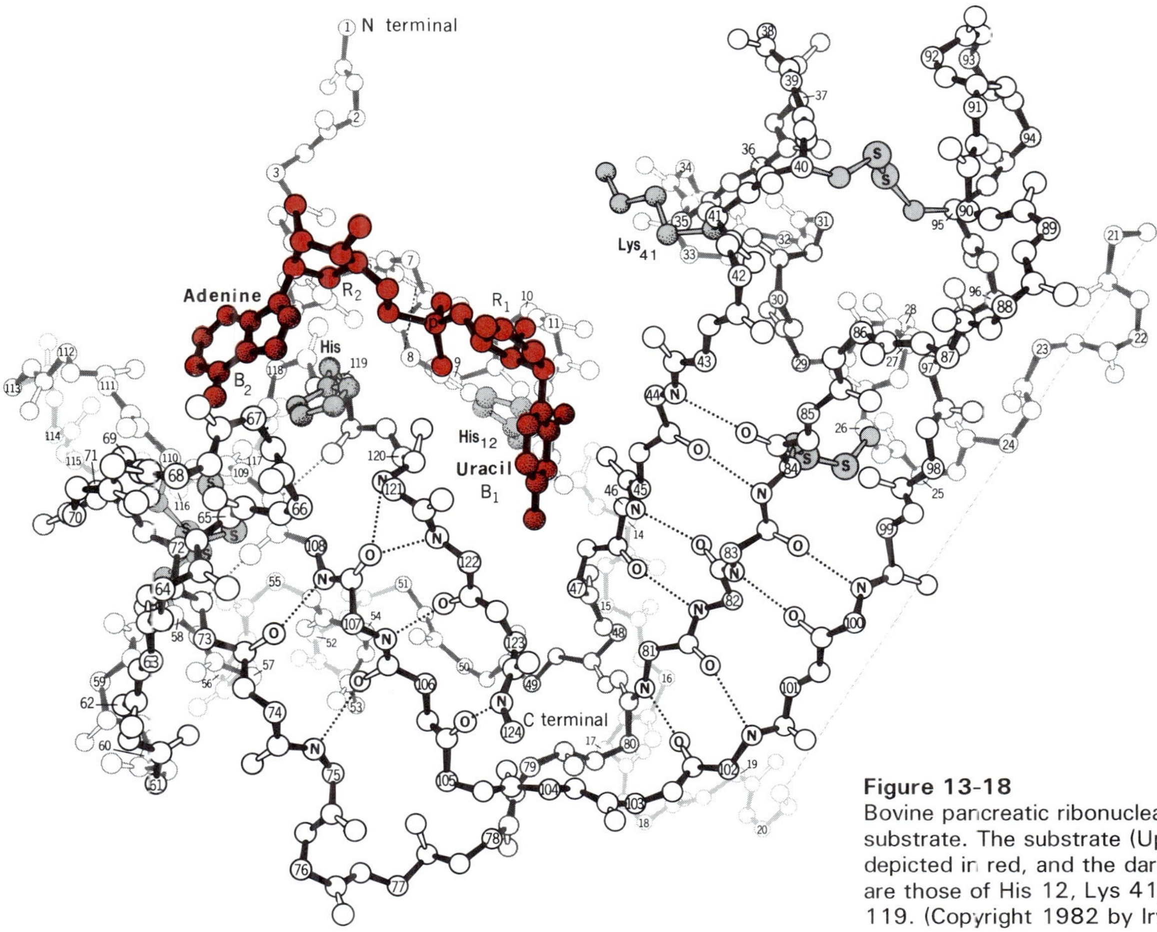

Figure 13-18
Bovine pancreatic ribonuclease with substrate. The substrate (UpA) is depicted in red, and the dark side chains are those of His 12, Lys 41, and His 119. (Copyright 1982 by Irving Geis)

Figure 13-19
Proposed mechanism of hydrolysis by ribonuclease A.

A. Production of first product

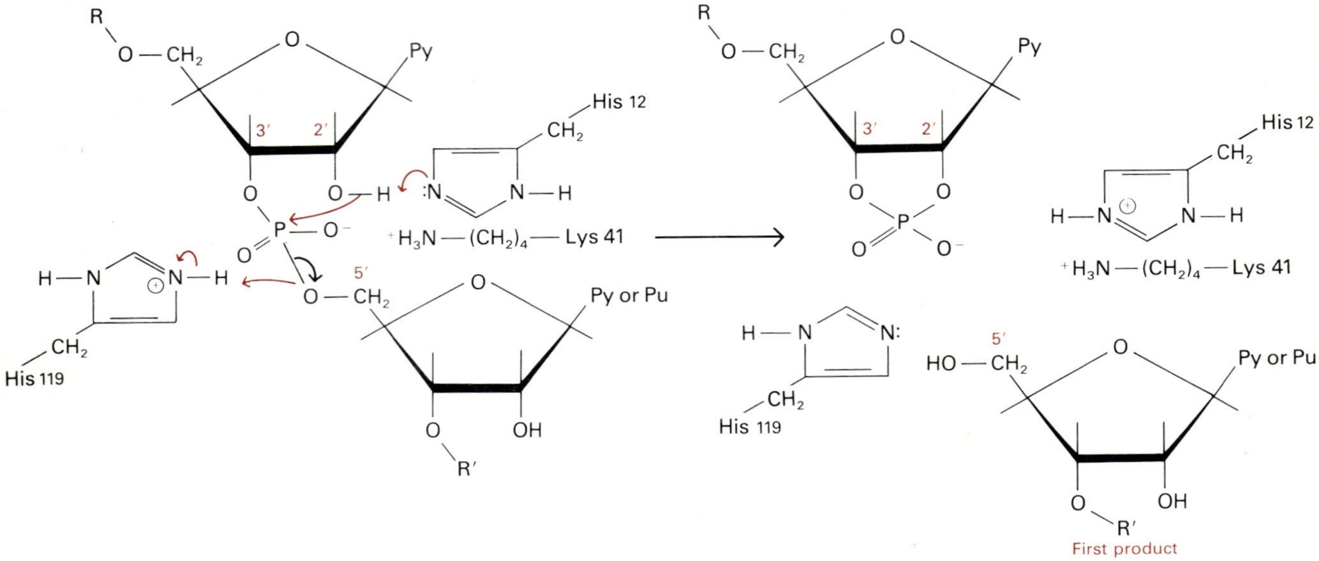

B. Production of second product

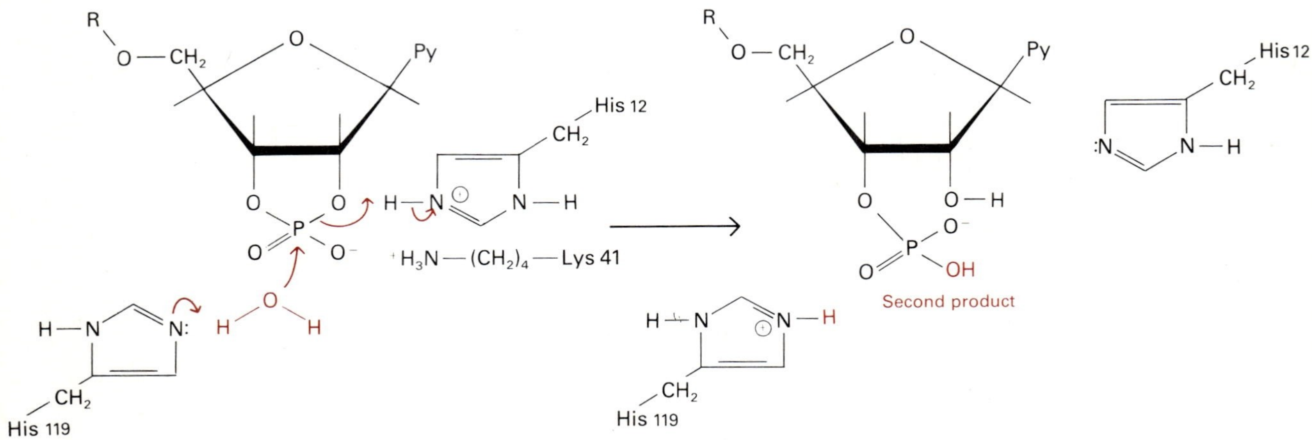

the binding pocket of the enzyme (analogous to the hydrophobic pocket described for chymotrypsin, p. 149). As diagrammed in Figure 13-19, the catalytic mechanism involves the ionic interaction of Lys41 with the scissile phosphodiester bond (the one on the 3′ side of the pyrimidine nucleotide), which serves to increase polarization of the phosphorus-oxygen bond and hence to render it more susceptible to nucleophilic attack. His12 behaves as a *general base catalyst* and abstracts the proton from the 2′-OH group, whereas His119 acts as a *general acid catalyst* and furnishes a proton to the 5′-oxygen as it is displaced from the phosphorus by the attacking nucleophile (the 2′-oxygen). Thus, a cyclic 2′,3′-cyclic phosphate is formed as an intermediate, and the first product of chain cleavage (with its 5′-OH group) leaves the active site. In the second step, the roles of the two histidyl residues are reversed, with His119 now serving as a conjugate base and His12 as a conjugate acid. Hence, the second step in the hydrolysis is the *enzyme-directed* hydrolysis of the 2′,3′-cyclic phosphate to yield the 3′-phosphomonoester. Being enzyme-directed, this cleavage is not random (unlike that of alkaline hydrolysis), and no 2′-phosphates are produced, i.e., protonation of the 2′-oxygen by His12 is directional (Figure 13-19).

2′,3′-Cyclic phosphodiester intermediates are formed by several endonucleases of varying specificities, which catalyze the *b* type of RNA hydrolysis. Ribonuclease T_1, from the mold *Aspergillus oryzae* (Table 13-3), is such an example; in this case, the 2′,3′-cyclic

Table 13-3
Specificities of some nucleases

Enzyme	Class	Specificity
Exonuclease		
Snake-venom phosphodiesterase	a*	Starts from 3′-terminus of RNA (or DNA)
Spleen phosphodiesterase	b**	Starts from 5′-terminus of RNA (or DNA)
Endonuclease		
Pancreatic ribonuclease A	b	5′-Linkages in which the 3′-linkage is to a pyrimidine nucleotide (RNA only)
Ribonuclease T_1 *(Aspergillus oryzae)*	b	5′-Linkage in which the 3′ -linkage is to a guanyl nucleotide (RNA only)

*Class a: hydrolyzes the ester linkage between C-3′ and the phosphate group.
**Class b: hydrolyzes the ester linkage between C-5′ and the phosphate group.

phosphate derivatives of guanosine residues are produced. Figure 13-20 summarizes the specific cleavage sites of an oligoribonucleotide by ribonuclease A, ribonuclease T_1, and the two exonucleases (spleen phosphodiesterase and snake-venom phosphodiesterase) listed in Table 13-3. Spleen phosphodiesterase sequentially cleaves 3′-mononucleotides from a free 5′-OH terminus, and snake-venom phosphodiesterase produces 5′-mononucleotides from a 3′-OH terminus.

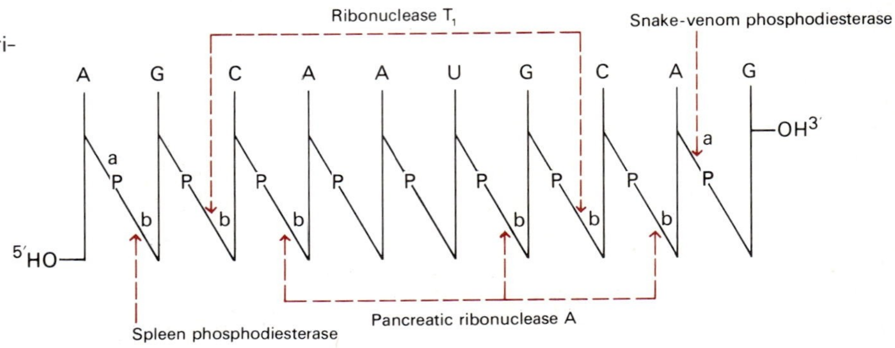

Figure 13-20
Schematic of the cleavage of an oligoribonucleotide by the various nucleases listed in Table 13-3.

STRUCTURE OF tRNAs

Figure 13-21 depicts the sequence drawn in a cloverleaf structure of yeast alanyl tRNA, proposed by Holley in 1964. Alanyl tRNA contains 77 nucleotide residues, ten of which have modified bases. The cloverleaf arrangement is a hypothetical structure, derived by maximizing the amount of *intramolecular* hydrogen bonding that can take place given the nucleotide sequence of the molecule. The model is characterized by double-stranded regions with single-stranded loops. The 3′-terminal adenyl nucleotide is the site of alanine attachment, and all tRNAs must have the specific pCpCpA-OH 3′-terminus to act as amino acid acceptors. The *anticodon* loop has a sequence of three nucleotides (pIpGpC) which hydrogen bonds with a coding unit *(codon)* for alanine within the structure of an mRNA. As a result of this specific relationship between codon (mRNA) and anticodon (tRNA), alanine is incorporated into its genetically designated positions during polypeptide synthesis (Chapter 23). The cloverleaf model is a two-dimensional representation and therefore lacks information about the actual foldings that a tRNA undergoes to attain its three-dimensional structure (Chapter 23). To date, over 70 different

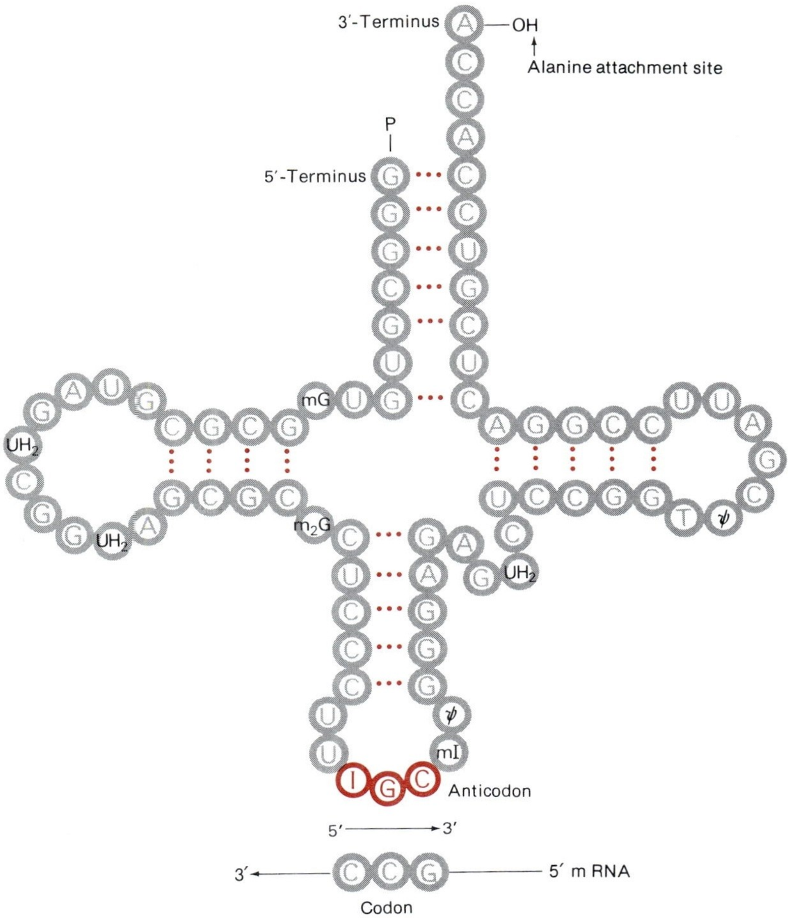

Figure 13-21
The nucleotide sequence of alanyl tRNA of yeast depicted in a cloverleaf arrangement. Abbreviations for nucleosides (in addition to A,G,U, and C) are: T = ribothymidine; I = inosine (hypoxanthine); mI = 1-methylinosine; mG = 1-methylguanosine; m_2G = N^2-dimethylguanosine; UH_2 = dihydrouridine; ψ = pseudouridine. The anticodon sequence is indicated in red and the hydrogen bonding by red dotted lines.

tRNAs have been sequenced, and a cloverleaf model similar to that of Figure 13-21 could be constructed with each of the unique sequences (p. 507). These findings emphasize the insight into structural relationships initially envisioned by Holley.

NUCLEIC ACIDS AS HEREDITARY MATERIAL

DNA serves as the genetic material of all procaryotes and eucaryotes. In the case of viruses, however, while many contain DNA as their genome, others contain RNA as their genetic material. Most DNAs are believed to occur naturally in double-helical structures, which affords cellular stability for the biomolecule carrying the genetic information of an organism. In a DNA molecule, only one

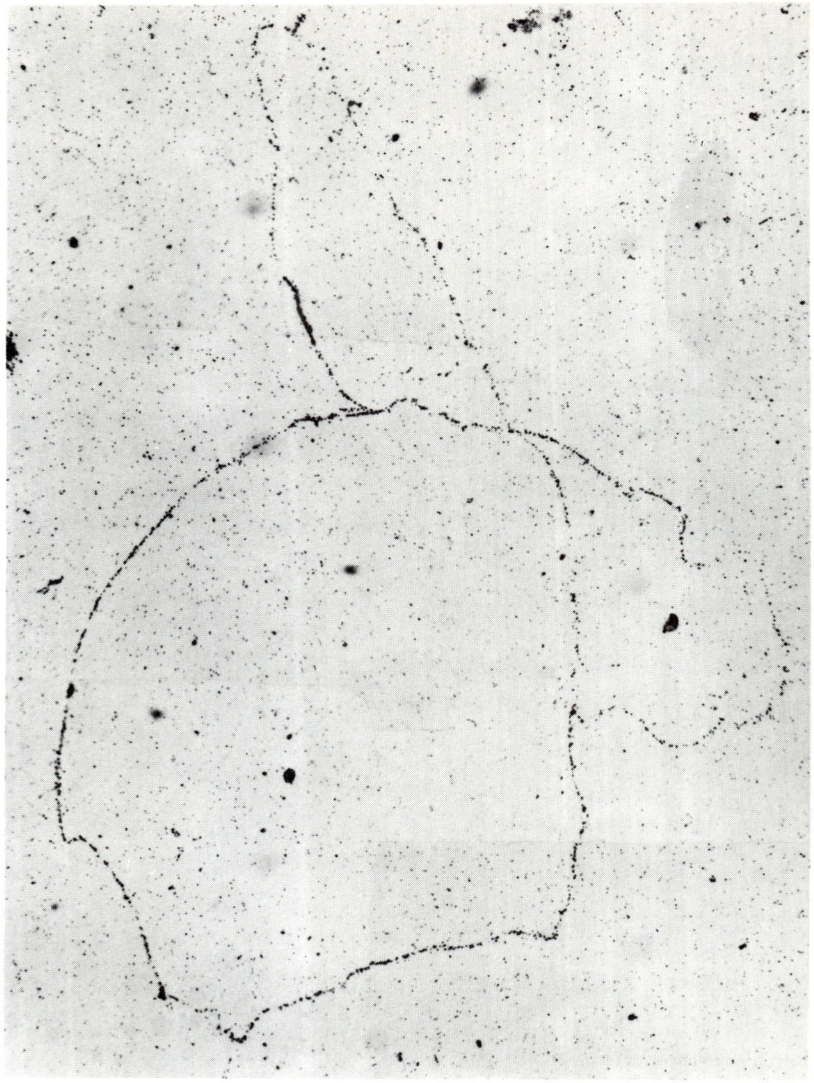

Figure 13-22
This classic autoradiograph shows the "circular" chromosome of *E. coli* in the process of replication. (Courtesy of Dr. John Cairns)

of the two complementary strands of a gene (specific sequence of nucleotides) contains the genic information; however, the informational content of different genes need not be located on the same strand. The *circular* form of a genome was first described in 1963 for the *E. coli* chromosome (Figure 13-22) by John Cairns and, for a while, was considered a peculiarity of bacteria. Subsequently, it has been demonstrated that many genetically active forms of DNA exist in circular structures, having no 3′- or 5′-termini, e.g., the DNAs of mitochondria (Figure 13-23) and chloro-

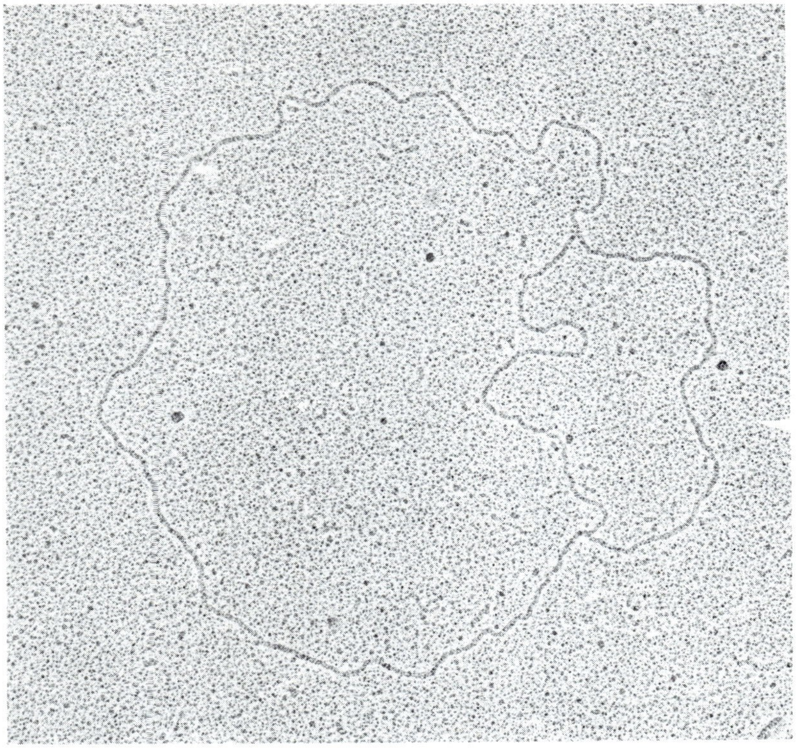

Figure 13-23
Electron micrograph of rotary-shadowed double-forked circular molecules of mitochondrial DNA from Novikoff rat ascites hepatoma cells. (Courtesy of Drs. Katsuro Koike and David R. Wolstenholme, *J. Cell Biology* (61), p. 14, 1974)

plasts in eucaryotes and that of bacterial plasmids. The lack of termini protects circular DNA from degradation by nucleases with specificities for 3′- or 5′-termini *(exonucleases).* Exceptions to the double-helical structure of genetic material are found in some bacteriophages; for example, phage ϕX174, which infects *E. coli,* has single-stranded circular (continuous) DNA (Chapter 23).

Certain bacterial viruses have linear double-stranded DNA that circularizes when in the host cell. For example, the linear form of bacteriophage λ of *E. coli* has a single-stranded sequence of 20-deoxynucleotides at the 5′-terminus of each strand (Figure 13-24). Because the two adhesive, or "sticky," ends have complementary base sequences, an *open-circle* structure can be produced when the linear molecule assumes a circular shape. Subsequent enzyme-catalyzed linkage (by DNA ligase) of the 3′- and 5′-terminal residues of each strand, by phosphodiester bond formation, converts the DNA molecule into a covalently *closed circle.*

Like DNA, single- and double-stranded RNAs are also able to carry genetic information. The utilization of RNA as primary hereditary material appears to be limited to viruses. Mammalian viruses

Figure 13-24
Schematic of the circularization of the linear DNA of bacteriophage λ.

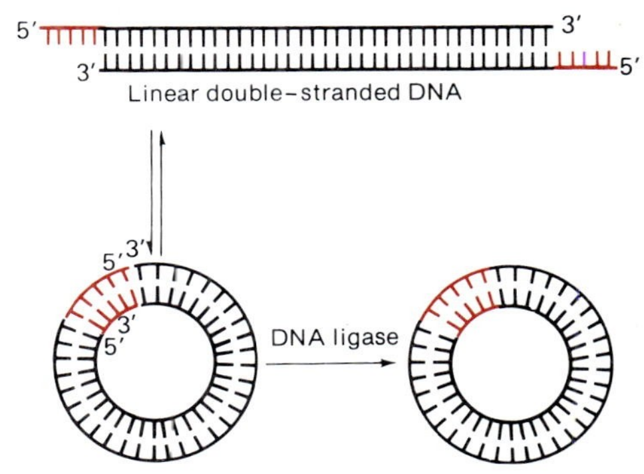

with single-stranded RNA are of research importance because of their medical and health implications. Viruses of this class include the influenza, polio, and oncogenic viruses, the last being responsible for certain cancers in animals. Animal reoviruses (associated with the respiratory tract in humans) are examples of viruses having double-stranded RNA.

The role of DNA and RNA as hereditary material provides an interesting contrast with regard to the biological functioning of these two classes of macromolecules. RNA exhibits the usual versatility of biomolecules by having distinct cellular and viral functions. DNA, on the other hand, is a rare example of a macromolecule performing the same singular function in viruses, procaryotes, and eucaryotes. As the bearer of genetic information of all cellular systems, DNA has maintained a purity of function throughout biological evolution. Nature's advantageous use of DNA for the propagation of the species apparently did not allow for a variety of functions for this macromolecule.

SUMMARY

A nucleic acid is a biomolecular polymer whose monomeric unit is a nucleotide, a molecule containing a nitrogenous base, a five-carbon sugar, and a phosphate. The two types of nucleic acid are ribonucleic acid (RNA) and deoxyribonucleic acid (DNA). In RNA and DNA, the pentoses are D-ribose and 2-D-deoxyribose, respectively. For nucleic acid formation, nucleotides are linked by phosphodiester bonds involving the C-3′ of one nucleotide and the C-5′ of its neighboring nucleotide. The four major nitrogenous bases of RNA are the two purines adenine and guanine and the two pyrimidines cytosine and uracil. DNA has the same major bases except that it has the pyrimidine thymine instead of uracil. Nucleic acids can also contain modified bases (various derivatives of the four major bases), e.g., methylated adenines, guanines, and cytosines.

A nucleoside is a purine or a pyrimidine linked to ribose (ribonucleoside) or to deoxyribose (deoxyribonucleoside); the *N*-β-glycosidic linkage is between C-1′ and N-9 in a purine nucleoside and between C-1′ and N-1 in a pyrimidine nucleoside. A nucleotide is a phosphate ester of a ribo- or deoxyribonucleoside; most naturally occurring nucleotides are 5′-monophosphoesters. The 5′-mono-, di-, and triphosphate nucleosides have a variety of functions in biological systems, e.g., in the conservation and utilization of chemical energy and as components of coenzymes.

The double-helical structure of DNA has two polydeoxyribonucleotide strands wound together about a common axis in a right-handed helix. This helix is maintained by the hydrogen bonding that occurs between complementary adenine and thymine nucleotides and between complementary guanine and cytosine nucleotides in the two antiparallel strands. The two forms of the double helix are referred to as DNA A and DNA B.

A nuclease (an enzyme that degrades nucleic acids) specifically hydrolyzes a phosphodiester bond either on the 3′ side (an *a* cleavage) or the 5′ side (a *b* cleavage). An *a* cleavage produces a free 5′-phosphate monoester, and a *b* cleavage yields a free 3′-phosphate monoester. Alkaline hydrolysis of RNA produces a mixture of 2′-,3′-mononucleotides because the 2′,3′-cyclic monophosphate intermediate generated by the base-catalyzed cleavage undergoes random hydrolysis. Ribonuclease A catalyzes a *b* cleavage on the 3′ side of pyrmidine ribonucleotides. Three residues at the active site, His21, Lys41, and His119, participate in the reaction. Lys41 interacts ionically with the substrate to increase polarization of the phosphorus-oxygen bond that is subject to nucleophilic attack. For this attack, His21 acts as a general base catalyst by abstracting the proton from the 2′-OH, and His119 behaves as a general acid catalyst by donating a proton to the leaving 5′-oxygen. In the second step of the catalytic mechanism (enzyme-directed hydrolysis), the two His residues reverse roles, and the 2′,3′-cyclic phosphate intermediate is converted into a 3′-phosphoester.

Three molecular species of cellular RNA are messenger RNA (mRNA), ribosomal RNA (rRNA), and transfer RNA (tRNA), and each is involved in protein synthesis. The first nucleotide sequence of an RNA was established for an alanyl tRNA of yeast. A cloverleaf arrangement (two-dimensional structure) was proposed for the tRNA, based on the maximal amount of intramolecular hydrogen bonding obtainable with the nucleotide sequence. Over 70 tRNAs have now been determined, and a cloverleaf arrangement can be constructed for each of the sequences.

DNA is the genetic material of procaryotes, animals, and plants, with the nucleic acid occurring as a linear or circular double-stranded molecule, e.g., linear chromosomal DNA and circular mitochondrial DNA. Some bacteriophage, e.g., ϕX174, have single-stranded DNA. For some viruses, single- or double-stranded RNA carries the genetic information. The polio virus, for example, has single-stranded RNA as its genetic material, whereas that of animal reoviruses is double-stranded RNA.

QUESTIONS

1. Draw the structures of the following modified bases found in nucleic acids: (a) 1-methyluracil, (b) 2-methylguanine, (c) 5-hydroxymethyluracil, and (d) N^4-acetylcytosine.

2. Which of the following statements are true about a nucleotide?
   (a) Refers only to 5′-mono-, di-, and triphosphate nucleosides
   (b) Has an *N*-β-glycosidic linkage
   (c) Can have either a D-ribosyl or a D-deoxyribosyl moiety
   (d) Is a phosphate ester
   (e) Serves only as a monomeric unit of nucleic acids

3. The hydrolysis of RNA with dilute alkali produces a mixture of 2′- and 3′-nucleoside monophosphates. Draw the structures of a 2′- and a 3′-nucleoside monophosphate (designate the nitrogenous bases as "B").

4. Which class(es) of RNA (mRNA, rRNA, and/or tRNA) best fits the following characterizations?
   (a) Smallest molecular species of RNA
   (b) Present in the largest amount in a cell
   (c) Considered cytoplasmic
   (d) Contains genetic information needed for protein synthesis
   (e) Recognizes codons in a genetic message

(f) Structural component of the site of protein synthesis
(g) Serves as the transporter of activated amino acids

5. What are the products of the complete hydrolysis of a polyribonucleotide (with a 3′- and 5′-hydroxyl groups) by (a) snake-venom phosphodiesterase and (b) spleen phosphodiesterase?

6. What would be the products of the hydrolysis if the following oligonribonucleotide

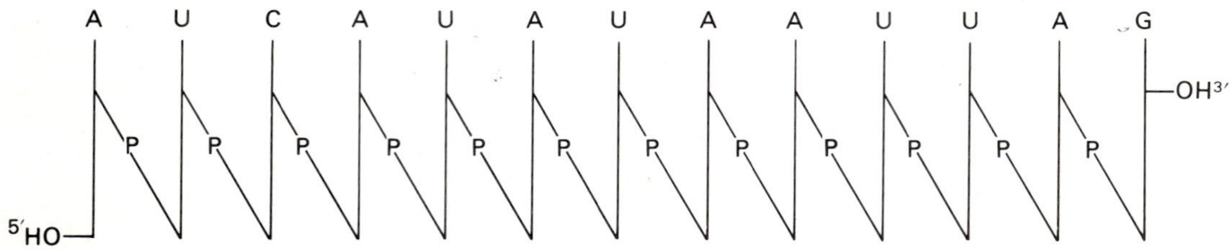

were treated with (a) ribonuclease T_1 or (b) pancreatic ribonuclease A?

7. Why does it require a higher temperature (more energy) to separate the two strands of a DNA molecule with a high G-C content than one with a high A-T content (refer to Figure 13-13)?

8. Which of the following statements about genetic material are considered true?
(a) No procaryote has DNA as its genetic material.
(b) Known circular DNA molecules are double stranded.
(c) Single-stranded RNA cannot serve as genetic material.
(d) Animals and plants, unlike *E. coli,* have no circular double-stranded DNA.
(e) No known animal virus has double-stranded RNA.
(f) The biological function of DNA appears to be limited to that of serving as genetic material.

9. Which of the following statements are true about the proposed catalytic mechanism of ribonuclease A?
(a) The enzyme catalyzes a *b*-type cleavage.
(b) The enzyme is specific for the phosphodiester bonds on the 3′ side of pyrmidine nucleotides.
(c) The enzyme cleaves only RNA.
(d) One Lys and two His residues of the enzyme participate in the hydrolysis mechanism.
(e) Only one of the His residues behaves as a general acid-base catalyst.
(f) The Lys residue interacts ionically with the phosphodiester bond to be cleaved.
(g) A cyclic 2′,3′-phosphate intermediate is formed.
(h) No 2′-phosphomonoesters are produced.

SUGGESTED READINGS

Davidson, J. N., *The Biochemistry of Nucleic Acids,* 7th ed. New York: Academic Press, 1972. An excellent paperback on the nucleic acids.

Fasman, G. D., ed., *Handbook of Biochemistry and Molecular Biology,* 3rd ed. Sec. B, *Nucleic Acids,* Vols. 1 and 2. Boca Raton, Fla: CRC Press, 1976. Good reference source for basic information on purines, pyrimidines, nucleotides, and DNAs.

Hurst, D. T., *An Introduction to the Chemistry and Biochemistry of Pyrimidines, Purines, and Pteridines.* New York: Wiley, 1980. The title aptly describes the book's contents.

Kochetkov, N. K., and E. I. Budovsky, eds., *Organic Chemistry of Nucleic Acids,* Parts A and B. New York: Plenum, 1972. An extensive and broad coverage of the chemistry of DNA and RNA.

Spencer, J. H., *The Physics and Chemistry of DNA and RNA.* Philadelphia: W. B. Saunders, 1972. A brief, in-depth study of the physical and chemical properties of DNA and RNA.

Stewart, P. R., and D. S. Letham, eds., *The Ribonucleic Acids,* 2nd ed. New York: Springer-Verlag, 1977. A comprehensive treatment of RNAs.

Watson, J. D., *Molecular Biology of the Gene,* 3rd ed. Menlo Park, Calif.: Benjamin/Cummings, 1976. An excellent presentation of the molecular biology of DNA and RNA.

Articles

Borst, P., "Structure and Function of Mitochondrial DNA." *TIBS,* 2:31, 1977.

Cairns, J., "The Bacterial Chromosome." *Scientific American,* 214(1):36 (offprint #1030), 1966.

Crick, F. H. C., "Nucleic Acids." *Scientific American,* 197(3):188 (offprint #54), 1957.

Goodenough, U. W., and R. P. Levine, "The Genetic Activity of Mitochondria and Chloroplasts." *Scientific American,* 223(5):22 (offprint #1203), 1970.

Holley, R. W., "The Nucleotide Sequence of a Nucleic Acid." *Scientific American,* 214(2):30 (offprint # 1033), 1966.

Horowitz, N. H., "The Gene." *Scientific American,* 195(4):78 (offprint #17), 1956.

Mirsky, A. E., "The Chemistry of Heredity." *Scientific American,* 188(2):47 (offprint #28), 1953.

Mirsky, A. E., "The Discovery of DNA." *Scientific American,* 218(6):78 (offprint #1109), 1968.

Younghusband, H. B., and R. B. Inman, "The Electron Microscopy of DNA." *Ann. Rev. Biochem.,* 43:605, 1974.

ENERGY AND LIFE

14

INTRODUCTION

The sun is the primary source of energy for life on earth. Photosynthesis (Chapter 18) is the unique mechanism employed by nature to transform radiant energy into chemical energy, the form of energy utilized by living organisms. Photosynthetic organisms convert solar energy into chemical energy through the production of biomolecules, which they then use as fuels to furnish the energy requirements for life processes. Since nonphotosynthetic life forms cannot utilize radiant energy to provide for their energy needs, they must rely on the chemical energy of biomolecular nutrients, e.g., sugars and lipids. The flow of biological energy within an organism encompasses the release, conservation, and utilization of chemical energy. The elucidation of the basic mechanisms governing the production and use of energy by living organisms *(bioenergetics)* has provided biological scientists with a set of unifying concepts, called *bioenergetic principles,* that effectively describe nature's use of energy in molecular terms.

Adenosine triphosphate, ATP, is the principal carrier of energy for all forms of life (Figure 14-1). The synthesis and hydrolysis of ATP (Figure 14-2) are metabolic pivotal points for the conservation and utilization, respectively, of chemical energy. ATP was first isolated from muscle in 1929 in the United States by Cyrus H. Fiske and Yellapragada SubbaRow and, independently, in Germany by Karl Lohman. It was not until a decade later, however, that the central role of ATP in energy transfer began to be recognized. In 1941, Fritz Lipmann (Nobel Prize, 1953), aided by the contributions of Herman Kalckar, hypothesized the cyclic nature of ATP's

Figure 14-1
Adenosine 5′-triphosphate (ATP).

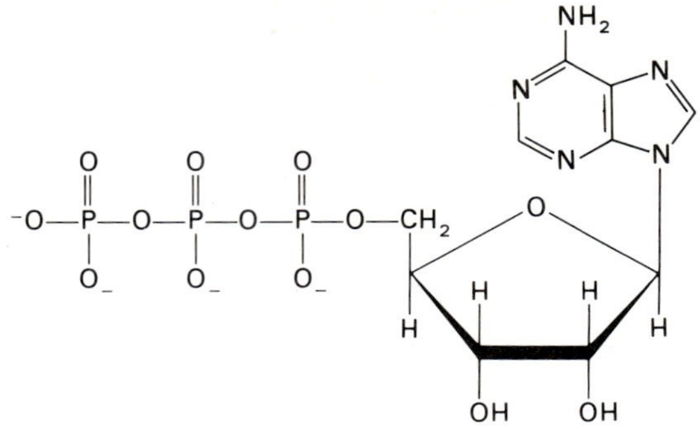

Fritz Lipmann.
(Nobel laureate, 1953)
In 1927 Lipmann, after obtaining his M.D. degree, started his research career as an unpaid graduate student of the famous biochemist Otto Meyerhof. In 1981, at age 82, Lipmann was still professionally active at Rockefeller University.

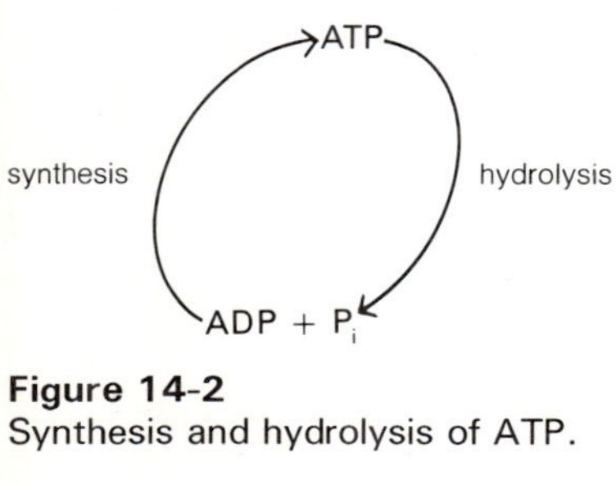

Figure 14-2
Synthesis and hydrolysis of ATP.

involvement in bioenergetic processes (Figure 14-3) and, in his proposal, wrote:

> Not very definite answers can be given to the question as to how high phosphate group potential operates as the promoter of various processes although a more or less loosely defined interconnection with phosphate turn-over is recognizable . . . the metabolic cycle [is] compared with a machine which generates electrical current. It seems, in fact, that in cellular organization the phosphate "current" plays a similar part as does electrical current in the life of human beings. It is also a form of energy utilized for all-around purposes.

To understand the above analogy between the chemical energy of ATP and electrical energy, it is necessary to appreciate the thermodynamic aspects of metabolic reactions which reveal nature's proficient use of the laws of thermodynamics.

FREE ENERGY CHANGE OF A CHEMICAL REACTION

The intrinsic energy contained in the molecular structure of a compound is referred to as its *free energy (G) content.* In the reaction $A \rightleftharpoons B$, the difference between the free energy content of product B (G_B) and that of reactant A(G_A) is called the *change in free energy,* or ΔG. The ΔG value is important to the basic concepts of bioenergetics because it represents the *maximum amount of energy that is released by or required for a reaction.* A reaction with a *negative* ΔG ($G_B < G_A$) occurs with a release of free energy, i.e., a decrease in free energy content. Reactions with negative values for ΔG are *exergonic* and are considered *spontaneous.* A reaction with a *positive* ΔG ($G_B > G_A$) is one that requires an input of free energy, and such reactions are *endergonic.*

STANDARD FREE ENERGY CHANGES

The formula for the calculation of ΔG of the reaction converting reactant A into product B is:

$$\Delta G = \Delta G^\circ + RT \ln [B]/[A]$$

The expression relates the free energy change to the equilibrium constant of the reaction. ΔG° is the standard change in free energy, which is ΔG determined under a specified set of experimental conditions, R is the universal gas constant, T is the absolute temperature, and [B] and [A] are the concentrations of B and A in moles per liter. ΔG is therefore a function of ΔG° and of the concentrations of the reactant and product. The free energy change of a reaction is most often cited as ΔG°, calculated at equilibrium and under standard conditions. ΔG is zero (no free energy change) at equilibrium because there is no net conversion of A into B and $[B]_{eq}/[A]_{eq}$ is equal to the equilibrium constant K_{eq}. Thus, at equilibrium, the above equation can be written:

$$0 = \Delta G^\circ + RT \ln K_{eq}$$

This can be arranged:

$$\Delta G^\circ = -RT \ln K_{eq} \qquad \text{or} \qquad \Delta G^\circ = -2.303RT \log K_{eq}$$

With the use of the above expression, ΔG° can be calculated if the concentrations of B and A at equilibrium are determinable.

The experimental conditions used in physical chemistry for the determination of ΔG° include reactant concentration value(s) of 1 molal, a temperature of 25°C (298 K), and a pressure of 1 atmosphere. In biochemical research, no distinction is usually made between molality (m) and molarity (M) of solutions, and the latter concentration is commonly used. Too, for bioenergetic considerations, the preferred ΔG° is that determined at pH 7.0, instead of pH 0.0 (the $[H^+]$ generally used in physical chemistry). The standard change in free energy, determined at pH 7.0, is identified as $\Delta G^{\circ\prime}$ to distinguish it from ΔG° (pH 0.0). R can be conveniently expressed in the energy units of calories (1.987 calories per mole per degree K). Although $\Delta G^{\circ\prime}$ is calculated as calories (cal) per mole, it is most often written as kilocalories (1,000 cal = 1 kcal) per mole. Recently, an international commission recommended that standard free energy be expressed in joules (J) or kJ per mole (1 cal is equivalent to 4.184 J).

Figure 14-3
Cyclic nature of the involvement of ATP in bioenergetic processes.

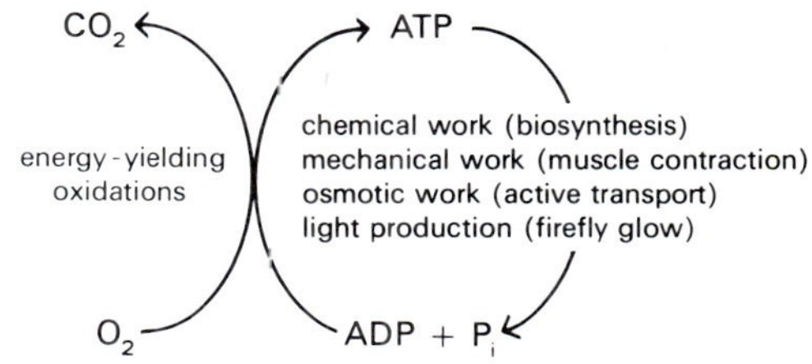

Molarity (M): the number of moles of solute per 1,000 ml (1 liter) of solution

Molality (m): the number of moles of solute per 1,000 g of solution

When the above information is incorporated into the general formula for $\Delta G^{\circ\prime}$ determination, the expression becomes:

$$\Delta G^{\circ\prime} = -(2.303)\,(1.987)\,(298)\,\log K_{eq}$$

$$\text{or } \Delta G^{\circ\prime} = -1{,}363 \log K_{eq} \text{ cal per mole (or cal mol}^{-1}\text{)}$$

Table 14-1
Relationship between equilibrium constants (K_{eq}) and changes in standard free energy ($\Delta G^{\circ\prime}$)

K_{eq}	$\Delta G^{\circ\prime}$ (cal mol^{-1})
0.001	+4,092
0.01	+2,728
0.1	+1,364
1.0	0
10.0	−1,364
100.0	−2,728
1000.0	−4,092

As illustrated in Table 14-1, $\Delta G^{\circ\prime}$ is a function of K_{eq}, with the ratio of $[B]_{eq}$ to $[A]_{eq}$ reflecting the exergonic or endergonic nature of a reaction. For example, a ratio of 1,000 (1,000 times as much product B as reactant A at equilibrium) would be indicative of a spontaneous (exergonic) reaction, whereas the reverse ratio (0.001), signifying very little reaction, would indicate an endergonic reaction. The $\Delta G^{\circ\prime}$ of a reaction, being a standardized value determined at equilibrium, can be very different from the ΔG of the reaction as it occurs under metabolic conditions, where molar concentrations of reactant (substrate) and other experimental parameters for determining $\Delta G^{\circ\prime}$ do not exist. Under physiological conditions, the initial concentrations of A *and* B, as well as the $\Delta G^{\circ\prime}$ of a reaction, determine whether or not a reaction will proceed and in which direction it will go. It is not unusual for a biological reaction to have a $\Delta G^{\circ\prime}$ value that is positive, e.g., phosphocreatine production from ATP and creatine (p. 259).

ATP AS AN ENERGY CARRIER

Living organisms use exergonic (energy-releasing) reactions to provide the energy requirements for the processes of life. As the predominant supplier of metabolic energy, ATP furnishes the chemical energy to drive endergonic (energy-requiring) reactions, perform mechanical work (movement), provide heat (help maintain body temperature), and produce light (the glow of fireflies). Ligases (p. 132) exemplify the exergonic/endergonic relationships found in liv-

Figure 14-4
Hydrolysis of ATP.

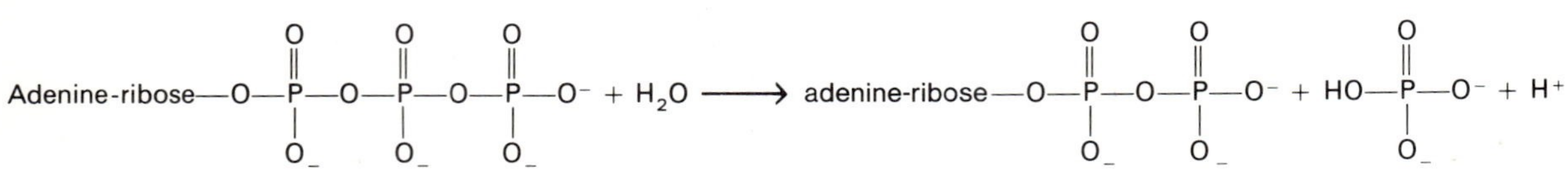

$$ATP + H_2O \longrightarrow ADP + P_i + H^+$$

$$\Delta G^{\circ\prime} = -7{,}300 \text{ cal mol}^{-1} \text{ (or } -7.3 \text{ kcal mol}^{-1}\text{)}$$

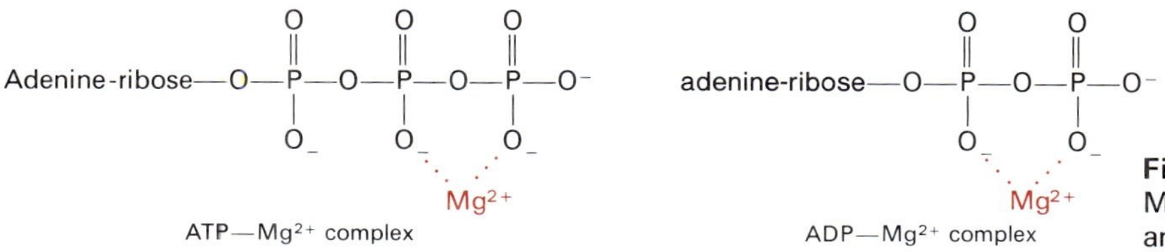

Figure 14-5
Magnesium ions complexed with ATP and ADP.

ing systems since the joining of two molecules (covalent bond formation), catalyzed by this class of enzymes, is an endergonic reaction and requires the energy released by an exergonic reaction, e.g., hydrolysis of ATP.

ATP's release of energy occurs principally by its hydrolysis (Figure 14-4) to yield ADP and inorganic phosphate (P_i), an *orthophosphate cleavage.* Under physiological conditions, both ATP and ADP exist in an electrostatic complex with Mg^{2+} (Figure 14-5), and therefore the metal ion is a cofactor for most reactions involving the two nucleotides. In certain ATP reactions, Mn^{2+} serves as the cofactor. The hydrolytic removal of the terminal (γ) phosphate group proceeds with a $\Delta G^{\circ\prime}$ of -7.3 kcal mol^{-1} and thereby releases molecular chemical energy which can be utilized for work. Hydrolysis of ADP to AMP and P_i (Figure 14-6) also has a $\Delta G^{\circ\prime}$ of -7.3 kcal mol^{-1}; however, in bioenergetic processes, this reaction is of limited importance compared to ATP hydrolysis. The $\Delta G^{\circ\prime}$ given for the ATP hydrolyses should be considered as *approximate* standard changes in free energy. Reported $\Delta G^{\circ\prime}$ values often differ because variations in experimental conditions, especially the Mg^{2+} concentration, highly influence the results obtained.

Lipmann introduced the "squiggle" notation ($\sim$) to designate the *energy-rich* bonds of biomolecules such as ATP and ADP (Figure 14-7). This bioenergetic concept of "high-energy" bonds, which expresses the *potential energy derivable from a bond* by

Figure 14-6
Hydrolysis of ADP.

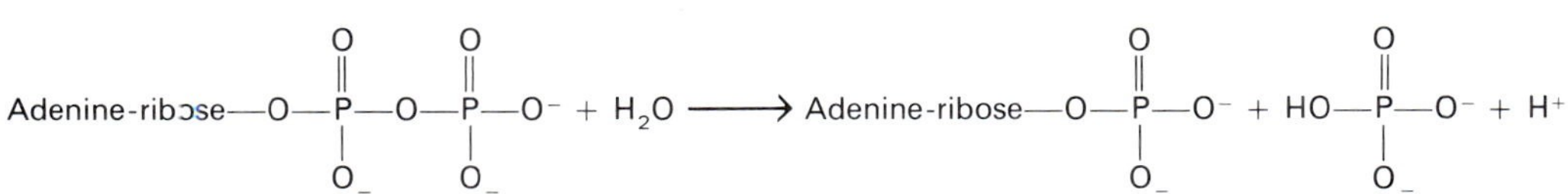

$$ADP + H_2O \longrightarrow AMP + P_i + H^+$$
$$\Delta G^{\circ\prime} = -7.3 \text{ kcal mol}^{-1}$$

Figure 14-7
ATP and ADP hydrolysis.

$$\text{Adenine-ribose-P} \sim \text{P} \sim \text{P} \xrightarrow{H_2O} \text{adenine-P} \sim \text{P} + P_i + H^+$$

$$\text{Adenine-ribose-P} \sim \text{P} \xrightarrow{H_2O} \text{adenine-ribose-P} + P_i + H^+$$

specific types of reactions, e.g., hydrolysis, differs from the chemist's definition of the term, which means the *energy expended to form (or to break) a bond.* To a chemist, high-energy bond generally refers to a *stable bond;* to a biochemist, the same term refers to an *unstable,* or *labile,* bond. Thus, the terminology and the "~" designation can be confusing if not clearly defined.

The other energetically important hydrolysis of ATP (Figure 14-8) is the reaction that, by *pyrophosphate cleavage,* produces AMP and inorganic pyrophosphate (PP_i); such hydrolysis has a $\Delta G°'$ of -8.0 kcal mol^{-1}. This particular hydrolytic reaction provides an

Figure 14-8
Pyrophosphate cleavage of ATP.

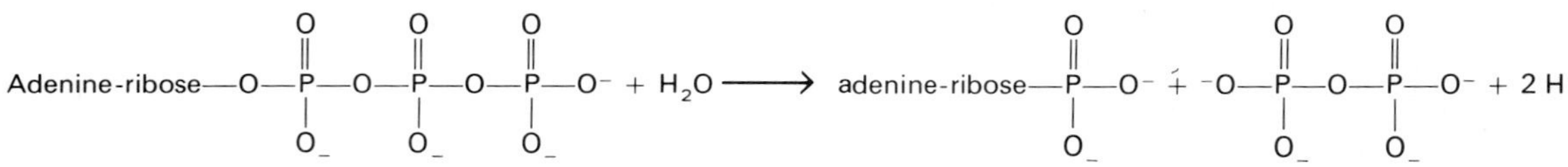

$$\text{ATP} + H_2O \longrightarrow \text{AMP} + PP_i + 2\,H^+$$
$$\Delta G°' = -8.0 \text{ kcal mol}^{-1}$$

effective mechanism for assuring the completeness of a metabolic reaction because of the subsequent rapid hydrolysis of the product PP_i by the enzyme *inorganic pyrophosphatase* (Figure 14-9). The hydrolysis of PP_i ($\Delta G°'$ of -6.5 kcal mol^{-1}) essentially ensures the irreversibility of the initial hydrolysis of ATP because it would require a large energy input of $+14.5$ kcal mol^{-1} (sum of the $\Delta G°'$ of the two hydrolytic reactions) to synthesize ATP from AMP and 2 P_i. The coupling of these two hydrolyses is a general biochemical mechanism used in many physiologically irreversible reactions catalyzed by ligases. The energy released by the hydrolysis

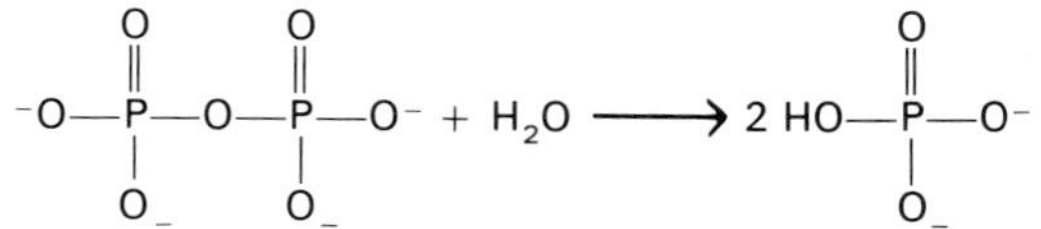

$$\text{Inorganic pyrophosphate } (PP_i) + H_2O \longrightarrow 2 \text{ inorganic phosphate } (P_i)$$
$$\Delta G°' = -6.5 \text{ kcal mol}^{-1}$$

Figure 14-9
Hydrolysis of inorganic pyrophosphate.

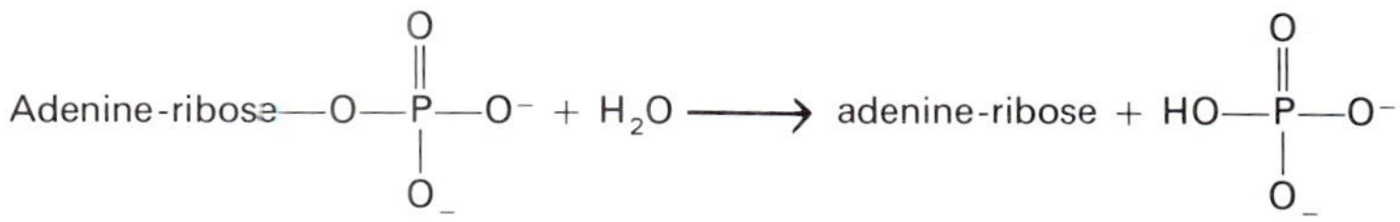

$$\text{AMP} + H_2O \longrightarrow \text{adenosine} + P_i$$
$$\Delta G^{\circ\prime} = -3.4 \text{ kcal mol}^{-1}$$

Figure 14-10
Hydrolysis of AMP.

of pyrophosphate, unlike that released by the ATP hydrolysis, is not generally utilized for metabolic work.

HYDROLYSIS OF AMP

The $\Delta G^{\circ\prime}$ of the hydrolysis of AMP (Figure 14-10) is significantly lower than those of ATP and ADP, i.e., -3.4 *vs.* -7.3 kcal mol^{-1}. The energy potentials of the two types of bonds cleaved in these nucleotides account for the difference in $\Delta G^{\circ\prime}$. The release of the phosphate (or pyrophosphate) group from ATP and ADP involves the hydrolysis of an *anhydride bond* (Figure 14-11), whereas that from AMP involves the hydrolysis of an *ester bond* between the C-5′ hydroxyl group of ribose and the phosphate moiety (Figure 14-12). Hydrolysis of an anhydride bond, in general, occurs with a greater release of energy than hydrolysis of an ester bond, as illustrated in Table 14-2 by the difference between the $\Delta G^{\circ\prime}$ of the hydrolysis of acetic anhydride, a four-carbon anhydride, and that of ethylacetate, a four-carbon ester (Figure 14-13). Nature evolved molecular mechanisms to use the energy potential of the anhydride bond as a common currency for energy production and utilization.

Figure 14-11
A phosphoanhydride bond of ATP and ADP.

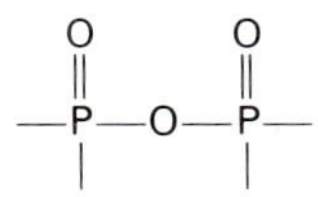

Figure 14-12
Ester bond of AMP.

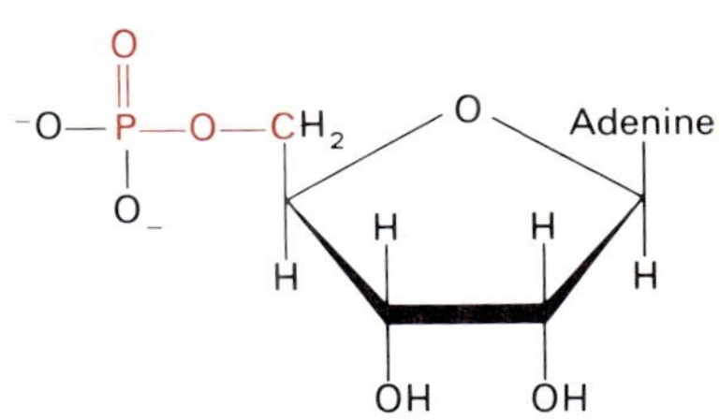

Table 14-2
Standard free energy changes of certain hydrolyses

Reaction	$\Delta G^{\circ\prime}$ (kcal mol^{-1})
Hydrolysis of an anhydride bond	
Acetic anhydride → 2 acetate	−21.8
ATP → ADP + P_i	−7.3
ADP → AMP + P_i	−7.3
Hydrolysis of an ester bond	
Ethyl acetate → ethanol + acetate	−4.7
Glucose 6-phosphate → glucose + phosphate	−3.3
Fructose 6-phosphate → fructose + phosphate	−3.3
sn-Glycerol 1-phosphate → glycerol + phosphate	−2.2
AMP → adenosine + phosphate	−3.4

Figure 14-13
Acetic anhydride and ethyl acetate.

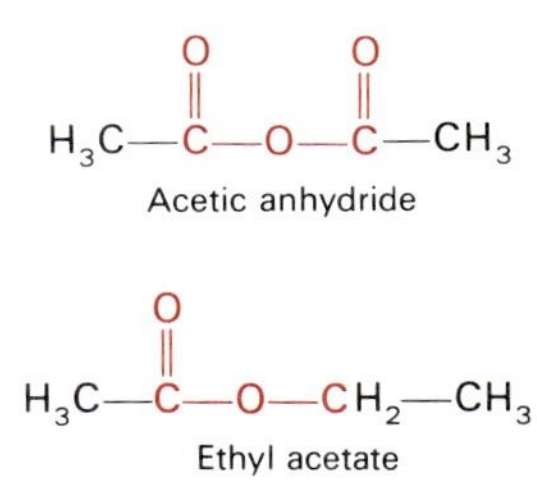

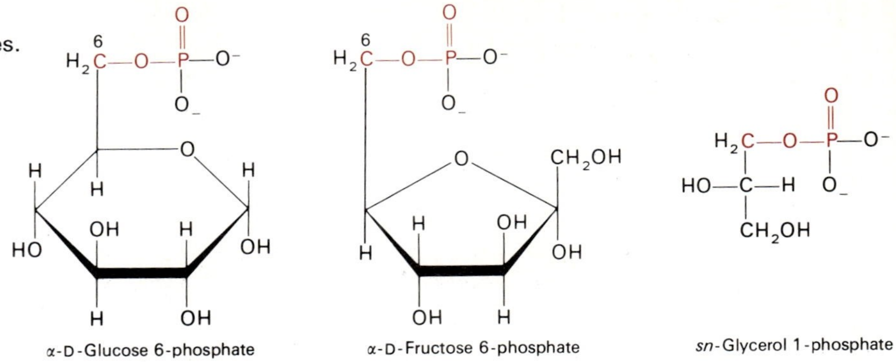

Figure 14-14
Some low-energy-yielding biomolecules.

OTHER PHOSPHORYLATED BIOMOLECULES

Not many of the numerous phosphorylated biomolecules yield the large negative free energy change upon hydrolysis that ATP does because of the chemical nature of the bond cleaved. Glucose and fructose 6-phosphate and *sn*-glycerol 1-phosphate (Table 14-3) are examples of such compounds, because the hydrolysis of each cleaves a low-energy-yielding ester bond (Figure 14-14); they are not considered high-energy biomolecules ($\Delta G^{\circ\prime} \approx -7.0$ kcal mol^{-1} or less). However, as seen in Table 14-3, certain biomolecules have negative $\Delta G^{\circ\prime}$ of hydrolysis that are equivalent to or larger than the -7.3 kcal mol^{-1} of ATP hydrolysis, and all of these compounds have important bioenergetic roles. Phosphoenolpyruvate and 1,3-bisphosphoglyceric acid (also called 1,3-diphosphoglyceric acid), as well as acetyl phosphate, in microorganisms (Figure 14-15), have well-defined metabolic roles which include the conservation and transfer of chemical energy. Phosphocreatine and phosphoarginine (Figure 14-16) are storage forms of metabolic energy. Phosphocreatine is particularly important for provid-

Table 14-3
Phosphorylated biomolecules with a large negative $\Delta G^{\circ\prime}$ of hydrolysis

Biomolecule	$\Delta G^{\circ\prime}$ (kcal mol^{-1})
Phosphoenolpyruvate	−14.8
1,3-Bisphosphoglyceric acid	−11.8
Phosphocreatine	−10.3
Acetyl phosphate	−10.1
Phosphoarginine	− 7.7

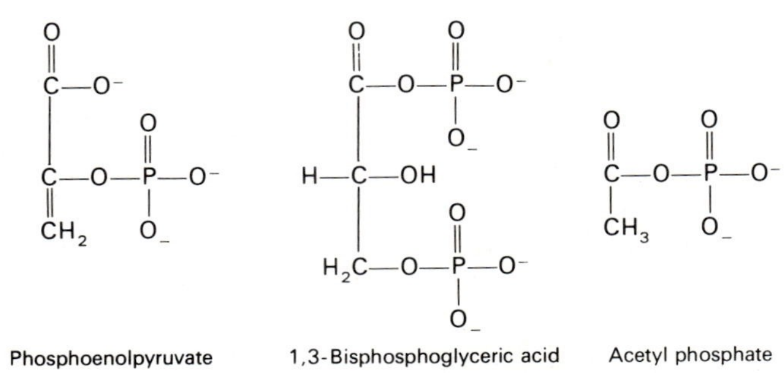

Figure 14-15
Some high-energy-yielding biomolecules.

ing energy to skeletal muscle by serving as a storage molecule for ~P used to produce ATP needed for contraction. When the concentration of muscular ATP is high, a transfer of the terminal phosphate group to creatine occurs, producing phosphocreatine (Figure 14-17). The reaction, which would be thermodynamically unfavorable under standard conditions ($\Delta G^{\circ\prime} = +3.0$ kcal mol^{-1}), is in fact favored and driven under physiological conditions by the high intracellular concentration of ATP. When the supply of ATP is being depleted by muscular work, the concomitant increase in ADP concentration promotes the reverse transfer of the phosphate group, i.e., synthesis of ATP. An analogous energy-storage function, utilizing phosphoarginine, is found in muscles of some invertebrates, e.g., crabs and lobsters.

Figure 14-16
Phosphocreatine and phosphoarginine.

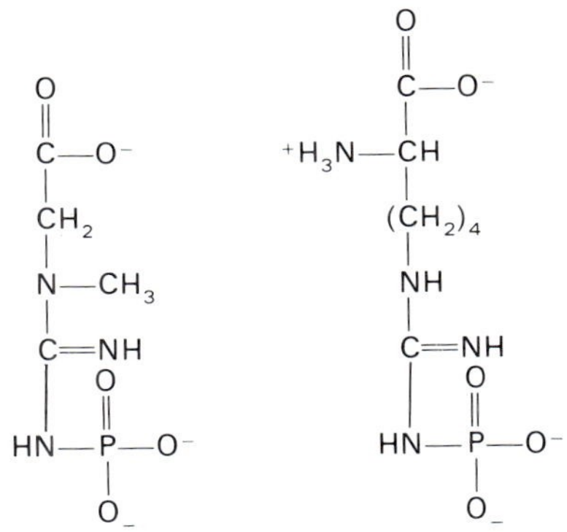

Phosphocreatine Phosphoarginine

$$\text{Creatine} + \text{ATP} \rightleftharpoons \text{phosphocreatine} + \text{ADP}$$
$$\Delta G^{\circ\prime} = +3.0 \text{ kcal mol}^{-1}$$

Figure 14-17
Production of phosphocreatine.

HYDROLYTIC REACTIONS YIELDING A LARGE NEGATIVE $\Delta G^{\circ\prime}$

The large negative change in free energy associated with the hydrolysis of high-energy phosphorylated biomolecules is determined by the chemical nature of the biological reaction. The instability of the reactant (substrate) and/or the stability and subsequent fate of the product(s) greatly influence the completeness of the reaction and, hence, the $\Delta G^{\circ\prime}$ determined.

ATP is an example of a reactant that contributes to the $\Delta G^{\circ\prime}$ of its hydrolysis. At physiological pH (Figure 14-18), the electronegative oxygen atoms in the P=O bonds of ATP attract electrons away from the phosphorus atoms, resulting in a partial positive ($\delta+$) and partial negative ($\delta-$) charge on each of the P and O atoms, respectively. The electrostatic repulsion thus established in the anhydride bonds imposes bond strain within the ATP molecule, which must have sufficient internal energy to overcome this repulsion of like charges. Upon hydrolysis, the internal molecular energy, required to maintain the strained anhydride bond that is cleaved, is released and makes a contribution to the observed negative $\Delta G^{\circ\prime}$ of the reaction. The hydrolysis of the ester bond in AMP proceeds with a smaller negative $\Delta G^{\circ\prime}$ than does that of ADP and ATP because there are fewer forces of electrostatic repulsion associated with the bond cleaved.

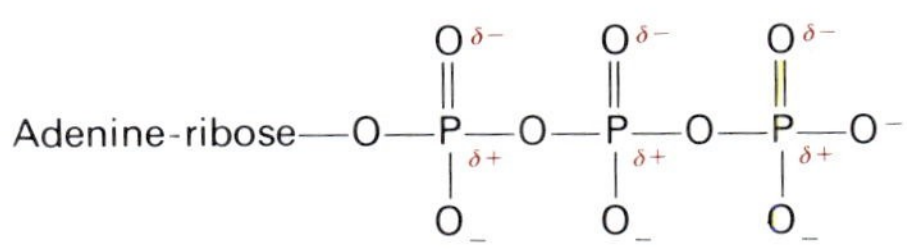

Figure 14-18
ATP at physiological pH.

In the hydrolysis of acetyl phosphate, both the reactant and the product contribute to the large $\Delta G^{\circ\prime}$ of -10.1 kcal mol^{-1}. Both the hydrolysis of the mixed anhydride bond $(-\overset{\displaystyle O}{\overset{\|}{C}}-O-\overset{\displaystyle O}{\overset{\|}{P}}-)$ and the *ionization of the product,* acetic acid, are thermodynamically important. The un-ionized COOH group (pK_a' of 4.7) of acetic acid, initial product of the hydrolysis, is instantaneously ionized at pH 7.0 (Figure 14-19). The ionization effectively drives the hydrolysis to completion by continuously decreasing the concentration of the reaction product, the undissociated acid. One pH unit difference between the pK_a' of an ionizable group of the product and the pH of the reaction environment contributes about -1.36 kcal mol^{-1} to the ΔG. Thus, the approximately -3.0 kcal mol^{-1} difference between the $\Delta G^{\circ\prime}$ of ATP hydrolysis and that of acetyl phosphate can be accounted for by the ionization of acetic acid. An analogous ionization also contributes to the $\Delta G^{\circ\prime}$ (-11.8 kcal mol^{-1}) of 1,3-bisphosphoglyceric acid, i.e., ionization of the product 3-phosphoglyceric acid to 3-phosphoglycerate.

In the case of phosphoenolpyruvate ($\Delta G^{\circ\prime}$ of -14.8 kcal mol^{-1}), a driving force of the hydrolysis is the *isomerization of the product.* The initial product of the reaction is the unstable *enol* form of pyruvate, which instantly tautomerizes to the stable *keto* form (Figure 14-20). The isomerization, as in the case of product ionization, drives the hydrolysis of phosphoenolpyruvate to completion by removal of the product, enol pyruvate. The isomerization contributes about -8.0 kcal mol^{-1} to the $\Delta G^{\circ\prime}$ of the hydrolysis. Thus, ionization and isomerization of the product make the hydrolysis of 1,3-bisphosphoglyceric acid and phosphoenolpyruvate, respectively, biologically important energy-yielding reactions.

Another contributing factor to a large negative $\Delta G^{\circ\prime}$ of hydrolysis is the *stabilization of the product(s) by resonance.* A product's

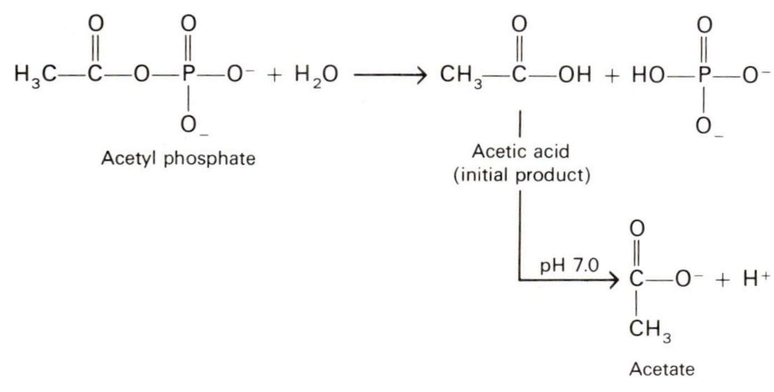

Figure 14-19
Hydrolysis of acetyl phosphate.

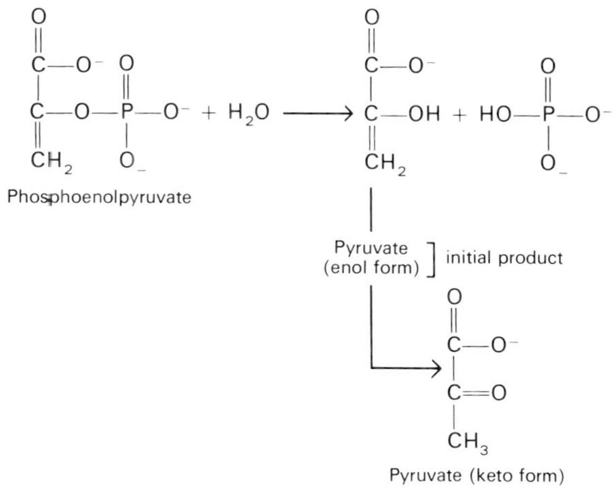

Figure 14-20
Hydrolysis of phosphoenolpyruvate.

stability is influenced by its number of resonance forms, i.e., stability increases with greater resonance. Very stable products decrease the probability of reversal of the reaction (resynthesis of the reactant) and, in this manner, contribute to the completion of the reaction. In the case of phosphocreatine, there are more possible resonance forms (18) for the products, creatine and inorganic phosphate (Figure 14-21), than there are for the reactant, phosphocreatine (12). Product stabilization by resonance also contributes to the $\Delta G^{\circ\prime}$ of ATP hydrolysis, i.e., a total of more resonance forms for ADP and P_i than for ATP.

Whereas the above discussion of negative $\Delta G^{\circ\prime}$ of hydrolysis of certain biomolecules considered the chemical nature of substrates and products, the role of the solvent (water) should be noted. The aqueous environment of living systems also contributes to the thermodynamics of these hydrolytic reactions, i.e., water, as a solvent, is not inert. Specifically, the *hydration (solvation)* of substrates and products that occurs in water (interaction between solutes and solvent) contributes to the observed free energy change of a reaction. Thus, in an analysis of a determined $\Delta G^{\circ\prime}$ value, various thermodynamic considerations of a reaction must be taken into account. For example, the $\Delta G^{\circ\prime}$ of the hydrolysis of ATP includes contributions from electrostatic repulsion, ionization, resonance, and solvation.

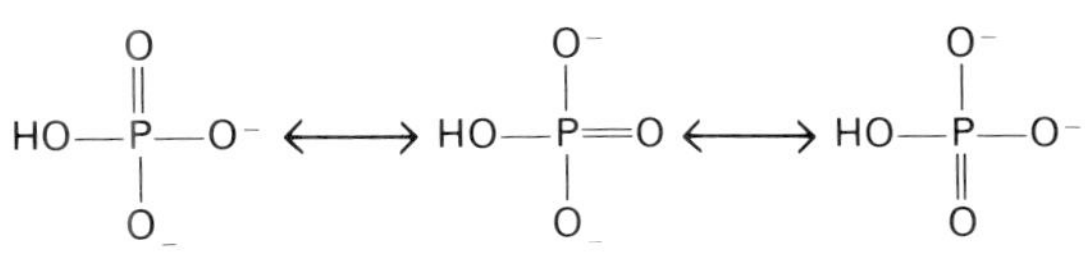

Figure 14-21
Resonance forms of inorganic phosphate at physiological pH.

SYNTHESIS OF NUCLEOSIDE TRIPHOSPHATES

Although ATP is the predominant donor of chemical energy in metabolic processes, the other nucleoside triphosphates are also used as energy-yielding biomolecules. For example, the substrates for RNA synthesis (ATP, GTP, UTP, and CTP) also provide the energy for the biosynthesis of these macromolecules. The synthesis of the various ribo- and deoxyribonucleoside triphosphates (all having the energy potential of ATP) relies primarily on the ability of ATP to transfer a phosphate group to other nucleotides. In the biosynthesis of ribonucleoside triphosphates (NTP), the nucleoside monophosphate (NMP) is the first nucleotide produced (p. 416). The acquisition of the additional two phosphate groups to attain the triphosphate state is accomplished by the broad specificity of two enzymes, *nucleoside monophosphate kinase* and *nucleoside diphosphate kinase.* The term *kinase* generally refers to an enzyme (class: transferase) that catalyzes the transfer of a phosphate group of a nucleoside triphosphate. As seen in Figure 14-22, nucleoside monophosphate kinase catalyzes the production of nucleoside diphosphates (NDP), which are then phosphorylated by nucleoside diphosphate kinase to yield the nucleoside triphosphates (NTP).

Figure 14-22
Biosynthesis of ribonucleoside triphosphates. NMP, NDP, and NTP represent nucleoside 5′-mono-, di-, and triphosphate, respectively.

Nucleoside monophosphate kinase: $\text{NMP} + \text{ATP} \rightleftharpoons \text{NDP} + \text{ADP}$

Nucleoside diphosphate kinase: $\text{NDP} + \text{ATP} \rightleftharpoons \text{NTP} + \text{ADP}$

Two biosynthetic routes have been described for the production of the deoxyribonucleoside triphosphates (dNTP), precursors of DNA. One pathway (Figure 14-23) involves the production of deoxyribonucleoside diphosphates (dNDP) from ribonucleoside diphosphates (NDP) by a multienzyme system and the subsequent phosphorylation of dNDP by nucleoside diphosphate kinase to yield deoxyribonucleoside triphosphates. The other pathway (Figure 14-23), found in some microorganisms, produces deoxyribonucleoside triphosphates directly from ribonucleoside triphosphates.

Figure 14-23
Biosynthesis of deoxyribonucleoside triphosphates. NDP and NTP represent nucleoside 5′-di- and triphosphates, respectively; dNDP and dNTP represent deoxyribonucleoside 5′-di- and triphosphates, respectively.

Multienzyme system: $\text{NDP} \longrightarrow \text{dNDP}$

Nucleoside diphosphate kinase: $\text{ATP} + \text{dNDP} \rightleftharpoons \text{ADP} + \text{dNTP}$

or

Some microorganisms: $\text{NTP} \longrightarrow \text{dNTP}$

The enzymatic versatility of nucleoside diphosphate kinase is broad enough to effect the transfer of a phosphate group from *any* NTP or dNTP to *any* NDP or dNDP. Hence, the enzyme provides a valuable mechanism for the metabolic flow of phosphate groups among all classes of nucleosides, particularly di- and triphosphates (Figure 14-24).

Figure 14-24
Some reversible reactions catalyzed by nucleoside diphosphate kinase.

$$\text{ATP} + \text{GDP} \rightleftharpoons \text{ADP} + \text{GTP}$$
$$\text{GTP} + \text{UDP} \rightleftharpoons \text{GDP} + \text{UTP}$$
$$\text{GTP} + \text{dADP} \rightleftharpoons \text{GDP} + \text{dATP}$$
$$\text{ATP} + \text{UDP} \rightleftharpoons \text{ADP} + \text{UTP}$$
$$\text{dCTP} + \text{ADP} \rightleftharpoons \text{ATP} + \text{dCDP}$$

SYNTHESIS OF ATP FROM AMP AND ADP

The hydrolysis of ATP by orthophosphate cleavage yields ADP and P_i and by pyrophosphate cleavage AMP and PP_i, with the PP_i being subsequently hydrolyzed to 2 P_i. ADP and P_i, being immediate precursors for much of the cellular synthesis of ATP, can be converted readily into ATP by a number of energy-yielding processes that catalyze the reaction ADP + P_i → ATP (Chapters 15, 17, and 18). The synthesis of ATP from AMP first requires the enzyme *adenylate kinase* (Figure 14-25), which catalyzes the transfer of a phosphate group from ATP to AMP to produce 2 ADP. Thus, AMP is converted into ADP at the expense of ATP, and then both ADP are made available for ATP synthesis. Adenylate kinase also provides an enzymatic mechanism to generate an ATP (and AMP) from two ADP molecules.

$$\text{AMP} + \text{ATP} \rightleftharpoons \text{ADP} + \text{ADP}$$

Figure 14-25
Adenylate kinase reaction.

The facile transfer of phosphate groups among the nucleotides establishes a metabolic network for the flow of biological energy (energy potential of the phosphoanhydride bond) that ensures adequate production of the various high-energy biomolecules other than ATP needed for a variety of biological functions, e.g., CTP, UTP, and GTP for lipid, polysaccharide, and protein synthesis, respectively. The fountainhead for this flow of chemical energy is the intracellular production of ATP (photosynthesis and/or anaerobic and aerobic oxidation), without which no biologically utilizable energy would be generated for secondary transfer to nucleotides.

SUMMARY

The maximum change in free energy of a reaction is called ΔG, which is expressed in calories per mole. A negative ΔG value is indicative of an exergonic (or spontaneous) reaction, which releases free energy. Conversely, a positive ΔG indicates an endergonic reaction, which requires an input of free energy. The thermodynamics of a reaction are often expressed as $\Delta G°'$, which is the free energy change determined at equilibrium under specific experimental conditions at pH 7. Living systems rely on exergonic reactions for their energy needs, with the hydrolysis of adenosine triphosphate (ATP) being the principal energy-yielding reaction ($\Delta G^{\circ\prime} = -7.3$ kcal mol^{-1}). ATP is hydrolyzed either to ADP and P_i or to AMP and PP_i; the latter hydrolysis is usually a physiologically irreversible reaction because of the subsequent hydrolysis of PP_i. Other biomolecules, e.g., phosphoenolpyruvate and 1,3-bisphosphoglyceric acid, besides ATP have a large negative $\Delta G^{\circ\prime}$ of hydrolysis, and they also have important roles in the bioenergetic processes of organisms. There are a number of factors that can account for the large negative $\Delta G^{\circ\prime}$ values obtained by the hydrolysis of certain biomolecules, including bond strain in the reactant because of internal electrostatic repulsion, ionization or isomerization of the product(s), stabilization of the product by resonance forms, and solvation effects.

The synthesis of ribonucleoside di- and triphosphates occurs by the phosphorylation of ribonucleoside mono- and diphosphates, respectively, with the enzymes nucleoside monophosphate and nucleoside diphosphate kinases catalyzing the reactions. ATP is usually the principal phosphate-group donor in these syntheses. For the production of the deoxyribonucleoside triphosphate precursors of DNA, most organisms convert ribonucleoside diphosphates into deoxyribonucleoside diphosphates, which are then phosphorylated by nucleoside diphosphate kinase.

QUESTIONS

1. Which of the following statements are true?
   (a) The product(s) of a reaction always has a higher free energy content than the reactant(s).
   (b) The change in free energy of a reaction is called ΔG.
   (c) ΔG is expressed in calories per liter.
   (d) Free energy is released in an exergonic reaction.
   (e) A spontaneous reaction is one that is exergonic.
   (f) Living systems rely on exergonic reactions to provide energy for biological processes.
   (g) Living systems use endergonic reactions to provide energy for the biosynthesis of biomolecules.
   (h) $\Delta G^{\circ\prime}$ values are determined at pH 0.0.
   (i) K_{eq} is the ratio of the concentration of product(s) to reactant(s) at equilibrium.
   (j) The $\Delta G^{\circ\prime}$ of hydrolysis of acetyl phosphate is less negative at pH 5 than at pH 7.
   (k) The ionization of a carboxyl group of a product contributes more to a negative $\Delta G^{\circ\prime}$ than does the ionization of a phosphate group.
   (l) ATP has no resonance forms.
   (m) The keto form of pyruvate is an initial product of phosphoenolpyruvate hydrolysis.
   (n) Hydrolysis of acetic anhydride yields an acid and an alcohol.
   (o) Glucose 6-phosphate is a monophosphate ester.

2. Would the hydrolysis of ATP to AMP and $2P_i$ depicted in the following two sets of reactions release approximately the same amount of energy?

   (a) $\mathrm{ATP} \xrightarrow{H_2O} \mathrm{ADP} + P_i$, then $\mathrm{ADP} \xrightarrow{H_2O} \mathrm{AMP} + P_i$

   (b) $\mathrm{ATP} \xrightarrow{H_2O} \mathrm{AMP} + PP_i$, then $PP_i \xrightarrow{H_2O} 2\,P_i$

3. Consider the hypothetical reaction (pH 7, 25°C, and 1 atmosphere) $A \rightleftharpoons B + C$. The initial concentration

of A is 0.2 *M*; at equilibrium, only 1 per cent of A remains. Calculate (a) the K_{eq} and (b) the $\Delta G^{\circ\prime}$ of the reaction. (c) What is the $\Delta G^{\circ\prime}$ for the synthesis of A from B and C?

4. (a) Using the $\Delta G^{\circ\prime}$ values for the hydrolysis of ATP (Table 14-2) and of phosphocreatine (Table 14-3), illustrate (by coupling the reactions) why the $\Delta G^{\circ\prime}$ for the production of phosphocreatine from ATP and creatine is $+3.0$ kcal mol^{-1}.
   (b) What is the $\Delta G^{\circ\prime}$ for the production of phosphoarginine from ATP and arginine? Use the $\Delta G^{\circ\prime}$ values for the hydrolysis of phosphoarginine (Table 14-3) and ATP (Table 14-2) to calculate the answer.
   (c) In the next chapter, the following reaction (which produces cellular ATP) is discussed:

   $$\text{Phosphoenolpyruvate} + \text{ADP} \rightarrow \text{pyruvate} + \text{ATP}$$

   Using the $\Delta G^{\circ\prime}$ of phosphoenolpyruvate and ATP hydrolysis (Tables 14-2 and 14-3), determine the $\Delta G^{\circ\prime}$ of the reaction.

5. It has been estimated that within a red blood cell [ATP], [ADP], and [P_i] are maintained under steady-state conditions at 1.85×10^{-3} *M*, 1.4×10^{-4} *M*, and 10^{-3} *M*, respectively. Calculate the ΔG of ATP hydrolysis under these steady-state conditions at pH 7, 25°C, and 1 atmosphere.

6. Devise a hypothetical scheme depicting a series of enzymatic reactions that could account for the following nucleotide interconversions

   2 ATP, 2 GMP, and 1 CDP →
   1 ATP, 1 AMP, 1 GDP, 1 GTP, and 1 CMP

   Utilize the following enzymes: nucleoside monophosphate kinase, nucleoside diphosphate kinase, and adenylate kinase (Figures 14-22, 14-24, and 14-25).

SUGGESTED READINGS

Bartley, W., and H. L. Kornberg, eds., *Essays in Cell Metabolism.* New York: Wiley-Interscience, 1970. Diversified topics on bioenergetics written in depth but, for the most part, understandable by the novice.

Fasman, G. D., ed., *Handbook of Biochemistry and Molecular Biology,* 3rd ed. Sec. D, *Physical and Chemical Data.* Vol. 1. Boca Raton, Fla.: CRC Press, 1976. Reference source for reliable data on the free energies of hydrolysis.

Kalckar, H. M., *Biological Phosphorylation: Development of Concepts.* Englewood Cliffs, N.J.: Prentice-Hall, 1969. A narrative whose theme is developed with the use of a collection of reprints encompassing the classic investigations in bioenergetics.

Krebs, H. A., and H. L. Kornberg, *Energy Transformations in Living Matter.* Berlin: Springer-Verlag, 1957. A classic on the energetics of life.

Lehninger, A. L., *Bioenergetics,* 2nd ed. Menlo Park, Calif.: Benjamin/Cummings, 1971. Highly recommended elementary treatment of the molecular basis of energy transformations in living systems.

Segel, I. H., *Biochemical Calculations,* 2nd ed. New York: Wiley, 1976. Good problem set on bioenergetics in the section on energy-yielding and -requiring reactions in Chapter 3.

Wood, W. B., J. H. Wilson, R. M. Benbow, and L. E. Hood, *Biochemistry—A Problems Approach,* 2nd ed. Menlo Park, Calif.: Benjamin/Cummings, 1981. Chapter 9 contains another good problem set on bioenergetics.

Articles

Lehninger, A. L., "Energy Transformation in the Cell." *Scientific American,* 202(5):102 (offprint #69), 1960.

Stumpf, P. K., "ATP." *Scientific American,* 188(4):85, 1953.

ANAEROBIC SYNTHESIS OF ATP (GLYCOLYSIS) AND PENTOSE PHOSPHATE PATHWAY

15

INTRODUCTION

One of nature's most primitive means of extracting chemical energy from organic compounds is *glycolysis,* the *anaerobic conversion of a molecule of glucose to two lactic acid molecules* (Figure 15-1). The series of chemical reactions constituting the glycolytic pathway is believed to have evolved in the reduced environment (no oxygen) in which primordial life arose. Glycolysis represents a successful attempt by early forms of life to transform some of the chemical energy of glucose, produced by photosynthetic processes, into other forms containing utilizable metabolic energy, i.e., solar energy →→ glucose →→ ATP. The continuing biological importance of this metabolic route for obtaining energy is evident by the fact that nearly all forms of life, aerobic and anaerobic, utilize glycolysis. In fact, the aerobic oxidation of carbohydrates, a subsequent evolutionary development, requires that sugars first be degraded anaerobically to pyruvate by glycolysis. All carbohydrates, from monosaccharides to polysaccharides, need to be routed metabolically through glycolysis to yield their chemical

Glycolysis is taken from the Greek words *glycos*, meaning sweet, and *lysis*, meaning *loosening* or *dissolution*.

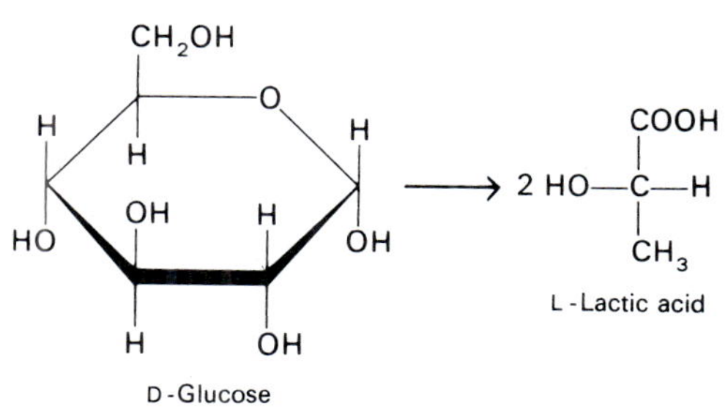

Figure 15-1
Glycolysis.

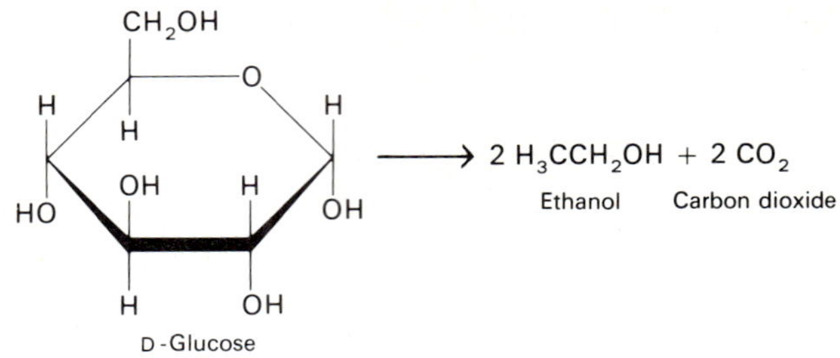

Figure 15-2
Alcoholic fermentation.

energy to biological systems. Starches, the polysaccharide-storage forms of glucose in plants, are the main dietary source of glucose for most humans.

Alcoholic fermentation, whereby a molecule of *glucose is converted into two ethanol and two carbon dioxide molecules* (Figure 15-2), is the same anaerobic catabolic process as glycolysis except for the final stages, which produce different end products. In glycolysis, the last metabolic intermediate, pyruvate, is reduced to lactate; in alcoholic fermentation, it is converted into ethanol and CO_2 (Figure 15-3). The biological studies on fermentation initiated early in the nineteenth century have culminated in the present century with a detailed biochemical description of the catabolic pathway for the anaerobic utilization of carbohydrates. By 1940, the accumulated results of interrelated studies on yeast (alcoholic fermentation) and muscle (glycolysis) provided biochemistry with its first major enzymatic pathway. Glycolysis is sometimes called the *Embden-Meyerhof pathway.* Gustav Embden and Otto Meyerhof (Nobel Prize, 1922) were German biochemists who, among a number of other distinguished scientists, made significant contributions toward the molecular elucidation of glycolysis. As a developing science, biochemistry accrued significant benefits from these pioneering studies on carbohydrate metabolism, because they introduced many new techniques and concepts into research on enzymes and intermediary metabolism.

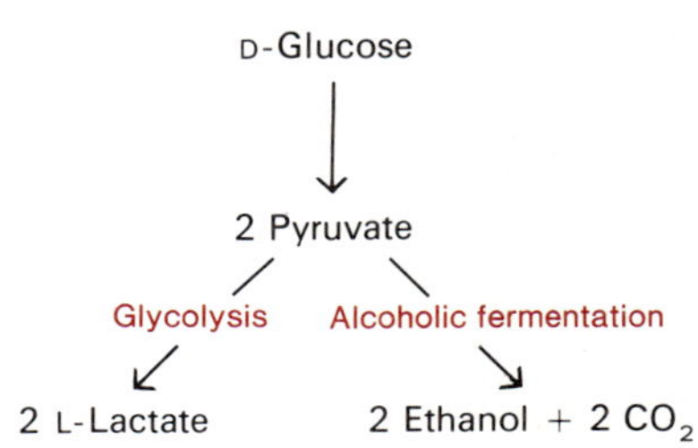

Figure 15-3
Difference between glycolysis and alcoholic fermentation.

REACTIONS OF THE GLYCOLYTIC PATHWAY

The conversion of D-glucose into two L-lactate molecules involves 11 enzymatic steps (Figure 15-4). If glycogen, the polysaccharide-storage form of carbohydrates in animals, supplies the glucose for glycolysis, then 12 reactions are involved. The elucidation of the pathway entailed the systematic identification and characterization of each enzyme to establish its role in the anaerobic metabolism of glucose. As the individual steps of glycolysis were described, they were fitted together, like pieces of a puzzle; this ultimately led to a biochemical description of the entire process.

Phosphorylation of Glucose

The entrance of D-glucose into glycolysis requires its phosphorylation to *D-glucose 6-phosphate* in a reaction usually catalyzed by *hexokinase* (Figure 15-5). Hexokinase, as implied by its name, can catalyze the transfer of a phosphate group from ATP to a variety of hexoses, e.g., D-glucose, D-fructose, and D-mannose. *Glucokinase,* a different kinase with a specificity for glucose, also phosphorylates glucose at the C-6 position; because of its high K_M for the sugar, this enzyme is believed to be of importance when the concentration of glucose in the liver is high, e.g., after a carbohydrate-rich meal. The priming of glucose for glycolysis, by phosphorylation, requires the orthophosphate cleavage of ATP ($\Delta G^{\circ\prime} = -7.3$ kcal mol^{-1}); hence, an input of chemical energy is needed to initiate the process for the anaerobic synthesis of ATP. (The phosphorylation of glucose also serves to trap glucose intracellularly.)

Derivation of Glucose 6-Phosphate from Glycogen

Glycogen of liver and muscle provides D-glucose 6-phosphate for glycolysis by a two-step process. The first reaction, catalyzed by *phosphorylase* a [an *α-(1-4)-glucan phosphorylase*], sequentially removes glucosyl residues from glycogen to yield *D-glucose 1-phosphate* (Figure 15-6). Phosphorylase *a* is capable of removing glucosyl residues bonded in α-(1-4) linkages but cannot cleave residues at the α-(1-6) branch points of glycogen. The regulation of glycolysis is highly influenced by hormonal *(epinephrine)* control of phosphorylase *a,* the enzyme that performs the important function of converting glucose from its storage form into a form readily utilizable for energy production. The release of epinephrine from the

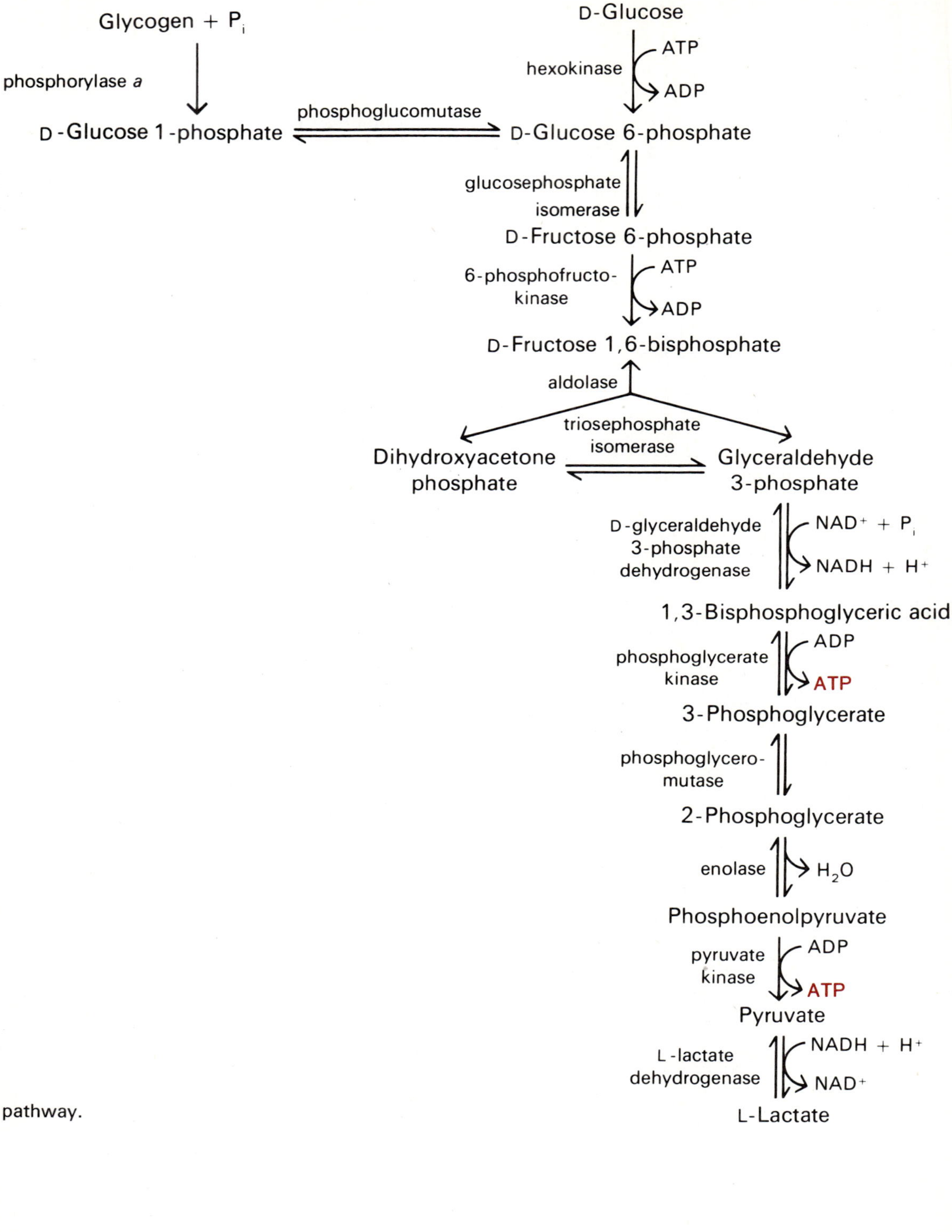

Figure 15-4
The glycolytic pathway.

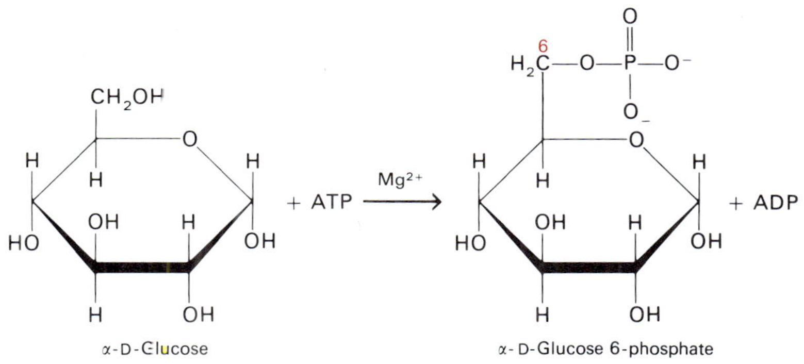

$\Delta G^{\circ\prime} = -4.0$ kcal mol^{-1}

Figure 15-5
Hexokinase reaction.

adrenal glands results in an enhancement of phosphorylase *a* activity in the liver and in muscle (p. 549), i.e., increases the rate of glycolysis.

In the second reaction, catalyzed by *phosphoglucomutase,* D-glucose 1-phosphate is converted into D-glucose 6-phosphate (Figure 15-7). Phosphoglucomutase is a *phosphoenzyme* which requires the phosphorylated form of a particular seryl residue (phosphoserine) for activity. The enzyme utilizes *glucose 1,6-bisphosphate* (Figure 15-7) as a cofactor. The proposed mechanism for phosphoglucomutase's catalytic action is of interest because the cofactor can actually be thought of as an intermediate in the reversible reaction (Figure 15-8). Although the overall reaction (Figure 15-7) indicates an intramolecular transfer of a phosphate group (C-1 ⇌ C-6), the interconversion of glucose 1- and 6-phosphate occurs by an exchange of the C-1 and C-6 phosphates with that of the enzyme.

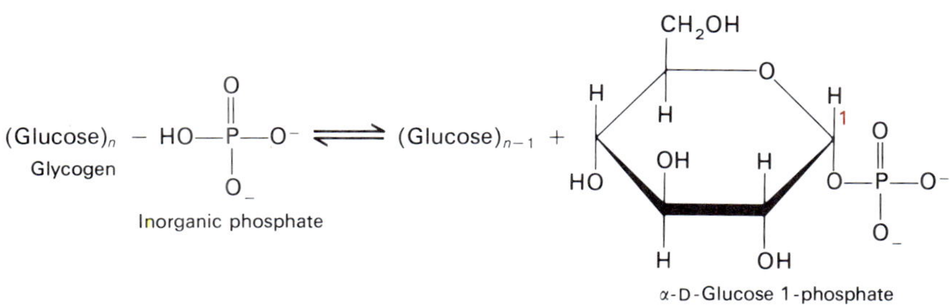

$\Delta G^{\circ\prime} = +0.73$ kcal mol^{-1}

Figure 15-6
Phosphorylase *a* reaction.

Figure 15-7
Phosphoglucomutase reaction.

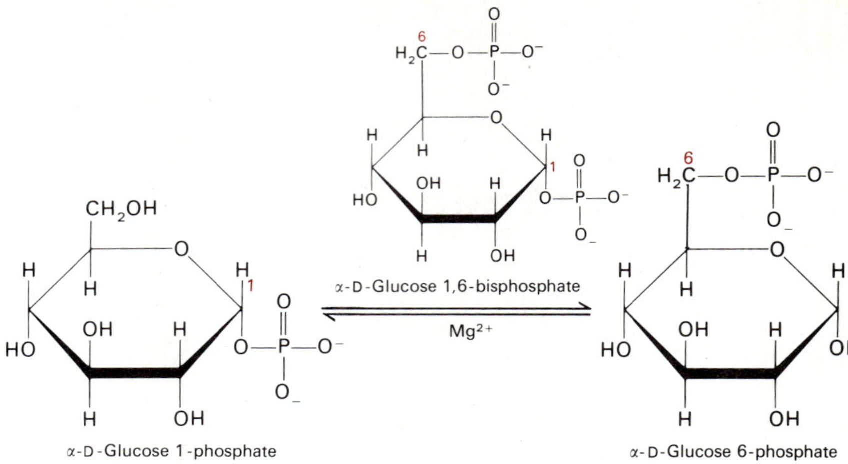

$\Delta G^{\circ\prime} = -1.74$ kcal mol^{-1}

Production of D-Fructose 1,6-Bisphosphate

The first reaction D-glucose 6-phosphate undergoes in glycolysis is an isomerization, catalyzed by *glucose phosphate isomerase,* to produce *D-fructose 6-phosphate* (Figure 15-9). D-Fructose 6-phosphate is then phosphorylated (Figure 15-10) in a reaction catalyzed by *6-phosphofructokinase* to yield *D-fructose 1,6-bisphosphate* (also called D-fructose 1,6-diphosphate). The formation of D-fructose 1,6-bisphosphate requires the expenditure of the energy potential of an anhydride bond of ATP. Hence, the initial reactions of glycolysis converting one molecule of D-glucose into D-fructose 1,6-bisphosphate require an input of the chemical energy of two ATP molecules. The synthesis of D-fructose 1,6-bisphosphate from glycogen, however, requires the hydrolysis of only one ATP. (It should be noted that in the overall scheme of carbohydrate metabolism, ATP utilization is balanced because an extra ATP is required for glycogen synthesis.)

Figure 15-8
Simplified depiction of the proposed mechanism of the phosphoglucomutase reaction.

Phosphoenzyme + glucose 1-phosphate

$\rightleftharpoons$

Dephosphoenzyme + glucose 1 6-bisphosphate

$\rightleftharpoons$

Phosphoenzyme + glucose 6-phosphate

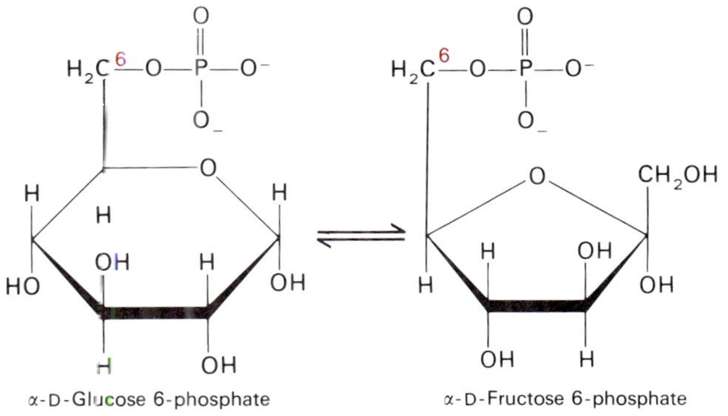

$\Delta G^{\circ\prime} = +0.4$ kcal mol^{-1}

Figure 15-9
Glucosephosphate isomerase reaction.

6-Phosphofructokinase as a Regulatory Enzyme

6-Phosphofructokinase is an allosteric enzyme and, as such, is critical for the metabolic regulation of the pathway. Whereas phosphorylase *a* responds to an extracellular stimulus (epinephrine), 6-phosphofructokinase is sensitive to the intracellular level of several allosteric effectors. *ADP* and *AMP,* as positive effectors, stimulate 6-phosphofructokinase activity; *ATP, citric acid,* and *long-chain fatty acids,* as negative effectors, inhibit the reaction. This mode of allosteric regulation is characteristic of the molecular mechanisms used to regulate intracellular synthesis of ATP. When supplies of ATP or of aerobically oxidizable fuels (citric acid and long-chain fatty acids) are plentiful, the high concentrations of these biomolecules serve as a cellular signal to decrease the rate of glycolysis, by inhibiting 6-phosphofructokinase. Conversely, when ATP is being rapidly utilized by cells, the increase in ADP and AMP

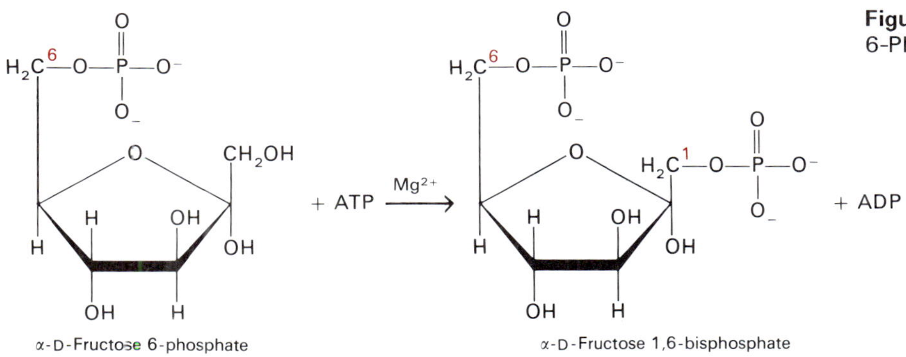

$\Delta G^{\circ\prime} = -3.40$ kcal mol^{-1}

Figure 15-10
6-Phosphofructokinase reaction.

concentrations triggers an increased synthesis of ATP, by stimulating 6-phosphofructokinase and other respiratory enzymes. The ratio of [ATP] to [ADP] and [AMP] is therefore an important monitoring system used by a cell to control the levels of readily available chemical energy (ATP). The role of citric acid and long-chain fatty acids as allosteric regulators of an enzyme which catalyzes a glycolytic reaction suggests that an evolutionary change in this anaerobic enzyme probably occurred after the evolution of aerobic systems.

Metabolic Production of D-Glyceraldehyde 3-Phosphate

The next series of reactions concerns the production of phosphorylated three-carbon sugars (trioses) and their utilization to synthesize ATP. The trioses are obtained by the catalytic cleavage of D-fructose 1,6-bisphosphate by *aldolase* to yield *dihydroxyacetone phosphate* and *D-glyceraldehyde 3-phosphate* (Figure 15-11), with the latter serving as the next intermediate in glycolysis (Figure 15-4). For glycolysis, dihydroxyacetone phosphate is readily converted into D-glyceraldehyde 3-phosphate by the enzyme *triosephosphate isomerase* (Figure 15-12). Thus, by means of these two reactions, one molecule of D-fructose 1,6-bisphosphate is converted into two molecules of D-glyceraldehyde 3-phosphate.

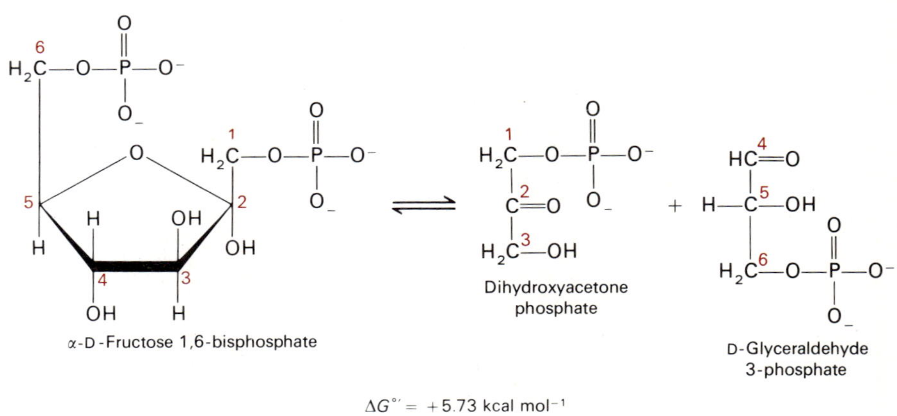

Figure 15-11
Aldolase reaction.

sn-Glycerol 3-Phosphate Dehydrogenase

Dihydroxyacetone phosphate is also the immediate precursor of sn-*glycerol 3-phosphate* (Figure 15-13); this reversible reaction, catalyzed by sn-*glycerol 3-phosphate dehydrogenase,* is an exam-

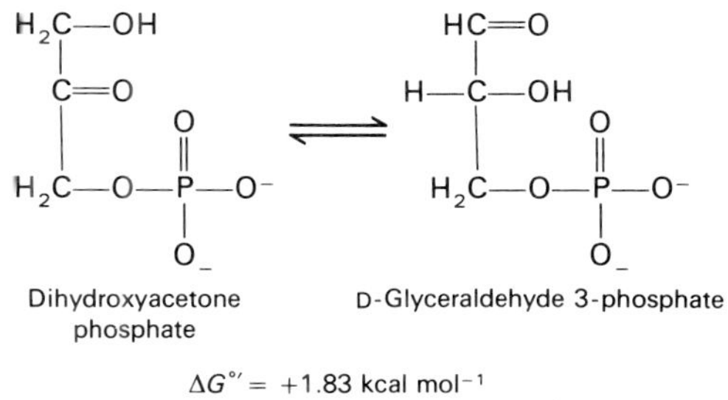

Figure 15-12
Triosephosphate isomerase reaction.

ple of a reaction that interconnects different biological processes. If required, dihydroxyacetone phosphate can be removed from the glycolytic pathway to furnish *sn*-glycerol 3-phosphate for lipid biosynthesis, i.e., glycolysis can provide the carbons of glucose for biosynthetic purposes. The reaction also constitutes the entry point of glycerol and *sn*-glycerol 3-phosphate (derived from dietary lipids) into glycolysis (glycerol → *sn*-glycerol 3-phosphate → dihydroxyacetone phosphate → D-glyceraldehyde 3-phosphate).

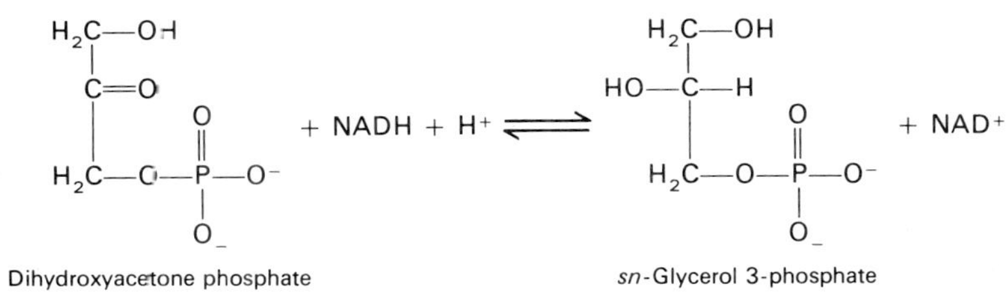

Figure 15-13
sn-Glycerol 3-phosphate dehydrogenase reaction.

Thermodynamics of the Aldolase and Triosephosphate Isomerase Reactions

For glycolytic purposes, both aldolase ($\Delta G^{\circ\prime}$ of $+5.73$ kcal mol^{-1}) and triosephosphate isomerase ($+1.83$ kcal mol^{-1}) appear to catalyze thermodynamically unfavorable reactions, i.e., the reverse reaction is energetically favored under standard conditions. The reaction catalyzed by triosephosphate isomerase, for example, greatly favors the production of dihydroxyacetone phosphate and, at equilibrium, about 93 per cent of the triosephosphate is dihydroxyacetone phosphate. The fact that these endergonic reactions proceed with efficiency in glycolysis is attributed to the energetically favorable subsequent utilization of D-glyceraldehyde 3-phosphate, end product of the two reactions, and to the concentration dependence of the ΔG of the aldolase reaction. At equilibrium, given a fructose 1,6-bisphosphate concentration of $0.1\,M$, approximately 97 per cent would remain in the hexose form at pH 7; however, if the concentration is 0.1 m*M*, about 40 per cent would exist as the hexose bisphosphate and 60 per cent as the triose-phosphate, at equilibrium. The intracellular concentration of fructose 1,6-bisphosphate is, in fact, less than 0.1 m*M*. Thus, the aldolase and triosephosphate isomerase reactions are illustrative of biological reactions whose laboratory-determined $\Delta G^{\circ\prime}$ values do not, at first glance, reflect their *in vivo* efficiency, which is determined by the immediate metabolic environment, e.g., by intracellular concentrations of substrates and products.

Oxidation of D-Glyceraldehyde 3-Phosphate

The next reaction in glycolysis generates 1,3-bisphosphoglyceric acid (Figure 15-14), whose high negative $\Delta G^{\circ\prime}$ of hydrolysis (-11.8 kcal mol^{-1}) is accounted for by the conservation (by phosphorylation) of the chemical energy released by the oxidation of D-glyceraldehyde 3-phosphate. The reaction is catalyzed by *D-glyceraldehyde 3-phosphate dehydrogenase* and requires inorganic phosphate and the coenzyme NAD^+, which is reduced to NADH. The overall reaction (Figure 15-14) is a summation of the coupling of an exergonic reaction (oxidation) to an endergonic reaction (phosphorylation). Figure 15-15 depicts a simplified description of the energetics of the reaction and illustrates how the energy released by the oxidation reaction ($\Delta G^{\circ\prime}$ of -10.3 kcal mol^{-1}) is used for the subsequent phosphorylation ($+11.8$ kcal mol^{-1}) of the oxidized product, *3-phosphoglycerate.* The small positive difference of $+1.5$ kcal mol^{-1} between the two $\Delta G^{\circ\prime}$ values indicates that the concentrations of substrates and products will determine the direction of the *in vivo* reaction.

Figure 15-14
D-Glyceraldehyde 3-phosphate dehydrogenase reaction.

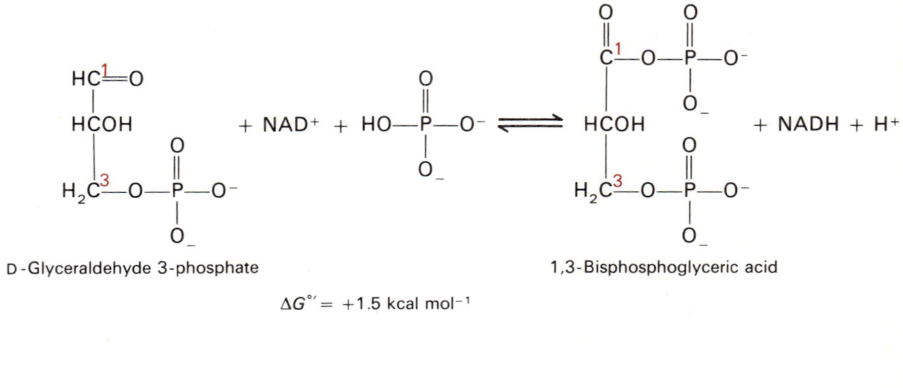

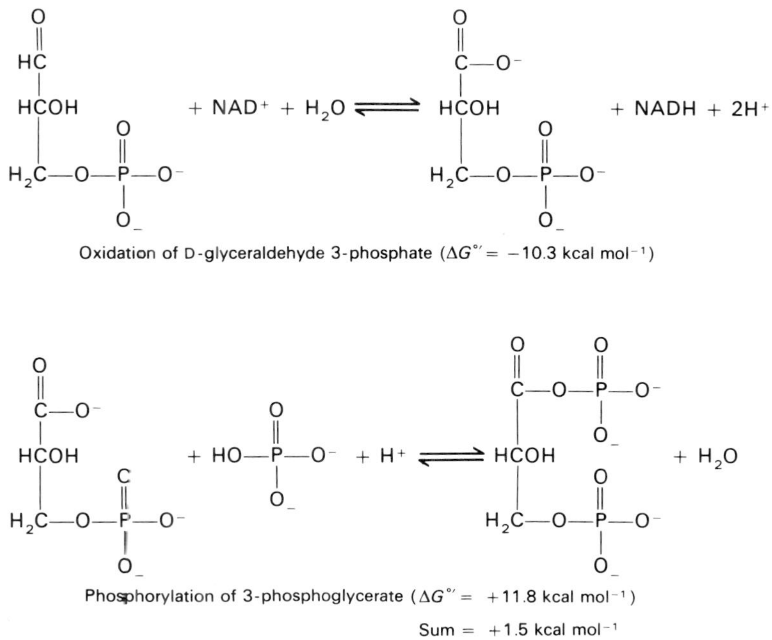

Figure 15-15
Simplified description of the energetics of the D-glyceraldehyde 3-phosphate dehydrogenase reaction.

First Synthesis of ATP

In the next reaction, catalyzed by *phosphoglycerate kinase,* the chemical energy of 1,3-bisphosphoglyceric acid is used to synthesize ATP from ADP and to produce *3-phosphoglycerate* (Figure 15-16). Thus, in this reaction, some of the oxidative energy released and conserved in the previous reaction is converted into a form that is readily utilizable by bioenergetic processes. This type of ATP production is called *substrate-level phosphorylation,* i.e., one high-energy substrate furnishing the energy to produce one

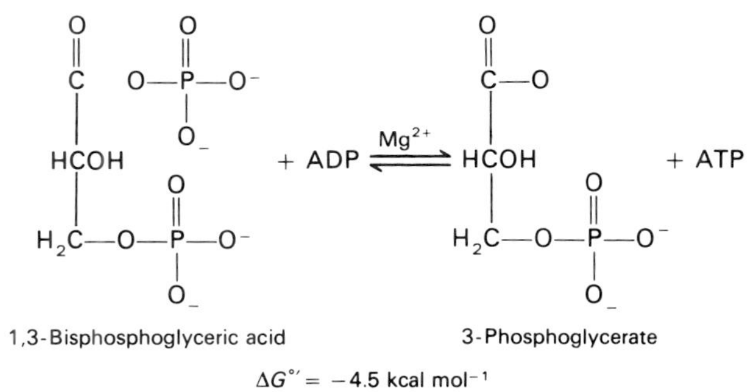

Figure 15-16
Phosphoglycerate kinase reaction.

nucleoside triphosphate from the corresponding diphosphate (another substrate).

Figure 15-17
Phosphoglyceromutase reaction.

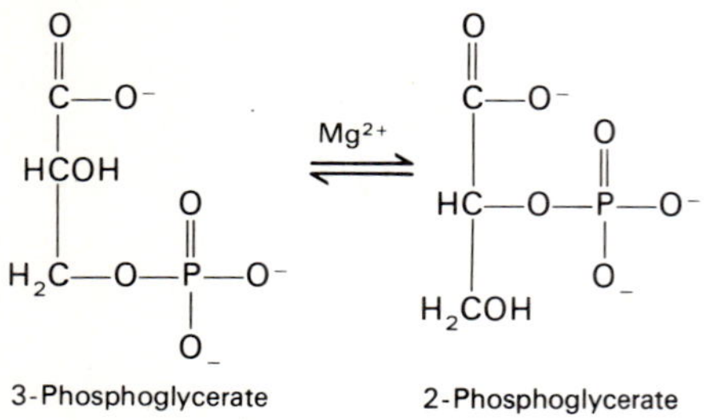

Second Synthesis of ATP

The next three reactions constitute the second synthesis of ATP in glycolysis. In the first reaction, catalyzed by *phosphoglyceromutase,* an intramolecular transfer of a phosphate group occurs when 3-phosphoglycerate is converted into *2-phosphoglycerate* (Figure 15-17). This reaction is analogous to the phosphoglucomutase reaction previously described (Figure 15-7). The second reaction (Figure 15-18) is a dehydration catalyzed by *enolase* and

Figure 15-18
Enolase reaction.

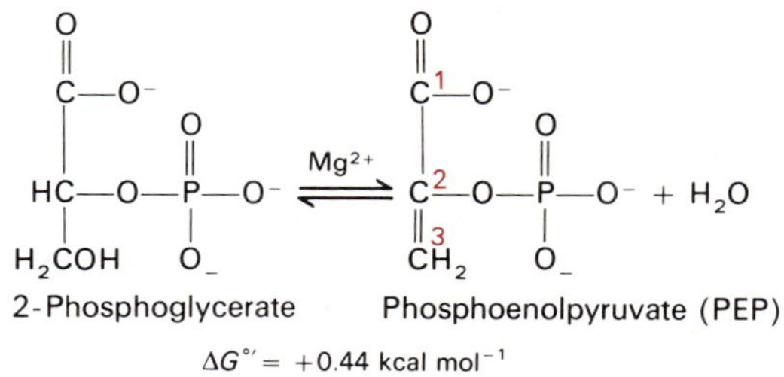

results in the production of *phosphoenolpyruvate (PEP),* a biomolecule with a very large negative $\Delta G^{\circ\prime}$ of hydrolysis (-14.8 kcal mol^{-1}). The enolase reaction can be regarded as an *intramolecular oxidoreduction,* with the C-2 and C-3 of 2-phosphoglycerate becoming more oxidized and reduced, respectively. The next reaction, catalyzed by *pyruvate kinase,* utilizes some of the chemical energy released by phosphoenolpyruvate's hydrolysis for the second synthesis of ATP (from ADP) and the production of *pyruvate* (Figure 15-19).

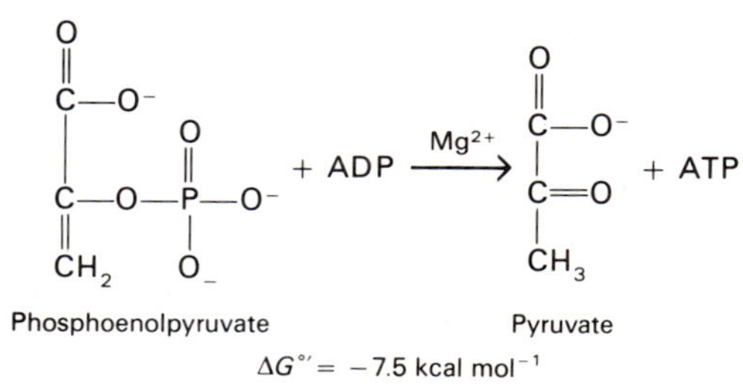

Figure 15-19
Pyruvate kinase reaction.

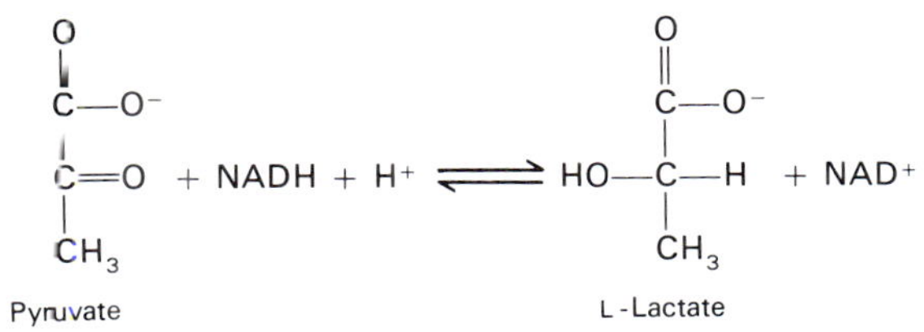

$\Delta G^{\circ\prime} = -6.0$ kcal mol^{-1}

Figure 15-20
L-Lactate dehydrogenase reaction.

Production of Lactate

The last step in glycolysis is the reduction of pyruvate to produce *lactate* (Figure 15-20). The reaction is catalyzed by *L-lactate dehydrogenase* (p. 144) and requires the reduced form of NAD^+, NADH, as a coenzyme. The utilization of NADH in the reaction establishes a *balance of the coenzyme* because it provides for the regeneration of NAD^+, required for the oxidation reaction catalyzed by D-glyceraldehyde 3-phosphate dehydrogenase (Figure 15-21). This balance within the glycolytic pathway assures that glycolysis will not cease because of a lack of the critical coenzyme, NAD^+.

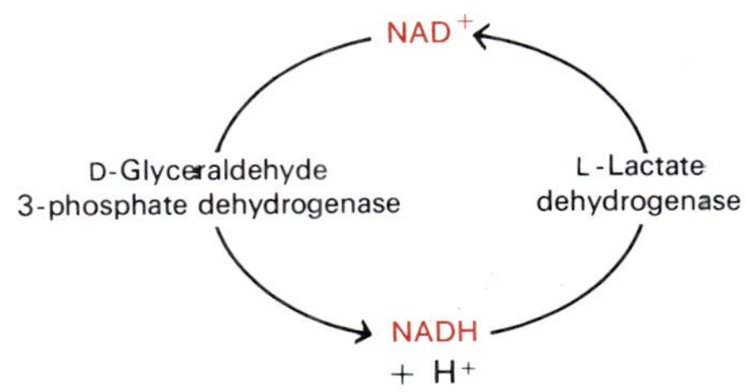

Figure 15-21
Balance of NAD^+ and NADH in glycolysis.

EFFICIENCY OF ENERGY CONSERVATION

The evolutionary development of glycolysis resulted in a biochemical mechanism whereby some of the chemical energy of glucose could be extracted by electron rearrangement for use in other biological processes. Chemically, the conversion of glucose into two lactate molecules can be abbreviated as follows:

$$\underset{(C_6H_{12}O_6)}{\text{Glucose}} + 2\ \text{ADP} + 2\ P_i \rightarrow 2\ \underset{(C_3H_6O_3)}{\text{lactate}} + 2\ \text{ATP} + 2\ H_2O$$

The above expression is derived by summarizing the total input (substrates) and output (products) of the pathway and then canceling the common molecules on both sides of the reaction equation (Table 15-1). The reactions accounting for the net production of two molecules of ATP are listed in Table 15-2.

The thermodynamics of the conversion of 1 mole of glucose to 2 moles of lactate are:

$$\text{Glucose} \rightarrow 2\ \text{lactate}$$
$$\Delta G^{\circ\prime} = -47.0\ \text{kcal mol}^{-1}$$

Table 15-1
Summary of the glycolytic pathway

Substrates: Glucose + 2 ATP + 4 ADP + 2 P_i + 2 NAD^+ + 2 NADH + 2 H^+
Products: 2 Lactate + 4 ATP + 2 ADP + 2 NAD^+ + 2 NADH + 2 H^+ + 2 H_2O
Net overall reaction: Glucose + 2 ADP + 2 P_i → 2 lactate + 2 ATP + 2 H_2O

Table 15-2
Net production of ATP in glycolysis

Reaction	ATP utilized (−) or produced (+)
Glucose → glucose 6-phosphate	−1
Fructose 6-phosphate → fructose 1,6-bisphosphate	−1
2 (1,3-Bisphosphoglyceric acid) → 2 (3-phosphoglycerate)	+2
2 Phosphoenolpyruvate → 2 pyruvate	+2
Net:	+2

On a molar basis, 47 kcal are released by glycolysis, of which 14.6 kcal are utilized for the synthesis of 2 moles of ATP (2 × 7.3 kcal). The *efficiency of energy conservation* is therefore *31 per cent.* If glycogen is the source of glucose for glycolysis, then the net synthesis of ATP is 3 moles because no input of ATP is needed to produce D-glucose 6-phosphate (Table 15-2); the hexokinase reaction is not required. Thus, 21.9 kcal (3 × 7.3 kcal) are conserved by the utilization of glycogen, and the efficiency of conservation is 49.8 per cent, since the $\Delta G°'$ of glycolysis starting with glycogen is −44 kcal mol^{-1}. These energy conservation estimates are only approximations, since the concentration of each cellular reactant is not molar, as required for the $\Delta G°'$ values used in the calculations. That fraction of chemical energy released by glycolysis that is not conserved in the formation of ATP is liberated as heat and, in warm-blooded animals, contributes to the maintenance of body temperature.

ALCOHOLIC FERMENTATION

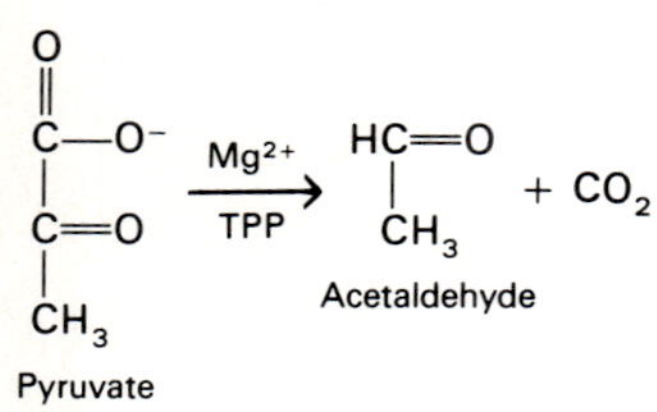

Figure 15-22
Pyruvate decarboxylase reaction.

The only difference between the pathways for alcoholic fermentation and glycolysis is the fate of the pyruvate that is produced. In alcoholic fermentation, pyruvate is first decarboxylated to *acetaldehyde* and CO_2 in a reaction catalyzed by *pyruvate decarboxylase.* This enzyme requires Mg^{2+} and TPP (thiamin pyrophosphate) as cofactors (Figure 15-22). The acetaldehyde produced is then reduced to *ethanol* by *alcohol dehydrogenase,* which utilizes NADH as a coenzyme (Figure 15-23). The reaction generates NAD^+ and, hence, the same coenzyme balance noted for glycolysis is established for alcoholic fermentation.

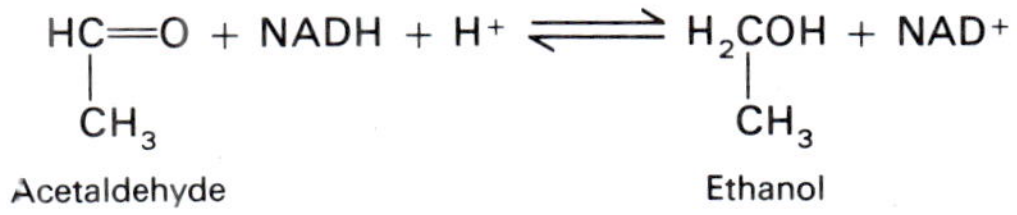

Figure 15-23
Alcohol dehydrogenase reaction.

The exergonic nature of the conversion of glucose into 2 moles of ethanol and 2 of CO_2 is expressed:

$$\text{Glucose} \rightarrow 2\ \text{ethanol} + 2\ CO_2$$

$$\Delta G^{\circ\prime} = -40.0\ \text{kcal mol}^{-1}$$

The net synthesis of ATP from glucose by alcoholic fermentation is two molecules, as in glycolysis. Therefore, on a molar basis, 14.6 kcal (2 × 7.3 kcal) of the 40.0 kcal released by alcoholic fermentation is conserved, a *36.5 per cent efficiency of energy conservation.*

2,3-Bisphosphoglycerate Synthesis and Degradation

1,3-Bisphosphoglyceric acid and 3-phosphoglycerate, in addition to their glycolytic functions, are key biomolecules in the metabolism of 2,3-bisphosphoglycerate (2,3-P_2-glycerate), the allosteric effector of hemoglobin (p. 115). Through a reaction catalyzed by *bisphosphoglycerate mutase* (Figure 15-24), 1,3-bisphosphoglyceric acid and 3-phosphoglycerate serve as precursors in the syn-

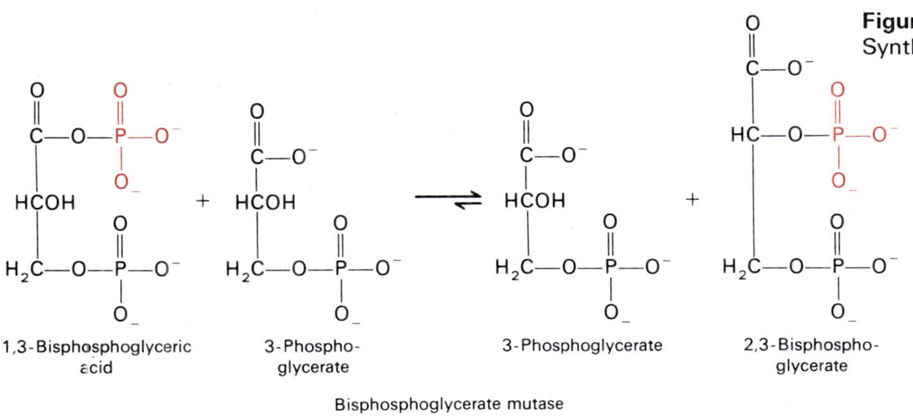

Figure 15-24
Synthesis of 2,3-bisphosphoglycerate.

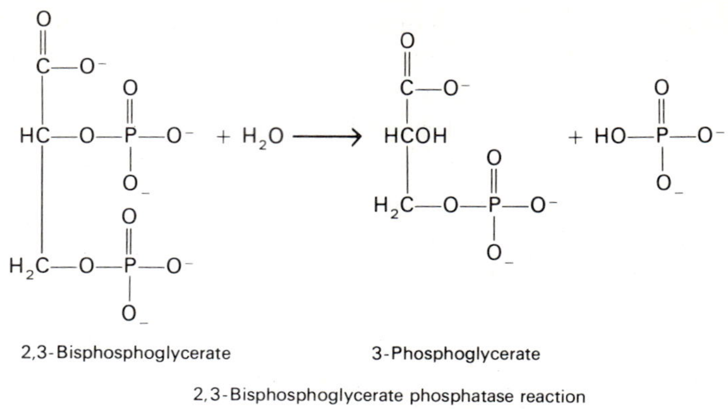

Figure 15-25
Degradation of 2,3-bisphosphoglycerate.

thesis of 2,3-P_2-glycerate. Degradation of 2,3-P_2-glycerate occurs by the hydrolysis of a phosphate ester to yield 3-phosphoglycerate, in a reaction catalyzed by *2,3-bisphosphoglycerate phosphatase* (Figure 15-25). Glycolytic production of 1,3-bisphosphoglyceric acid and 3-phosphoglycerate is therefore directly linked to the physiological functioning of hemoglobin. Metabolic disorders involving glycolysis that affect the synthesis of 2,3-P_2-glycerate result in changes in the effective affinity of hemoglobin for oxygen.

GLUCONEOGENESIS FROM LACTATE

Strenuous muscle activity requires the continuous production of ATP by glycolysis. The lactate produced, upon rapid accumulation, contributes to the painful ache associated with vigorous, prolonged use of skeletal muscle. Lactate is released into the bloodstream and is transported to the liver, where it can be converted back into glucose; this synthesis of glucose is called *gluconeogenesis.* [As will be discussed (p. 306), biomolecules other than lactate can serve as precursors for glucose synthesis.] Extensive glycolysis in skeletal muscles and concomitant gluconeogenesis in the liver exemplify a physiological interrelationship whereby, with blood serving as the transport system, the liver regenerates the fuel (glucose) needed for the continuous production of energy (ATP) in active skeletal muscle (Figure 15-26).

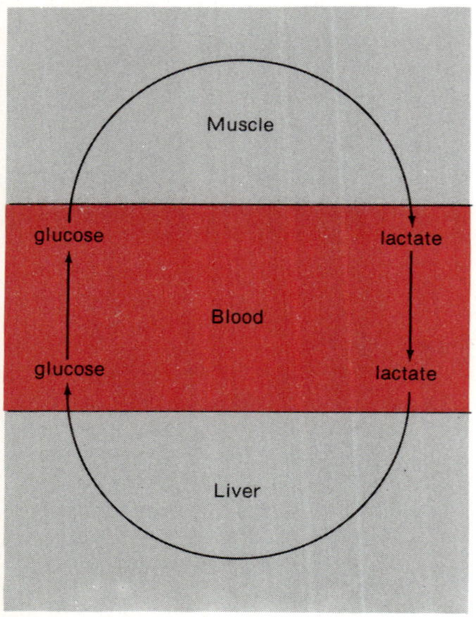

Figure 15-26
Schematic of the interrelationship between glycolysis in muscle and gluconeogenesis in liver.

Physiologically Irreversible Reactions of Glycolysis

Although most of the reactions of glycolysis are physiologically reversible, three are not; gluconeogenesis is therefore not merely a reversal of glycolysis. The three irreversible steps in glycolysis are the production of (1) pyruvate from phosphoenolpyruvate, (2) D-fructose 1,6-bisphosphate from D-fructose 6-phosphate, and (3) D-glucose 6-phosphate from D-glucose. The chemical energy required to drive these endergonic reactions cannot be furnished by reversal of the glycolytic steps involved. Note that the $\Delta G°'$ for the production of phosphoenolpyruvate from pyruvate and ATP (Figure 15-19) would be $+7.5$ kcal mol^{-1} and that the synthesis of both D-glucose and D-fructose 6-phosphate, by reversal of glycolysis (Figures 15-5 and 15-10), would require the synthesis of ATP from ADP. In the early decades of this century, it was generally believed that the anabolic and catabolic processes pertaining to a certain biomolecule utilize the same reversible pathway. This was a mistaken belief and, in fact, many biomolecules are synthesized and degraded by pathways that are distinctly different.

Synthesis of Phosphoenolpyruvate from Pyruvate

The generation of phosphoenolpyruvate (PEP) from pyruvate requires an input of the chemical energy of two nucleoside triphosphates, ATP and GTP. The first energy-utilizing reaction is catalyzed by *pyruvate carboxylase* and, in addition to pyruvate, requires ATP, Mg^{2+}, biotin, and acetyl-CoA (p. 258) to produce *oxaloacetate,* ADP, and P_i (Figure 15-27). The acetyl-CoA requirement is understandable in view of the reaction's relationship to the tricarboxylic acid (TCA) cycle, i.e., aerobic oxidation of glucose (Chapter 16). The acetyl portion of acetyl-CoA is a two-carbon moiety derived from glucose that is oxidizable by the TCA cycle, and oxaloacetate is a key intermediate in this oxidation cycle. For the pyruvate carboxylase reaction, acetyl-CoA acts as a *positive allosteric effector,* i.e., high levels of the oxidizable effector prime the TCA cycle by increasing oxaloacetate synthesis.

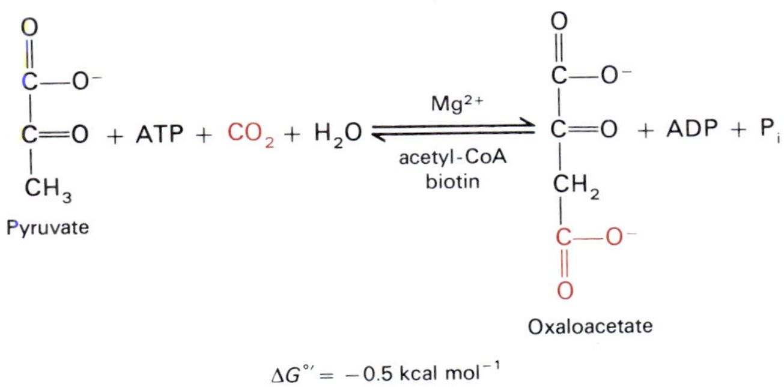

Figure 15-27
Pyruvate carboxylase reaction.

Figure 15-28
Phosphoenolpyruvate carboxykinase reaction.

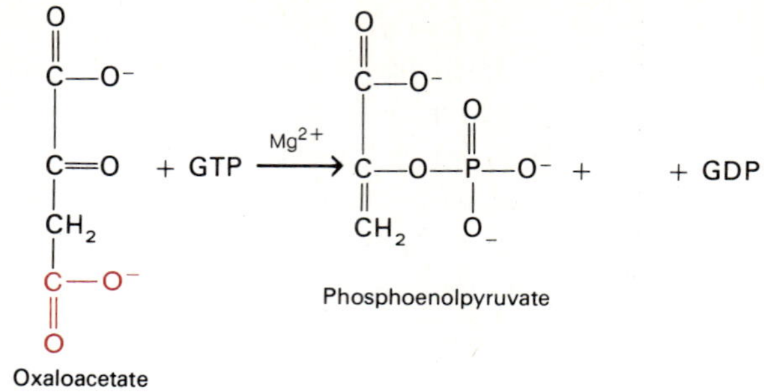

$\Delta G^{\circ\prime} = +1.0$ kcal mol^{-1}

The second reaction, catalyzed by *phosphoenolpyruvate carboxykinase,* utilizes oxaloacetate and GTP to produce PEP, CO_2, and GDP (Figure 15-28). The production of PEP is summarized in Figure 15-29). The *in vivo* production of PEP from pyruvate, however, is more complex than accounted for by these two reactions,

Figure 15-29
Summation of the production of phosphoenolpyruvate from pyruvate.

$$\text{Pyruvate} + \text{ATP} + \text{GTP} \rightleftharpoons \text{phosphoenolpyruvate} + \text{ADP} + \text{GDP} + P_i$$

$\Delta G^{\circ\prime} = +0.5$ kcal mol^{-1}

because the pyruvate carboxylase reaction occurs within the mitochondria and the phosphoenolpyruvate carboxykinase reaction takes place in the cytosol. Because of the selective permeability of the mitochondrial membrane, oxaloacetate must be indirectly transported out into the cytosol as *L-malate,* where it is then reoxidized to oxaloacetate for PEP synthesis (Figure 15-30). Once PEP is synthesized, the reactions leading the production of D-fructose 1,6-bisphosphate are physiologically reversible.

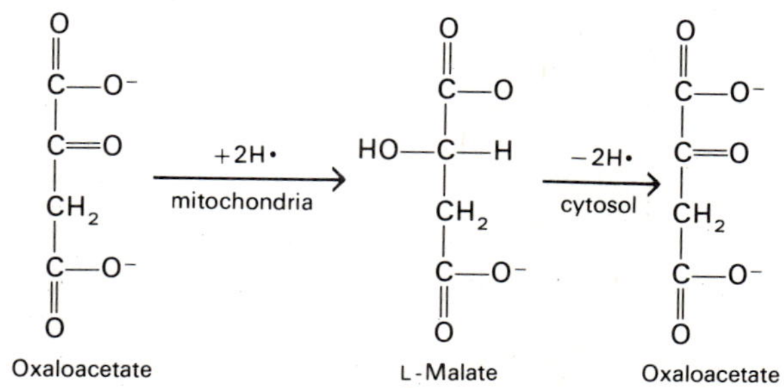

Figure 15-30
Biochemical mechanism for the transfer of mitochondrial oxaloacetate to the cytosol.

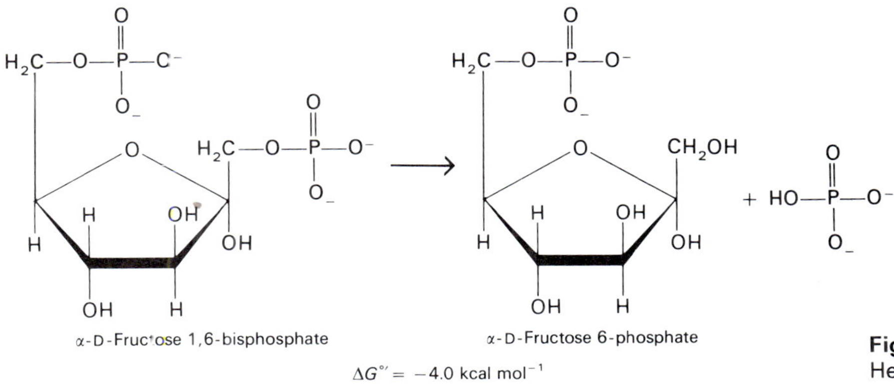

Figure 15-31
Hexose bisphosphatase reaction.

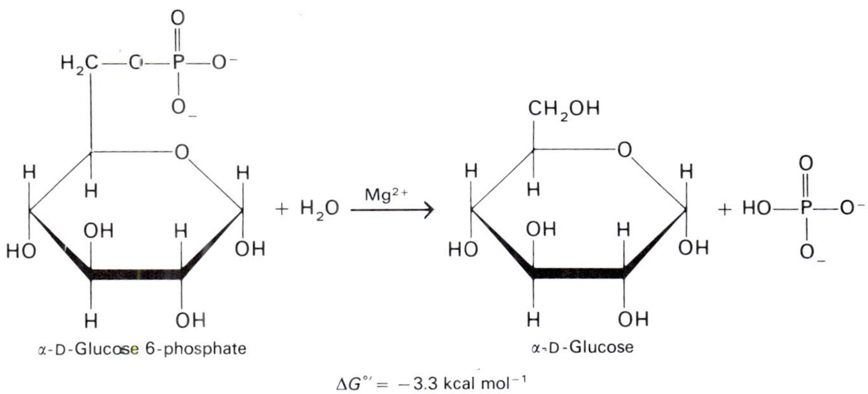

Figure 15-32
D-Glucose 6-phosphatase reaction.

Production of D-Fructose 6-Phosphate and D-Glucose

In gluconeogenesis the production of *D-fructose 6-phosphate* is accomplished by the hydrolytic removal of the C-1 phosphate from D-fructose 1,6-bisphosphate by *hexose bisphosphatase* (Figure 15-31). Similarly, *D-glucose* is produced from D-glucose 6-phosphate in a reaction catalyzed by *D-glucose 6-phosphatase* (Figure 15-32). Because hexose bisphosphatase is an allosteric enzyme, the reaction it catalyzes serves as a control point for gluconeogenesis.

UTILIZATION OF FRUCTOSE AND GALACTOSE

Two monosaccharides readily utilized by humans for energy are fructose (from fruits and sucrose) and galactose (from lactose of milk). In vertebrates, D-fructose enters glycolysis as *D-glyceralde-*

1. D-Fructose + ATP $\xrightarrow{\text{fructokinase}}$ D-fructose 1-phosphate + ADP

2. D-Fructose 1-phosphate $\xrightarrow[\text{aldolase}]{\text{fructose 1-phosphate}}$ D-glyceraldehyde + dihydroxyacetone phosphate

3. D-Glyceraldehyde + ATP $\xrightarrow{\text{kinase}}$ D-glyceraldehyde 3-phosphate + ADP

Figure 15-33
Utilization of D-fructose.

hyde 3-phosphate and *dihydroxyacetone phosphate* by a series of reactions that occurs in the liver (Figure 15-33). Note that, as with glucose, complete fructose utilization requires an expenditure of two ATP molecules. In infants, utilization of D-galactose requires first its phosphorylation to *D-galactose 1-phosphate* and then its conversion into *D-glucose 1-phosphate,* with the latter reactions utilizing *uridine diphosphate (UDP)* as the carrier of glucose and galactose (Figure 15-34). Note that although *hexose 1-phosphate uridyltransferase* appears to convert D-galactose 1-phosphate into D-glucose 1-phosphate, the actual production of glucose from galactose occurs in the reaction catalyzed by *UDP-glucose 4-epimerase.*

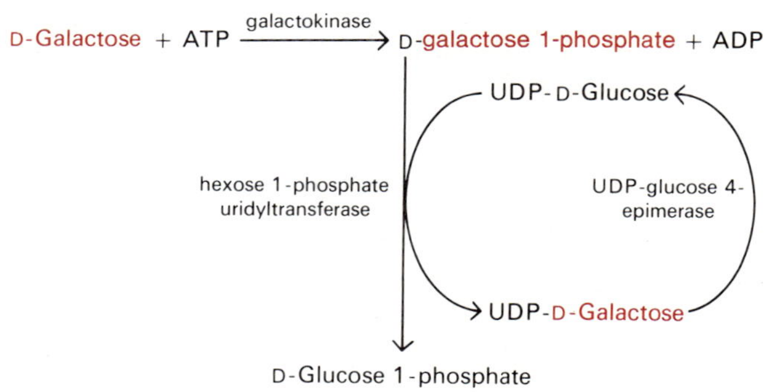

Figure 15-34
Utilization of D-galactose.

ABNORMAL CARBOHYDRATE METABOLISM

There are many well-documented inborn errors of metabolism involving genetic defects that affect the utilization of carbohydrates. The array of disorders includes the defective utilization of glucose, galactose, and fructose, as well as enzymatic deficiencies in glycolysis and gluconeogenesis.

Galactosemia

Galactosemia was one of the first inborn errors to be investigated on a biochemical basis, and the definitive studies by Herman Kalckar, like those on sickle-cell anemia (p. 120), were instrumental in generating the current research interest in molecular diseases. Galactosemia, the *inability to utilize dietary galactose,* is an autosomal recessive trait characterized by a deficiency in *hexose 1-phosphate uridyltransferase* (Figure 15-34). Because of the lactose content of milk, galactosemia afflicts the child homozygous for the altered gene early in life. The main clinical features are hepatomegaly, cataracts, and mental retardation. If detected early, a diet low in galactose provides an effective treatment for the disorder.

Fructose Intolerance (Fructosemia)

Fructose intolerance is also an autosomal recessive trait that becomes apparent soon after an afflicted child is weaned, i.e., when dietary fructose and/or sucrose is increased. The severely ill infant experiences *hypoglycemia* (low blood glucose) and vomiting and suffers from malnutrition. The child cannot utilize fructose because of a deficiency in *fructose 1-phosphate aldolase* activity (Figure 15-33). Interestingly, fructosemic children often have an aversion to sweets and fruits (and, also, remarkably few dental caries). The syndrome is treated by placing the patient on a low-fructose diet.

Hexokinase Deficiency of Hemolytic Anemia

An example of an inborn error involving the utilization of glucose is the deficiency of *hexokinase of red blood cells* (autosomal recessive trait), which results in hemolytic anemia. Glycolysis, the energy pathway for red blood cells, is impaired by insufficient production of *D-glucose 6-phosphate* (Figure 15-5), and the metabolic defect results in a premature breakdown of the cells. This *biochemical block* also affects the production of 2,3-bisphosphoglycerate (Figure 15-24); because of a *lowered concentration of* *2,3-P_2-glycerate,* an afflicted individual's hemoglobin has an unusually *high oxygen affinity.*

Pyruvate Kinase Deficiency of Erythrocytes

This metabolic error also involves glycolysis in red blood cells because of a deficiency in *pyruvate kinase* (Figure 15-19). The reduced synthesis of pyruvate from phosphoenolpyruvate results in an accumulation of glycolytic intermediates, including 2,3-P_2-glycerate. Because of the *increase in 2,3-P_2-glycerate concentration,* the affected individual's hemoglobin has a *low affinity for oxygen.* Pyruvate kinase deficiency in red blood cells (autosomal recessive trait) produces a severe anemia that results in death during the first years of life if not treated by blood transfusions and removal of the spleen (to retard red blood cell destruction). Generally, inborn errors of glycolysis create anemic conditions—a reflection of the importance of the pathway for normal maintenance of red blood cells.

Glycogen Storage Disease I

Glycogen storage disease I is a syndrome that arises from a deficiency in *D-glucose 6-phosphatase,* an enzyme associated with gluconeogenesis (Figure 15-32). This metabolic error is of special interest because of the various medical problems created by the lack of normal enzyme activity. The inability of glucose 6-phosphate to be converted into D-glucose results in hypoglycemia and increased levels of glycogen (glycogen storage) since a portion of the phosphorylated glucose is routed to glycogen synthesis. *Lipidemia* (high lipid levels in the blood) also occurs and, as a consequence of high blood levels of cholesterol, *xanthoma* formation (yellowish tumors under the skin) occurs. Also, some patients develop *gout,* caused by excess uric acid production (p. 425). Survival to adulthood is rare. These multiple *(pleiotropic)* effects of a single mutation stress the general metabolic importance of glucose 6-phosphate in normal metabolism.

PENTOSE PHOSPHATE PATHWAY

Another fundamentally important pathway of carbohydrate metabolism in most forms of life is the *pentose phosphate pathway.* The physiological importance of this pathway does not include the generation of ATP (like glycolysis) but, rather, the production of essential NADPH and the synthesis and/or interconversion of various sugars, including the production of glycolytic intermediates. Also known as the hexose monophosphate or phosphogluconate pathway, the cytosolic pentose phosphate pathway (Figure 15-35) was elucidated about the middle of this century by several noted biochemists, including Fritz Lipmann, Efraim Racker, Bernard Horecker, and their colleagues.

The pathway commences with the oxidation of *glucose 6-phosphate* to *6-phosphoglucono-δ-lactone;* the reaction is catalyzed by *glucose 6-phosphate dehydrogenase,* which requires $NADP^+$ and Mg^{2+} as cofactors. The enzyme was the first $NADP^+$-linked dehydrogenase described (by Otto Warburg in the early 1930s, p. 166). In the second reaction, *lactonase* cleaves the lactone ring to yield the oxidized derivative of *glucose 6-phosphogluconate.* In the next reaction, *6-phosphogluconate dehydrogenase,* which also requires $NADP^+$ and Mg^{2+}, catalyzes the production of CO_2 and the pentose *D-ribulose 5-phosphate.* The pentose can serve as a substrate for two different reactions, one of which is catalyzed by *phosphopentose isomerase* to produce *D-ribose 5-phosphate.* (Figure 15-35).

In some metabolic systems, the pathway's principal function is to produce NADPH and ribose; hence, it terminates after the syn-

Warburg, a student of Emil Fischer, is considered by many to be the greatest biochemist of the first half of this century. His first publication (with Fischer) appeared in 1904 and his last in 1970, the year of his death at age 87.

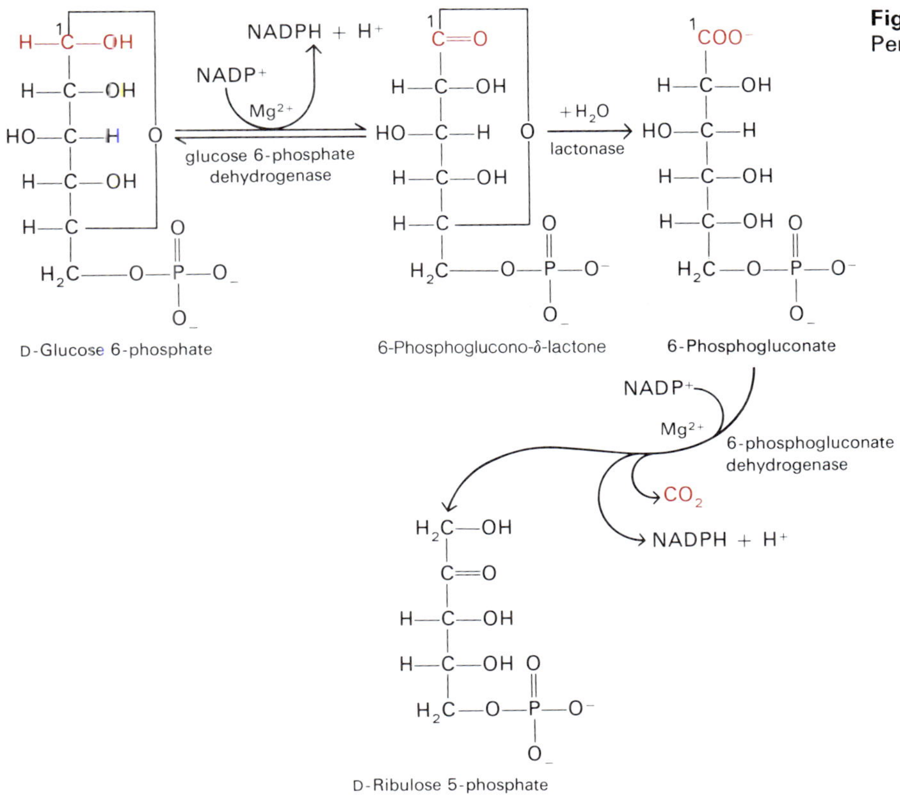

Figure 15-35
Pentose phosphate pathway.

Figure 15-35, continued

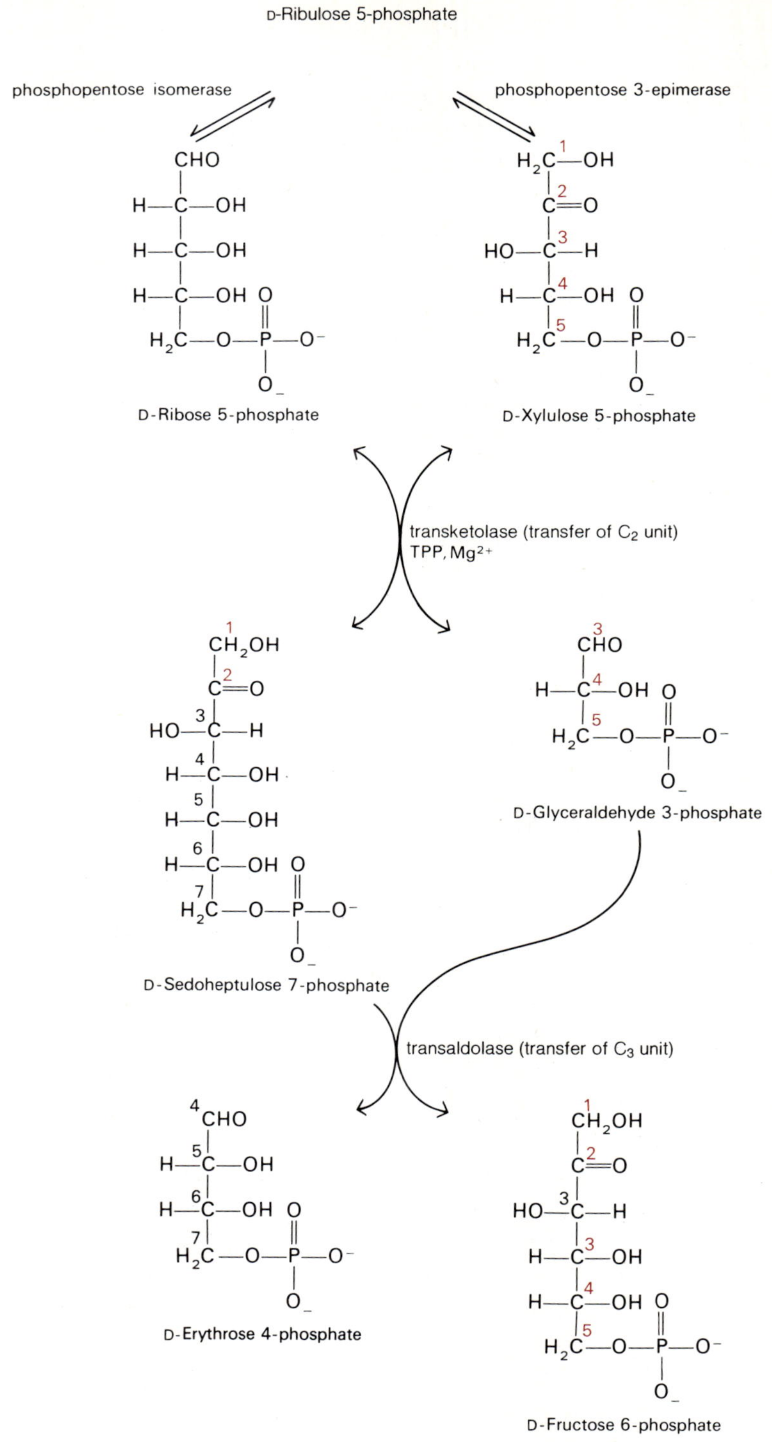

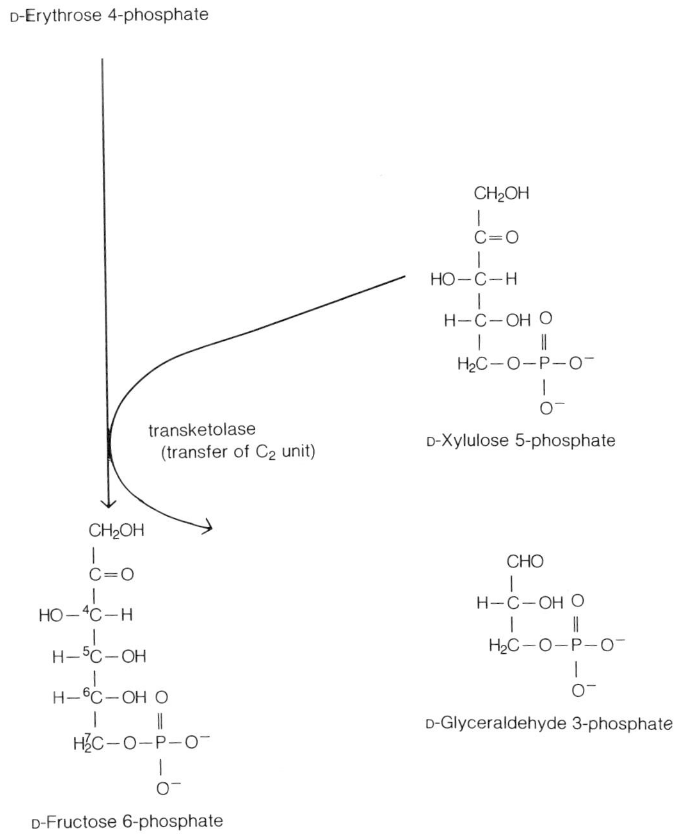

Figure 15-35, continued

thesis of the pentose. The ribose synthesized is used for the production of nucleotides which, in turn, serve as components of RNA and DNA and of coenzymes. The production of NADPH is also a key metabolic aspect of the pathway, since the coenzyme is selectively used for reductive biosynthesis and for furnishing reducing power in biological systems. For example, in humans, NADPH is needed for the normal maintenance of red blood cells, where it serves to generate *reduced glutathione,* a deficiency of which results in hemolysis of the cells. The most prevalent human inborn error is *glucose 6-phosphate-dehydrogenase deficiency* (first reaction of the pathway); this impaired ability to synthesize NADPH is usually manifested clinically as acute hemolytic anemia, induced by drugs (e.g., pamaquine, an antimalarial agent) or by toxic substances (e.g., accidental ingestion of mothballs by an afflicted child). Glucose 6-phosphate-dehydrogenase deficiency is an X-linked recessive trait, most frequently observed in regions of the

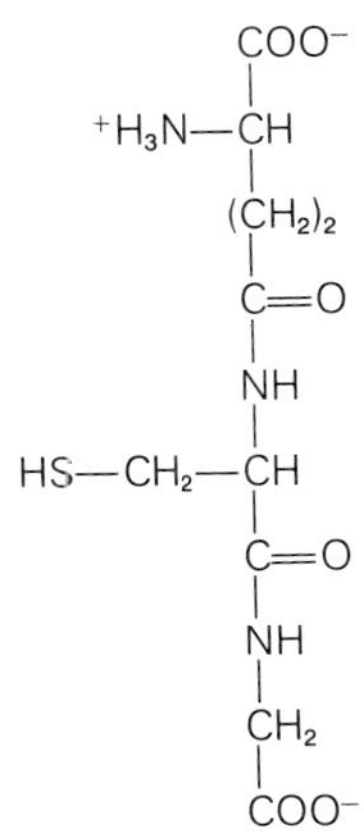

Reduced glutathione
(γ-Glutamylcysteinylglycine)

world where malaria is endemic. It is estimated that 100,000,000 people in the world have the deficiency. Like sickle-cell anemia, a deficiency of the dehydrogenase's activity is believed to protect an individual against malaria and to be another example of an inborn error that, from a populational viewpoint, can be advantageous in a particular environment.

The three other enzymes of the pathway (Figure 15-35) are *phosphopentose 3-epimerase, transketolase,* and *transaldolase.* The epimerase, the second reaction that utilizes D-ribulose 5-phosphate as a substrate, converts the pentose (by epimerization of C-3) to *D-xylulose 5-phosphate.* The transketolase reaction, using *D-ribose 5-phosphate* and *D-xylulose 5-phosphate* as substrates, catalyzes the transfer of the glycoaldehyde ($CH_2OH—CO—$) unit from xylulose to ribose to produce *D-sedoheptulose 7-phosphate* (seven-carbon sugar) and *D-glyceraldehyde 3-phosphate* (three-carbon intermediate of glycolysis). Thiamin pyrophosphate (TPP) and Mg^{2+} are required for the transketolase reaction. In the transaldolase reaction, the dihydroxyacetone moiety ($CH_2OH—CO—CHOH—$) of sedoheptulose is transferred to glyceraldehyde to yield *D-fructose 6-phosphate* and *D-erythrose 4-phosphate* (a tetrose). If D-xylulose 5-phosphate then donates two carbons to D-erythrose 5-phosphate, via the transketolase reaction, *D-fructose 6-phosphate* and *D-glyceraldehyde 3-phosphate* are generated. Thus, as summarized in Figure 15-36, the pentose phosphate pathway can serve to convert pentoses into glycolytic intermediates for anaerobic and/or aerobic respiration. As noted, *one glyceraldehyde* and *two fructose* intermediates of glycolysis can be produced from *three phosphorylated pentoses.*

As seen in Figure 15-35, the pentose phosphate pathway provides a biochemical route by which three-, four-, five-, six-, and seven-carbon carbohydrates can be synthesized or interconverted.

Figure 15-36
Production of three- and six-carbon intermediates of glycolysis from pentoses.

1. D-Xylulose 5-PO_3^{2-} + D-ribose 5-PO_3^{2-} $\rightleftharpoons$ D-sedoheptulose 7-PO_3^{2-} + D-glyceraldehyde 3-PO_3^{2-}
2. D-Sedoheptulose 7-PO_3^{2-} + D-glyceraldehyde 3-PO_3^{2-} $\rightleftharpoons$ D-fructose 6-PO_3^{2-} + D-erythrose 4-PO_3^{2-}
3. D-Xylulose 5-PO_3^{2-} + D-erythrose 4-PO_3^{2-} $\rightleftharpoons$ D-fructose 6-PO_3^{2-} + D-glyceraldehyde 3-PO_3^{2-}

Net overall: 2 D-Xylulose 5-PO_3^{2-} + D-ribose 5-PO_3^{2-} $\rightleftharpoons$ 2 D-fructose 6-PO_3^{2-} + D-glyceraldehyde 3-PO_3^{2-}

2 (5C) + 5C → 2 (6C) + 3C

The principal enzymes in this versatile exchange of carbon atoms between carbohydrates are transaldolase (transfer of a C_3 unit) and transketolase (transfer of a C_2 unit) with the latter enzyme being able to utilize various 2-keto and aldose sugar phosphates as donor and acceptor substrates, respectively. The pathway is also of prime importance in the photosynthetic process (p. 354).

SUMMARY

Glycolysis and alcoholic fermentation represent anaerobic pathways developed by primitive biological systems to extract some of the chemical energy of D-glucose for subsequent use in the processes of life. In glycolysis, a glucose molecule is converted into two pyruvate molecules, which are then reduced to produce two lactate molecules, the end product of the process. In alcoholic fermentation, pyruvate is also produced from glucose; in this pathway, however, the pyruvate is converted by a two-step process into ethanol and carbon dioxide. Both pathways involve six-carbon (hexoses) and three-carbon (trioses) sugars and proceed by the following general scheme. In the first part of the pathway, D-glucose is first converted into D-fructose 1,6-bisphosphate by a series of three reactions, a conversion that requires the chemical energy released by two ATP hydrolyses. In the second part of the pathway, fructose 1,6-bisphosphate is cleaved into two phosphorylated trioses, which are then converted into pyruvate. In this second phase, which entails seven reactions, two ATP are produced for each pyruvate produced (a total production of four ATP). Hence, the conversion of glucose into two pyruvate molecules yields a net total of two ATP. If glycogen serves as the source of glucose, then the net yield is three ATP, since only one ATP is required for the production of fructose 1,6-bisphosphate. The efficiency of energy conservation (percentage of the free energy released from glucose that is used for ATP synthesis) of glycolysis and alcoholic fermentation is 31 and 36.5 per cent, respectively. Other hexoses besides glucose, e.g., fructose and galactose, can also be utilized by organisms for the anaerobic synthesis of ATP.

Two of the enzymes that help to regulate glycolysis are phosphorylase *a* and 6-phosphofructokinase. Phosphorylase *a*, which sequentially removes D-glucosyl residues from glycogen, is under hormonal control, i.e., epinephrine release enhances the enzyme's activity and, hence, the rate of glycolysis. 6-Phosphofructokinase, which converts D-fructose 6-phosphate into D-fructose 1,6-bisphosphate, is an allosteric enzyme, with AMP and ADP being positive effectors and ATP, citric acid, and long-chain fatty acids being negative effectors.

In the glycolytic pathway, the following three reactions are not physiologically reversible: phosphoenolpyruvate → pyruvate; D-fructose 6-phosphate → D-fructose 1,6-bisphosphate; and D-glucose → D-glucose 6-phosphate. However, the synthesis of glucose from lactate (a pathway that is seemingly a reversal of glycolysis) does occur. This biosynthetic process, called gluconeogenesis, relies on the reversible enzymes of glycolysis but requires nonglycolytic enzymes to produce phosphoenolpyruvate, fructose 6-phosphate, and glucose.

Another metabolic pathway of carbohydrates is the pentose phosphate pathway, which is of critical physiological importance for several reasons. The pentose phosphate pathway produces D-ribose 5-phosphate (pentose of RNA and precursor of the pentose of DNA) and NADPH (essential coenzyme for physiological reductive purposes). The versatile pathway also provides a biochemical means by which three-, four-, five-, six-, and seven-carbon sugars can be synthesized or interconverted, which can include the production of the glycolytic intermediates D-fructose 6-phosphate and D-glyceraldehyde 3-phosphate.

A number of inborn errors involving the metabolism of carbohydrates are known. Among the disorders are the defective utilization of galactose and fructose and enzyme deficiencies in glycolysis, gluconeogenesis, and the pentose phosphate pathway.

QUESTIONS

1. Complete each of the reactions listed below with the appropriate reaction from the following list:
   $NADH + H^+ \rightarrow NAD^+$
   $NAD^+ \rightarrow NADH + H^+$
   $ATP \rightarrow ADP + P_i$
   $ATP \rightarrow ADP$
   $ADP \rightarrow ATP$
   (a) Pyruvate → lactate
   (b) D-Glucose → D-glucose 6-PO_4
   (c) Pyruvate + CO_2 + H_2O → oxaloacetate
   (d) Acetaldehyde → ethanol
   (e) Phosphoenolpyruvate → pyruvate
   (f) D-Glyceraldehyde 3-PO_4 + P_i → 1,3-bisphosphoglycerate
   (g) D-Fructose 6-PO_4 → D-fructose 1,6-bisphosphate
   (h) 1,3-Bisphosphoglyceric acid → 3-phosphoglycerate
2. With regard to Table 15-1, which enzymatic reaction or reactions of the glycolytic pathway account for the (a) 2 ATP, (b) 2 P_i, and (c) 2 H^+ listed as substrates and which reaction or reactions yield the (d) 4 ATP, (e) 2 NADH, and (f) 2 H_2O as products?
3. Select the answers to the statements below from the following list:

D-Glucose 1,6-bisphosphate	TPP
Acetyl-CoA	ATP
Biotin	GTP

   (a) Coenzyme for the transketolase reaction
   (b) Coenzyme for the pyruvate carboxylase reaction
   (c) Coenzyme for the pyruvate decarboxylase reaction
   (d) Cofactor for the phosphoglucomutase reaction
   (e) Required for the phosphoenolpyruvate carboxykinase reaction
   (f) Negative allosteric effector for 6-phosphofructokinase reaction
   (g) Positive allosteric effector for the pyruvate carboxylase reaction
4. Select your answers to the questions below from the following list of enzymes:

Hexokinase	Phosphoglycerate kinase
Lactate dehydrogenase	Pyruvate decarboxylase
6-Phosphofructokinase	Pyruvate dehydrogenase
Phosphoglucomutase	Pyruvate kinase
Glyceraldehyde 3-phosphate dehydrogenase	

   (a) Which three catalyze the physiologically irreversible reactions of glycolysis?
   (b) Which three are oxidoreductases?
   (c) Which five are transferases?
   (d) Which four are *not* required for the conversion of D-glucose into two pyruvate molecules?
5. (a) If yeast were supplied with glucose labeled with the radioisotope ^{14}C at the C-1 position (^{14}C-1) to produce ethanol and CO_2 anaerobically, would the ^{14}C appear in

the CO_2? (b) Answer the same question asked in part a for glucose with ^{14}C at the C-3 position.

6. (a) Postulate what effect a deficiency of enolase activity in red blood cells would have on the oxygen affinity of the hemoglobin of such an afflicted person. (b) Answer the same question with regard to a deficiency in phosphoglyceromutase activity.

7. Write the general reaction depicting the role of NADH in the normal maintenance of reduced glutathione (G-SH) in red blood cells.

8. Which enzymes in the pentose phosphate pathway catalyze each of the three reactions listed in Figure 15-36?

SUGGESTED READINGS

Bondy, P. K., and L. E. Rosenberg, eds., *Molecular Control and Disease,* 8th ed. Philadelphia: W. B. Saunders, 1980. Chapter 6 contains an extensive coverage of the disorders of carbohydrate metabolism.

Dickens, F., P. J. Randle, and W. L. Whelan, eds., *Carbohydrate Metabolism and Its Disorders,* 2 vols. New York: Academic Press, 1968. Reviews about various aspects of carbohydrate metabolism and its regulation.

Florkin, M., and E. Stotz, eds., *Carbohydrate Metabolism,* Vol. 17 of *Comprehensive Biochemistry.* New York: American Elsevier, 1967. The detailed articles provide a thorough review of glycolysis.

Fruton, J. S., *Molecules and Life.* New York: Wiley-Interscience, 1972. An excellent account of the scientific history of glycolysis and its influence on the development of enzymology.

Greenberg, D. M., ed., *Metabolic Pathways,* 3rd ed., Vol. 1. New York: Academic Press, 1967. This book contains an authoritative chapter on glycolysis by B. Axelrod.

Howald, H., and J. R. Poortmans, eds., *Metabolic Adaptation to Prolonged Physical Exercise.* Basel: Kirkhauser Verlag, 1975. An article by R. W. McGilvery on the use of fuels for muscular energy has a good discussion of regulation by ADP and AMP.

Stanbury, J. B., J. B. Wyngaarden, and D. S. Fredrickson, eds., *The Metabolic Basis of Inherited Disease,* 4th ed. New York: McGraw-Hill, 1978. Chapters 4 through 9 are devoted to disorders of carbohydrate metabolism.

Articles

Hofmann, E., "Phosphofructokinase—A Favourite of Enzymologists and of Students of Metabolic Regulation." *TIBS,* 3:145, 1978.

Nordlie, R. C., "Multifunctional Hepatic Glucose-6-Phosphatase and the 'Tuning' of Blood Glucose Levels." *TIBS,* 1:199, 1976.

Tejwani, G. A., "The Role of Phosphofructokinase in the Pasteur Effect." *TIBS,* 3:30, 1978.

AEROBIC SYNTHESIS OF ATP I: TRICARBOXYLIC ACID CYCLE

16

INTRODUCTION

The aerobic oxidation of glucose to CO_2 and H_2O, called *respiration*, was a critically important evolutionary advancement because it allowed organisms to greatly enhance their ability to extract the chemical energy of the sugar for use in life processes. Although the anaerobic pathway of glycolysis is successful in releasing some of the free energy content of glucose, the amount is only a small fraction of the total energy that can be made available by complete oxidation of the hexose to CO_2 and H_2O. The $\Delta G^{\circ\prime}$ values for the two modes of glucose oxidation are:

Anaerobic: Glucose $\rightarrow$ 2 lactate ($\Delta G^{\circ\prime} = -47.0$ kcal mol^{-1})

Aerobic: Glucose + 6 O_2 $\rightarrow$ 6 CO_2 + 6 H_2O ($\Delta G^{\circ\prime} = -686.0$ kcal mol^{-1})

Thus, the two molecules of lactate (end product of glycolysis) still contain about 93 per cent of the free energy content of the glucose molecule.

The aerobic oxidation of glucose requires the participation of three metabolically interrelated processes (*tricarboxylic acid cycle, electron transport,* and *oxidative phosphorylation*), all of which take place in the *mitochondria*. Whereas glycolysis occurs in the cytosol of eucaryotic cells, aerobic oxidation is a highly specialized cellular phenomenon limited to mitochondria (or the plasma membrane of procaryotes). Mitochondria are often referred to as the energy factories of the cell. All biomolecular fuels (carbohydrates, lipids, and amino acids) are oxidized in the mitochondria, where the bioenergetic processes for the generation of chemical energy are concentrated.

In aerobic oxidation, the pyruvate produced by glycolysis is not

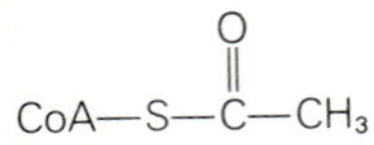

Acetyl-CoA

converted into lactate but, rather, into *CO_2 and acetyl-CoA*; the latter is then utilized by the *tricarboxylic (TCA) cycle*. The TCA cycle completes the oxidation of the four remaining carbon atoms of a glucose molecule, i.e., two acetyl-CoA to four CO_2. The subsequent production of water, to complete the production of six H_2O per oxidized glucose, involves *electron transport*, which consists of a series of reactions that transfer the electron pairs derived from the oxidation of the sugar to oxygen to form water. The flow of electrons through electron transport is an evolutionary development that ensures a release of electron energy, some of which is conserved through the synthesis of ATP by *oxidative phosphorylation*, a process that is physically coupled to the electron transport system (Chapter 17).

The elucidation of the TCA cycle was one of the classic achievements of biochemistry during the first half of this century. As was true for the studies on glycolysis, a number of distinguished scientists, beginning in the 1910s, aided in uncovering the molecular details of an oxidative process in animal tissues that proved to be the central oxidative pathway for energy production. The accumulated findings about the oxidation of specific di- and tricarboxylic acids to CO_2 by molecular oxygen prompted the English biochemist Hans A. Krebs (Nobel Prize, 1953) to study this process of aerobic oxidation. One intriguing observation, first noted by Albert Szent-Györgyi (Nobel Prize, 1937), was that when small amounts of added organic acids were oxidized by minced tissues, the consumption of oxygen was much larger than the quantity needed to oxidize the exogenous acid. Szent-Györgyi correctly surmised that some endogenous molecule was also being oxidized. Krebs brilliantly deduced and established the relationship between the utilization of endogenous pyruvate and the oxidation process and, in 1937, proposed the *citric acid cycle*, the original name for the TCA cycle. The cycle, including later modifications, is diagrammed in Figure 16-1. As illustrated, the "catalytic nature" of the cycle is maintained by the continuous regeneration of oxaloacetate. Theoretically, only one molecule of oxaloacetate (or of another intermediate of the cycle) is required for the oxidation of many molecules of pyruvate.

When Krebs' original manuscript on the TCA cycle was submitted for publication, it was rejected because of a lack of publishing space.

PRODUCTION OF ACETYL-CoA FROM PYRUVATE

The formation of acetyl-CoA is a prerequisite for the entrance of the carbon atoms of pyruvate, produced by all carbohydrates being

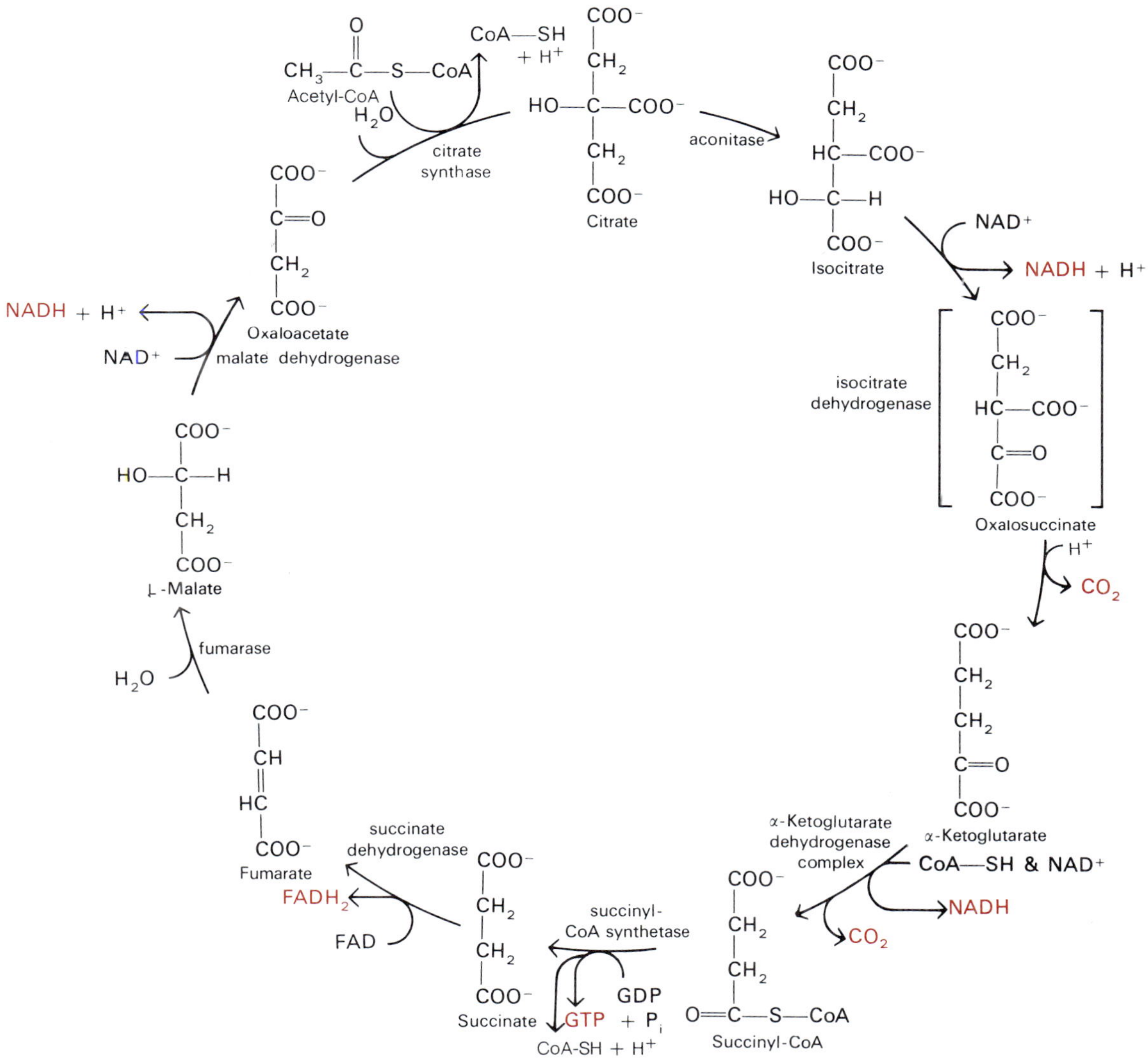

Figure 16-1
The tricarboxylic acid cycle (TCA).

utilized for energy, into the TCA cycle. As implied by the requirements of six cofactors (Figure 16-2), acetyl-CoA production is a complicated enzymatic process catalyzed by an aggregate of enzymes called the *pyruvate dehydrogenase complex*. The comprehensive studies of Lester J. Reed and his associates have provided molecular details of the enzyme complex and of the individual reactions involved.

The overall reaction is a summation of a sequential five-step mechanism which requires three different enzymes (Figure 16-3).

Figure 16-2
Overall reaction for acetyl-CoA production from pyruvate.

$$\text{Pyruvate} + \text{NAD}^+ + \text{CoA—SH} \xrightarrow[\text{lipoic acid, FAD}]{\text{Mg}^{2+},\ \text{TPP}} \text{acetyl-CoA} + CO_2 + \text{NADH} + H^+$$

$$\Delta G^{\circ\prime} = -8.0 \text{ kcal mol}^{-1}$$

The first step, catalyzed by *pyruvate dehydrogenase*, decarboxylates pyruvate to yield *CO_2* and the *α-hydroxyethyl derivative* of the enzyme-bound coenzyme, thiamine pyrophosphate. In the second reaction, involving *dihydrolipoyl transacetylase*, the hydroxyethyl group is dehydrogenated to an acetyl group, which is transferred to the sulfur atom of C-6 of lipoic acid (transacetylase's prosthetic group). Lipoic acid is contained in the *lipoyllysyl side chain* of the transacetylase which, as indicated in Figure 16-3, functions as a "swinging arm." The third reaction transfers the acetyl group from the lipoyllysyl arm to the thiol group of CoA, producing *acetyl-CoA* and *dihydrolipoic acid* (reduced form of the prosthetic group). The fourth reaction restores lipoic acid to its oxidized state with the concomitant reduction of FAD, the prosthetic group of *dihydrolipoyl dehydrogenase;* FAD is then reoxidized in the last reaction by the reduction of NAD^+. The exergonic nature of the first reaction (decarboxylation of pyruvate) establishes the irreversibility of the overall reaction ($\Delta G^{\circ\prime}$ of -8.0 kcal mol^{-1}) in animal systems.

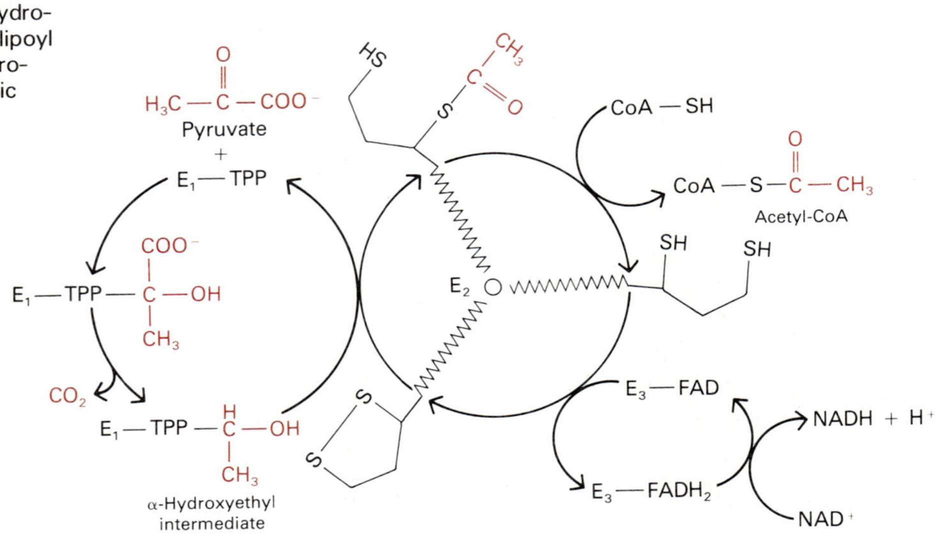

Figure 16-3
Schematic of the pyruvate dehydrogenase complex, depicting the long lipoyllysyl side chain of dihydrolipoyl transacetylase ( ), which functions as a "swinging arm." E_1 = pyruvate dehydrogenase, E_2 = dihydrolipoyl transacetylase, E_3 = dihydrolipoyl dehydrogenase, TPP = thiamin pyrophosphate, and S–S = oxidized lipoic acid.

Molecular Composition of the Complex

The studies by Reed and his associates also provided detailed information on the physical structure of the pyruvate dehydrogenase complex (Figure 16-4), which has a molecular weight of 4.6 million. The core of the complex consists of 24 dihydrolipoyl transacetylase (MW $\approx$ 70,000) molecules, to which are attached 12 pyruvate dehydrogenase dimers (MW of monomer $\approx$ 96,000) and six dihydrolipoyl dehydrogenase dimers (MW of monomer $\approx$ 56,000). In addition to containing the three enzymes required for the production of acetyl-CoA, the complex also possesses the two regulatory enzymes pyruvate dehydogenase kinase (MW $\approx$ 62,000) and *pyruvate dehydrogenase phosphatase* (MW $\approx$ 100,000).

Figure 16-4
Drawing of the pyruvate dehydrogenase complex of *Escherichia coli*, prepared from a photograph of a model provided by Dr. Lester J. Reed. White spheres represent the dihydrolipoyl transacetylase core; black spheres are pyruvate dehydrogenase dimers; red spheres are dihydrolipoyl dehydrogenase dimers.

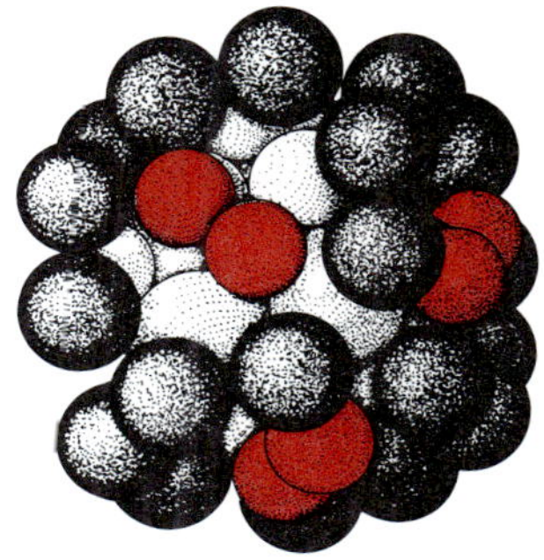

Regulation by Covalent Modification

Cellular production of acetyl-CoA is inhibited and stimulated by ATP and Ca^{2+}, respectively. The regulatory mechanism involves *covalent modification, viz.,* phosphorylation and dephosphorylation of the complex. When levels of ATP are high, acetyl-CoA production is *decreased* as a consequence of *phosphorylation* of proteins in the complex by the constituent regulatory enzyme, pyruvate dehydrogenase kinase. The inhibition diminishes the rate of ATP production by aerobic oxidation. When ADP concentrations are high and pyruvate is available, the catalytic activity of the complex is *stimulated* by *dephosphorylation* of the phosphorylated proteins in the complex, catalyzed by the other regulatory enzyme, pyruvate dehydrogenase phosphatase. The dephosphorylation reaction is markedly stimulated by Ca^{2+}. Increased acetyl-CoA synthesis subsequently results in enhanced ATP production. Covalent modification, like allosteric regulation, is effectively employed by biological systems to adjust biochemical processes to respond to immediate metabolic demands.

PRODUCTION OF THE TWO TRICARBOXYLIC ACIDS OF THE CYCLE

The initial reaction of the TCA cycle is the synthesis of the first tricarboxylic acid, *citrate*, by the condensation of *oxaloacetate* and the *acetyl moiety* of acetyl-CoA (Figure 16-5). The reaction is catalyzed by *citrate synthase,* formerly called condensing enzyme, and is highly exergonic because of the energy released by the

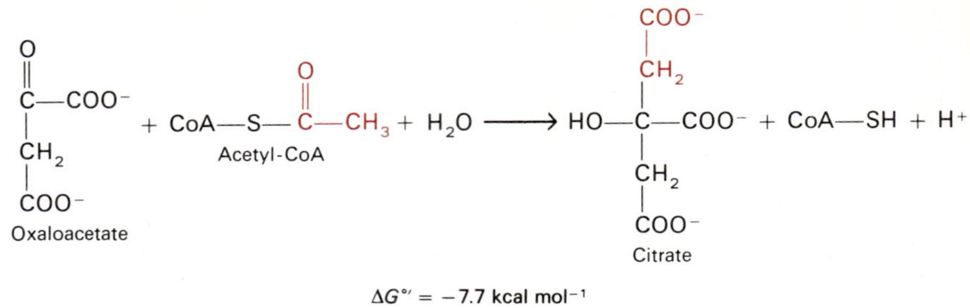

Figure 16-5
Citrate synthase reaction.

hydrolysis of the *thioester* $(-\overset{\displaystyle O}{\overset{\|}{C}}-S)$ *bond* of acetyl-CoA ($\Delta G°'$ of -7.7 kcal mol^{-1}). The levels of the two substrates greatly influence the rate of the reaction and, hence, are considered a major regulatory feature of the TCA cycle.

Citrate is converted into the second tricarboxylic acid of the cycle, *isocitrate*, in a reaction catalyzed by *aconitase* (Figure 16-6), whose metal cofactor, Fe^{2+}, is contained in the enzyme's structure. This particular step has been of considerable research interest because of the question regarding the status of cis-*aconitate,* formerly believed to be an intermediate of the cycle (Figure 16-7). Illustrations of the TCA cycle often include *cis*-aconitate as an enzyme-bound intermediate of the aconitase reaction. There is evidence, however, that the true intermediate is a *carbonium ion* in equilibrium with citrate, isocitrate, and *cis*-aconitate, three tricarboxylic acids that are interconvertible by aconitase (Figure 16-8). Thus, although *cis*-aconitate may appear to be a tricarboxylic acid of the cycle, it is now considered by many to be an artifact of the aconitase reaction.

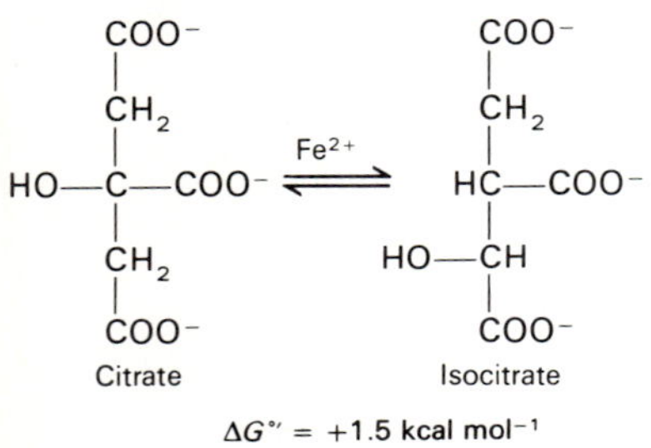

Figure 16-6
Aconitase reaction.

An interesting example of a poison that affects the TCA cycle is *fluoroacetate* (FCH_2COO^-), found in some toxic plants of South Africa. When ingested, fluoroacetate is converted into *fluoroacetyl-CoA*, which effectively competes with acetyl-CoA as a substrate for citrate synthase, resulting in the endogenous production of *fluorocitrate*. As a potent inhibitor of aconitase, fluorocitrate is a lethal poison, one that has been used as a rodenticide. Hence, fluoroacetate is not the physiological toxin but, instead, is its precursor.

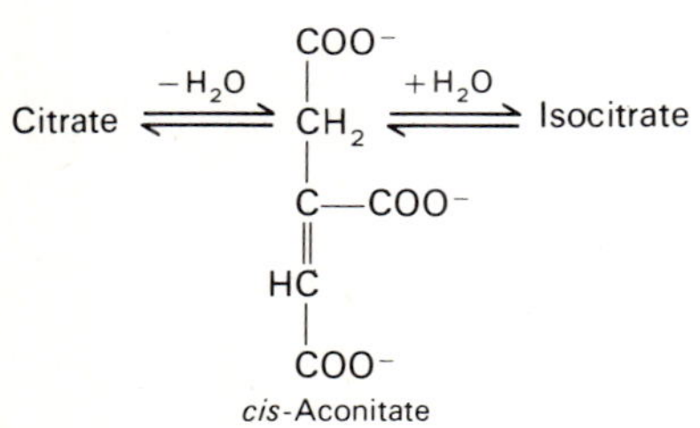

Figure 16-7
cis-Aconitate as a proposed intermediate of the aconitase reaction.

OXIDATIVE DECARBOXYLATION OF ISOCITRATE

The next reaction of the cycle involves the oxidation of isocitrate to yield *oxalosuccinate* which, as an enzyme-bound product, rapidly decarboxylates (β-carboxyl group) to produce the dicarboxylic acid *α-ketoglutarate* and *CO_2* (Figure 16-9). In the reaction, catalyzed by *isocitrate dehydrogenase*, the coenzyme NAD^+ is reduced.

Isocitrate dehydrogenase is allosterically regulated by *ATP* and *NADH* (negative effectors) and by *ADP* (positive effector). Allosteric control by NADH (or NAD^+), like that by ATP, ADP, and AMP, is another mechanism for regulating ATP production. High intracellular [NADH] is indicative of an active TCA cycle and therefore of ample aerobic production of ATP; conversely, high [NAD^+] reflects a need for additional ATP synthesis. Thus, the [NADH] and/or [NAD^+] are utilized by certain regulatory enzymes as molecular indicators of the energy status of the cell.

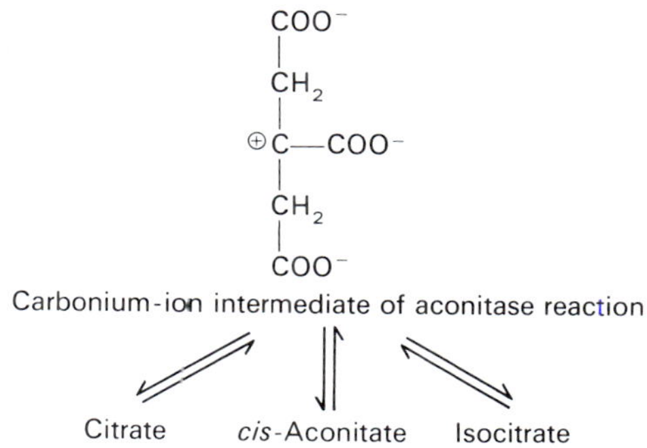

Figure 16-8
Possible interconversion of citrate, *cis*-aconitate, and isocitrate.

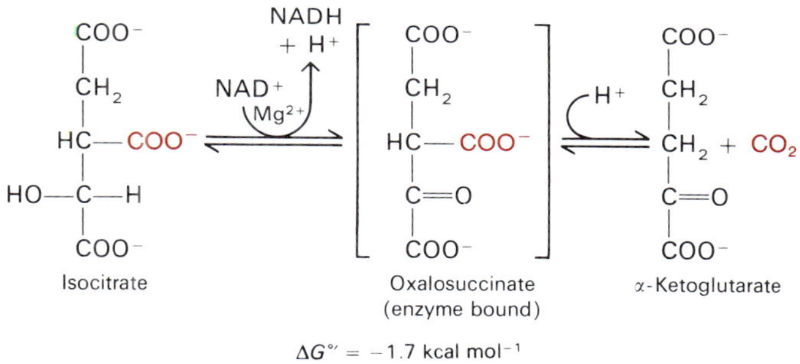

Figure 16-9
Isocitrate dehydrogenase reaction.

PRODUCTION OF SUCCINATE AND GTP

In a reaction analogous to that catalyzed by the pyruvate dehydrogenase complex, α-ketoglutarate is decarboxylated by the *α-ketoglutarate dehydrogenase complex*, which requires the same six cofactors needed for the production of acetyl-CoA. The reaction catalyzes the second *release of CO_2* in the cycle and the production of *succinyl-CoA* (Figure 16-10). The overall reaction, like that of

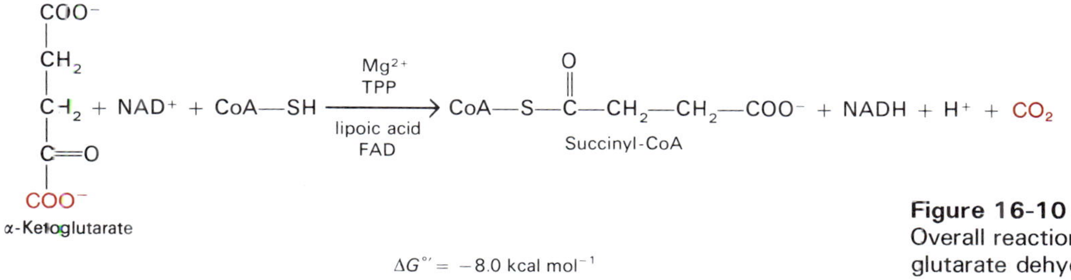

Figure 16-10
Overall reaction catalyzed by the α-ketoglutarate dehydrogenase complex.

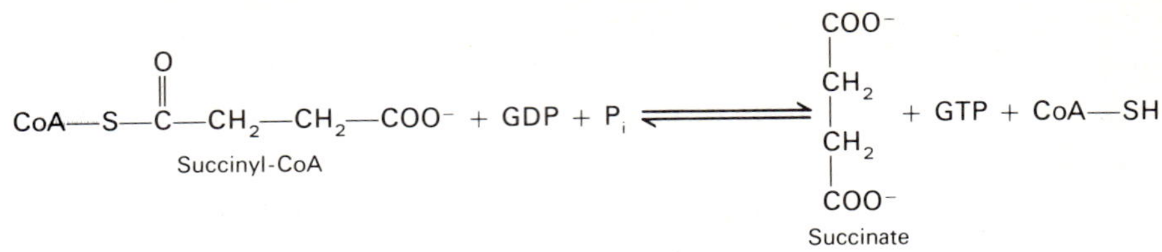

$\Delta G^{\circ\prime} = -0.8$ kcal mol^{-1}

Figure 16-11
Succinyl-CoA synthetase reaction.

the pyruvate dehydrogenase complex, is physiologically irreversible.

In the next reaction, the free energy released by the hydrolysis of the thioester bond of succinyl-CoA, to produce succinate and CoA-SH, is utilized to synthesize GTP from GDP and P_i (Figure 16-11). *Succinyl-CoA synthetase* catalyzes this substrate-level phosphorylation (p. 277). The GTP produced by the hydrolysis of succinyl-CoA can readily transfer its terminal phosphate to ADP to produce an ATP molecule (nucleoside diphosphate kinase, p. 262).

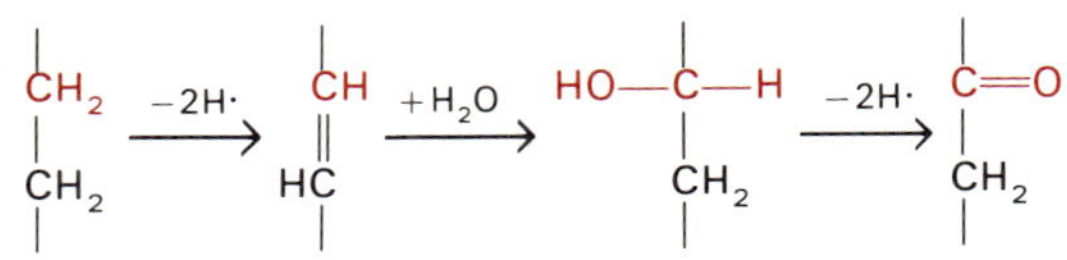

Figure 16-12
Biological oxidation of a methylene group to a carbonyl group.

REGENERATION OF OXALOACETATE

The last three reactions of the cycle catalyze the oxidative synthesis of oxaloacetate from succinate. Basically, the reactions involve the oxidation of a *methylene* to a *carbonyl group* (Figure 16-12). In biological systems, the process can be accomplished by the (1) removal of a hydrogen atom from each of two carbons to create a double bond, followed by (2) addition of water to the double bond and, finally, (3) removal of two hydrogen atoms from the hydroxylated carbon to yield the appropriate α-keto group.

In the TCA cycle, the first reaction oxidizes succinate to produce *fumarate*, in a reaction catalyzed by *succinate dehydrogenase* (Figure 16-13). The acceptor of the two hydrogen atoms is the prosthetic group FAD, which is covalently bonded to the enzyme. The addition of water to the double bond of fumarate is then catalyzed by *fumarase*, which produces *L-malate* (Figure 16-14). The second oxidation, catalyzed by *malate dehydrogenase* and requiring NAD^+ as the coenzyme, converts malate into *oxaloacetate* (Figure 16-

Figure 16-13
Succinate dehydrogenase reaction.

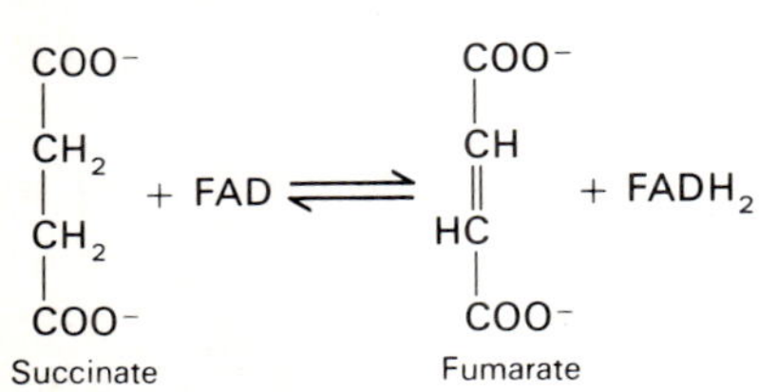

$\Delta G^{\circ\prime} = \approx 0$ kcal mol^{-1}

15). The cycle is now complete, and another molecule of oxaloacetate is available to reinitiate the cycle.

Malonate, as a competitive inhibitor of the succinate-dehydrogenase reaction (p. 139), was advantageously used by early investigators of the TCA cycle to "poison" (inhibit) the oxidative process. This experimental approach of using malonate as an inhibitor greatly aided in the deciphering of the cyclic nature of pyruvate oxidation (see Problem 9 at the end of the chapter).

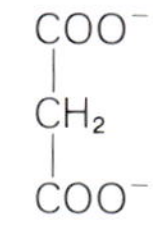

Malonate

SUMMARY OF THE OXIDATION OF PYRUVATE

The overall reaction for the oxidation of one molecule of pyruvate via the pyruvate dehydrogenase complex and the TCA cycle (Figure 16-16) includes the production of *three molecules of* CO_2, *a nucleoside triphosphate (GTP)*, and *five molecules of reduced coenzymes (4 NADH and 1* $FADH_2$*)*. The only CO_2 derived directly from the pyruvate molecule is produced in the formation of acetyl-CoA; the other two, as indicated in Figure 16-1, are obtained from the oxaloacetate that condenses with the acetyl-CoA molecule.

Figure 16-14
Fumarase reaction.

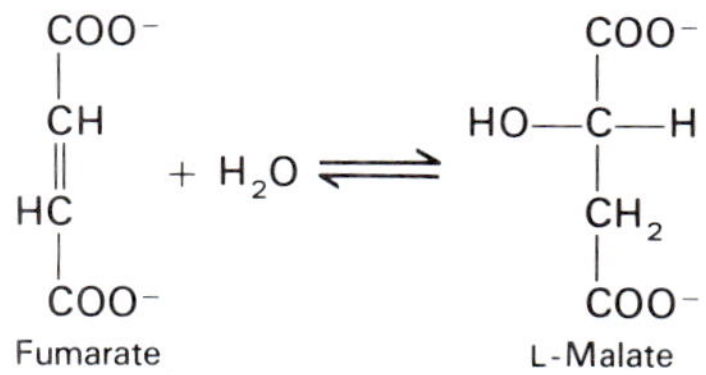

$\Delta G^{\circ\prime} = -0.9\ \text{kcal mol}^{-1}$

Figure 16-15
Malate dehydrogenase reaction.

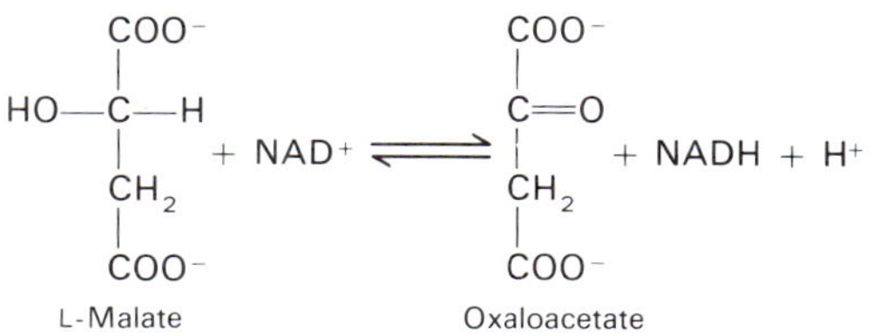

$\Delta G^{\circ\prime} = +7.0\ \text{kcal mol}^{-1}$

However, the cyclic nature of the oxidative process ensures the rapid subsequent conversion of the acetyl carbon atoms into CO_2. Since GTP is the single nucleoside triphosphate produced, the TCA cycle supplies only a very limited amount of the chemical energy

Figure 16-16
Summation of the oxidation of pyruvate to carbon dioxide.

I. Pyruvate dehydrogenase complex:
Pyruvate + CoA-SH + NAD^+ $\longrightarrow$ acetyl-CoA + NADH + H^+ + CO_2

II. Tricarboxylic acid cycle:
Acetyl-CoA + 3 NAD^+ + FAD + GDP + P_i + 2 H_2O $\longrightarrow$ 2 CO_2 + CoA-SH + 3 NADH + 3 H^+ + $FADH_2$ + GTP

Net overall: Pyruvate + 4 NAD^+ + FAD + GDP + P_i + 2 H_2O $\longrightarrow$ 3 CO_2 + 4 NADH + 4 H^+ + $FADH_2$ + GTP released by the aerobic oxidation of glucose. The five reduced coenzyme molecules, because they possess the electron pairs released by the oxidative process, contain a significant percentage of the free energy released by glucose oxidation. The two other components of aerobic oxidation succeed in freeing a substantial portion of the chemical energy stored in these reduced coenzymes (electron transport) and conserving some of it in ATP structure (oxidative phosphorylation).

UTILIZATION OF AMINO ACIDS FOR ENERGY

As stated previously, the TCA cycle is a central metabolic system for the aerobic production of ATP, regardless of the class of biomolecules being utilized for fuel. For organisms subsisting on a high-protein diet, e.g., carnivores, the catabolic pathways of amino acids assure the entry of their carbon atoms into the TCA cycle. As depicted in Figure 16-17, all 20 amino acids are degradable to intermediates of the cycle. Some are catabolized to a single intermediate, e.g., alanine and glutamate, and others to two, e.g., phenylalanine and leucine. (The catabolic pathways for a number of the amino acids are presented in Chapter 20.) Figure 16-17 also indicates how the carbon atoms of amino acids, via the TCA cycle,

Figure 16-17
Points of entry of the carbon atoms of the 20 amino acids into the TCA cycle.

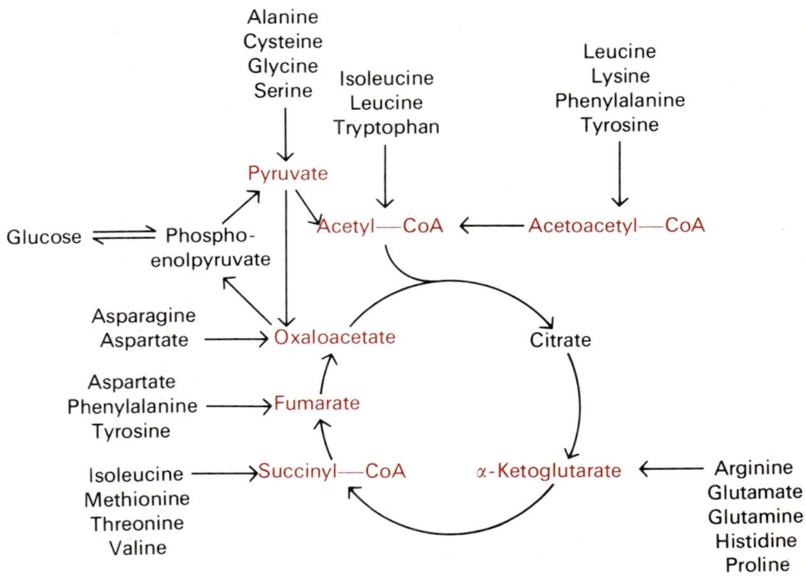

Degradation products of the amino acids are in red.

can be metabolically routed to produce glucose, i.e., oxaloacetate → phosphoenolpyruvate →→ glucose. The enzymatic system for gluconeogenesis from amino acids includes most of the enzymes, e.g., phosphoenolpyruvate carboxykinase, required for the synthesis of glucose from lactate (p. 282).

GLYOXYLATE CYCLE

Although higher animals can utilize many of the amino acids for the net synthesis of glucose (gluconeogenesis), they cannot use fatty acids (derived from lipids) for the same purpose. As illustrated in Figure 16-17, most degradation pathways of amino acids yield four- or five-carbon intermediates of the TCA cycle, which can be oxidized by the cycle for energy or used as precursors for glucose production. However, in the case of fatty acids, their degradation product, acetyl-CoA (p. 334), is oxidized to CO_2 by the TCA cycle (1 C_2 → 2 CO_2); thus, fatty acid oxidation does not result in a net synthesis of TCA cycle intermediates which can serve as glucose precursors.

In contrast to higher animals, most plants and microorganisms are able to use fatty acids or acetate, in the form of acetyl-CoA, as a sole or major carbon source; this is accomplished by a modified TCA cycle, called the *glyoxylate cycle*, which was elucidated by Krebs and Hans R. Kornberg. In plants, the glyoxylate cycle occurs in glyoxysomes (p. 62), organelles present in germinating *oil-storage seeds*, e.g., castor beans, peanuts, cotton, and corn. A key role of the cycle is to convert acetyl-CoA produced from fatty acids (obtained from the seed oil) into succinate, which is then utilized as a precursor of glucose (principal product) and other biomolecules needed by a growing seedling.

The glyoxylate cycle is illustrated in Figure 16-18. As in the TCA cycle, acetyl-CoA condenses with oxaloacetate to produce citrate, which is then converted into isocitrate. The next two reactions are unique features of the cycle. In the first reaction, *isocitrate lyase* cleaves isocitrate, yielding *succinate* and *glyoxylate*. Then, in a reaction catalyzed by *malate synthase*, a second *acetyl-CoA* condenses with glyoxylate, producing *L-malate*. The cycle is completed by the oxidation of L-malate to oxaloacetate. This cycle, which bypasses the two decarboxylation steps characteristic of the TCA cycle, succeeds in converting *two* molecules of *acetyl-CoA* into *one* molecule of *succinate*. The overall reaction of the cycle is expressed as follows:

$$2\ \text{Acetyl-CoA} + NAD^+ + H_2O \rightarrow \text{succinate} + 2\ \text{CoA} + NADH + H^+$$

The utilization of fatty acids by seedlings to synthesize glucose and other necessary biomolecules entails a metabolic interplay between glyoxysomes and mitochondria. Whereas the glyoxysomes convert acetyl-CoA generated from fatty acids into succinate, the biomolecule's role in glucose synthesis transpires in the mitochondria. In mitochondria, the succinate is converted by the TCA cycle into oxaloacetate, which serves as a substrate for the phosphoenolpyruvate carboxykinase reaction (p. 284) to produce phosphoenolpyruvate (PEP). The PEP synthesized then serves as a

Figure 16-18
The glyoxylate cycle.

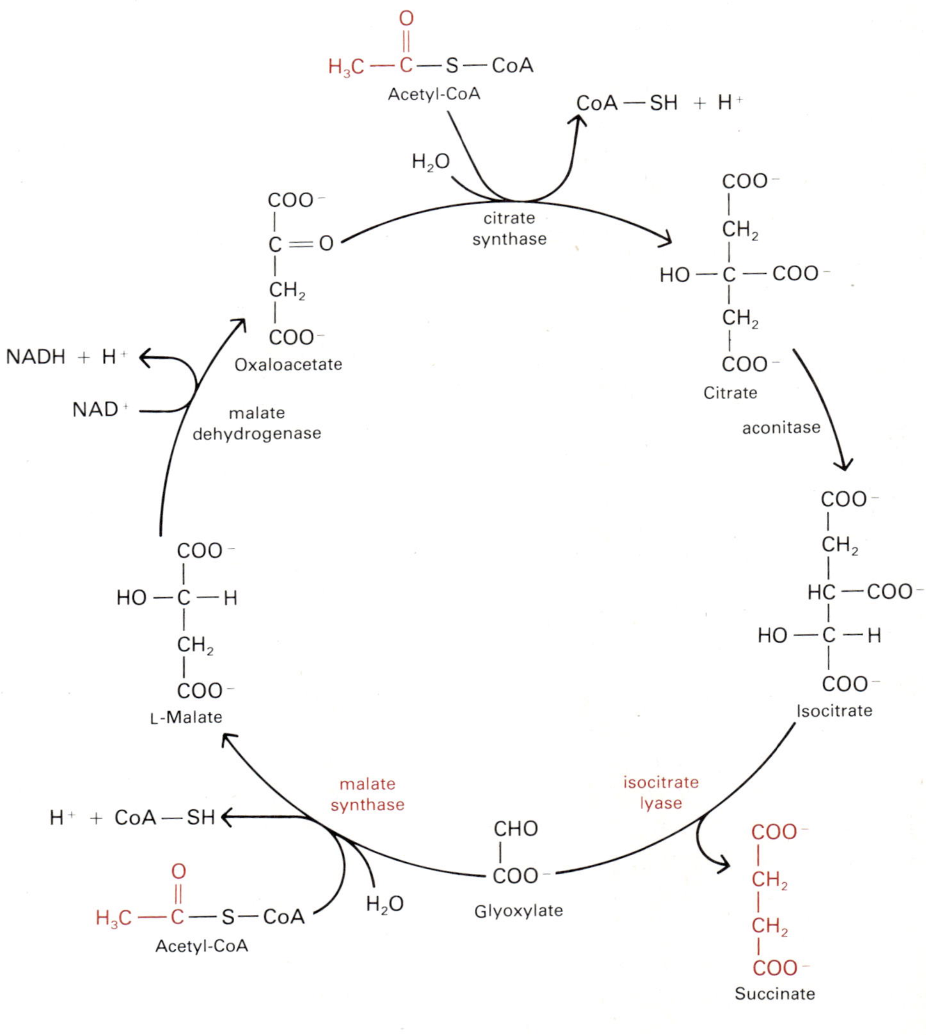

key intermediate for glucose production. Thus, a three-carbon intermediate (PEP) of gluconeogenesis is derived from the four carbons originally furnished to the glyoxylate cycle by two acetyl-CoA molecules.

AMPHIBOLIC NATURE OF THE TCA CYCLE

Although the TCA cycle is a *catabolic pathway*, it also participates in *anabolism* by providing intermediates for biosynthetic purposes. The TCA cycle is therefore *amphibolic*, a term used to designate the dual catabolic-anabolic nature of some pathways. In addition to occupying a major position in carbon metabolism, which has been emphasized in this chapter, the cycle is also of prime importance in *nitrogen metabolism* because of its anabolic functions.

L-Glutamate Dehydrogenase

One of the important anabolic functions of the TCA cycle is to furnish the intermediate *α-ketoglutarate* as a precursor of the synthesis of *L-glutamate* (Figure 16-19). Many organisms rely on this reaction, catalyzed by *L-glutamate dehydrogenase*, for the synthesis of glutamate. The enzyme, in animals, can use either NADH (NAD^+) or NADPH ($NADP^+$) as the coenzyme, with the former being preferred generally. In plants, NADPH ($NADP^+$) is often required as the coenzyme.

The synthesis of glutamate by L-glutamate dehydrogenase is a significant reaction in nitrogen metabolism because it is one of the few that convert *inorganic nitrogen* (NH_4^+) into *organic nitrogen* (carbon-bound amino group), i.e., constitutes an entry point for nitrogen into biological systems. The reversibility of the reaction also ensures that, for catabolic purposes, the carbon atoms of excess glutamate are made directly available to the TCA cycle.

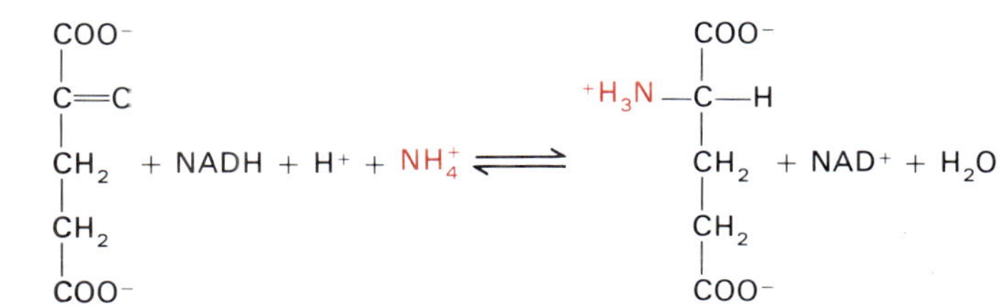

α-Ketoglutarate L-Glutamate

Figure 16-19
L-Glutamate dehydrogenase reaction.

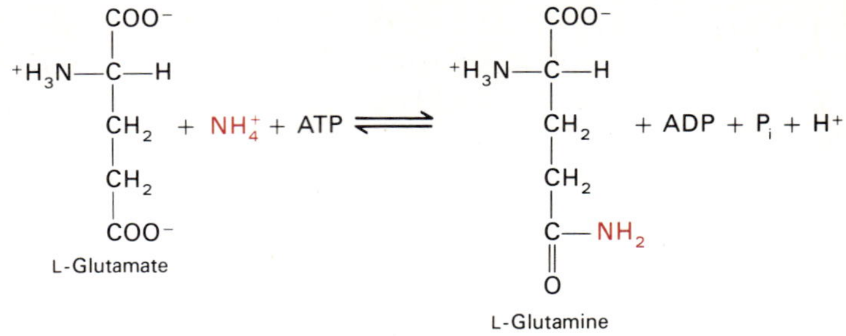

Figure 16-20
Glutamine synthetase reaction.

Glutamine Synthetase

The synthesis of the amino acid L-glutamine is readily accomplished enzymatically through the use of glutamate and NH_4^+ as precursors (Figure 16-20). The reaction, catalyzed by *glutamine synthetase*, requires the energy of ATP hydrolysis and is another of the few biological reactions that incorporate inorganic nitrogen into an organic molecule. For many bacterial species, the synthesis of glutamine is the principal enzymatic reaction for the incorporation of inorganic nitrogen (p. 614). The synthesis of both glutamate and glutamine, utilizing the five carbon atoms of α-ketoglutarate, exemplifies the anabolic character of the TCA cycle.

Glutamine synthetase of *E. coli*, like the pyruvate dehydrogenase complex, is subject to regulation by covalent modification. Studies by Earl Stadtman reveal that the enzyme's activity can be regulated by *adenylylation* (inhibition) and *deadenylylation* (restoration of activity). The covalent modification occurs with the enzymatic transfer of the adenosine 5′-monophosphate (AMP) moiety of ATP to a specific tyrosyl residue in each of the 12 protomers of the enzyme (Figure 16-21); deadenylylation relieves the inhibition of activity. This control of the activity of glutamine synthetase is of particular interest because the same two proteins, *adenylyltransferase (ATase)* and *P_{II}*, catalyze the adenylylation and deadenylylation of the enzyme. The specific reaction catalyzed by the two

Figure 16-21
Adenylylation of glutamine synthetase *(Escherichia coli)*.

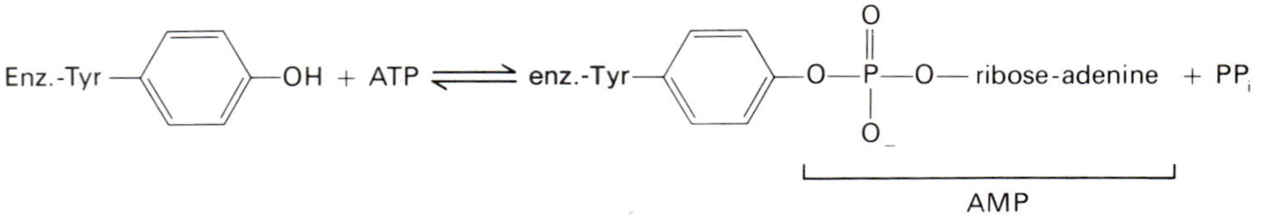

proteins depends on the form of P_{II}. *Uridylylation* (attachment of uridine 5′-monophosphate) of P_{IIA}, the form that activates *adenylylation*, converts the protein to another form, called P_{IID}, which activates *deadenylylation*.

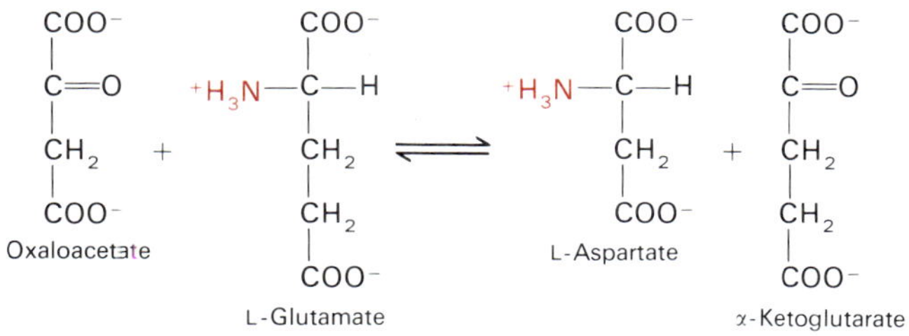

Figure 16-22
Aspartate aminotransferase (transamination) reaction.

Aspartate Aminotransferase

Another intermediate of the TCA cycle that serves as a direct precursor of an amino acid is *oxaloacetate*, which, by transamination, is converted into *L-aspartate* (Figure 16-22). The reaction is catalyzed by *aspartate aminotransferase (transaminase)* and utilizes glutamate as the *amino donor* and oxaloacetate as the *amino acceptor*. The reaction is illustrative not only of another amphibolic role of the TCA cycle but also of a major mechanism for the flow (transfer) of nitrogen in biological systems, i.e., transamination. Glutamate and aspartate synthesis emphasizes the key role of glutamate and the TCA cycle in nitrogen metabolism. As illustrated in Figure 16-23, when glutamate donates its amino group in transamination, α-ketoglutarate is regenerated and another inorganic NH_3 can be incorporated by glutamate synthesis. Aspartate aminotransferase is only one of a variety of transaminases that utilize glutamate as the amino donor; the synthesis of this amino acid and the routes of transamination for the distribution of its amino group among other biomolecules are critically important aspects of nitrogen metabolism.

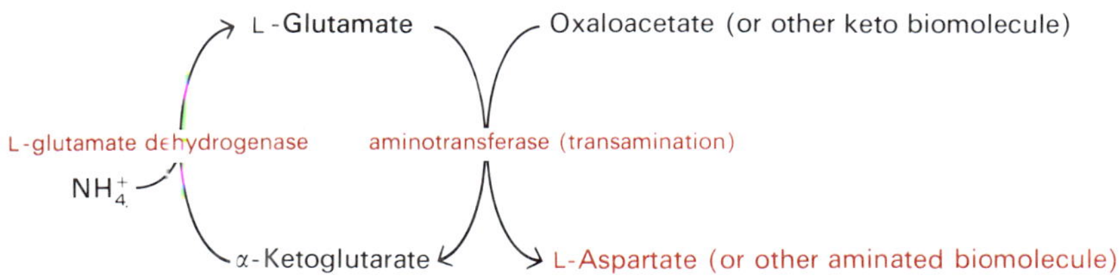

Figure 16-23
Cyclic nature of the role of glutamate as a principal nitrogen donor.

SUMMARY

The tricarboxylic acid (TCA) cycle, which takes place in the mitochondria of eucaryotic cells, is used for the aerobic oxidation of biomolecular fuels, i.e., carbohydrates, lipids, and amino acids. With regard to the aerobic oxidation of D-glucose, the pyruvate produced by glycolysis is converted into CO_2 and acetyl-CoA, and it is the latter two-carbon derivative of glucose that is utilized by the TCA cycle to complete the oxidation of the sugar to CO_2. The production of acetyl-CoA from pyruvate is catalyzed by a series of three enzymatic reactions; the enzymes involved are clustered in an aggregate called the pyruvate dehydrogenase complex, and the following six cofactors are required: TPP, Mg^{2+}, lipoic acid, CoA, FAD, and NAD^+.

The first reaction of the TCA cycle is the condensation of the acetyl moiety of acetyl-CoA with oxaloacetate (four-carbon dicarboxylic acid) to produce citrate (six-carbon tricarboxylic acid). Citrate is converted into isocitrate, which is then decarboxylated to yield α-ketoglutarate (five-carbon dicarboxylic acid). In the next reaction, α-ketoglutarate is decarboxylated and succinyl-CoA is produced; this reaction is catalyzed by the α-ketoglutarate dehydrogenase complex. The production of succinyl-CoA proceeds by a series of reactions analogous to those of acetyl-CoA production. Succinyl-CoA is then cleaved to produce succinate (four-carbon dicarboxylic acid) and CoA; the free energy released by the cleavage of the thioester bond of succinyl-CoA is used to synthesize GTP from GDP and P_i. The last three reactions of the cycle result in the regeneration of oxaloacetate from succinate. Overall, the oxidation of the two acetyl carbons, derived from glucose, by the TCA cycle results in the production of two CO_2, a GTP, and four reduced coenzymes (3 NADH and 1 $FADH_2$).

Degradation of all 20 amino acids produces TCA cycle intermediates which can then be used to produce energy via aerobic respiration. Many amino acids produce TCA cycle intermediates which, if not needed for energy production, can be used for gluconeogenesis. In higher animals, acetyl-CoA derived from fatty acids (lipids) cannot be used for net glucose synthesis because the two acetyl carbons furnished to the TCA cycle are converted into two CO_2. However, because many plants and bacteria possess the glyoxylate cycle, they can utilize acetyl-CoA (lipids) for gluconeogenesis. In the glyoxylate cycle, two acetyl-CoA are converted into succinate, which is used by the mitochondrial TCA cycle to produce oxaloacetate. The oxaloacetate is then converted into phosphoenolpyruvate, a key three-carbon intermediate of gluconeogenesis.

The TCA cycle is amphibolic, since it is utilized metabolically for both catabolic and anabolic purposes. The TCA cycle intermediate α-ketoglutarate is a direct precursor of the amino acid L-glutamate which, in turn, can be used to synthesize L-glutamine. The reactions for the synthesis of glutamate and glutamine (catalyzed by L-glutamate dehydrogenase and glutamine synthetase, respectively) are of particular importance to nitrogen metabolism, since both convert inorganic nitrogen (NH_4^+) into organic nitrogen (carbon-bound amino group). The activity of glutamine synthetase is regulated by covalent modification, with adenylylation of the enzyme exerting an inhibitory effect and deadenylylation restoring activity. L-Glutamate, by means of transaminase (aminotransferase) reactions, is a principal donor of amino groups to other biomolecules. For example, in the aspartate aminotransferase reaction, glutamate donates its amino group to oxaloacetate (TCA cycle intermediate) to synthesize L-aspartate and to regenerate α-ketoglutarate; the latter product can then be reutilized, via the L-glutamate dehydrogenase reaction, to incorporate inorganic nitrogen.

QUESTIONS

1. Which reaction(s) in the aerobic oxidation of pyruvate to CO_2
   (a) produces CO_2?
   (b) produces NADH?
   (c) produces $FADH_2$?
   (d) has Fe^{2+} as a cofactor?
   (e) adds the elements of H_2O across the double bond of the substrate?
   (f) catalyzes substrate phosphorylation?
   (g) is physiologically irreversible?
   (h) is a condensation reaction that uses a four-carbon dicarboxylic acid as a substrate?
2. Which of the following statements are true?
   (a) Isocitrate dehydrogenase is an allosteric enzyme.
   (b) L-Glutamate dehydrogenase catalyzes a reversible reaction.
   (c) α-Ketoglutarate is the direct precursor of L-aspartate.
   (d) L-Glutamine synthesis requires ATP.
   (e) L-Glutamate synthesis requires ATP.
   (f) The TCA cycle is limited to the aerobic oxidation of pyruvate derived from glucose.
   (g) The activity of L-glutamate dehydrogenase is regulated by covalent modification.
   (h) The activity of pyruvate dehydrogenase is regulated by covalent modification.
   (i) Fumarase is a *cis* isomer.
   (j) Intermediates of the TCA cycle can be utilized for amino acid synthesis but not for gluconeogenesis.
3. Which B-vitamins are involved in the oxidation of pyruvate to CO_2 via the pyruvate dehydrogenase complex and the TCA cycle?
4. What is the $\Delta G^{\circ\prime}$ of (a) the oxidation of pyruvate by pyruvate dehydrogenase and the TCA cycle? (b) of the oxidation of the acetyl moiety of acetyl-CoA by the TCA cycle?
5. Based on the $\Delta G^{\circ\prime}$ of the succinyl-CoA synthetase reaction, Figure 16-11, what is the approximate $\Delta G^{\circ\prime}$ of the hydrolysis of the thioester bond of succinyl-CoA?
6. Which amino acids can be catabolically degraded to produce two different tricarboxylic acid intermediates of the TCA cycle?
7. If the methyl group of pyruvate is labeled with radioactive carbon-14 ($H_3{}^{14}C-\overset{O}{\overset{\|}{C}}-COO^-$), where would the label appear
   (a) after the oxidation of the pyruvate by one turn of the TCA cycle?
   (b) after two turns of the TCA cycle?
8. Which of the following enzymes catalyze a reaction in the TCA cycle (T), glyoxylate cycle (G), both (B), or neither (N)?
   (a) Aconitase
   (b) Citrate synthase
   (c) Fumarase
   (d) Glutamate dehydrogenase
   (e) Isocitrate dehydrogenase
   (f) Isocitrate lyase
   (g) Malate dehydrogenase
   (h) Malate synthase
   (i) Succinate dehydrogenase
   (j) Succinyl-CoA synthetase
9. Krebs astutely used malonate-poisoned muscle preparations in his studies elucidating the details of the TCA cycle. Questions a-f concern some of his experiments and the conclusions derived from them (refer to Figure 16-1).
   (a) What enzyme in the cycle does malonate inhibit?
   (b) What biological process does malonate specifically affect in these muscle preparations?

(c) What intermediate of the cycle was found to accumulate when Krebs added citrate, isocitrate, or α-ketoglutarate to muscle suspensions containing malonate?
(d) What intermediate of the cycle was found to accumulate when he added fumarate, malate, or oxaloacetate to malonate-poisoned muscle suspensions?
(e) Did the results of the experiments referred to in parts c and d of the question support Krebs' postulate about the cyclic nature of the sequence of reactions involved in pyruvate oxidation?
(f) When the cycle was not inhibited, one molecule of added oxaloacetate accounted for the utilization of many molecules of pyruvate; however, when inhibited, one molecule of oxaloacetate accounted for the utilization of only one molecule of pyruvate. How did Krebs explain these observations?

SUGGESTED READINGS

Goodwin, T. W., ed., *The Metabolic Roles of Citrate*. New York: Academic Press, 1968. Eight lectures delivered in a symposium held in honor of Sir Hans A. Krebs, sponsored by the Biochemical Society of England.

Greenberg, D. M., *Metabolic Pathways*, 3rd ed., Vol. 1. New York: Academic Press, 1967. This book has an excellent review by J. M. Lowenstein on the tricarboxylic acid cycle, with an emphasis on stereochemical relationships.

Mehlman, M. A., and R. W. Hanson, eds., *Energy Metabolism and the Regulation of Metabolic Processes in Mitochondria*. New York: Academic Press, 1972. Very good articles on the regulation of the tricarboxylic acid cycle.

Reed, L. J., "Pyruvate Dehydrogenase Complex," in *Current Topics in Cellular Regulation*, Vol. 1. New York: Academic Press, 1969. A review of the structure, function, and regulation of this multienzyme system.

(Additional references concerning the TCA cycle are found in the Suggested Readings for Chapter 17).

Articles

Krebs, H. A., "The History of the Tricarboxylic Acid Cycle." *Perspect. Biol. Med.,* 14:154, 1970.

Magasanik, B., "Regulation of Bacterial Nitrogen Assimilation by Glutamine Synthetase." *TIBS,* 2:9, 1977.

Randle, P. J., "Pyruvate Dehydrogenase Complex—Meticulous Regulator of Glucose Disposal in Animals." *TIBS,* 3:217, 1978.

Reed, L. J., "Multienzyme Complexes." *Accounts Chem. Res.,* 7:40, 1974.

AEROBIC SYNTHESIS OF ATP II: ELECTRON TRANSPORT AND OXIDATIVE PHOSPHORYLATION

17

INTRODUCTION

The generation of NADH and $FADH_2$ by the TCA cycle is a necessary prelude for the next phase, *electron transport*, in the aerobic synthesis of ATP from glucose and other carbohydrates. The reoxidation of these reduced coenzymes (carrying the electron pairs derived from the oxidation of pyruvate) initiates electron transport, a mitochondrial process designed for the transfer of electrons through a series of acceptor molecules until the last acceptor, O_2, is reduced to water. This transport system results in the release of large amounts of energy, a significant percentage of which is conserved through ATP synthesis by the last component of aerobic respiration, *oxidative phosphorylation*.

Detailed molecular analysis of electron transport and oxidative phosphorylation has proved to be one of the most demanding research challenges in biochemistry. Whereas most enzymes of the TCA cycle are located in the soluble environment of the *matrix* of the mitochondrion (Figure 17-1), the protein systems for electron transport and oxidative phosphorylation are associated with the *inner membrane*, with many of the proteins integrated into the membrane's structure and therefore essentially undissociable. Thus, unlike studies on the TCA cycle and glycolysis, isolation and analysis of most individual components of the electron transport chain and oxidative phosphorylation have not been possible. To compound the experimental difficulties, the two systems are *coupled*, i.e., existing in a functionally critical physical arrangement. Hence, integrity of the membrane is essential for ATP synthesis

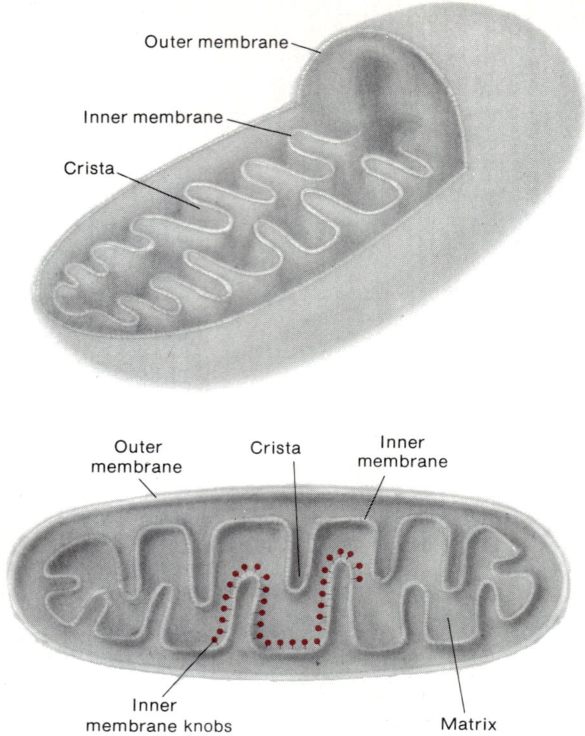

Figure 17-1
Drawings of a mitochondrion.

because fragmentation results in *uncoupling*, i.e., electron transport but no oxidative phosphorylation. This problem illustrates the complexities encountered when an interdependent relationship exists between a membrane's function and its structure, since disruption of the structure destroys the biological phenomenon being studied. However, in spite of the experimental obstacles, a number of distinguished biochemists have painstakingly but successfully elucidated many molecular aspects of mitochondrial ATP synthesis. Their pioneering efforts over the past 30 years have made the inner membrane of the mitochondrion one of the best understood biological membrane systems. The accumulated knowledge about the mitochondrial membrane has also made valuable contributions to current studies on other types of membranes.

ELECTRON TRANSPORT SYSTEM

Standard Reduction Potentials

The electron transport chain is a sequence of oxidation-reduction *(redox)* reactions, each of which involves the *oxidation* of one molecule (the *electron donor*) accompanied by the reduction of another (the *electron acceptor*). Just as individual Brönsted-Lowry acids and bases differ in their ability to donate or accept protons, biological redox molecules also vary in their tendency to donate or accept electrons, i.e., some are stronger oxidizing or reducing agents than others. In biochemistry, the tendency for a molecule to gain or lose electrons is given by its *reduction potential* (E, expressed in volts) and the E of a reaction is most often expressed as the *standard reduction potential* (E_o). Like a $\Delta G°$ value, E_o refers to a value pertaining to specified experimental conditions. By convention, the oxidation-reduction of H_2

$$2H^+ + 2e^- \rightleftharpoons H_2$$

was arbitrarily assigned a standard reduction potential of 0.0 V (volts) when the temperature is 25°C (298 K), $[H^+]$ is 1.0 *M* (pH = 0.0), and pressure of the H_2 gas is 1.0 atm. When corrected to pH 7.0 for biochemical purposes, the standard reduction potential (E_o') for hydrogen becomes −0.42 V. The prime (′) designation identifies an E_o value as one for pH 7.0.

Table 17-1 lists a number of biological reactions (written as reductions) and their E_o' values. Note that, depending on the reaction, electrons can be accepted (or donated) in the form of a hydrogen atom ($H^+ + e^-$). Each *half reaction* in Table 17-1 identifies a specific electron acceptor, e.g., NAD^+ or pyruvate, but not an electron donor, and therefore represents *half of an oxidation-reduction reaction*. With regard to reduction potentials, molecules with the *more positive* E_o' *are stronger oxidizing agents* (have a

Table 17-1
Reduction potentials of certain oxidation-reduction half reactions

Half reactions (written as reductions)	E_o' (V)
$2\ H^+ + 2\ e^- \rightarrow H_2$	−0.42
Ferredoxin-Fe^{3+} + 1 e^- → Ferredoxin-Fe^{2+}	−0.39
α-Ketoglutarate + CO_2 + 2 H^+ + 2 e^- → Isocitrate	−0.38
$NAD^+ + 2\ H^+ + 2\ e^- \rightarrow NADH + H^+$	−0.32
Riboflavin + 2 H^+ + 2 e^- → Riboflavin-H_2	−0.20
Dihydroxyacetone phosphate + 2 H^+ + 2 e^- → Glycerol 3-phosphate	−0.19
Pyruvate + 2 H^+ + 2 e^- → Lactate	−0.19
Oxaloacetate + 2 H^+ + 2 e^- → L-Malate	−0.17
Cytochrome *b*-Fe^{3+} + 1 e^- → Cytochrome *b*-Fe^{2+}	−0.04
Fumarate + 2 H^+ + 2 e^- → Succinate	+0.03
Cytochrome *c*-Fe^{3+} + 1 e^- → Cytochrome *c*-Fe^{2+}	+0.25
Cytochrome a_3-Fe^3 + 1 e^- → Cytochrome a_3-Fe^{2+}	+0.38
½ O_2 + 2 H^+ + 2 e^- → H_2O	+0.82

greater tendency to accept electrons and become reduced); conversely, molecules with the *more negative* E'_o *are stronger reducing agents (have a greater tendency to donate electrons and become oxidized). Of the compounds listed in Table 17-1, oxygen* (E'_o). $=$ $+0.82$ V) and hydrogen (E'_o $-$ 0.42 V) are the best oxidizing and reducing agents, respectively.

Coupling of Two Half Reactions

When two half reactions are coupled to establish an oxidation-reduction reaction, the one with the *more positive reduction potential* will proceed as a *reduction* (as written in Table 17-1) and the other as an *oxidation* (reversal of the reduction reaction). Coupling of the half reaction for O_2 ($E'_o = +0.82$ V) and NAD^+ (-0.32 V), for example, is written as follows:

Reduction (more positive E'_o):	$\frac{1}{2}O_2 + 2H^+ + 2e^- \rightarrow H_2O$
Oxidation (less positive E'_o):	$NADH + H^+ \rightarrow NAD^+ + 2H^+ + 2e^-$
Redox reaction:	$\frac{1}{2}O_2 + NADH + H^+ \rightarrow H_2O + NAD^+$

In this reaction, oxygen accepts electrons donated by NADH and, thus, oxygen is reduced and NADH is oxidized.

Calculation of $\Delta G^{\circ\prime}$ of an Oxidation-Reduction Reaction

The E'_o values of two half reactions can be used to calculate the $\Delta G^{\circ\prime}$ of the redox reaction by employing the following expression: $\Delta G^{\circ\prime} = -nF\Delta E'_o$, where n is the number of electrons transferred, F is Faraday's constant (23,063 cal per V equiv.), and $\Delta E'_o$ is the difference between the standard reduction potentials of the two half reactions. For the above reaction involving the reduction of O_2 by NADH, $n = 2$ and $\Delta E'_o = 0.82 - (-0.32)$, or 1.14, V and therefore

$$\Delta G^{\circ\prime} = -\ (2)\ (23{,}063)\ (1.14)$$

$$\Delta G^{\circ\prime} = -\ 52{,}583 \text{ cal mol}^{-1}\ (-52.6 \text{ kcal mol}^{-1})$$

The large negative $\Delta G^{\circ\prime}$ means that significant amounts of energy can be released by oxidation-reduction reactions; however, the calculation provides no information about the rate (kinetics) of the reaction. In fact, the reduction of O_2 by NADH at ambient temperatures would occur only if catalyzed by the appropriate enzyme.

The $\Delta G^{\circ\prime}$ of the O_2-NADH reaction summarizes the energetics of electron transport, whereby the initial re-oxidation of NADH subsequently results in the reduction of O_2 and, in the process, enough energy is released to synthesize several ATP molecules. The amount of energy released through the electron transport chain by the oxidation of one mole of NADH ($\Delta G^{\circ\prime} = -52.6$ kcal) is comparable to that freed by the anaerobic conversion (glycolysis) of one mole of D-glucose to two moles of L-lactate ($\Delta G^{\circ\prime} = -47.0$ kcal).

Components of the Electron Transport System

The complexities encountered when attempts are made to clarify the structural and functional details of electron transport continue to beset investigators. However, although a complete molecular description of the system is not yet possible, several distinct classes of protein redox carriers are now known to participate in the respiratory chain. These molecular species of biomolecules are *pyridine-linked dehydrogenases, flavin-linked dehydrogenases, iron-sulfur proteins, ubiquinones,* and *cytochromes*.

Pyridine-Linked Dehydrogenases

These enzymes require NAD^+ or $NADP^+$ (or their reduced forms, p. 167) as loosely bound *(dissociable)* coenzymes. They catalyze the following general reaction in which $X \cdot H_2$ and X represent reduced and oxidized forms of a specific substrate, respectively:

$$X \cdot H_2 + NAD^+ (NADP^+) \rightleftharpoons X + NADH\ (NADPH) + H^+$$

Of the 200 or more known pyridine-linked dehydrogenases, most *NAD^+-linked dehydrogenases* are associated with biological redox reactions, such as those involved in aerobic respiration. About 60 percent of the liver's NAD^+ is in the mitochondria, reflecting the major involvement of various mitochondrial NAD^+-linked dehydrogenases in aerobic oxidation, e.g., isocitrate and malate dehydrogenases. *$NADP^+$-linked dehydrogenases*, on the other hand, are primarily used for biosynthetic purposes.

Flavin-Linked Dehydrogenases

These enzymes also oxidize and/or reduce substrates but require FMN or FAD (or their reduced forms, p. 168) as tightly bound coenzymes or prosthetic groups. Flavin-linked dehydrogenases catalyze the following general reactions:

$$X\cdot H_2 + \text{E-FMN} \rightleftharpoons X + \text{E-FMNH}_2$$

$$X\cdot H_2 + \text{E-FAD} \rightleftharpoons X + \text{E-FADH}_2$$

NADH dehydrogenase, containing FMN as its prosthetic group, is the flavin-linked dehydrogenase that transfers electrons from NADH to the next acceptor in respiratory electron transport. In mitochondria, there are additional flavin-linked dehydrogenases, e.g., succinate dehydrogenase, that are important in the aerobic oxidation of biomolecular fuels.

Iron-Sulfur Proteins

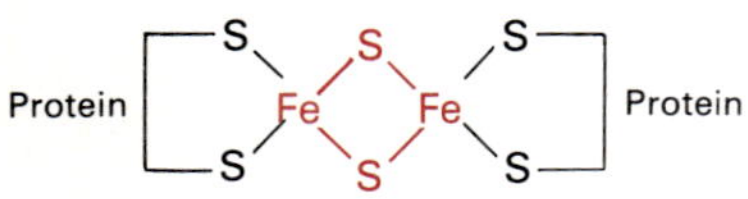

Proposed two iron-sulfur (Fe·S) centers of chloroplast ferredoxin.

These components are proteins characterized by their stoichiometric content of iron and acid-labile sulfur. Since the discovery of the first iron-sulfur protein, *ferredoxin*, in an anaerobic bacterium, others have been isolated from other microorganisms, plants, and animals. As electron carriers ($Fe^{3+} \rightleftharpoons Fe^{2+}$), proteins of this type are associated with photosynthetic transport as well as with respiration (p. 350). A number of iron-sulfur proteins have been identified in the electron transport chain, complexed with other respiratory carriers. For example, the flavoproteins NADH dehydrogenase and succinate dehydrogenase have four and two *iron-sulfur centers* (Fe·S) per molecule, respectively. Although the electron transport functions of some of the iron-sulfur centers are now fairly well established, the specific roles of others are still uncertain.

Ubiquinones (CoQ)

Ubiquinones are fat-soluble coenzymes that serve as electron carriers. The prefix *ubi* refers to the ubiquitous occurrence of this class of quinones, found in microorganisms, plants, and animals. A ubiquinone, also called *coenzyme Q (CoQ),* is a quinone derivative with a long, nonpolar side chain composed of isoprenoid (branched five-carbon) units (Figure 17-2). Several types of CoQ are known; they differ in the number of isoprene units in their side chains. Mitochondria of mammals have a CoQ with ten isoprenoid units in its side chain *(CoQ_{10})*, whereas some microorganisms have a CoQ with six *(CoQ_6)*. As seen in Figure 17-2, when CoQ participates in electron transfer, two hydrogen atoms ($2H^+ + 2e^-$) are gained or lost.

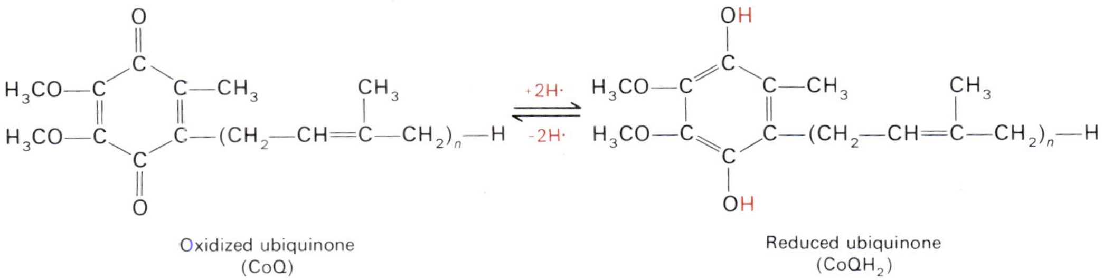

Figure 17-2
Oxidized and reduced ubiquinone.

Cytochromes

The cytochromes are a class of proteins that, like iron-sulfur proteins, rely on the redox potential of the iron atom ($Fe^{3+} \rightleftharpoons Fe^{2+}$) to transfer electrons. In eucaryotes, a variety of cytochromes are found in the mitochondria (respiratory chain) and the endoplasmic reticulum (hydroxylation reactions). Cytochromes are classified as families, known as *a, b,* or *c,* designations that refer to the distinctive characteristic of the α absorbance band that each type, in the reduced state, exhibits in the green-red range of the spectrum (Figure 17-3).

All cytochromes have *iron-porphyrin groups* and are therefore *heme proteins.* The *b* and *c* cytochromes, like myoglobin and hemoglobin (p. 107), have *iron-protoporphyrin IX* as their heme group, which is covalently bound to the protein in the *c* cytochromes but not in the *b* cytochromes. In cytochrome *c* and c_1 (cyt *c* and cyt c_1) of electron transport, the heme is covalently bonded by thioether linkages to two cysteinyl residues (Figure 17-4). The iron-porphyrin group of cytochrome *a* and a_3 (cyt *a* and cyt a_3) is noncovalently bound heme A (Figure 17-5), a molecule that differs from protoporphyrin IX in that a *long hydrocarbon chain replaces one of the vinyl groups and a formyl group one of the methyl groups.* These differences in heme structure or heme attachment account for the distinctive absorption spectra of the reduced cytochromes.

Five cytochromes have been identified in the electron transport chain of mammals (*b, c,* c_1, *a,* and a_3). Recent evidence reveals that cytochrome *b* (cyt *b*) has two components, cyt b_k and cyt b_t, and data suggest that cyt b_k is the redox carrier and cyt b_t is involved in energy conservation, i.e., coupling of electron transport to oxidative phosphorylation. Cytochrome *c*, one of the few readily extractable electron carriers of the chain, has been purified from many sources, and comparisons of the amino acid sequences of

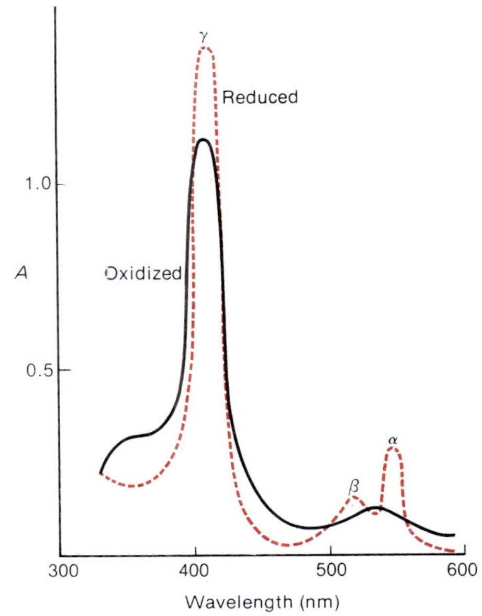

Figure 17-3
Absorption spectra of oxidized and reduced cytochrome *c* at pH 7.0, with the α, β, and γ peaks identified.

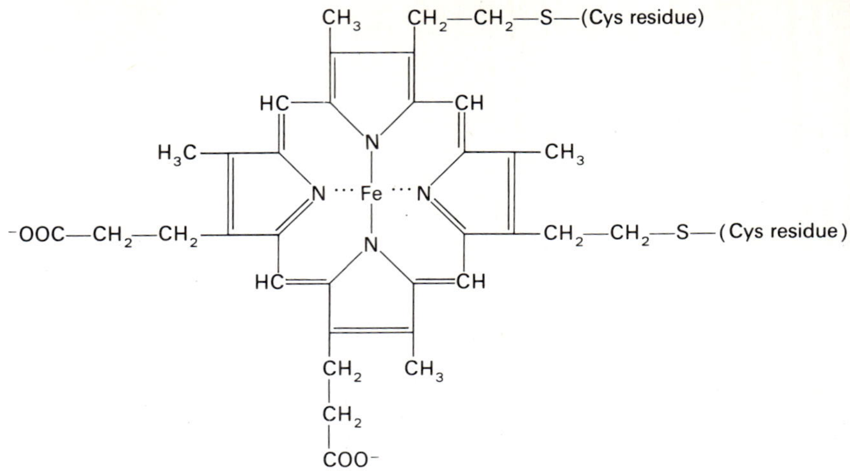

Figure 17-4
Heme of cytochrome *c* and c_1, which is covalently bonded to the sulfhydryl groups of two cysteines.

the protein of many species have contributed greatly to the molecular approach to the study of the biochemical evolution of macromolecules (p. 153).

Cytochromes *a* and a_3 exist as an oligomeric protein, called *cytochrome oxidase* (cyt aa_3) and are the terminal cytochrome acceptors in transferring electrons to O_2. The cyt aa_3 complex has *two heme A* (Fe) and *two atoms of copper*, and both metal ions function in the transfer of electrons. The transfer of electrons in

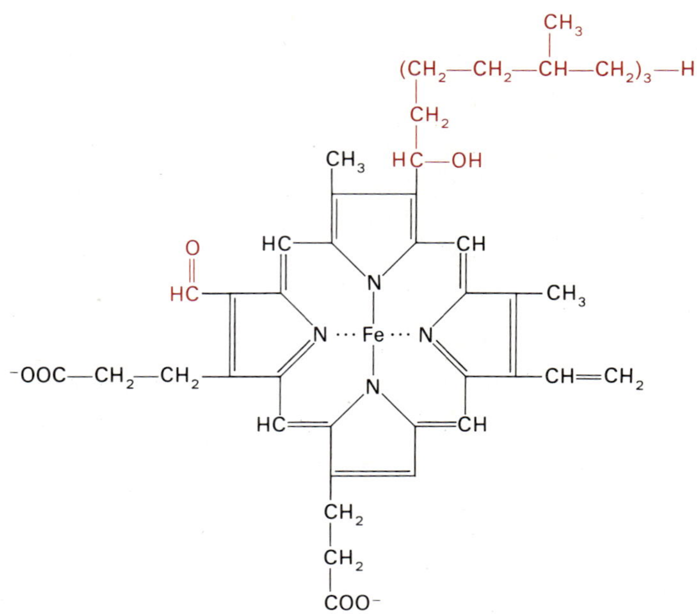

Figure 17-5
Heme A of cytochromes *a* and a_3. Groups in red are the two that differ between heme A and protoporphyrin IX.

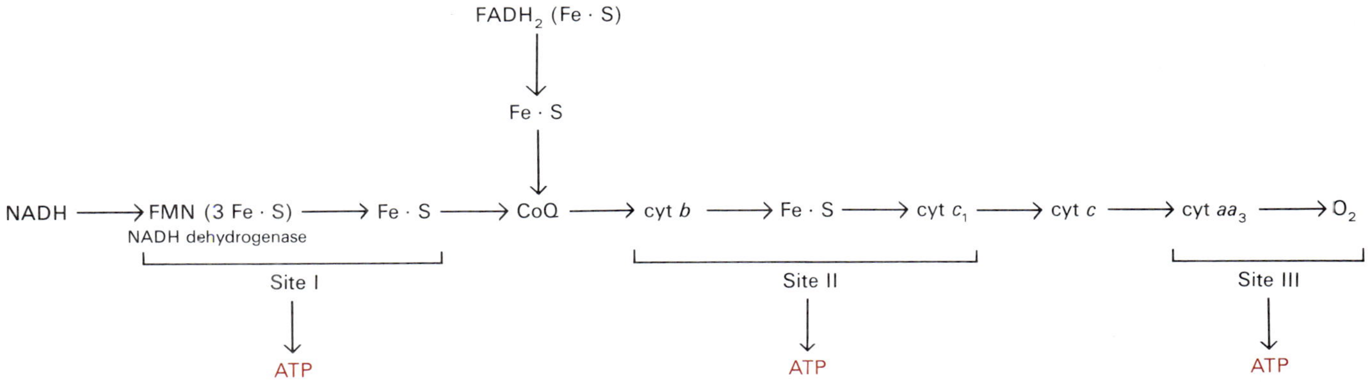

Figure 17-6
Schematic of the transport of electrons from NADH and $FADH_2$ to oxygen, identifying the three regions (sites) of the chain where the amount of free energy released is large enough to allow for the synthesis of an ATP molecule. (Fe·S and CoQ represent an iron-sulfur center and the ubiquinone coenzyme Q, respectively.)

the final stages of transport is believed to proceed as follows: cyt $a \rightarrow$ cyt $a_3 \rightarrow O_2$, with copper ($Cu^{2+} \rightleftharpoons Cu^+$) participating in the transfer of electrons from cyt *a* to cyt a_3.

Electron Transport System of Mammals

The matrix of a mitochondrion, where oxidation of the carbon atoms of fuel molecules transpires, serves as a rich source of NADH, the major electron donor to respiration, and also of $FADH_2$. The sequence of reactions that transfers electrons to the terminal acceptor O_2 in the inner membrane of a mammalian mitochondrion is summarized in Figure 17-6. Note that whereas the two electrons of NADH enter into the transport chain by their transfer to the FMN moiety of NADH dehydrogenase, the two of $FADH_2$ enter at a subsequent site, by transfer to CoQ. The passage of electrons through the cytochromes proceeds from cyt *b* to cyt c_1 and *c* and then to cyt aa_3. The iron-sulfur centers now identified as respiratory electron carriers are those complexed with NADH dehydrogenase [FMN (3 Fe·S) → Fe·S], succinate dehydrogenase [FAD (Fe·S) → Fe·S] and cytochrome *b* (cyt *b* → Fe·S). Since each cytochrome and iron-sulfur center can transfer only one electron at a time, two molecules of each are needed to convey the two electrons donated by NADH or $FADH_2$.

Another way of depicting electron transport, one that involves an initial transfer of electrons from a substrate to NAD^+, is illustrated in Figure 17-7. This version emphasizes the cyclic nature of the individual redox reactions and also those electron transfers in which one or two protons (H^+) are taken up ($2\ Fe^{2+}\cdot S + CoQ + 2\ H^+ \rightarrow 2\ Fe^{3+}\cdot S + CoQH_2$) or ejected ($CoQH_2 + 2\ cytb\text{-}Fe^{3+} \rightarrow CoQ + 2\ cytb\text{-}Fe^{2+} + 2\ H^+$).

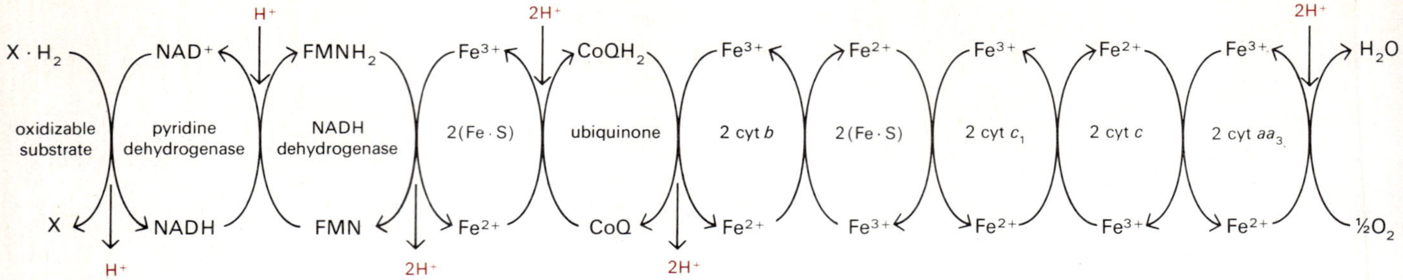

Figure 17-7
Schematic of electron transport, emphasizing the cyclic nature of the system and the input and output of protons from the mitochondrial matrix to the outside of the inner membrane. (Fe·S represents an iron-sulfur center.)

Energetics of Electron Transport and ATP Synthesis

Respiratory electron transfer produces a continuous decline in free energy as electron pairs are passed from NADH to O_2 (−52.6 kcal mol^{-1}). Figure 17-8 diagrams this sequential decrease in free energy and identifies the three regions of the chain where the amount of energy released is large enough to account for the synthesis of an ATP molecule (−7.3 kcal mol^{-1}) by the coupling of oxidative phosphorylation to electron transport. These regions of the electron transport system are also indicated in Figure 17-6 as Sites I, II, and III of ATP synthesis. As implied, a pair of electrons donated by *NADH* or by *FADH*$_2$ results in the production of *three*

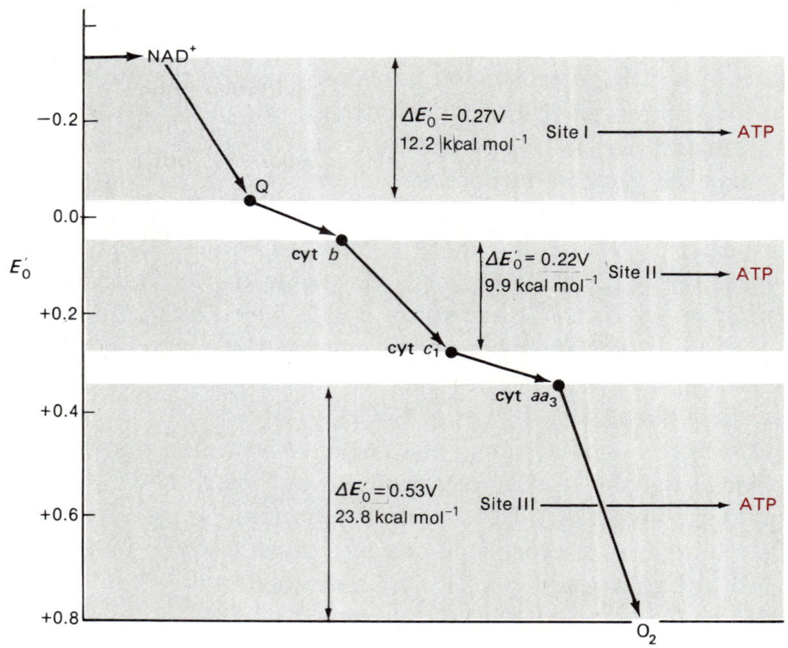

Figure 17-8
Schematic of the decline in free energy as an electron pair is transported to oxygen, depicting the amounts of free energy released by the three regions of the chain that result in the subsequent synthesis of ATP.

or *two ATP molecules,* respectively. Thus, whereas substrate-level phosphorylation (p. 277) transfers the phosphate from a phosphorylated substrate to the terminal phosphate of a nucleoside diphosphate, oxidative phosphorylation succeeds in producing two or three molecules of ATP from ADP and P_i, using the free energy released by the transfer of an electron pair.

OXIDATIVE PHOSPHORYLATION

The molecular mechanisms by which the energy released by electron transport is conserved by ATP synthesis remain a scientific challenge. However, during the last two decades, a number of investigators have succeeded in providing insights into the mitochondrial production of ATP which, in turn, have generated several hypotheses in an attempt to explain this phenomenon.

Reconstitution of Oxidative Phosphorylation

One of the most enlightening contributions to the understanding of the complex molecular systems of aerobic ATP synthesis is that made by Efraim Racker and his associates. By appropriate disruption of mitochondria (Figure 17-9), these investigators fractionated the inner membrane into *membranous vesicles* and a *soluble protein fraction*, the latter being the *inner membrane knobs.* The vesicles catalyze electron transport but not oxidative phosphorylation, i.e., no ATP production from ADP and P_i. The soluble fraction catalyzes ATP hydrolysis, i.e., has *ATPase activity,* but not electron transport. When the two fractions are combined, *submitochondrial vesicles* (with inner membrane knobs) are reconstituted (Figure 17-9) and a significant percentage of oxidative phosphorylation is restored. This successful reconstitution of ATP synthesis provides evidence that the *inner membrane* (membranous vesicles) contains the *enzymes of electron transport* and that the *inner membrane knobs,* called *F_1-ATPase,* contain the *protein-coupling factors* which link electron transport to oxidative phosphorylation. Racker and his colleagues have established that F_1-ATPase is a large oligomeric complex (MW $\approx$ 380,000) composed of ten protomers and containing five or six different types of proteins. The *in vitro* ATPase activity of the complex is thought to represent the reverse of the normal function (ADP + $P_i \rightarrow$ ATP), i.e., F_1-ATPase is ATP synthetase.

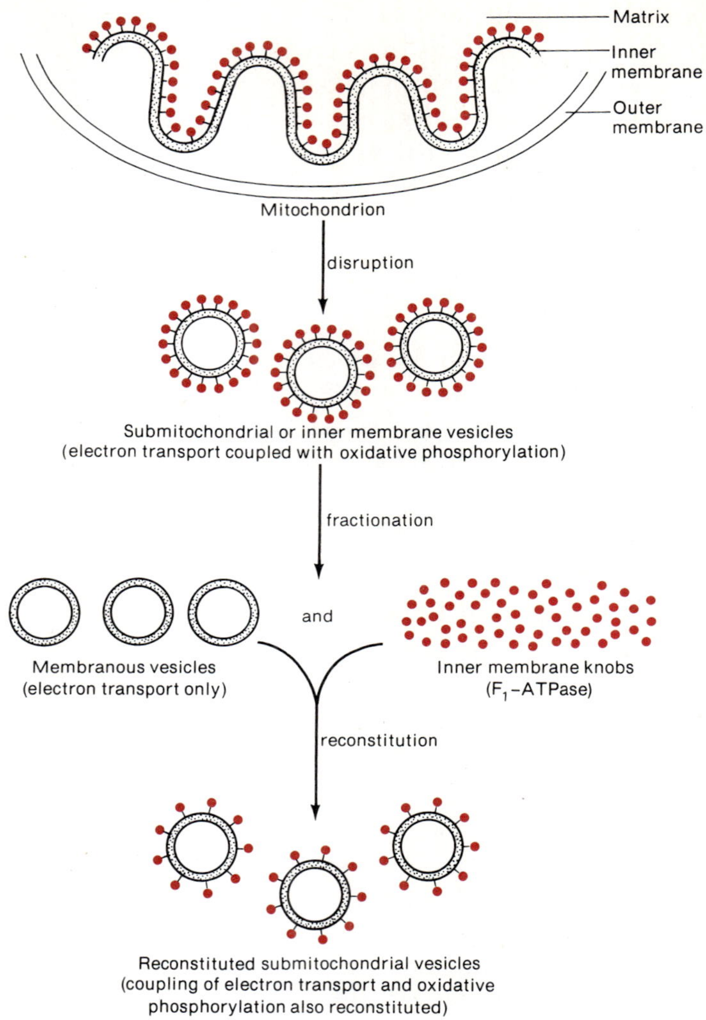

Figure 17-9
Schematic of the experimental treatment of mitochondrial inner membranes to demonstrate the uncoupling and subsequent recoupling of electron transport and oxidative phosphorylation.

PROPOSED MECHANISMS OF OXIDATIVE PHOSPHORYLATION

The molecular events that transpire between the release of energy by electron transport and its conservation by phosphorylation of ADP to yield ATP continue to be scientifically intriguing. Three hypotheses have been proposed over the past 25 years, and each has served as a focal point, at one time or another, for much of the research on mitochondrial ATP synthesis.

The Chemical-Coupling Hypothesis

In the *chemical-coupling hypothesis*, the earliest proposed, one scheme (Figure 17-10) postulates that the energy released in the transfer of electrons from carrier A to carrier B (spanning an ATP-synthesis site) in electron transport is conserved by the production of A~I and is subsequently transferred to X~P, with the intermediate formation of X~I. Both the energy and the phosphate group for the synthesis of ATP from ADP are provided by X~P. The involvement of *coupling factors* is represented by the hypothetical molecules I and X, neither of which has been chemically identified. This hypothesis has its analogy in the chemical-coupling reaction catalyzed by glyceraldehyde 3-phosphate dehydrogenase in glycolysis and the subsequent substrate-level phosphorylation (p. 276). Because no high-energy intermediates have been detected after several decades of intensive research, interest in the hypothesis has diminished.

1. $AH_2 + B + \text{I-OH} \rightleftharpoons A{\sim}I + BH_2 + OH^-$

2. $A{\sim}I + \text{X-H} + OH^- \rightleftharpoons X{\sim}I + A + H_2O$

3. $X{\sim}I + \text{P-OH} \rightleftharpoons X{\sim}P + \text{I-OH}$

4. $X{\sim}P + ADP \rightleftharpoons ATP + \text{X-H}$

Overall: $AH_2 + B + ADP + \text{P-OH} \rightleftharpoons A + BH_2 + ATP + H_2O$

Figure 17-10
Summary of the types of reactions proposed for the chemical-coupling hypothesis. A and B are electron carriers; I-OH and X-H are hypothetical coupling factors; P-OH is inorganic phosphate; and A~I, X~I, and X~P are high-energy intermediates.

The Conformational-Coupling Hypothesis

The *conformational-coupling hypothesis* proposes that energy released by electron transport is conserved by a conformational change in a protein, e.g., coupling factor, resulting in the macromolecule's attainment of a high-energy conformational state. The release of this energy when the protein returns to its original lower-energy conformational state provides the driving force for the phosphorylation of ADP. The basic difference between the chemical- and conformational-coupling hypotheses is that, in the former, the energy is conserved by formation of covalent bonds, e.g., X~P, and in the latter, a large number of weak, noncovalent bonds (needed to bring about the high-energy conformational change) collectively conserve the energy. Observations of physical changes in the inner membranes during active transport of electrons offer one of the lines of suggestive evidence that support the hypothesis.

The Chemiosmotic-Coupling Hypothesis

This hypothesis is the most novel mechanism advanced to explain oxidative phosphorylation. Proposed by Peter Mitchell (Nobel Prize, 1978), the *chemiosmotic-coupling hypothesis* abandons the concept of high-energy molecular intermediates. Instead, Mitchell proposes that an *electrochemical potential difference of protons and/or electrons* created across the inner membrane of a mitochondrion, as a result of electron transport, is the energetic driving force in oxidative phosphorylation. In the proposal, electron transport is responsible for translocating protons from within the mitochondrial matrix across the inner membrane (or electrons in the opposite direction); this results in the creation of a *H^+-concentration gradient*, as well as an electropotential, because of the membrane's impermeability to protons (Figure 17-11). The free energy stored in this gradient is called a *protonmotive force*. This *electrochemical gradient* drives the synthesis of ATP by causing the dehydration of ADP and P_i. As illustrated in Figure 17-12, ATP synthesis is catalyzed by F_1-ATPase (ATP synthetase), which utilizes the free energy difference of the proton gradient.

Mitchell is a rare example of the truly independent scientist since, in his native England, he is not affiliated with a university, industry, or government.

The molecular mechanism of proton ejection presupposes a specific vectorial arrangement of the electron carrier components of electron transport in a series of *oxidation-reduction loops*, which couple hydrogen carriers to electron carriers (Figure 17-13). As a result of the flow of electrons through this sequence of *proton-translocating loops*, an outward translocation of two protons (or an inward translocation of two electrons) is believed to occur per pair of electrons transferred in each loop. It is also believed that *two protons* must pass through F_1-ATPase for the synthesis of *one ATP*. Thus, the production of three ATP and two ATP for each NADH and $FADH_2$, respectively, donating electrons to the trans-

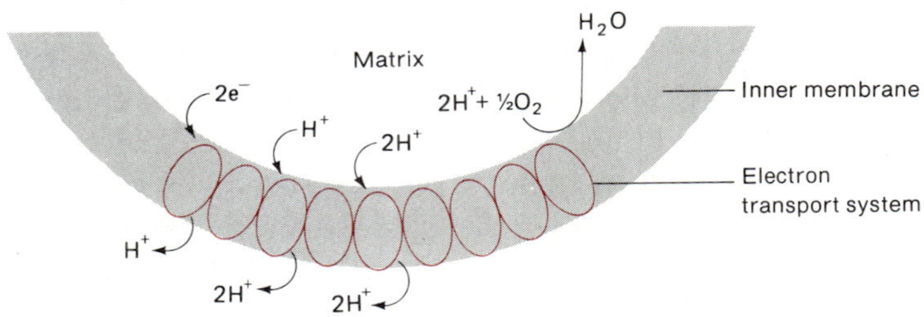

Figure 17-11
Schematic of the creation of a proton (H^+) concentration gradient across the mitochondrial inner membrane by electron transport.

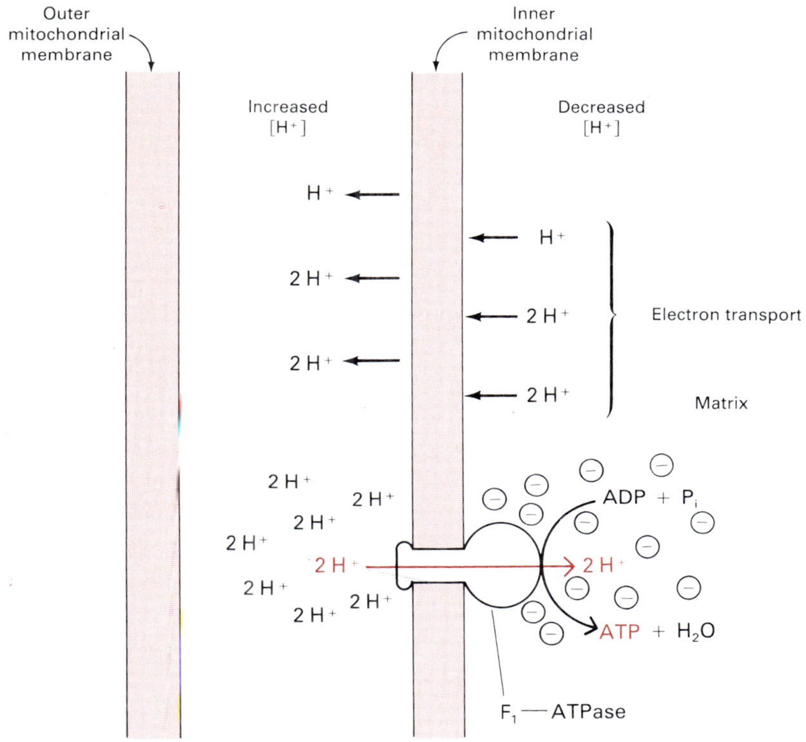

Figure 17-12
Simplified diagram of the production of ATP with the H^+ gradient that is generated by electron transport serving as the driving force.

port system is still accepted as generally valid. Verification and identification of the proposed loops, as well as the obtainment of more detailed information about actual proton (electron) translocation, are currently the subject of intensive research.

Interest in the chemiosmotic hypothesis extends beyond the question of oxidative phosphorylation because of its broad biological implications. The basic concept of energy being stored in the form of an electrochemical gradient across a membrane is now incorporated into hypotheses about photorespiration and respiration-linked transport across bacterial membranes, e.g., Ca^{2+} flowing inward to replace two H^+ flowing outward. Bacterial membrane systems have been especially useful in providing evidence that the H^+ gradient generated by electron transport (as well as by an artificially generated electropotential gradient) can drive the active transport systems that concentrate specific molecules against a gradient. The thorough molecular studies by Walther Stoeckenius and Racker on the purple membrane of a halobacterium (microorganisms requiring high salt concentrations for growth) have established that the light energy captured by these photosynthetic organisms is conserved as a H^+ gradient, which is then used for ATP synthesis. These findings support the concept of the involvement of the chemiosmotic phenomenon in photosynthetic processes, as discussed on p. 351.

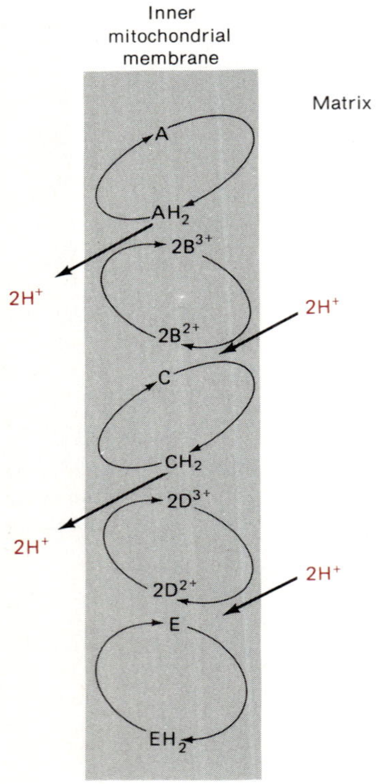

Figure 17-13
Diagram of a series of proposed proton-translocating loops arranged vectorically in the inner membrane of a mitochondrion to ensure the ejection of protons during electron transport. (A, B, C, D, and E represent hypothetical electron carriers.)

ATP PRODUCTION FROM PYRUVATE

The production of H_2O by respiration completes the oxidation of glucose to CO_2 and H_2O. When the oxidation of pyruvate is coupled to electron transport and oxidative phosphorylation, the *four NADH* and *one FADH₂* produced by pyruvate dehydrogenase and the TCA cycle (Table 17-2) are responsible for the generation of *12* (4 × 3) and *2 ATP molecules*, respectively. The additional production of GTP (energetically equivalent to ATP) by substrate-level phosphorylation in the TCA cycle brings to 15 the total number of nucleoside triphosphates produced by complete mitochondrial oxidation of one pyruvate. The net reaction for the oxidation of pyruvate to three CO_2 ($\Delta G^{\circ\prime} = -273$ kcal mol^{-1}) is

$$\text{Pyruvate} + 2\tfrac{1}{2}\,O_2 + 15\,P_i + 15\,\text{ADP} \rightarrow 3\,CO_2 + 15\,\text{ATP} + 17\,H_2O$$

The production of 15 ATP accounts for the conservation of 109.5 kcal mol^{-1} (15 × 7.3) of the 273 kcal mol^{-1} released; hence, the *efficiency of energy conservation is 40 per cent.*

Table 17-2
Reduced coenzymes produced by the aerobic oxidation of pyruvate

Reacton catalyzed by	Reduced coenzyme generated
Pyruvate dehydrogenase	1 NADH
Isocitrate dehydrogenase	1 NADH
α-Ketoglutarate dehydrogenase	1 NADH
Succinate dehydrogenase	1 $FADH_2$
Malate dehydrogenase	1 NADH
	Total: 4 NADH + 1 $FADH_2$

ATP PRODUCTION BY THE AEROBIC OXIDATION OF GLUCOSE

The equations for the anaerobic (glycolysis) and aerobic oxidation of glucose, which together account for the complete oxidation of the sugar to CO_2 and H_2O are:

$$\text{Anaerobic: Glucose} + 2\ \text{ADP} + 2\ \text{P}_i + 2\ \text{NAD}^+ \rightarrow 2\ \text{pyruvate} + 2\ \text{ATP} + 2\ \text{NADH} + 2\ \text{H}^+ + 2\ \text{H}_2\text{O}$$

$$\text{Aerobic: } 2\ \text{Pyruvate} + 5\ \text{O}_2 + 30\ \text{ADP} + 30\ \text{P}_i \rightarrow 6\ \text{CO}_2 + 30\ \text{ATP} + 34\ \text{H}_2\text{O}$$

$$\text{Sum: Glucose} + 5\ \text{O}_2 + 32\ \text{ADP} + 32\ \text{P}_i + 2\ \text{NAD}^+ \rightarrow 6\ \text{CO}_2 + 32\ \text{ATP} + 2\ \text{NADH} + 2\ \text{H}^+ + 36\ \text{H}_2\text{O}$$

ENZYMATIC SHUTTLES

Note that in the above summation the two NADH generated in glycolysis have not been reoxidized because the two pyruvate produced in the anaerobic portion of the process are not reduced to lactate but, instead, are routed through the TCA cycle for aerobic oxidation. These NADH molecules, being produced in the cytosol, are not directly available for mitochondrial oxidation because the organelle's membrane is impermeable to NADH. To utilize the electron energy stored in these cytosolic NADH molecules for ATP production (and to regenerate cytosolic NAD^+), eucaryotic cells employ *enzymatic shuttle systems* to transport *reducing equivalents* (electrons) across the mitochondrial membrane. Such shuttles capitalize on the limited variety of biomolecules that can transverse the mitochondrial membrane.

Glycerol Phosphate Shuttle

One such system is the *glycerol phosphate shuttle* (Figure 17-14), which first uses *cytosolic glycerol 3-phosphate dehydrogenase* to produce *glycerol 3-phosphate* from *dihydroxyacetone phosphate* and cytosolic NADH; the product, glycerol 3-phosphate, can diffuse into mitochondria. On the outer surface of the inner membrane of mitochondria, there is a *mitochondrial glycerol 3-phosphate dehydrogenase,* having FAD as a prosthetic group, which regenerates dihydroxyacetone phosphate from glycerol 3-phosphate. The electrons transferred to FAD in the reaction result in the production of two ATP molecules by oxidative phosphorylation. The shuttle is completed by the diffusion of dihydroxyacetone phosphate into the cytosol, where it is again available for enzymatic reduction. The shuttle is critically important for sustained flight by those insects that rely on sugars for their energy, as evidenced by the observations that mutants of the fruitfly, *Drosophila melanogaster*, which lack cytosolic glycerol 3-phosphate dehydrogenase cannot maintain sustained flight.

Figure 17-14
The glycerol phosphate shuttle.

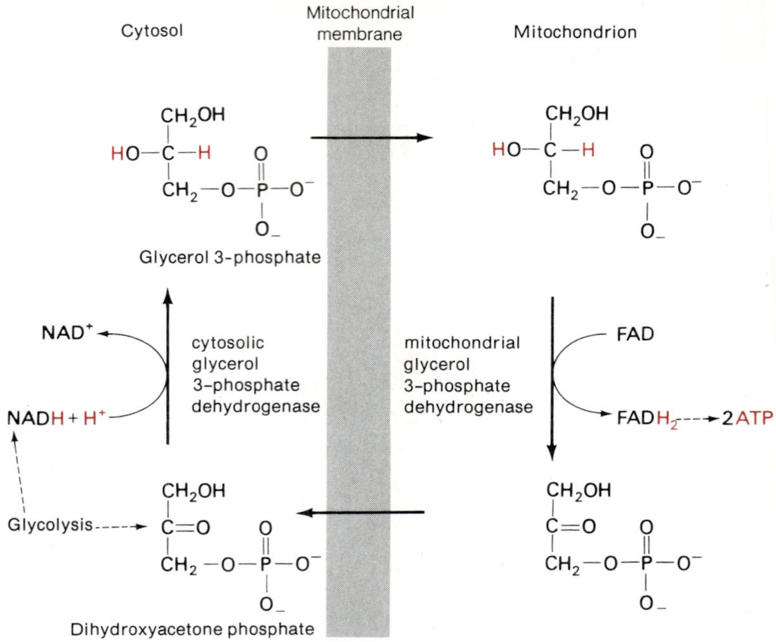

Malate-Aspartate Shuttle

A second shuttle, predominant in mammalian liver and heart, is the *malate-aspartate shuttle*, which involves the participation of *cytosolic* and *mitochondrial forms of both malate dehydrogenase and aspartate aminotransferase* (Figure 17-15). In the shuttle, the reducing equivalents of cytosolic NADH are transferred to oxaloacetate to produce *L-malate*. Malate can then be transported into the mitochondria, where it is reoxidized to oxaloacetate, and the NADH generated in the reaction results in the production of three ATP. To re-establish the cytosolic phase of the shuttle, oxaloacetate (which cannot transverse the membrane) is converted into *aspartate (mitochondrial aspartate aminotransferase),* which is transported out of the mitochondria. Once in the cytosol, the reverse transamination reaction occurs *(cytosolic aspartate aminotransferase),* and oxaloacetate is regenerated. Glutamate and α-ketoglutarate serve as NH_2-donor and -acceptor, respectively, in these transamination reactions. The malate-aspartate shuttle is more complex and versatile than the glycerol phosphate shuttle because of the reversible transport systems involving aspartate, glutamate, α-ketoglutarate, and malate. For example, in gluconeogenesis (p. 284), mitochondrial malate dehydrogenase reduces

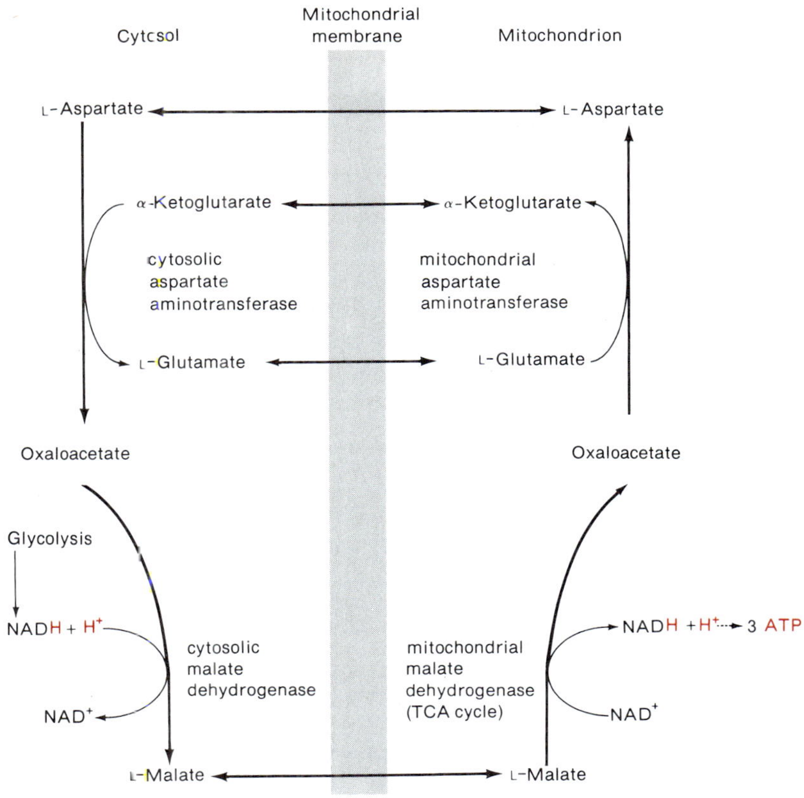

Figure 17-15
The malate-aspartate shuttle.

oxaloacetate (product of the pyruvate carboxylase reaction) to produce malate, which is then transported to the cytosol and re-oxidized to oxaloacetate by cytosolic malate dehydrogenase.

Efficiency of Energy Conservation

Oxidation of a molecule of NADH produced by glycolysis, therefore, can result in the synthesis of either *two* or *three ATP* molecules, depending on the shuttle used, as follows:

$$NADH + H^+ + \tfrac{1}{2} O_2 + 2\ ADP + 2\ P_i \rightarrow NAD^+ + 2\ ATP + 3\ H_2O$$

or

$$NADH + H^+ + \tfrac{1}{2} O_2 + 3\ ADP + 3\ P_i \rightarrow NAD^+ + 3\ ATP + 4\ H_2O$$

When this oxidation of NADH is incorporated into the overall oxidation of glucose, the reaction becomes

1) Glucose $+ 5\ O_2 + 32$ ADP; $+ 32\ P_i + 2\ NAD^+ \rightarrow$
$6\ CO_2 + 32$ ATP $+ 2$ NADH $+ 2\ H^+ + 36\ H_2O$

2) 2 NADH $+ 2\ H^+ + O_2 + 4(6)$ ADP $+ 4(6)\ P_i \rightarrow$
$2\ NAD^+ + 4(6)$ ATP $+ 6(8)\ H_2O$

Sum: Glucose $+ 6\ O_2 + 36(38)$ ADP $+ 36(38)\ P_i \rightarrow$
$6\ CO_2 + 36(38)$ ATP $+ 42(44)\ H_2O$

If *36 ATP* is selected as the total number produced by oxidation of glucose, then 263 (36 × 7.3) kcal of the 686 kcal released by the oxidation of one mole of glucose is conserved, a *38 per cent efficiency of energy conservation*. Production of *38 ATP* results in the conservation of 277 kcal mol^{-1}, i.e., *40 per cent efficiency of conservation*. Under physiological conditions, the conservation of energy is considered to be even more efficient. When a cell is viewed as an "engine," nature's biochemical pathways for extracting utilizable energy from glucose and other biomolecular fuels have an efficiency that is far superior to that attained by any engine yet designed.

β-OXIDATION CYCLE

In previous discussions, the utilization of carbohydrates as a major fuel source has been emphasized. The catabolism of amino acids to yield intermediates of the TCA cycle has also been illustrated (p. 306). Fatty acids derived from lipids are a third important source of biomolecular fuel, and, like pyruvate and catabolized amino acids, they are oxidized in the mitochondrial matrix. The oxidative degradation of *saturated fatty acids* (with even numbers of carbon atoms) proceeds by *β-oxidation* (Figure 17-16) and provides products that are utilizable in respiration.

Activation of a Fatty Acid

The first step in the oxidative process, which occurs in the cytosol, is the *activation* of a fatty acid. In a reaction catalyzed by an *acyl-CoA synthetase,* a fatty acid is covalently linked to a CoA-SH molecule (Figure 17-16). Synthesis of an *acyl-CoA* requires ATP, and the reaction is an example of a pyrophosphate cleavage of ATP (p. 256) that assures the physiological irreversibility of the reaction because of the subsequent hydrolysis of pyrophosphate. In the transfer of a fatty acyl-CoA derivative into the mitochondria, *car-*

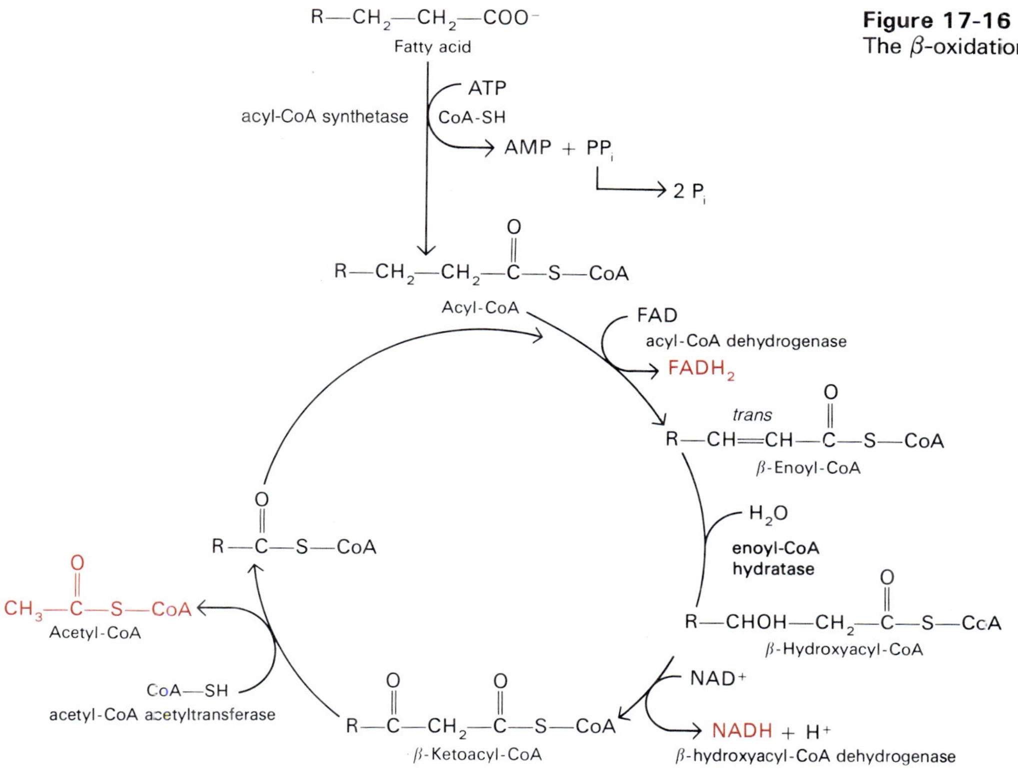

Figure 17-16
The β-oxidation cycle for fatty acids.

nitine (Figure 17-17) serves as the carrier. The acyl moiety of the fatty acid is transferred to carnitine, and *acyl carnitine* (Figure 17-17) is transported into the mitochondria, where the reverse reaction, using mitochondrial CoA-SH, regenerates the acyl-CoA derivative.

Figure 17-17
Transfer of the acyl moiety of an acyl-CoA into the mitochondria by carnitine.

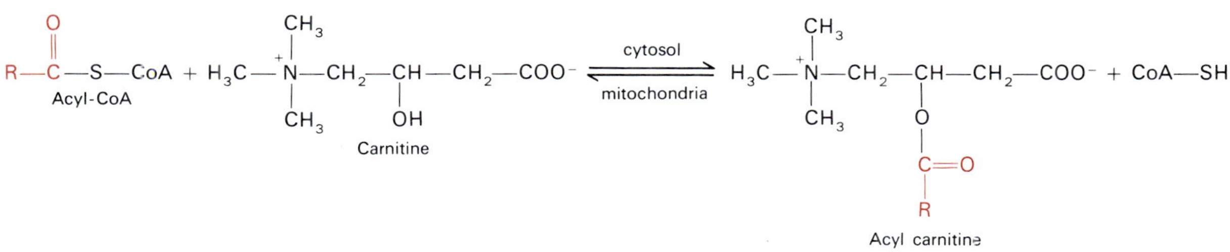

Oxidation of the β-Carbon

The next three reactions in the scheme result in the oxidation of the β-carbon of a fatty acid and are analogous to the last three steps in the TCA cycle, which convert succinate into oxaloacetate (p. 304).

In the first reaction, catalyzed by *acyl-CoA dehydrogenase* (Figure 17-16), dehydrogenation occurs at the α and β carbon atoms, creating a double bond with a *trans* configuration and forming a β *-enoyl-CoA.* The enzyme's prosthetic group, FAD, is reduced to $FADH_2$ in the reaction. In the second reaction, *enoyl-CoA hydratase* catalyzes the addition of H_2O across the double bond to synthesize the L-isomer of a *β-hydroxyacyl-CoA.* The third reaction, a second dehydrogenation, requires NAD^+ and produces a *β-ketoacyl-CoA and NADH;* this reaction is catalyzed by *β-hydroxyacyl-CoA dehydrogenase.*

Completion of a Turn of the Cycle

The last reaction, requiring CoA-SH, is the cleavage of a β-ketoacyl-CoA to release the two carbons from the carboxyl terminus of the fatty acid as *acetyl-CoA* (Figure 17-16) and to regenerate an *acyl-CoA derivative of the fatty acid,* now shorter by two carbon atoms. The reaction, catalyzed by *acetyl-CoA acetyltransferase (thiolase),* completes a *turn of the cycle* and furnishes an acyl-CoA derivative for the next turn. Thus, by repeated turns of the cycle, a fatty acid is degraded to acetyl-CoA molecules, with *one* being produced *every turn* until the last cycle, wherein two are produced.

Energy Conservation of Fatty Acid Oxidation

As illustrated in Figure 17-16, the β-oxidation cycle produces *one* molecule of *acetyl-CoA, $FADH_2$,* and *NADH* per turn. In the case of *palmitic acid* (p. 207), a 16-carbon saturated fatty acid, *seven* turns of the cycle would be required to degrade palmitoyl-CoA to eight acetyl-CoA [seventh turn: butyryl-CoA (four-carbon acyl-CoA) → 2 acetyl-CoA], as expressed in the following equation:

$$\text{Palmitoyl-CoA} + 7\ \text{CoA} + 7\ \text{FAD} + 7\ \text{NAD}^+ + 7\ \text{H}_2\text{O} \rightarrow 8\ \text{acetyl-CoA} + 7\ \text{FADH}_2 + 7\ \text{NADH} + 7\ \text{H}^+$$

When the β-oxidation of palmitic acid is coupled to the TCA cycle and the respiratory chain, each *acetyl-CoA* yields *12 ATP* and, hence, a total of *96* (12 × 8) ATP are produced. An additional *35 ATP* are generated by the $FADH_2$ (2 × 7 = 14) and NADH (3 × 7 = 21) molecules; thus, an overall production of *131 ATP* is obtained by the oxidation of palmitoyl-CoA. Because of the energy equivalent of 2 ATP (ATP and PP_i hydrolysis) required initially for the activation of palmitic acid, *viz.*, synthesis of palmitoyl-CoA, the *net overall production of ATP is 129.*

The $\Delta G^{\circ\prime}$ equation for the oxidation of palmitic acid is expressed:

$$\text{Palmitic acid} + 23\ O_2 \rightarrow 16\ CO_2 + 16\ H_2O$$

$$\Delta G^{\circ\prime} = -2{,}340\ \text{kcal mol}^{-1}$$

The 129 ATP produced account for a conservation of *942* kcal of the 2,340 kcal released by the oxidation of one mole of palmitic acid, again a *40 per cent efficiency* of energy conservation.

When compared, the $\Delta G^{\circ\prime}$ of palmitic acid ($-2{,}340$ kcal mol^{-1}) and that of glucose (-686 kcal mol^{-1}) reveal that *341 per cent* more energy is released by the aerobic oxidation of one mole (255 g) of palmitic acid than by a mole (180 g) of glucose. On a mass-per-mass basis, therefore, palmitic acid produces about *240 per cent* (2.4 times) more calories than does glucose (341 × 180/255). If a mole of a triacylglycerol, containing three palmitic acid residues, was oxidized by biological means, the fatty acid residues would account for a conservation of 2,826 kcal (942 × 3). These very large amounts of energy involved in the utilization of lipids for energy are the basis for the statement that a 20-mile walk would be required to lose one pound of fat. The high caloric content of lipids per unit weight is effectively used by migratory birds, who rely on high-fat diets (or lipids stored in their bodies) to supply the large amounts of energy required for long, sustained flight.

Metabolic Water

Another important biological feature of fatty acid oxidation (and of aerobic respiration in general) is the production of metabolic water, e.g., 16 molecules of H_2O per molecule of palmitic acid oxidized. This metabolic production of water is of significant importance to many organisms. In the case of a camel, for example, the lipids stored in its humps serve both as a source of energy and as a source of the water needed to help sustain the animal for extended periods of time in the desert.

SUMMARY

To complete the aerobic oxidation of D-glucose, the five reduced coenzymes (4 NADH and 1 $FADH_2$) derived from the oxidation of pyruvate are utilized in the electron transport system. This system of electron carriers constitutes a series of oxidation-reduction reactions that effect the transfer of electrons donated by NADH (as a hydride ion) and $FADH_2$ (as two hydrogen atoms) to molecular O_2, to produce H_2O. As a result of the free energy released by this flow of electrons, ATP is produced by the oxidative phosphorylation system, which is physically coupled to electron transport. Electron transport and oxidative phosphorylation, like the TCA cycle, occur in the mitochondria of eucaryotic cells and are intimately associated with the inner membrane of the organelle.

The five classes of redox carriers known to be involved in electron transport are the pyridine-linked dehydrogenases (coenzyme: NAD^+), flavin-linked dehydrogenases (coenzyme: FAD or FMN), iron-sulfur (Fe·S) proteins, ubiquinones (CoQ), and cytochromes (*a, b,* and *c*). When an electron pair, obtained from NADH or $FADH_2$, is transferred by electron transport, there is a continuous and significant decline in its free energy; it is this released free energy that is utilized by oxidative phosphorylation to synthesize ATP from ADP. The energy released in electron transport by the flow of an electron pair donated by NADH or $FADH_2$ results in the synthesis of three ATP or two ATP, respectively.

The molecular details of oxidative phosphorylation remain unclear; however, the following three hypotheses have been proposed to account for the phenomenon: chemical coupling, conformational coupling, and chemiosmotic coupling. The chemiosmotic-coupling hypothesis is an innovative proposal that is proving to be of broad biological significance, e.g., may also account for active transport of molecules and ions across membranes.

The complete oxidation of two molecules of pyruvate (8 NADH, 2 $FADH_2$, and 2 GTP) derived from a molecule of glucose results in the synthesis of 30 ATP molecules. To calculate the total amount of ATP produced by the aerobic oxidation of a molecule of glucose, the two ATP and the two NADH produced by glycolysis must also be taken into account. The NADH produced by glycolysis furnishes reducing equivalents (electrons) to the oxidative process by means of enzymatic shuttle systems, which transport the coenzyme's electrons from the cytosol into the mitochondria. The glycerol phosphate and the malate-aspartate shuttle are two such enzymatic systems; the former accounts for the production of two ATP per cytosolic NADH and the latter for three ATP. Thus, a total of 36 or 38 ATP are produced per glucose molecule by aerobic respiration, representing a 38 or 40 per cent efficiency of energy conservation, respectively.

Fatty acids, another important biomolecular fuel, are oxidized in the matrix of mitochondria by the β-oxidation cycle. This cyclic pathway sequentially degrades a fatty acid, which participates as an acyl-CoA derivative, by removing a two-carbon unit (as acetyl-CoA) from the molecule with each turn of the cycle. Every acetyl-CoA produced by β-oxidation results in the synthesis of three NADH, one $FADH_2$, and one GTP when oxidized by the TCA cycle. Collectively, the NADH and $FADH_2$ generated by the oxidation of a fatty acid account for the production of a large number of ATP molecules.

QUESTIONS

1. Select your answers to the statements below from the following list:

Cytochrome aa_3	Cytochrome c_1
Cytochrome *b*	Coenzyme *Q*
Cytochrome *c*	NADH dehydrogenase

A redox carrier (or carriers) in electron transport that
   (a) accepts and donates two hydrogen atoms
   (b) accepts and donates an electron
   (c) transfers electrons to O_2
   (d) has FMN as a coenzyme
   (e) utilizes iron as an electron carrier
   (f) utilizes copper as an electron carrier
   (g) accepts the electrons donated by NADH
   (h) has a covalently bound iron-protoporphyrin IX group
   (i) has a noncovalently bound iron-protoporphyrin IX group
   (j) has heme A

2. Which of the following statements about the chemiosmotic-coupling (Mitchell) hypothesis are currently considered true?
   (a) The function of mitochondrial electron transport is to translocate protons across the inner membrane into the mitochondrial matrix.
   (b) The free energy released by mitochondrial electron transport is stored in an electrochemical gradient.
   (c) The inner membrane knobs of mitochondria are F_1-ATPase complexes.
   (d) F_1-ATPase catalyzes the *in vivo* synthesis of ATP from ADP and P_i.
   (e) The components of electron transport are arranged in a series of oxidation-reduction loops linking hydrogen carriers to electron carriers.
   (f) The chemiosmotic-coupling hypothesis does not support the contention that each NADH and $FADH_2$ donating electrons to electron transport produces three ATP and two ATP, respectively.
   (g) ATP production by chemiosmotic coupling occurs only in mitochondria.

3. In Figure 17-8, the decline in free energy of an electron pair transferred from cyt aa_3 complex to O_2 is listed as 23.8 kcal mol^{-1}.
   (a) Theoretically, how many ATP molecules could be produced by the release of this amount of free energy (assuming $\Delta G^{\circ\prime}$ conditions)?
   (b) In reality, how many ATP are produced?
   (c) What percentage of the released free energy is conserved in ATP structure?
   (d) With the use of the E_0' values listed in Table 17-1, calculate the $\Delta G^{\circ\prime}$ for the transfer of an electron pair from cyt a_3 to O_2.
   (e) What percentage of the free energy released by the transfer of an electron pair from the cyt aa_3 complex to O_2 can be accounted for by their transfer from cyt a_3 to O_2?

4. In Figure 17-8, the diagram indicates that the amount of free energy released by the transfer of an electron pair from cyt c_1 to the cyt aa_3 complex is not large enough to account for the production of an ATP molecule by oxidative phosphorylation.
   (a) Using the E_0' values listed in Table 17-1 for the reduction of cyt *c* and cyt a_3, what is the approximate $\Delta G^{\circ\prime}$ for the transfer of an electron pair between cyt *c* and the cyt aa_3 complex? (b) Could this amount of free energy account for the production of an ATP molecule?

5. Calculate (a) the $\Delta G^{\circ\prime}$ for the production of glycerol 3-phosphate from dihydroxyacetone phosphate and NADH in the glycerol phosphate shuttle and (b) the $\Delta G^{\circ\prime}$ for the regeneration of glycerol 3-phosphate from dihydroxyacetone phosphate and FAD in the mitochondria. (Use the E_0' value for riboflavin reduction for that of FAD reduction in the latter calculation.)

6. (a) How much more chemical energy, expressed in kcal, is obtained by the aerobic oxidation of one mole of D-glucose than by its anaerobic conversion into two moles of lactate by glycolysis? (Assume an aerobic production of 38 ATP molecules per glucose molecule and the $\Delta G^{\circ\prime}$ of ATP hydrolysis.) (b) The total amount of free energy conserved by glycolysis is what percentage of the total amount conserved by aerobic oxidation?

7. (a) In the β-oxidation of the *three* 20-carbon fatty acid residues of a triacyglycerol, how many acetyl-CoA, NADH, and $FADH_2$ molecules would be produced?
   (b) If the acetyl-CoA produced is then oxidized by the TCA cycle, how many NADH, $FADH_2$, and GTP molecules would be produced?

(c) If the combined NADH and $FADH_2$ produced by the β-oxidation and TCA cycles are used for aerobic respiration, how many ATP would be produced?
(d) What would be the net overall production of ATP (GTP $\approx$ ATP)?
(e) If a mole of this triacylglycerol is used as a biomolecular fuel (assuming $\Delta G^{\circ\prime}$ conditions), how many kcal would be conserved in ATP structure by the aerobic oxidation of its three fatty acids?

SUGGESTED READINGS

Fruton, J. S., *Molecules and Life.* New York: Wiley-Interscience, 1972. This text includes an excellent essay on the history of the scientific studies on cellular respiration.

Keilin, D., *The History of Cell Respiration and Cytochromes.* New York: Cambridge University Press, 1966. This book relates the historical development of this area of the biological sciences.

King, R. E., and M. Klingenberg, eds., *Electron and Coupled Energy Transfer in Biological Systems,* 2 vols. New York: Marcel Dekker, 1971. Excellent volumes—concise, yet thorough, in their coverage of the processes of energy transformation in a cell.

Lehninger, A. L., *The Mitochondrion: Molecular Basis of Structure and Function.* New York: W. A. Benjamin, 1964. A recommended text about past research on aerobic respiration.

Lemberg, R., and J. Barrett, *Cytochromes.* New York: Academic Press, 1973. A thorough presentation of information on these electron carriers.

Lovenberg, W., ed., *Iron-Sulfur Proteins,* 3 vols. New York: Academic Press, 1973–1977. Chemical and biochemical reviews of iron-sulfur proteins in the electron transport system.

Mitchell, P., *Chemiosmotic Coupling and Energy Transduction.* Bodmin, England: Glynn Research, 1968. A presentation of chemiosmotic coupling published by the scientist who proposed the hypothesis.

Racker, E., *Mechanisms in Bioenergetics.* New York: Academic Press, 1965. A recommended monogram on mitochondria and oxidative phosphorylation.

Racker, E., *A New Look at Mechanisms in Bioenergetics.* New York: Academic Press, 1976. A well-written sequel to his 1965 book, emphasizing the chemiosmotic hypothesis.

Articles

Bremer, J., "Carnitine and Its Role in Fatty Acid Metabolism." *TIBS,* 2:207, 1977.

Dickerson, R. E., "The Structure and History of an Ancient Protein (Cytochrome *c*), *Scientific American,* 226(4):58 (offprint #1245), 1972.

Dickerson, R. E., "Cytochrome *c* and the Evolution of Energy Metabolism." *Scientific American,* 242(3):136 (offprint #1464), 1980.

Eventoff, W., and M. G. Rossmann, "The Structure of Dehydrogenases." *TIBS,* 1:227, 1976.

Fillingame, R. H., "The Proton-Translocating Pumps of Oxidative Phosphorylation." *Ann. Rev. Biochem.,* 49:1079, 1980.

Hinkle, P. C., and R. E. McCarty, "How Cells Make ATP." *Scientific American,* 238(3):104, 1978.

Lehninger, A. L., "How Cells Transform Energy." *Scientific American,* 205(3):62 (offprint #91), 1961.

Mitchell, P., "Keilin's Respiratory Chain Concept and Its Chemiosmotic Consequences" (Nobel lecture). *Science,* 206:1148, 1979.

Racker, E., "The Membrane of the Mitochondrion." *Scientific American,* 218(2):32 (offprint #1101), 1968.

Schmidt-Nielson, K., "The Physiology of the Camel." *Scientific American,* 201(6):140 (offprint #1096), 1959.

Siekevitz, P., "Powerhouse of the Cell." *Scientific American,* 197(1):131 (offprint #36), 1957.

Stoeckenius, W., "The Purple Membrane of Salt-Loving Bacteria." *Scientific American,* 234(6):38 (offprint # 1340), 1976.

PHOTOSYNTHESIS: SINE QUA NON OF LIFE

18

INTRODUCTION

The quotation "all flesh is grass" (Isaiah 40:6) succinctly summarizes the vital importance of photosynthesis to all animals, since they depend on the photosynthetic conversion of light energy into chemical energy for their continued existence. The use of radiant energy from the sun to synthesize carbohydrates from CO_2 and H_2O is the ultimate basic biochemical process which allows life, as we know it, to be expressed in its many different forms. The overall reaction that accounts for photosynthesis in eucaryotic organisms and cyanobacteria (blue-green algae) is generally written:

$$6\ CO_2 + 6\ H_2O \xrightarrow{\text{light}} C_6H_{12}O_6\ \text{(glucose)} + 6\ O_2$$

In the process, H_2O (as an electron donor) is oxidized to produce O_2, with the subsequent reduction of CO_2 (electron acceptor). In the biological world, photosynthesis is the only process whereby H_2O is oxidized, i.e., a reverse of the aerobic respiratory process in which O_2 is reduced to H_2O. Eucaryotic organisms that utilize sunlight as the primary energy source to generate their chemical energy (ATP) to satisfy their biomolecular needs include *higher green plants, algae* (brown, green, and red), and the *euglenoids, dinoflagellates,* and *diatoms.* Photosynthetic procaryotes include the cyanobacteria, the *brown* and *red Rhodospirillaceae,* and the sulfur utilizers *brown* and *green Chlorobiaceae* and *brown, purple,* and *red Chromatiaceae.* Photosynthetic bacteria use electron donors other than H_2O and therefore *do not produce O_2;* most of these species are strict anaerobes for whom O_2 is, in fact, toxic.

The green and purple sulfur bacteria, found in sulfur springs, use H_2S as an electron source and form inorganic sulfur as a product:

$$2\ H_2S + CO_2 \xrightarrow{\text{light}} (CH_2O) + H_2O + 2\ S$$

Photosynthesis appears to have evolved when the earth's atmosphere was a reducing one and is therefore an ancient life process closely aligned with the early stages of biological evolution. The biological development of photosynthesis and the concomitant accumulation of its by-product O_2 in the atmosphere made possible the evolution of aerobic organisms. As pointed out in previous chapters, the ability of living systems to use O_2 to accept electrons greatly enhances the amounts of chemical energy obtainable from biomolecular fuels. Photosynthetic organisms still produce the oxygen of the atmosphere and, although higher green plants supply a significant amount of O_2, the major contributors are the *phytoplankton* (microscopic algae, diatoms, and dinoflagellates) of the seas and oceans.

Ingen-Housz, trained as a physician, was inoculating people against smallpox (using live virus) decades before Jenner introduced his vaccine (1798). In 1768, Ingen-Housz inoculated the Austrian royal family and, in grateful appreciation, Empress Maria Theresa appointed him court physician with a lifelong annual income of 5,000 gulden. He used his financial independence to pursue his research interests. He published his experiments on photosynthesis (done in England) in a book entitled *Experiments upon Vegetables, Discovering Their Power to Purify the Common Air in the Sunshine and of Injuring It in the Shade and at Night* (1779).

Joseph Priestley established the production of oxygen by plants in the 1770s and also recognized that the process seemingly represented the reverse of animal respiration, which consumes O_2. However, it was Jan Ingen-Housz who, in 1779, correlated O_2 production with the need for light and with the green parts of a plant. (Oxygen, at that time, was identified as *dephlogisticated air* and carbon dioxide as *fixed air.*) By the middle of the 1800s, it had been deduced that plant photosynthesis requires CO_2, H_2O, and the energy of light to produce organic matter and O_2. Today, after years of intensive research, it is known that photosynthesis in plants occurs in two distinguishable phases, called the *primary (light reactions)* and *secondary* (sometimes called *dark reactions*) *processes,* and that the primary process consists of *two separate light reactions* (I and II). In the primary process, solar energy is first trapped by *light-absorbing pigments* (*photoreceptor* molecules); some of this energy is then utilized to *cleave H_2O molecules to yield O_2, protons* (H^+), and *electrons,* with the last two products being used to *produce reducing power* (NADPH from $NADP^+$) and *chemical energy* (ATP from ADP and P_i). The general reaction for the primary process is written:

$$\text{Water} + NADP^+ + ADP + P_i \xrightarrow{\text{light}} NADPH + H^+ + ATP + \text{oxygen}$$

In the secondary process, the NADPH and ATP generated by the primary process serve as energy sources for the *fixation and reduc-*

tion of CO_2 to the level of carbohydrate, as expressed in the following general equation:

$$CO_2 + NADPH + H^+ + ATP \rightarrow (CH_2O) + NADP^+ + ADP + P_i$$

The secondary process also regenerates the precursors for NADPH and ATP synthesis. The principal carbohydrate produced by photosynthesis is D-glucose, which can be stored in plants as starch.

The structural and biochemical complexities associated with photosynthesis, especially those of the primary process, make a detailed step-by-step analysis of the system an extremely difficult task. Research on photosynthesis shares the experimental obstacles that are encountered in studies on electron transport and oxidative phosphorylation. Because of the fundamental importance of photosynthesis to human survival, there is a recognized need for a better molecular understanding of the process and its relationship(s) to plant physiology. Such knowledge would furnish scientific insights that could be advantageously applied to the worldwide problem of food production.

SITES OF PHOTOSYNTHESIS

In eucaryotic plants, photosynthesis takes place in specialized organelles called *chloroplasts,* discussed in Chapter 5 (p. 62). A cell of a higher plant may contain as many as 40 chloroplasts, but some photosynthetic algae may have only one. The primary energy-conserving processes (light reactions) take place in the *membranes of the thylakoids and intergranal* (stromal) *lamellae* of chloroplasts (Figure 18-1), where the light-absorbing pigments are located. The enzymes of the secondary process (which converts CO_2 into carbohydrate structure) are found in the soluble environment, the *stroma,* of chloroplasts. Thus, the specialized functions of chloroplasts, like those of mitochondria, occur in distinct compartments of the organelle, and it is this intimate association with membranes that makes molecular elucidation of the primary process a challenging task.

Chloroplasts are not found in procaryotic photosynthetic bacteria and cyanobacteria; instead, the photosynthetic apparatus in these organisms is situated in the cellular membrane, which folds inward and extends across the cell to form *photosynthetic lamellae* (Figure 18-2). *Chromatophores,* vesicular structures derived from

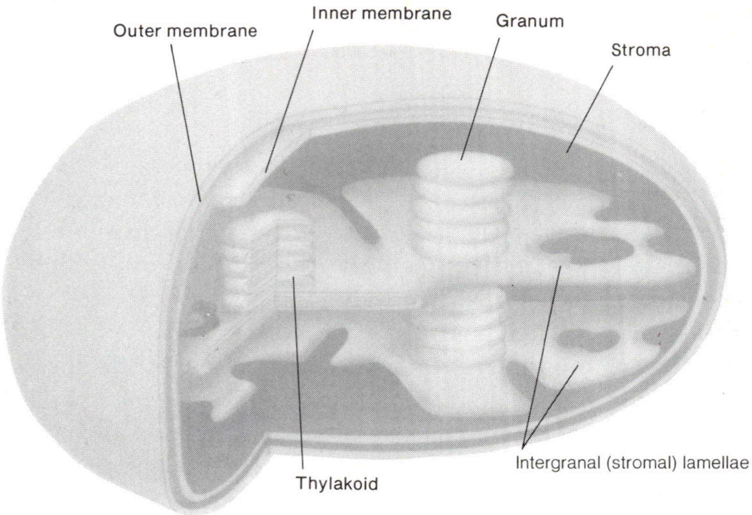

Figure 18-1
Drawing of a chloroplast.

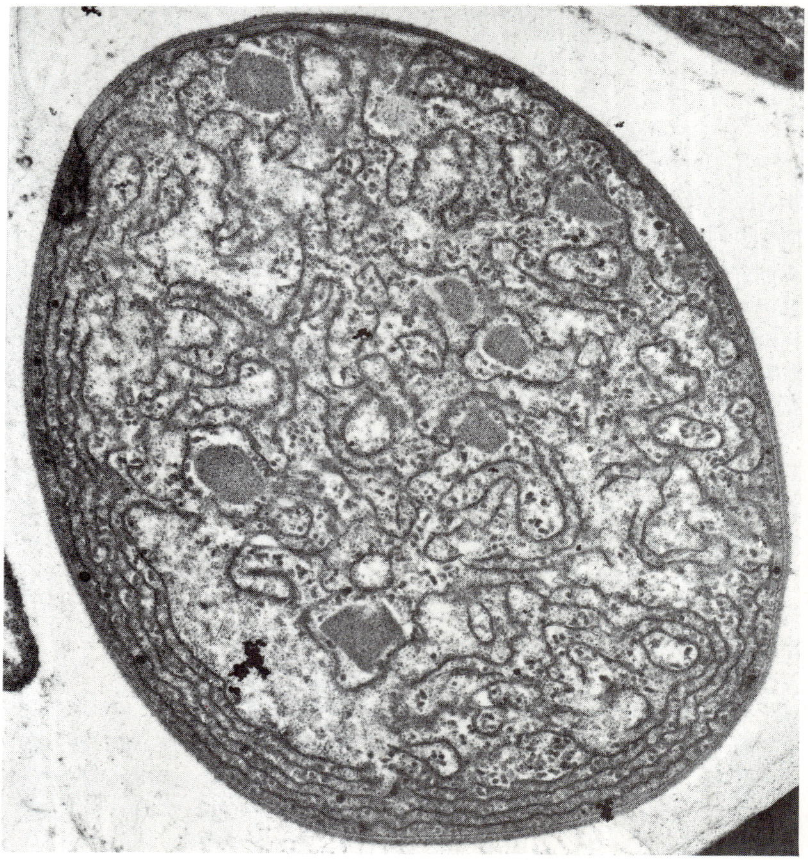

Figure 18-2
Electron micrograph of the cyanobacterium *Anabaena azollae.* Membranes that constitute the photosynthetic lamellar system pervade the cytoplasm. (Courtesy of Dr. Norma Lang)

lamellae, constitute an alternative site of photosynthesis in certain of these procaryotes.

LIGHT-ABSORBING PIGMENTS

The familiar green color of plants and the colors of some algae and bacteria (brown, purple, and red) are associated with a photosynthetic mode of life. The pigments responsible for these colors are used by these organisms to absorb the energy of the sun. One class of light-absorbing pigments common to all photosynthetic cells is the *chlorophylls,* which are usually (but not always) green. Other classes are the *carotenoids* and the *phycobilins,* called accessory pigments, which participate in conjunction with chlorophylls in trapping solar energy.

Chlorophylls

Chlorophylls *a* and *b* (Figure 18-3) are the two major photoreceptor pigments found in higher plants. The two forms are identical except that chlorophyll *a* has a methyl group at C-3 of ring II and chlorophyll *b* has a formyl group at that position. Chlorophylls are *magnesium porphyrins,* and their organic structures differ from that of an iron porphyrin (heme, p. 107) in the types and positions of the substituted groups on the tetrapyrrole ring and also in having fused cyclopentanone rings (V). The hydrophobic side chain (composed of four isoprenoid units) of chlorophyll is *phytol,* an alcohol esterified to the propionic-acid group of ring IV (Figure 18-3).

Chlorophylls *a* and *b* absorb certain wavelengths of visible light (400 to 700 nm) effectively because of their conjugated double-bond systems. The similar but distinct absorption spectra of chlorophylls *a* and *b*, depicted in Figure 18-4, illustrate how the two types of chlorophyll complement each other in absorbing light to increase the total amount of solar energy trapped in photosynthesis. Chlorophyll *a* is always present in oxygen-producing photosynthetic cells; it is found with chlorophyll *b* in all green plants and with chlorophyll *c* in brown algae, diatoms, and dinoflagellates. Cyanobacteria contain only chlorophyll *a*. For their role in photosynthesis, chlorophylls are conjugated to specific proteins. Procaryotic photosynthetic systems, which do not produce oxygen, have no chlorophyll *a* but, rather, contain either *bacteriochlorophyll a* or *b*, which are characterized by a reduced ring II and by their absorbance at longer wavelengths (up to 1100 nm).

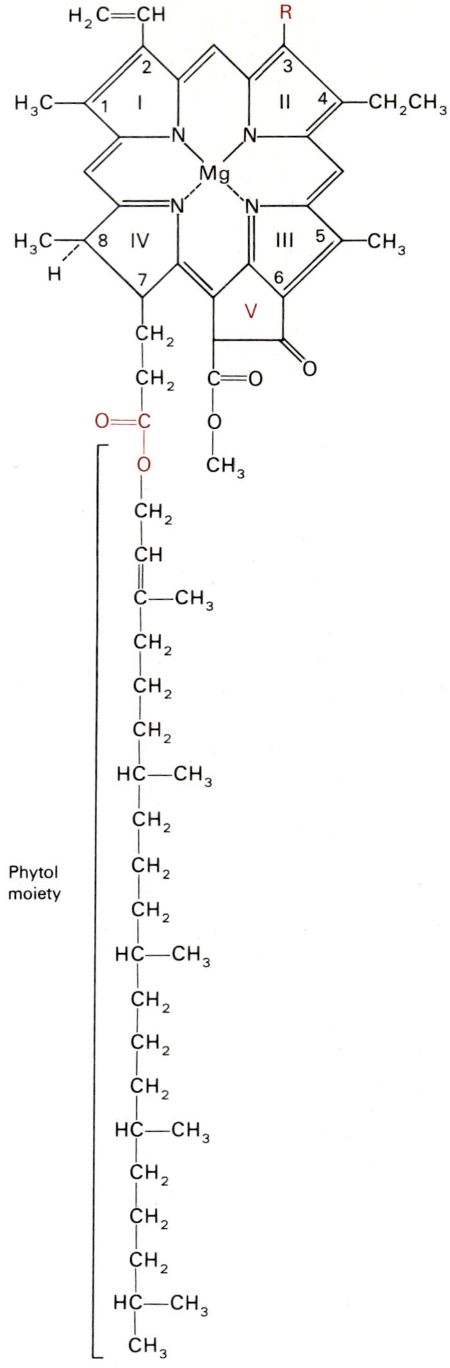

Figure 18-3
The structure of chlorophylls *a* and *b*. In *a*, R = CH_3; in *b*, R = —CHO.

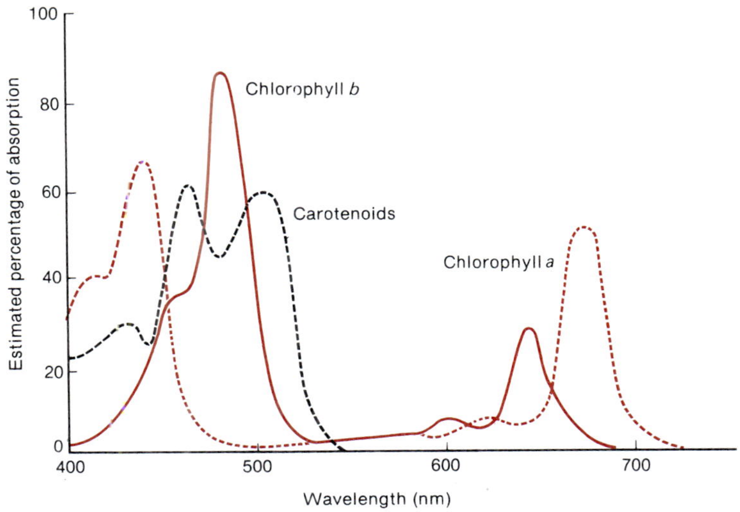

Figure 18-4
Spectra of chlorophylls *a* and *b* and of carotenoids.

Accessory Pigments

The purple, red, or yellow carotenoids and the blue or red phycobilins represent the two major classes of accessory pigments. Because the absorption maxima of carotenoids (Figure 18-4) differ from those of chlorophylls, photosynthetic organisms having this type of accessory pigment can absorb a broader range of visible light. For photosynthesis to occur, the solar energy absorbed by the accessory pigments must be transferred to chlorophylls, which are the critical pigments for donating light energy to the primary photochemical process.

In chloroplasts there are two major types of carotenoids, *carotenes* (precursors of vitamin A for animals, p. 577) and *xanthophylls* (Figure 18-5). Both types of carotenoids are composed of

Figure 18-5
Two classes of carotenoids.

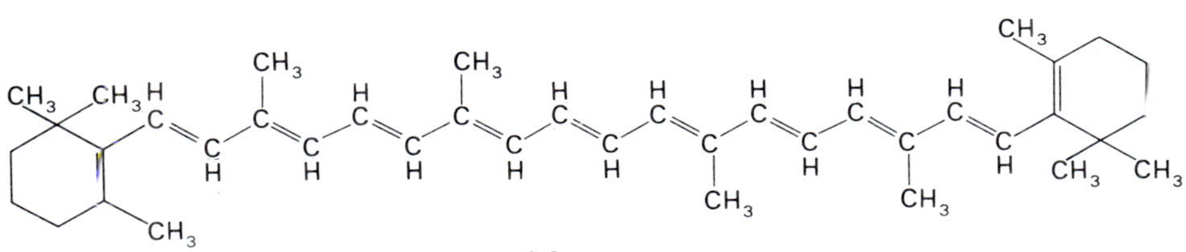

β-Carotene

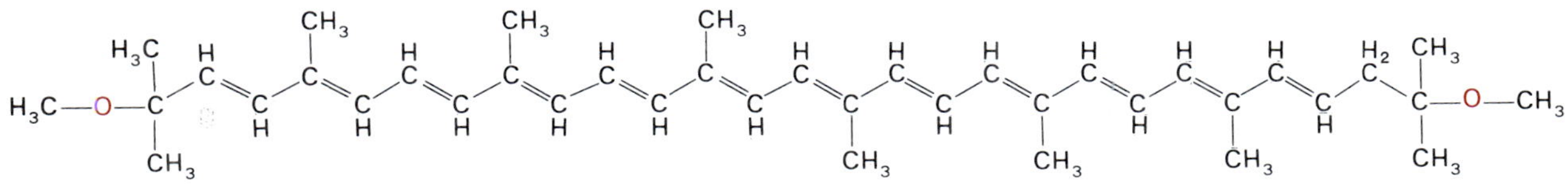

Spirilloxanthin
(a xanthophyll)

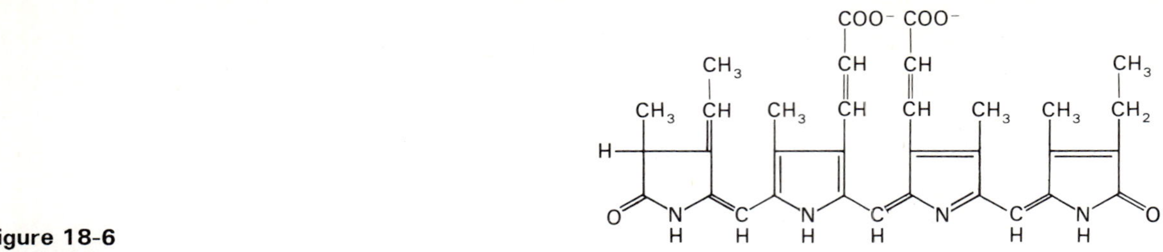

Phycocyanobilin

Figure 18-6
A phycobilin of cyanobacteria.

isoprenoid units and contain conjugated double bonds; the chemical distinction is that xanthophylls have oxygen as a molecular component but carotenes do not. In the case of phycobilins, which are found in red algae *(phycoerythrobilin)* and in cyanobacteria *(phycocyanobilin),* the pigments are *linear tetrapyrroles* (Figure 18-6). Like chlorophylls, phycobilins are conjugates of specific proteins and, being the major photoreceptors, furnish these algae with their distinctive colors.

PRIMARY PROCESS (LIGHT REACTIONS): GENERATION OF NADPH, ATP, AND O_2

The light reactions of all algae and of green plants involve *two photosystems* (*I* and *II*), which are structurally distinct photosynthetic units in chloroplasts, even though both function similarly by trapping the energy of light ($h\nu$) and transferring electrons to acceptor molecules. The photosynthetic activities of the two systems, however, are interconnected, in tandem, by an *electron transport system* (Figure 18-7), and it is by this mechanism that the overall function of the primary process to produce NADPH and ATP is efficiently coordinated.

Although both photosystems I and II are excited primarily by wavelengths of light less than 680 nm, photosystem I can be activated by longer wavelengths, up to 700 nm. Photosystems I and II are responsible for the production of NADPH, ATP, and O_2, respectively, and *both* systems must be excited for NADPH production and O_2 evolution. (Photosynthetic bacteria, which do not produce O_2, have only one photosystem.) Each photosystem also has a particular light-absorbing pigment: P_{700} in system I and P_{680} in system II (P stands for pigment, and the subscripts refer to absorption maxima). The pigments are located at the *reaction center* of their respective photosystems.

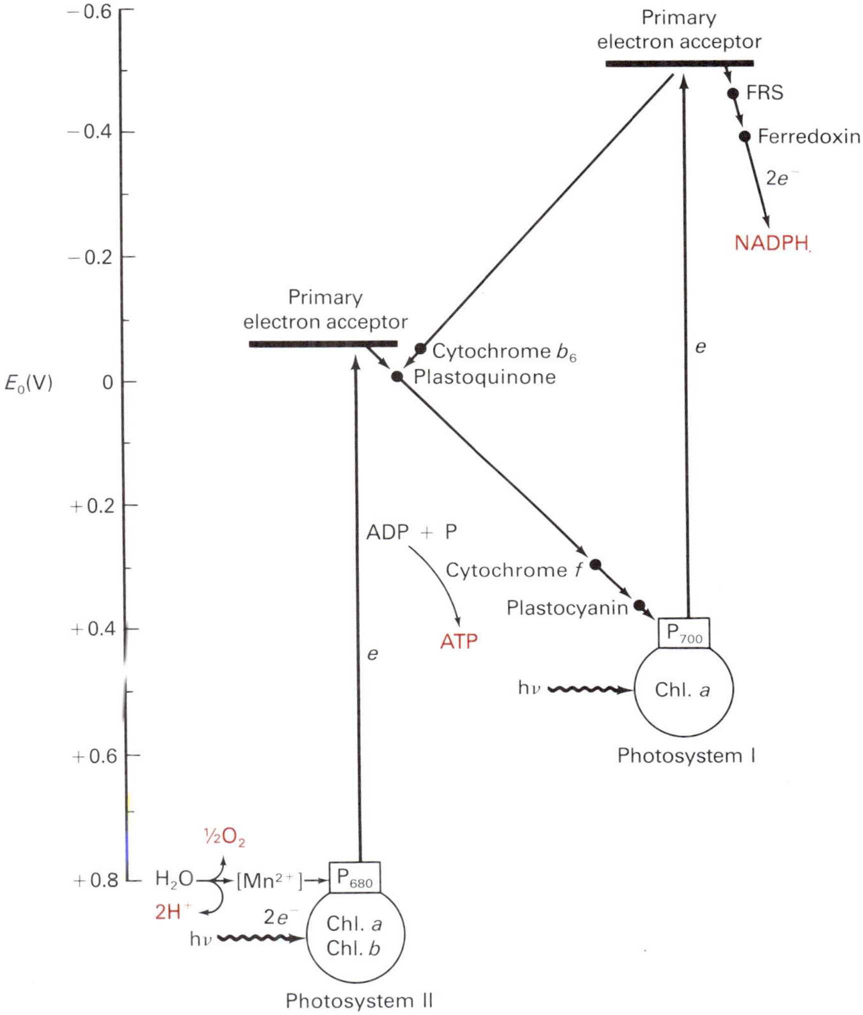

Figure 18-7
Diagram (Z scheme) of photosynthetic transport by photosystems I and II. Approximate standard reduction potentials E_0') of the individual electron carriers were used to plot the diagram. Attainment of the excited state by P_{680} and P_{700}, as a result of the absorption of light energy, is indicated by red arrows. Production of NADPH, ATP, H^+, and ½ O_2 is also indicated in red. (hν = energy of a photon; FRS = ferredoxin-reducing substrate; and chls. *a* and *b* = chlorophylls *a* and *b*.) (After an illustration prepared by Dr. Joseph S. Kahn)

The concentration of chlorophyll in chloroplasts (and in photosynthetic lamellae and chromatophores) is much greater than any of the other intermediates of the photosystem. Between 80 and 400 chlorophyll molecules are packed tightly in a highly organized manner around the reaction center for each of the light reactions. When a light-harvesting chlorophyll (LHC) captures a photon of light, it becomes excited, i.e., one of its electrons is raised to an excited state. This excitation is transferred very rapidly to other chlorophyll molecules in its vicinity until, by "random walk," it reaches P_{700} or P_{680} at a reaction center. Since both P_{700} and P_{680} need less energy for excitation, the transmitted energy is sufficient

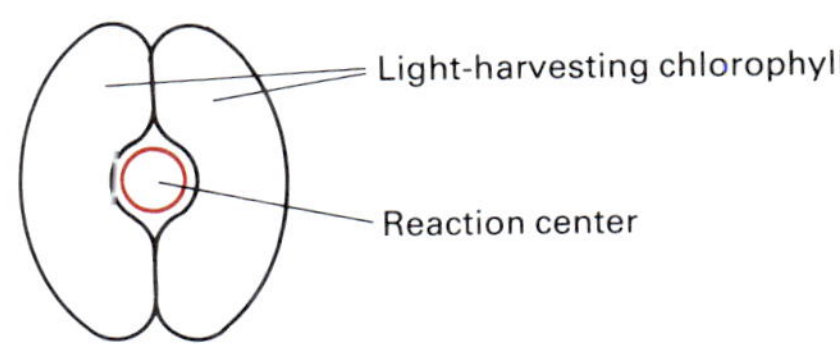

Schematic of a photosystem.

to excite either of these molecules to a state in which one electron can be removed. Thus, the excited pigments become oxidized (ionized), a state expressed as $P_{700\ ox.}$ (P^{+}_{700}) and $P_{680\ ox.}$ (P^{+}_{680}). In photochemistry, the instant that P_{700} or P_{680} becomes excited, the *photo-* ends and the *-chemistry* begins.

Photosystem I (Light Reaction I)

Photosystem I (Figure 18-7) relies primarily on specific chlorophyll *a*-protein complexes, which absorb the energy of photons up to the limits of the visible spectrum (700 nm). Activation of the system is brought about, as stated above, by the raising of the energy level of P_{700}, which in the ground state has a standard reduction potential of about +0.4 V and therefore has little tendency to lose electrons. However, by excitation, the reduction potential changes to about −0.5 V, producing an energy-rich state which allows P_{700} to donate a high-energy electron to a primary electron acceptor (as yet unidentified). By means of an electron transport system, the electron is made available for NADPH formation. The transport chain includes a *ferredoxin-reducing substance,* called *FRS* (perhaps the primary electron acceptor from P_{700}), which donates an electron to *ferredoxin,* a protein (MW = 11,600) containing two iron-sulfur centers (p. 320). NADPH is produced when each of two ferredoxin molecules furnishes an electron to reduce $NADP^+$; the reaction is catalyzed by *ferredoxin-$NADP^+$ oxidoreductase,* a flavoprotein (FAD). In photosystem I, NADPH synthesis leaves P_{700} in an oxidized (and colorless) state, and the pigment must regain an electron before the system can be activated again.

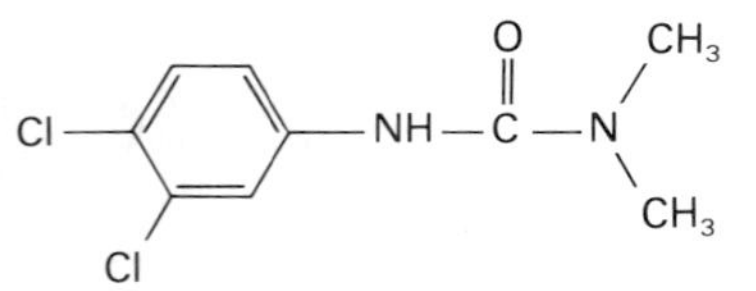

Diuron

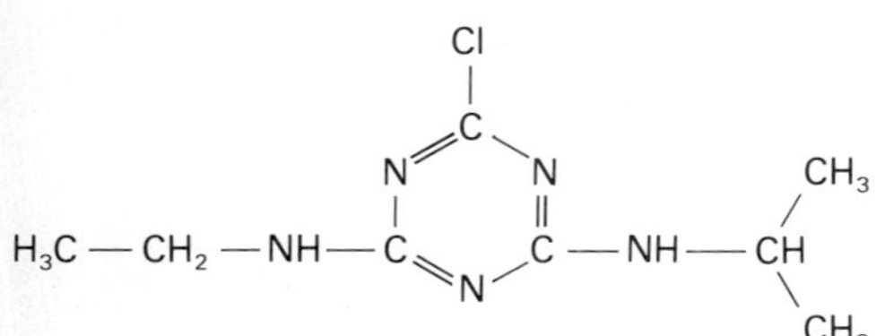

Atrazine

The herbicides diuron and atrazine, each of which binds to a different membrane protein, kill plants by disrupting the flow of electrons from photosystem II to photosystem I. Diuron, it is now known, binds to a protein associated with plastoquinone.

Photosystem II (Light Reaction II)

A critical function of photosystem II is to restore oxidized P_{700} of photosystem I to its ground state, a process that utilizes the electron transport system that links the two systems. This transfer of an electron to P_{700} also results in ATP formation. Thus, prior formation of oxidized P_{700} by photosystem I is a prerequisite for the excitation of photosystem II.

Upon illumination, photosystem II is activated by the collection of excitation energy from chlorophylls *a* and *b* by P_{680}, a mechanism similar to that described for photosystem I. Chlorophyll *b*, it should be noted, transmits its excitation energy to P_{680} via chlorophyll *a*. Upon excitation, the standard reduction potential of P_{680} changes from the highly positive value of about +0.8 V to −0.05 V. In its elevated energy state, P_{680} transfers an electron to a pri-

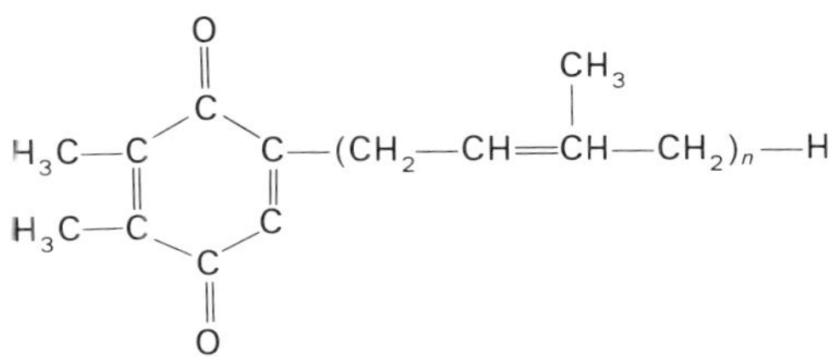

Figure 18-8
Plastoquinone.

mary electron acceptor (which may be a quinone) and, as seen in Figure 18-7, the electron is then shuttled through a series of electron carriers to P_{700}. The three known components of this electron transfer chain are a *plastoquinone* (Figure 18-8), the plant *cytochrome f*, and a blue-copper protein, *plastocyanin.* Plastoquinones are structural analogues of CoQ (p. 320), an electron carrier in mitochondrial electron transport.

Oxygen Production

The production of O_2 by photosystem II results in the transfer of electrons from H_2O to oxidized P_{680} which, like P_{700}, needs to gain an electron to replace the one transferred to the primary electron acceptor. The biochemical mechanisms by which two molecules of H_2O donate four electrons (to produce a molecule of O_2) to the very electropositive P_{680} are still unclear. Because the intermediate reduction states between H_2O and O_2 are so unstable and thus transient, it has been difficult to detect and identify them; it is known, however, that the reaction requires Mn^{2+}.

Photophosphorylation

Noncyclic

The passage of electrons from the primary electron acceptor of photosystem II to P_{700}, as well as the splitting of water, results in ATP production from ADP and P_i. This photoinduced synthesis of ATP is called *photophosphorylation* and, as illustrated in Figure 18-9, occurs by molecular mechanisms analogous to those described in Chapter 17 for mitochondrial oxidative phosphorylation. As shown, during photosynthesis, protons accumulate in the interthylakoid space because of the transport of electrons to photosystem I and the splitting of water by photosystem II. The net result is that a proton gradient is generated across the thylakoid membrane, and it is this electrochemical force that drives the synthesis of ATP from ADP and P_i. ATP production by photosynthesis is illustrated schematically in the lower part of Figure 18-9 and, as

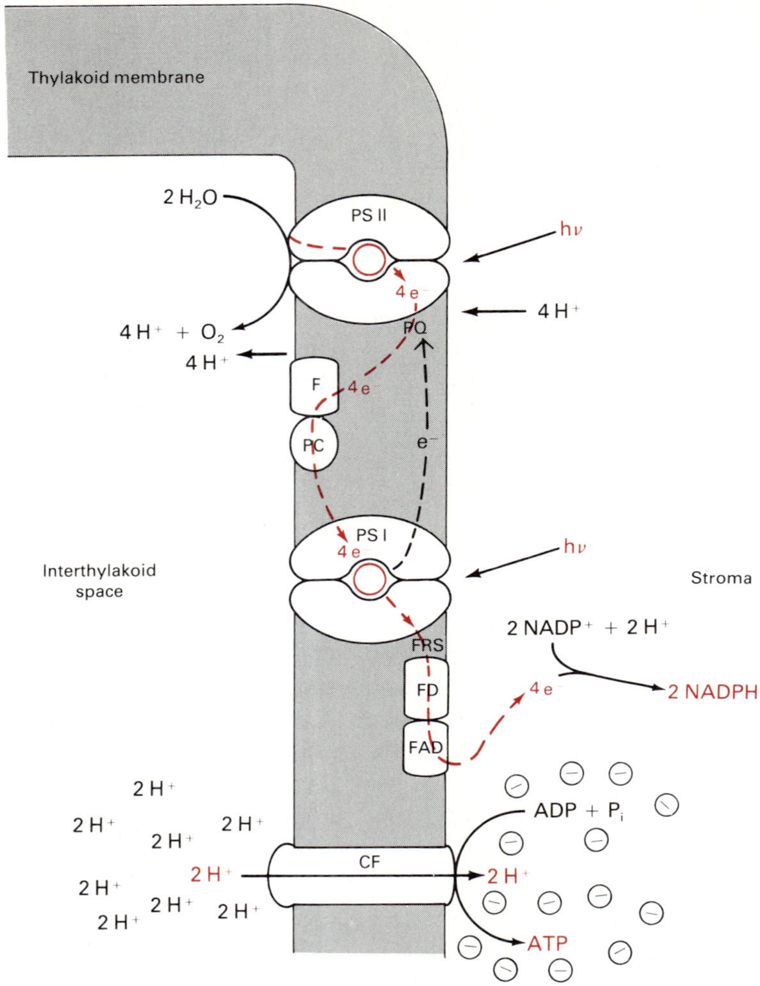

Figure 18-9
Schematic of photosystems I and II and photophosphorylation. PS I and II = photosystems I and II; hν = energy of a photon; PQ = plastoquinone; F = cytochrome *f*; PC = plastocyanin; FRS = ferredoxin-reducing substances; FD = ferredoxin; FAD = ferredoxin-NADP$^+$ oxidoreductase; and CF = coupling factor.

depicted, phosphorylation occurs basically in the same manner as that proposed by the chemiosmotic hypothesis (p. 328). The thylakoid coupling factor performs a function similar to that described for the F_1-ATPase of the mitochondrial system (p. 325). ATP formation involving *both* photosystems I and II is referred to as *noncyclic photophosphorylation* to distinguish it from *cyclic photophosphorylation,* a photosystem I event.

Cyclic

As Figure 18-7 illustrates, the electron donated to the primary electron acceptor by excited P_{700} in photosystem I can be utilized in an electron transfer system other than the one leading to NADPH production. This alternative route of transfer is cyclic in nature, since the electron subsequently returns to its original donor, P_{700}. Apparently, in the process, *cytochrome b_6* accepts an electron from the primary acceptor and donates it to plastoquinone of the electron transport chain, which then transfers it to P_{700} via cytochrome *f* and plastocyanin. At least one other intermediate carrier is required for this electron transfer, but its identity is as yet unknown. Some of the energy released in this cyclic transfer is utilized for ATP production (cyclic photophosphorylation). This mode of ATP formation is governed by the amounts of $NADP^+$ available for reduction, being stimulated when there is insufficient $NADP^+$ (high levels of NADPH) to accept the electrons generated by photosystem I. Photosystem II does not participate in cyclic photophosphorylation.

Summary of ATP, NADPH, and O_2 Production

As shown in Figure 18-9, the production of *one oxygen* molecule from two of H_2O (which also releases four electrons and four protons) accounts for the generation of *two NADPH.* With regard to ATP production, it is currently estimated that every *two H_2O* utilized by photosystem II result in the synthesis of *three ATP.* Since it is also estimated that about *one ATP* is produced for every *three H^+* generated in the interthylakoid space, protons derived from the splitting of water molecules, from electron transport from photosystems II to I, *and* from cyclic photophosphorylation (indicated by the black dashed lines between PS I and PQ in Figure 18-9) are needed to account for the observed ATP production. Note that both NADPH and ATP production occur in the chloroplast stroma, site of the secondary process.

Summary of Electron Transfer in the Primary Process

In the primary process in algae and green plants, the combined transfer systems of photosystems I and II succeed in transferring electrons from H_2O to $NADP^+$, i.e., $H_2O + NADP^+ \rightarrow NADPH + H^+ + \frac{1}{2} O_2$. Photosynthetic production of NADPH, therefore, constitutes a reversal of electron transport as it occurs in mitochondria. Although it may appear that photosynthetic light reactions do not obey the maxim that, in redox reactions, the reactant with the more positive standard reduction potential is reduced (E_o' of $\frac{1}{2}$ O_2 and $NADP^+$ are $+0.82$ and -0.32 V, respectively), it is the input of light energy into photosynthetic systems that drives this seemingly thermodynamically unfavorable reduction of $NADP^+$. Within each electron transport segment of the overall pathway, as indicated in Figure 18-7, the components do obey the above maxim.

SECONDARY PROCESS

Whereas the primary process utilizes biological reactions that are unique to photosynthetic organisms, the secondary process (which fixes CO_2 and produces sugars) employs a number of enzymes common to most heterotrophic species, e.g., enzymes of glycolysis and of the pentose phosphate pathway. This secondary process, relying on the ATP and NADPH produced by the primary process, accounts for a production of glucose by the plant kingdom that makes the carbohydrate (as the polymer cellulose) the most abundant biomolecule on earth. The biochemical answers explaining how CO_2 is fixed photosynthetically and subsequently incorporated into hexoses were provided by Melvin Calvin (Nobel Prize, 1961) and his associates. Their studies culminated with the elucidation of the Calvin cycle, a series of reactions that interrelates many enzymes of carbohydrate metabolism to accomplish the incorporation of CO_2. The cycle, commencing with the fixation of CO_2, includes the pathway for (1) the synthesis of glucose containing photosynthetically fixed CO_2 and (2) the regeneration of ribulose 1,5-bisphosphate, the biomolecular acceptor of CO_2. In contrast to the usual reactions of the pentose phosphate pathway and glycolysis, which occur in the cytosol, the reactions involved in photosynthetic CO_2 fixation are localized in the stroma of chloroplasts.

Fixation of CO_2

Figure 18-10 illustrates the reaction established by Calvin as the primary fixation step. The incorporation of CO_2 is catalyzed by *ribulose 1,5-bisphosphate carboxylase,* which is associated with the thylakoid membranes. As depicted, prior to CO_2 fixation, enolization occurs between C-2 and C-3 of *D-ribulose 1,5-bisphosphate,* producing the carbanion center at C-2 that then serves as the attachment site for electrophilic CO_2. The enzyme-bound product, *2-carboxy-3-ketoribitol 1,5-bisphosphate,* instantly hydrolyzes to yield *two 3-phosphoglycerate* molecules, with the fixed CO_2 appearing as the carboxyl group of one of the products.

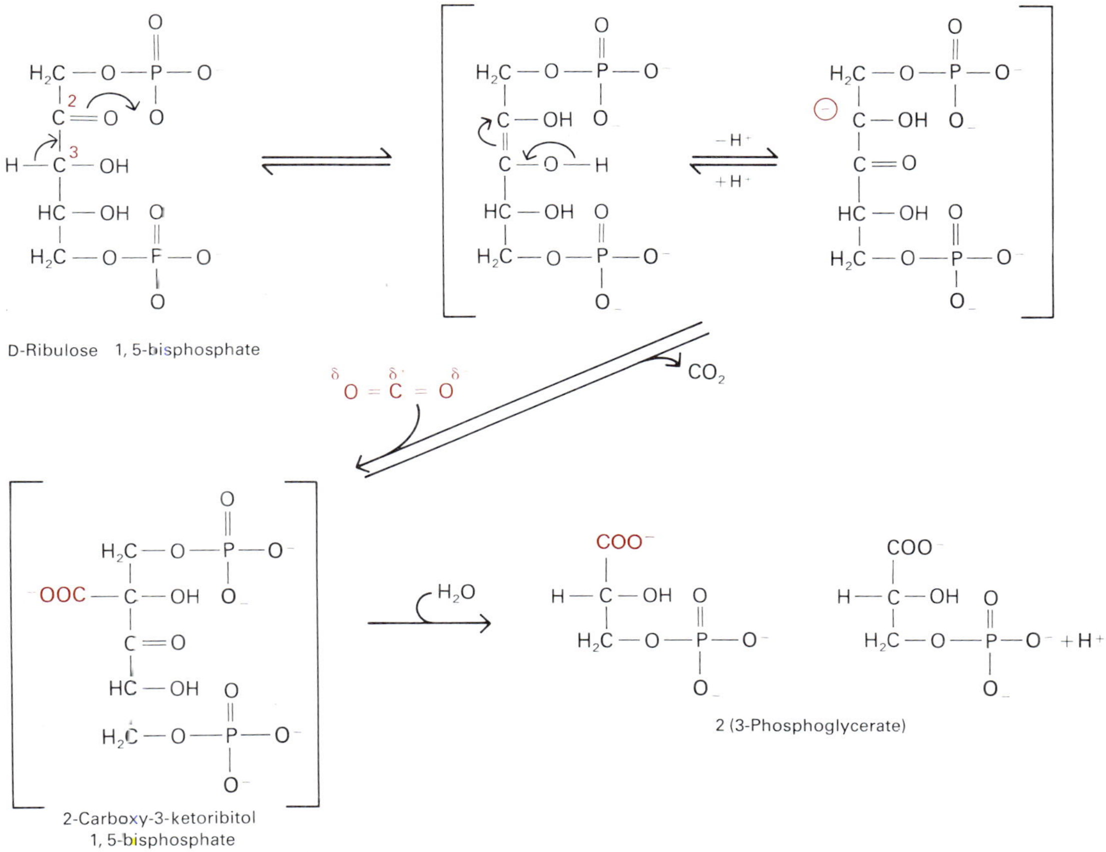

Figure 18-10
Photosynthetic fixation of CO_2 by ribulose 1,5-bisphosphate carboxylase.

Ribulose 1,5-Bisphosphate Carboxylase

Its Role as a Monooxygenase

This enzyme is unusually interesting because it is critical not only for photosynthesis but also for photorespiration. In addition to fixing CO_2, the enzyme is also the *internal monooxygenase* that catalyzes the first reaction in photorespiration. In the reaction, D-ribulose 1,5-bisphosphate and *molecular oxygen* react to produce 3-phosphoglycerate and *2-phosphoglycolate.* The latter product is the immediate precursor of at least 90 per cent of the glycolate that serves as the principal substrate for photorespiration (p. 62). Although the biochemical details of the O_2 reaction remain elusive, there is evidence that CO_2 and O_2 compete for the same site on the enzyme. This CO_2-O_2 competition is believed to determine the relative rates of photosynthesis and photorespiration, which, in turn, influence the production of plant mass.

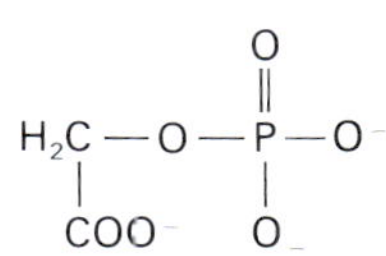

2-Phosphoglycolate

As the major soluble protein of plant leaves, ribulose, 1,5-bisphosphate carboxylase is considered the most abundant enzyme in nature.

Molecular Structure

The carboxylase, which has a molecular weight of about 550,000, is composed of two types of protomers, called the large (MW ≈ 55,000) and small (MW ≈ 14,000) subunits. The active enzyme consists of 16 protomers (eight each of the small and large subunits). An interesting genetic feature of the enzyme is that the *large subunit* is encoded for by a *chloroplast gene* (chloroplast DNA) and the small subunit by a *nuclear gene.* The carboxylase is a pioneering example of the genetic interplay that has evolved between the nuclear and organelle genomes of eucaryotic cells.

The Carboxylase as a Nutrient

Since ribulose 1,5-bisphosphate carboxylase is so plentiful in the plant world, attempts are now under way to utilize the enzyme as a protein supplement in diets. From 2 to 3 mg of the carboxylase (which is odorless, colorless, and tasteless) can be easily isolated from 1 g of tobacco leaf (fresh weight). Animal experiments show that rats fed a diet containing tobacco carboxylase as the sole protein source grow faster than those fed a comparable diet containing casein. With regard to its amino acid content, tobacco carboxylase is considered a good-quality plant protein because it contains adequate amounts of all the essential amino acids except methionine (p. 572). The possibility of using an enzyme as a nutrient is a novel concept that, until recently, neither an enzymologist nor a nutritionist would have considered feasible. Of course, it is the natural abundance of the carboxylase that makes the possibility a reality.

Production of D-Glucose from 3-Phosphoglycerate

In the Calvin cycle, the synthesis of hexoses from the product of the fixation step, 3-phosphoglycerate (a glycolytic intermediate), and from the primary process products (ATP + NADPH) proceeds by reactions previously described for gluconeogenesis (Chapter 15), as outlined in Figure 18-11. Note that the production of D-glyceraldehyde 3-phosphate utilizes *NADPH,* rather than NADH, as a coenzyme; glyceraldehyde 3-phosphate dehydrogenase of chloroplasts, unlike the analogous enzyme in glycolysis, can utilize NADPH. As implied in Figure 18-11, two 3-phosphoglycerate molecules are needed to synthesize one molecule of glucose; thus, *two ATP and two NADPH* are required.

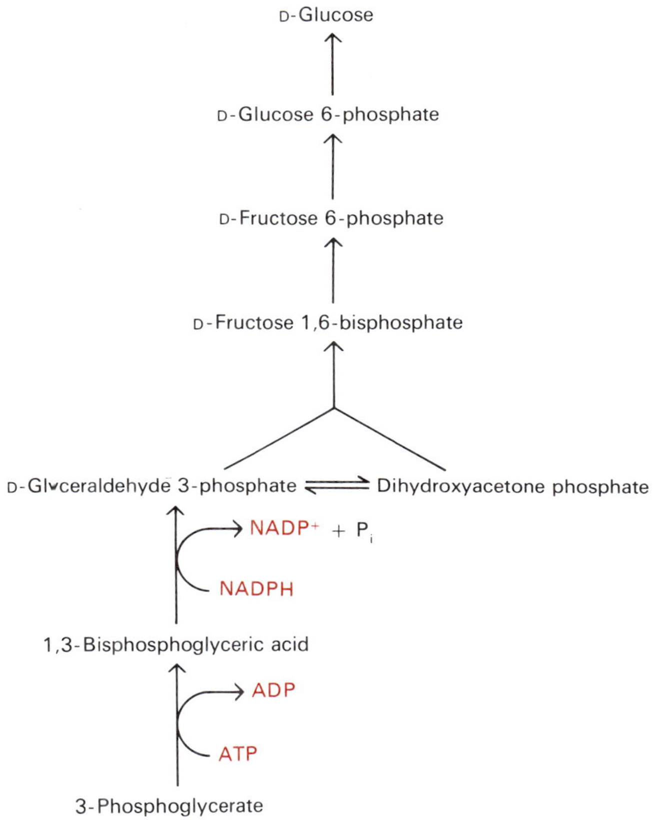

Figure 18-11
Photosynthetic pathway for the synthesis of D-glucose from 3-phosphoglycerate, utilizing NADPH and ATP generated by photosynthesis.

Regeneration of D-Ribulose 1,5-Bisphosphate

To regenerate D-ribulose 1,5-bisphosphate and complete the cycle, three- and six-carbon intermediates in the synthesis of glucose (Figure 18-11) serve as precursors for the production of the five-carbon sugar. The reactions leading to ribulose 1,5-bisphosphate synthesis are listed in Figure 18-12. The initial reaction, catalyzed by transketolase, converts fructose 6-phosphate (6 C) and glyceraldehyde 3-phosphate (3 C) into erythrose 4-phosphate (4 C) and the pentose xylulose 5-phosphate. In the second reaction, aldolase (a glycolytic enzyme), which normally utilizes dihydroxyacetone phosphate but which can use various phosphorylated aldoses as substrates, produces *sedoheptulose 1,7-bisphosphate* (7 C) from the phosphorylated forms of erythrose and dihydroxyacetone. A phosphatase then hydrolytically removes a phosphate group to produce *sedoheptulose 7-phosphate,* which serves as a substrate,

1. $\underset{6C}{\text{Fructose 6-PO}_3^{2-}} + \underset{3C}{\text{glyceraldehyde 3-PO}_3^{2-}} \xrightarrow{\text{transketolase}} \underset{4C}{\text{erythrose 4-PO}_3^{2-}} + \underset{5C}{\text{xylulose 5-PO}_3^{2-}}$

2. $\underset{4C}{\text{Erythrose 4-PO}_3^{2-}} + \underset{3C}{\text{dihydroxyacetone PO}_3^{2-}} \xrightarrow{\text{aldolase}} \underset{7C}{\text{sedoheptulose 1,7-bisPO}_3^{2-}}$

3. $\text{Sedoheptulose 1,7-bisPO}_3^{2-} \xrightarrow{\text{hexose bisphosphatase}} \text{sedoheptulose 7-PO}_3^{2-} + P_i$

4. $\underset{7C}{\text{Sedoheptulose 7-PO}_3^{2-}} + \underset{3C}{\text{glyceraldehyde 3-PO}_3^{2-}} \xrightarrow{\text{transketolase}} \underset{5C}{\text{ribose 5-PO}_3^{2-}} + \underset{5C}{\text{xylulose 5-PO}_3^{2-}}$

5. $\text{Ribose 5-PO}_3^{2-} \xrightarrow{\text{phosphopentose isomerase}} \text{ribulose 5-PO}_3^{2-}$

$\text{2 Xylulose 5-PO}_3^{2-} \xrightarrow{\text{phosphopentose 3-epimerase}} \text{2 ribulose 5-PO}_3^{2-}$

6. $\text{3 Ribulose 5-PO}_3^{2-} + \text{3 ATP} \xrightarrow{\text{phosphoribulokinase}} \text{3 ribulose 1,5-bisPO}_3^{2-} + \text{3 ADP}$

Net: $\underset{6C}{\text{Fructose 6-PO}_3^{2-}} \underset{+}{+} \underset{2(3C)}{\text{2 glyceraldehyde-3PO}_3^{2-}} \underset{+}{+} \underset{3C}{\text{dihydroxyacetone PO}_3^{2-}} + \text{3 ATP} \longrightarrow$

$\underset{3(5C)}{\text{3 ribulose 1,5-bisPO}_3^{2-}} + \text{3ADP}$

Figure 18-12
Photosynthetic production of D-ribulose 1,5-bisphosphate.

with glyceraldehyde 3-phosphate, in a transketolase reaction to produce the two pentoses *ribose 5-phosphate* and *xylulose 5-phosphate.* The isomerase and epimerase of the pentose phosphate pathway (Chapter 15) convert ribose 5-phosphate and xylulose 5-phosphate, respectively, into *ribulose 5-phosphate.* The last step, catalyzed by phosphoribulokinase and requiring ATP, produces *ribulose 1,5-bisphosphate* and regenerates the substrate needed for CO_2 incorporation. In this scheme, one fructose (6 C), one dihydroxyacetone (3 C), and two glyceraldehyde (2×3 C) produce three ribulose bisphosphates (3×5 C), with the synthesis of each of the three requiring hydrolysis of an ATP molecule.

Overall Requirements for ATP and NADPH in the Calvin Cycle

The synthesis of a D-glucose molecule by the Calvin cycle requires 2 ATP and 2 NADPH, and the generation of D-ribulose 1,5-bisphosphate utilizes one ATP. The overall reaction, per CO_2 equivalent, is written:

$$CO_2 + 3\text{ ATP} + 2\text{ NADPH} + 2\text{ H}^+ + 2\text{ H}_2O \rightarrow (CH_2O) + 3\text{ ADP} + 3\text{ P}_i + 2\text{ NADP}^+$$

The secondary process that synthesizes glucose, therefore, requires 3 ATP and 2 NADPH for each molecule of CO_2 reduced.

Because six turns of the Calvin cycle are required to account for all the carbon atoms of glucose, the overall reaction can be summed up as follows:

$$6\ CO_2 + 18\ ATP + 12\ NADPH + 12\ H^+ + 12\ H_2O \rightarrow C_6H_{12}O_6 + 18\ ADP + 18\ P_i + 12\ NADP^+$$

C-4 Dicarboxylic Acid Pathway for CO_2 Fixation

In the 1960s it was discovered that, in some tropical plants, four-carbon dicarboxylic acids (C_4) are among the first intermediates in CO_2 fixation. This deviation from the Calvin cycle, studied by George O. Burns, Marshall (Hal) D. Hatch, and C. Roger Slack, proved to be a *second mode of CO_2 incorporation,* used by plants of tropical origin. Known as the *C-4 dicarboxylic acid pathway,* the biochemical reactions and physiological mechanisms of this scheme accommodate photosynthetic processes that occur in intracellular environments where the CO_2 concentration is low. In such an environment, ribulose 1,5-bisphosphate carboxylase (which has a low affinity for CO_2) functions inefficiently. In hot climates, the stomata of plants often must remain closed to avoid excessive loss of water, and, as a consequence, the inflow of atmospheric CO_2 into leaves is limited. Plants in these environments rely on *phosphoenolpyruvate carboxylase* (an enzyme with a high affinity for CO_2) in *mesophyll cells,* located near the leaf's epidermis, to incorporate CO_2 into *phosphoenolpyruvate* to produce *oxaloacetate* (Figures 18-13 and 18-14). This fixation reaction is followed by conversion of oxaloacetate into *L-malate,* catalyzed by an $NADP^+$- requiring malate dehydrogenase (Figure 18-13). The malate produced is then transported into *bundle-sheath cells,* photosynthetic cells surrounding the vascular bundle in these plants. In these cells, CO_2 is released in the chloroplasts by a reaction, catalyzed by the *malic enzyme,* which produces *pyruvate*

A. In the mesophyll cells:

1. $\text{Phosphoenolpyruvate} + CO_2 + H_2O \xrightarrow[\text{carboxylase}]{\text{phosphoenolpyruvate}} \text{oxaloacetate} + P_i$

2. $\text{Oxaloacetate} + NADPH + H^+ \xrightarrow[\text{malate dehydrogenase}]{NADP^+\text{-linked}} \text{L-malate} + NADP^+$

B. In the bundle-sheath cells:

3. $\text{L-Malate} + NADP^+ \xrightarrow{\text{malic enzyme}} CO_2 + \text{pyruvate} + NADPH + H^+$

4. $\text{Ribulose 1,5-bis}PO_3^{2-} + CO_2 \xrightarrow[\text{carboxylase}]{\text{ribulose 1,5-bisphosphate}} 2\ (\text{3-phosphoglycerate})$

Figure 18-13
C-4 dicarboxylic acid pathway for CO_2 fixation, biochemically depicting how CO_2 is fixed in mesophyll cells and is subsequently released in bundle-sheath cells for use in the Calvin cycle.

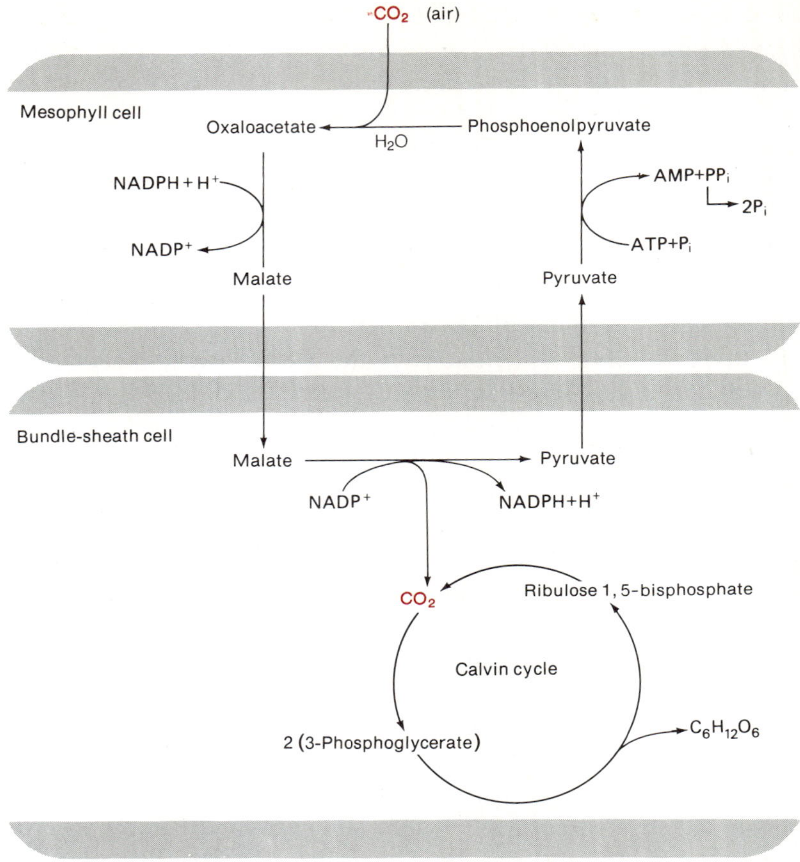

Figure 18-14
Schematic of the functioning of the C-4 dicarboxylic acid pathway in C_4 plants.

from malate (Figures 18-13 and 18-14). By these reactions and the transport process, CO_2 is effectively concentrated in the chloroplasts of the bundle-sheath cells, where enzymes of the Calvin cycle utilize the CO_2 to produce hexoses. The pyruvate generated in bundle-sheath cells is returned to the mesophyll cells, where it is converted, at the expense of two high-energy bonds of ATP, into phosphoenolpyruvate to complete the cycle. This production of phosphoenolpyruvate is catalized by *pyruvate phosphate dikinase,* present in the chloroplasts of mesophyll cells. Plants that rely on the C-4 dicarboxylic acid pathway are referred to as *C_4 plants* (for four-carbon dicarboxylic acid carriers of fixed CO_2) and include commercial crops such as corn, sorghum, and sugar cane (as well as most weeds). Temperate-climate plants that use only the Calvin cycle are called *C_3 plants* (for 3-phosphoglycerate).

In another version of the C-4 dicarboxylic acid pathway found in some C_4 plants, the oxaloacetate produced by the fixation reaction is transaminated to produce *L-aspartate* (oxaloacetate + glutamate → aspartate + α-ketoglutarate), and the amino acid is transported into bundle-sheath cells, where it is deaminated and decarboxylated to produce CO_2.

The C_4 plants require two more ATP per CO_2 equivalent in glucose production than do C_3 plants because of the synthesis of phosphoenolpyruvate from pyruvate (Figure 18-14). The overall equation for C_4 plant synthesis of glucose is:

$$6\ CO_2 + 30\ ATP + 12\ NADPH + 12\ H^+ + 24\ H_2O \rightarrow C_6H_{12}O_6 + 30\ ADP + 30\ P_i + 12\ NADP^+$$

The requirement for extra ATP in C_4 plants does not hamper CO_2 fixation and glucose formation because light intensity is not a limiting factor in tropical areas. Indeed, C_4 plants grow faster and synthesize hexoses faster per unit leaf area than do C_3 plants whenever the plants grow in strong sunlight and at high temperatures.

PHOTOSYNTHETIC PRODUCTION OF BIOMASS AND ITS EFFICIENCY

When the amount of photosynthesis continuously occurring on earth is considered, it is readily apparent that this fundamental process for the synthesis of biomolecules produces enormous quantities of plant biomass (organic material). It is estimated that approximately 5×10^{16} g, or 5×10^{10} metric tons, of carbon are photosynthetically fixed yearly. This amount of fixed carbon is equivalent to a biomass cube with dimensions of $1 \times 1 \times 100$ km, or $1 \times 1 \times 24$ miles. The quantity of captured solar energy accounting for this remarkable synthesis of biomass can be calculated by determining the number of calories needed to fix and reduce the CO_2 that gives rise to the biomass. Since about 4.2×10^{15} moles of CO_2 are fixed yearly and 114,000 cal are required to reduce each mole to (CH_2O), a total of approximately 4.8×10^{20} (480,000,000,000,000,000,000) calories are stored yearly in biomass.

The percentage of the light energy reaching the earth stored in this yearly production of plant biomass also can be calculated. The energy in the light that the earth receives from the sun is about 1.34×10^{24} calories per year. Since the biomass synthesized yearly contains a total of 4.8×10^{20} cal, only about *0.036 per cent* of radiant energy is conserved in biomolecular structures. It should be realized, however, that only a portion of the energy available in the sun's light is trapped by photosynthetic pigments; for example, as seen in Figure 18-4, absorption of photon energy between 500 and 600 nm in the visible spectrum is deficient. Also, not all of the absorbed energy is converted into chemical energy, since some is released in the form of heat. If these factors are taken into consideration, then the caloric content of the biomass represents about *1.6 per cent* of the energy absorbed annually. It should also be noted that this figure is a worldwide estimate; for cultivated croplands, the average quantity of solar energy conserved chemically is *3 per cent* and can be as high as *7 per cent.* When the current attempts to improve food production by increasing the efficiency of photosynthesis are considered, it is evident that neither light quality nor intensity is a limiting factor. Improved photosynthetic efficiency can best be studied by stressing other, more important variables, such as duration of illumination, plant nutrition, soils, H_2O, and temperature.

SUMMARY

All forms of life on earth ultimately rely on photosynthesis for their continued existence since the process is nature's unique way of capturing radiant energy and converting it into the type of chemical energy needed for life processes. Photosynthesis is generally written as follows:

$$6\ CO_2 + 6\ H_2O \xrightarrow{\text{light}} C_6H_{12}O_6 + 6\ O_2$$

This reaction summarizes the oxidation of water to produce O_2 and the fixation and reduction of CO_2 to the level of carbohydrate (D-glucose). In eucaryotic plants, photosynthesis occurs in specialized organelles called chloroplasts, and the biochemical phenomenon requires two distinct phases, namely, the primary (light reactions) and secondary processes. The primary process (which requires light) is responsible for trapping solar energy and for using some of this energy to cleave water molecules to yield O_2, protons, and electrons. The electrons are used for the production of NADPH (which also requires an input of protons) and ATP. The secondary process (which does not require light) fixes and reduces CO_2, using the NADPH and ATP generated by the primary process.

The primary process consists of two light reactions, called photosystems I and II, each of which is a distinct structural unit. In photosystem I, when a light-harvesting pigment, chlorophyll *a*, or an accessory light-harvesting pigment, a carotenoid or xanthophyll, absorbs the energy of a photon, the molecule becomes excited. This energy of excitation is eventually transferred to a particular pigment, called P_{700}, located at the reaction center of the system. Accessory pigments must transfer their excitation energy to P_{700} via chlorophyll *a*. In its excited state, P_{700} transfers an electron to a primary electron acceptor, which then donates the electron to an electron transport system that terminates with the production of NADPH from $NADP^+$. Two known electron carriers in this transport chain are ferredoxin-reducing substance and ferredoxin. Two electrons, released by P_{700} excitation, are required for the production of an NADPH molecule.

As a result of the activation of photosystem I, P_{700} is left in an oxidized state, and activation of photosystem II is needed to restore P_{700} to its ground state. Photosystem II utilizes two types of chlorophyll (*a* and *b*) to absorb light energy and, at its reaction center, has the pigment P_{680}. The transfer of excitation energy to P_{680} occurs via chlorophyll *a*. When excited, P_{680} loses an electron to the primary electron acceptor of photosystem II, which then initiates the transfer of the electron, by means of an electron transport system, to P_{700} in photosystem I. The three known components of this transfer chain are plastoquinone, plastocyanin, and cytochrome *f*. To restore oxidized P_{680} to its ground state, a molecule of water is cleaved into ½ O_2, 2 H^+, and 2 e^-, with the last product being transferred to P_{680}.

Another transport system, which involves only photosystem I, exists in which an electron donated to the photosystem's primary electron acceptor is routed back to its original donor, P_{700}. This electron transport system includes cytochromes b_6 and *f* and plastocyanin.

With regard to ATP production (photophosphorylation), protons generated by electron transport (1) from photosystem II to I (noncyclic photophosphorylation) and (2) by photosystem I (cyclic photophosphorylation) and also (3) by the splitting of water create a proton gradient across the thylakoid membrane that drives the synthesis of ATP from ADP and P_i.

The secondary process, called the Calvin cycle, commences with the fixation of CO_2 by its incorporation at C-2 of D-ribulose 1,5-bisphosphate. The reaction is catalyzed by ribulose 1,5-bisphosphate carboxylase, and the products are two 3-phosphoglycerate molecules. The carboxylase also serves as an internal monooxygenase which catalyzes the reaction between ribulose 1,5-bisphosphate and molecular oxygen to yield 3-phosphoglycerate and 2-phosphoglycolate (principal substrate for photorespiration).

The 3-phosphoglycerate produced by CO_2 fixation is used for gluconeogenesis. In this production of D-glucose, the ATP and NADPH generated by the primary process are utilized. The final phase of the Calvin cycle involves the regeneration of ribulose 1,5-bisphosphate, with three- and six-carbon intermediates of glycolysis serving as precursors. With the use of enzymes of the glycolytic and pentose phosphate pathways, a fructose 6-phosphate, a dihydroxyacetone phosphate, and two glyceraldehyde 3-phosphate molecules can account for the production of three ribulose 1,5-bisphosphate molecules.

A second mechanism for CO_2 fixation and reduction, called the C-4 dicarboxylic acid pathway, occurs in many tropical plants. In this pathway, CO_2 is fixed in mesophyll cells by the production of oxaloacetate from phosphoenolpyruvate. Oxaloacetate is then converted into L-malate (or L-aspartate), which is then transported into bundle-sheath cells; in these cells, malate is converted into CO_2 and pyruvate (or aspartate is transaminated and decarboxylated to produce CO_2 and pyruvate). The CO_2 is then fixed and reduced by the Calvin cycle, and the pyruvate is transported to mesophyll cells, where it is used to regenerate phosphoenolpyruvate. Plants that utilize the C-4 dicarboxylic acid pathway are called C_4 plants and those that use only the Calvin cycle, C_3 plants.

Excerpt from *The Web of Life* by John H. Storer

"Air, rock, water, and sunlight—these are the four sources from which come all living things and their environment. On the bare sands of the desert the sun's rays strike in tiny units of energy moving with atomic speed. Some of them we can feel as heat or see as light. These speeding units impart some of their energy to the dead sands, which temporarily store it in the form of heat, but when the sun sinks, this newly acquired energy is radiated back into the air and lost. The sand becomes as cold and dead as ever. But chlorophyll in the leaves of green plant exists as an agent for garnering these units of solar energy. It makes of the green leaf a laboratory in which nature creates food for living creatures and carries on unceasingly the magic of building life.

Like the sand, a field of grass absorbs the sun's rays; but when night comes the grass does not give back this newly gained energy. In its green laboratory the chlorophyll blends the sun's captured radiance together with elements taken from the air, the water, and the soil, and builds these dead materials into organized living form to make new blades of grass.

This grass is cool and quiet, giving no hint of the sunlight stored within its framework. But dry it out and touch a match to it. The blades of grass—these tiny bits of organized gas and sunlight—blaze up with a flame hot enough to kill a man. All of that fierce heat is merely a release of the same energy that the cool, moist plants have been quietly gathering from the sunlight and storing for later use.

If the grass is not burned, the energy will remain stored within its substance. If it is eaten by an animal, its life force is transferred with it into the body of the animal to sustain the spark that we call life."

QUESTIONS

1. Select the answers to the statements below from the following list:

P_{680}	Mn^{2+}	Chlorophyll *b*	Plastoquinone
P_{700}	Chlorophyll *a*	H_2O	Plastocyanin
		Cytochrome b_6	Ferredoxin
		Cytochrome *f*	

(a) Components of photosystem I
(b) Components of photosystem II
(c) Associated with the electron transport system linking photosystems I and II
(d) Donates two electrons directly to $NADP^+$
(e) Donates an electron directly to P_{700}

2. (a) What is the $\Delta G^{o\prime}$ for the transfer of an electron from the primary electron acceptor of photosystem II ($E_o' = -0.05$ V) to P_{700} ($E_o' = +0.40$ V) of photosystem I?
(b) Is the free energy released sufficient to account for the synthesis of an ATP from ADP and P_i?
(c) How many ATP molecules are produced per oxygen molecule generated by photosystem II?

3. (a) Calculate the $\Delta G^{o\prime}$ for the transfer of an electron in cyclic phosphorylation from the primary electron acceptor of photosystem I ($E_o' = -0.50$ V) to cytochrome b_6 ($E_o' = -0.05$ V).
(b) Is the free energy released sufficient to account for the synthesis of an ATP from ADP and P_i?

4. Calculate the $G^{o\prime}$ for the photosynthetic reduction of $NADP^+$ ($E_o' = -0.32$ V) in spinach, whose ferredoxin has a standard reduction potential of -0.42 V.

5. In Figure 18-14, the C-4 dicarboxylic acid pathway is illustrated using L-malate as the CO_2 carrier into the bundle-sheath cells. Draw the analogous cycle of reactions when L-aspartate, instead of L-malate, is the carrier.

6. In the chapter it was stated that +114 kcal are required to reduce 1 mole of CO_2 to (CH_2O). Based on the information given in Chapter 17 on the complete aerobic oxidation of glucose, show how the $\Delta G^{o\prime}$ for the fixation of 1 mole of CO_2 is derived.

7. Write the general reaction for the production of 3-phosphoglycerate and 2-phosphoglycolate from D-ribulose, 1,5-bisphosphate and O_2 by ribulose 1,5-bisphosphate carboxylase.

SUGGESTED READINGS

Calvin, M., and J. A. Bassham, *The Photosynthesis of Carbon Compounds.* New York: W. A. Benjamin, 1962. Written by two of the principal scientists who elucidated the Calvin cycle.

Campbell, P. N., and G. D. Greville, eds., *Essays in Biochemistry,* Vol. 1. New York: Academic Press, 1965. Article entitled "The Biochemists' Green Mansions: The Photosynthetic Electron Chain in Plants," written by R. Hill (a pioneer in the field), is highly recommended.

Rabinowitch, E. I., and Govindjee, *Photosynthesis,* 2nd ed. New York: Wiley, 1969. A dated but comprehensive treatise on all aspects of photosynthesis.

Reinhold, L., and Y. Liwschitz, eds., *Progress in Phytochemistry,* Vol. 2. London: Interscience, 1970. Contains an account of the photosynthetic process in C_4 plants written by M. D. Hatch and C. R. Slack, investigators who helped elucidate the pathway.

San Pietro, A., *Photosynthesis and Nitrogen Fixation.* Part C, *Methods Enzymol,* Vol. 69. New York: Academic Press, 1980. An excellent reference on the methods used to study photosynthesis.

Articles

Bassham, J. A., "The Path of Carbon in Photosynthesis." *Scientific American,* 206(6):88 (offprint #122), 1962.

Björkman, O., and J. Berry, "High-Efficiency Photosynthesis." *Scientific American,* 229(4):80 (offprint # 1281), 1973.

Blankenship, R. E., and W. N. Parson, "The Photochemical Electron Transfer Reactions of Photosynthetic Bacteria and Plants." *Ann. Rev. Biochem.,* 47:635, 1978.

Ellis, R. J., "The Most Abundant Protein in the World: RUBPcase." *TIBS,* 4:241, 1979.

Govindjee and R. Govindjee, "The Primary Events of Photosynthesis." *Scientific American,* 231(6):68 (offprint #1310), 1974.

Hatch, M. D., ''C_4 Pathway Photosynthesis: Mechanism and Physiological Function.'' *TIBS,* 2:199, 1977.

Levine, R. P., ''The Mechanisms of Photosynthesis.'' *Scientific American,* 221(6):58 (offprint #1163), 1969.

Miller, K. R., ''The Photosynthetic Membrane.'' *Scientific American,* 241(4):102, 1979.

Rabinowitch, E. I., and Govindjee, ''The Role of Chlorophyll in Photosynthesis.'' *Scientific American,* 213(1):74 (offprint #1016), 1965.

Radmer, R., and B. Kok, ''Energy Capture in Photosynthesis: Photosystem II.'' *Ann. Rev. Biochem.,* 44:409, 1975.

Shavit, N., ''Energy Transduction in Chloroplasts: Structure and Function of the ATPase Complex.'' *Ann. Rev. Biochem.,* 49:111, 1980.

Thronber, J. P., and J. P. Markwell, ''Photosynthetic Pigment-Protein Complexes in Plant and Bacterial Membranes.'' *TIBS,* 6:122, 1981.

Zelitch, I. ''Pathways of Carbon Fixation in Green Plants.'' *Ann. Rev. Biochem.,* 44:123, 1975.

LIPID BIOSYNTHESIS

19

INTRODUCTION

The topic of lipids was introduced in Chapter 12, in which the roles of these biomolecules as storage forms of energy and as components of biological membranes were emphasized. In Chapter 17 the metabolic mechanism (β-oxidation cycle) by which saturated fatty acids of triacylglycerols, a major storage form of energy (p. 334), and other lipids can be used for energy production was discussed. Although both previous discussions were concerned primarily with the functions and metabolism of saponifiable lipids (p.205), those of nonsaponifiable lipids are equally as important. Among the nonsaponfiable lipids are the fat-soluble vitamins (discussed in Chapter 25) and cholesterol, which plays a pivotal role in lipid metabolism. In addition to cholesterol's function as a component of animal membranes (p. 208), this lipid is also the parent compound for the various classes of steroid hormones, and its catabolic products provide the bile salts necessary for normal digestion of lipids. With regard to hormone production, certain fatty acids also serve as precursors for a class of hormones called the prostaglandins. Thus, lipid metabolism is interassociated with that of many hormones.

Studies on cholesterol biosynthesis were instrumental in revealing the metabolic interrelationship among the nonsaponifiable lipids since the isoprenoid units generated for cholesterol production are also the building blocks for other lipids, e.g., fat-soluble vitamins. Cholesterol production, as well as that of fatty acids, from a common two-carbon (C_2) precursor, acetate, constitutes an instructive lesson in biological synthesis. Most biological synthetic processes are characterized by the construction of large and often complex biomolecules from simple precursors.

FATTY ACID BIOSYNTHESIS

Because the carbon atoms of fatty acids are derived from acetyl-CoA, it would appear at first approximation that the synthesis of these lipid biomolecules occurs by reversal of β-oxidation, which degrades fatty acids to acetyl-CoA (p. 334). However, whereas β-oxidation is localized in mitochondria, fatty acid synthesis takes place in the endoplasmic reticulum membranes (p. 57) and is catalyzed by a different set of enzymatic activities. Hence, the metabolism of fatty acids is a well-documented example of the independent and distinct enzymatic processes generally associated with the synthesis and catabolism of a particular biological compound or class of biomolecules—a feature of metabolism that was not established until the biochemistry of biosynthetic processes was understood (p. 283).

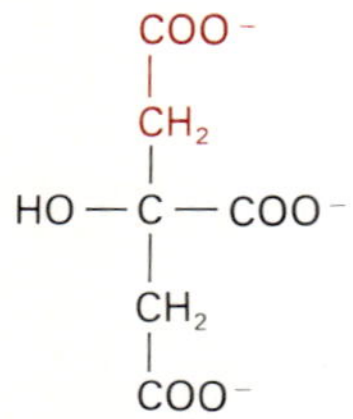
Citrate

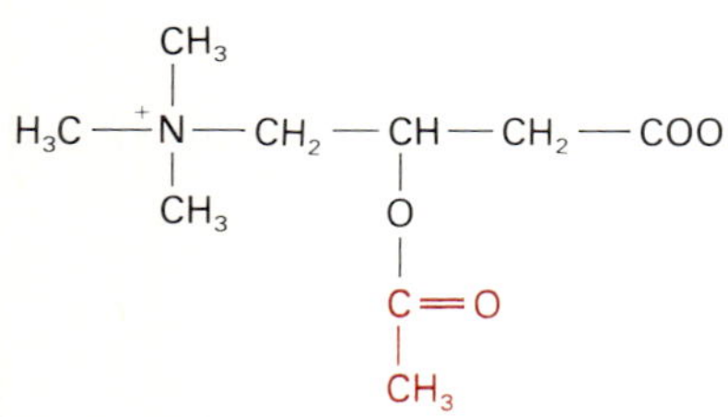
Acetylcarnitine

Acetyl-CoA Transport into the Cytosol

For the production of fatty acids, acetyl-CoA (which is produced in mitochondria) must first be transported across the organelle's membrane into the cytosol. Since acetyl-CoA itself cannot transverse the membrane, this transfer relies on transport of the acetyl moiety as citrate (produced from acetyl-CoA and oxaloacetate, p. 301). After citrate is transferred via the tricarboxylate transport system from mitochondria into the cytosol, it is cleaved by *ATP-citrate lyase* to produce acetyl-CoA by the following reaction:

$$\text{Citrate} + \text{CoA} + \text{ATP} \rightarrow \text{acetyl-CoA} + \text{oxaloacetate} + \text{ADP} + P_i$$

Although carnitine has been implicated as a carrier of acetyl groups, as well as of fatty acids (p. 335), across the mitochondrial membrane, current evidence supports the contention that citrate and not acetylcarnitine is the principal source of cytosolic acetyl-CoA.

Production of Malonyl-CoA — The Initial Reaction

A key observation that subsequently provided an important clue to the biochemical mechanisms involved in fatty acid synthesis was made by Salih Wakil and his associates, who noted that CO_2

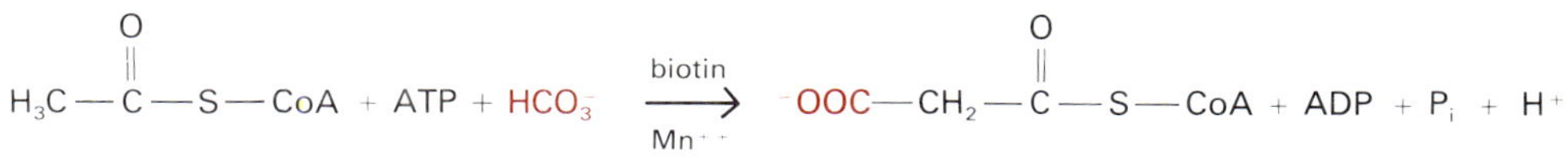

Acetyl-CoA Malonyl-CoA

Figure 19-1
Acetyl-CoA carboxylase reaction.

greatly stimulates the incorporation of acetyl-CoA into fatty acid structures. Their studies on the phenomenon revealed that acetyl-CoA must be converted into *malonyl-CoA* prior to its utilization for fatty acid synthesis. The enzyme catalyzing this reaction is *acetyl-CoA carboxylase* and, as shown in Figure 19-1, the *ATP-dependent carboxylation* requires Mn^{2+} and biotin as cofactors. Production of malonyl-CoA from acetyl-CoA is considered the first committed step in fatty acid synthesis and, as would be expected, it is also the primary rate-limiting reaction of the process. As an allosteric enzyme, acetyl-CoA carboxylase is activated by its positive allosteric effectors, *citrate* and *isocitrate,* and inhibited by its negative effectors, the *acyl-CoA derivatives of long-chain fatty acids.* In avian liver, it has been demonstrated that the enzyme exists as an *inactive monomer* (MW $\approx$ 400,000) in the absence of citrate; however, when citrate is present, acetyl-CoA carboxylase becomes an active polymer, i.e., a long, filamentous structure of about 20 monomeric units and about 400 nm in length. Thus, as a positive effector, citrate (or isocitrate) is required for production of the catalytically active conformation of the enzyme which leads to its polymerization.

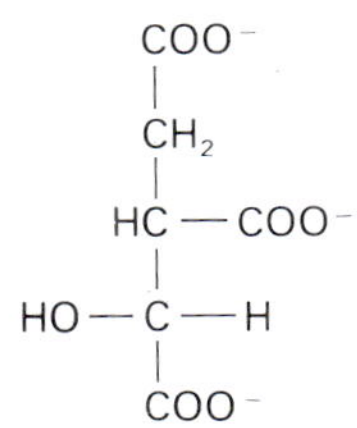

Isocitrate

The regulation of acetyl-CoA carboxylase activity and, hence, of fatty acid synthesis is believed to perform an important *in vivo* function. When the energy demands of a cell are high, ATP synthesis is accelerated; this, in turn, requires that acetyl-CoA generated in the mitochondria be utilized in the TCA cycle (generating ATP while being oxidized to CO_2). Under these conditions, the available acetyl-CoA is metabolized to produce reduced nucleotide coenzymes (NADH and $FADH_2$). No *net* synthesis of TCA cycle intermediates occurs, and intracellular levels of citrate and other intermediates remain low. When the need for enhanced ATP production is relieved, as indicated by increased ratios of ATP/ADP and NADH/NAD^+, allosteric inhibition of isocitrate dehydrogenase (TCA cycle enzyme) by its negative effectors, ATP and NADH (p. 303), results in increased levels of isocitrate and its thermodynamically favored precursor, citrate. When in excess, these two TCA cycle intermediates are transported from mitochondria to the cytosol, where they act as positive effectors for acetyl-CoA carboxylase. Thus, as a result of this series of events, acetyl-CoA not immediately needed for energy production is routed into fatty acid synthesis for energy storage.

Palmitate Synthetase (Fatty Acid Synthetase)

The primary product of fatty acid synthesis is *palmitate* (16-carbon saturated fatty acid). Much of the information about the synthesis of palmitate was obtained from studies on the six individual enzymes of *E. coli* that catalyze the process. A striking difference between palmitate synthesis in *E. coli* and in mammalian species is that, in the case of the latter, the analogous six enzymatic activities are contained in a *single* polypeptide chain. Two of these multienzymatic polypeptides, in dimer formation, constitute *palmitate synthetase (fatty acid synthetase),* which has a MW of about 500,000. Both polypeptides of the dimer independently catalyze palmitate synthesis.

$H_3C—(CH_2)_{14}—COO^-$

Palmitate

A nonenzymatic but critical protein component of palmitate synthetase is the *acyl carrier protein* (ACP), first isolated from *E. coli* by P. Roy Vagelos and co-workers. Their studies showed that ACP is a relatively small protein containing 77 amino acid residues and has a prosthetic group, 4′-phosphopantetheine, that is covalently linked to the hydroxyl group of Ser36 of the protein (Figure 19-2). In higher animals, ACP appears to be an integral part of the single polypeptide chain. The function of ACP in fatty acid synthesis, as implied by its name, is to serve as a carrier of the intermediates, whose acyl groups are covalently bonded in thioester linkage to the sulfhydryl terminus of the 4′-phosphopantetheine moiety. As illustrated in Figure 19-4 (p. 374), the phosphopantetheine unit of ACP serves as an extended, pivoting arm that carries the intermediates to each of the active sites on palmitate synthetase. This proposed role of the phosphopantetheine arm is analogous to that previously described for the lipoyllysyl component of the pyruvate dehydrogenase complex (p. 300).

Both the synthesis and the catabolism (β-oxidation) of fatty acids utilize acyl intermediates and, in both systems, the acyl

Figure 19-2
Prosthetic group of acyl carrier protein (ACP). Fatty acid binds to the prosthetic group by forming a thioester bond with the sulfhydryl group (in red).

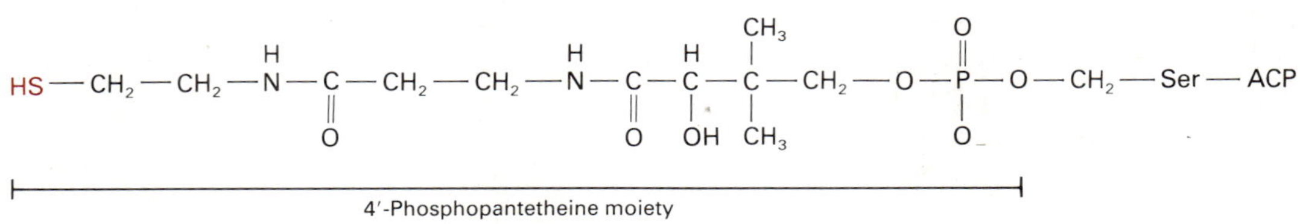

groups are attached by thioester linkage to 4′-phosphopantetheine. In the case of palmitate synthetase, phosphopantetheine is the prosthetic group of ACP and, in β-oxidation, the acyl groups of the acyl-CoA intermediates are bonded to the phosphopantetheine portion of the CoA molecule (p. 173).

First Cycle of Palmitate Synthesis

Figure 19-3 lists the series of reactions that constitute the first cycle in fatty acid synthesis. The cycle commences with the prim-

Figure 19-3
First cycle of palmitate synthesis.

ing of the system with acetyl-CoA and involves the reaction catalyzed by *ACP-acyltransferase,* one of the six enzymatic activities associated with palmitate synthetase. In the reaction, acetyl-CoA reacts with ACP to produce *acetyl-ACP,* which then transfers the acetyl group to a specific cysteinyl residue of *β-ketoacyl-ACP synthase,* another enzymatic activity of palmitate synthetase. With the attachment of the acetyl unit to the synthase *(acetyl-synthase)* and ACP free to accept an acyl group, the system is now primed to initiate the first cycle of reactions that will add a two-carbon unit (furnished by malonyl-CoA).

The entrance of malonyl-CoA into the system is catalyzed by a third enzymatic activity, *ACP-malonyltransferase,* and entails the transfer of the malonyl moiety of malonyl-CoA to a Ser residue of ACP-malonyltransferase and then to ACP, producing *malonyl-ACP.* In the next reaction, the acetyl group bound to a sulfhydryl group of β-ketoacyl-ACP synthase condenses with malonyl-ACP to produce *acetoacetyl-ACP* and CO_2. The reaction is catalyzed by the synthase. Of biological significance is the fact that decarboxylation of the malonyl residue, which makes the reaction highly exergonic, ensures that the equilibrium is far to the right, i.e., toward acetoacetyl-ACP synthesis. This thermodynamic feature of the reaction probably explains why, in biological systems, a C_3 unit (malonyl) is used to provide the C_2 units for fatty acid synthesis. In the following enzymatic synthesis of acetoacetyl-CoA from C_2 units (two acetyl-CoA),

$$\text{Acetyl-S-CoA} + \text{acetyl-S-CoA} \rightleftharpoons \text{acetoacetyl-S-CoA} + \text{CoA-SH}$$

the equilibrium lies to the left since the reaction is endergonic; thus, the reaction is unlikely to occur in biological systems. Incorporation of a C_2 unit into fatty acid structure via the use of a C_3 residue (that is decarboxylated in the process) is therefore a thermodynamically superior way of assuring the elongation of fatty acid chains.

The next three reactions of the cycle (IV-VI in Figure 19-3) convert acetoacetyl-ACP into butyryl-ACP, i.e., the keto group at C-3 of acetoacetyl-ACP is reduced to a methylene group. In the first reaction, catalyzed by *β-ketoacyl-ACP reductase,* acetoacetyl-ACP is reduced to produce *D-β-hydroxybutyryl-ACP.* NADPH serves as the coenzyme for the reaction. (The analogous reaction in β-oxidation produces the L-isomer of the hydroxy intermediate, p. 336.) The second reaction is catalyzed by *β-hydroxyacyl-ACP dehydratase* and involves dehydration of *D-β-hydroxybutyryl-ACP* to yield *crotonyl-ACP,* a *trans*-Δ^2-enoyl-ACP. The last reaction, catalyzed by *β-enoyl-ACP reductase,* is a second reduction, which converts crotonyl-ACP into *butyryl-ACP* and, again, NADPH is the required coenzyme. Thus, after the first cycle of elongation, the C_4 (butyryl) precursor of palmitate has been synthesized from a C_2 (acetyl) and a C_3 (malonyl) unit, with the acetyl group constituting the two terminal carbons of the growing fatty acid chain.

Synthesis of Odd-Numbered Fatty Acids

The cycle described above initiates synthesis of a fatty acid with an even number of carbon atoms. In the synthesis of fatty acids with odd numbers of carbons, a process carried out by many marine organisms, the system is primed by propionyl-CoA instead of by acetyl-CoA. Such a priming is possible because ACP-acyltransferase (but not ACP-malonyltransferase) is capable of transferring acyl groups other than the acetyl residue. The subsequent condensation of a propionyl and a malonyl group during the initial cycle yields a C_5 intermediate and CO_2.

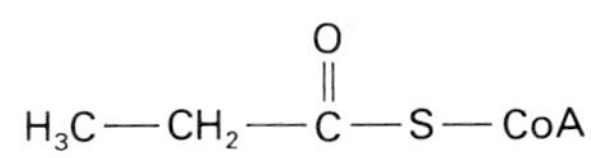

Propionyl-CoA

Second Cycle of Palmitate Synthesis

Figure 19-4 depicts the second cycle in palmitate synthesis, and the illustration emphasizes the carrier role assigned to the pivoting 4′-phosphopantetheine arm of the ACP portion of palmitate synthetase. In the second cycle, the butyryl group of butyryl-ACP (product of the first cycle) is first transferred by ACP-acyltransferase to the thiol group of β-ketoacyl-ACP synthase—a reaction analogous to the transfer of the acetyl residue to prime the system (Figure 19-3). The transfer of the malonyl unit of malonyl-CoA to ACP by ACP-malonyltransferase then takes place. What follows next is the condensation of the butyryl moiety of butyryl-synthase with malonyl-ACP to produce the C_6 intermediate, *β-ketocaproyl-ACP,* and CO_2. To complete the cycle, β-ketocaproyl-ACP is first reduced (β-ketoacyl-ACP reductase), then dehydrated (β-hydroxyacyl-ACP dehydratase), and finally reduced again (β-enoyl-ACP reductase) to yield *caproyl-ACP* (not shown in Figure 19-4). Thus, at the end of two cycles, a C_6-fatty acid intermediate has been synthesized.

Completion of Palmitate Synthesis

After *seven* turns of the cycle, fatty acid synthesis terminates with the production of palmitate (C_{16} saturated fatty acid). It is inter-

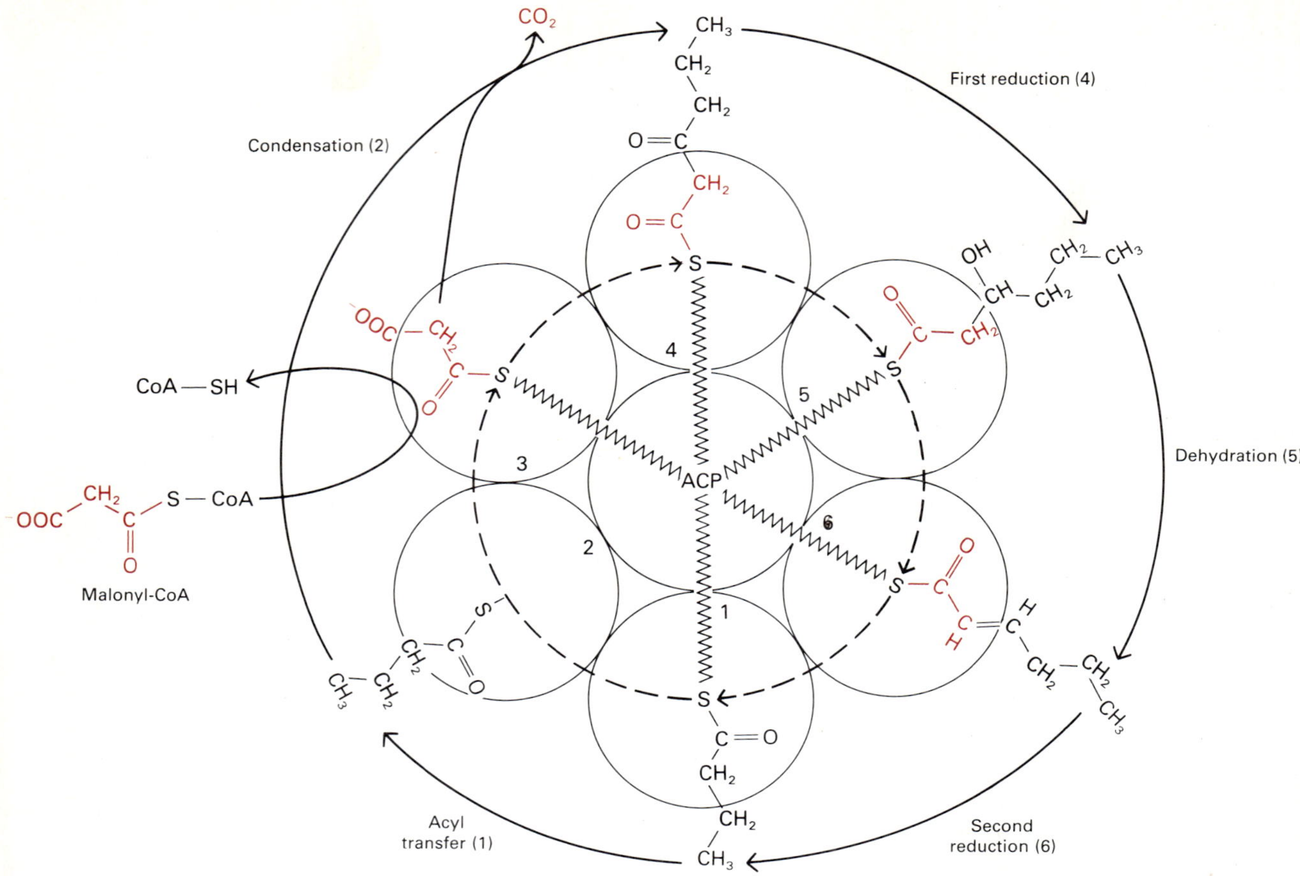

Figure 19-4
Schematic of palmitate synthetase, depicting the second cycle of synthesis (C_4 fatty acid → C_6 fatty acid). ACP = acyl carrier protein; ~SH = 4′-phosphopantetheine side chain of ACP; 1 = ACP-acyltransferase; 2 = β-ketoacyl-ACP synthase; 3 = ACP-malonyltransferase; 4 = β-ketoacyl-ACP reductase; 5 = β-hydroxyacyl-ACP dehydratase; and 6 = β-enoyl-ACP reductase.

esting that, in most organisms, palmitate is the major product of the synthetic process [about 85 per cent palmitate and 15 per cent myristate (C_{14})]. After synthesis, palmitate is released hydrolytically from ACP by *palmitoyl-ACP deacylase* or transferred to CoA.

Another point of interest is that palmitoyl-CoA inhibits palmitate synthetase by dissociating it into two inactive monomers, i.e., the product of the synthetase controls its own production. The system is also specific for acetyl-CoA and malonyl-CoA and will not accept any of the acyl-CoA derivatives of fatty acid intermediates. When the above features of fatty acid synthesis are considered, it is apparent that palmitate synthetase represents a highly integrated and controlled system that is programmed primarily for palmitate synthesis.

Stoichiometry of Palmitate Synthesis

The overall reaction for palmitate synthesis is written:

$$\text{Acetyl-CoA} + 7\ \text{malonyl-CoA} + 14\ \text{NADPH} + 7\ \text{H}^+ \rightarrow \text{palmitate} + 7\ \text{CO}_2 + 14\ \text{NADP}^+ + 8\ \text{CoA} + 6\ \text{H}_2\text{O}$$

Production of the seven malonyl-CoA from acetyl-CoA is written:

$$7\ \text{Acetyl-CoA} + 7\ \text{CO}_2 + 7\ \text{ATP} \rightarrow 7\ \text{malonyl-CoA} + 7\ \text{ADP} + 7\ \text{P}_i + 7\ \text{H}^+$$

Thus, for palmitate synthesis, the overall stoichiometry is expressed as follows:

$$8\ \text{Acetyl-CoA} + 7\ \text{ATP} + 14\ \text{NADPH} \rightarrow \text{palmitate} + 8\ \text{CoA} + 14\ \text{NADP}^+ + 6\ \text{H}_2\text{O} + 7\ \text{ADP} + 7\ \text{P}_i$$

Note that the CO_2 utilized (formation of malonyl-CoA) and the CO_2 produced (condensation reaction) cancel each other when the overall stoichiometry is tabulated.

Comparison of Synthesis and Catabolism

As stated at the beginning of this section, fatty acid synthesis and catabolism represent independent cellular mechanisms; however, as was made apparent in the discussion about synthesis, the two systems share many similarities. As illustrated in Figure 19-5, the last three steps in synthesis are the biochemical reversal of the three key catabolic reactions of β-oxidation. The principal difference in this series of reactions, as catalyzed by the two systems, is that the configurations of the β-hydroxyacyl intermediates differ. Also, β-oxidation uses FAD and NAD^+ as redox coenzymes, whereas synthesis uses $NADP^+$. The latter difference is an excellent example of the concept that NAD^+ and $NADP^+$ coenzymes are used in catabolic and biosynthetic reactions, respectively (p. 319). The one consistent feature of all synthetic and catabolic reactions is that they have a strict specificity for fatty acyl derivatives (either those of CoA or those of ACP).

Elongation and Unsaturation of Fatty Acids

Mammals have the ability to synthesize fatty acids that are longer than palmitate, and the production of these biomolecules occurs

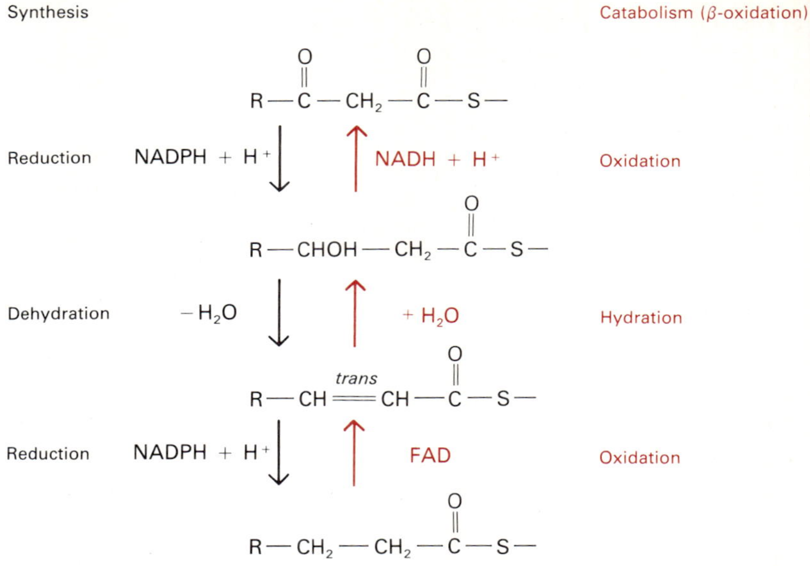

Figure 19-5
Comparison of fatty acid synthesis and catabolism.

by either mitochondrial or microsomal (endoplasmic reticulum) mechanisms. In mitochondria, elongation of pre-formed fatty acids (both saturated and unsaturated) occurs by the addition of acetyl-CoA groups (C_2 units), which is catalyzed by an enzyme complex. The complex, which is enzymatically distinct from palmitate synthetase, is believed to utilize acyl-CoA and not acyl-ACP derivatives. The elongation process is similar to that described for palmitate synthetase and includes (1) reduction (NADH), (2) dehydration, and (3) a second reduction (NADPH).

The microsomal system also catalyzes elongation of saturated and unsaturated fatty acids and requires acyl-CoA derivatives as substrates. However, unlike the mitochondrial process, the microsomal mechanism employs malonyl-CoA as the C_2 donor; thus, except for the lack of ACP-bound intermediates, it is similar to palmitate synthetase.

The microsomes are also the major site for the production of unsaturated fatty acids. The desaturation systems for fatty acids are molecularly complex and require *NADH* and *molecular oxygen,* as well as microsomal *cytochrome* b_5 as an electron carrier. The overall reaction for the conversion of palmitate into its monoenoic derivative, palmitoleate (16:1 *cis*-Δ^9 or 9-$C_{16:1}$), is written:

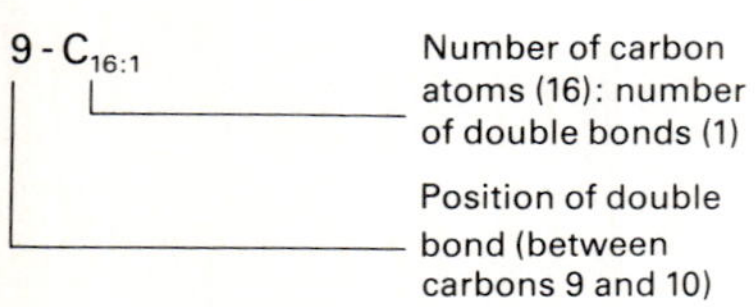

Number of carbon atoms (16): number of double bonds (1)
Position of double bond (between carbons 9 and 10)

$$\text{Palmitoyl-CoA} + \text{NADH} + \text{H}^+ + \text{O}_2 \rightarrow$$
$$\text{palmitoleoyl-CoA} + \text{NAD}^+ + 2\ \text{H}_2\text{O}$$

Since elongation also takes place in microsomes, addition of a C_2 unit(s) can follow desaturation or vice versa. For example, following the synthesis of palmitoleate, the fatty acid can be converted into *vaccenate* (18:1 *cis*-Δ^{11} or 11-$C_{18:1}$) by addition of a C_2 unit. Thus, by use of the microsomal elongation and desaturation systems, a number of saturated and unsaturated fatty acids of various lengths can be generated.

Essential Fatty Acids

Although an animal can synthesize many of the polyunsaturated (two or more double bonds) fatty acids found in its tissues, there are some that must be obtained from the diet. This class of *essential fatty acids,* which cannot be synthesized *de novo,* are the polyunsaturates containing *one or more double bonds within the terminal seven carbon atoms* of the molecule, *viz., linoleate, γ-linolenate,* and *arachidonate* (Figure 19-6). Because linoleate (C_{18}) can be converted into γ-linolenate (C_{18}) and arachidonate (C_{20}) by the microsomal systems, it is considered the true essential fatty acid. As outlined in Figure 19-6, γ-linolenate is produced from linoleate by a desaturation reaction that creates a double bond

Figure 19-6
Arachidonate synthesis from linoleate.

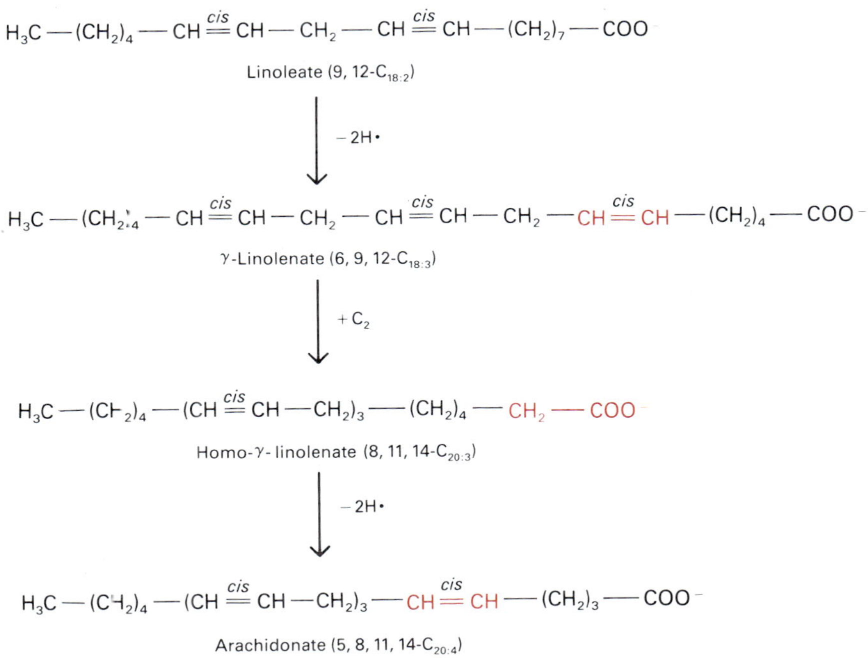

between carbons 6 and 7. To synthesize arachidonate, a C_2 unit is added to γ-linolenate to produce *homo-γ-linolenate,* which is then desaturated (C-5 and C-6) to generate the tetraenoic fatty acid. Currently, there is immense scientific and medical interest in arachidonate metabolism since this fatty acid is the precursor of a class of potent hormones, the *prostaglandins.* These hormones perform critical physiological functions and, as discussed in Chapter 24, are rapidly becoming the newest "miracle drugs" in medicine.

CHOLESTEROL SYNTHESIS

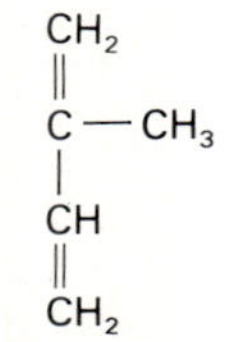
Isoprene

Studies by Konrad Bloch and associates in the early 1940s were the first to reveal that all the carbon atoms of cholesterol (a nonsaponifiable lipid) are derived from acetate (Figure 19-7). This notable finding stimulated subsequent investigations by a number of researchers who, in time, succeeded in elucidating the enzymatic complexities of cholesterol biosynthesis. These studies were also valuable in providing new and important information about the metabolism of many other natural products, especially about the terpenes (biomolecules constructed from two or more isoprene units). Bloch and Theodor Lynen, another eminent biochemist, were awarded Nobel prizes in 1964 for their substantial and influential contributions to this area of biochemistry.

Mevalonate Synthesis

The first two reactions in cholesterol synthesis lead to the production of *mevalonate* from three acetates (Figure 19-8). In the first reaction, *3-hydroxy-3-methylglutaryl-CoA* is synthesized by condensation of acetyl-CoA and acetoacetyl-CoA; the reaction is cat-

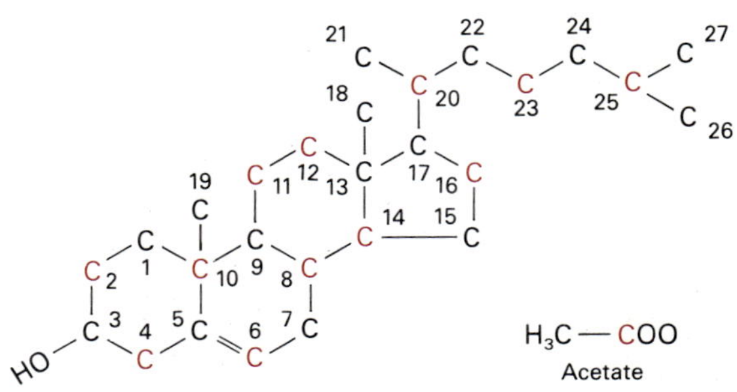

Figure 19-7
Carbon atoms of cholesterol derived from a methyl (C) or carboxyl (C) group of acetate.

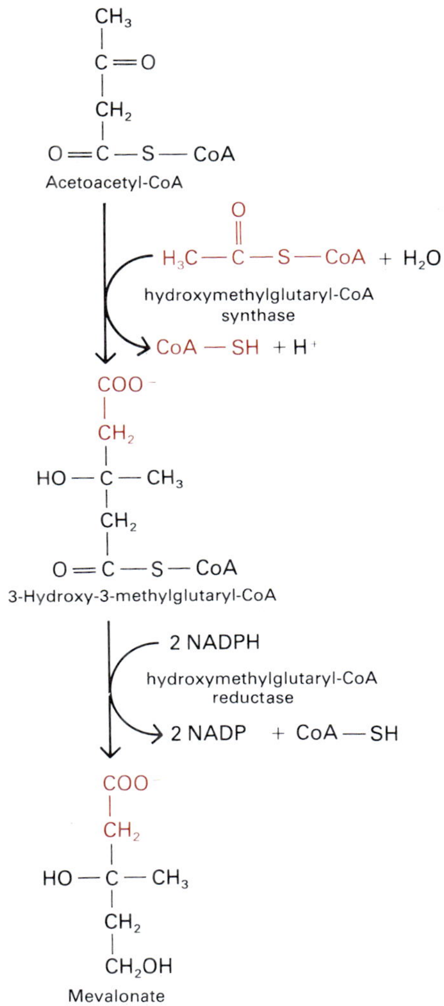

Figure 19-8
Mevalonate synthesis from acetate.

alyzed by *hydroxymethylglutaryl-CoA synthase.* 3-Hydroxy-3-methylglutaryl-CoA can be either cleaved in mitochondria to yield acetyl-CoA and acetoacetate for energy production or, as depicted in Figure 19-8, reduced in the cytosol to produce mevalonate. The reduction step, catalyzed by *hydroxymethylglutaryl-CoA reductase,* is the committed step in cholesterol synthesis (and, as will be discussed, is subject to regulation, p. 385).

Isoprenoid Synthesis

In the next series of reactions (Figure 19-9), mevalonate is utilized for the synthesis of two key isoprenoid precursors of cholesterol.

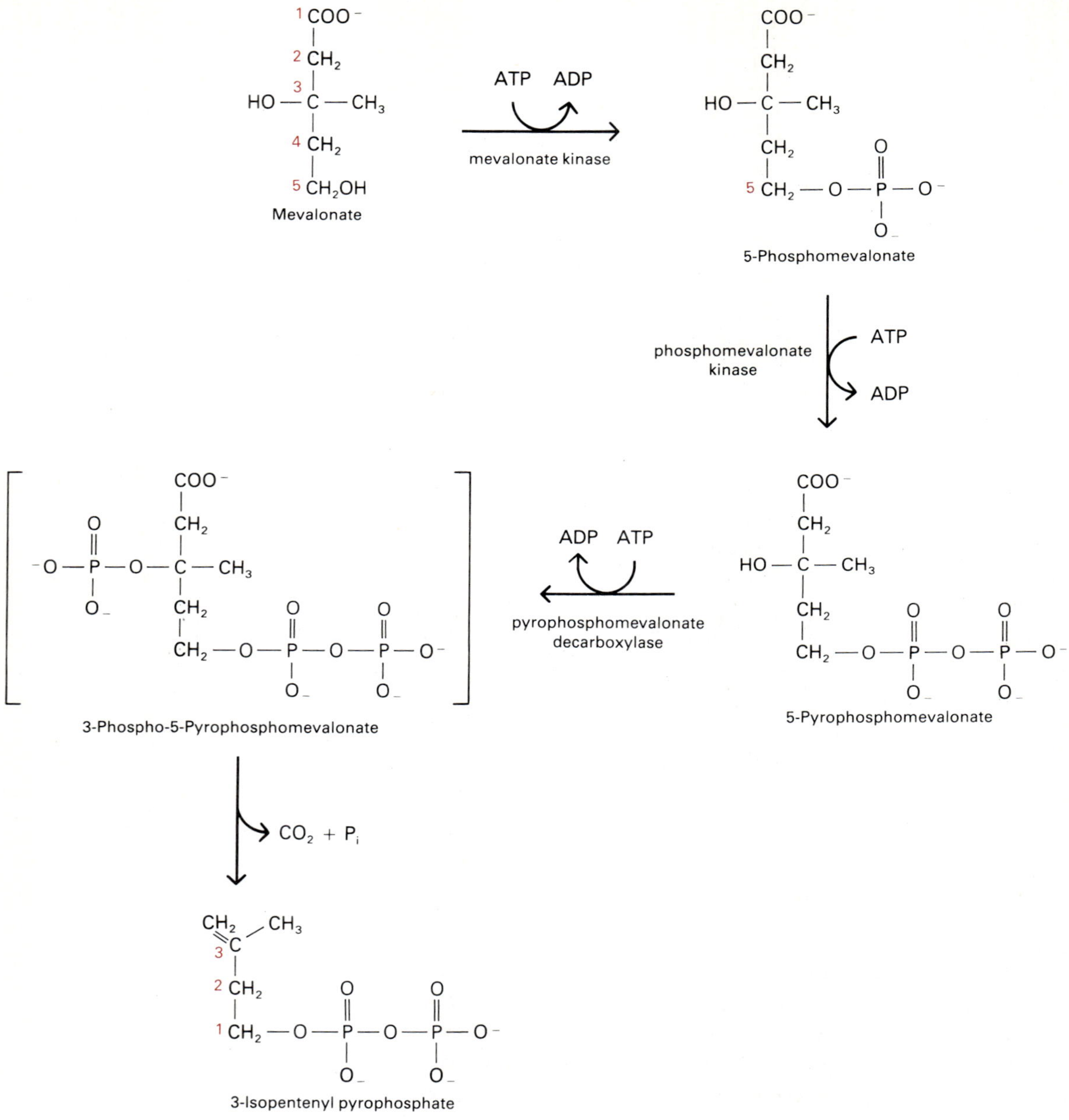

Figure 19-9
Isopentenyl pyrophosphate synthesis from mevalonate.

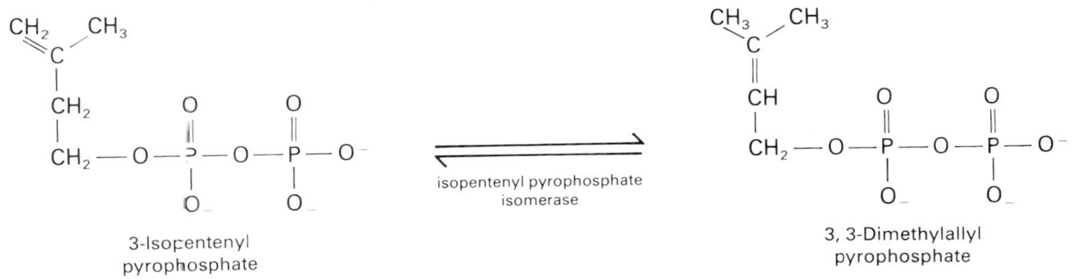

Figure 19-10
Isomerization of isopentenyl pyrophosphate.

The first two reactions are phosphorylations which require ATP; the first is catalyzed by *mevalonate kinase* to produce *5-phosphomevalonate,* and the second by *phosphomevalonate kinase* to yield *5-pyrophosphomevalonate.* What follows next is decarboxylation by the enzyme *phosphomevalonate decarboxylase* to furnish the isoprenoid derivative *3-isopentenyl pyrophosphate.* It is proposed that in this complex reaction the hydroxyl group of pyrophosphomevalonate is first phosphorylated (ATP as phosphate donor) to produce the intermediate *3-phospho-5-pyrophosphomevalonate,* which then undergoes decarboxylation and dephosphorylation to yield 3-isopentenyl pyrophosphate. The production of the other key isoprenoid compound, *3,3-dimethylallyl pyrophosphate,* is accomplished by an isomerase reaction catalyzed by *isopentenyl pyrophosphate isomerase,* an enzyme that interconverts 3-isopentenyl pyrophosphate and 3,3-dimethylallyl pyrophosphate (Figure 19-10).

Squalene Synthesis

The next phase of synthesis includes a series of reactions involving the five-carbon units of the two isoprenoid intermediates to produce *squalene,* a C_{30} intermediate (Figure 19-11). In two reactions catalyzed by *dimethylallyl transferase,* 3,3-dimethylallyl pyrophosphate and 3-isopentenyl pyrophosphate condense to form *geranyl pyrophosphate* (C_{10}), which is then transferred to isopentenyl pyrophosphate to synthesize *farnesyl pyrophosphate,* a C_{15} intermediate. Two farnesyl pyrophosphates then condense, in a two-step mechanism, to produce squalene (C_{30}). The product of the first reaction, catalyzed by *presqualene synthase,* is presqualene (an asymmetric C_{30} condensation product having a cyclopropane ring), which is then reduced and rearranged by *squalene synthase* (NADPH as coenzyme) to yield the symmetrical squalene molecule.

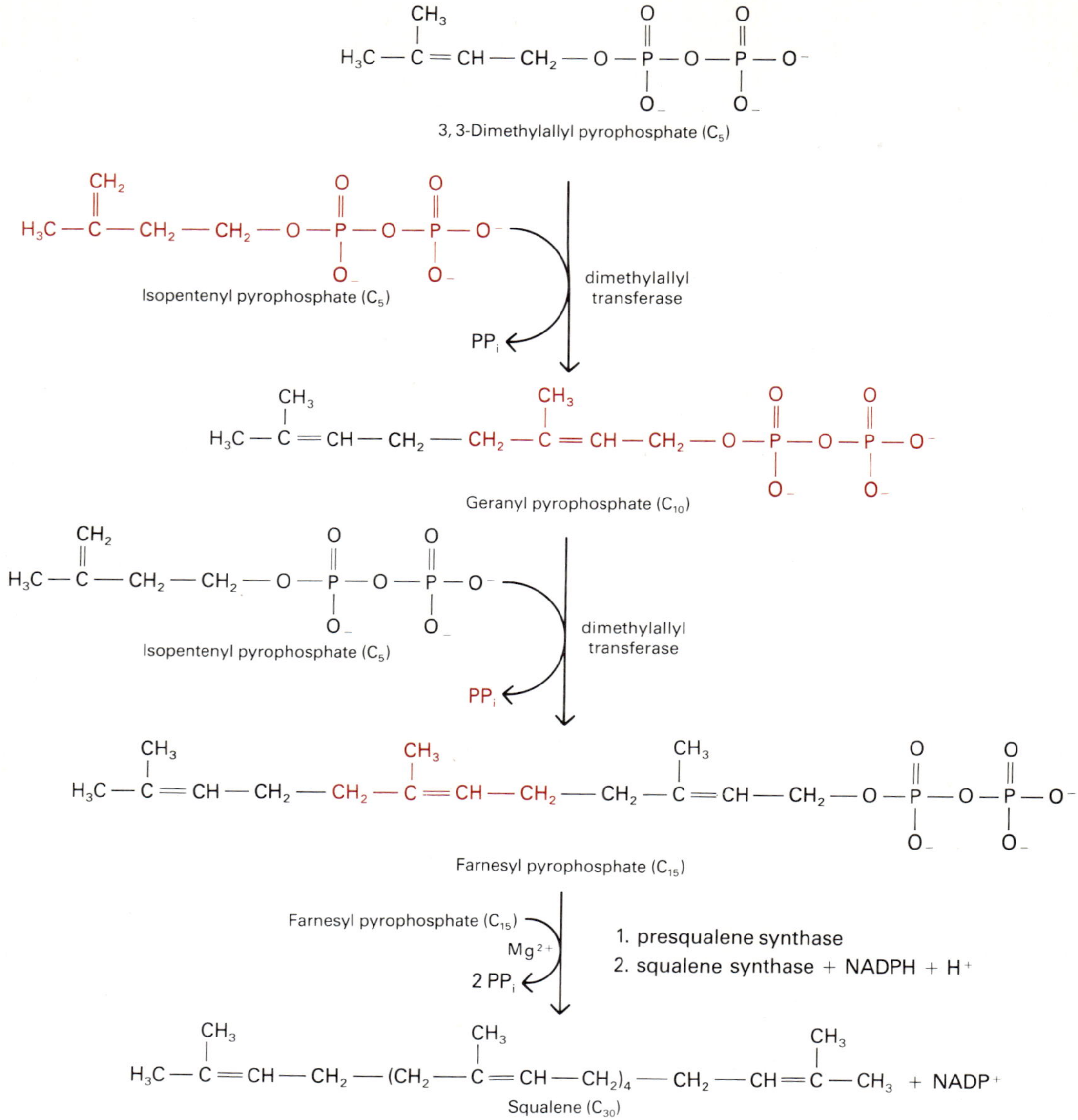

Figure 19-11
Squalene synthesis.

Final Steps in Cholesterol Synthesis

In the next two reactions, squalene is converted into a tetracyclic steroidal configuration (Figure 19-12). The first enzyme, *squalene monooxygenase*—in a reaction requiring NADPH and molecular oxygen—produces *squalene 2,3-epoxide,* which then undergoes cyclization to yield *lanosterol,* the first sterol to be formed. The cyclization, catalyzed by *squalene epoxide cyclase,* is a most unusual reaction because, in addition to bringing about a concerted movement of electrons, it also involves migration of two methyl groups.

The series of reactions synthesizing cholesterol from lanosterol (Figure 19-13) catalyzes (1) the saturation of the double bond in the side chain, (2) the removal of three methyl groups (two from C-4 and one from C-14), and (3) the shift of the double bond from C-8,9 to C-5,6. The conversion does not necessarily proceed by a set of ordered reactions, i.e., saturation of the double bond in the side chain can occur at any stage of the synthesis. However, the principal direct precursors of cholesterol are *7-dehydrocholesterol*

A substituent group *above* the plane containing the four rings of a steroid is *β-oriented,* indicated by a *solid line.* Conversely, a group *below* the plane is *α-oriented,* shown by a *dotted or dashed line.*

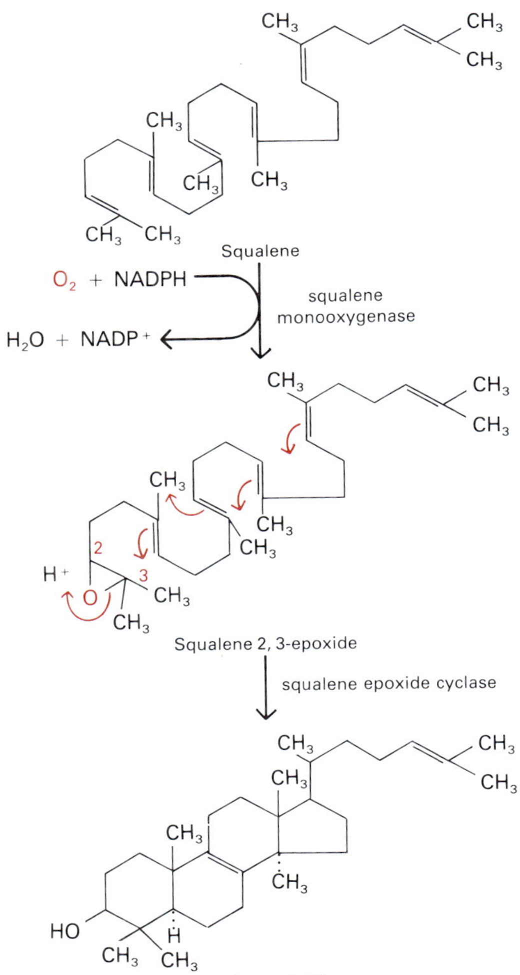

Figure 19-12
Lanosterol synthesis from squalene.

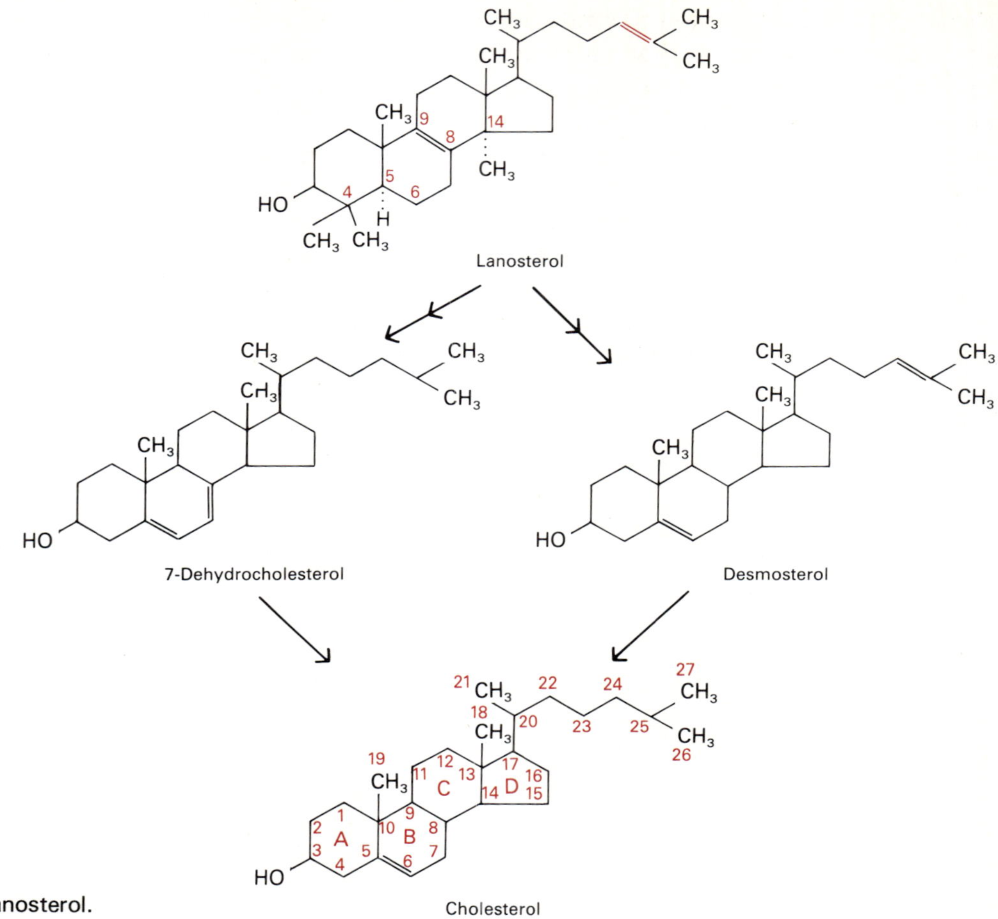

Figure 19-13
Cholesterol synthesis from lanosterol.

and *desmosterol.* Cholesterol synthesis from 7-dehydrocholesterol entails reduction of the C-7,8 double bond and, from desmosterol, reduction of the C-24,25 double bond. (In skin, 7-dehydrocholesterol also serves as a precursor of vitamin D, a subject discussed in Chapter 25.)

CONTROL OF CHOLESTEROL SYNTHESIS

Although the liver is the major site of cholesterol synthesis, other tissues also produce this sterol, e.g., intestine, skin, nervous tissue, adrenal, and gonads. It is estimated that an adult can synthesize approximately 800 mg of cholesterol daily. In addition to this *de novo* synthesis, an individual's diet also provides substantial amounts of cholesterol. To control internal levels of the sterol (being furnished by synthesis and diet), there exists a mechanism whereby cholesterol regulates its own biosynthesis. This regulatory process, recently elucidated, is initiated when the plasma carrier of cholesterol, called *low-density lipoprotein,* or *LDL,* binds to its specific receptor sites on membranes of cholesterol-producing cells. The carrier complex, which transports cholesterol as an ester of linoleate *(cholesteryl linoleate),* is referred to as the *LDL-cholesterol complex.* After binding to the membranes, LDL is internalized by endocytosis and delivered to the lysosomes (p. 59), where it is degraded. The *cholesterol* liberated by LDL degradation accumulates in the cells and represses both the activity and the synthesis of *hydroxymethylglutaryl-CoA reductase,* the enzyme that catalyzes the committed step in cholesterol synthesis (mevalonate production, p. 378). The free cholesterol is also available for entry into cellular membranes.

Endocytosis is taken from the Greek words *endo,* meaning *within* or *inner,* and *kytos,* a *hollow vessel* (cell), and refers to the process whereby something is taken into a cell.

FAMILIAL HYPERCHOLESTEROLEMIA

In certain families, there is a history of high plasma levels of cholesterol and LDL and of premature atherosclerosis, a heart disease characterized by plaque formation on the inner walls of blood vessels, often called "hardening of the blood vessels." Recent studies have revealed that the condition results from an inborn error called *familial hypercholesterolemia,* which affects both heterozygous and homozygous individuals. Somewhat less than 1 per cent of the U.S. population is heterozygous for the disorder. The molecular basis for the syndrome was identified as a *deficiency of LDL receptors.* Because of a lack of membrane receptors, entry of the LDL-cholesterol complex into cells is impaired and, as a result, loss of normal regulation of cholesterol synthesis results in high plasma levels of cholesterol and, with time, early onset of atherosclerosis. Heterozygous individuals often develop atherosclerosis or a more severe heart disease by age 50, and homozygous persons usually die before age 20. Xanthomas (lipid deposits rich in cholesterol) of the skin and bone are a clinical manifestation of the homozygous conditions and often of the heterozygous.

STEROID HORMONE PRODUCTION FROM CHOLESTEROL

A key metabolic function of cholesterol is to serve as the precursor of the following five classes of steroid hormones: *progestagens, glucocorticoids, mineralocorticoids, androgens* (male hormones), and *estrogens* (female hormones). A general scheme for the synthesis of six of these hormones is outlined in Figure 19-14, and a description of their functions is presented in Chapter 24. In the production of steroid hormones, hydroxylation reactions play a prominent role and, as in the case of lanosterol synthesis from squalene (Figure 19-12), they are catalyzed by *monooxygenases* (mixed-function oxidases of microsomes) and require *molecular oxygen* and *NADPH.* A principal intermediate of the synthetic pathways is *pregnenolone,* which is produced by the cleavage of six carbons from the side chain of cholesterol. The removal of the carbons from C_{27} cholesterol yields the C_{21} structure common to many steroid hormones. Pregnenolone is then converted into *progesterone* (a progestagen) in a two-step process that includes oxidation of the hydroxyl group of C-3 to a keto group and isomerization of the double bond from C-5,6 to C-4,5. Progesterone synthesis occurs primarily in the corpus luteum.

The remains of the ovarian follicle, after ovulation, develop into the *corpus luteum* (yellow body), which produces and secretes progesterone for a period of 13 to 14 days, after which it becomes a fibrous mass of scar tissue *(corpus albicans).*

The three main steroid hormones made in the adrenal cortex are the glucocorticoids *cortisol* and *corticosterone* and the mineralocorticoid *aldosterone.* Hydroxylation of C-11, C-17, and C-21 converts progesterone into the major corticosteroid cortisol (Figure 19-14) and, in this reaction series, hydroxylation of C-17 must occur prior to that of C-21. The C-11 hydroxylation can take place at any stage of the conversion. For aldosterone synthesis, progesterone is first hydroxylated at C-21 and then at C-11 to produce corticosterone, whose C-18 methyl group is then oxidized to an aldehyde in the final step, which yields aldosterone.

Progesterone is also the precursor of *androgens,* which in turn are precursors of *estrogens* (Figure 19-14). Production of *testosterone* (C_{19} androgen) occurs mainly in the testis, and synthesis involves loss of C-20 and C-21 from progesterone (C_{21}); the steroid product of these reactions is *androstenedione* (an androgen), which has a keto group at C-17. This keto group is then reduced to a hydroxyl group to produce testosterone. With regard to estrogen (C_{18}) synthesis from testosterone, loss of the C-19 methyl group occurs, and the A ring is converted into an aromatic structure. *Estrone,* another estrogen, is produced in an analogous manner from androstenedione.

CONGENITAL ADRENAL HYPERPLASIA

Inborn errors of adrenal hormone synthesis are categorized as *congenital adrenal hyperplasia* because enlargement of the adrenal

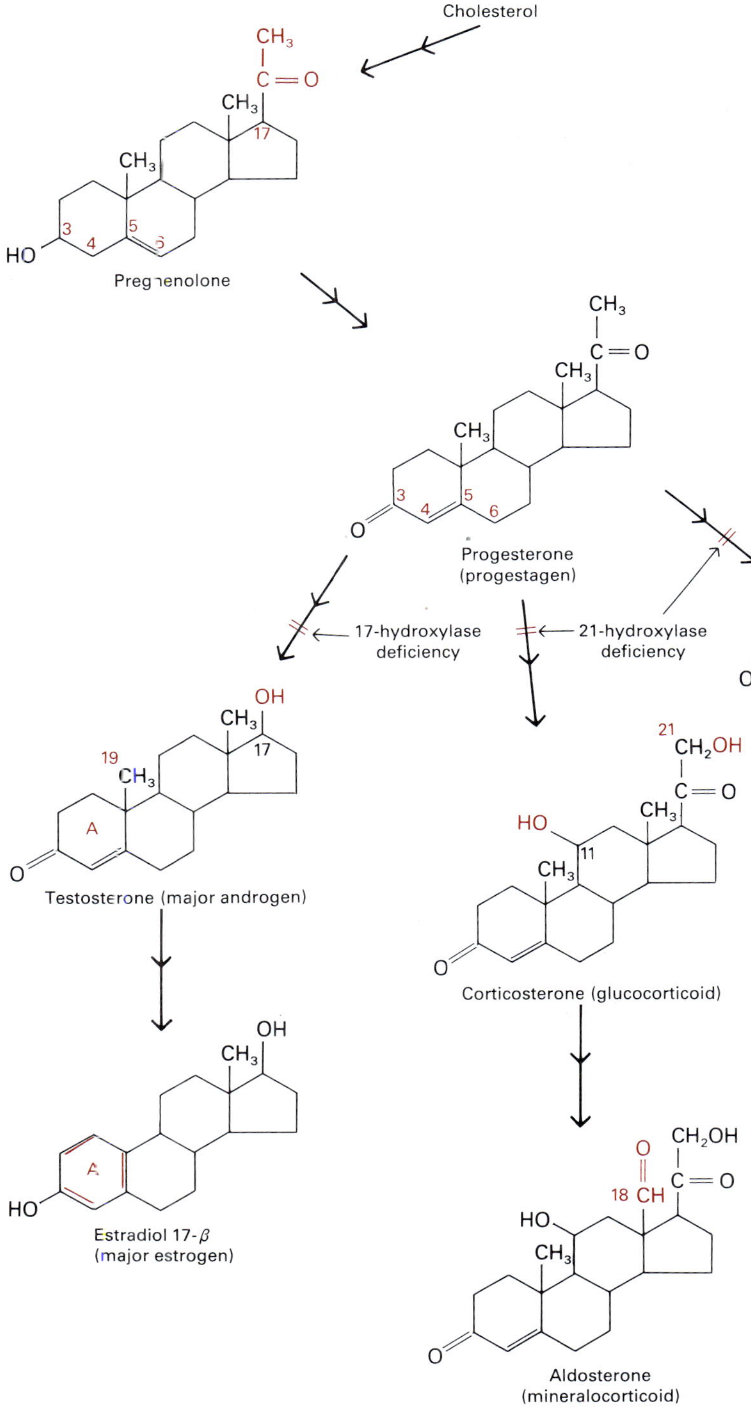

Figure 19-14
Steroid hormone production from cholesterol.

gland is common to all the disorders. The most prevalent form of congenital adrenal hyperplasia (about 90 per cent of the cases) is *21-hydroxylase deficiency* (a Mendelian recessive trait), characterized by *low amounts of cortisol* and often of *aldosterone,* whose syntheses are most affected by the biochemical block (Figure 19-14). A serious metabolic consequence of the deficiency is elevated levels of *ACTH (adrenocorticotropin hormone)* which stimulates hyperplasia of the adrenal (p. 547). Because hyperplasia enhances adrenal hormone synthesis, excess *progesterone* and *17-hydroxyprogesterone* (substrate for 17-hydroxylase) that cannot be used for cortisol and aldosterone production are converted into *androstenedione,* which is then transformed into *testosterone.* This increase in androgens brings about *virilization,* a pronounced clinical manifestation of the disorder. In the fetal development of afflicted females, masculization of the external genitalia occurs. Although afflicted males appear normal at birth, sexual precocity (exceptionally early sexual development) is noted within a few months. Since these individuals prematurely undergo rapid growth and bone maturation during childhood, they are usually short in stature as adults. If aldosterone synthesis is severely affected, there is an additional problem of a persistent *loss of sodium* in the urine (p. 544), which can induce shock and sometimes death. It is important that the syndrome be diagnosed early so that treatment with corticosteroids can be initiated to alleviate the hormonal imbalance. In the case of females, surgical remedy of the effects of virilization on the genitalia may also be required.

Virilization is not always associated with congenital adrenal hyperplasia. For example, in cases of *17-hydroxylase deficiency,* male infants are almost always characterized by female external genitalia, and females are not virilized. As illustrated in Figure 19-14, because lack of 17-hydroxylase activity affects all androgen production and hence also all estrogen production, no sex hormones are produced.

BILE SALT PRODUCTION FROM CHOLESTEROL

Bile salts are metabolic products of cholesterol that are needed for the normal digestion of lipids. Because *bile salts* are polar derivatives of nonpolar cholesterol, they are amphipathic molecules (p. 28) which function in the small intestine to *solubilize dietary lipids.* Solubilization of lipids greatly enhances both their hydrolysis by intestinal lipases and their absorption. Approximately 80 per cent

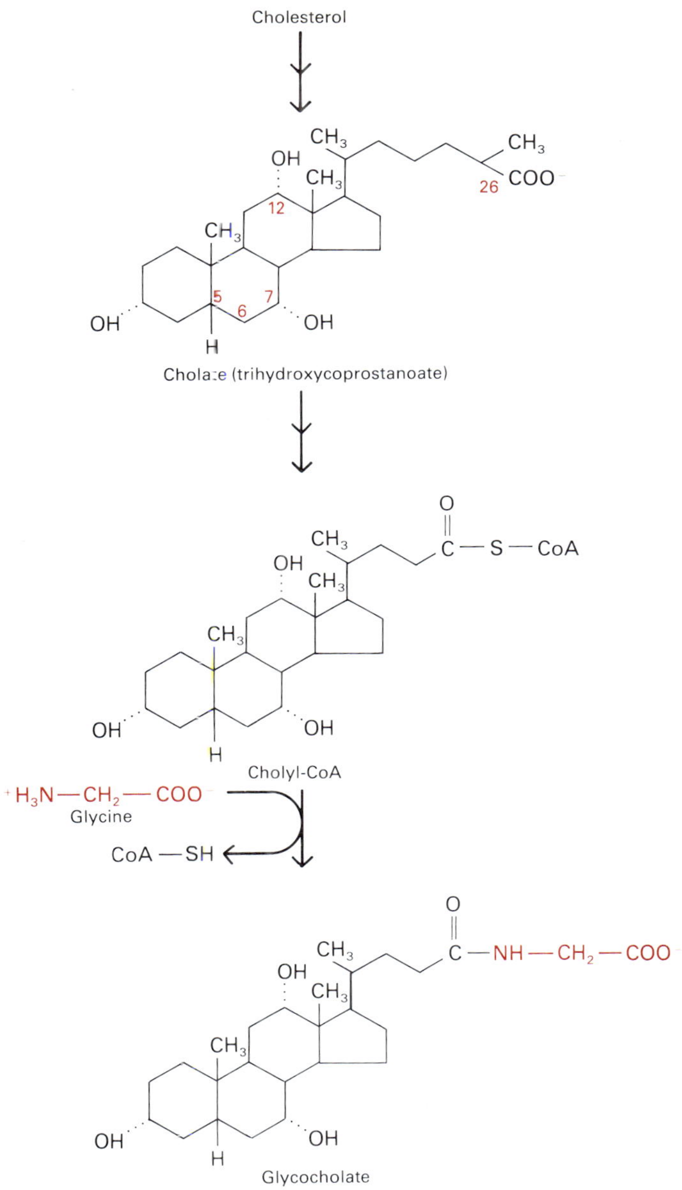

Figure 19-15
Glycocholate synthesis from cholesterol.

of the body's cholesterol is converted by the liver into various *bile acids,* which then conjugate with *glycine* or *taurine* to produce bile salts. *Glycocholate,* the major bile salt, is produced by condensation of cholate, a bile acid, and *glycine* (Figure 19-15). As shown in Figure 19-15, *cholate (trihydroxycoprostanoate)* is an oxidized

(C-26), hydroxylated (C-7 and C-12), and reduced (C-5,6 double bond) derivative of cholesterol. (Another major bile salt is *chenodeoxycholate,* which lacks the hydroxyl group at C-12.) The CoA derivative of cholate, *cholyl-CoA,* is generated prior to the condensation reaction with glycine that produces glycocholate. If, instead of glycine, taurine is the conjugate, the bile salt made is *taurocholate.* Bile salts, after synthesis, are stored in the gall bladder, from where they are transported by a duct into the small intestine for digestive purposes.

$^{+}H_3N—CH_2—CH_2—SO_3^-$

Taurine

SUMMARY

For fatty acid synthesis, which occurs in the endoplasmic reticulum membrane, acetyl-CoA is transported out of the mitochondria as citrate, which is cleaved in the cytosol to regenerate acetyl-CoA and oxaloacetate. Prior to its utilization for fatty acid synthesis, acetyl-CoA is converted by a carboxylation reaction into malonyl-CoA. In mammalian species, the six enzymatic activities required for palmitate (principal fatty acid produced) synthesis are contained in a single polypeptide which, in dimer formation, is called palmitate synthetase. A nonenzymatic protein component of palmitate synthetase is the acyl carrier protein (ACP), whose prosthetic group, 4′-phosphopantetheine, serves as the carrier of the intermediates during fatty acid synthesis.

In the first cycle of synthesis, acetyl-CoA primes the system by transferring its acetyl group to ACP, which then donates it to β-ketoacyl-ACP synthase. Malonyl-CoA then enters the system by transferring its malonyl moiety to ACP-malonyltransferase, which then passes the C_3 unit to ACP. Next, the acetyl group bound to β-keto-ACP synthase condenses with malonyl-ACP to produce acetoacetyl-ACP and CO_2. In the next three reactions (a reduction, a dehydration, and a second reduction), acetoacetyl-ACP is converted into butyryl-CoA. Thus, at the end of the first cycle, the C_4 precursor of palmitate is produced. In those species that can synthesize fatty acids with an odd number of carbons, the system is primed by propionyl-CoA instead of by acetyl-CoA, and the initial cycle yields a C_5 intermediate and CO_2. It requires seven turns of the cycle, eight acetyl-CoA, seven ATP, and 14 NADPH to synthesize one palmitate molecule. Palmitoyl-CoA inhibits the system.

In mammals, elongation of fatty acids occurs either in the mitochondria or in the endoplasmic reticulum (microsomes). The mitochondrial and microsomal systems elongate fatty acids by the addition of acetyl (C_2) units furnished by acetyl-CoA and malonyl-CoA, respectively. The microsomes are also the major site for the production of unsaturated fatty acids. The three essential fatty acids are the polyunsaturates linoleate, γ-linolenate, and arachidonate; however, since the microsomal system can produce γ-linolenate and arachidonate from linoleate, only linoleate is considered to be truly essential.

Cholesterol biosynthesis commences with mevalonate production from three acetates. Mevalonate is then converted into two key isoprenoid (C_5)

precursors of cholesterol, 3-isopentenyl pyrophosphate and 3,3-dimethylallyl pyrophosphate. The two isoprenoids then condense to form geranyl pyrophosphate (C_{10}) which then bonds with 3-isopentenyl pyrophosphate (C_5) to produce farnesyl pyrophosphate (C_{15}). Two farnesyl pyrophosphates condense and, in a two-step process, form squalene (C_{30}), which is then converted into lanosterol (C_{30}), the first sterol produced. Next, lanosterol is converted into 7-dehydrocholesterol (C_{27}) and desmosterol (C_{27}), either of which can be used directly to produce cholesterol (C_{27}).

Cellular production of cholesterol is controlled by the low-density lipoprotein (LDL)-cholesterol carrier complex, which is internalized (by endocytosis) after attaching to its specific cellular membrane receptor. Within the cell, the LDL-cholesterol complex is degraded in lysosomes, and the freed cholesterol represses both the activity and the synthesis of the enzyme that catalyzes mevalonate production (first committed step in cholesterol synthesis). Individuals deficient in LDL membrane receptors suffer from familial hypercholesterolemia, a disease that produces premature atherosclerosis.

Cholesterol is the precursor of the following five classes of steroid hormones: progestagens (e.g., progesterone), glucocorticoids (cortisol), mineralocorticoids (aldosterone), androgens (testosterone), and estrogens (estradiol 17-β). Congenital adrenal hyperplasia, an inborn error, results from an enzyme deficiency that brings about a reduced production of cortisol (and often of aldosterone) and elevated levels of ACTH, which stimulates adrenalhyperplasia. As a consequence, testosterone synthesis is increased and virilization occurs.

Cholesterol is also the precursor of bile salts, needed for the normal digestion of lipids. About 80 per cent of the body's cholesterol is converted by the liver into bile salts. Two major bile salts are glycocholate (condensation product of cholate and glycine) and taurocholate (condensation product of cholate and taurine). Bile salts are stored in the gall bladder, from where they are transported by a duct into the small intestine for digestive purposes.

QUESTIONS

1. Select your answers to the statements below about fatty acid synthesis from the following list:

Acetyl-CoA carboxylase	β-Enoyl-ACP reductase
ACP-acyltransferase	β-Hydroxy-ACP dehydratase
ACP-malonyltransferase	β-Keto-ACP reductase
Acyl carrier protein	β-Keto-ACP synthase
ATP-citrate lyase	

(a) Catalyzes acetyl-CoA production in the cytosol
(b) Catalyzes malonyl-CoA production
(c) Its prosthetic group is 4′-phosphopantetheine

First round of synthesis
(d) Enzyme that transfers the acetyl group of acetyl-CoA to β-ketoacyl-ACP synthase
(e) Enzyme that transfers the malonyl group of malonyl-CoA to ACP
(f) Catalyzes acetoacetyl-ACP production
(g) Catalyzes D-β-hydroxybutyryl-ACP production
(h) Catalyzes crotonyl-ACP production
(i) Catalyzes butyryl-ACP production
(j) Two enzymes that utilize NADPH as a coenzyme
(k) The six enzymatic activities associated with palmitate synthetase

2. Select your answers to the questions below on steroid biosynthesis from the following list.

Acetate	Farnesyl pyrophosphate
3,3-Dimethylallyl pyrophosphate	Geranyl pyrophosphate

3-Isopentenyl pyrophosphate — Mevalonate
Lanosterol — Squalene

Fill in the blanks:

In cholesterol biosynthesis, the first C_5 intermediate, (a) ________, is produced from three (b) ________. This C_5 intermediate then serves as the precursor for the key isoprenoid precursors, (c) ________ and (d) ________, which condense to form the C_{10} intermediate (e) ________. This C_{10} intermediate then combines with (f) ________ to produce the C_{15} intermediate (g) ________. Two C_{15} precursors then combine to yield the C_{30} compound (h) ________, which undergoes transformation to produce the first sterol, (i) ________.

3. In the conversion of squalene 2,3-epoxide into lanosterol (Figure 19-12), which two methyl groups migrate?

4. Which of the following statements are true about the steroid hormones?
   (a) Cholesterol is the parent compound for the steroid hormones.
   (b) Progesterone is the precursor of glucocorticoids but not of mineralocorticoids.
   (c) Progesterone is the precursor of the sex hormones.
   (d) Cortisol is the precursor of pregnenolone.
   (e) Corticosterone is the precursor of aldosterone.
   (f) Androgens are produced from estrogens.
   (g) Androgens and estrogens are C_{19} steroids.
   (h) Cortisol and aldosterone are C_{21} steroids.

5. Draw the structure of (a) androstenedione, (b) estrone, (c) chenodeoxycholate, and (d) taurocholate.

6. Which of the following statements are true about the inborn errors discussed in the chapter?
   (a) Familial hypercholesterolemia
      (1) is due to decreased production of LDL.
      (2) affects individuals heterozygous for the condition.
      (3) affects cholesterol absorption from the gut.
      (4) results in defective brain development.
   (b) 21-Hydroxylase deficiency
      (1) is characterized by low amounts of cortisol.
      (2) is characterized by increased androgen production.
      (3) can affect aldosterone production.
      (4) produces virilization in male, but not female, infants.
   (c) 17-Hydroxylase deficiency
      (1) primarily affects cortisol and aldosterone production.
      (2) stimulates androgen and estrogen production.
      (3) brings about virilization in both male and female infants.
      (4) is corrected by lowering the salt content of the diet.

SUGGESTED READINGS

Bondy, P. K., and L. E. Rosenberg, eds., *Metabolic Control and Disease,* 8th ed. Philadelphia: W. B. Saunders, 1980. Chapter 7 discusses lipid metabolism and diseases.

Fasman, G. D., ed., *Handbook of Biochemistry and Molecular Biology,* 3rd ed. Sec. C, *Lipids, Carbohydrates and Steroids,* Vol. 1. Boca Raton, Fla.: CRC Press, 1976. Good reference source for information on lipids and steroids.

Heftman, E., *Steroid Biochemistry.* New York: Academic Press, 1970. A short summary of the biosynthesis of many of the steroid hormones; also covers the structure, function, clinical significance, and analogues of these hormones.

Hobkisk, R., *Steroid Biochemistry,* Vols. 1 and 2. Boca Raton, Fla.: CRC Press, 1979. Comprehensive presentation of the synthesis and biological functions of steroid hormones.

Nes, W. R., and M. L. McKean, *Biochemistry of Steroids and Other Isoprenoids.* Baltimore, Md.: University Park Press, 1977. A recommended text.

Snyder, F., ed., *Lipid Metabolism in Mammals,* Vols. 1 and 2. New York: Plenum Publishing, 1977. A comprehensive treatment of the subject.

Wakil, S. J., ed., *Lipid Metabolism.* New York: Academic Press, 1970. An advanced and thorough treatment of lipid metabolism.

Articles

Bloch, K., and D. Vance, "Control Mechanisms in the Synthesis of Saturated Fatty Acids." *Ann. Rev. Biochem.,* 46:263, 1977.

Bosch, H. van den, "Phosphoglyceride Metabolism." *Ann. Rev. Biochem.,* 43:243, 1974.

Dempsey, M. E., "Regulation of Steroid Biosynthesis." *Ann. Rev. Biochem.,* 43:967, 1974.

Fulco, A. J., "Metabolic Alterations of Fatty Acids." *Ann. Rev. Biochem.,* 43:215, 1974.

Numa, S., "Two Long-Chain Acyl Coenzyme A Synthetases: Their Different Roles in Fatty Acid Metabolism and Its Regulation." *TIBS,* 6:113, 1981.

Saggerson, D., "The Role of Insulin in Fatty Acid Metabolism." *TIBS,* 4:33, 1979.

NITROGEN METABOLISM

20

INTRODUCTION

In the discussion on the amphibolic nature of the TCA style (Chapter 16), attention was given to the synthesis of glutamate and aspartate from α-ketoglutarate and oxaloacetate, respectively, and of glutamine from glutamate. These three syntheses, it was noted, are key reactions interconnecting carbon metabolism with that of nitrogen. Of particular importance is that glutamate and glutamine syntheses constitute two major reactions for converting inorganic nitrogen (NH_4^+) into an organic form ($-NH_2$). Aspartate synthesis from glutamate also demonstrates how amino nitrogen, by transamination reactions, can be transferred from one biomolecule to another.

This chapter complements and expands that introduction to nitrogen metabolism. The presentation commences by exploring the unique biological mechanisms for converting nitrogen (N_2) and nitrate (NO_3^-) into NH_4^+, which subsequently becomes incorporated into biomolecules. The chapter then continues with descriptions of the biosynthetic and catabolic pathways of various classes of nitrogenous compounds, including amino acids, purines, and pyrimidines. Such pathways, which permit the tracing of the flow of nitrogen through biological systems, are the essence of nitrogen metabolism.

NITROGEN FIXATION

Fixation of the 175 million tons of atmospheric nitrogen estimated to occur annually can be carried out only by bacteria and cyanobacteria (p. 344). However, even among the numerous bacterial species, nitrogen fixation is limited to anaerobic soil bacteria, a few aerobes, e.g., *Azotobacter,* and the *Rhizobium*-legume symbiotic relationship. From a thermodynamic viewpoint, reduction of nitrogen to ammonia is a chemical conversion that is difficult to accomplish because the N≡N bond (bond energy of 225 kcal mol^{-1}) is exceedingly stable and therefore unreactive. For example, commercial production of NH_3 from N_2 requires a catalyst (iron), a temperature of about 500° C, and a pressure of 300 atm. That some bacteria and cyanobacteria perform the same feat *in vivo* is an impressive example of the catalytic (enzymatic) capabilities of living organisms.

Nitrogenase Complex

The metabolic machinery responsible for nitrogen fixation is called the *nitrogenase complex,* an aggregate of enzymatically active polypeptides that harness reducing power via an electron transport system; this reducing power is then used to reduce N_2 to NH_4^+. (The nitrogen-fixing process, it should be noted, is poisoned by oxygen.) There are two major components of the complex. *Component I,* also called the *nitrogenase component,* has a MW of about 220,000. It contains four protomers (each having an Fe·S center, p. 320) and a small polypeptide which also possesses an Fe·S center, as well as two molybdenum (Mo) atoms. The importance of Mo in nitrogen fixation was suggested before the nitrogenase complex was isolated as a result of the observation that soils deficient in molybdenum are unable to support the growth of legumes. To remedy such a situation, Mo compounds are added to legume seeds at planting time. *Component II,* or the *reductase component,* has a MW of about 60,000 and is composed of two identical protomers, each having an Fe·S center. There is one component II for every two component I in the nitrogenase complex.

A soluble ferredoxin, an Fe·S protein (p. 320), serves as the electron source for N_2 reduction. Recent evidence suggests that the flow of electrons proceeds as follows:

$$\text{Ferredoxin} \rightarrow \text{component II} \rightarrow \text{Fe-Mo polypeptide (component I)}$$

Ferredoxin transfers electrons to component II, which then transfers them to the small Fe-Mo polypeptide of component I, after which they are used to reduce N_2 bound to component I. Biochemical details of the enzymatic reduction of N_2 are vague since the intermediates are enzyme bound and hence have not been isolated.

The electron transport requires *ATP,* which is hydrolyzed in the process. The role of ATP is to bind to component II, an event that alters the dimer's conformation. This structural change lowers the redox potential of component II from −0.29 to −.040 V, which enhances the transfer of electrons to the Fe-Mo polypeptide. The stoichiometry of the *six-electron reduction* of a molecule of N_2 to NH_4^+ is:

$$N_2 + 6\,e^- + 12\,ATP + 12\,H_2O \rightarrow 2\,NH_4^+ + 12\,ADP + 12\,P_i + 4H^+$$

Hydrogen Production by the Nitrogenase Complex

Nitrogen fixation is an energy-expensive process, e.g., about 20 per cent of the ATP provided by photosynthesis in a pea plant (legume) is used by its *Rhizobium*-containing nodules. Such large expenditures of ATP are incurred because, in addition to reducing N_2, the nitrogenase complex also produces hydrogen. In most legumes, about 50 per cent of the electrons processed by the complex are used to produce H_2. This feature of nitrogen fixation is illustrated in Figure 20-1 where, as shown, an input of 12 e^- and 12 H^+ into the system results in the reduction of a N_2 molecule (6 e^- and 6 H^+) and the production of three hydrogen molecules (6 e^- and 6 H^+). Hydrogen production is considered a side reaction of the complex (which does not have a strict substrate specificity), and its significance remains to be ascertained. As also indicated in Figure 20-1, some bacteria have a *hydrogenase* that oxidizes H_2 to H^+ ions and electrons, which are then utilized for nitrogen reduction and for ATP synthesis.

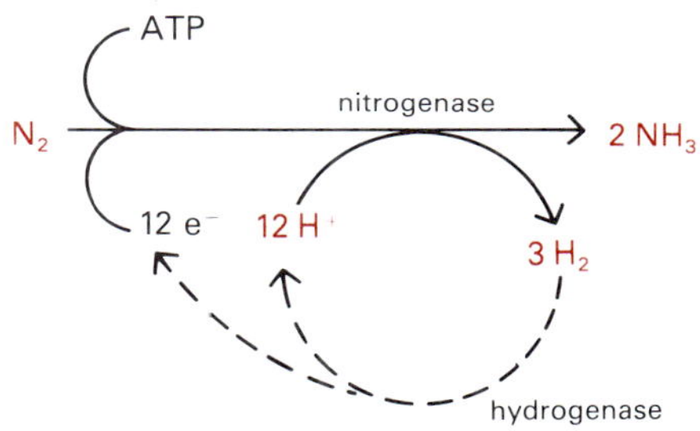

Figure 20-1
Summation of the nitrogenase and hydrogenase reactions.

Hydrogen production by most legumes results in substantial losses of energy. For example, it is estimated that the hydrogen gas produced annually by the 50 million acres of soybeans grown in the United States has an energy equivalent of 300 billion cubic feet of natural gas. In view of these losses, research efforts are currently being made to select for nitrogen-fixing bacteria of agricultural importance that exhibit reduced hydrogen production. Such strains utilize ATP more efficiently for nitrogen fixation—a metabolic function that translates into more efficient nitrogen reduction and, as a consequence, better crop production.

NITRATE REDUCTION

Whereas nitrogen-fixing plants can obtain their nitrogen requirements from N_2, nonfixing plants acquire most of their nitrogen from nitrate (NO_3^-), the most abundant form of nitrogen in soil. Most anaerobic and facultative aerobes (those that can also grow anaerobically) and algae are also capable of converting NO_3^- into biological nitrogen. This utilization of NO_3^- entails its reduction to ammonia, which is then incorporated into nitrogenous biomolecules. Reduction is accomplished by the catalytic actions of two enzymes, *nitrate reductase* and *nitrite reductase.*

Nitrate Reductase

This enzyme is a tetrameric aggregation of four dimers, each possessing one protomer of MW $\approx$ 150,000 and another of MW $\approx$ 50,000. The mechanism of nitrate reduction is analogous to that of nitrogen fixation in that nitrate reductase also contains an electron pathway, with an FAD, a Mo atom, and several Fe·S centers (per dimer) serving as electron carriers. The flow of electrons, derived from NADH in plants, that results in NO_3^- reduction to NO_2^- (nitrite), appears to proceed as follows:

$$\text{NADH} \rightarrow \text{FAD} \rightarrow \text{Fe·S} \rightarrow \text{Mo} \rightarrow NO_3^- \rightarrow NO_2^-$$

The overall reaction is written:

$$NO_3^- + \text{NADH} + H^+ \rightarrow NO_2^- + \text{NAD}^+ + H_2O$$

The nitrate reductase of plants is induced by nitrate and is believed by some to be repressed by ammonia.

Nitrite Reductase

After conversion of NO_3^- to NO_2^-, the second enzyme, *nitrite reductase,* completes the reduction to NH_4^+. Few molecular details have been elucidated about the *six-electron reduction* catalyzed by the enzyme since none of the intermediates (presumably enzyme bound) are detected. The nitrite reductase of plants has an Fe·S center and *siroheme,* the immediate reductant of nitrogen, as carriers of the electrons furnished to the enzyme by reduced ferredoxin. Some fungal reductase also possess an FAD or FMN as an electron carrier. The overall reaction for the reduction of NO_2^- to NH_4^+ is written:

$$NO_2^- + 6\ e^- + 8\ H^+ \rightarrow NH_4^+ + 2\ H_2O$$

AMINO ACID METABOLISM

As will be discussed in Chapter 25, humans are able to synthesize their requirements for 11 of the 20 amino acids needed for protein synthesis. The production of three of the 11 *nonessential* amino acids (glutamate, glutamine, and aspartate) has already been described (p. 309), and their key roles in nitrogen metabolism have also been noted. The following section discusses the biosynthesis and catabolism of the remaining eight (alanine, arginine, asparagine, serine, glycine, cysteine, proline, and tyrosine).

Alanine Synthesis and Catabolism

Pyruvate is readily converted into *alanine* by transamination, catalyzed by *alanine aminotransferase.* The synthesis utilizes glutamate as the NH_2 donor and occurs as follows:

$$\text{Pyruvate} + \text{glutamate} \rightleftharpoons \text{alanine} + \alpha\text{-ketoglutarate}$$

For alanine catabolism, the reversible nature of the reaction assures the amino acid's conversion back into pyruvate. Thus, alanine, glutamate, and aspartate are synthesized directly from intermediates of carbohydrate metabolism (pyruvate, α-ketoglutarate, and oxaloacetate, respectively) by a reversible, one-step mechanism.

Asparagine Synthesis and Catabolism

Whereas in microorganisms, asparagine is produced from aspartate and NH_4^+ by a reaction analogous to that catalyzed by glutamine synthetase (p. 310), mammals synthesize the amino acid principally from *aspartate* and *glutamine.* As illustrated in Figure 20-2, glutamine donates its amide nitrogen to aspartate to produce *asparagine* and *glutamate;* the reaction is catalyzed by *glu-*

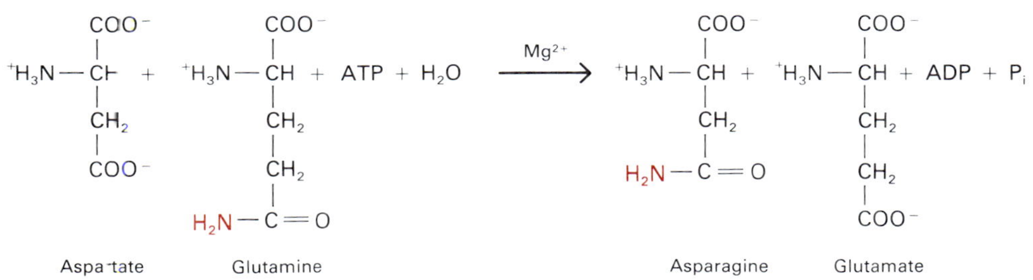

Figure 20-2
Glutamine-dependent asparagine synthetase reaction.

Asparagine was isolated in 1813 from asparagus, for which it was named.

tamine-dependent asparagine synthetase. The energy for this reaction is provided by ATP. Asparagine catabolism, catalyzed by *asparaginase,* involves its hydrolysis by the following reaction:

$$\text{Asparagine} + H_2O \rightarrow \text{aspartate} + NH_4^+$$

The aspartate produced can then undergo transamination to yield oxaloacetate, a TCA cycle intermediate.

Serine Synthesis and Catabolism

The name *serine* is derived from the word *serum.*

An intermediate of glycolysis, *3-phosphoglycerate,* serves as a major source of *serine* in mammals. In the first of three reactions (Figure 20-3), 3-phosphoglycerate is oxidized by *3-phosphoglycerate dehydrogenase* to produce *3-phosphohydroxypyruvate.* The enzyme requires NAD^+ as a coenzyme. In the next step, catalyzed by *phosphoserine aminotransferase,* 3-phosphohydroxypyruvate is converted into *3-phospho-L-serine,* which is then hydrolyzed by *phosphoserine phosphatase* to produce serine. The pathway is regulated by serine, which is a negative effector for both the 3-phosphoglycerate dehydrogenase and the phosphoserine phosphatase reactions.

For catabolic purposes, serine is readily converted into *pyruvate* by *serine-threonine dehydratase* (the two hydroxylated amino acids are substrates for the mammalian enzyme), as illustrated in Figure 20-4. In the reaction, an initial *α,β-dehydration* produces unstable *dehydroalanine,* which then reacts with water to yield pyruvate and NH_4^+.

Glycine Synthesis and Catabolism

Glycine, because of its sweet taste, is named for the Greek word *glycos,* meaning *sweet.*

The primary source of glycine is *serine,* and in the production of the two-carbon amino acid (Figure 20-5), serine transfers its β-carbon to tetrahydrofolate (FH_4), converting the coenzyme into N^5,N^{10}-methylene-FH_4 (p. 175). The reaction is catalyzed by *serine transhydroxymethylase,* which also requires pyridoxal phosphate as a coenzyme. Another enzyme that synthesizes glycine is *glycine synthase,* which catalyzes the following reaction:

$$CO_2 + NH_4^+ + N^5,N^{10}\text{-methylene-}FH_4 + NADH + H^+ \rightleftharpoons \text{glycine} + FH_4 + NAD^+$$

This tetrahydrofolate reaction also requires pyridoxal phosphate.

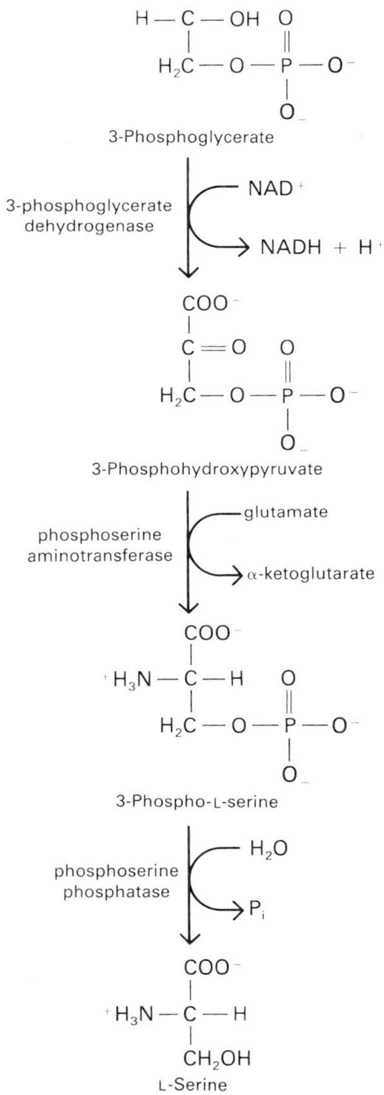

Figure 20-3
Serine synthesis.

The catabolism of glycine involves the reversal of its synthetic pathways. Because both glycine synthase and serine transhydroxymethylase catalyze physiologically reversible reactions, glycine can be converted into CO_2 and NH_4^+ or into serine. Serine can then be catabolized to pyruvate. Note that the catabolic transformation of glycine into serine is also an alternative enzymatic route for serine synthesis.

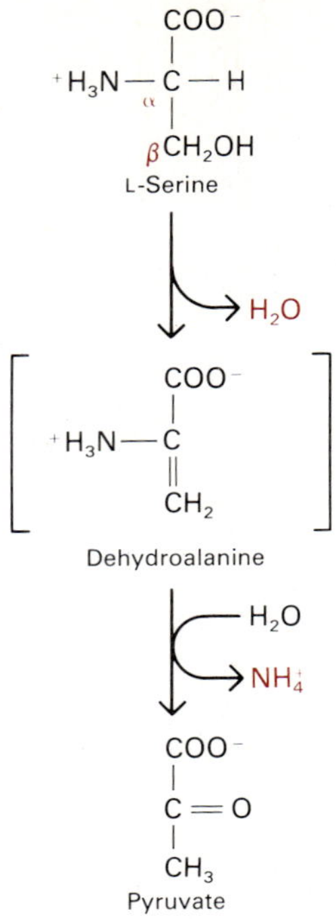

Figure 20-4
Serine-threonine dehydratase reaction.

Cysteine Synthesis and Catabolism

Cystine, from which the name *cysteine* was later derived, was isolated in 1843 from urinary stones and was named for the Greek word *cystos*, meaning *bladder*.

Cysteine, one of the two sulfur-containing amino acids, is synthesized from serine and methionine (the sulfur amino acid required by humans). As depicted in Figure 20-6, cysteine production initially entails formation of S-*adenosylmethionine* in a reaction that is catalyzed by *methionine adenosyltransferase* and utilizes *methionine* and *ATP* as substrates. As a substrate for *transmethylation* reactions, *S*-adenosylmethionine is a key methyl-group donor and is therefore of general metabolic importance. This "active form" of methionine (its sulfonium form) has a $\Delta G^{o\prime}$ equivalent to that of a pyrophosphate bond of ATP. As indicated in Figure 20-6, when *S*-adenosylmethionine transfers its methyl group, it is converted into S-*adenosylhomocysteine,* which is then hydrolyzed to produce *L-homocysteine* and adenosine.

Figure 20-5
Serine transhydroxymethylase reaction.

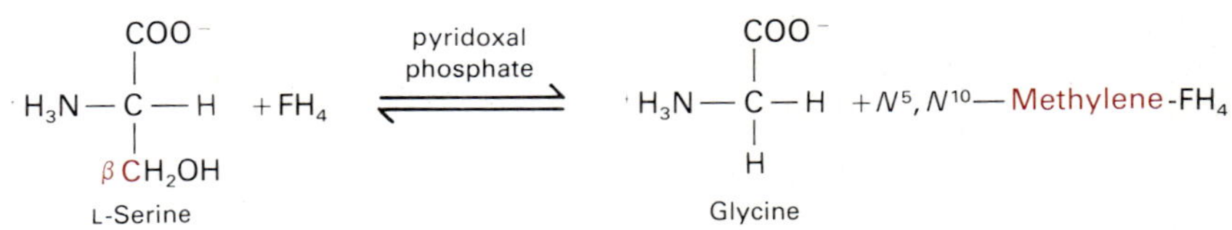

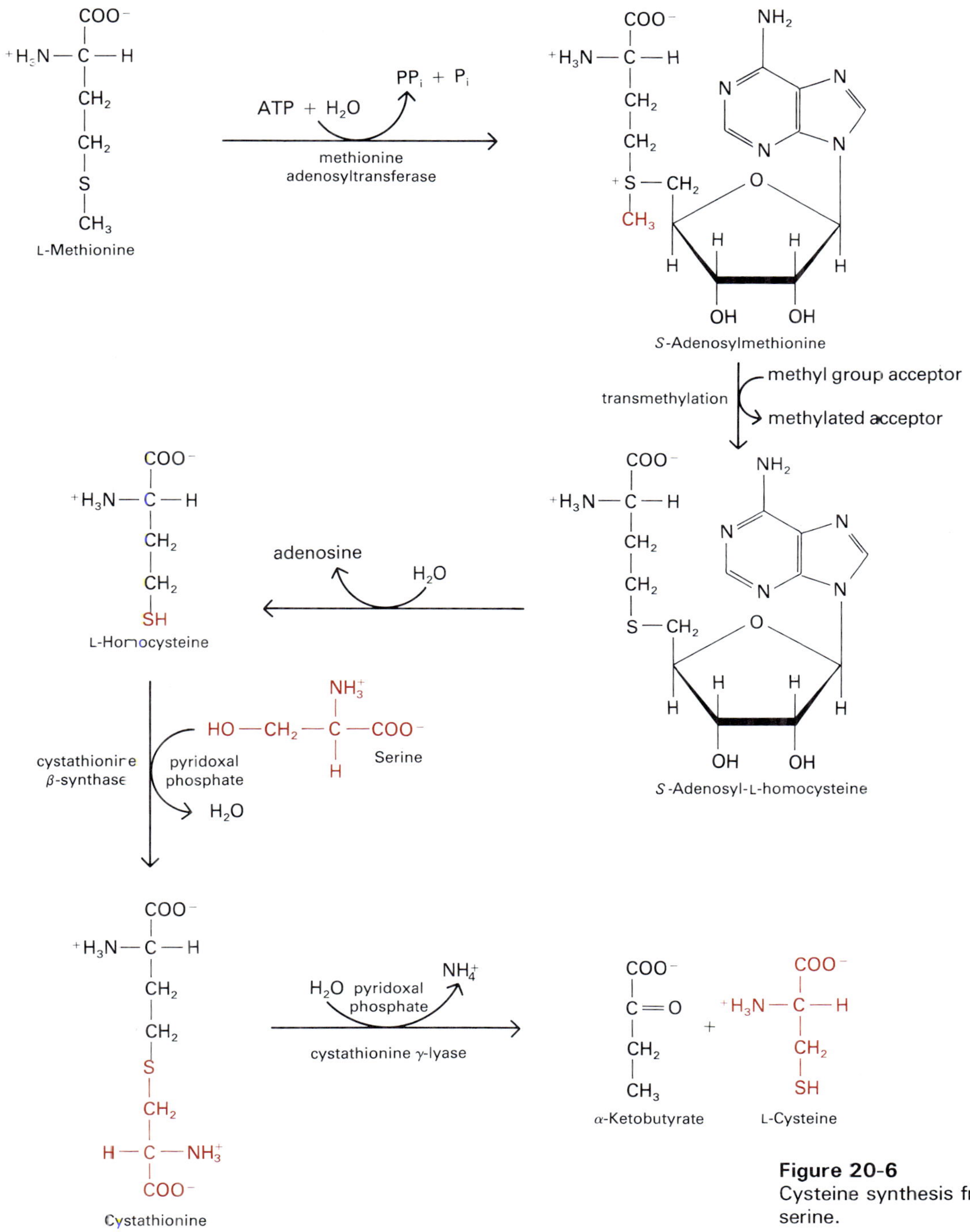

Figure 20-6
Cysteine synthesis from methionine and serine.

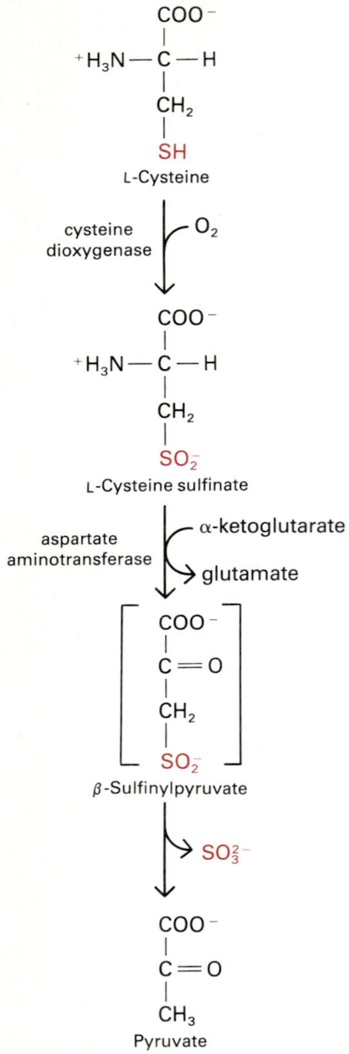

Figure 20-7
Oxidative catabolism of cysteine.

Figure 20-8
Cysteine catabolism by transamination.

In the next step, *serine* conjugates with homocysteine to form *cystathionine;* the reaction is catalyzed by *cystathionine β-synthase* and requires pyridoxal phosphate. The final step, catalyzed by *cystathionine γ-lyase,* brings about deamination and cleavage of cystathionine to produce *cysteine, α-ketobutyrate,* and NH_4^+. Cysteine, as a negative effector, is an allosteric inhibitor of cystathionine γ-lyase. With regard to cysteine synthesis, the amino acid's carbon atoms are derived from serine and its sulfur from methionine, i.e., the net effect of the pathway is to exchange the hydroxyl group of serine for the sulfhydryl of methionine.

The oxidative pathway shown in Figure 20-7 is the main catabolic route for cysteine in animals. Cysteine is first converted into *L-cysteine sulfinate* by oxidation of its sulfhydryl group to a *sulfinic acid* moiety. The reaction is catalyzed by *cysteine dioxygenase,* an enzyme that utilizes oxygen. Then, in a transamination reaction catalyzed by *aspartate aminotransferase,* cysteine sulfinate is converted into *pyruvate* and *sulfite* (SO_3^{2-}). Another mechanism by which cysteine is transformed into pyruvate is illustrated in Figure 20-8. In this catabolic scheme, cysteine is transaminated by *cysteine aminotransferase* to produce *β-mercaptopyruvate,* which is then desulfurated to yield *pyruvate* and H_2S.

Inborn Errors of Cysteine Synthesis

Two genetic defects (Mendelian recessive traits) associated with cysteine synthesis are *homocystinuria* and *cystathionuria,* and both inborn errors are characterized by mental retardation. In homocystinuria, cystathionine synthesis is defective and, as a result, high levels of *homocystine* (oxidized product of homocysteine) are excreted in the urine. In cystathionuria, there is an imbalance between cystathionine synthesis and degradation, i.e., either augmented synthesis or decreased catabolism, and the condition results in high levels of cystathionine being excreted in the urine. Because the normal concentrations of cystathionine in the brain

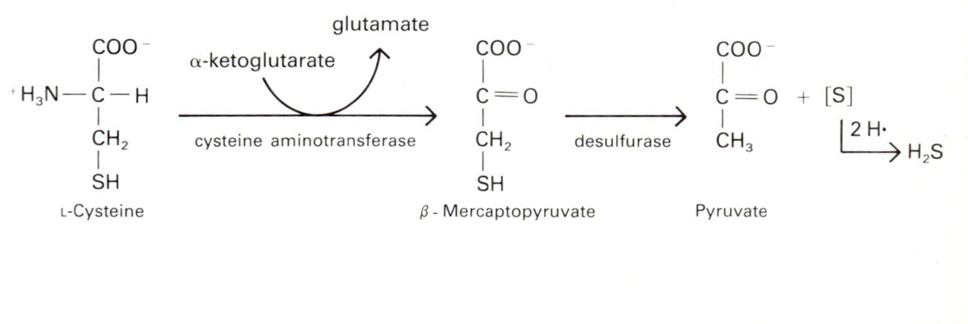

are unusually high (about 300 μg per g of brain tissue *vs.* about 7 μg per g of other tissues), it is suspected that cystathionine has some, as yet unknown, function(s) in the physiology of the nervous system. It is not surprising, therefore, that abnormally low or high levels of cystathionine could affect brain development and/or functions.

Proline Synthesis and Catabolism

As illustrated in Figure 20-9, proline is synthesized from *glutamate.* The initial step involves conversion of the γ-carboxyl group of glutamate into an aldehyde to produce *glutamate γ-semialdehyde.* The reaction, catalyzed by *glutamate γ-semialdehyde dehy-*

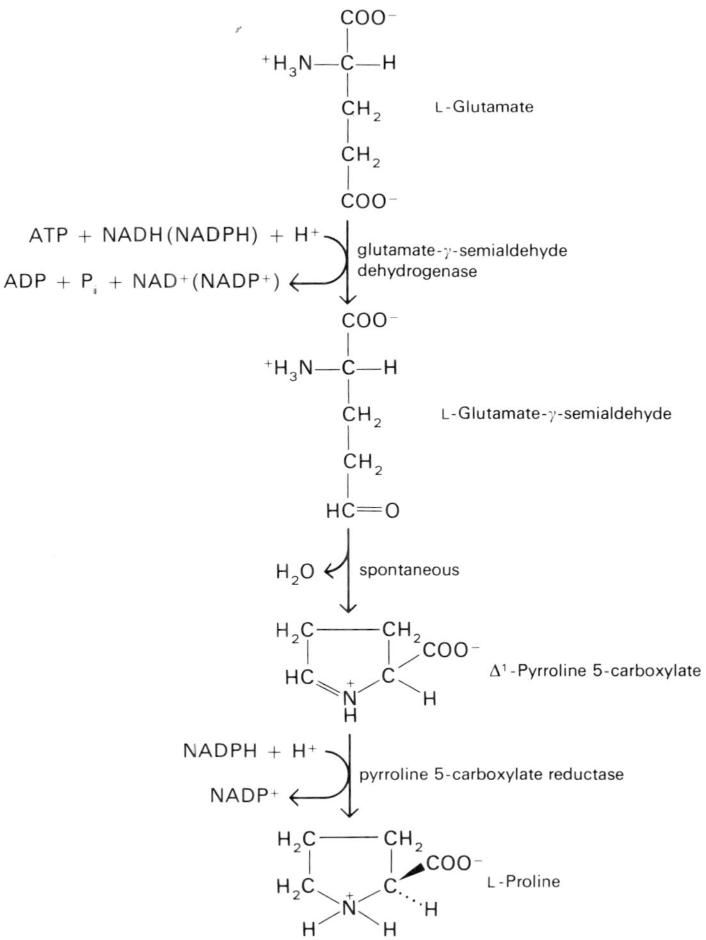

Figure 20-9
Proline synthesis.

drogenase, requires ATP and NADH (or NADPH); the carboxyl group is believed to be first phosphorylated, then reduced by NADH (NADPH). The next step is spontaneous and entails cyclization of glutamate semialdehyde (involving loss of a water molecule) to produce *Δ^1-pyrroline 5-carboxylate.* The last reaction is catalyzed by *pyrroline 5-carboxylate reductase* and uses NADPH to reduce Δ^1-pyrroline 5-carboxylate to proline.

For proline catabolism, the amino acid is reconverted into glutamate, which is then oxidatively deaminated by glutamate dehydrogenase (p. 309), to produce α-ketoglutarate, a TCA cycle intermediate. This production of glutamate from proline proceeds by a series of reactions that mimic reversal of the synthetic pathway depicted in Figure 20-9. Proline is first *oxidized* by mitochondrial *proline oxidase* to Δ^1-pyrroline 5-carboxylate, which is then *hydrolyzed* to produce L-glutamate-γ-semialdehyde. *Oxidation* of glutamate-γ-semialdehyde to glutamate completes the conversion.

Tyrosine Synthesis and Catabolism

When tyrosine was isolated from cheese in 1857, it was named for the Greek word *tyri,* meaning *cheese.*

Tyrosine is synthesized by the first step in phenylalanine catabolism, and thus the degradation pathway of phenylalanine (a required amino acid) illustrated in Figure 20-10 encompasses both synthesis and catabolism of the hydroxylated amino acid. Tyrosine production is catalyzed by *phenylalanine 4-monooxygenase* (also called *phenylalanine hydroxylase*) in a reaction that requires an unusual coenzyme, *tetrahydrobiopterin* (Figure 20-11). The role of NADPH in the synthesis is to reduce dihydro- to tetrahydrobiopterin. Tyrosine then loses its NH_2 group in a transamination reaction, catalyzed by *tyrosine aminotransferase,* to produce *4-hydroxyphenylpyruvate.* Next, a carbon atom is lost as CO_2 and a hydroxyl group is added to the phenyl ring to yield *homogentisate;* the reaction is catalyzed by *4-hydroxyphenylpyruvate dioxygenase.* What follows is the oxidative cleavage of the phenyl ring by *homogentisate 1,2-dioxygenase* to produce *4-maleylacetoacetate,* which is then isomerized by *4-maleylacetoacetate isomerase* to produce its *trans* isomer, *4-fumarylacetoacetate.* In the last step, catalyzed by *fumaryl acetoacetase,* 4-fumarylacetoacetate is cleaved to generate *fumarate* and *acetoacetate.* The latter product can subsequently be activated by *3-ketoacid CoA transferase* (with succinyl-CoA serving as the second substrate) to produce *acetoacetyl-CoA,* which is aerobically oxidizable.

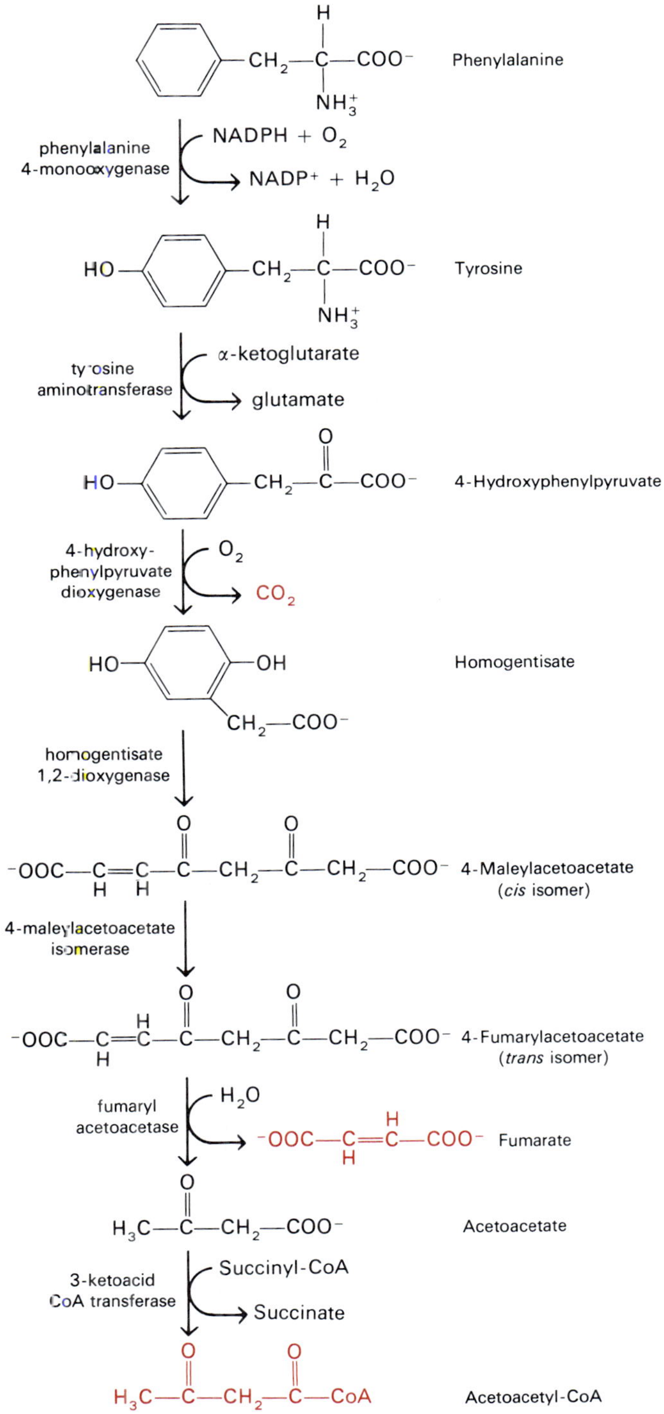

Figure 20-10
Catabolic pathway of phenylalanine and tyrosine.

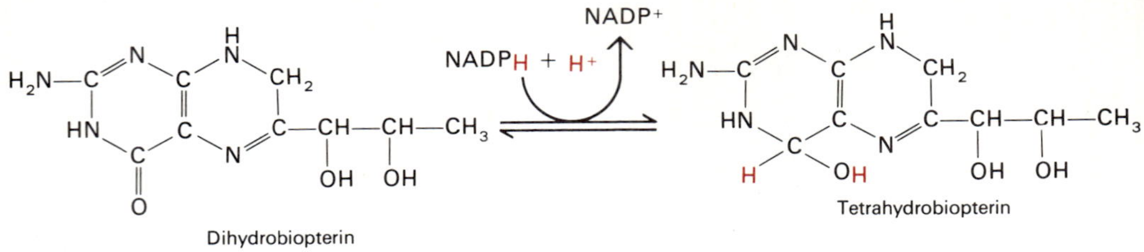

Figure 20-11
Dihydrobiopterin reductase reaction.

Inborn Errors of Phenylalanine and Tyrosine Metabolism

Alkaptonuria

The catabolic pathway of phenylalanine was one of the first studied by early biochemists and the first to be associated with an inborn error. At the turn of the century, Archibald Garrod, an English pediatrician, recognized that *alkaptonuria,* a clinical syndrome, showed a pattern of familial inheritance. At that time, when Gregor Mendel's genetic studies on garden peas were being rediscovered, Garrod astutely associated alkaptonuria with human genetics. A predominant clinical manifestation of alkaptonuria (a benign condition) is the darkening of urine upon its standing open to the atmosphere. The phenomenon is due to an accumulation in the urine of *homogentisate* (Figure 20-10), a compound that oxidizes to produce a dark polymer. Garrod, knowing that homogentisate was an intermediate of phenylalanine catabolism, postulated that alkaptonuria was the result of a genetic defect (Mendelian recessive trait), expressed as an individual's inability to cleave the phenyl ring of homogentisate to produce 4-maleylacetoacetate (Figure 20-10). Homogentisate accumulation in the urine, therefore, is the result of a biochemical block.

Garrod, often called the father of biochemical genetics, is an example of a scientist ahead of his time. Although Garrod had developed his hypothesis by 1902, it is of interest to note that the word *gene* was not coined until 1911 and the first enzyme was not crystallized until 1926.

In 1909 Garrod developed, in his book *Inborn Errors of Metabolism* (original use of the term), the proposal that alkaptonuria and certain other metabolic disorders are due to genetic defects, each of which results in the lack of the activity of a particular metabolic enzyme. In essense, he proposed the one gene-one enzyme hypothesis, which Beadle and Tatum later elucidated in 1940. Forty-five years later, liver biopsy analysis confirmed Garrod's diagnosis for the cause of alkaptonuria, *viz.,* alkaptonurics lack homogentisate 1,2-dioxygenase (Figure 20-10).

Albinism

Another inborn error associated with phenylalanine and tyrosine metabolism is *albinism,* an autosomal recessive trait. The biochemical defect involves *melanin* (the pigment that gives skin and

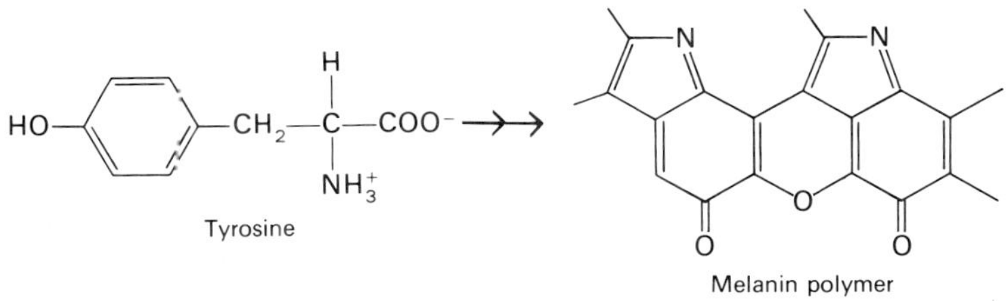

Figure 20-12
Production of melanin.

hair their color) production, for which tyrosine is the precursor (Figure 20-12). One type of albinism is believed to be due to a deficiency of tyrosinase (tyrosine 3-monooxygenase), an enzyme needed for melanin synthesis. A variation of albinism is noted in Siamese cats, whose distinctive pattern of pigmentation results from a *temperature-sensitive form* of tyrosinase. At regular body temperature, the enzyme is relatively inactive, i.e., very little pigment produced; however, in parts of the body, e.g., paws, tail, and ears, which have lower temperatures and/or are more exposed to the environment, the enzyme is active and consequently the fur is pigmented.

Phenylketonuria

Studies on *phenylketonuria (PKU),* like those on galactosemia, were instrumental in alerting the medical world to the serious physiological consequences of many inborn errors and in bringing genetics into the mainstream of medical research. The biochemical block in PKU is a deficiency of *phenylalanine 4-monooxygenase* activity (Figure 20-10), which results in impaired phenylalanine metabolism. Continued accumulation of phenylalanine in the blood leads to the amino acid's transamination to produce abnormal amounts of phenylpyruvate (Figure 20-13), which damages the developing brain of afflicted infants. Severe mental retardation (a mean IQ of 20) is a clinical feature of PKU. It is estimated that about 1 per cent of institutionalized children are phenylketonuric.

Very fair skin and light blonde hair are two characteristics of

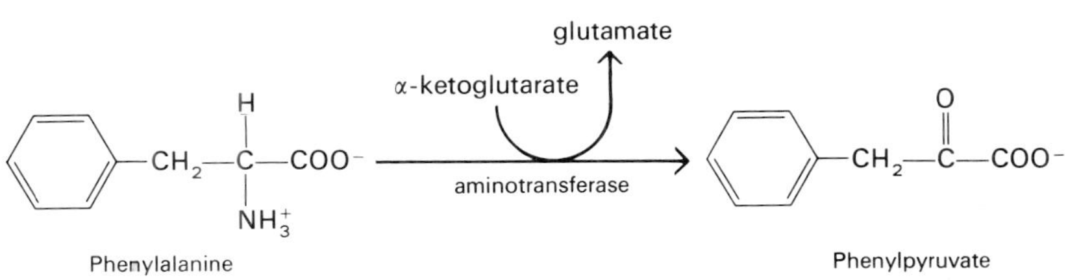

Figure 20-13
Production of phenylpyruvate.

PKU. At first approximation, it might be assumed that the inability to synthesize tyrosine, the precursor of melanin, is responsible for the observed lack of pigmentation. However, tyrosine is not lacking, since food proteins, e.g., casein of milk, furnish adequate amounts of the amino acid. Melanin production is impaired because high levels of phenylalanine effectively compete with tyrosine as a substrate, i.e., competitive inhibition.

The medical treatment of PKU centers on early diagnosis in order to place the afflicted infant on a low-phenylalanine diet and, thus, to avoid brain damage. Today, newborns are tested for PKU in several ways. One test involves exposing an infant's urine to $FeCl_3$, a compound that reacts with excreted phenylketone compounds (e.g., phenylpyruvate) to produce a greenish color. Another test is to assay a blood sample for phenylalanine 4-monooxygenase activity. A child diagnosed as a phenylketonuric is maintained on the restricted diet until at least age three, by which time brain development is complete. Although severe mental retardation can be prevented by this treatment, phenylketonurics exhibit other physical and/or emotional problems, many of which remain biochemically unexplained.

Medical and Genetic Studies on PKU

The numerous medical investigations on PKU (an autosomal recessive trait) reveal the complexities inherent in attempting to understand an inborn error thoroughly. One puzzling case was a clincally diagnosed phenylketonuric (mental retardation, high level of phenylalanine in the blood and of phenylketone compounds in the urine) who possessed normal levels of phenylalanine 4-monooxygenase activity. The anomaly was resolved when it was discovered that the deficient enzyme is *dihydrobiopterin reductase* (Figure 20-11) and that conversion of phenylalanine into tyrosine is blocked because the tetrahydrobiopterin coenzyme cannot be regenerated.

As more patients are being studied, it is becoming apparent that high blood levels of phenylalanine (10 to 20 mg per cent) are not necessarily associated with PKU. Many individuals are classified as *phenylalaninemics,* i.e., high levels of phenylalanine because of unknown reasons. A number of infants originally diagnosed as PKU (persistent high blood levels of phenylalanine) were observed, after a period of several months, to have normal blood levels of the amino acid ($\approx$ 1 mg per cent) when fed regular diets. This phenomenon is referred to as *transitory phenylalaninemia* and probably reflects an infant's late development of the enzyme.

It is now known that particular genetic makeups found in some families mask the classic PKU symptom of mental retardation. Indeed, evidence exists that 20 to 20 per cent of the phenylketonurics have near-normal or normal intelligence. In one family studied, although two male offspring were phenylketonuric (as determined by blood and urine analyses), one was severely mentally retarded and the other had normal intelligence. Further study established that although several members of the family were phenylketonuric, this defect was not expressed in an incapacitating manner. The answer was found to be linked to the aminotransferase (transaminase) that produces phenylpyruvate from phenylalanine. The unaffected phenylketonurics possess an altered aminotransferase, one that has a higher K_M for phenylalanine than does the normal enzyme. Hence, although phenylalanine blood levels are high in these people, they are never high enough (because of the altered affinity of the aminotransferase) to allow production of phenylpyruvate levels that threaten brain development. In genetic terminology, these individuals have a *supressor mutation,* a mutation (altered aminotransferase) that supresses the phenotypic expression(s) of another mutation (altered phenylalanine 4-monooxygenase).

These disguised cases of phenylketonuria, however, pose another serious problem. Children born to "normal" phenylketonuric mothers almost always suffer from mental impairments and often have other birth defects. The continuous high levels of phenylalanine and phenylketone compounds in the mother's blood are blamed for the damage to the developing fetus. Thus, continued research on PKU has revealed a previously unsuspected medical problem associated with the inborn error.

GLUCOGENIC AND KETOGENIC AMINO ACIDS

Based on their catabolic products, amino acids are classified as glucogenic or ketogenic. Those amino acids that generate *precursors of glucose,* e.g., pyruvate or a TCA cycle intermediate, are referred to as *glucogenic,* and those that are degraded to *ketone bodies,* e.g., acetoacetate, are called *ketogenic.* Overproduction of ketone bodies, as discussed in Chapter 3 (p. 37), is a major cause of acidosis. Table 20-1 lists the 20 amino acids according to their glucogenic or ketogenic status. Among the 11 nonessential amino acids discussed, the following are glucogenic: alanine (pyruvate), arginine (α-ketoglutarate), aspartate (oxaloacetate), asparagine (oxaloacetate), cysteine (pyruvate), glutamate (α-ketoglutarate),

Table 20-1
Classification of amino acids as glucogenic or ketogenic

Glucogenic	
Alanine	Histidine
Arginine	Methionine
Asparagine	Proline
Aspartate	Serine
Cysteine	Threonine
Glutamate	Tryptophan
Glutamine	Valine
Glycine	
Ketogenic	
Leucine	
Glucogenic and ketogenic	
Isoleucine	Phenylalanine
Lysine	Tyrosine

glutamine (α-ketoglutarate), glycine (pyruvate), proline (α-ketoglutarate), and serine (pyruvate). Note that the tenth amino acid, tyrosine (and its precursor phenylalanine), is both glucogenic and ketogenic because its catabolic products are fumarate (TCA cycle intermediate) and acetoacetate (ketone body). However, as described above, acetoacetate can be converted into acetoacetyl-CoA, which can then furnish acetyl-CoA to the TCA cycle (p. 301) for the production of ATP (but not of glucose, p. 309).

NITROGEN EXCRETION

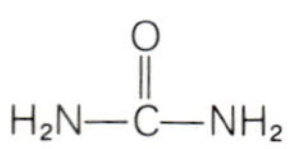

Urea

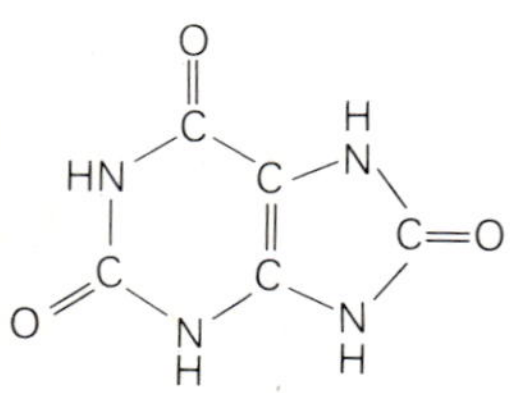

Uric acid (keto form)

NH_4^+
Ammonium ion

The three principal waste products of nitrogen excreted by animals are *NH_3, urea,* and *uric acid* (a purine). The major type of nitrogen compound excreted by an animal is believed to reflect a meaningful relationship between the toxicity and solubility of the molecule and the organism's environment. Most fish are *ammoniotelic:* they *excrete NH_3* which, although very toxic when concentrated, is not dangerous to them since it is very water soluble and rapidly dilutes in the aqueous environment. Terrestrial animals, because of the toxicity of NH_3, rely on either urea (much less toxic yet appreciably water soluble) or uric acid (nontoxic because of its water insolubility) as their major excretory nitrogen product. The English biochemist Joseph Needham postulated that the environment of the embryo is correlated with whether a terrestrial animal is *ureotelic (excretes urea)* or *uricotelic (excretes uric acid).* According to the proposal (which is of a teleological nature), birds and terrestrial reptiles are uricotelic and their embryos develop in eggs, which have only the limited amount of water needed for the embryos to develop and hatch. If NH_3 or urea were the excretory product, accumulation of either toxic compound would be lethal. On the other hand, uric acid is harmless to the developing embryo because its insolubility ensures that the nitrogen wastes will precipitate (in a small sac) in the egg. Terrestrial mammals are ureotelic, since the solubility of urea produced by their embryos assures its efficient and continuous removal by the mother's circulatory and excretory systems. For terrestrial mammals, urea spares the fetus and mother from the danger of ammonia toxicity and avoids problems that the insolubility of uric acid might create.

Many adult amphibia are also ureotelic. In the case of frogs, adults are ureotelic and aquatic tadpoles ammoniotelic. It is during metamorphosis that the tadpole begins to excrete urea rather than ammonia and, by the end of the transition, the newly developed frog is predominantly ureotelic. (These patterns of nitrogen excretion in animals are a classic and historically interesting example of *comparative biochemistry,* a field that examines phylogenetic schemes at a biochemical level.) Some animals excrete nitrogen in other forms; spiders excrete guanine instead of uric acid, and some fish form trimethylamine oxide. (In plant systems, asparagine and glutamine serve as the storage forms of nitrogen.)

Teleology is taken from the Greek word *telos,* meaning *end.* In biology the term refers to explaining biological phenomena in terms of utility or purpose, i.e., they are directed toward definite ends. If divine purpose is invoked, teleological reasoning is undistinguishable from vitalism.

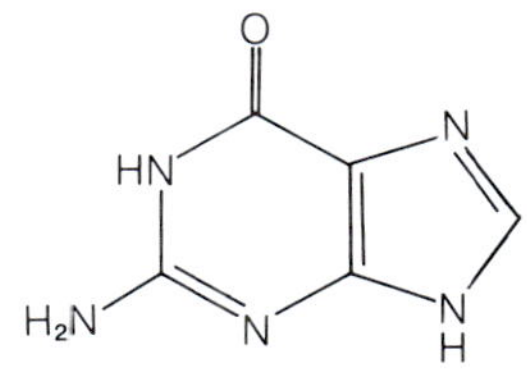

Guanine

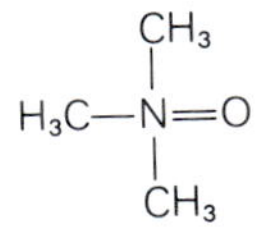

Trimethylamine oxide

THE UREA CYCLE

In ureotelic animals, synthesis of urea occurs via the urea cycle (Figure 20-14), a metabolic cycle proposed by Hans A. Krebs and Kurt Henseleit in 1932. Urea is produced in the liver, then carried by the blood to the kidneys for excretion. Because ureotelic vertebrates are exceedingly sensitive to NH_3 (which rapidly affects brain metabolism and causes convulsions), the cycle performs the vital function of continuously removing this toxic substance to avoid its accumulation in the body.

Ammonia (as ammonium ion) is a substrate for the synthesis of *carbamoyl phosphate* (Figure 20-15), which furnishes the first of two amino groups removed by urea synthesis. The reaction is catalyzed by *carbamoyl phosphate synthetase (ammonia)* in the mitochondrial matrix and also requires CO_2, 2 ATP, H_2O, and a positive effector, *N-acetylglutamate.* This form of carbamoyl phosphate synthetase is designated "(ammonia)" to distinguish it from the cytosolic synthetase that utilizes glutamine instead of ammonia as a substrate (p. 428). These reactions synthesizing carbamoyl

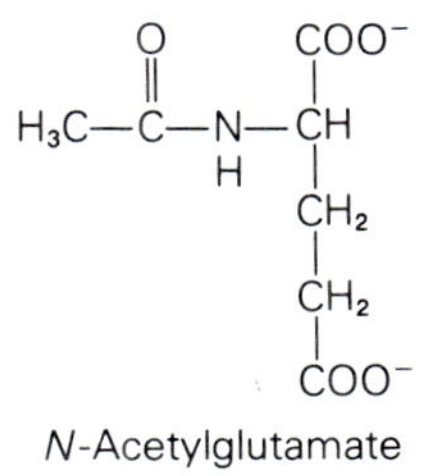

N-Acetylglutamate

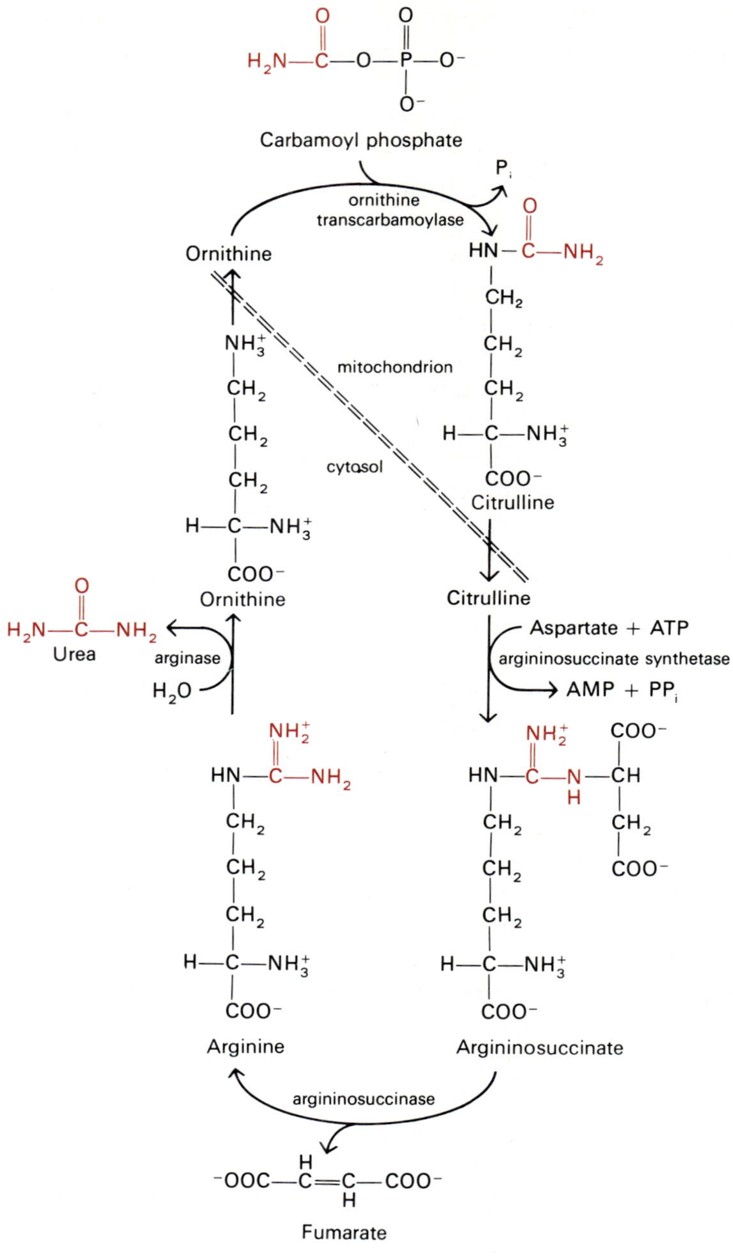

Figure 20-14
The urea cycle.

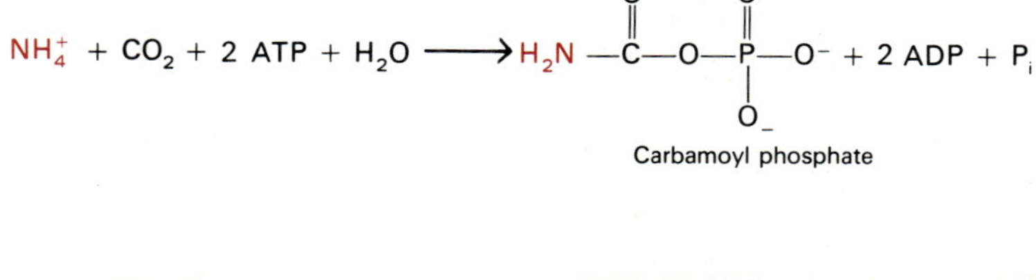

Figure 20-15
Carbamoyl phosphate synthetase (ammonia) reaction.

phosphate and those producing glutamate and glutamine (Chapter 16) are the three major biochemical routes for the incorporation of inorganic nitrogen into an organic form.

Urea synthesis (Figure 20-14) commences in the mitochondrial matrix with the condensation of L-*ornithine* and carbamoyl phosphate, catalyzed by *ornithine transcarbamoylase*, to produce L-*citrulline,* which enters into the cytosol, where the remainder of the reactions of the cycle occur. In the next two reactions, L-*aspartate* donates its amino group to L-citrulline. The first step, catalyzed by *argininosuccinate synthetase,* requires a pyrophosphate cleavage of ATP (followed by subsequent PP_i hydrolysis) to produce *argininosuccinate,* which, in the second step, is cleaved by *argininosuccinase* to yield L-*arginine* and *fumarate.* This is the enzymatic route by which organisms synthesize arginine. In ureotelic animals, which have large amounts of *arginase,* the cycle is completed by the synthesis of *urea* from arginine, a reaction that regenerates *ornithine.* The synthesis of urea, commencing with carbamoyl phosphate synthesis, requires the hydrolysis of four high-energy phosphate bonds (three ATP and one PP_i).

As diagrammed in Figure 20-16, a close metabolic relationship exists between the urea and the TCA cycles. The production of arginine in the urea cycle generates the TCA intermediate fumarate, which can be used to synthesize oxaloacetate. Transamination involving oxaloacetate and glutamate synthesizes aspartate, which can re-enter the urea cycle, and α-ketoglutarate, which can incorporate NH_3 (L-glutamate dehydrogenase) to regenerate glutamate. This particular set of interrelating reactions illustrates how the NH_2 group donated by aspartate to the urea cycle can readily be derived from free NH_3. Aspartate can also be produced from amino acid donors other than glutamate in transamination reactions.

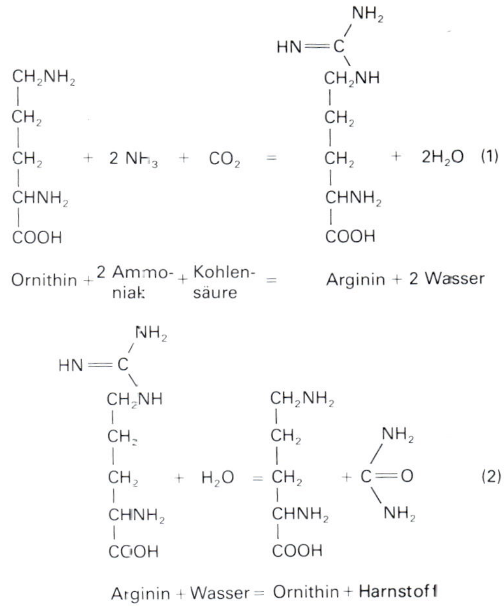

"The Ornithine Cycle"—first version of the urea cycle, published in 1932 by Krebs and Henseleit. At the time of its publication, Krebs was waiting for a sample of the novel and rare amino acid *citrulline,* which was isolated from watermelons (genus: *Citrullis*) in 1914 and characterized chemically in 1930. From his studies, he had deduced that citrulline should be an intermediate in the cycle. The 20-mg gift of citrulline was used by Krebs and Henseleit to confirm the amino acid's role in the cycle, which was reported in their second 1932 publication.

INBORN ERRORS INVOLVING UREA SYNTHESIS

Metabolic impairments caused by defective urea synthesis include known deficiencies of each of the enzymes directly associated with

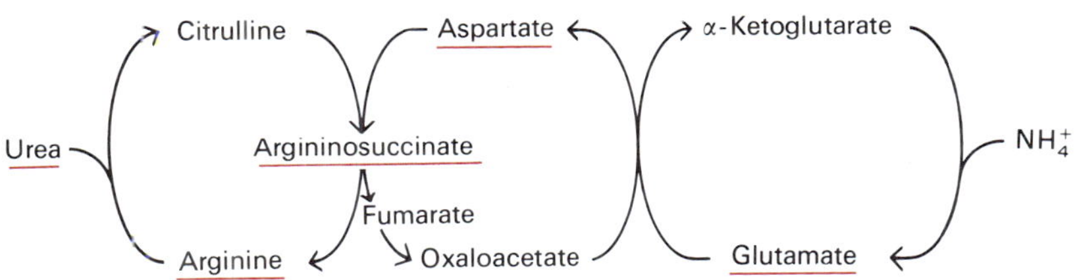

Figure 20-16
Metabolic interrelationship between the incorporation of the amino group of aspartate into urea and the incorporation of inorganic nitrogen. Biomolecules underlined in red trace the metabolic flow of incorporated nitrogen to urea.

the cycle. The disorders (and their deficient enzymes) are *hyperammonemia Type I* (carbamoyl phosphate synthetase) and *Type II* (ornithine transcarbamoylase), *citrullinemia* (argininosuccinate synthetase), *argininosuccinate acidurea* (argininosuccinase), and *hyperargininemia* (arginase). All share the tragedy of mental retardation. Convulsions and hepatomegaly are also clinical symptoms of argininosuccinate aciduria; coma, vomiting, lethargy, and irritability are characteristics of the other four inborn errors.

Medical therapy of these disorders relies on protein-restricted diets. Many afflicted individuals have an aversion for foods rich in proteins. Lowered protein intake diminishes the quantity of α-amino groups that need to be removed by urea synthesis. The α-keto acid analogues of essential amino acids can also be included in the diet, since they are converted by the body into needed amino acids (by accepting NH_2 groups in transamination reactions). This utilization of amino groups by the keto acids aids in alleviating NH_3 accumulation.

PURINE NUCLEOTIDE METABOLISM

Uricotelic animals rely on the *biosynthetic (anabolic) and catabolic pathways for adenine and guanine* to produce uric acid. Thus, uricotelic animals effectively use the pathway that synthesizes the purine nucleotides (needed for RNA and DNA synthesis) to serve also for nitrogen excretion. The catabolism of purines to uric acid, it should be noted, is not limited to uricotelic animals. Uric acid and other purine catabolites (p. 425) occur in most animal species; however, in ureotelic animals, they are not the major excretory form of nitrogen. Purine synthesis, in addition to its pertinent dual role in uricotelic animals, is an instructive example of biosynthesis because it aptly illustrates (1) the synthesis of a large biomolecule from smaller ones, (2) a biosynthetic use of cellular ATP, (3) different biochemical routes of nitrogen incorporation, and (4) regulatory mechanisms of biomolecular synthesis.

One of the major biosynthetic pathways elucidated in the 1950s was that for the production of purines, accomplished by John M. Buchanan and G. Robert Greenberg and their colleagues. The first breakthrough came when Buchanan identified the origin of the atoms of the purine molecule by chemical-degradation studies on uric acid obtained from pigeons (uricotelic organisms) fed isotopic compounds. As seen in Figure 20-17, the evidence obtained established that N-1 is derived from asparatate, N-3 and N-9 from

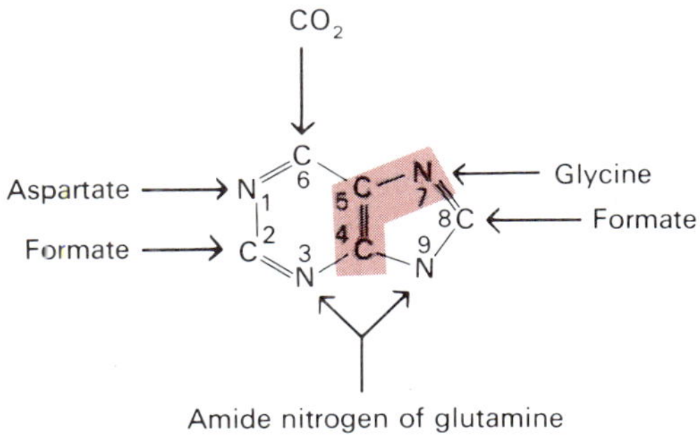

Figure 20-17
Precursors of the atoms of the purine molecule.

the amide nitrogen of glutamine, C-2 and C-8 from formate, C-6 from CO_2, and C-4, C-5, and N-7 from glycine. A second significant advance toward understanding the pathway was the discovery that purines are not synthesized as free nitrogenous bases, as previously believed, but in their nucleotide forms. Subsequent research succeeded in unraveling the following series of reactions by which the purine nucleotide molecule is synthesized.

Synthesis of 5-Phosphoribose 1-Pyrophosphate (PRPP)

To initiate the synthesis of purine, D-ribose 5′-phosphate (p. 290) first needs to be activated by pyrophosphorylation. In a reaction catalyzed by *ribosephosphate pyrophosphokinase* (Figure 20-18), a pyrophosphate group is transferred from ATP to C-1 of *α-D-ribose 5-phosphate* to produce *5-phospho-α-D-ribose 1-pyrophosphate,* abbreviated *PRPP,* and AMP. This kinase reaction is uncommon since it catalyzes the transfer of an intact pyrophosphate group. Because PRPP is also required for the synthesis of pyrimidine nucleotides, its production is catalyzed by an allosteric enzyme whose inhibition by *ADP* and *GDP* provides regulation of all nucleotide biosynthesis.

Figure 20-18
Ribosephosphate pyrophosphokinase reaction.

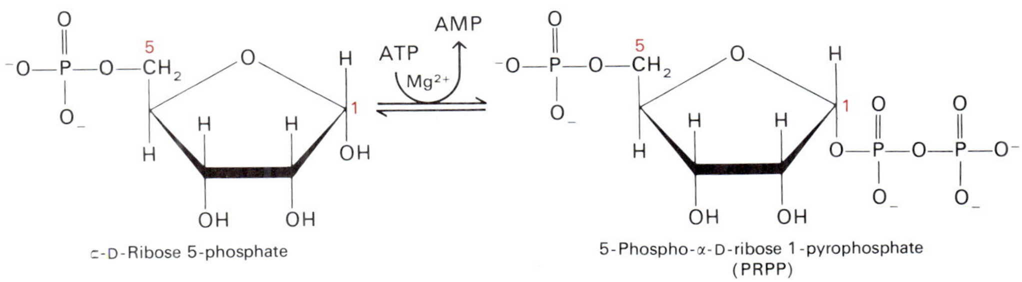

Synthesis of the Purine Molecule

Figure 20-19 depicts the ten sequential enzymatic reactions that utilize the precursors needed to furnish the six carbon and three nitrogen atoms of a purine molecule. As shown, the first purine nucleotide synthesized is *inosine 5′-phosphate (IMP),* or *inosinic acid,* which has *hypoxanthine* as its nitrogenous base; IMP serves as the precursor for the nucleotides AMP and GMP. The following presentation is a brief summary of each of the steps involved in the production of IMP from PRPP, with minimal attention given to the names of enzymes and intermediates.

The first atom incorporated into the purine structure is *N-9,* donated by *glutamine.* In the reaction, the amino group of the amide moiety of glutamine replaces the pyrophosphate group of C-1 of PRPP, and the substitution is driven by the hydrolysis of the product PP_i. Note that the reaction results in an inversion of the anomeric C-1 of ribose to a β configuration, which is maintained throughout the subsequent formation of the nucleotide. This first committed step of the pathway is a second site of regulation, since the enzyme is inhibited by purine nucleotides (p. 422). Allosteric control of the first enzyme of a biosynthetic pathway is a general feature of metabolic regulation.

The second reaction results in the incorporation of the nitrogen and two carbon atoms of *glycine* (as *C-4, C-5,* and *N-7* of the purine molecule) by amide bond formation between the carboxyl group of the amino acid and the amino sugar. The reaction requires the energy derived from the hydrolysis of ATP to ADP and P_i. In the third step, a *formyl group* (C_1 *unit*), furnished by the N^{10}-*formyl-coenzyme derivative* of *tetrahydrofolate* (FH_4), is attached to the free amino group of the substrate to incorporate *C-8.* Another nitrogen atom *(N-3)* is then introduced by the fourth reaction, with *glutamine* serving as an amino donor (its amide nitrogen). Unlike the first step, which also utilizes glutamine but relies on the energy furnished by PP_i hydrolysis, this incorporation depends on ATP hydrolysis to drive the reaction.

The fifth reaction catalyzes not an incorporation but, instead, the removal of a molecule of water to form the *imidazole ring portion* of the purine structure, yielding *5′-phosphoribosyl-5-aminoimidazole.* (Note that the numbering system of this and subsequent

---

Figure 20-19
The ten enzymatic steps involved in the biosynthesis of inosinic acid (IMP), commencing with 5-phospho-α-D-ribose 1-pyrophosphate (PRPP).

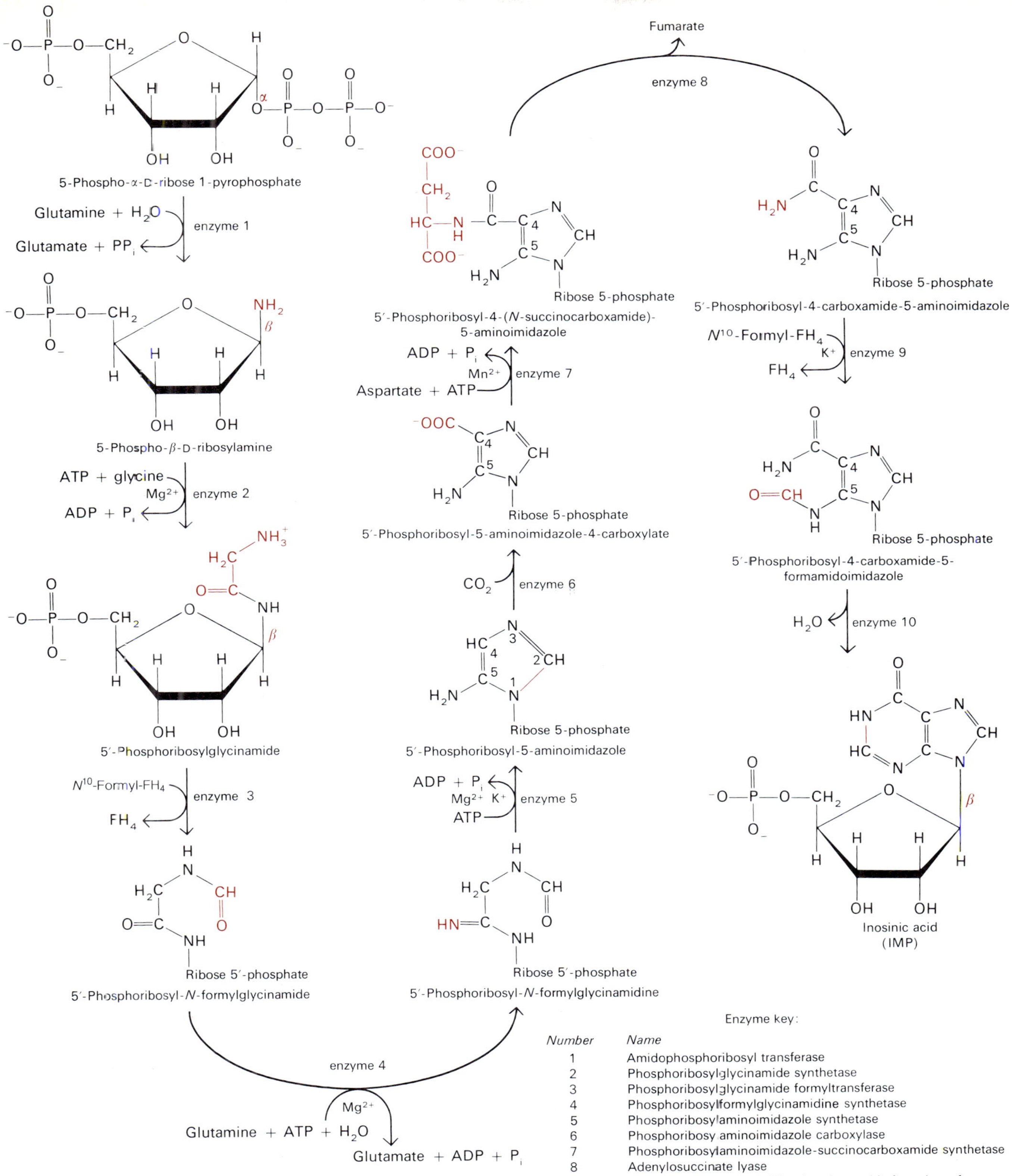

Enzyme key:

Number	*Name*
1	Amidophosphoribosyl transferase
2	Phosphoribosylglycinamide synthetase
3	Phosphoribosylglycinamide formyltransferase
4	Phosphoribosylformylglycinamidine synthetase
5	Phosphoribosylaminoimidazole synthetase
6	Phosphoribosylaminoimidazole carboxylase
7	Phosphoribosylaminoimidazole-succinocarboxamide synthetase
8	Adenylosuccinate lyase
9	Phosphoribosylaminoimidazolecarboxamide formyltransferase
10	IMP cyclohydrolase

intermediates is that for an imidazole.) The ring closure requires ATP hydrolysis as an energy source. The sixth step is a carboxylation reaction that incorporates CO_2 as *C-6* of the purine by its attachment to C-4 of the imidazole ring. In the seventh and eight steps, an amino group *(N-1)* is bonded to the newly attached carbon atom. The amino group is furnished by aspartate in a two-step process. The first step, with an expenditure of an ATP, catalyzes the attachment of aspartate through its amino group (in an amide linkage) to the intermediate; the second step eliminates fumarate, leaving the amino group of aspartate incorporated in the nucleotide intermediate. Thus, unlike the transfer of nitrogen by glutamine in previous reactions (steps 1 and 4), nitrogen donation by aspartate is not accomplished in a single step.

Figure 20-20
AMP synthesis from IMP.

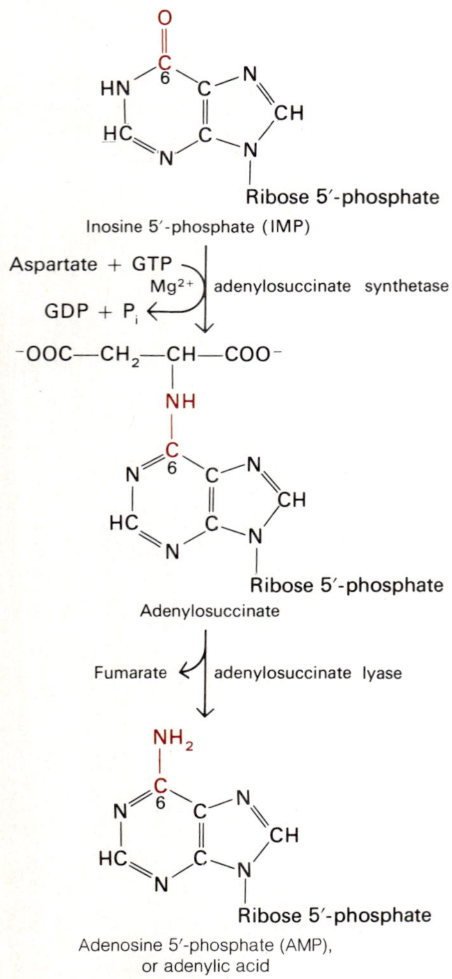

The ninth step incorporates the last carbon atom *(C-2)* into the *pyrimidine portion* of the purine ring, and the C_1 unit is derived from *N^{10}-formyl-FH_4*. The last step closes the pyrimidine structure by elimination of water to produce the purine nucleotide *inosine 5′-phosphate (IMP),* or *inosinic acid.* The closure of the pyrimidine ring, unlike that of the imidazole ring, does not require concomitant ATP hydrolysis.

AMP and GMP Synthesis

The conversion of IMP into *adenosine 5′-phosphate (AMP),* or *adenylic acid,* involves the transfer of the amino group of aspartate to C-6 of the purine and, as previously noted for the addition of N-1 of the purine molecule, the process requires two steps (Figure 20-20). The addition of aspartate to C-6 of IMP utilizes the hydrolysis of *GTP,* and the subsequent removal of fumarate produces AMP. One of the enzymes, *adenylosuccinate lyase,* is the same one that incorporates the nitrogen of aspartate in the synthesis of IMP (step 8).

To synthesize *guanosine 5′-phosphate (GMP),* or *guanylic acid,* from IMP, C-2 is first oxidized, then aminated (Figure 20-21). In the first reaction, catalyzed by *IMP dehydrogenase,* the nucleotide product is *xanthosine 5′-phosphate (xanthylic acid),* which has *xanthine* as its base. NAD^+ is the coenzyme for the reaction. The amination of xanthosine 5′-phosphate to produce GMP is catalyzed by *GMP synthetase* and requires glutamine as the amino donor and, again, a cleavage of ATP to AMP and PP_i. The subsequent production of the purine di- and triphosphonucleosides from AMP and GMP utilizes the enzymes *nucleoside mono-* and *diphosphate kinases,* which catalyze the transfer of phosphate groups in nucleotides (p. 262).

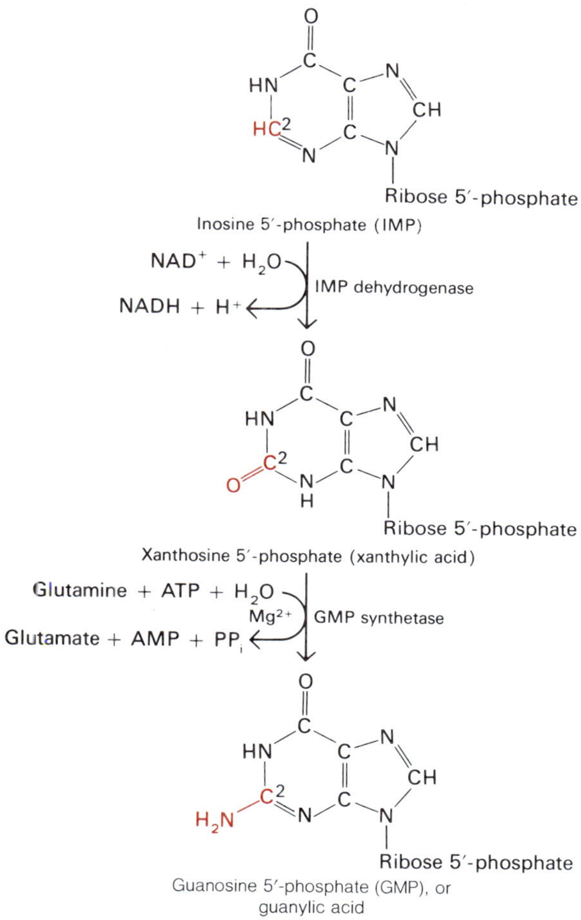

Figure 20-21
GMP synthesis from IMP.

Energy Requirements for AMP and GMP Synthesis

The synthesis of purine nucleotides is an example of how some of the chemical energy produced by an organism is used to synthesize the biomolecules needed for life processes. To produce IMP from D-ribose 5-phosphate (a pathway that represents a step-by-step synthesis involving covalent-bond formation), the hydrolysis of six high-energy bonds (five ATP and one PP_i) is required. The synthesis of AMP from IMP utilizes an additional nucleoside triphosphate (GTP); hence, a *total of seven high-energy phosphate groups* are needed for *AMP synthesis from ribose 5-phosphate.* For the comparable *synthesis of GMP,* the *total required is eight* because the amination of xanthosine 5′-phosphate requires the hydrolysis of two high-energy bonds (ATP and PP_i). An anaerobic organism, generating two ATP per glucose molecule oxidized, needs to oxidize four glucose molecules to produce an amount of chemical energy equivalent to that required to synthesize AMP or GMP from ribose 5-phosphate. The aerobic production of 36 or 38 ATP per glucose molecule suffices to provide the energy requirements for the synthesis of about five AMP or four GMP. The production of purine nucleotides also illustrates the energy costs for the production of uric acid by uricotelic organisms.

Regulation of Purine Nucleotide Biosynthesis

Studies on regulation of purine biosynthesis in *Escherichia coli* reveal a system of controls which operates at different steps in the pathway (Figure 20-22). This seemingly complex control mechanism is best understood if viewed as regulation at two particular levels of synthesis, the first at the *initial stages* of synthesis and the second at the *branch reactions* leading to the synthesis of AMP and GMP from IMP.

In the early stages of synthesis, as stated previously, allosteric inhibition of the production of PRPP from D-ribose 5-phosphate

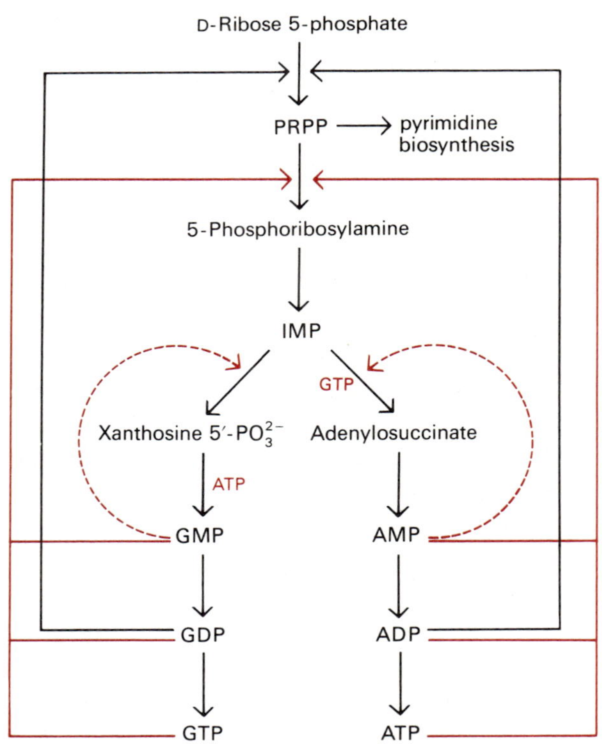

Figure 20-22
Regulation of purine nucleotide biosynthesis in *Escherichia coli.*

(Figure 20-18) by ADP and GDP constitutes a control over the synthesis of both purine and pyrimidine nucleotides. Specific regulation of purine nucleotide synthesis is effected by inhibition of the production of 5-phosphoribosylamine (Figure 20-19) by all nucleotides of adenine and guanine. The amidotransferase that catalyzes the reaction is a *bivalent regulatory enzyme,* meaning that it possesses separate allosteric sites for the two classes of purine nucleotides. Hence, although each of the two types of nucleotides inhibits enzymatic activity, maximum inhibition is observed when high levels of both adenine and guanine nucleotides are present, i.e., inhibition is cumulative or additive.

Each of the two pathways leading from IMP to the synthesis of either AMP or GMP is allosterically controlled by its product. High levels of GMP and AMP inhibit the production of xanthosine 5′-phosphate and adenylosuccinate, respectively. An interesting mode of nonallosteric control, which also affects the synthesis of AMP and GMP, is the requirement of ATP for GMP production and of GTP for AMP synthesis (Figure 20-22). Because of these particular requirements, excess ATP stimulates GMP synthesis and, conversely, GTP accelerates AMP production, i.e., there is a balanced rate of synthesis. Too, at the nucleotide level, neither AMP nor GMP synthesis from IMP depends directly on its homologous triphosphonucleoside, e.g., AMP synthesis does not require ATP hydrolysis. Together, the two control systems constitute a mechanism whereby the rate of synthesis of AMP and GMP from IMP can be increased, decreased, or maintained—as dictated by cellular conditions.

Catabolism of Purine Nucleotides

Through the dynamics of metabolism, purine nucleotides are continuously being synthesized and catabolized, with the greater percentage of the degradative products being metabolically reutilized and the remainder being excreted. Generally, *nucleotidases* degrade *nucleotides* to *nucleosides* by hydrolysis, and then *nucleosidases* catalyze the phosphorylytic cleavage of *nucleosides* to *free bases* and *ribose 1-phosphate:*

$$\text{Base-ribose-PO}_3^{2-} + H_2O \rightarrow \text{base-ribose} + P_i$$

$$\text{Base-ribose} + P_i \rightarrow \text{base} + \text{ribose 1-PO}_3^{2-}$$

There are additional pathways for AMP catabolism (Figure 20-23). Further degradation of the purines, regardless of the route, leads

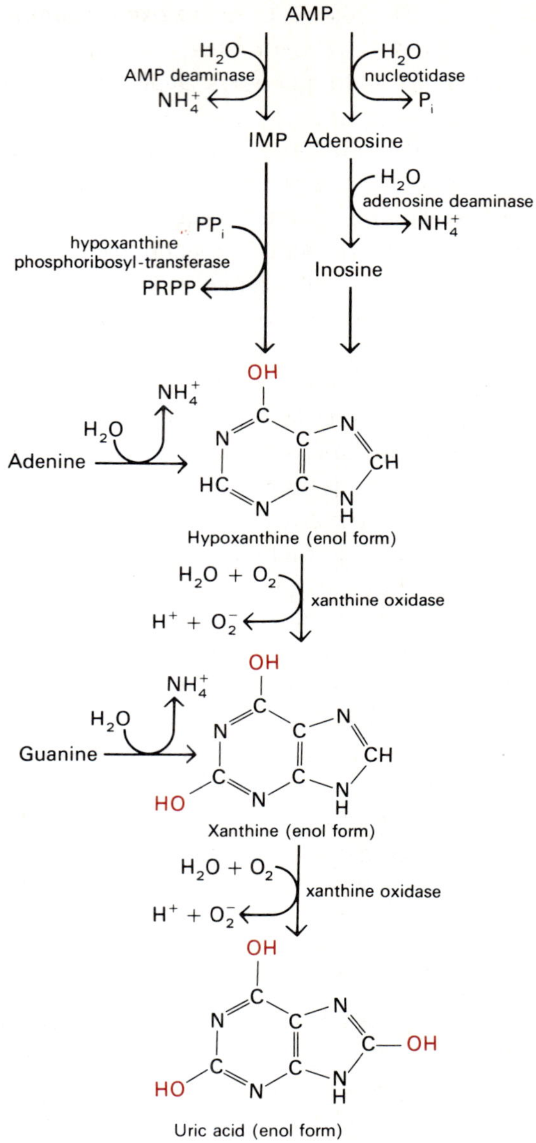

Figure 20-23
Major steps in the production of uric acid.

to the production of *uric acid* (Figure 20-23), which is the end product of purine catabolism in humans, primates, and some reptiles.

The production of uric acid from hypoxanthine or xanthine is catalyzed by *xanthine oxidase,* a complex flavoprotein containing iron and molybdenum. The *superoxide radicals,* O_2^-, generated by the reactions are rapidly converted, by *superoxide dismutase,* into O_2 and *hydrogen peroxide,* H_2O_2, with the latter then being decomposed to H_2O and ½ O_2 by *catalase.* Superoxide dismutase and catalase are critically important since O_2^- and H_2O_2 being very reactive molecules, are extremely dangerous to life processes (e.g., oxidative inactivation of sensitive enzymes) if allowed to accumulate.

For the majority of animal species, the degradation of purines does not terminate with uric acid production but proceeds further (Figure 20-24). *Allantoin* is excreted by most mammals and by turtles, mollusks, and certain reptiles. Among species of fish, *allantoate* and *urea* are the end products of purine metabolism. Most amphibia excrete urea, and certain marine invertebrates degrade purines to NH_4^+ and CO_2. Apparently, during the course of evolution, primates lost the genetic information (DNA) to code for the enzymes needed to degrade uric acid.

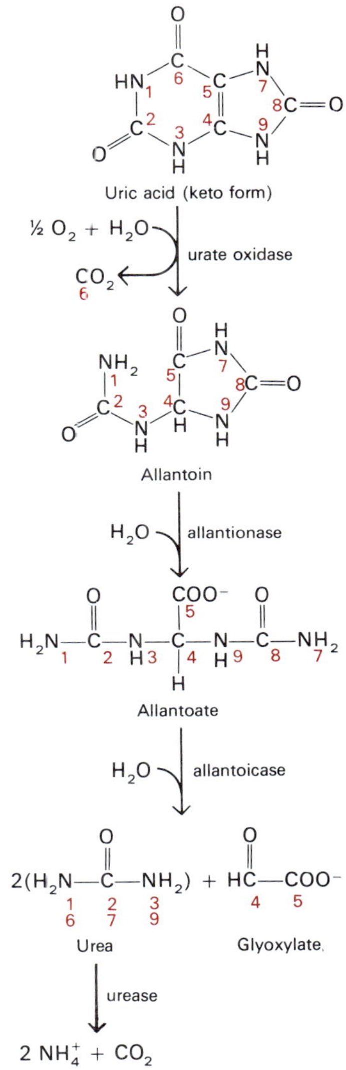

Figure 20-24
Metabolic degradation of uric acid.

Disorders Involving Purine Metabolism

Gout

One of the oldest known afflications of humans, *gout,* is the result of imparied purine metabolism. The underlying cause of gout is *overproduction of uric acid.* Whereas the water insolubility of uric acid may be advantageous to a developing embryo in an egg, the same property is responsible for the clinical symptoms of gout. Excess production of uric acid results in circulatory levels of *monosodium urate* (physiological form), which tends to precipitate in the joints of the body's digital regions, where the normal temperature (a degree or two lower than 37° C) enhances the salt's insolubility. Gout is characterized by acute and painful arthritis and sometimes *tophi,* deposits of sodium urate and other material, which grossly disfigure hands and feet. Kidney stones of sodium urate can be another serious medical consequence of gout.

Dietary restrictions are imposed on gouty patients, with special emphasis on avoiding foods high in nucleic acids (to decrease purine intake). The old adage that gout is "a rich person's disease" stems from the fact that the diets of the affluent in the past were high in meats (nucleic acids) and alcohol (which aggravates the gouty condition), in contrast to a poor person's diet of cereals and grains (low in nucleic acids). *Colchicine* (Figure 20-25) is one drug that successfully controls gout; another effective drug is *allopurinol* (Figure 20-25), which acts as a *competitive inhibitor of xanthine oxidase* to reduce the production of uric acid.

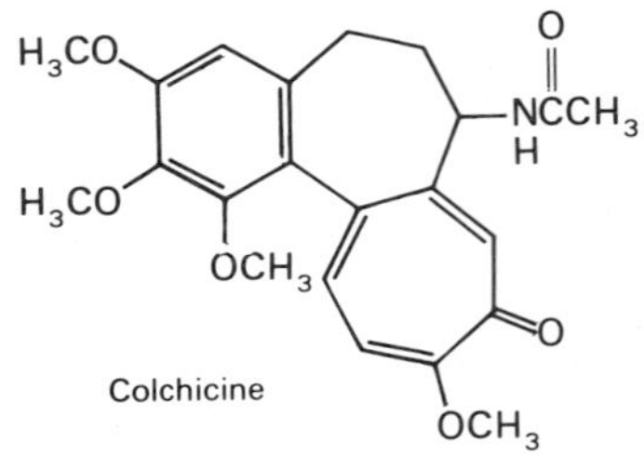

Colchicine

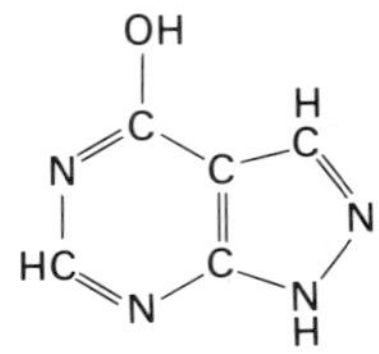

Allopurinol

Figure 20-25
Two compounds used to treat gout.

Molecular Basis of Gout

Although an excess production of uric acid is the underlying cause of gout, the biochemical basis for this overproduction is often not known, since the disease can be a manifestation of numerous metabolic disorders (most of which are unknown). Thus, in spite of the impressive medical advances made toward understanding gout as an inborn error of metabolism, much still remains to be learned about the disease.

One molecular basis for gout is believed to be an overproduction of purine nucleotides due to inborn errors that affect the allosteric regulation of the amidotransferase (Figure 20-19) and, therefore, that abolish the controlled production of purines. The consequence of this loss of normal control is an excessive synthesis of uric acid. This type of gout is an example of an inborn error that *alters the normal regulatory mechanism of a metabolic pathway.* The following two inborn errors (glucose 6-phosphatase deficiency and Lesch-Nyham syndrome) are not caused by defects in the purine pathway *per se,* but illustrate why gout is a difficult medical problem to comprehend fully on a molecular basis and, more generally, the interdependence of metabolic pathways.

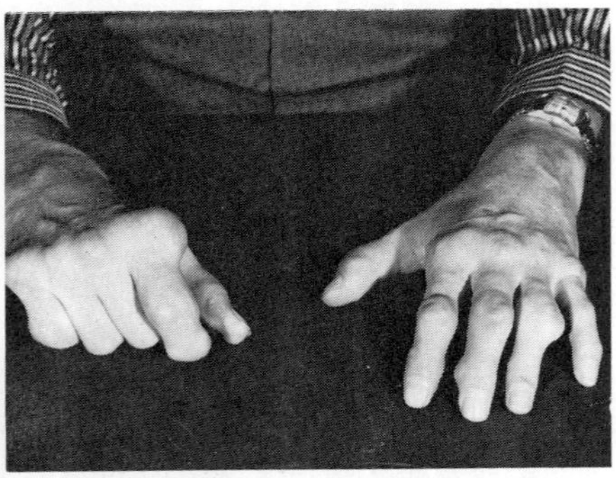

Chronic gouty arthritis of the hands. Note that extensive deposits of urate have created joint enlargements and the large soft tissue tophi. (Courtesy, National Institute of Arthritis, Metabolism, and Digestive Diseases)

Gout and Glucose 6-Phosphatase Deficiency

Among the inborn errors that affect carbohydrate metabolism is glucose 6-phosphatase deficiency (p. 288), and one of the clinical symptoms noted in some afflicted individuals is gout. Although this relationship between carbohydrate and purine metabolism may appear puzzling, it is explainable by the metabolic interrelatedness of the two classes of biomolecules. The relationship also illustrates the complexities of metabolism that must always be considered when physiological conditions are being studied. An inability to desphosphorylate glucose 6-phosphate results in an accumulation of this phosphorylated carbohydrate, some of which is metabolically shunted into the pentose phosphate pathway (Chapter 15), resulting in increased production of D-ribose 5-phosphate. The higher levels of ribose 5-phosphate promote PRPP synthesis (p. 417), which, in turn, stimulates production of purine nucleotides. Enhanced rates of synthesis of purines result in excessive production of uric acid, which explains why some glucose 6-phosphatase-deficient patients develop gout. The key point in this biochemical analysis is that the primary biochemical lesion in these patients is over 20 enzymatic steps removed from the reaction producing the uric acid that causes the gouty condition.

Lesch-Nyhan Syndrome

One of the most biochemically interesting yet bizarre inborn errors is the Lesch-Nyhan syndrome which, until 1964, was diagnosed as a rare form of cerebral palsy. The clinical symptoms include muscle spasticity, mental retardataion, and a compulsion for self-mutilation, e.g., to bite away the lips, tongue, and fingertips. To avoid permanent disfigurement, afflicted children must wear arm restraints to prevent them from biting off their fingertips. In severe cases, removal of teeth is required to prevent loss of parts of the lips and tongue. Although the children often test as mentally retarded by standardized tests, this condition is not often apparent since they are generally very cheery and have bright personalities.

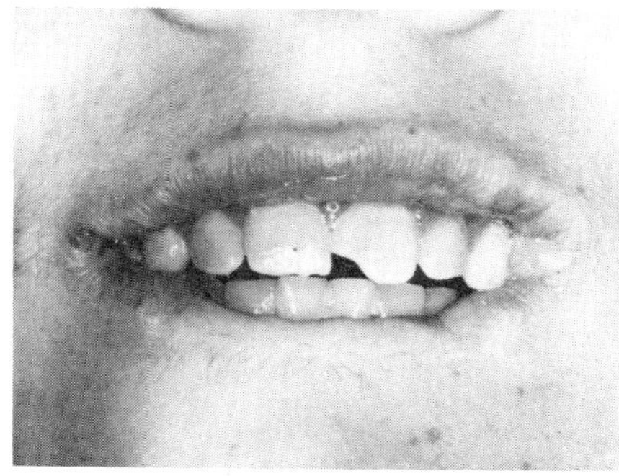

Self-mutilation of the lips of a child with Lesch–Nyhan syndrome. (Courtesy of Dr. J. Edwin Seegmiller and by permission of HP Publishing Co., Inc. *Medical Genetics,* McKusick and Claiborne, eds., 1973)

The *Lesch-Nyhan syndrome* is a *recessive X-linked trait* and therefore is observed in male offspring of women heterozygous for the defect. The condition was recognized as a defect in purine metabolism when uric acid crystals were noted in a patient's urine. The Lesch-Nyhan patient characteristically excretes more uric acid ($\approx$50 mg/kg body wt/24 hr) than do gouty individuals ($\approx$9 mg/kg body wt/24 hr). The deficient enzyme activity associated with purine metabolism is *hypoxanthine-guanine phosphoribosyltransferase (HGPRT),* which catalyzes the *synthesis of IMP (or GMP) from hypoxanthine (or guanine) and PRPP* (Figure 20-26). The enzyme has been known for years as a *salvage enzyme,* one that aids (by direct nucleotide synthesis) in the conservation of about 90 per cent of the free purine bases metabolically produced. Because of the severe and unusual medical aspects of the syndrome, biochemists now realize that HGPRT-catalyzed reactions have physiological importances that hitherto were unrecognized.

The relationship between HGPRT and brain metabolism (cerebral palsy) is still unclear. With regard to excess production of uric acid, it is believed that the salvage role of the enzyme is especially important for the maintenance of appropriate intracellular levels of PRPP and that a deficiency of HGPRT activity results in elevated concentrations of PRPP, which stimulates purine synthesis. One of

Figure 20-26
Hypoxanthine-guanine phosphoribosyltransferase (HGPRT) reaction.

$$\text{Hypoxanthine} + \text{PRPP} \rightleftharpoons \text{IMP} + \text{PP}_i$$

$$\text{Guanine} + \text{PRPP} \rightleftharpoons \text{GMP} + \text{PP}_i$$

the most revealing facts, though, is that the inborn error results in a *specific form of aberrant behavior.* This relationship between an inborn error and an emotional problem raises the possibility that other types of psychological conditions may be due to biochemical blocks. The Lesch-Nyhan syndrome also illustrates how limited may be the present understanding of the true *in vivo* physiological functioning of many known enzymes.

PYRIMIDINE NUCLEOTIDE METABOLISM

Synthesis

Carbamoyl phosphate, which is required for the urea cycle (p. 413), is also an initial precursor for pyrimidine synthesis. However, the cytosolic reaction producing carbamoyl phosphate for pyrimidine production is distinct from that catalyzed by mitochondrial carbamoyl phosphate synthetase (ammonia) because the enzyme requires glutamine, and not NH_3, as its nitrogenous substrate. The latter enzyme, *carbamoyl phosphate synthetase (glutamine),* catalyzes the following reaction:

$$\text{Glutamine} + HCO_3^- + 2\ \text{ATP} + H_2O \rightarrow$$
$$\text{carbamoyl phosphate} + \text{glutamate} + 2\ \text{ADP} + 2\ P_i$$

As illustrated in Figure 20-27, the first step is catalyzed by *aspartate transcarbamoylase* and, in the reaction, carbamoyl phosphate condenses with *aspartate* to produce N-*carbamoylaspartate.* The next reaction, catalyzed by *dihydroorotase,* involves a dehydration and brings about ring closure to yield *L-dihydroorotate,* which is then reduced by *dihydroorotate dehydrogenase* to *orotate.* At this stage in pyrimidine nucleotide production, a *phosphoribosyl* moiety, derived from 5-phosphoribose 1-pyrophosphate (PRPP), is added to orotate to generate the nucleotide *orotidine 5′-phosphate* (Figure 20-28). The reaction is catalyzed by *orotate phosphoribosyl transferase.* Once orotidine 5′-phosphate is produced, it is decarboxylated by *oritidine 5′-phosphate decarboxylase* to yield *uridine 5′-monophosphate* (UMP). Note that whereas *de novo* synthesis of the purine molecule occurs with the phosphoribosyl group attached (Figure 20-19), the pyrimidine molecule is synthesized prior to its transformation into a nucleotide structure.

Cytosolic synthesis of uridine 5′-monophosphate is an interesting metabolic phenomenon because the fourth reaction of the six-step process occurs in mitochondria. As depicted in Figure 20-29,

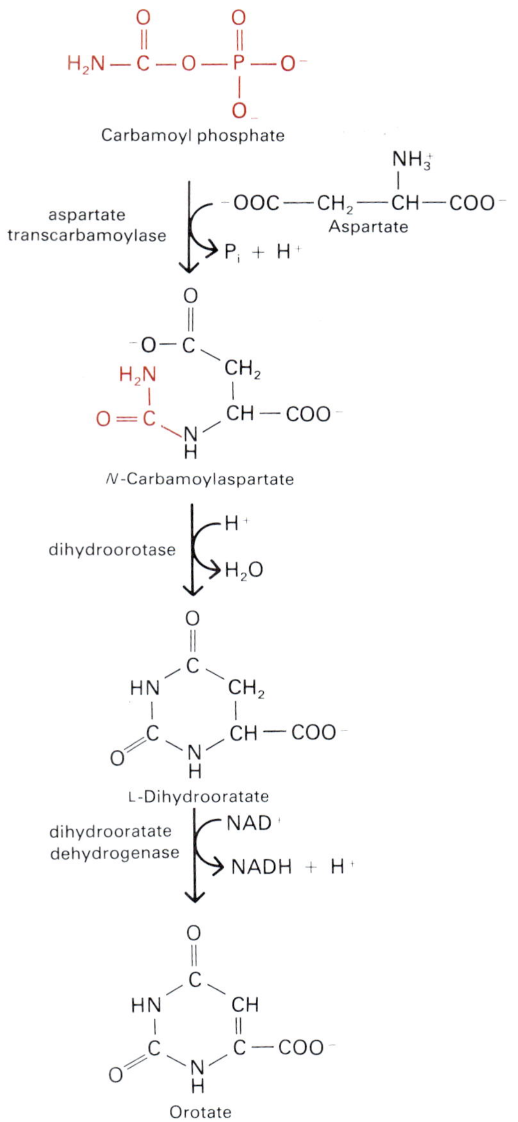

Figure 20-27
Orotate synthesis.

the first three enzymatic activities of pyrimidine synthesis are contained in a multienzymic polypeptide (MW ≈ 200,000), a structural feature reminiscent of that of palmitate synthetase (p. 370). Dihydroorotate, which is produced by the polypeptide, enters into the mitochondria, where its conversion into orotate is catalyzed by dihydroorotate dehydrogenase, an enzyme associated with the outer surface of the inner mitochondrial membrane. Because of the

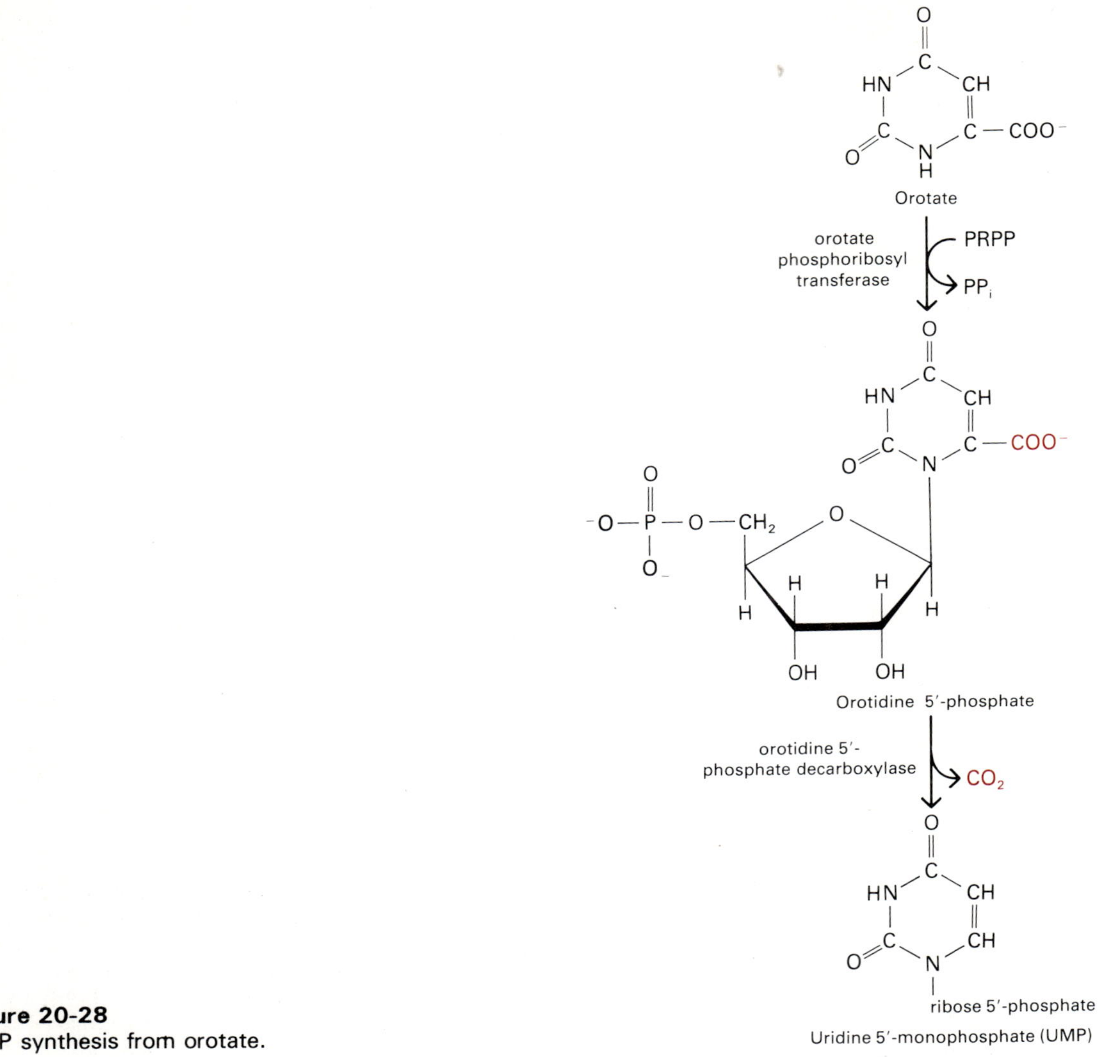

Figure 20-28
UMP synthesis from orotate.

enzyme's mitochondrial location, the NADH produced by the reaction is available for use in aerobic respiration. Orotate then enters into the cytosol where a second multienzymic polypeptide (MW ≈ 55,000), containing the remaining two enzymes of the pathway, completes the synthesis of uridine 5′-monophosphate. These findings of polypeptides containing two or more enzymatic activities (pyrimidine and fatty acid synthesis) pose the novel speculation that such multifunctional proteins may be relatively common in eucaryotic biosynthetic processes.

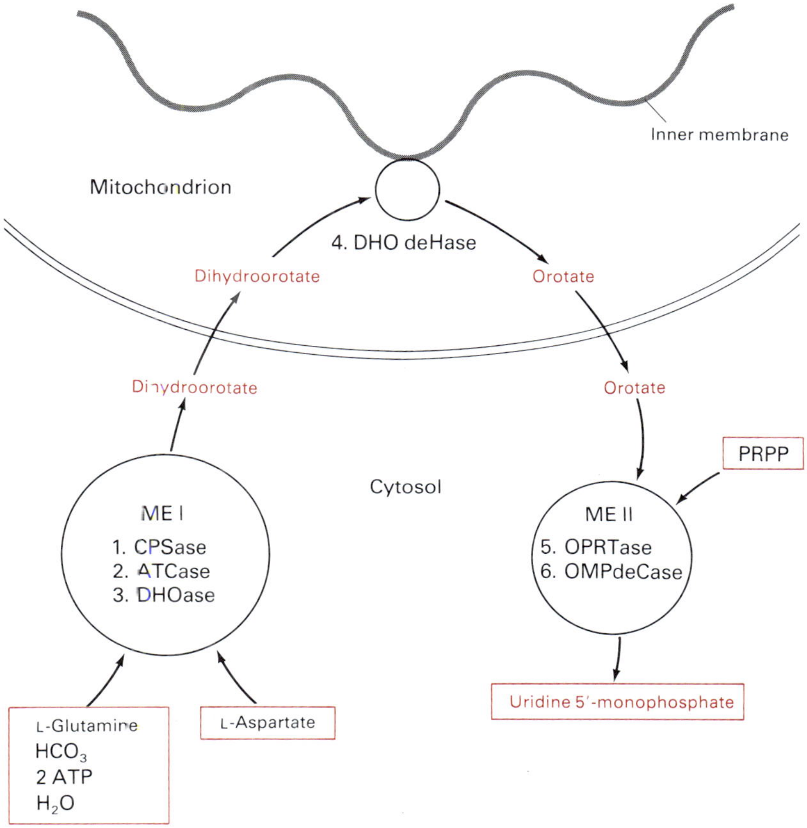

Figure 20-29
Schematic of the intracellular location of the six enzymes for uridine 5′-monophosphate (UMP) biosynthesis in animals. ME I is multienzyme I, consisting of carbamoyl phosphate synthetase (CPSase), aspartate transcarbamoylase (ATCase), and dihydroorotase (DHOase); DHOdeHase is dihydroorotate dehydrogenase, and ME II is multienzyme II, consisting of orotate phosphoribosyltransferase (OPRTase) and orotidine 5′-monophosphate decarboxylase (OMPdeCase). (After M. E. Jones, *Ann. Rev. Biochem.*, 49:253, 1980).

Cytosine synthesis does not proceed by direct conversion of UMP into CMP. Instead, UMP undergoes two phosphorylation reactions (p. 262) to produce uridine 5′-triphosphate (UTP), which is then converted into cytidine 5′-triphosphate (CTP) by *CTP synthetase* in the following reaction:

$$\text{UTP} + \text{glutamine} + \text{ATP} \rightarrow \text{CTP} + \text{glutamate} + \text{ADP} + P_i + H_2O$$

In the reaction, the carbonyl oxygen at C-4 of uracil is replaced by an amino group furnished by glutamine.

Catabolism

The general route for catabolism of cytosine and uracil is outlined in Figure 20-30. As depicted, cytosine is *deaminated* in a hydrolytic reaction to produce uracil, which is then *reduced* to N-*car-*

bamoylproprionate. In the final step, *N*-carbamoylpropionate is hydrolyzed to yield *β-alanine,* NH_4^+, and CO_2.

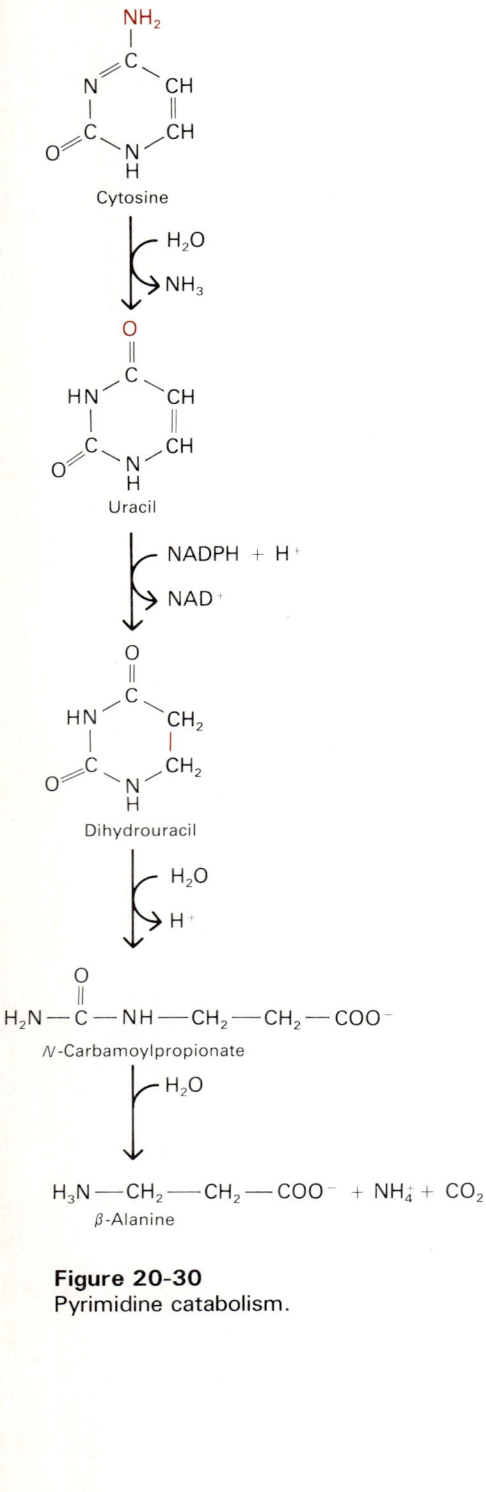

Figure 20-30
Pyrimidine catabolism.

Orotic Aciduria

In pyrimidine biosynthesis, there is an inborn error that is manifested by the accumulation and excretion of orotate. The syndrome is called *orotic aciduria,* and afflicted children do not grow normally and suffer from megaloblastic anemia (abnormal formation of red blood cells). The biochemical block has been identified as reduced activity of the phosphoribosyl transferase and decarboxylase associated with the second multienzymic polypeptide (Me II in Figure 20-29). The disorder is treated with a diet high in uridine and cytidine, which alleviates the clinical manifestations of the disease.

SUMMARY

Nitrogen fixation, which reduces N_2 to NH_4^+, is limited to anaerobic soil bacteria, a few aerobes, and the *Rhizobium*-legume relationship. Fixation occurs via the nitrogenase complex, which has two major components (I and II), and requires six electrons, 12 ATP, and 12 H_2O for the reduction of one N_2 molecule to two NH_4^+. Hydrogen production is a significant side reaction of the complex. Nonfixing plants obtain most of their nitrogen by the two-step conversion of NO_3^- into NH_4^+. In the first reaction, NO_3^- is reduced to NO_2^- by nitrate reductase, and in the second reaction, NO_2^- is reduced to NH_4^+ by nitrite reductase.

In the biosynthesis of amino acids, alanine is synthesized from pyruvate in a reaction catalyzed by alanine aminotransferase; the reverse reaction catabolizes alanine. For asparagine synthesis, glutamine donates its amide group to aspartate; for asparagine catabolism, its amide group is lost by hydrolysis to produce aspartate. Serine is synthesized from 3-phosphoglycerate, a glycolytic intermediate, and is catabolized by conversion into pyruvate. Glycine is produced from serine in a reaction that converts tetrahydrofolate (FH_4) into N^5,N^{10}-methylene-FH_4. Glycine can also be synthesized from CO_2, NH_4^+, N^5,N^{10}-methylene-FH_4, and NADH. Catabolism of glycine involves reversal of its synthetic pathways. Cysteine is produced from serine and methionine, with its carbon atoms being derived from serine and its sulfur atom from methionine. The main degradation pathway for cysteine involves its conversion into pyruvate and sulfite (SO_3^{2-}). Homocystinuria and cystathionuria are two inborn errors associated with cysteine synthesis. Proline is synthesized by the conversion of glutamate into Δ^1-pyrroline 5-carboxylate, which is then reduced to yield proline. Catabolism of proline occurs by a series of reactions that mimic the synthetic pathway.

In the catabolism of phenylalanine, the first step is the production of tyrosine; in the next five steps, tyrosine (as well as phenylalanine) is degraded to fumarate and acetoacetate. The first inborn error described was alkaptonuria, which results from an inability to cleave the ring of homogentisate, an intermediate in phenylalanine (tyrosine) catabolism. Another well-studied inborn error of phenylalanine metabolism is phenylketonuria, which results from an inability to synthesize tyrosine from phenylalanine.

Glucogenic amino acids are those degraded to precursors of glucose, e.g., TCA cycle intermediates, and ketogenic amino acids are those catabolized to ketone bodies, e.g., acetoacetate. Leu is the only true ketogenic amino acid; Ile, Lys, Phe, and Tyr are both glucogenic and ketogenic, and the remaining 15 amino acids are glucogenic.

Animals excrete nitrogen in the form of urea, uric acid, or ammonia. Most fish are ammoniotelic (excrete ammonia), and most terrestrial animals are either ureotelic (excrete urea) or uricotelic (excrete uric acid). Ureotelic animals, e.g., humans, rely on the urea cycle to produce their excretory form of nitrogen (urea). The cycle commences with the covalent linkage of ornithine and carbamoyl phosphate (first donor of nitrogen) to produce citrulline, which then bonds with aspartate (second donor of nitrogen) to form argininosuccinate. In the next step, argininosuccinate is cleaved to produce arginine (synthesis of this amino acid) and fumarate. In the last reaction of the cycle, arginine is hydrolyzed to produce urea (containing the two donated nitrogens) and to regenerate ornithine. The synthesis of urea requires an input of four high-energy phsophate groups.

Uricotelic animals incorporate disposable nitrogen into the structure of purines, which are then catabolized to uric acid. The biosynthesis of purines, which occurs in their nucleotide form, first requires the activation of D-ribose 5-phosphate by its conversion into 5-phospho-D-ribose 1-pyrophosphate (PRPP). Then, in a series of ten reactions, the purine molecules is synthesized attached to the phosphorylated pentose. The first nucletotide produced is inosine 5′-monophosphate (IMP). AMP and GMP are produced from IMP; AMP and GMP synthesis requires GTP and ATP as an energy source, respectively. Seven high-energy bonds are needed for AMP production, and eight are required for GMP production. The purine pathway is allosterically regulated at the initial stages of synthesis and also at the branch reactions leading to AMP and GMP synthesis from IMP.

In catabolism, adenine and guanine are first converted into hypoxanthine, which is then utilized by the xanthine oxidase reaction to produce uric acid. In many animal species, uric acid is further degraded to allantoin, allantoate, or glyoxylate and urea (or ammonia and carbon dioxide).

The illness gout is brought about by an excessive production of uric acid, which can be due to a variety of metabolic disorders. D-Glucose 6-phosphatase and hypoxanthine-guanine phosphoribosyltransferase (HGPRT) deficiencies are two inborn errors that result in an enhanced production of purine nucleotides and hence of uric acid. HGPRT deficiency (Lesch–Nyhan syndrome) is of particular interest because of the aberrant behavioral problems associated with the inborn error.

For pyrimidine nucleotide synthesis, orotate is first produced in a series of three reactions from carbamoyl phosphate and aspartate. The phosphoribosyl moiety of PRPP is then transferred to orotate, which is decarboxylated to produce uridine 5′-monophosphate. The first three enzymatic activities involved in uridine 5′-monophosphate synthesis (including carbamoyl phosphate synthesis) are contained in a cytosolic multienzymic polypeptide. The fourth step, orotate synthesis, occurs in the mitochondria. Orotate then enters into the cytosol, where a second multienzymic polypeptide (possessing two enzymatic activites) completes the synthesis of uridine 5′-monophosphate. Cytosine synthesis occurs at the triphosphate level by the conversion of UTP into CTP. The pyrmidines are catabolized to β-alanine, NH_4^+, and CO_2. Orotic aciduria is an inborn error of pyrimidine metabolism associated with the reduced activity of the two enzymes of the second multienzymic polypeptide. Afflicted individuals accumulate and excrete orotate and are treated with a diet high in uridine and cytidine.

QUESTIONS

1. With regard to amino acid metabolism, which of the following statements are true?
   (a) Pyruvate is the direct precursor of alanine.
   (b) Alanine, aspartate, and glutamate are produced by transamination reactions.
   (c) In mammals, asparagine is produced from asparate and NH_4^+.
   (d) In mammals, asparagine is catabolized to aspartate and NH_4^+.
   (e) 3-Phosphoglycerate, a glycolytic intermediate, is the major source of serine in animals.
   (f) Serine-threonine dehydratase converts serine into threonine.
   (g) Glycine synthase produces glycine from serine.
   (h) In cysteine synthesis, the hydroxyl group of serine is exchanged for the sulfhydryl group of methionine.
   (i) Cysteine is catabolized to yield pyruvate.
   (j) Proline is synthesized from glutamate.
   (k) Proline is catabolized to yield glutamate.
   (l) Tyrosine is a catabolic intermediate of phenylalanine.
   (m) In humans, phenylalanine can be synthesized from tyrosine.

2. Match the amino acids in the left-hand column with the appropriate classification in the right-hand column:

____ (a) Tyrosine	(1) Glucogenic
____ (b) Alanine	(2) Ketogenic
____ (c) Glutamine	(3) Glucogenic and ketogenic
____ (d) Glutamate	
____ (e) Leucine	
____ (f) Phenylalanine	

3. Select the answers to the following statements about the urea cycle from the following list:

Arginase	Citrulline
Arginine	Ornithine
Argininosuccinase	Ornithine transcarbamoylase
	Argininosuccinate synthetase

   (a) Catalyzes a mitochondrial reaction
   (b) Catalyzes the entrance of aspartate into the cycle
   (c) Catalyzes the entrance of carbamoyl phosphate into the cycle
   (d) Catalyzes the production of urea
   (e) Catalyzes the production of a TCA intermediate
   (f) Catalyzes the production of ornithine
   (g) Catalyzes the hydrolysis of ATP to AMP and PP_i
   (h) Transported from mitochondria into the cytosol
   (i) Transported from the cytosol into the mitochondria
   (j) Immediate precursor of urea

4. (a) How many α-amino acids participate directly in the urea cycle?
   (b) How many of these amino acids can be used for protein synthesis?

5. Select the answers to the following statements about nucleotide biosynthesis from the following list:

Aspartate	Glycine
ATP	GTP
CO_2	N^{10}-Formyl-FH_4
Glutamine	

   Furnishes the following atom to the purine ring:
   (a) N-1
   (b) N-3
   (c) N-7

(d) N-9
(e) C-2
(f) C-6
(g) C-8

Nucleoside triphosphate used in the synthesis of:
(h) IMP
(i) AMP from IMP
(j) GMP from IMP

6. Write the two general reactions catalyzed by superoxide dismutase and catalase that convert dangerous superoxide radicals (O_2^-) into O_2 and H_2O.

7. How many high-energy bonds are saved by the intracellular synthesis of GMP from guanine and PRPP (HGPRT reaction), rather than from the pathway that leads to IMP production?

8. If the α-carbonyl carbon of α-ketoglutarate used for proline biosynthesis was labeled with ^{14}C, which carbon atom of proline would be labeled?

9. Select your answers to the statements below from the following list of enzymes involved in pyrimidine nucleotide biosynthesis:

Aspartate transcarbamoylase
Carbamoyl phosphate synthetase
Dihydroorotase
Dihydroorotate dehydrogenase
Orotate phosphoribosyl transferase
Orotidine phosphate decarboxylase

(a) Utilizes HCO_3^- as a substrate
(b) Utilizes carbamoyl phosphate as a substrate
(c) Catalyzes the production of UMP
(d) Catalyzes the first reaction in the pathway
(e) Catalyzes a mitochondrial reaction
(f) Adds a phosphoribosyl group to orotate
(g) Are components of multienzyme I
(h) Are components of multienzyme II
(i) Reduced activities associated with orotic aciduria

10. For the curious student: List the reactions that account for the excess production of uric acid in some patients deficient in D-glucose 6-phosphatase (because of an accumulation of D-glucose 6-phosphate). Are there at least 20 enzymatic steps involved?

SUGGESTED READINGS

Baldwin, E., *An Introduction to Comparative Biochemistry,* 4th ed. New York: Cambridge University Press, 1964. A classic biochemical text discussing the comparative aspects of nitrogen-excreting organisms.

Bender, D. A., *Amino Acid Metabolism.* New York: Wiley, 1975. A good presentation on the biochemistry of amino acids.

Bondy, P. K., and L. E. Rosenberg, eds., *Metabolic Control and Disease,* 8th ed. Philadelphia: W. B. Saunders, 1980. Chapters 11 and 12 have complete and informative discussions on the inborn errors involving amino acid, purine, and pyrimidine metabolism.

Florkin, M., and H. S. Mason, eds., *Free Energy and Biological Function of Comparative Biochemistry,* Vol. 2. New York: Academic Press, 1960. The article by P. P. Cohen and G. W. Brown Jr., "Ammonia Metabolism and Urea Biosynthesis," is recommended.

Garrod, A. E., *Inborn Errors of Metabolism.* New York: Oxford University Press, 1963. A reprint of the 1909 prophetic book on inborn errors, with a supplement by the noted human geneticist H. Harris.

Greenberg, D. M., *Metabolic Pathways,* Vol. 3. New York: Academic Press, 1969. A volume containing excellent reviews on the biosynthesis and catabolism of amino acids.

Henderson, J. F., and A. R. P. Paterson, *Nucleotide Metabolism: An Introduction.* New York: Academic Press, 1972. A good presentation of the subject.

McKusick, V. A., and R. Claiborne, eds., *Medical Genetics.* New York: HP Publishing, 1973. Chapter 10, by J. C. Seegmiller, has an excellent discourse on the Lesch–Nyhan syndrome.

Meister, A., *Biochemistry of the Amino Acids,* 2nd ed., 2 vols. New York: Academic Press, 1965. A highly respected and comprehensive treatise on amino acid metabolism.

San Pietro, A., *Photosynthesis and Nitrogen Fixation.* Part C, *Methods Enzymol,* Vol. 69. New York: Academic Press, 1980. The last section of the volume is devoted to research techniques used to study nitrogen fixation.

Stanbury, J. B., J. B. Wyngaarden, and D. S. Fredrickson, eds., *The Metabolic Basis of Inherited Diseases,* 4th ed. New York: McGraw-Hill, 1978. A good reference source for the inborn errors discussed in the chapter.

Articles

Brill, W. J., ''Biological Nitrogen Fixation.'' *Scientific American,* 236(3):68, 1977.

Jones, M. E., ''Pyrmidine Biosynthesis in Animals: Genes, Enzymes, and Regulation of UMP Biosynthesis.'' *Ann. Rev. Biochem.,* 47:533, 1978.

Mortenson, L. E., and R. N. F. Thorneley, ''Structure and Function of Nitrogenase.'' *Ann. Rev. Biochem.,* 48:387, 1979.

Umbarger, H. E., ''Amino Acid Biosynthesis and Its Regulation.'' *Ann. Rev. Biochem.,* 47:533, 1978.

Yates, M. G., ''Nitrogen Fixation.'' *TIBS,* 1:17, 1976.

MOLECULAR INSIGHTS INTO DNA

21

INTRODUCTION

Since the announcement of the double-helical model of DNA in 1953 by Watson and Crick, molecular studies on macromolecules have made unprecedented contributions to basic knowledge in biology. As mentioned in Chapter 1, an intellectual contribution that markedly influenced the earlier molecular studies was the Central Dogma, proposed by Crick in the late 1950s. The dogma singularly stresses a fundamental characteristic of living processes—*the flow of biological information.* Based on accumulated scientific evidence and emerging research findings, Crick surmised that a molecular understanding of genetic information and its replication and transmission within a living system resides primarily in the biochemical functionings of the following three classes of macromolecules: DNA, RNA, and protein. He envisioned, as is schematically diagrammed in Figure 21-1, nine hypothetical possible modes of informational transfer: each of the three types of macromolecules could replicate itself, and each could transfer its informational content to the other two. The later proposal, in biochemical terms, implies that a class of macromolecules can transmit its information by serving physically as a structure-specifying template for the synthesis of the other two. The available evidence at the time of Crick's postulate identified only the replication of DNA and of viral RNA and the transfer of genetic information to protein structure (one gene–one enzyme hypothesis) as documented avenues for the flow of biological information.

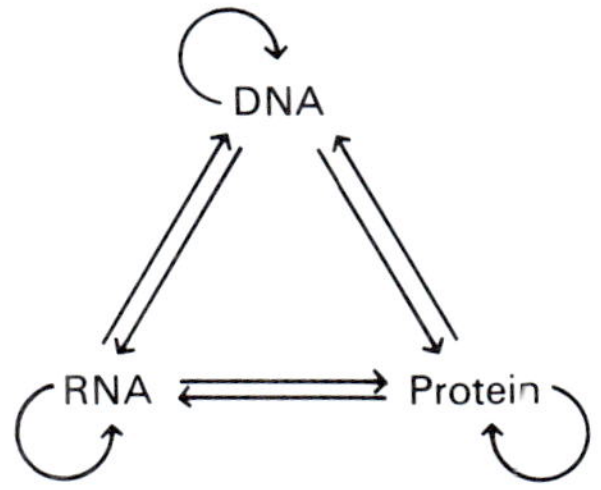

Figure 21-1
Hypothetical routes for the flow of biological information, as postulated by Crick.

Because of the rapid and exciting advances being made during that period toward unraveling the biochemical mechanisms of protein synthesis and the chemical nature of the genetic code, the

Figure 21-2
Biochemical route for the transfer of genetic information to protein structure.

scientific world focused its attention on the transfer of information as it occurs in gene expression (Figure 21-2). The DNA-protein relationship thereby assumed a dominant role in the Central Dogma, and the one-way flow of information associated with the relationship became entrenched in scientific thinking. When Howard Temin and David Baltimore (shared Nobel Prize, 1975) independently established by 1970 that *oncogenic* (tumor-causing) *RNA viruses* have an enzyme, *reverse transcriptase,* that uses the single-stranded RNA viral genome as a template to synthesize DNA, i.e., RNA → DNA, the finding was received with skepticism by many scientists. The postulate that the flow of biological information could proceed from RNA to DNA was considered by some as akin to a heretical proposal. However, as Crick subsequently pointed out, his original thoughts about the flow of biological information (Figure 21-1) did not exclude such a possibility, and the discovery of reverse transcriptase *(RNA-directed DNA polymerase)* does not violate but, rather, expands the fundamental concepts of the dogma. With the incorporation of this newer evidence, the known flow of biological information is now abbreviated as shown in Figure 21-3. DNA and viral RNA are *replicons,* i.e., molecules capable of self replication; both classes of nucleic acids can serve as templates for the synthesis of the other, and proteins are synthesized from information contained in RNA. The solid arrows in Figure 21-3 denote general transfers that occur in all cells, and dashed-line arrows show specialized transfers that are known to occur only in RNA virus infected cells. Because there is no evidence of biochemical mechanisms that allow proteins to replicate or to serve as templates for DNA or RNA synthesis, the dogma postulates that the informational content of proteins is not transferable.

Figure 21-3
Known mechanisms for the transfer of biological information. Solid and dashed arrows represent general and special transfers, respectively.

The effect that the discovery of reverse transcriptase in oncogenic viruses has had on cancer research also deserves comment. Although it was known that some DNA and RNA viruses are capable of causing certain types of cancer in experimental animals, there was no unifying model that satisfactorily incorporated a cellular basis for the cancer-causing action of both types of viruses. Studies on the intracellular relationships between bacteria and their DNA bacteriophages (bacterial viruses) provided a sound scientific foundation for the following rationale by which researchers could account for the transformation of normal mammalian cells to a cancerous state by certain DNA viruses. After entrance of the virus into a host cell, the double-stranded viral genome *(provirus)* undergoes recombination with a chromosome(s) of the infected cell, and the genetic influence of the incorporated provirus results in cellular

Three oncogenic RNA viruses are the Rous sarcoma (cancer in chickens), avian myeloblastosis (leukemia in birds), and Rauscher leukemia (leukemia in mice). Two oncogenic DNA viruses are SV40 (cancer in monkeys) and polyoma (cancer in mice).

transformation. This model does not encompass oncogenic RNA viruses because it would need to assume recombination between RNA (viral genome) and DNA (host-cell genome)—a biological phenomenon that is not known to occur. As illustrated in Figure 21-4, the reverse transcriptase reaction initiates a biochemical means by which a viral RNA can transfer its genetic information to a double-stranded DNA proviral structure that can then incorporate into a host cell's genome. Thus, the discovery of reverse transcriptase provided a biochemical link that greatly aided the proponents of the viral theory of cancer. (Although viruses are implicated in certain animal cancers, some researchers do not consider them to be significant causative agents in human cancer; instead, they postulate that exposure to environmental factors, e.g., smoking, diets, and pollutants, is the major underlying cause of most cancers.)

MOLECULAR RESEARCH ON DNA

Of the molecular studies on the three classes of informational macromolecules, those on DNA continue to exert the greatest impact on research trends in biology. The advances made toward elucidating molecular details about the carrier of the genetic blueprint and its biological functions transcend DNA's role in the flow of biological information (replication and gene expression), since research has also uncovered other biochemical phenomena involv-

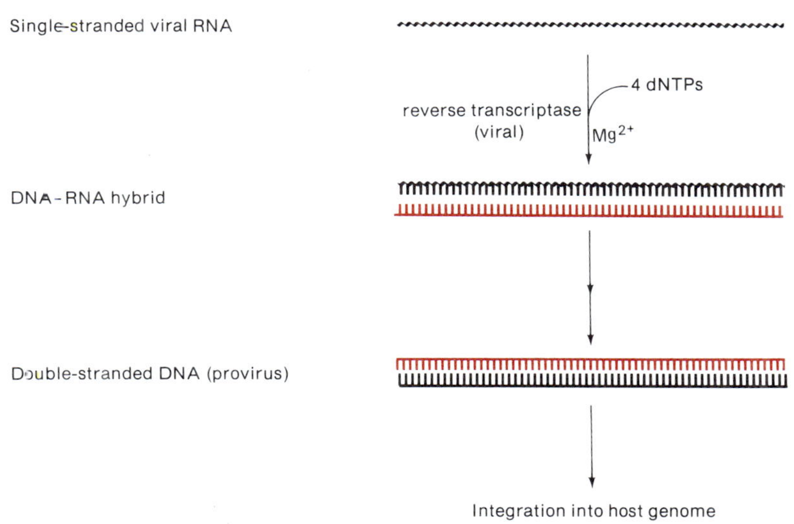

Figure 21-4
Simplified diagram of the intracellular conversion of viral RNA into double-stranded DNA. Details of cellular enzymatic events leading to the synthesis of provirus are uncertain. Deoxyribonucleoside triphosphates needed for DNA synthesis are represented by four dNTPs.

ing the nucleic acid. It has been discovered, for example, that a cell is capable of repairing its damaged DNA (repair mechanism, p. 447) and that certain procaryotes degrade foreign DNA that enters into a cell (modification and restriction mechanisms, p. 448). These types of mechanisms involve specific biochemical interactions between proteins (enzymes) and DNA and exemplify some of the types of molecular information that are now known to be contained in the nucleotide sequence of a DNA molecule.

The molecular studies on DNA emphasized an increasing need for techniques by which to sequence DNA molecules and thereby allow for the deciphering of the various sets (base sequences) of molecular information that collectively allow DNA to participate in a variety of cellular functions. By the early months of 1977, two effective and relatively simple sequencing techniques were available, and the first nucleotide sequence of a complete genome, that of the bacteriophage ϕX174, was announced (Chapter 23). These landmark accomplishments have issued in yet another dynamic era of DNA research.

The most controversial molecular studies on DNA were those which provided scientists with the techniques to construct recombinant-DNA molecules (composed of DNA from two different sources, e.g., a bacterial and an animal recombinant) and to propagate them in bacteria. This area of biological research, because of its implied social implications, became a matter of public concern and debate (p. 597).

REPLICATION OF DNA

Although the double-helical structure of DNA immediately suggested a semiconservative mode of replication for the molecule to Watson and Crick (p. 11), a comprehensive understanding of the molecular details of the replicative process has proved to be an unusually difficult and, at times, puzzling task. In fact, it has only been within recent years that the complexities of DNA replication have been fully appreciated.

DNA Polymerase (Kornberg's Enzyme)

In the 1950s, the discovery of and subsequent studies on the enzyme *DNA-directed DNA polymerase* of *Escherichia coli* by Arthur Kornberg (Nobel Prize, 1959) and associates, offered, so it then seemed, the catalytic mechanism by which DNA replication occurs. The DNA polymerase (Figure 21-5) was found to require a DNA template and all four deoxyriboside triphosphate substrates; the enzyme was also shown to prefer single-stranded DNA for a template, from which a complementary strand is synthesized. Collectively, these findings fulfilled the expected requirements for the enzyme that catalyzes the *in vivo* replication of DNA. As molecular and genetic studies on DNA replication continued, it became apparent that the views on DNA replication were perhaps too simplistic, and even the role of DNA polymerase became a matter of debate. Indeed, if catalogued, the pros and cons about the status of Kornberg's enzyme in DNA replication over the past several decades provide an ordered, historical view of the gradual evolvement of an accurate understanding of DNA replication. Few enzymes have ever been so closely scrutinized with respect to their biological function as has DNA polymerase.

$$\begin{matrix}\text{dATP}\\ \text{dGTP}\\ \text{dCTP}\\ \text{dTTP}\end{matrix} \xrightarrow[\text{Mg}^{2+}]{\text{DNA template}} \text{nascent DNA}$$

Figure 21-5
General features of the DNA polymerase reaction.

Some Molecular Features of DNA Replication

The following findings are some of the salient molecular aspects of DNA replication that have been progressively revealed. In semiconservative replication, the two strands of the parent double helix are synthesized simultaneously and, because of the antiparallelism of the two strands, 5′ → 3′ and 3′ → 5′ syntheses (as indicated in Figure 21-6) were postulated to proceed concomitantly. This hypothesis cast doubt about the proposed replication role of Kornberg's enzyme, which can synthesize only in a 5′ → 3′ direction. Initiation of DNA replication, as determined from studies on *E. coli,* requires a particular site *(origin of replication)* on the molecule, i.e., synthesis occurs in a specified, ordered manner. DNA replication was also discovered to involve *discontinuous synthesis* (Figure 21-7). Whereas synthesis of a complementary

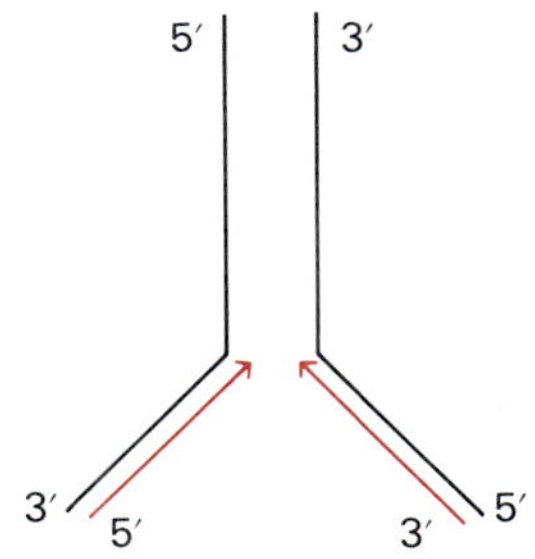

Figure 21-6
Simplified scheme of the semiconservative replication of DNA.

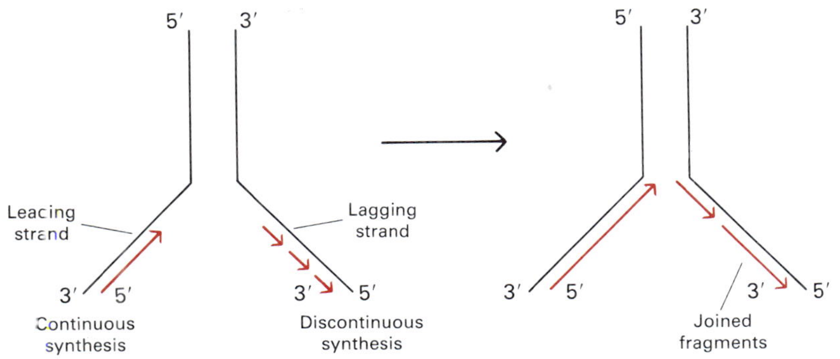

Figure 21-7
Simplified diagram of DNA replication.

strand to the 3′ → 5′ parent strand (called the *leading strand*) takes place in a *continuous* manner, synthesis occuring on the 5′ → 3′ parent strand *(lagging strand)* proceeds in a *discontinuous* fashion. Discontinuous synthesis entails the production of DNA fragments which subsequently, by covalent linkage to one another, yield an intact DNA strand. These fragments (about 1,000 to 2,000 nucleotides long in *E. coli*) are often called *Okazaki fragments,* named for Rieji Okazaki, who discovered them in the late 1960s. Note that discontinuous synthesis occurs in a 5′ → 3′ direction, a finding that restored the postulated role for Kornberg's enzyme in DNA replication.

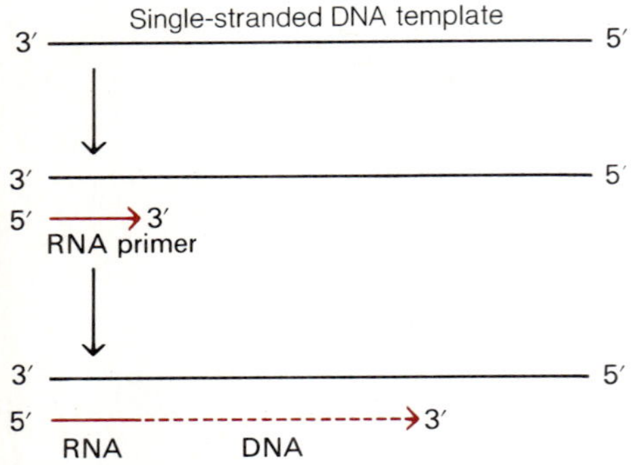

Figure 21-8
Simplified illustration of RNA and DNA synthesis in DNA replication.

A starting revelation was that RNA synthesis, which uses single-stranded DNA as a template, is a prerequisite for DNA replication. As illustrated in Figure 21-8, a small *RNA primer* (about 20 to 30 nucleotides long) is produced first and then deoxyribonucleotides are *added on* to the 3′-terminus of the primer to continue 5′ → 3′ synthesis. This inclusion of a 5′-RNA terminus in nascent DNA fragments revealed that the final steps in DNA replication, especially in discontinuous synthesis, are more complicated than previously suspected. The RNA portions of the fragments must be removed and replaced by analogous DNA sequences before the DNA fragments can be linked to produce the completed DNA strand.

The Three DNA Polymerases of *E. coli*

As the molecular details of DNA replication were progressively unfolding, genetic and enzymatic studies were also uncovering new facets about the *in vivo* synthesis of polydeoxyribonucleotides. To summarize, three DNA polymerases are now identified in *E. coli,* and Table 21-1 lists some of the properties of each. The

Table 21-1
DNA polymerases of *Escherichia coli*

	Polymerase I	Polymerase II	Polymerase III
Functions			
5′ → 3′ polymerization	+	+	+
5′ → 3′ exonuclease	+	−	−
3′ → 5′ exonuclease	+	+	+
Molecular weight	109,000	120,000	180,000
Molecules/cell	400	100	10
Nucleotides polymerized at 37°/min/molecule	≈1,000	≈50	≈15,000

(Data from *DNA Synthesis,* A. Kornberg, San Francisco: Freeman Publishing, 1974, p. 127.)

original polymerase discovered by Kornberg is now called *DNA polymerase I,* and the more recently described enzymes are *DNA polymerase II* and *III.* All three enzymes polymerize deoxyribonucleotides in a 5′ → 3′ direction. They also exhibit 3′ → 5′ exonuclease activity, i.e., degrade DNA, one nucleotide at a time, from its 3′-terminus. DNA polymerase I has a third catalytic activity: it can also degrade DNA (or RNA) from the 5′-terminus (5′ → 3′ exonuclease activity). As a single polypeptide, DNA polymerase I is a remarkable enzyme, capable of catalyzing three distinct reactions. DNA polymerase II is also a single polypeptide, whereas catalytically active DNA polymerase III is composed of two identical protomers. One biochemical feature that all three enzymes share is that *none synthesizes DNA de novo* but, instead, can only add on DNA sequences to the 3′-terminus of a pre-existing polynucleotide (primer).

The biological functions of the three polymerases have been a puzzle, particularly with respect to their possible roles in DNA replication. Studies on various classes of bacterial DNA-polymerase mutants support the contention that membrane-bound DNA polymerase III is the true DNA-replicating enzyme. (*DnaE* mutants of *E. coli* that cannot replicate their chromosome at an elevated temperature also have a DNA polymerase III, but not polymerase I and II, that is similarly temperature sensitive.) DNA polymerase I is postulated to be required for DNA repair mechanisms (p. 447), and the enzyme is also believed to have a role in DNA replication. DNA polymerase II is still an enigma, since no specific cellular function has been identified for the enzyme. Continued research, however, will undoubtedly establish a specific cellular role for each of the polymerases.

Current Model for DNA Replication

Through the years, Kornberg and his colleagues have steadfastly continued to probe the molecular mechanisms of DNA replication. From studies on the *in vitro* replication of the single-stranded DNA genomes of bacteriophages ϕX174, G4, and M13, Kornberg's group succeeded in identifying many of the individual molecular components required for the synthesis of the complementary strand of these bacteriophages' DNAs. Their findings, as well as those of other investigators, provide the basis for a general model of the stepwise replication of DNA, which requires a *multienzyme DNA replicase* system. The model, proposed by Kornberg, recognizes the following four stages in the replication process: *prepriming* (preparation for replication), *priming* (synthesis of RNA primers), *elongation* (addition of DNA sequences to the 3′-termini of primers), and *termination* (removal of the RNA primers and their replacement with analogous DNA sequences and the subsequent covalent joining of the DNA fragments).

Prior to the following discussion of DNA replication, it should be noted that DNA replication occurs only during a particular phase of a cell's cycle. Since circular DNA exists as a compact, *supercoiled* molecule in a cell, it is necessary that the molecule be untwisted *(relaxed)* for its replication. The enzyme *DNA gyrase* (a *topoiomserase)* performs this function by cleaving phosphodiester bonds *(''nicking'')* and then re-forming these bonds. Thus, in this manner, supertwisted DNA is relaxed. (The enzyme, with the energy provided by ATP hydrolysis, also catalyzes the reverse phenomenon, i.e., twisting relaxed circular DNA into a supercoiled structure.)

Prepriming and Priming

Prepriming, the initial stage of replication, prepares the two DNA strands for replication, and at least a half dozen proteins participate in the process. As illustrated in Figure 21-9, a *helicase* is required for the unwinding of the DNA at the *replication fork,* and subsequent stabilization of the unwound DNA is provided by molecules of *DNA-binding protein (DBP),* which coats the two separated strands. The particular regions of DNA that serve as initiation sites for priming are established by *dnaB protein* molecules, which also attach to the DNA strands. The positioning of a dnaB protein requires the involvement of three additional proteins, called the *dnaC, i,* and *n proteins;* these latter proteins are believed to func-

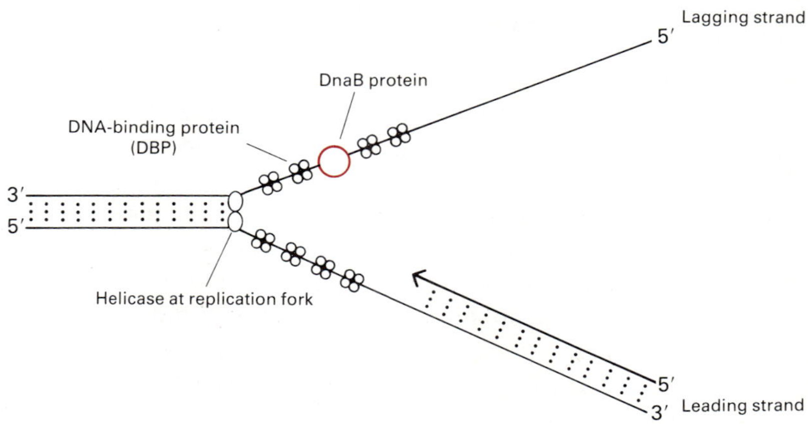

Figure 21-9
Prepriming for DNA replication.

tion as an "escort service" for the dnaB protein. (The dna proteins are so named because they are the products of genes, e.g., *dnaB* and *dnaC,* which were previously determined by genetic and biochemical studies to be required for DNA replication.) The function of the dnaB protein is to open a region on the DNA strand that allows the enzyme *primase,* also called the *dnaG protein,* to commence the 5′ → 3′ synthesis of an RNA primer (Figure 21-10). An additional role of the dnaB protein may be to travel along the DNA strand opening other initiation sites for the primase, as indicated by the priming of the second fragment in Figure 21-10.

Elongation and Termination

Synthesis of an RNA primer is followed by the addition of a DNA sequence to its 3′-terminus *(chain elongation).* The elongation process is catalyzed by the *DNA polymerase III holoenzyme* (Figure 21-10). For the completion of the synthesis of an intact DNA strand, two additional enzymes are required—*DNA polymerase I* and *DNA ligase.* The termination process, as illustrated in Figure 21-10, requires the joining of one fragment to another. As indicated, DNA polymerase I has the dual role of removing RNA primers (5′ → 3′ exonuclease activity) and extending the DNA sequence on the 3′-terminus of fragments (5′ → 3′ polymerase activity). Note that the degradation and elongation processes, catalyzed by two distinct active sites on DNA polymerase I, occur concurrently and result in the production of fragments containing only DNA. The final step, catalyzed by DNA ligase (Figure 21-11), establishes a phosphodiester bond between the two fragments by joining the 3′-OH terminus of one to a phosphate group at the 5′ -terminus of the other. In animal cells and for the replication of bacteriophage, the energy for the reaction is provided by ATP hydrolysis (ATP → AMP + PP_i); in bacteria, NAD^+ serves as the energy source (NAD^+ → AMP + NMN, nicotinamide mononucleotide).

Additional Comments on DNA Replication

The diagrams in Figures 21-9 and 21-10 and the discussion of DNA replication are simplified versions of a biological process whose molecular complexities are yet to be fully understood. It is now apparent that DNA replication is a highly programmed and dynamic mechanism with previously unsuspected molecular features, which ensure efficient and faithful copying of genetic information. The new insights gained from the recent studies on DNA

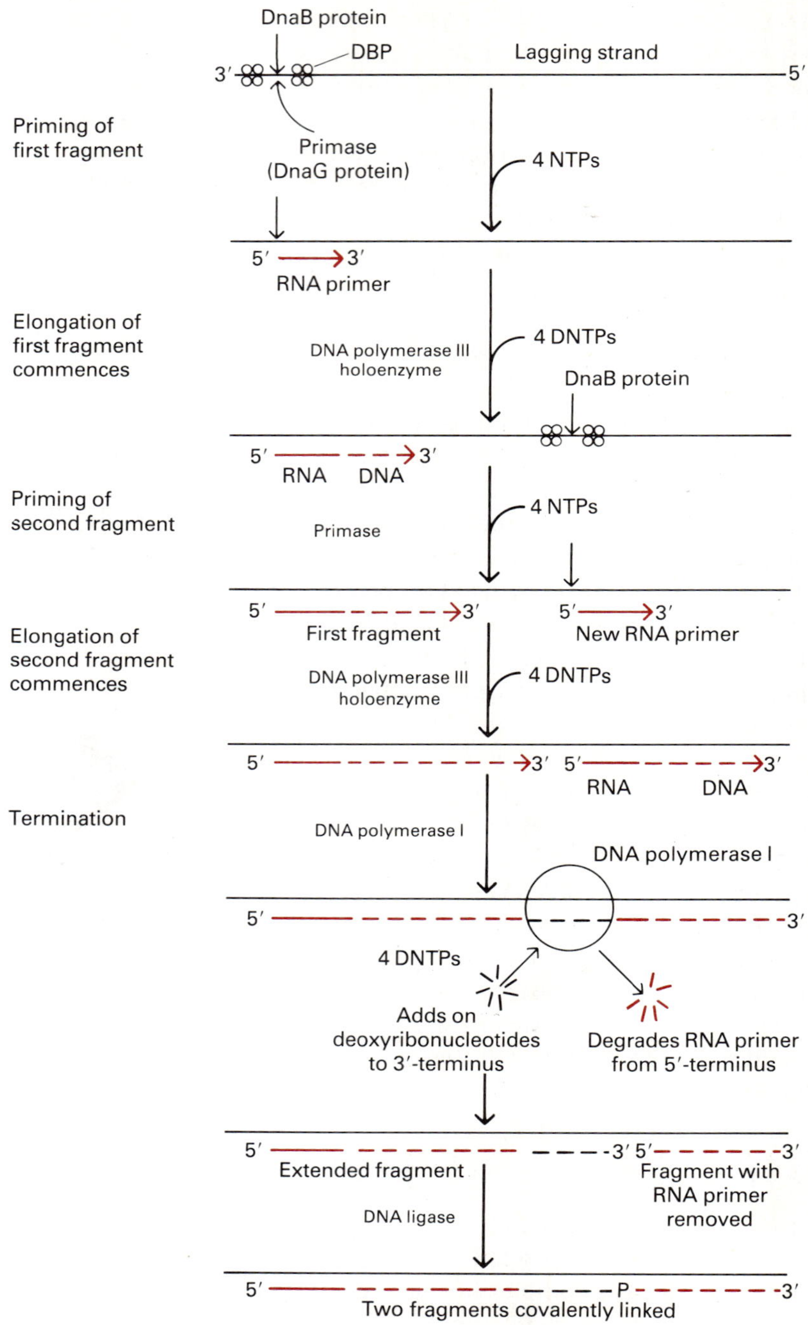

Figure 21-10
Schematic of DNA replication on lagging strand.

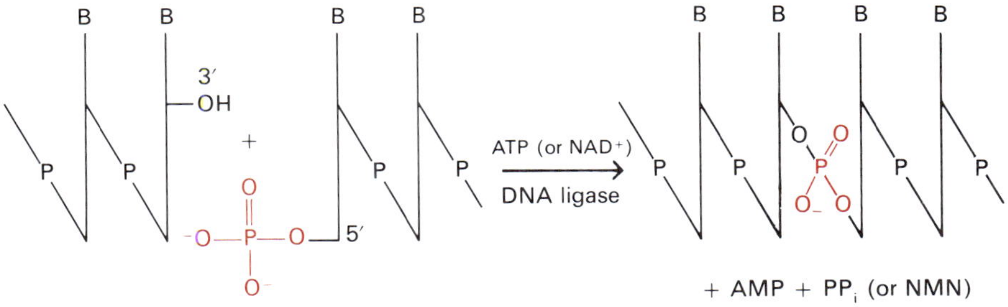

Figure 21-11
DNA ligase reaction.

replication have, in turn, furnished new questions about the molecular mechanisms involved in the process. For example, the enzyme that synthesizes the RNA primers is not the well-defined RNA polymerase responsible for the production of cellular RNAs (Chapter 22). Also, the functions of many of the seven different components of the DNA polymerase III holoenzyme (elongation enzyme) remain to be elucidated. The molecular details of DNA replication that continue to unfold are reminiscent of a statement once made by Albert Einstein (Nobel Prize, 1921) that, in science, for every milestone reached there is a signpost pointing to yet another.

DNA REPAIR MECHANISMS

In the 1960s, Richard B. Setlow and Paul Howard-Flanders, studying UV (ultraviolet light) sensitivity and resistance in *E. coli,* discovered that resistant strains are enzymatically capable of repairing their DNA when it is damaged by UV irradiation. Exposure to UV is lethal to sensitive strains, since they cannot repair their damaged DNA. The irradiation damage principally involves the formation of *thymine dimers* (Figure 21-12), although other pyrimidine dimers can also be produced. The covalent linkage of two thymine residues (T⌒T) on a strand of DNA creates a distortion in that region of the double helix (Figure 21-13). Such a localized conformational change in a DNA molecule is often lethal, since it interferes with normal DNA replication by causing faulty strand synthesis and eventually cellular death.

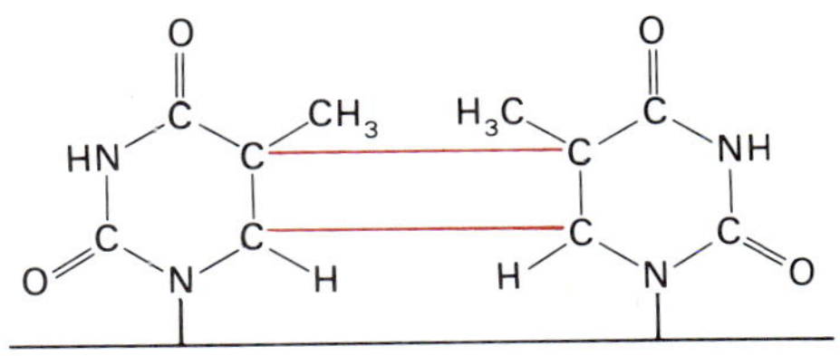

Figure 21-12
Thymine dimer.

Excision-Repair Mechanism

The proposed steps that account for the repair process, called *excision repair,* are depicted in Figure 21-13. The first reaction is

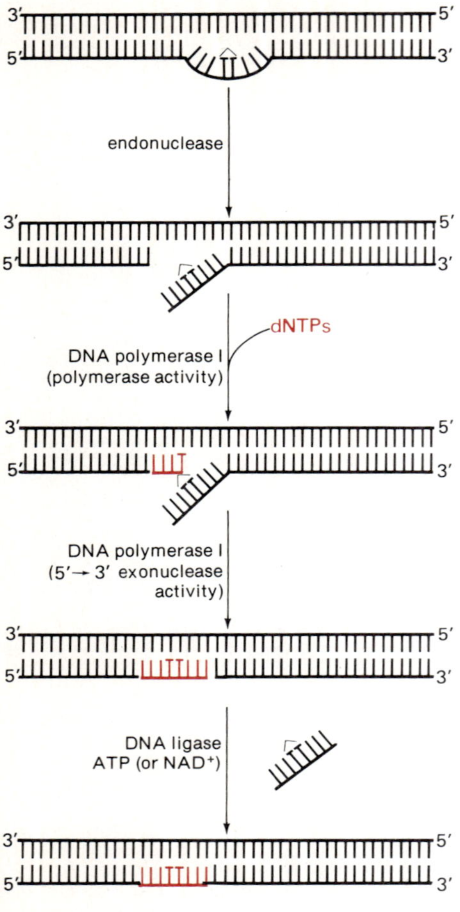

Figure 21-13
Schematic illustration of the excision-repair mechanism.

catalyzed by a specific *endonuclease* (perhaps DNA polymerase I), which recognizes the structural deformity in the DNA molecule and cleaves a phosphodiester bond in the DNA strand several bases away from the dimer, creating a "nick" in the strand. *DNA polymerase I,* using the other DNA strand as a template, then initiates a repair process by elongating the 3′-OH terminus generated by the endonuclease *("nicking" enzyme).* The 5′ → 3′ *exonuclease* activity of DNA polymerase I is responsible for removing the DNA fragment containing the thymine dimer. The newly synthesized replacement for the lost fragment is then covalently linked to tbe DNA strand by *DNA ligase,* and the DNA region is now restored to an undamaged state.

Essentiality of Repair Mechanisms

All cells have DNA repair mechanisms (including others besides excision repair), which are vital for the maintenance of the undamaged genomes that must be passed on intact to progeny. The repair mechanism uncovered by the studies on UV damage to DNA emphasizes the harmful effects that sunlight, with its UV rays, has on living systems and the essentiality of biochemical mechanisms which cope with the problem. In humans, there is a rare Mendelian recessive inborn error, *xeroderma pigmentosum,* a deficiency in excision repair. Because the cells of an afflicted individual (believed to lack the initial endonuclease activity) cannot repair their DNA, the continuous death of skin cells results in coarse deposits of pigmented tissue (atrophic scarring) appearing on exposed areas of the body, e.g., arms and face. These deposits of pigmented tissue can become so extensive on the face that a patient may have trouble keeping the eyelids open. Skin cancer is an inevitable consequence of the condition. It is sometimes necessary to operate several times a year to remove new skin cancers; however, the continued appearance of malignancies which *metastasize* (spread to other parts of the body) eventually leads to the patient's death.

MODIFICATION AND RESTRICTION ENZYMES

A novel revelation obtained from molecular studies on bacteria is that some procaryotes, e.g., *E. coli* and *Haemophilus* species, have specific enzymatic mechanisms to degrade foreign DNA that enters a cell, e.g., bacteriophage DNA. A procaryote's ability to destroy foreign DNA, without harming its own genome, involves

Table 21-2
Double-stranded base sequences recognized by two modification and restriction enzyme systems

Source	Restriction enzyme	Sequence recognized
Escherichia coli KY13	EcoR I	5′—C↓T—T—A*—A—G—3′ 3′—G—A—A—T—T↑C—5′
Haemophilus influenzae Rd	Hind III	5′—A↓A—G—C—T—T—3′ 3′—T—T—C—G—A↑A—5′
		Axis of symmetry

*Adenines (A) shown in red denote those methylated by the modification methylases. Arrows indicate the location of phosphodiester bond cleavage by the restriction endonucleases.

an unusual interplay between two enzymatic reactions. One enzyme, a *methylase,* modifies a cell's DNA by introducing a particular pattern of methylated bases into the structure which, in turn, assures protection against attack by a other enzyme, an *endonuclease.* This latter enzyme cleaves (restricts) any DNA molecule lacking that particular methylated identity. The two enzymatic reactions are interrelated because they share a requirement for the same sequences of bases (generally four to eight) in double-stranded DNA (Table 21-2). For example, both the *restriction endonuclease* EcoR I and its corresponding *modification methylase* of *E. coli* recognize the specific base sequence depicted in Table 21-2. The role of the methylase is to ensure that the two adenyl residues, in such sequences in a cell's DNA, are methylated (Figure 21-14). Sequences bearing that methylation pattern are not recognized by the endonuclease as a substrate and, hence, are spared cleavage. The methyl donor for the methylase reaction is *S-adenosyl methionine* (p. 402). It is of interest to note that newly replicated DNA, having a methylated parent strand and a nonmethylated daughter strand, is readily methylated.

Foreign DNA, lacking the identifying methylated bases of the cell's DNA, have the specific restriction sites recognized by the endonuclease, which cleaves both strands of the molecule. This two-strand cleavage cannot be corrected by DNA ligase and renders the molecule vulnerable to other cellular nucleases. The two restriction endonucleases listed in Table 21-2 produce *staggered cuts* in the double strand; others, such as Hpa I and Alu I (Table 21-3), make *blunt cuts.*

Me
5′—C—T—T—A—A—G—3′
3′—G—A—A—T—T—C—5′
Me

Figure 21-14
Methylated sequence of double-stranded DNA that is protected from cleavage by endonuclease EcoRI.

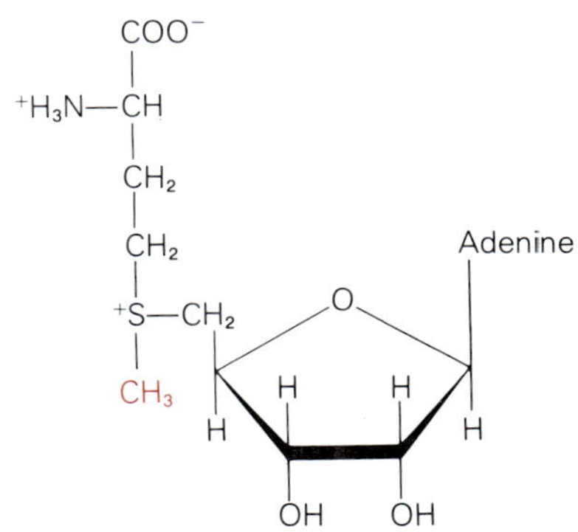

S-Adenosyl methionine

Table 21-3
Two restriction endonucleases that cleave phosphodiester bond opposite each other

Source	Restriction enzyme	Sequence recognized
Haemophilus parainfluenzae	Hpa I	5′—G—T—T↘A—A—C—3′
		3′—C—A—A↖T—T—G—5′
Arthrobacter luteus	Alu I	5′—A—G↘C—T—3′
		3′—T—C↖G—A—5′
		Axis of symmetry

Red arrows indicate the location of phosphodiester bond cleavage by the restriction endonucleases.

A MAN A PLAN A CANAL PANAMA
A palindrome

5′—R—O—M—A—3′
3′—A—M—O—R—5′

A palindrome illustrated as a double-stranded sequence

An unusual feature of the region of the DNA molecule recognized by restriction and modification enzymes is the twofold symmetry in the base sequence. (The axis of symmetry is indicated for the sequences listed in Tables 21-2 and 21-3.) This type of serial arrangement of bases in double-stranded DNA is called a *palindrome,* or an inverted repeating sequence, characterized by having the same (or nearly the same) order of bases when the two strands are read in opposite directions. In the case of the EcoR I sequence (Figure 21-14), the base order is C-T-T-A-A-G when read from the 5′-terminus of either strand, i.e., 180° rotational symmetry. Palindromic sequences are currently of great interest in DNA research, since many serve as distinguishing sets of molecular information in the DNA sequences involved in DNA-protein interactions.

RESTRICTION MAPPING

The pioneers of restriction endonuclease research, Hamilton O. Smith and Daniel Nathans of the United States and Werner Arber of Switzerland, were awarded Nobel prizes in 1978 for their discovery of these enzymes, which have proved to be of considerable importance to many areas of biological research. One valuable technique that relies on restriction endonucleases is *restriction mapping,* a procedure being used successfully to construct physical maps of DNA molecules. The technique basically involves cleavage of the DNA under investigation with a series of endonucleases and subsequent determination of fragment sizes by electrophoresis through either an Agarose or an acrylamide gel matrix (Figure 21-15).

In Agarose-gel or acrylamide-gel electrophoresis, the samples being analyzed migrate in an electrical field through a porous matrix (Agarose gel or a cross-linked acrylamide gel).

A hypothetical example of the use of restriction endonucleases to construct a genomic structure is illustrated in Figure 21-16. In the example, a 5,300-base-pair (bp) circular, double-stranded genome (DNA) is cleaved by a variety of treatments with three endonucleases (EcoR I, Hpa I, and Alu I) and, by analysis of the DNA fragments obtained, a map of the DNA molecule is generated. When the DNA is treated with EcoR I or Hpa I, it is converted from a *circular* into a *linear* molecule, revealing that the molecule possesses a *single cleavage site* for each of the two endonucleases. The production of *three fragments* by Alu I treatment identifies *three cleavage sites* for this enzyme. The restriction analysis continues with subsequent cleavages of the EcoR I product (5,300-bp, linear DNA molecule) with Alu I and Hpa I. When cleaved with Alu I, four fragments (identified as Alu I-A, -B, -C, and -D) are obtained. When these fragments are compared with the three produced by Alu I treatment (no cleavage by EcoR I), it is apparent that the 2,000-bp fragment from Alu I cleavage can be accounted for by the 1,500- and 500-bp fragments of the double treatment (EcoR I and Alu I). From this comparison, it is concluded that the EcoR I cleavage site is located within the 2,000-

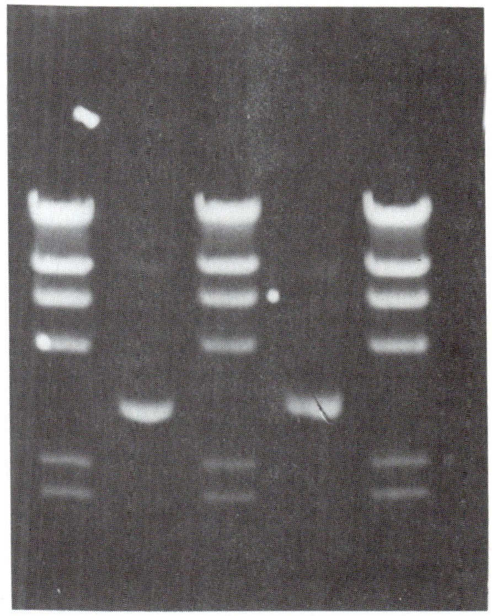

Figure 21-15
Gel electrophoresis patterns of pBR322 (channels 2 and 4) and fragments of bacteriophage λ produced by Hind III (Table 21-2) digestion (channels 1, 3, and 5). To detect the bands, the gel is first exposed to ethidium bromide, which intercalates with DNA, and then viewed under a long-wave length UV lamp. The fluorescence of ethidium bromide identifies the location of the DNA bands on the gel. The sizes of the six Hind III fragments, in descending order, are 23,000 bp (23 "kilobases"), 9,800 bp, 6,800 bp, 4,500 bp, 2,500 bp, and 2,200 bp. The predominant band in the pBR322 channels is the supercoiled form of the plasmid; the two trace bands are contaminating DNA. (Courtesy of D. W. Niesel)

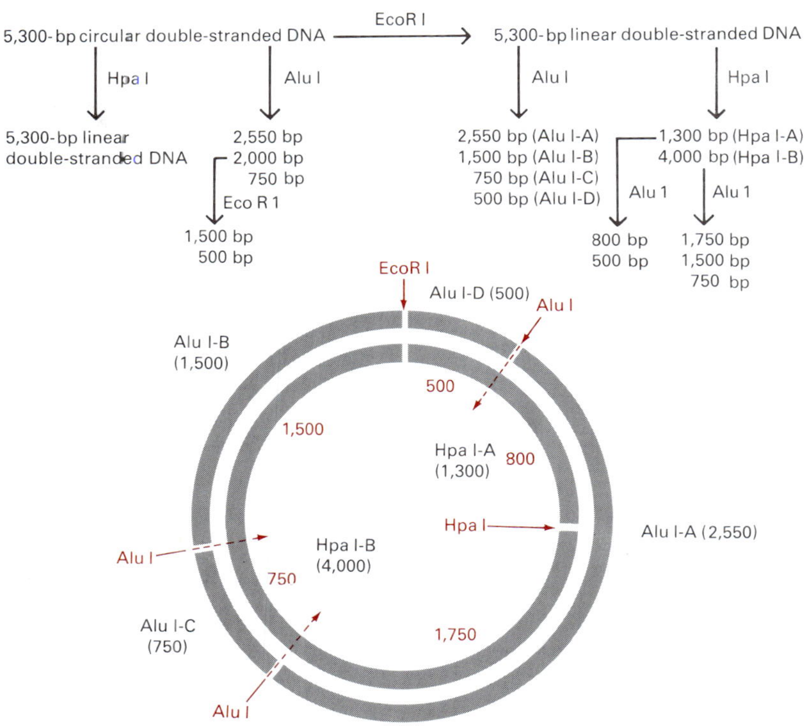

Figure 21-16
Hypothetical example of restriction mapping.

bp Alu I segment. This conclusion is confirmed by treating the 2,000-bp fragment with EcoR I, which yields two fragments the same sizes as Alu I-B and -D, i.e., 1,500 and 500 bp, respectively. This information, as depicted in Figure 21-16, is used to initiate construction of the DNA molecule, using the EcoR I site as a reference point from which the remainder of the molecule will be assembled.

When the EcoR I linear DNA molecule is cleaved with Hpa I, two fragments (1,300 and 4,000 bp) are obtained—an expected result since each endonuclease recognizes a single cleavage site. These two products are then cleaved with Alu I, and the 1,300- and 4,000-bp segments yield two and three products, respectively. This collection of five fragments represents the total number of segments that can be generated from the original 5,300-bp circular DNA by treatment with the three enzymes, and it is the relatedness of these fragments with those obtained by the previous treatments that allows for completion of the genomic structure. Note that Hpa I-A and -B yield the 500- and 1,500-bp fragments, respectively, that were used to initiate construction of the molecule. This information, as depicted in the figure, allows for the ordering of the 1,500-bp Hpa I-B segment and also of Hpa I-A with respect to Alu I-D (1,300-bp Hpa I-A ≈ 500-bp Alu I-D + 800-bp Hpa I-A fragment).

With the above information regarding the ordering of certain of the fragments, the remainder of the circular DNA molecule can be ascertained by continuing to compare the sizes of the fragments listed below:

EcoR I/Hpa I/Alu I	EcoR I/Alu I
1,750 bp (Hpa I-B)	2,550 bp (Alu I-A)
1,500 bp (Hpa I-B)	*1,500 bp (Alu I-B)*
800 bp (Hpa I-A)	750 bp (Alu I-C)
750 bp (Hpa I-B)	*500 bp (Alu I-D)*
500 bp (Hpa I-A)	
5 300 bp total	5,300 bp total

In the above listing, those written in italics have been assigned positions in the partial structure already constructed. Thus, the remaining segments whose placements remain to be identified are the 1,750- and 750-bp products of Hpa I-B cleavage, Alu I-A (2,550 bp), and Alu I-C (750 bp). Both sets of segments have a 750-bp fragment, and an association of the 1,750-bp fragment (Hpa I-B) with Alu I-A (2,550 bp) can be established as follows.

Since the 500-bp cleavage product of 1,300-bp Hpa I-A is the same as Alu I-D, the remaining 800 bp of Hpa I-A can be added to the 1,750-bp product of Hpa I-B cleavage to obtain the 2,550-bp Alu I-A. Thus, based on this rationale, the 1,750-bp Hpa I-B fragment can be linked to 1,300-bp Hpa I-A, and the Alu I-A fragment can be placed adjacent to Alu I-D. The one remaining unassigned segment is Alu I-C (750 bp) which, in view of the positions of all the other fragments, can be situated only between Alu I-A and Alu I-B. As illustrated in Figure 21-16, the circular DNA molecule derived accommodates all the data obtained by the different fragmentation procedures.

This hypothetical example of restriction mapping illustrates how the physical structure of a DNA molecule can be constructed by comparing its products after various endonuclease digestions to establish segment overlaps. When this type of biochemical analysis is coupled with genetic studies on the molecule, the combined results provide an accurate chemical and biological evaluation of a genetic system.

RECOMBINANT DNA MOLECULES

Procaryotic restriction endonucleases also proved to be a key laboratory tool in the development of recombinant DNA research. The feasibility of constructing a recombinant DNA molecule composed of genetic material from two different sources capitalizes on the fact that a restriction enzyme cleaves *any* double-stranded DNA not bearing a particular methylation pattern. How this information is used to construct DNA-recombinant molecules, utilizing endonucleases that produce staggered cuts, is schematically depicted in Figure 21-17.

In Vitro Construction of a Recombinant DNA Molecule

One method employed to produce recombinant DNA molecules relies on the use of a *bacterial plasmid* as one source of DNA. These plasmids are *circular extrachromosomal DNA molecules,* which are harbored by many species of bacteria. The well-studied plasmids of *E. coli* have been adopted for use in much of the research. The second source of DNA is the investigator's choice, since it can be genetic material from any organism. The use of eucaryotic DNA, e.g., fruitfly, mouse, or human, is illustrated in Figure 21-17. As indicated, the two DNAs are independently

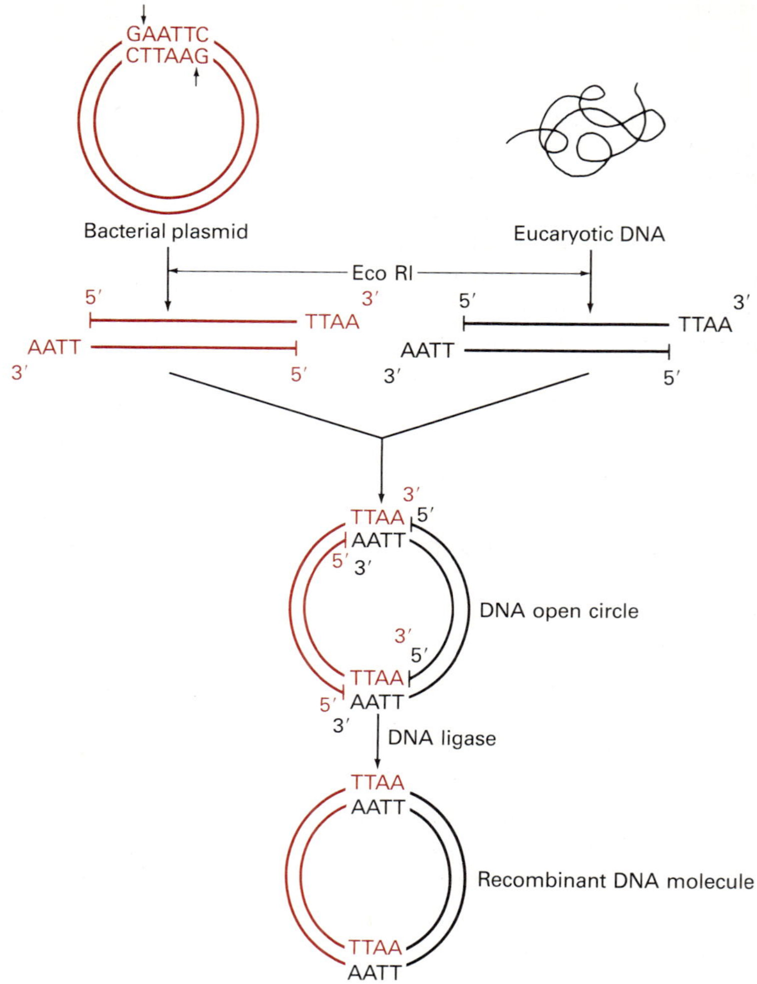

Figure 21-17
Schematic of the *in vitro* construction of a recombinant DNA molecule.

treated with an endonuclease (EcoR I), which produces identical staggered cuts in both molecules. The specific cleavage produces linear double-stranded DNAs with "sticky ends," i.e., single-stranded 3′-termini on each of the cleaved products. With regard to large-molecular-weight eucaryotic DNA, a variety of DNA fragments are produced, since the molecule probably has a number of nonmethylated endonuclease-specific sites. The linear double-stranded plasmid DNA and an isolated eucaryotic fragment (or the collection of fragments) are mixed under appropriate conditions, and the complementary "sticky ends" of the linear DNAs base-pair with one another to form an *open-circle DNA molecule.* With the use of the DNA ligase reaction, the four hydrogen-bonded 3′- and 5′-termini of the open-circle molecule are covalently linked to produce a *closed-circle recombinant DNA molecule* with genetic information from two different sources.

In Vivo Propagation of Recombinant DNA Molecules

Once a recombinant DNA molecule, also called *chimeric DNA,* is constructed, it is then propagated *in vivo* by experimentally introducing it into an *E. coli* cell, where it is maintained (Figure 21-18). Acquisition of a chimeric DNA molecule by *E. coli* requires that the bacteria first be treated with Ca^{2+} to induce *competency,* i.e., a physiological state during which a cell is capable of taking up exogenous DNA (recombinant DNA molecule in this case). The recombinant DNA molecule, having the genetic information of a bacterial plasmid (a replicon), undergoes replication that is independent of that of the cell's chromosome (but linked with cell duplication). Hence, because of the molecule's replication and the rapid cellular division of *E. coli,* the genetic material originally incorporated into a recombinant DNA molecule is amplified, i.e., numerous copies are made and distributed among many billions of bacterial cells. The bacterial plasmid is called a *vector,* since its genetic information allows for the harboring and replication of the recombinant molecule in an *E. coli* cell.

Chimera—a fire-breathing monster of Greek mythology, having a lion's head, a goat's body, and a serpent's tail.

In a second experimental approach, circular DNA of a bacteriophage substitutes for that of the bacterial plasmid in the construction of a recombinant DNA molecule. In these experiments, the bacteriophage serves as the vector, since its genetic information is responsible for the infection of an *E. coli* cell and the subsequent replication of the recombinant molecule. The topic of recombinant DNA research is considered in more detail in Chapter 26.

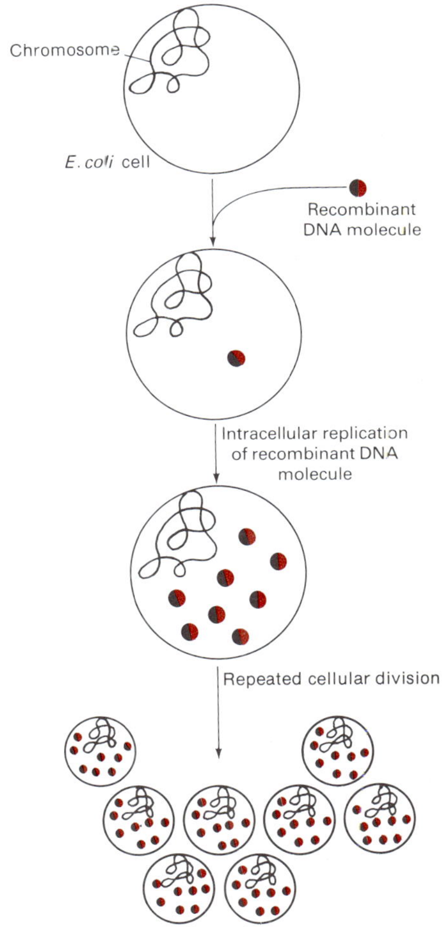

Figure 21-18
Schematic of the *in vivo* propagation of a recombinant DNA molecule.

DETERMINATION OF DNA SEQUENCES

By the late 1970s, two simple and rapid methods for determining the nucleotide sequence of single-stranded DNA had been published. Although Robert Holley (Nobel Prize, 1968) succeeded in providing a breakthrough for the sequencing of tRNAs (≈75 nucleotides), it remained an extremely laborious task to obtain the complete nucleotide sequences of larger RNA and DNA molecules (hundreds or thousands of nucleotides). The first relatively simple method (plus-and-minus method) for sequencing single-stranded DNA was announced in 1975 by Frederick Sanger (Nobel Prize, 1958) and Alan R. Coulson (*Journal of Molecular Biology,*

94:441); it uses enzymatic analysis of DNA fragments. This outstanding achievement was followed by an equally impressive contribution by Allan Maxam and Walter Gilbert, who in 1977 published a convenient chemical method for sequencing single-stranded DNA (*Proceedings of the National Academy of Science, U.S.A.,* 74:560). The scientific significance of their contributions was recognized in 1980 when Gilbert and Sanger were awarded Nobel prizes—the second time this honor was bestowed on Sanger.

Plus-and-Minus Method of Sanger and Coulson

For DNA-sequencing studies, electrophoresis is performed in the presence of 8 *M* urea which, by denaturing the single-stranded DNA fragments, allows for their separation on the basis of size.

Sanger and Coulson's method incorporates an ingenious use of DNA replication and DNA degradation to produce mixtures of DNA fragments, each of which is then resolved by acrylamide gel electrophoresis to determine oligonucleotide sequences.

Partial Synthesis of a Complementary Strand

The first step involves the initiation of DNA replication, using a single-stranded DNA as a template. The single-stranded DNA is obtained from a double-stranded fragment produced by restriction endonuclease digestion. As indicated in Figure 21-19, a *synthetic primer* (a sequence of ten nucleotides in Sanger's work) and *DNA polymerase I* are used to commence the synthesis of a DNA strand complementary to the template. Since the use of a primer neces-

Figure 21-19
Plus-and-minus method I: initiation of DNA replication using a single-stranded DNA template.

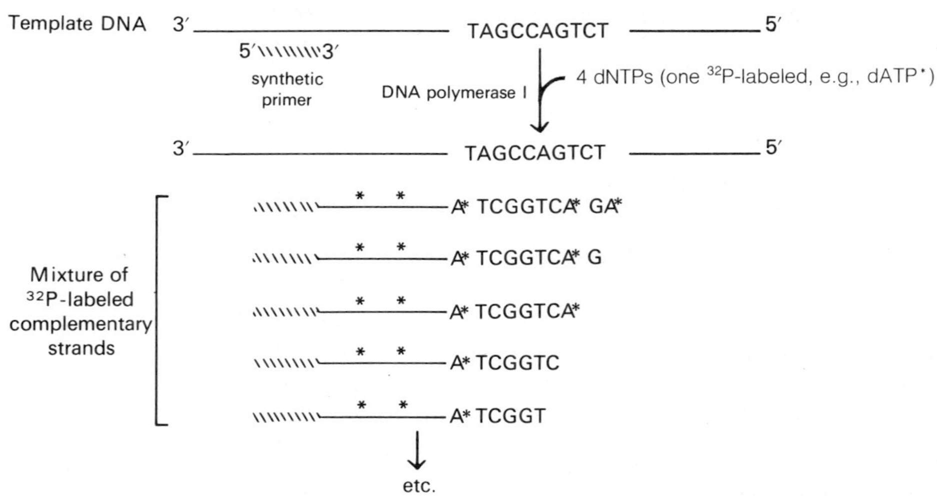

sitates prior knowledge of a complementary sequence in the templates, a small complementary strand (produced from double-stranded DNA by separate digestions with two endonucleases) can be substituted. With the four deoxyribonucleoside triphosphates present, DNA polymerase I catalyzes chain elongation of the 3′-terminus of primers. Note that one of the nucleoside triphosphate substrates is α-^{32}P-labeled, ensuring the production of complementary strands that are radioactivally "tagged." This step in the method is limited to partial replication of the DNA template and, ideally, a mixture of fragments of varying lengths is produced (Figure 21-19). The collection of heterogeneous fragments is then used in the plus-and-minus systems.

The Minus System

For the minus system, a sample of the mixture of newly synthesized strands (still hydrogen bonded to their templates) is used to reinitiate DNA replication. Four different replicating conditions are used, with each system lacking *(minus)* one of the four nucleotide substrates. The experimental design permits replication to proceed on the strands in each system until there is a need to incorporate the missing nucleotide, at which point chain elongation terminates. Figure 21-20 depicts a hypothetical minus system using a nucleotide sequence of ten bases located in the interior of the single-stranded DNA template illustrated in Figure 21-19. Reinitiation of DNA replication, using DNA polymerase I, is demonstrated with ten complementary strands that terminate within the identified sequence of the template.

In the minus system lacking dATP (Figure 21-20), DNA replication of fragment 10 proceeds until an adenyl nucleotide is to be incorporated (denoted by an arrow, since termination is at a point

Figure 21-20
Plus-and-minus method II: minus system (chain elongation using DNA polymerase).

		Template DNA: 3′-TAGCCAGTCT-5′	Deoxyribonucleoside triphosphate lacking (−): dATP	dGTP	dCTP	dTTP
Mixture of ten fragments of varying lengths	10	5′-ATCGGTCAGA-3′	⟶	⟶	⟶	⟶
	9	-ATCGGTCAG	-ATCGGTCAG(9)	⟶	⟶	⟶
	8	-ATCGGTCA	-ATCGGTCAG	-ATCGGTCA(8)	⟶	⟶
	7	-ATCGGTC	-ATCGGTC(7)	-ATCGGTCA	⟶	⟶
	6	-ATCGGT	-ATCGGTC	-ATCGGTCA	-ATCGGT(6)	⟶
	5	-ATCGG	-ATCGGTC	-ATCGGTCA	-ATCGGT	-ATCGG(5)
	4	-ATCG	-ATCGGTC	-ATCG(4)	-ATCGGT	-ATCGG
	3	-ATC	-ATCGGTC	-ATC(3)	-ATCGGT	-ATCGG
	2	-AT	-ATCGGTC	-ATC	-AT(2)	-ATCGG
	1	-A	-ATCGGTC	-ATC	-AT	-A(1)

beyond the identified sequence). Fragment 9 cannot elongate its DNA chain, since dATP is required for the next nucleotide incorporation; however, fragment 8 incorporates one nucleotide before replication ceases. Fragment 7 is also unable to reinitiate DNA synthesis, but all remaining fragments elongate their chains to the length of that of fragment 7, since no dATP is required. Hence, within the limits of the illustrated example, the minus system lacking dATP (−dATP) produces two classes of fragments, one the length of fragment 9 and the other of fragment 7. The same rationale, when applied to the particular conditions of the other three minus systems, produces three classes of fragments for the −dGTP system and two each for the −dCTP and −dTTP systems.

The four incubation mixtures of the minus system are then denatured to free the newly synthesized strands from their templates, and the strands are then subjected to acrylamide gel electrophoresis in the presence of 8 *M* urea. An electrophoretic procedure was developed to allow for the separation of DNA strands of different sizes, even if the length difference between two strands is only one nucleotide. After electrophoresis, an autoradiograph of each gel is prepared by placing the gel in contact with photographic film, which detects the ^{32}P radioactivity of the DNA strands. The radioactive DNA strands in the gels blacken the emulsion on the film and, as a result, a series of dark bands is produced which identifies the location of the various strands. As illustrated in Figure 21-21, when the four autoradiographs are compared, the order of the base sequence (in a 5′ → 3′ direction) can be read upward starting from the bottom bands. Note in the illustrated example that the terminal 5′-base (A) is not identified by the minus analysis. Each fragment identifies which base is next in the 5′ → 3′ sequence, e.g., band 1 (−A in dTTP column) identifies thymine as the next base toward the 3′-terminus. The order of the original single-stranded DNA obtained from a restriction-enzyme fragment (Figure 21-19) is an antiparallel complementary copy of the sequence determined:

Determined sequence	5′—T—C—G—G—T—C—A—G—A—3′
DNA template	3′—A—G—C—C—A—G—T—C—T—5′

The Plus System

The plus system, which verifies the sequence derived by the minus system, also makes use of a portion of the mixture of newly replicated strands and their templates produced in the first step (Figure 21-19). In the system, the fragments are subjected to degra-

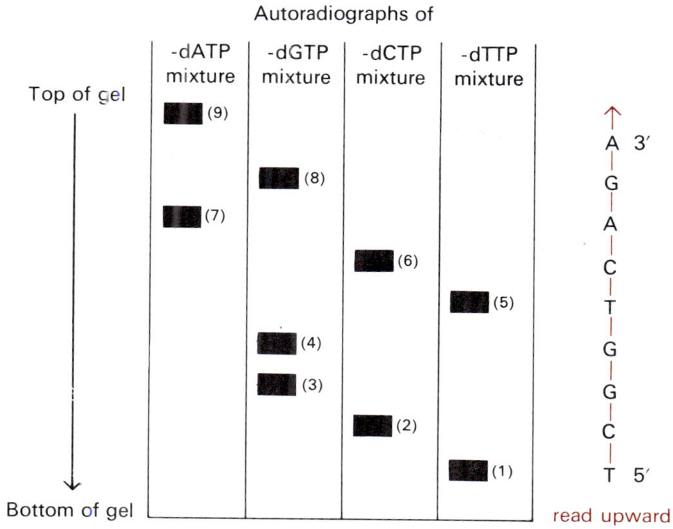

Number in parentheses beside each autoradiographic band refers to the fragment so numbered in Figure 21-20.

Figure 21-21
Plus-and-minus method III: schematic of a comparison of the autoradiographs obtained from the polyacrylamide gels following the electrophoresis of each of the four incubation mixtures of the minus system illustrated in Figure 21-20.

dation by the 3′-exonuclease activity of *DNA polymerase of bacteriophage T4.* An unusual feature of this exonuclease is that, in the presence of a deoxyribonucleoside triphosphate, the enzyme degrades double-stranded DNA from its 3′-terminus but terminates its activity at residues corresponding to the nucleotide present. For example, if dATP is present in the reaction mixture, degradation ceases when an adenyl residue in the sequence is to be cleaved next.

In the plus system, four different mixtures are used, with one of the four deoxyribonucleoside triphosphates being added *(plus)* to each system. A hypothetical system is illustrated in Figure 21-22

Figure 21-22
Plus-and-minus system IV: plus system (3′-exonuclease degradation).

Template DNA 3′-TAGCCAGTCT-5′			Deoxyribonucleoside triphosphate added (+) dATP	dGTP	dCTP	dTTP
Mixture of ten fragments of varying lengths	10	5′-ATCGGTCAGA 3′	-ATCGGTCAGA(10)	-ATCGGTCAG(9)	-ATCGGTC(7)	-ATCGGT(6)
	9	-ATCGGTCAG	-ATCGGTCA(8)	-ATCGGTCAG	-ATCGGTC	-ATCGGT
	8	-ATCGGTCA	-ATCGGTCA	-ATCGG(5)	-ATCGGTC	-ATCGGT
	7	-ATCGGTC	-A(1)	-ATCGG	-ATCGGTC	-ATCGGT
	6	-ATCGGT	-A	-ATCGG	-ATC(3)	-ATCGGT
	5	-ATCGG	-A	-ATCGG	-ATC	-AT(2)
	4	-ATCG	-A	-ATCG(4)	-ATC	-AT
	3	-ATC	-A	←	-ATC	-AT
	2	-AT	-A	←	←	-AT
	1	-A	-A	←	←	←

with the same oligonucleotide sequence and collection of partially replicated fragments used to demonstrate the minus system (Figure 21-20). In the reaction containing dATP, the 3′-exonuclease activity cannot degrade fragment 10 with its adenyl 3′-terminus; however, one nucleotide can be removed from fragment 9. Fragment 8 is also unaffected by the exonuclease, but the remaining seven fragments are all degraded to the length of fragment 1. An analogous rationale explains the strand degradation that occurs in each of the other three plus systems. (The arrows in the figure denote that degradation of those particular fragments extends beyond the base sequence depicted.) The four incubation mixtures of the plus system are then subjected to the denaturation, electrophoretic, and autoradiographic procedures described for the minus system. When the four autoradiographs of the gels (Figure 21-23) are compared, reading the order of bases upward from the bottom of the gels reveals a 5′ → 3′ sequence, the identical order derived from the minus system (plus the 5′-terminal adenyl nucleotide). With the use of this elegant technique, Sanger and his collaborators sequenced the first complete genome, the order of 5,375 nucleotides of the DNA of bacteriophage ϕX174 (p. 527).

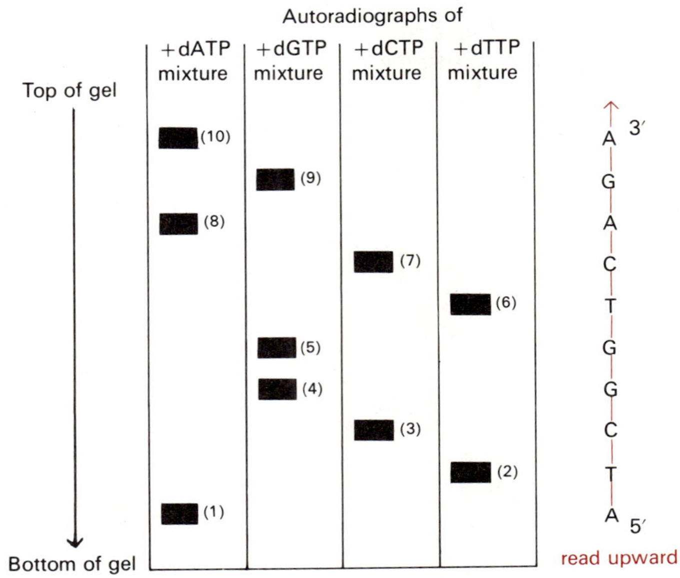

Number in parentheses beside each autoradiographic band refers to the fragment so numbered in Figure 21-22.

Figure 21-23
Plus-and-minus system V: schematic of a comparison of the autoradiographs obtained from polyacrylamide gels following the electrophoresis of each of the four incubation mixtures of the plus system illustrated in Figure 21-22.

Chain-Termination System

Since the plus-and-minus method was introduced, Sanger and his associates have developed a second DNA-sequencing protocol, one that is more rapid and accurate than the first. This newer procedure, which has replaced the older method, makes use of analogues of the deoxyribonucleoside 5′-triphosphate substrates required for DNA synthesis (chain elongation). These nucleotide analogues have either 2,3-dideoxy-D-ribose or D-arabinose instead of 2-deoxy-D-ribose as their carbohydrate component, e.g., 2′,3′-dideoxy analogue of dGTP or arabinosyl analogue of dATP. When such an analogue is added to an elongation system that utilizes *E. coli* DNA polymerase I (similar to the conditions described for the minus system), its incorporation into the DNA sequence terminates elongation - hence, the name *chain-termination system.* In the case of the incorporation of a 2′,3′-dideoxy analogue, its lack of a 3′-hydroxyl group precludes the addition of the next nucleotide; likewise, the presence of arabinose (a stereoisomer of ribose) at the 3′-terminus of a growing chain prevents additional incorporation. Currently, the analogues of choice for these studies are the dideoxy compounds.

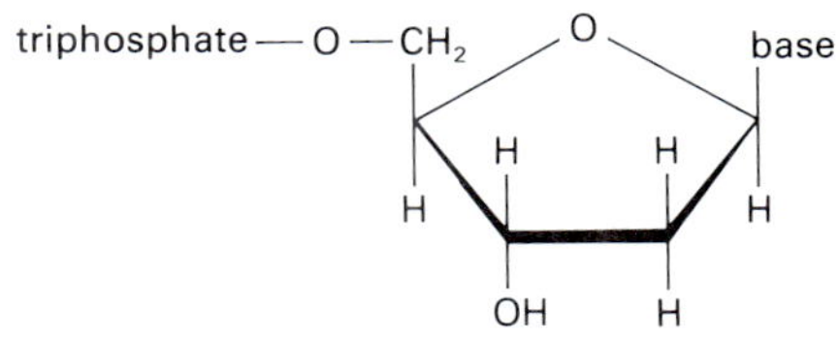

2′-Deoxy-D-ribonucleoside 5′-triphosphate

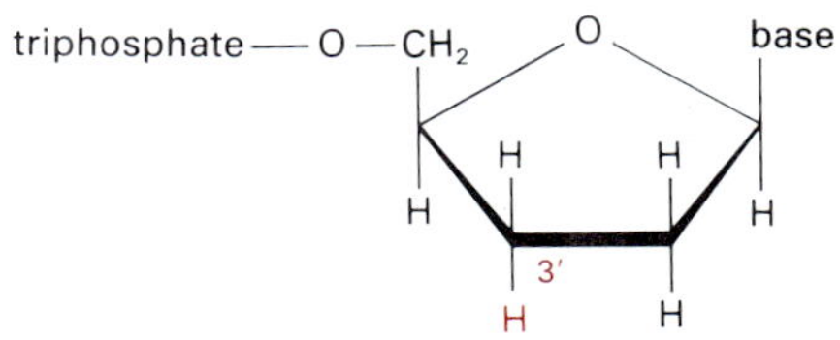

2′, 3′-Dideoxy-D-ribonucleoside 5′-triphosphate

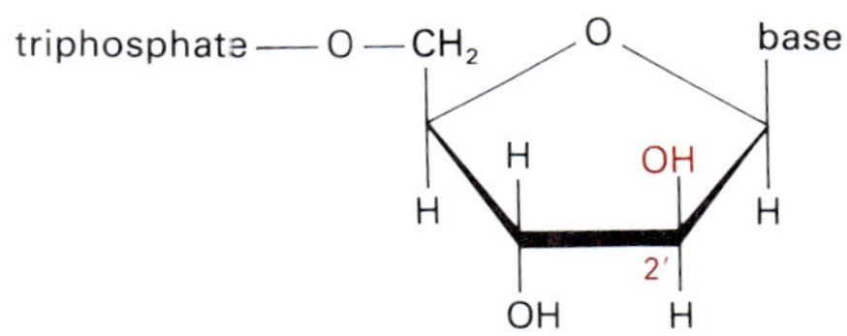

D-Arabinonucleoside 5′-triphosphate

One advantage of the chain-termination system over the first method is that no preliminary or partial synthesis (p. 456) is required. Instead, synthesis of the complementary strand to be analyzed proceeds by the addition of nucleotides to a primer (a short sequence obtained by a restriction enzyme cleavage). A hypothetical example of the chain-termination system is presented in Figure 21-24. As indicated, the 2′,3′-dideoxy analogue of dGTP (ddGTP) is added to an elongation system containing DNA polymerase I and the four natural deoxyribonucleoside triphosphates, one of which is α-^{32}P labeled to provide for the radioactive "tagging" of the DNA fragments. (The DNA sequence being extended is the same as that used for the plus-and-minus system, Figure 21-20.) Under these conditions, ddGTP terminates chain elongation at those sites in the sequence at which guanyl nucleotides are incorporated. In the example, a mixture of three fragments (4, 5, and 9) are obtained as products. A critical feature of this system

Elongation system substrates: dGTP, dCTP, dTTP, [α − ^{32}P]dATP, and ddGTP(G)	
DNA template: 3′ − TAGCCAGTCT − 5′	
Chain termination fragments produced:	5′ − ATCGGTCAG − 3′ (9)
	− ATCGG (5)
	− ATCG (4)

Figure 21-24
Illustration of the results of the chain-termination system using the 2′,3′-dideoxy analogue of GTP (ddGTP).

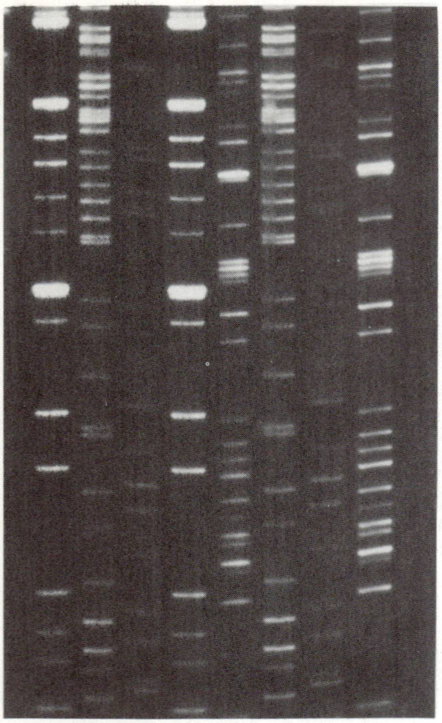

C T A C G T A G

Figure 21-25
Section of an eight-channel gel pattern (two channels for each of the four bases) obtained by the chain-termination system. The DNA being sequenced is that of maize mitochondria, sections of which were inserted into the single-stranded bacteriophage M13 (vector) and cloned in *E. coli* (host). (Courtesy of Dr. C. S. Levings III)

is that the ratio of the terminating nucleotide to the natural nucleotide, e.g., ddGTP to dGTP, is adjusted so that incorporation of the terminating analogue occurs in a random manner (to ensure obtaining a collection of fragments representing all guanyl nucleotide incorporation sites). The ratios are 100 and 5,000 for the dideoxy and arabinosyl analogues, respectively. In three similar experimental systems and with each containing the appropriate analogue, the adenyl, cytidyl, and thymidyl sites can also be located. Subsequent analyses of the four reaction mixtures follow the electrophoretic and autoradiographic protocols described for the plus-and-minus system. The 5′ → 3′ nucleotide sequence is read from gel patterns (Figure 21-25) in the same manner as that described for the plus-and-minus method (p. 459).

The Chemical Method of Maxam and Gilbert

The second method for determining deoxyribonucleotide sequences relies on the selective chemical cleavage of single-stranded DNA that, as in the Sanger and Coulson method, is obtained from double-stranded fragments produced by restrictive endonuclease digestion. For DNA sequencing, endonucleases which catalyze blunt cuts (Table 21-3) are utilized. Prior to chemical treatment, a dephosphorylated (alkaline-phosphatase treated) DNA strand is labeled with ^{32}P at its 5′-terminus (Figure 21-26) to produce a terminally labeled strand. (The labeling of the 3′-terminus is an alternative way of "tagging" a strand with radioactivity and requires [γ-^{32}P] ATP and terminal transferase.) Double-stranded fragments can also be terminally labeled by the same enzymes, after which the strands are separated for analysis. Selective fragmentation of a labeled DNA strand is then accomplished with the use of four separate chemical treatments, each designed to produce specific cleavage and radioautographic patterns.

The Two Treatments for Selective Cleavage at Purine Positions
For the cleavage at purine positions, a DNA strand is treated with *dimethylsulfate,* which methylates *adenyl* and *guanyl* residues at

Figure 21-26
Labeling of the 5′-terminus of a single-stranded DNA fragment by polynucleotide kinase.

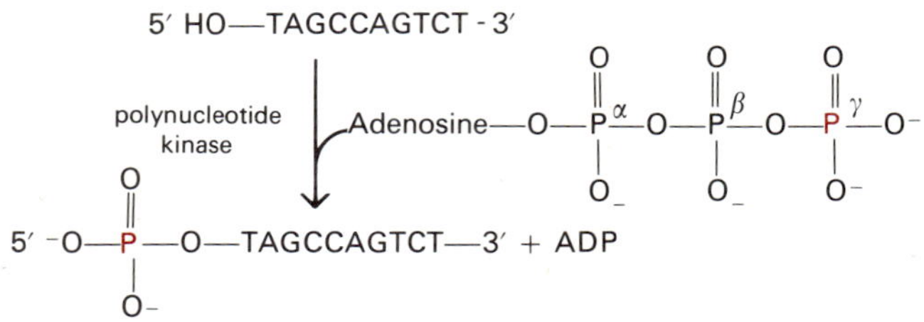

N-3 and *N-7*, respectively (Figure 21-27). The reaction is a limited one which methylates only one purine for every 50 to 100 bases in a strand, ensuring a random distribution of methylated purines in the strands. Because the glycosidic bond of a methylated purine is unstable, the bases are easily removed by heating the strands at neutral pH, as illustrated in Figure 21-27 by cleavage treatment 1 of the strand with a methylated guanine. Subsequent treatment with mild alkali at 90° C frees depurinated deoxyribose residues completely from their 3′- and 5′-phosphate linkages (Figure 21-28), producing a variety of labeled fragments of the DNA strand. When the fragments are resolved by polyacrylamide gel electrophoresis and autoradiographs are prepared, a pattern of dark and light bands is obtained (*dark G-light A* column in Figure 21-29). The dark bands are fragments produced by the loss of guanyl nucleotides since, in the methylation reaction, guanine is methylated at a rate five times faster than adenine. Hence, the greater number of fragments derived from guanine cleavages produces more intense (darker) bands in an autoradiograph than those created by the loss of adenine.

Although the results of the first treatment identify the positions of purines in a strand, a second sample of methylated strands is cleaved in a different manner to confirm the findings. This second procedure was developed to enhance the removal of methylated adenines, whose glycosidic bonds are less stable than those of methylated guanines. Hence, by *gentle treatment with dilute acid,* methylated adenyl residues are preferentially released (cleavage treatment 2 in Figure 21-28). The autoradiograph subsequently produced from methylated strands treated in this mild manner (*dark A-light G* column in Figure 21-29) reveals darker bands for the strands cleaved at adenyl positions. The two selective treatments of a DNA strand with methylated purines identify and verify all purine positions.

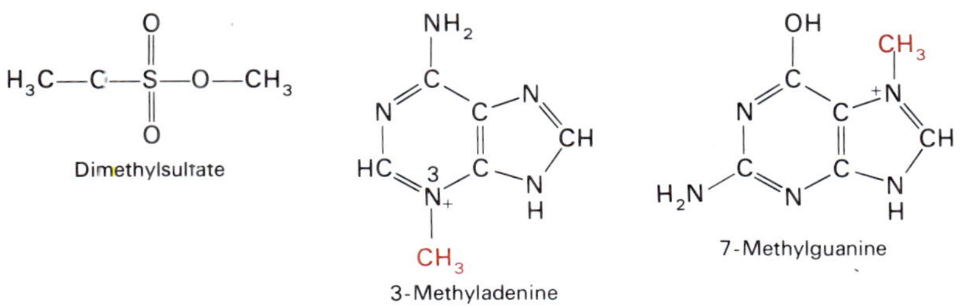

Figure 21-27
Dimethylsulfate, 3-methyladenine, and 7-methylguanine.

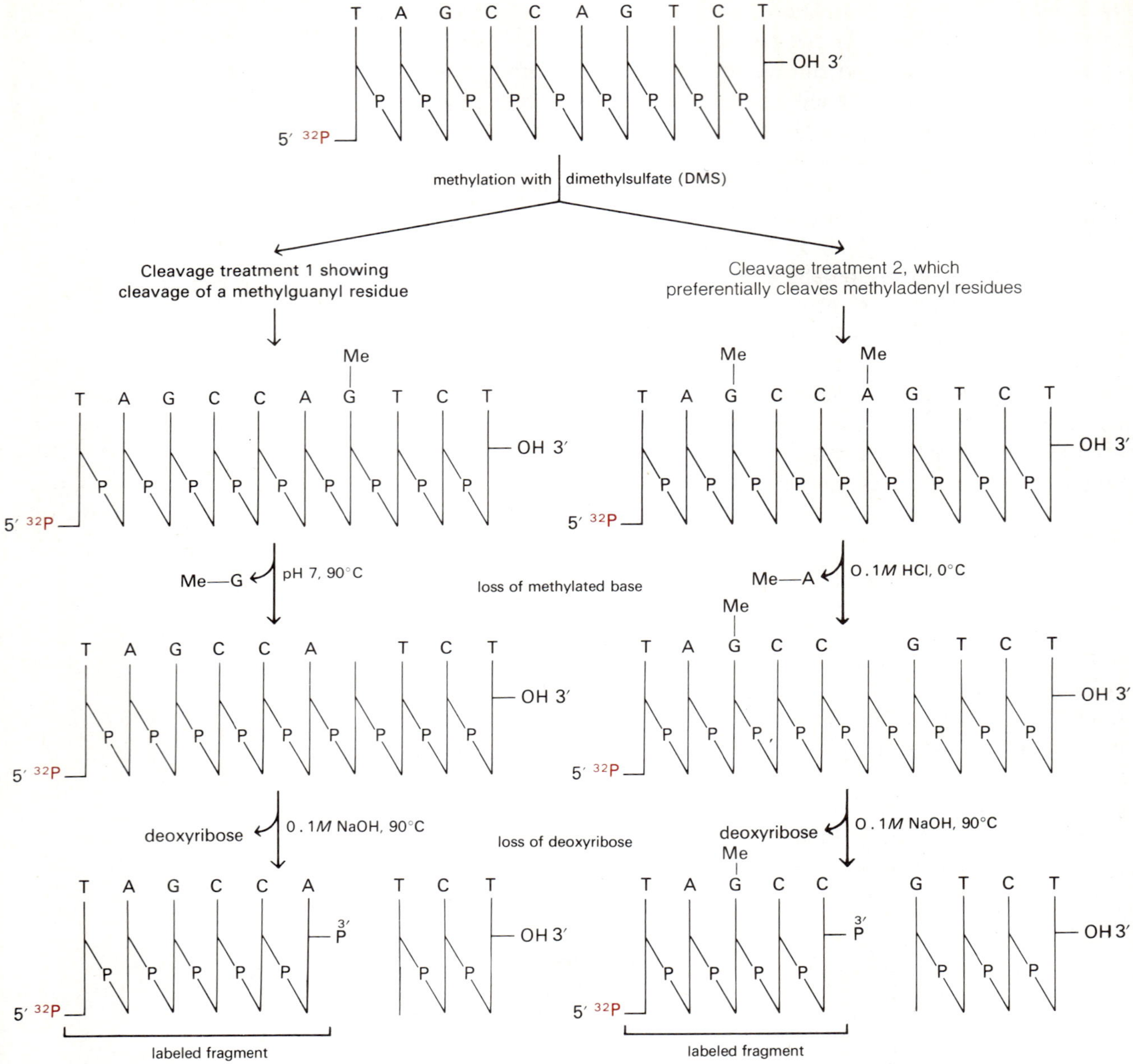

Figure 21-28
Schematic of the cleavage of a hypothetical DNA strand at purine residues.

The Two Treatments for Selective Cleavage at Pyrimidine Positions An analogous chemical approach is utilized to fragment a DNA strand selectively at pyrimidine positions. A strand is treated with hydrazine, which reacts with and destroys thymine and cytosine residues. After hydrazinolysis, the strand is cleaved with aqueous piperidine at 90° C, with frees the deoxyribosyl residues previously linked to pyrimidines, yielding fragments of the DNA strand. The

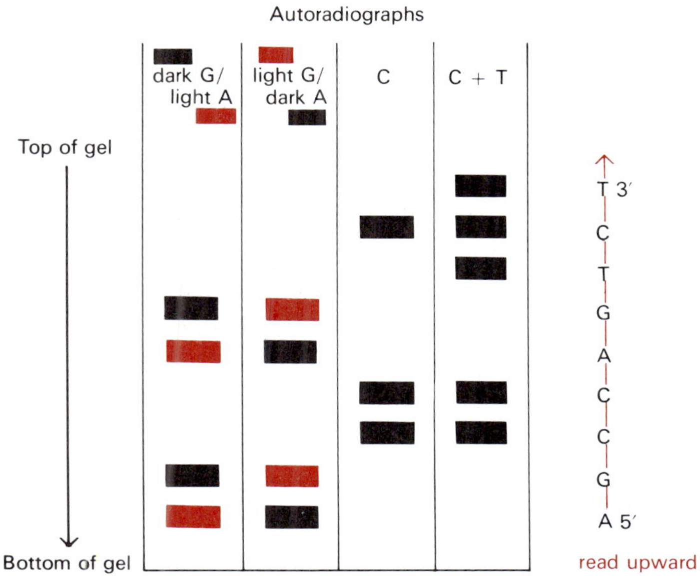

Compare the sequence derived with that depicted in Figure 21-28.

Figure 21-29
Patterns of bands in the autoradiographs obtained from polyacrylamide gels following the electrophoresis of a DNA fragment (Figure 21-28) that underwent the four chemical treatments of the Maxam and Gilbert method.

autoradiographs obtained after electrophoretic resolution of these fragments produce bands of similar intensity from cleavage at both cytosine and thymine positions (C + T column in Figure 21-29).

H_2NNH_2
Hydrazine

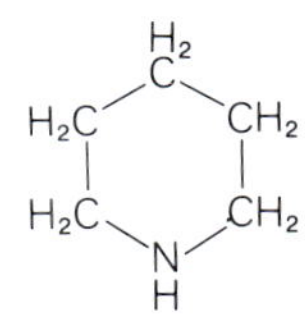

Piperidine

The hydrazine reaction is repeated in the presence of *2 M* NaCl on another sample of the terminally labeled strand. The salt suppresses the hydrazine reaction with thymine and, as a result, only fragments cleaved at cytosyl positions (by piperidine treatment) are eventually detected in an autoradiograph (column C in Figure 21-29). When the C and C + T autoradiograph patterns are compared, a band appearing in both autoradiographs denotes cleavage at a cytosine, whereas a band present only in the C + T pattern represents breakage at a thymine.

When the autoradiographic patterns derived from the four chemical treatments are compared, the sequence read from the bottom upward is the 5′ → 3′ order of the original DNA strand analyzed (Figure 21-28). Only the 5′-terminal base (T) is not identified.

Newer Method for Purine Determinations

As with the plus-and-minus system, the chemical method of Maxam and Gilbert has also undergone revision. The major changes involve the chemical procedures for identifying purine positions in a DNA sequence. Whereas the original protocol relies on both the faster rate of guanine methylation by dimethylsulfate

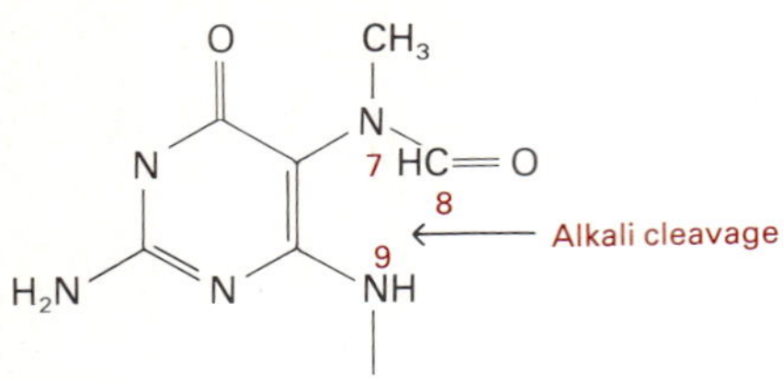

Cleavage of a 7-methylguanine residue

(G > A) and the more sensitive lability of the glycosidic bond of a methylated adenine (A > G), the two newer methods utilize chemical treatments that result in the loss of (1) guanyl nucleotides (G) and (2) guanyl and adenyl nucleotides (A + G). These newer protocols are easier and faster to perform than the older ones.

For the specific removal of guanyl residues, advantage is taken of the fact that at an alkaline pH of 10 or higher, the bond between C-8 and N-9 of 7-methylguanine is cleaved. The cleaved methylated purine and its deoxyribosyl residue can then be sequentially removed by treatment with piperidine. Thus, following limited purine methylation, the DNA fragments are treated with 1 *M* piperidine which, as a free base, brings about not only ring cleavage of 7-methylguanine residues but also their release from the DNA sequences and elimination of their deoxyribose moieties. The loss of the guanyl and deoxyribosyl residues by piperidine treatment is analogous to that previously described for pyrimidine identification. Scissions in the DNA fragments are specific for guanyl sites because the just-mentioned three-reaction effect of piperidine does not take place with 3-methyladenine.

For determination of guanyl and adenyl positions, use is made of *acid depurination,* i.e., loss of purine residues by acid treatment of DNA. The use of acid to produce *apurinic acids* (purineless DNA) is an old technique in nucleic acid chemistry. Loss of the purines occurs because their protonation by acid, like methylation, destabilizes their glycosidic bonds. In the Maxam and Gilbert procedure, limited acid depurination, to assure random cleavage, is accomplished by mild acid treatment of the DNA fragments with 4 per cent formate, adjusted to pH 2 with piperidine (piperidinium formate). After depurination, the fragments are treated with 1 *M* piperidine to free the deoxyribosyl residues, i.e., to cleave the fragments. In this procedure, both adenyl and guanyl sites are identified.

Thus, the two newer methods for the purines (G and A + G) are currently coupled with the two described previously for the pyrimidines (C and C + T) to obtain the information needed to determine DNA sequences. Figure 21-30 illustrates the autoradiographic patterns that would be obtained if the sequence determined by the original method (Figures 21-28 and 21-29) was analyzed by substituting the newer procedures for the purines.

Comments on DNA Sequencing Methods

These two relatively simple, but remarkable, sequencing techniques are limited only by the resolving capabilities of the electro-

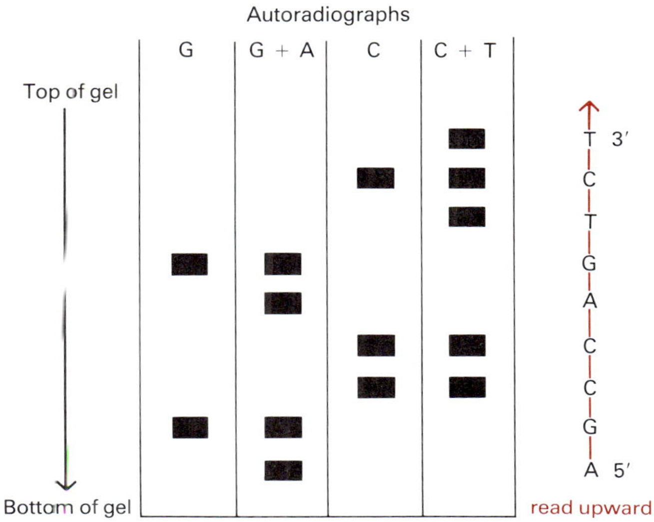

Figure 21-30
Patterns of bands in the autoradiographs obtained from polyacrylamide gels following the electrophoresis of a DNA fragment (Figure 21-28) which underwent the four chemical treatments of the newer, modified Maxam and Gilbert method.

phoretic procedures used. It is now electrophoretically possible to resolve more than 200 fragments of different lengths. Now that a DNA sequence can be determined with relative convenience, many amino acid sequences of proteins and hormones are being inferred by elucidation of the nucleotide sequences of the genes that code for the individual polypeptides.

SUMMARY

The Central Dogma of molecular biology, which stresses the flow of biological information, has been revised in recent years to accommodate the discovery of reverse transcriptase, which catalyzes RNA-directed DNA synthesis. The known modes for the transfer of biological information now include the self-replication of DNA and viral RNA, DNA-directed RNA synthesis, RNA-directed DNA synthesis (in RNA virus-infected cells), and RNA-directed protein synthesis.

Studies on DNA replication have revealed that a series of complex molecular events is involved. In DNA replication, complementary synthesis of the 3′ → 5′ parent strand (leading strand) takes place in a continuous manner; however, on the 5′ → 3′ strand (lagging strand), synthesis occurs in a discontinuous fashion. In discontinuous synthesis, complementary fragments are produced; these are then covalently linked to one another to yield an intact daughter strand of DNA. Replication occurs in four stages. Prepriming, the first stage, prepares the two DNA strands for replication, and at least a half dozen proteins participate in the process. The second stage, priming, entails the 5′ → 3′ synthesis of a small RNA primer (complementary to the parent DNA strand) by dnaG protein. The synthesis of additional RNA primers at different sites along the lagging strand subsequently follows. In elongation, the third stage, DNA polymerase III initiates synthesis of DNA sequences complementary to the parent strand by adding deoxyribonucleotides to the 3′-terminus of RNA primers. At this point in the replication process, each replicated fragment along the lagging strand has both an RNA (primer) and a DNA sequence. In the fourth stage, termination, DNA polymerase I performs the dual role of removing the ribonucleotides (RNA primer) from the fragments and replacing these primers with deoxyribonucleotide sequences. It is in this manner that the replicated fragments are converted into deoxyribonucleotide sequences, whose covalent linkage to one another (by phosphodiester bonds) is then catalyzed by DNA ligase.

Cells are capable of repairing their damaged DNA. In the case of UV irradiation, the damage is primarily due to the formation of thymine dimers. In the excision-repair mechanism, the distortion created in the DNA double helix by a thymine dimer is recognized by an endonuclease, which cleaves a phosphodiester bond near the thymine dimer. DNA polymerase I then proceeds to replace the DNA sequence in the region containing the thymine dimer by adding dexoyribonucleotides to the 3′-terminus created at the site where the strand was cleaved. DNA polymerase I utilizes the other DNA strands as a template. DNA polymerase I also cleaves the second phosphodiester bond that releases the region containing the thymine dimer from the DNA strand. Once the sequence of the removed section has been resynthesized, DNA ligase catalyzes the linkage of the newly regenerated region to the existing strand.

Some bacteria are also capable of recognizing and degrading foreign DNA, e.g., that of a bacteriophage, which enters their cellular environment. Two enzymes are involved in this phenomenon that allows a bacterium to destroy foreign DNA but not its own. One enzyme (a methylase) is responsible for providing a bacterium's DNA with a distinctive methylation pattern by recognizing and methylating a particular deoxyribonucleotide sequence wherever it occurs in the genome. The other enzyme (an endonuclease) cleaves the same DNA sequence if it is not methylated and, in this way, restricts foreign DNA. The sequence recognized by both enzymes is usually a series of four to eight deoxyribonucleotides that has a twofold symmetry, i.e., the same order of bases when the two strands are read in opposite directions. Such sequences are also called palindromes and are now known to serve as distinctive sets of molecular information in DNA sequences involved in DNA-protein interactions.

Restrictive endonucleases have provided scientists with the means of constructing recombinant DNA molecules, so named because they contain genetic material from two species, e.g., DNA derived from a bacterium and from a mouse. These recombinant DNA molecules can be propagated in bacteria and amplified, i.e., numerous copies can be produced. Restrictive endonucleases are also widely used in restriction mapping, a technique used to construct physical maps of DNA molecules. In this technique, the DNA under investigation is cleaved with a series of endonucleases, and the sizes of the fragments obtained by the different cleavage treatments are determined by gel electrophoresis and then compared to determine a physical map.

By the late 1970s, two simple yet elegant procedures for sequencing DNA have been developed. One method makes use of the enzymatic synthesis (by DNA polymerase I) and degradation (by an exonuclease) of single-stranded DNA to elucidate a deoxyribonucleotide sequence; the other method utilizes chemical treatments of single-stranded DNA to selectively cleave the fragment. Both procedures rely on acrylamide gel electrophoresis to separate DNA fragments so that a base-by-base identification can be established.

Since their introduction, both methods have been modified. The enzymatic procedure now utilizes analogues of the deoxyribonucleotide 5′-triphosphates required for DNA synthesis. When such an analogue is incorporated into a DNA strand, elongation terminates. This chain-termination system is similar to the original synthetic (minus) system. In the chemical method, the two systems for determining purine positions in a DNA sequence have been replaced with newer methods, which are easier and faster to perform. In one system, guanyl sites are specifically identified and, in the other, guanyl and adenyl positions are determined.

QUESTIONS

1. Based on the current knowledge about the flow of biological information, which of the following macromolecules can serve as templates for the syntheses indicated?

Structure-specifying template for the	*synthesis of*
(a) DNA	DNA
(b) DNA	RNA
(c) DNA	Protein
(d) RNA	DNA
(e) RNA	RNA
(f) RNA	Protein
(g) Protein	DNA
(h) Protein	RNA
(i) Protein	Protein

2. What would be the products of a conservative mode of DNA replication?

3. In bacteria the DNA-ligase reaction (Figure 21-11) relies on the hydrolysis of NAD^+ to provide the energy needed. (a) Draw the structures in the reaction $NAD^+ \xrightarrow{H_2O} AMP + NMN$. (b) What type of bond in NAD^+ is hydrolyzed?

4. Complete the following palindromic sequence:

   5′—A—G—A—T—?—?—?—?—3′
   3′—?—?—?—?—?—?—?—?—5′

5. (a) Which enzymatic activities are now believed to be involved in DNA replication?
   (b) Which of these enzymes are also involved in the excision-repair mechanism?

6. If the following hypothetical patterns of autoradiographic bands were obtained by the minus system (Sanger and Coulson method), what is the probable 5′ → 3′ sequence of the single-stranded DNA fragment that was used as the template?

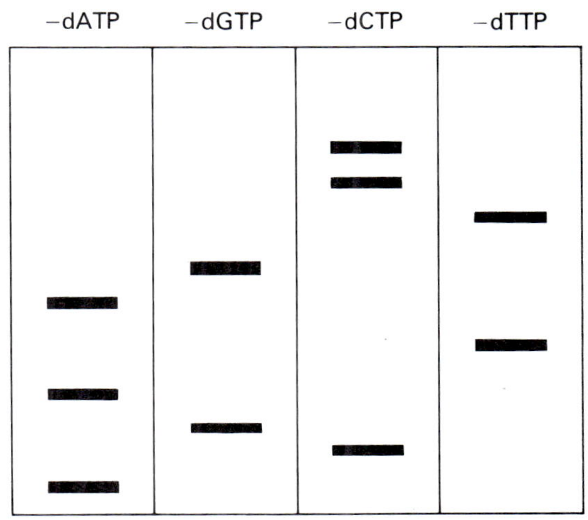

7. In Figure 21-24 an example of Sanger's chain-termination system is given using the analogue of 2′,3′-dideoxyGTP. If the same DNA template is used in chain-termination systems with the 2′,3′-dideoxy analogues of (a) dATP, (b) dCTP, and (c) dTTP, what length fragments would be obtained in each system?

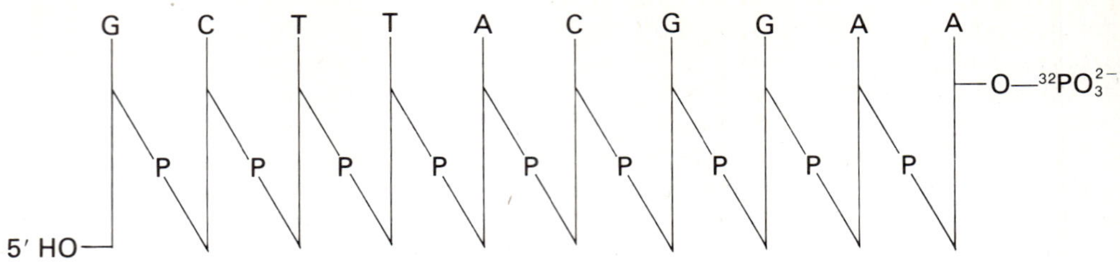

8. (a) What four patterns of bands would be expected on the autoradiographs if the sequence of the following single-stranded DNA fragment were determined by the original Maxam and Gilbert method (Figure 21-29)?

(b) With the newer method for determining purines by the Maxam and Gilbert system, what would be the patterns of bands for the two gels determining the location of the adenyl and guanyl nucleotides in the sequence of the DNA fragment?

SUGGESTED READINGS

Kornberg, A., *DNA Synthesis.* San Francisco: W. H. Freeman, 1974. An excellent, highly readable account of the molecular studies on DNA.

Kornberg, A., *DNA Replication.* San Francisco: W. H. Freeman, 1980. The highly recommended sequel to *DNA Synthesis.*

Lewin, B., *Gene Expression.* Vol. 1, *Bacterial Genomes.* New York: Wiley, 1974. This book and its two companion volumes are an authoritative and comprehensive review of the advances made toward understanding DNA and its biological function at the molecular level. The three volumes are highly recommended as references for this chapter and for Chapters 22 and 23. The volumes are available in paperback.

Lewin, B., *Gene Expression.* Vol. 2, *Eucaryotic Chromosomes,* 2nd ed. New York: Wiley, 1980.

Lewin, B., *Gene Expression.* Vol. 3, *Plasmids and Phages.* New York: Wiley, 1977.

Stent, G. S., and R. Calendar, *Molecular Genetics: An Introductory Narrative,* 2nd ed. San Francisco: W. H. Freeman, 1978. A well-written and interesting text on the historical development of molecular biology and the advances in scientific thinking this area of research has provided.

Taylor, J. H., ed., *Selected Papers on Molecular Genetics.* New York: Academic Press, 1965. A paperback collection of reprints containing many of the outstanding studies on DNA.

Watson, J. D., *Molecular Biology of the Gene,* 3rd ed. Menlo Park, Calif.: Benjamin/Cummings, 1976. Another highly recommended text on molecular biology, written by one of the leading scientists in the field.

Zubay, G. L., and J. Marmur, eds., *Papers in Biochemical Genetics,* 2nd ed. New York: Holt, Rinehart and Winston, 1973. Another valuable paperback containing reprints of the key studies on DNA.

Articles

Beadle, G. W., "The Genes of Man and Molds." *Scientific American,* 179(3):30 (offprint #1), 1948.

Cohen, S. N., "The Manipulation of Genes." *Scientific American,* 233(3):24 (offprint #1324), 1975.

Fiddes, J. C., "The Nucleotide Sequence of a Viral DNA." *Scientific American,* 237(6):54 (offprint #1374), 1977.

Grossman, L., A. Braun, R. Feldberg, and I. Mahler, "Enzymatic Repair of DNA." *Ann. Rev. Biochem.,* 44:19, 1975.

Hanawalt, P. C., and R. H. Haynes, "The Repair of DNA." *Scientific American,* 216(2):36, 1967.

Harland, R., "Initiation of DNA Replication in Eukaryotic Chromosomes." *TIBS,* 6:71, 1981.

Kasamatsu, H., and J. Vinograd, "Replication of Circular DNA in Eucaryotic Cells." *Ann. Rev. Biochem.,* 43:695, 1974.

Kornberg, A., "The Synthesis of DNA." *Scientific American,* 219(4):65 (offprint #1124), 1968.

Luria, S. E., "The Recognition of DNA in Bacteria." *Scientific American,* 222(1):88 (offprint #1167), 1970.

Nathans, D., and H. O. Smith, "Restriction Endonucleases in the Analysis and Restructuring of DNA Molecules." *Ann. Rev. Biochem.,* 44:273, 1975.

Nu, R., "DNA Sequence Analysis." *Ann. Rev. Biochem.,* 47:607, 1978.

Ogawa, T., and T. Okazaki, "Discontinuous DNA Replication." *Ann. Rev. Biochem.,* 49:421, 1980.

Temin, H. M., "RNA-Directed DNA Synthesis." *Scientific American,* 226(1):24 (offprint #1239), 1972.

Wickner, S., "DNA Replication Proteins in *Escherichia coli.*" *Ann. Rev. Biochem.,* 47:1163, 1978.

Wintersberger, E., "DNA-Dependent DNA Polymerases from Eukaryotes." *TIBS,* 2:58, 1977.

GENETIC INFORMATION: TRANSCRIPTION AND ITS CONTROL

22

INTRODUCTION

Since the early 1950s, molecular studies on DNA have provided information about the physical form by which genetic information is stored and how it is maintained and perpetuated. Within the last two decades, equally impressive contributions have also been made toward understanding the meaning of the informational content of a genome and how genes are regulated and expressed. These revelations about the transfer of genetic information within a cell and the mechanisms that control this flow of biological information are of prime importance in the quest to understand many biological phenomena, e.g., differentiation and development, at the molecular level. To comprehend the ordered translation of genetic information to the biological information used by a cell is the key to understanding the molecular programming of life.

Gene expression and its regulation became amenable to molecular analysis because of the discoveries made about the biochemical details of protein synthesis (latter half of the 1950s) and the proposal of a molecular model to explain procaryotic gene regulation, made in 1961 by François Jacob and Jacques Monod (joint Nobel Prize recipients in 1965). The proposals by Crick and his colleagues in the late 1950s concerning the chemical nature of the genetic code were a third influential contribution that greatly stimulated interest in molecular studies on the transcription of genetic information. Their brilliant deductions suggested how genetic information could be stored in the nucleotide sequence of a gene (the code) and how the information is used to specify the primary structure (amino acid sequence) of a polypeptide. Their innovative ideas, which immediately caught the imagination of the scientific community, generated the molecular hypothesis that guided experimental efforts which, within a decade, revealed the biochemical identity of the code. The deciphering of the genetic code ranks among the greatest scientific contributions of the century because, as expressed by Crick, ''It is, in a sense, the key to molecular biology because it shows how the two great polymeric languages, the nucleic acid language and the protein language, are linked together.''

The Jacob-Monod model offered provocative molecular insights into gene regulation in procaryotes and sparked the initiation of analogous studies in the regulation of eucaryotic genomes. The feasibility of probing the molecular mechanisms that switch genes on and off brought biochemical research on differentiation and development into the mainstream of molecular biology, since these biological processes rely on an ordered and selective use of an organism's genome.

Examples of genes involved in developmental processes have been indirectly alluded to in previous information presented on hemoglobin (p. 109) and urea exretion by frogs (p. 413). Embryonic, fetal, and adult hemoglobins are the oxygen carriers used by humans (and other higher vertebrates) and, during the lifetime of an individual, each form is specifically utilized in a particular stage of development, as implied by its name. The differences between the various hemoglobins, as previously described, reside in the two types of polypeptides, e.g., α, ϵ, and γ chains, that make up the biologically active tetramer. At the genetic level, the selective occurrence of each hemoglobin means, for example, that during the earliest periods of embryonic development ($\zeta_2\epsilon_2$) the genes responsible for the α, γ, and β chains of fetal ($\alpha_2\gamma_2$) and adult ($\alpha_2\beta_2$) hemoglobin are not expressed, i.e., are turned off. During the early postnatal period, on the other hand, only the α and β genes are expressed, and α and β chains continue to be produced for the remainder of a human's life. Conversely, the genes responsible for the production of the ζ, ϵ, and γ chains remain switched off permanently.

In the case of urea excretion by a frog, the genes possessing the information for the synthesis of the urea cycle enzymes are present in a tadpole's genome, but they are turned off until metamorphosis occurs. There are also many genes that are expressed only in a certain organ or tissue, e.g., the gene responsible for insulin production in the pancreas or the genes for the hormones synthesized by the pituitary (p. 547).

Studies on differentiation and development represent one of the most promising areas of research, since it is likely that humankind will accrue many benefits from the knowledge gleaned about the molecular mechanisms governing eucaryotic gene regulation, e.g., improved plant and animal breeding. An explanation of the molecular basis of cancer is another scientific reward that may be gained from such studies. The recognition that cancerous tissue is composed of cells that have lost the normal control systems that regulate cellular division (and, hence, cannot cease dividing) implicates faulty gene expression and/or regulation as primary causes of cancer. Among the various hypotheses to account for the induction of the cancerous state of a cell, a defect(s) in the molecular control of certain genes critical for division is one distinct possibility. Such a loss of control could occur in several ways, e.g., by mutation (either spontaneous or induced by carcinogens) or by the genetic influence of a provirus. Possibly the turning on of genes that are not expressed in normal adult cells, e.g., embryonic genes, may be another underlying cause of cancer. A defective control mechanism (caused for the same reasons listed above) could also explain this type of cellular malfunction. Thus, as the molecular patterns that regulate eucaryotic genes are uncovered, the information gained will undoubtedly be utilized astutely in cancer research.

SYNTHESIS OF RNA

The cellular production of RNA (Figure 22-1) proceeds by a biochemical mechanism similar to that described for DNA replication. The synthesis of single-stranded RNA, i.e., probably all molecular species of cellular RNA, depends on a DNA template (chromosomal DNA), occurs in a 5′ → 3′ direction, and produces an anti-

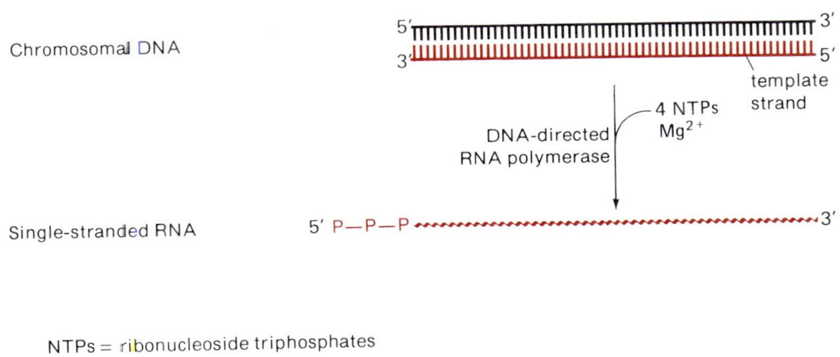

Figure 22-1
General features of RNA synthesis.

parallel, complementary copy of one of the two strands of a gene. The 5′-triphosphate group is a typical feature of nascent RNA. Like DNA replication, RNA synthesis also requires Mg^{2+} and the triphosphate forms of the nucleotide substrates and proceeds by pyrophosphate cleavage. The uracil of RNA replaces the thymine residues of DNA.

DNA-Directed RNA Polymerase

It was in 1959 that *DNA-directed RNA polymerase* was discovered. In *Escherichia coli,* the enzyme is responsible not only for the synthesis of mRNA (transcribed genetic information specifying a protein) but also for that of tRNAs and rRNAs. At least three different RNA polymerases (found in the nucleoplasm, nucleolus, and mitochondria) occur in animal cells; chloroplasts in plants also have the enzyme. Hence, the DNA of mitochondria and chloroplasts are probably each transcribed by a specific RNA polymerase in each of the organelles. The basic biochemical features of the reaction catalyzed by all RNA polymerases are those given in the preceding paragraph. The RNA polymerase of *E. coli* is a large, noncovalently bonded complex with a MW of about 490,000. The protomeric composition of the *holoenzyme* (Table 22-1) includes two α protomers and one protomer of β, β', and σ.

Table 22-1
Molecular composition of RNA polymerase of *Escherichia coli*

Protomer	Number	MW
α	2	40,000
β	1	155,000
β'	1	165,000
σ	1	90,000

General Features of RNA Synthesis in *E. coli*

As diagrammed in Figure 22-2, the holoenzyme ($\alpha_2\beta\beta'\sigma$) is required for the binding of RNA polymerase to a specific site on double-stranded DNA (preferred template). The role of the σ (sigma) factor is to aid the complex in recognizing the particular base sequence that constitutes the binding site. The β and β' chains are believed to function in the binding of the complex to DNA.

Molecular analysis of a number of *RNA polymerase binding sites* in procaryotes, called *promoters* (p. 486), reveals that they vary somewhat in sequence and length but are usually located about ten base pairs (bp) from the start of transcription, i.e., about ten bp *upstream.* The following A·T-rich sequence of seven base pairs is generally inlcuded in a promoter:

5′—T—A—T—Pu—A—T—G—3′

3′—A—T—A—Py—T—A—C—5′

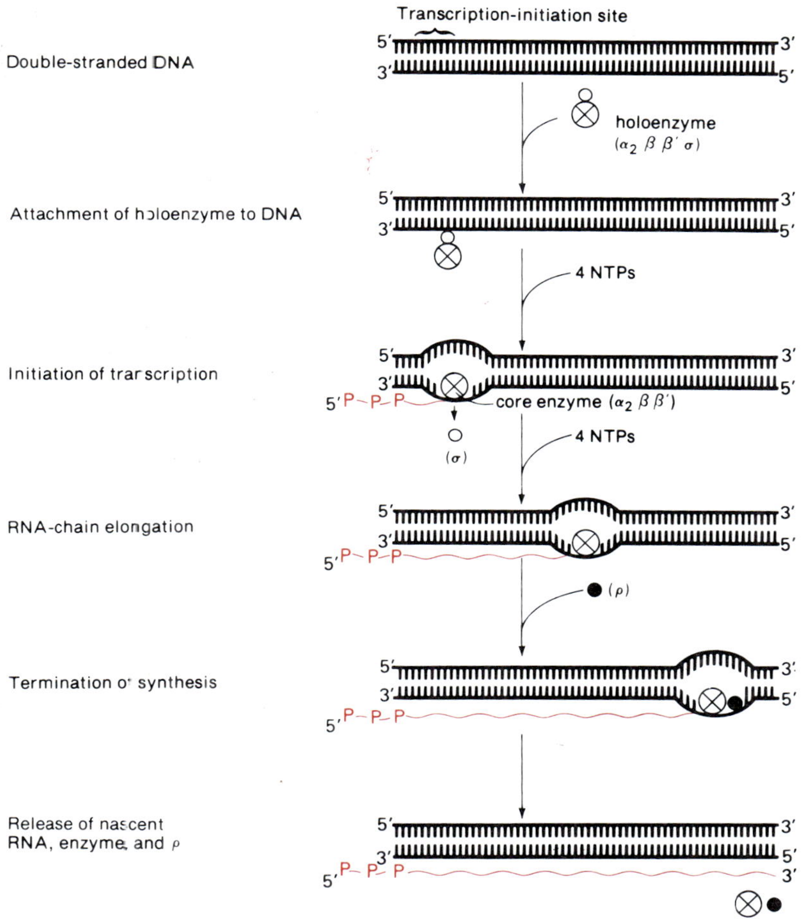

Figure 22-2
Schematic of RNA synthesis. ⊗ = RNA polymerase; ○ = sigma (σ) factor; ● = rho (ρ) factor; NTPs = ribonucleoside triphosphates.

The sequence in procaryotes is called the *Pribnow box.* In eucaryotes, a similar promoter sequence is located about 20 to 30 bp from the start of transcription, i.e., is 20 to 30 bp upstream, and it is known as the *Goldberg-Hogness box,* or the *TATA region* (p. 611).

Soon after the initiation of transcription of one of the chromosomal strands (after incorporation of about ten nucleotides), the σ factor dissociates from the complex (Figure 22-2), and the continued synthesis of RNA is catalyzed by the core enzyme ($\alpha_2\beta\beta'$). For the termination of transcription at the end of a gene, the ρ (rho) factor is involved. The ρ factor (protein of MW $\approx$ 50,000) functions in a multimeric state and is believed to interact with the DNA. After synthesis ends, the nascent RNA molecule, core enzyme, and ρ factor are released from the DNA template. Another type of termination of transcription has been described, one that involves the recognition by RNA polymerase of a terminal sequence UUUUUU(A); the ρ factor is not required for termination in this case.

Another similarity between the synthesis of the macromolecules DNA and RNA is that both processes include initiation, elongation, and termination. The singularly important advantage that molecular research on RNA synthesis had over comparable studies on DNA replication is that the complete synthesizing complex (holoenzyme) was initially isolated, and thus the molecular complications of the process were recognized early in the studies. As discussed previously, the isolation of DNA polymerase I and the subsequent attention focused on this one component of the DNA replicase system made the molecular studies on DNA replication, at times, an enigmatic task.

THE GENETIC CODE PROPOSED BY CRICK

As research on the biochemical analysis of transcription was in progress, companion studies deciphering the genetic code were also under way. An intriguing proposal of a *commaless code,* first made in 1957 by Crick, Sidney Brenner, and associates, was based primarily on the assumption that genetic information is contained in a message that relies on a four-letter alphabet (A, G, C, and T of DNA) and which, in transcribed form, provides the instructions for the incorporation of the 20 amino acids (''magic twenty'') used by all living systems for protein synthesis. Based on a critical evaluation of available data, Crick concluded that genetic information is contained in a *triplet, nondegenerate, nonoverlapping, commaless code.* The term *triplet* means that the basic coding unit for an amino acid *(codon)* is a sequence of *three nucleotide bases* in a DNA strand; *nondegenerate* refers to the idea that, of the possible 64 triplet arrangements derivable from the four bases (4^3), only 20 code for the 20 amino acids (one *sense* codon for each amino acid) and the other 44 do not code, i.e., are *nonsense* codons. (This original concept of nondegeneracy is now defunct, since subsequent research established that the *code is degenerate,* with all but two of the amino acids having more than one codon.) *Commaless* and *nonoverlapping* are terms that specify how the code is read.

As illustrated in Figure 22-3, the code contained in mRNA

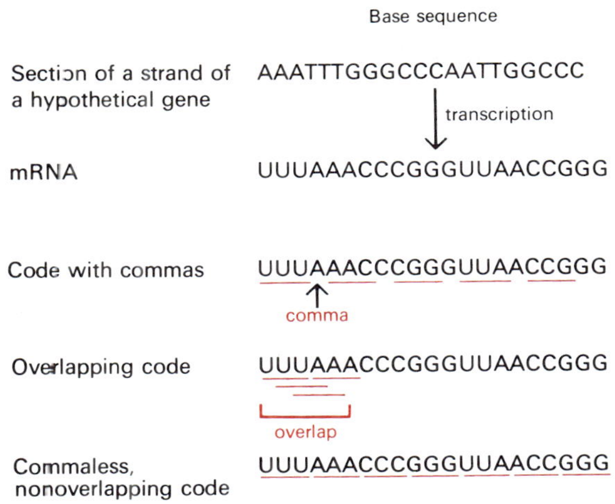

Figure 22-3
Various postulated ways of reading a genetic message.

(whose transcribed genetic information is translated into polypeptide structure during protein synthesis) could be read several ways. A code with commas has one or more bases situated between two codons; the intermittent base(s) is not in a codon sequence and hence can be thought of as a "comma" in the message. The illustrated example of an overlapping code indicates that within the first six bases of the genetic message, both codons UUU and AAA and the two other triplet sequences (UUA and UAA) occurring within the UUUAAA sequence can code for amino acids, i.e., overlapping reading of codons. The proposed commaless, nonoverlapping code has neither of the above features; instead, the genetic information is stored in a manner that allows for an uninterrupted reading of the codons, triplet after triplet, until the termination of the message.

THE CODON MESSAGE

In 1961 Marshall Nirenberg (Nobel Prize, 1968) surprised the scientific world by proposing that the triplet codon UUU codes for phenylalanine. His study made use of a cell-free system for protein synthesis and a synthetic mRNA, *poly U* (RNA with a base composition of uracil), and showed that the artificial message coded for *polyphenylalanine* (polypeptide composed of phenylalanyl residues). Nirenberg's heralded announcement provided an exciting incentive for the subsequent biochemical research that, in a matter of a few years, successfully "cracked the code," i.e., revealed the

Table 22-2
The genetic code. The 64 triplet codons are listed in the 5′ → 3′ direction in which they are read. The third nucleotide of each codon is indicated in red, as are the three termination (Term.) codons.

	U		C		A		G	
U	UUU	Phe	UCU	Ser	UAU	Tyr	UGU	Cys
	UUC	Phe	UCC	Ser	UAC	Tyr	UGC	Cys
	UUA	Leu	UCA	Ser	UAA	Term.	UGA	Term.
	UUG	Leu	UCG	Ser	UAG	Term.	UGG	Trp
C	CUU	Leu	CCU	Pro	CAU	His	CGU	Arg
	CUC	Leu	CCC	Pro	CAC	His	CGC	Arg
	CUA	Leu	CCA	Pro	CAA	Gln	CGA	Arg
	CUG	Leu	CCG	Pro	CAG	Gln	CGG	Arg
A	AUU	Ile	ACU	Thr	AAU	Asn	AGU	Ser
	AUC	Ile	ACC	Thr	AAC	Asn	AGC	Ser
	AUA	Ile	ACA	Thr	AAA	Lys	AGA	Arg
	AUG	Met	ACG	Thr	AAG	Lys	AGG	Arg
G	GUU	Val	GCU	Ala	GAU	Asp	GGU	Gly
	GUC	Val	GCC	Ala	GAC	Asp	GGC	Gly
	GUA	Val	GCA	Ala	GAA	Glu	GGA	Gly
	GUG	Val	GCG	Ala	GAG	Glu	GGG	Gly

informational content of the 64 triplets (Table 22-2). The accolades for this remarkable achievement belong to Nirenberg, Severo Ochoa (Nobel Prize, 1959), H. Gobind Khorana (Nobel Prize, 1968), and their colleagues.

Of the 64 possible triplets of mRNAs (Table 22-2), 61 code for amino acids and the remaining three (*UAG, UAA,* and *UGA*) serve as *termination signals* for polypeptide synthesis, i.e., are the "periods" at the end of a message. The code is *degenerate,* with 14 amino acids having either a two- or fourfold degeneracy, i.e., have two or four codons, respectively. Isoleucine has a threefold degeneracy, and that of arginine, leucine, and serine is sixfold. Only methionine and tryptophan have a single codon each. Note that the degeneracy of the code (except for arginine, leucine, and serine) involves only the third base (3′-end) of a codon. The general pattern is that the third base (position 3) can be either of the two purines (e.g., GAA or GAG for glutamic acid), either pyrimidine (e.e., UGU or UGC for cysteine), or any of the four bases (e.g., GCU, GCC, GCA, and GCC for alanine). Positions 1 and 2 of a codon are evidently the significant determinants of its coding specificity; position 3, because of its variability, is referred to as the *wobble* base (p. 510).

MITOCHONDRIAL GENETIC CODE

After the genetic code was deciphered, numerous studies provided evidence that all biological systems utilize the same code, and thus the concept of a *universal code* became generally accepted. However, recent biochemical and molecular investigations on mitochondria have established that the code used in these eucaryotic organelles (which possess their own DNA and protein-synthesizing system) differs from the established code used for cytoplasmic protein synthesis. The differences between the mammalian mitochondrial and cytoplasmic codes are listed in Table 22-3. Whereas the cytoplasmic code has only one codon *(AUG)* for *methionine,* the mitochondrial code has two (*AUG* and *AUA*). For cytoplasmic protein synthesis, the triplet *AUA* codes for *isoleucine.* In the case of *tryptophan,* both *UGA* and *UGG* code for the amino acid in mitochondria; hence, the cytoplasmic termination codon, UGA, is used as a coding triplet in mitochondria. Note that the amino acids coded for by three codons (Ile) and by one (Met and Trp) in the cytoplasm have two codons each in the mitochondrial system. A third difference is that two of the six cytoplasmic codons for arginine (*AGA* and *AGG*) serve as *termination codons* for mitochondrial protein synthesis.

The differences between the mitochondrial and cytoplasmic codes are believed by some to be the evolutionary result of a coding economy dictated by the small size of the organelle's genome

Table 22-3
Differences between the mammalian cytoplasmic and mitochondrial genetic codes

Codon (5′ → 3′)	Cytoplasmic code	Mitochondrial code
AUU	Ile	Ile
AUC	Ile	Ile
AUA	Ile	Met
AUG	Met	Met
UGA	Term.	Trp
UGG	Trp	Trp
AGA	Arg	Term.
AGG	Arg	Term.

(MW of about 1×10^7 for mammalian mitochondrial DNA *vs.* about 2×10^9 for the *E. coli* chromosome). Because of its diminutive size, the mitochondrial genome can accommodate only a limited number of genes (about 35), which includes those specifying the ribosomal RNAs and the transfer RNAs (p. 234) used for mitochondrial protein synthesis. If mitochondrial protein synthesis relied on the cytoplasmic coding system, a minimum of 32 transfer RNAs (tRNAs) would be required (see Question 6 of Chapter 23). However, because of the modified code and the unique tRNAs of mitochondria, all codons for the 20 amino acids can be recognized by a total of 22 tRNAs. The codons, as recognized by the anticodons of the tRNAs (p. 509), are read in a simpler fashion than those used for cytoplasmic protein synthesis. In mitochondria, each tRNA recognizes a set of four codons or a set of two. For example, valine has a set of four codons (5′-G-U-X-3′), and the mitochondrial valyl tRNA recognizes all four, i.e., position 3 (-X-3′) can be one of the four nucleotides. Isoleucine and methionine have a set of two codons each; for isoleucine, the set is 5′-A-U-Py-3′ (pyrimidine in position 3) and for methionine the set is 5′-A-U-Pu-3′ (purine in position 3). Hence, the isoleucyl tRNA recognizes the set with a 3′-pyrimidine and methionyl tRNA the set with a 3′-purine. Although the cytoplasmic coding system follows the same general pattern, it is not as stringent as that of mitochondria. Thus, because of this modified code and tRNA species that are specific for sets of four codons (cytoplasmic tRNAs can read a maximum of three, p. 510), protein synthesis in mitochondria requires 10 fewer tRNAs than does the analogous cytoplasmic system—a reduced requirement that abolishes the need for ten additional genes in the mitochondrial genome.

JACOB–MONOD MODEL OF GENE REGULATION

The molecular understanding of transcription and of the language of the message carried by the RNA product was complemented by Jacob and Monod's explanation of the control of gene expression. Monod began his studies on gene regulation in *E. coli* by investigating the inducibility of the enzyme β-galactosidase, which catalyzes the hydrolysis of the disaccharide lactose (p. 189) into its component monosaccharides, glucose and galactose (carbon and energy sources for the bacterium). Figure 22-4 illustrates β-galactosidase inducibility. When glucose is the carbon and energy source, levels of β-galactosidase activity are very low. Upon deple-

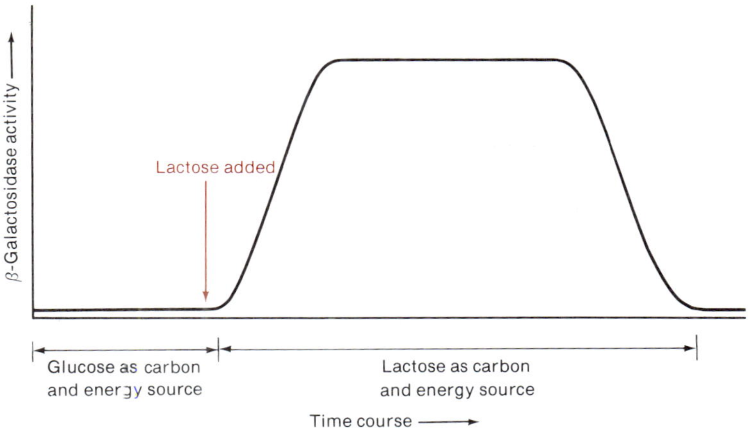

Figure 22-4
Simplified diagram of β-galactosidase induction in *Escherichia coli.*

tion of glucose and addition of lactose to the medium, a marked induction of the enzyme's activity soon occurs. The bacteria, during this growth period, rely on lactose for carbon and energy. Once the lactose is utilized and its intracellular level declines, the β-galactosidase activity returns to its original low level.

In the 1950s, the phenomenon of β-galactosidase induction was explained by either of two rationales. One explanation was that the gene responsible for β-galactosidase production is turned off when the preferred energy source, glucose, is available (an enzyme to degrade lactose is not needed) but is turned on by lactose (inducer) when a cell is forced to substitute the disaccharide for its primary carbon and energy source. The other explanation was that regulation of the β-galactosidase gene is not involved but, rather, that the enzyme is continuously produced in an inactive form. The function of lactose is to activate the zymogen, as reflected by the observed increase in β-galactosidase activity. Thus, one rationale explained enzyme induction by a genetic mechanism and the other by a biochemical one.

Monod conclusively proved that β-galactosidase induction entails *de novo* synthesis of the enzyme, i.e., the responsible gene is turned on. As a result of a collaborative effort, Jacob and Monod eventually developed a molecular model to explain the control of inducible enzyme systems, as deduced from the studies on the β-galactosidase and related genes (*lac* system). Their monumental paper in 1961 (*Journal of Molecular Biology,* 3:318) became a classic of molecular biology the day it was published.

The Jacob-Monod model (Figure 22-5) introduced the concept of two classes of genes: *structural genes* (SG) are responsible for

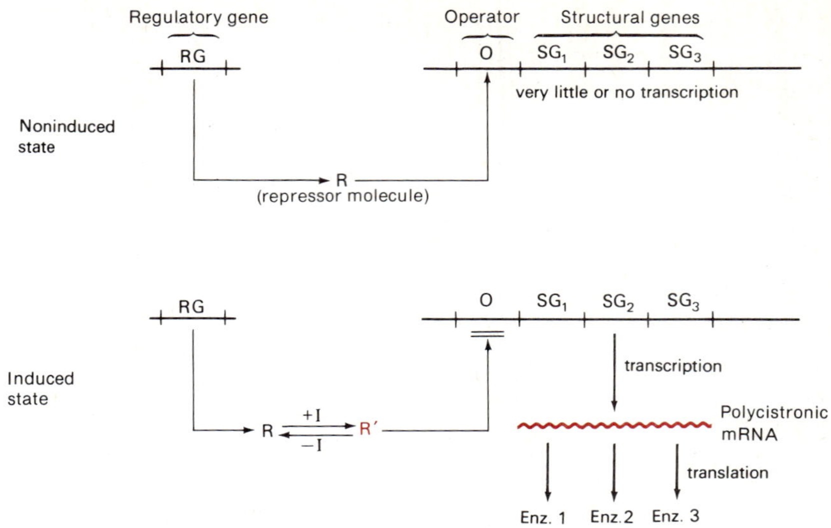

Figure 22-5
Jacob–Monod model of enzyme induction.

the synthesis of proteins (one gene-one enzyme) and other cytoplasmic products, and *regulatory genes* (RG and O in the figure) control the expression of structural genes. An inducible system has two regulatory genetic elements; the first is a specific regulatory gene (RG) that codes for a *repressor molecule* (R), which interacts with the second regulatory element, the *operator* (O) gene or region of that system. The operator is physically linked to the structural gene(s) of the inducible enzyme system and was originally believed to contain the RNA polymerase binding site needed for the transcription of the structural gene(s). The term *operon* was coined to describe an operator and the structural genes under its control; one or more structural genes may be contained in an operon, e.g., the *lac* operon has three. Note that the regulatory gene is set apart from the operon in Figure 22-5, indicating that the regulatory gene may or may not be linked to the operon it controls. A regulatory gene may also control one or more operons.

In a noninduced state (Figure 22-5), the repressor (a protein) binds to the operator and, as a result, very little or no transcription occurs. Such a situation exits, for example, when *E. coli* uses glucose, and the observed β-galactosidase activity is very low (Figure 22-4). Induction is triggered when an inducer (I) for a particular system, e.g., lactose or a derivative of the disaccharide for the *lac* operon, noncovalently binds to the repressor ($R + I \rightarrow R'$) and alters the molecule's high affinity for the operator. This loss of affinity results from a conformational change in the repressor molecule. The release from inhibition of transcription allows for the expression of the operon's structural gene(s) and for the subsequent synthesis of the inducible enzymes, e.g., induced levels of β-galactosidase activity in Figure 22-4. Note that an operon with more than one structural gene is transcribed into a single mRNA molecule and that the individual enzymes are synthesized from the instructions contained in this *polycistronic message.*

Cistron is a molecular redefinition of a gene and its function, i.e., one cistron–one polypeptide replacing one gene–one enzyme.

When the inducer is depleted and its cellular concentration decreases, the repressor returns to its original state ($R' \rightarrow R + I$) of high affinity for the operator and binds to the region. Transcription of the operon is prevented once again, and the levels of inducible enzyme(s) eventually decline as the enzyme molecules are catabolized.

The basic molecular features of the model for control of the synthesis of enzymes explain how a cell can effectively regulate its metabolism. When the regulation of enzymatic synthesis at a genic level is superimposed on the regulation of enzymes already produced, e.g., by positive and negative allosteric effectors, it is apparent that these two levels of control provide a cell with the flexibility needed to maintain optimal control of its metabolic activities in its response to environmental changes.

THE *LAC* OPERON TODAY

Since the announcement explaining genetic regulation of inducible enzymes, continued analysis of the *lac* operon of *E. coli* has added new dimensions to the molecular understanding of control systems. Today, as in 1961, the *lac* operon remains a model regulatory system because of the thorough insights it provides into a cell's regulation of its genome.

The Physical Structure of the *Lac* Operon

The *lac* operon of *E. coli* is contained in a contiguous section of chromosomal DNA and has six genetic elements (Figure 22-6); three are structural genes that code for the following enzymes: β-

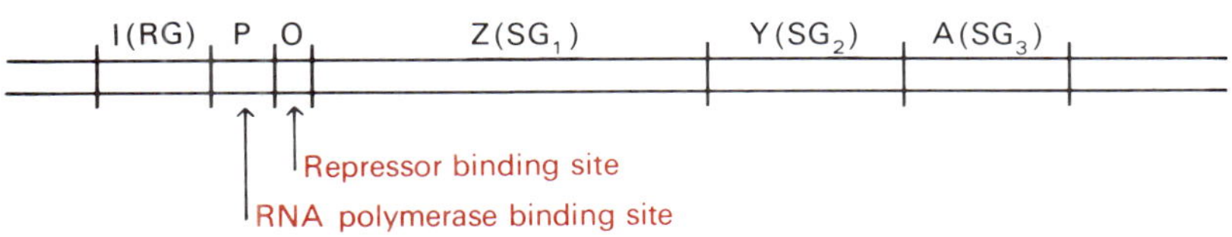

Figure 22-6
Simplified diagram of the *lac* operon of *E. coli*. I = regulatory gene (RG); P = promoter; O = operator; Z = structural gene for β-galactosidase; Y = structural gene for galactoside permease; A = structural gene for galactoside acetylase.

galactosidase (gene Z), *galactoside permease* (gene Y), and *galactoside acetylase* (gene A). Galactoside permease controls the rate of uptake of β-galactosides, e.g., lactose, into *E. coli;* the function of galactoside acetylase remains unknown. The occurrence of the three genes in the same operon ensures that the enzymes needed for β-galactoside metabolism are coordinately synthesized and repressed. A single control mechanism is believed to be an advantage derived from the evolutionary development of procaryotic operons. The I gene is the regulatory gene and is responsible for the synthesis of the repressor molecule (a protein tetramer with a MW of about 148,000), which, as stated previously, interacts with the operator (O). Since Jacob and Monod's initial proposal, it has been discovered that RNA polymerase does not bind within the operator as originally thought but, instead, interacts with a region, called the *promoter,* located between the operator and the regulatory gene.

cAMP Involvement in Enzyme Induction

An illuminating discovery about the control of the *lac* operon was made when a role for cyclic AMP (cAMP) was established in the induction process. After the inhibition of transcription by the repressor is relieved (R → R′), cAMP complexed with a specific receptor protein (CAP, for cAMP protein) must first bind to the promoter before RNA polymerase can attach to the promoter site and commence mRNA synthesis. (The abbreviation *CRP* is also used for the cyclic AMP receptor protein.) The regulatory region of the operon (Figure 22-7), therefore, contains specific binding site for three proteins, i.e., separate attachment sites for the cAMP-CAP complex (CAP site), RNA polymerase in the promoter, and the binding site for the repressor in the operator. With the discovery of the cAMP involvement, it became apparent that the *lac*

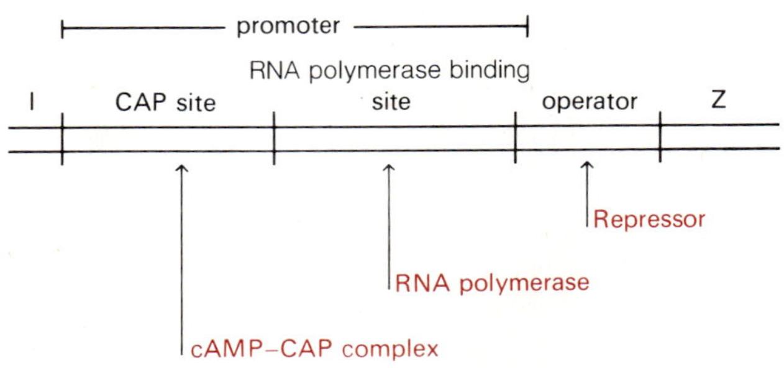

Figure 22-7
Diagram of the regulatory region of the *lac* operon in *E. coli.* I = regulatory gene; CAP = cyclic AMP protein; Z = structural gene for β-galactosidase.

operon (and probably all *E. coli* inducible enzyme systems) is under both positive and negative control. The function of the repressor is a *negative control* because, by binding to the operator, *transcription is prevented.* Conversely, the attachment of the CAP complex to the promoter is a *positive control,* since its presence *permits* the *initiation of transcription.*

The discovery of the critical importance of cAMP in regulatory phenomena of procaryotes is of broad biological interest, since the same nucleotide is the second messenger that initiates the intracellular metabolic responses induced by many hormones (Chapter 24). The biological antiquity of the role of cAMP as a metabolic regulator supports the contention that the endocrine system of vertebrates represents an evolutionary extension of intracellular homeostasis.

Nucleotide Sequence of the Promoter and Operator

A more recent outstanding advancement furnished by the studies on the *lac* operon is the identification of the nucleotide sequence of the promoter and operator, a feat that offers definitive insights into the chemical nature of the binding sites of the three proteins. When the base-pair sequence is correlated with the substantial quantity of accumulated data on the operon, an expanded understanding is obtained of the molecular features of this region of the operon. As indicated in Figure 22-8, the operator and promoter sections overlap, since the site where transcription starts (promoter) is contained within the region covered by the bound repressor (operator). This conclusion substantiates the proposal that the repressor sterically prevents transcription. Note that mRNA synthesis is initiated prior to the beginning of the first structural gene (Z) of the operon and that the 21-base-pair sequence binding site of the operator (Figure 22-9) is a palindrome.

In the 85-base-pair sequence of the promoter (Figure 22-10), the CAP binding site is thought to be the palindromic sequence that precedes the left G·C region in the figure. The *entry site* of RNA polymerase is postulated to be the high-A·T sequence located between the two regions with high G·C content, since the two strands of an A·T-rich sequence separate more easily (and, hence, more frequently) than those of a G·C-rich sequence. How binding of the CAP complex enhances that of RNA polymerase is unknown, but it has been suggested that binding of the complex (Figure 22-11) destabilizes the helical conformation of the entry

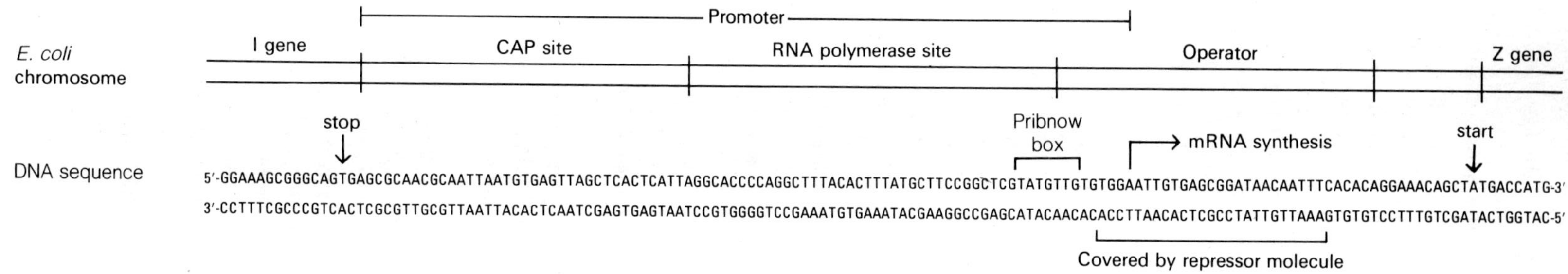

Figure 22-8
The nucleotide sequence of the regulatory region of the *lac* operon in *E. coli.*

site (A·T-rich region) which, in turn, facilitates the attachment of the enzyme. Since the entry site of RNA polymerase is a number of base pairs removed from the point of transcription initiation, it is also assumed that the enzyme "drifts" until it becomes firmly bound, after which mRNA synthesis begins. With these new molecular intrigues, studies on the *lac* operon re-emphasize the penetrating powers of analysis that molecular studies provide toward understanding biological phenomena.

5′—AATTGTGAGCGGATAACAATT—3′
3′—TTAACACTCGCCTATTGTTAA—5′

Figure 22-9
Nucleotide sequence of the *lac* operator region in *E. coli.* Bases in red denote those that give the operator its palindromic character.

REGULATION OF EUCARYOTIC GENES

Unlike the simple double-helical structure of the *E. coli* chromosome, a eucaryotic chromosome is a much larger and more complex structure, containing protein, e.g., histones, in addition to DNA. Because of this difference and because of other complexities associated with eucaryotic systems, studies on the regulation of eucaryotic genes have not been as amenable to the types of detailed molecular analysis successfully pursued with *E. coli.* However, as studies on the regulation of eucaryotic genomes continue, it is increasingly apparent (but not unexpected) that the molecular mechanisms vary from those described for procaryotes. In eucaryotes, for example, there is little evidence of operons, i.e., physically linked groups of related structural genes that are synchronously controlled. Also, the principal pattern of regulation appears to be one of positive control, and often a single regulatory effector (e.g., a steroid hormone) initiates a cascade of events, i.e., activates a battery of genes (p. 556).

To account for the emerging evidence about eucaryotic genes and their molecular control, Roy Britten and Eric Davidson formulated a working hypothesis to explain eucaryotic gene expression. In eucaryotic research, their model performed the same role as the Jacob-Monod model did when the latter was introduced in 1961 to explain the control of procaryotic genes. Although in light of subsequent research the Britten-Davidson model does not enjoy the popularity it once did, it nevertheless remains a good introduction to the scientific thinking that has evolved concerning eucaryotic gene regulation. The answers to the questions the model seeks to resolve continue to remain elusive; however, as will be discussed in Chapter 26, combined recombinant DNA research and DNA-sequencing analyses are rapidly furnishing new and provocative information about eucaryotic genes and their expression.

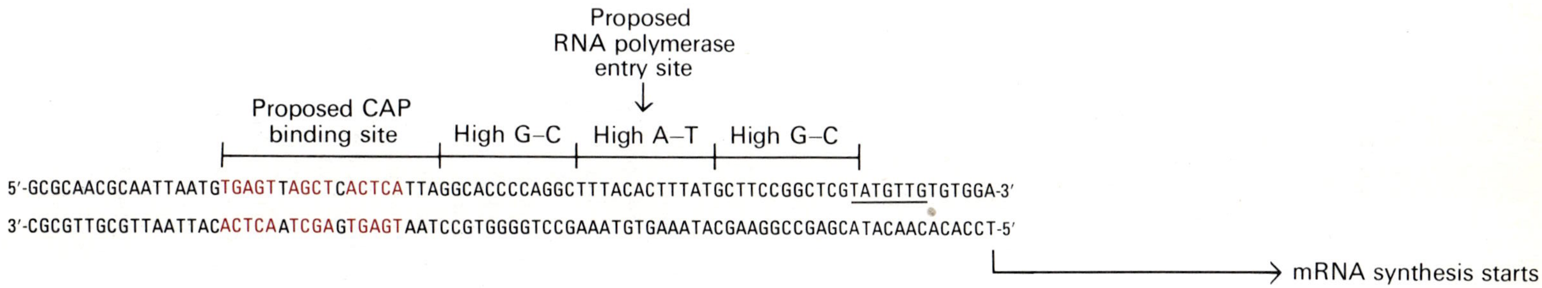

Figure 22-10
Nucleotide sequence of the promoter region of the *lac* operon in *E. coli.* Palindromic sequence proposed for the CAP binding site is indicated in red, and the Pribnow box is underlined.

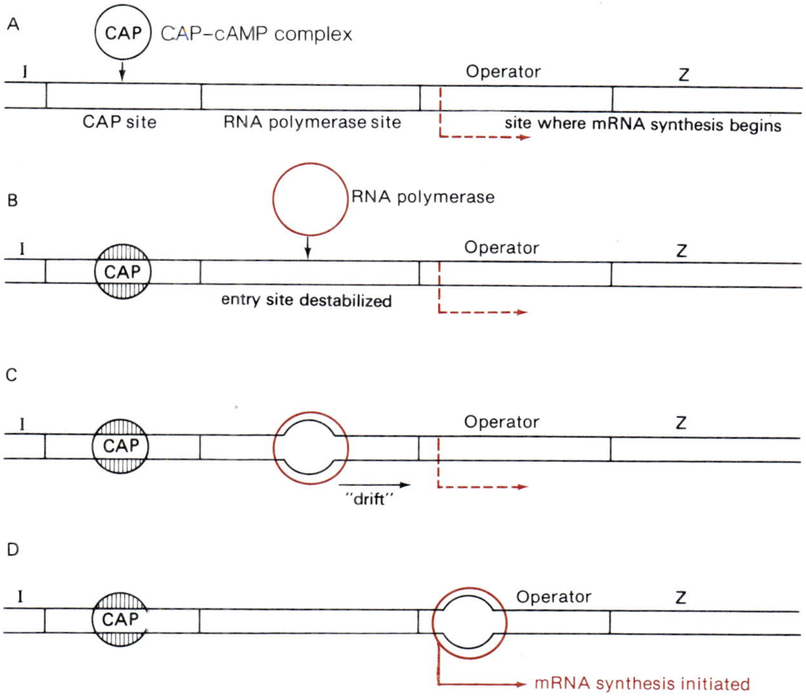

Figure 22-11
Diagram of the initiation of transcription of the *lac* operon *(E. coli).* (A) Binding of the CAP–cAMP complex; (B) destabilization of RNA polymerase entry site; (C) binding of RNA polymerase and subsequent "drift"; (D) initiation of RNA synthesis. (After R. C. Dickerson et al., *Science,* 187:27, 1975. Copyright 1975 by the American Association for the Advancement of Science)

BRITTEN–DAVIDSON MODEL

The model postulates a two-tier system of positive control (Figure 22-12) requiring two classes of regulatory proteins. The first level of control is initiated when a primary regulatory protein, a *sensor protein* (S^{α}), after interacting with its specific *effector* (α), acts on a *sensor gene* that controls the transcription of a designated *set of integrator genes* (I^{α}). A cytoplasmic hormonal-receptor protein qualifies as a sensor protein and its specific hormone as an effector (p. 00). The turning on of the appropriate sensor gene results in the production of an RNA *activator message set,* which codes for the secondary regulatory proteins, *activator proteins.* (The transcription of the integrator gene set is reminiscent of synthesis of the polycistronic message of the *lac* operon.) At the second level of control, the set of activator proteins interacts with specific *receptor sequences* to promote the expression of structural genes. Note that both levels of control are positive. An activator protein in Figure 22-12 is identified by a superscript that indicates the integrator gene set that coded for it and by a subscript that specifies the sequence receptor it recognizes. Hence, protein PA_b^{α} is

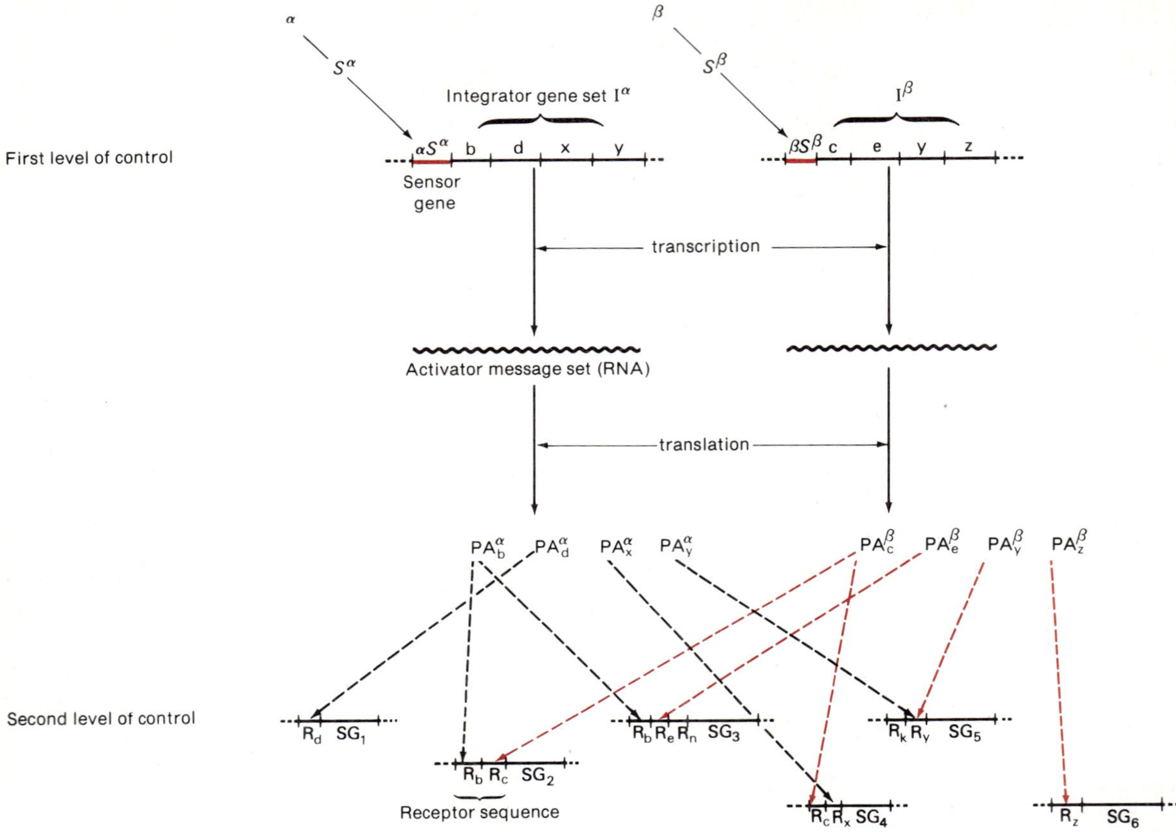

Figure 22-12
Diagram of the Britten–Davidson model for the regulation of eucaryotic genes. α and β = effectors; S^{α} and S^{β} = sensor proteins; PA = activator protein; SG = structural gene. (After E. H. Davidson and R. J. Britten, *Quart. Rev. Biol.*, 48:565, 1973)

coded for by integrator set I^{α} and recognizes the receptor sequence R_b (SG_2 and SG_3). As indicated, an effector is responsible for the ultimate expression of a battery of genes (SG_1 through SG_5). If the same rationale is applied to effector β, the activated battery is SG_2 through SG_6. A structural gene (SG_2 through SG_5) may have more than one receptor sequence and thus can be activated by various effectors. Also, activator proteins produced by the transcription of different integrator sets (PA_y^{α} and PA_y^{β}) can recognize the same receptor sequence (R_y). Note, too, that a particular receptor sequence can be associated with more than one structural gene, e.g., R_c. The Britten-Davidson model, therefore, provides for a very flexible control of structural genes and explains how the regulatory signal of a single effector is amplified into a cascading effect.

POST-TRANSCRIPTIONAL MODIFICATION OF RNA

Accumulated evidence reveals that many RNAs are synthesized as larger precursor molecules which are then *processed,* by nuclease and other enzymatic activities, to yield the functional cytoplasmic products. The molecular details of the processing of ribosomal and transfer RNAs (rRNAs and tRNAs, respectively) exemplify the general phenomenon of *post-transcriptional modification.*

The 28 S and 18 S rRNAs of the 80 S ribosome of mammalian cells are synthesized together in a 45 S precursor molecule, called *pre-rRNA* (Figure 22-13). In its processing, the pre-rRNA first undergoes methylation of about 1 per cent of the 2′- hydroxyl groups of its ribosyl residues and is then cleaved into 41 S and 24 S fragments. Two additional cleavages of the 41 S molecule are required to produce the 28 S and 18 S rRNA components of ribosomes.

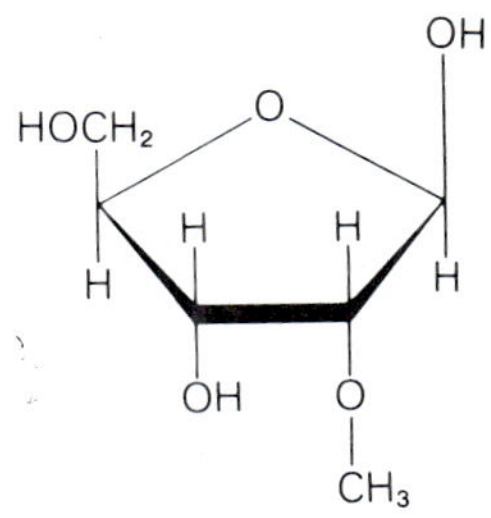

2-*O*-Methylribose

The processing of tRNAs entails not only nuclease trimming of a larger precursor (pre-tRNA) but also enzymatic modification of particular bases of the processed forms of the tRNAs (Figure 22-14), which endows the molecules with their distinctive sets of modified bases, e.g., methylated adenines and guanines and pseudouridine (p. 229). The final products are the functionally *mature tRNAs.*

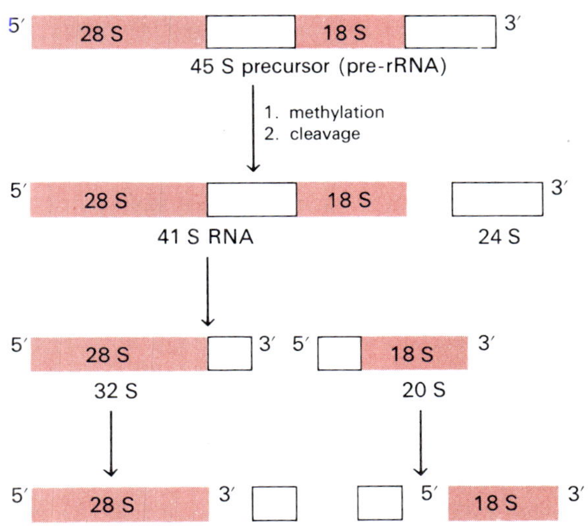

Figure 22-13
Processing of the precursor molecule of mammalian 18 S and 28 S ribosomal RNA.

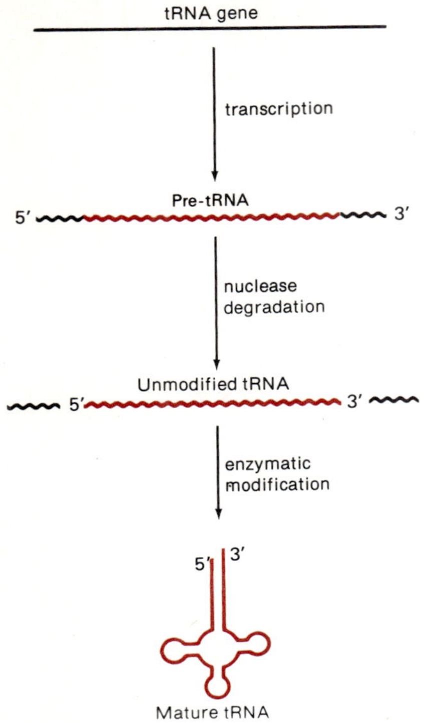

Figure 22-14
Processing of a transfer RNA precursor molecule.

LEADER SEQUENCES OF PROCARYOTIC mRNAs

The previous notion that the size of an mRNA conforms closely to the length of the polypeptide for which it codes, i.e., three nucleotides (codon) per amino acid residue, is no longer valid since it is now known that procaryotic mRNAs have *leader sequences* at their 5′-termini, e.g., the *lac* message (Figure 22-8). These sequences vary in length from one mRNA to another, e.g., the *E. coli gal* operon mRNA has 26 bases in its leader sequence and the *E. coli trp* operon mRNA has 140. One function of a leader sequence is to provide a recognition sequence, called the *Shine-Dalgarno sequence,* that allows for the binding of ribosomes to an mRNA (*ribosomal binding site,* p. 517). Many of these sequences in procaryotic mRNAs also participate in the control of protein synthesis at the transcription level. This type of control entails initiation of transcription within the leader sequence and its termination prior to transcription of the structural genes of the operon. The name *attenuation* has been given to this control phenomenon.

Attenuation

In recent years, Charles Yanofsky and his associates have studied attenuation of the *E. coli trp* operon, which contains five genes coding for the enzymes that synthesize tryptophan, and have developed a hypothesis to explain the molecular details of this type of control. Attenuation is not limited to the *trp* operon since the same mechanism of transcriptional control has been identified for several *E. coli* operons containing genes for amino acid biosynthesis, e.g., *his* and *leu* operons. A key feature of the attenuation hypothesis is that transcription and translation proceed concomitantly in bacteria (which have no nuclei). Hence, while an RNA polymerase is producing an mRNA (transcription), ribosomes are attaching to the nascent message and protein synthesis is taking place (translation). For attenuation, initiation of transcription and translation begins in the section of the leader sequence containing the code for a *leader peptide.* In the case of the *trp* operon, the leader peptide has 14 amino acid residues and, in its sequence, there are two tryptophyl residues in tandem (an unusual feature for such a small peptide, since tryptophyl residues are rare in *E. coli* proteins). How the incorporation of these two residues into the leader peptide during translation influences transcription of the *trp* operon is schematically illustrated in Figure 22-15.

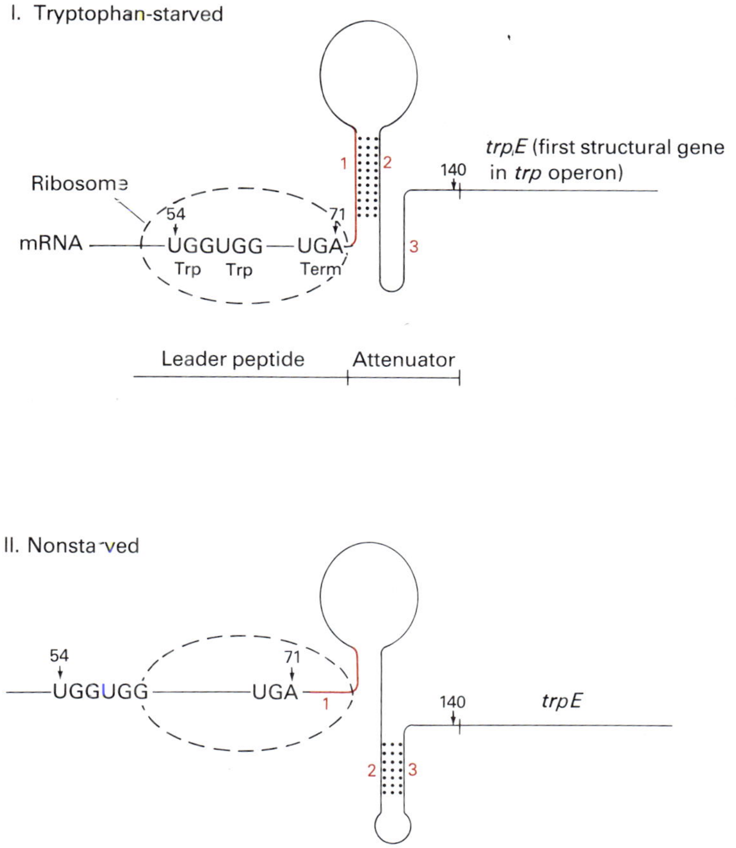

Figure 22-15
Simplified diagram of attenuation of the *trp* operon of *E. coli.*

Within the RNA sequence (called the *attenuator*) between the termination codon for the leader peptide and the beginning of the first structural gene *(trpE),* two types of secondary structures that regulate transcription are possible. These structures are produced by intramolecular hydrogen bonding occurring between complementary segments of the RNA molecule (similar to the cloverleaf arrangements in a tRNA molecule, p. 243). It is the secondary structure (represented as 1 and 2 or 2 and 3) generated by translation of the leader sequence that determines whether transcription will terminate at the end of the leader sequence or continue on into the structural genes of the operon. This decision about the transcriptional fate of the operon is determined by intracellular levels of tryptophan (end product of the pathway).

When the intracellular level of tryptophan is low (tryptophan-starved in Figure 22-15), only a limited number of tryptophyl tRNAs bearing tryptophan are available for protein synthesis. Because of this shortage of tryptophyl-bearing tRNAs, synthesis of the leader sequence is delayed when the ribosome arrives at the two tryptophyl codons (UGGUGG) in the mRNA. This ``stalling'' of the ribosomes, which serves as a regulation signal, results in the formation of the 1-2 secondary structure because region 1 is available for hydrogen bonding with complementary region 2 as the latter is synthesized by RNA polymerase. (Recall that transcription by RNA polymerase is immediately preceding the translational process occurring on the ribosome.) Formation of the 1-2 secondary structure ensures that the polymerase will continue its transcription of the entire operon. Thus, it is by this mechanism that low levels of intracellular tryptophan enhance transcription of the *trp* operon, which ultimately results in increased synthesis of tryptophan.

On the other hand, when the intracellular level of tryptophan is adequate (nonstarved), there is no ``stalling'' of the ribosome at the Trp codons since there is no shortage of tryptophyl-bearing tRNAs. As indicated in Figure 22-15, because of this lack of translational hesitation, a majority of region 1 is not available for hydrogen bonding because of the more rapidly moving ribosome. However, as region 3 is synthesized by the polymerase, it base pairs with region 2 to produce the 2-3 secondary structure. This 2-3 secondary structure acts as a termination signal recognized by RNA polymerase, which terminates mRNA synthesis at the end of the leader sequence, i.e., *trp* structural genes are not transcribed. Thus, an ample supply of intracellular tryptophan reduces transcription of the *trp* operon.

The attenuation hypothesis represents an intricate control system that relies on the close physical and dynamic interrelationship between transcription and translation, the influence of the coding sequence for the leader peptide on the rate of translation, and the use of secondary RNA structures to accomplish its purposes. The fact that one section of the attenuator (region 2 in Figure 22-15) can base-pair with a sequence that precedes it (region 1) or with one that follows it (region 3) is a versatile and critical molecular feature of the system. The occurrence of codons in the leader peptide sequence that code for the amino acid synthesized by the enzymes specified by the operon has been verified in other attenuation systems. For example, in the leader peptide for the *his* operon, there are seven His residues in tandem and, for the *leu* operon, the peptide has four Leu residues in tandem.

MOLECULAR DIFFERENCES BETWEEN EUCARYOTIC AND PROCARYOTIC RNAs

As more eucaryotic mRNAs are isolated and examined, several unusual molecular features of their 3′- and 5′-termini, not observed in procaryotic mRNAs, have been noted. One difference is the *3′-polyA* (sequence of adenylic residues) *terminus* that eucaryotic mRNAs usually have. The adenyl nucleotides making up the polyA sequences are added as a tail to the 3′-termini of newly synthesized mRNAs by the enzyme *polyA polymerase.* The polyA tracts of different mRNAs vary in length, generally ranging from 50 to 200 nucleotides. Although it has been proposed that 3′-polyA endings function in the transport of mRNAs from the nucleus (through nuclear membrane pores) to the cytoplasm or to retard degradation of mRNAs, either role remains to be verified.

At the 5′-termini of many eucaryotic mRNAs, there is another curious molecular modification. After transcription, a *capping* of the 5′-termini occurs by the addition of *7-methylguanosine* (Figure 22-16), via a 5′-to-5′ condensation, to the triphosphate group of the 5′-terminal nucleotide of the nascent mRNA. Note that the 2′-hydroxyl group of the true 5′-terminal nucleotide and sometimes that of the penultimate nucleotide are methylated. It is believed that capping is required for translation of the RNA message (pro-

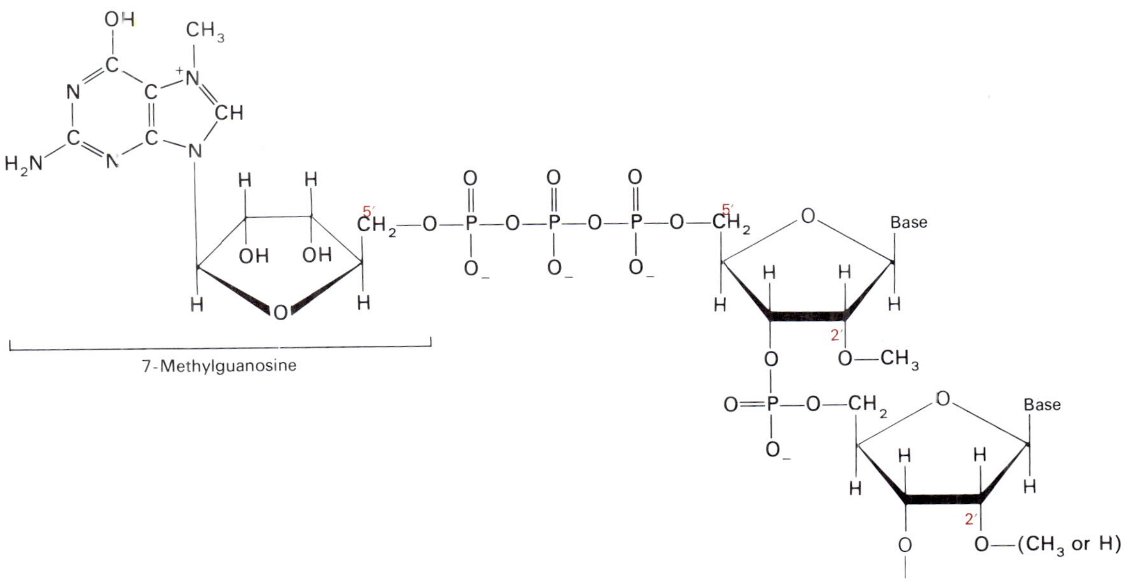

Figure 22-16
Schematic of post-transcriptional modification of the 5′-terminus (capping) of a eucaryotic messenger RNA.

tein synthesis). However, although capping is needed for the translation of many eucaryotic mRNAs, there are several known uncapped eucaryotic messages that function well as templates for protein synthesis. Additional information on eucaryotic mRNAs is presented in Chapter 26.

SUMMARY

Cellular RNA synthesis is catalyzed by DNA-directed RNA polymerase, which uses one of the two strands of genic DNA as a template to synthesize an antiparallel, complementary copy of that strand. The synthesis occurs in a $5' \rightarrow 3'$ direction. The initiation of synthesis in *E. coli* requires RNA polymerase holoenzyme, composed of five protein molecules (α_2, β, β', and σ). However, the sigma (σ) factor dissociates from the holoenzyme after RNA synthesis commences, and the core enzyme (α_2, β, and β') continues RNA chain elongation along the template. For the termination of RNA synthesis, an additional protein, the rho (ρ) factor, is usually required.

The genetic information used for protein synthesis is stored in a code that is triplet, degenerate, nonoverlapping, and commaless. *Triplet* refers to a sequence of three nucleotide bases that serve as a coding unit for an amino acid. *Degenerate* means that an amino acid may be coded for by more than one of the possible 64 triplet codons, e.g., isoleucine and arginine have three and six codons, respectively. *Commaless* and *nonoverlapping* describe how the code is read (translated into protein structure), which is by an uninterrupted reading of each triplet codon in an RNA message. The code used in mammalian mitochondria differs somewhat from that elucidated for cytoplasmic protein synthesis. Because of this difference, cytoplasmic and mitochondrial protein synthesis requires a minimum of 32 and 22 tRNAs, respectively.

Studies on the *lac* operon of *Escherichia coli* have provided detailed molecular insights into how a genome is regulated, i.e., how genes are turned on and off. In the case of the *lac* operon (representative of an inducible enzyme system), a regulatory gene (I) is responsible for the production of a repressor molecule. This repressor molecule interacts with an operator region (O) that controls the transcription of three structural genes (Z, Y, and A). The operator region and the three structural genes under its control constitute an operon. The binding of the repressor prevents the transcription of the operon, i.e., no enzyme synthesis. In the presence of the inducer lactose, the operon is expressed (enzymes are produced), since the inducer specifically binds to the *lac* repressor and alters the latter's affinity for the operator region. The turning on of the *lac* operon also requires the interaction of a cyclic AMP receptor protein (cAMP–CAP) complex at a site in the regulatory region of the operon. This interaction allows for the binding of RNA polymerase to initiate transcription. The regulatory region of the *lac* operon, therefore, has the following three specific binding sites for protein: operator (for repressor binding), CAP site (for cAMP–CAP complex binding), and RNA polymerase binding site. The interaction of the repressor with the operon is a negative control since it prevents transcription. The cAMP–CAP interaction is a positive control because it allows transcription to occur.

In eucaryotes, regulation of genes appears to occur mainly by positively controlled mechanisms. Although the molecular mechanisms involved in eucaryotic gene regulation remain to be elucidated, a model has been proposed that offers some intriguing speculations about how such controls could operate. The model postulates the following two levels of control. The first level involves the activation of a sensor gene by an effector, e.g., steroid hormone, complexed with a specific sensor protein, e.g., hormonal receptor protein. The sensor gene controls the transcription of a set of integrator genes, which code for a variety of activator proteins. Each of the activator proteins produced by the activation of the first level of control then interacts with a specific receptor sequence(s) to activate the second level of control. This activation of the receptor sequences can promote the transcription of many genes. The proposed model effectively explains how the regulatory signal of a single effector can be amplified to activate a battery of structural genes.

Molecular studies have revealed that many RNAs are synthesized as large precursor molecules, which are processed to produce the functional cytoplasmic RNAs, i.e., are post-transcriptionally modified. The precursor molecule of mammalian 18 S and 28 S ribosomal RNA (rRNA) is a 45 S RNA that undergoes a process of selective methylations and cleavages to produce the RNA components of ribosomes. The processing of precursor molecules of transfer RNA (tRNA), in addition to nuclease cleavages, also includes the modification of particular nitrogenous bases to furnish the tRNAs with their distinctive pattern of modified bases, e.g., methylated adenines and guanines.

Studies on the *trp* operon and other biosynthetic operons have revealed an intricate mode of transcriptional regulation in procaryotes, called attenuation, that involves the leader sequence of an mRNA. After the leader sequence (which precedes that of the first structural gene) is transcribed, its translation to produce a leader peptide begins immediately. The rate of translation of the leader sequence determines if transcription of the operon will occur. In the case of the *trp* operon, low intracellular levels of Trp cause a delay in translation of the leader sequence, which has two Trp codons in tandem. This hesitation allows transcription of the operon to take place because of the secondary structure it permits the mRNA to assume as it is being synthesized. Conversely, adequate amounts of Trp (that allow for uninterrupted translation of the leader sequence) cause transcription to cease before the first structural gene is transcribed. Once again, it is the secondary structure (different from that formed under conditions of tryptophan starvation) of the mRNA being made that decides the fate of the transcriptional process.

One of the two post-transcriptional modifications frequently observed with eucaryotic messenger RNAs is the addition of a tract of adenyl nucleotides, i.e., a polyA tail, to the 3′-terminus of an mRNA molecule. The other is the capping of the 5′-terminus by the addition of 7-methylguanosine, via a 5′-to-5′ condensation, to the triphosphate group of the 5′-terminal nucleotide. In a capped mRNA, the 2′-hydroxyl group of the true 5′-terminal nucleotide is methylated.

QUESTIONS

1. Which of the following statements are true about cellular production of RNA by DNA-directed RNA polymerase? (Statements (a) through (e) apply to the discussion of RNA synthesis in *E. coli.*)
   (a) Synthesis has no metal ion requirement.
   (b) Holoenzyme consists of five protein molecules.
   (c) Core enzyme lacks the sigma (σ) factor.
   (d) Elongation of the RNA chain requires the holoenzyme.
   (e) Termination of synthesis always requires the rho (ρ) factor.
   (f) Product has a triphosphate group at the 3′-terminus.
   (g) An antiparallel, complementary copy of one of the two DNA strands is produced.
   (h) Animal cells have at least three different DNA-directed RNA polymerases.
   (i) Cells of higher plants have at least four different DNA-directed RNA polymerases.

2. Why did Crick and his colleagues, in their proposal about the genetic code, rule out the possibility that a codon is a doublet (sequence of two nucleotide bases) or a sequence of four nucleotide bases?

3. What effect would the following mutations in *E. coli* probably have on the expression of the *lac* operon, which is normally an inducible enzyme system (Figure 22-5)?
   (a) Mutation in the I gene
   (b) Mutation in the O region
   (c) Mutation affecting the synthesis of cyclic AMP
   (d) Mutation affecting the production of CAP

4. Based on current knowledge, which of the following characteristics apply to messenger RNAs of procaryotes (P), of eucaryotes (E), or of both (B)?
   (a) Carry the genetic information needed for protein synthesis
   (b) Can possess a polycistronic message coding for two or more enzymes
   (c) Are synthesized in a 5′ → 3′ direction
   (d) Have 3′-polyA termini
   (e) Have capped 5′-termini
   (f) Are synthesized by DNA-directed RNA polymerase

5. Select the answers to the following statements about the Britten–Davidson proposal for the regulation of eucaryotic genes from the following list:

Activator message set	Sensor gene
Activator protein	Sensor protein
Effector	Set of integrator genes
Receptor sequence	

   (a) Steroid hormone
   (b) Hormonal receptor protein
   (c) Controls transcription of a set of integrator genes
   (d) Produces an activator message set
   (e) Has the code for the production of activator proteins
   (f) Interacts with a specific receptor sequence(s)
   (g) Activation initiates the first level of control
   (h) Activation initiates the second level of control
   (i) Controls the transcription of a structural gene

6. Based on the nucleotide sequence presented in Figure 22-8, (a) what three amino acids are coded for by the terminal codons of the message of the I gene, (b) what is the termination codon of the message, (c) what are the first three amino acids incorporated into the polypeptide specified by the Z gene, and (d) what proteins are coded for by the I and Z genes of the *lac* operon in *E. coli?*

7. Which of the following statements apply to the genetic code as used in mammalian mitochondria (M) or cytoplasm (C), in both (B), or in neither (N).
   (a) Has one codon for tryptophan
   (b) Has two codons for methionine
   (c) Has two codons for phenylalanine
   (d) Has two codons for cysteine
   (e) Has three codons for isoleucine
   (f) Has four codons for arginine
   (g) Has four termination codons
   (h) Has six codons for leucine
   (i) Has six codons for threonine
   (j) Has 61 "sense" codons

8. In the chapter, control of the *trp* operon by attenuation was described. Transcription of the operon is also controlled by *end product repression,* a regulatory mechanism proposed by Jacob and Monod for operons possessing genes for biosynthetic pathways. Their model was patterned after the one proposed for the *lac* operon (Figure 22-5) and, by using the same regulatory elements (R gene and O region), explains how a cell can regulate its biosynthetic pathways. Describe briefly the molecular aspects of regulation of the *trp* operon by end product repression.

SUGGESTED READINGS

Grossman, L., and K. Moldave, eds., *Nucleic Acids.* Part I, *Methods Enzymol,* Vol. 65. New York: Academic Press, 1980. A volume devoted to articles on restriction endonucleases and restriction mapping techniques.

Losick, R., and M. Chamberlin, eds., *RNA Polymerases.* Cold Spring Harbor, N.Y.: Cold Spring Harbor Laboratory, 1976. Comprehensive review and research articles.

Stewart, P. R., and D. S. Letham, eds., *The Ribonucleic Acids,* 2nd ed. New York: Springer-Verlag, 1977. A comprehensive treatment of the various classes of RNA.

Vogel, H. J., ed., *Metabolic Regulation.* Vol. 5 of *Metabolic Pathways,* 3rd ed. New York: Academic Press, 1971. This volume has good, authoritative reviews on the regulation of various biosynthetic and catabolic pathways, including the *lac* operon.

Woese, C. R., *The Genetic Code.* New York: Harper and Row, 1967. A good reference book summarizing the knowledge about the genetic code.

Ycas, M., *The Biological Code.* Amsterdam: North-Holland, 1969. (U.S. distributor: Wiley-Interscience, New York). Another well-written story about the code.

The following references listed at the end of Chapter 21 are also recommended: Lewin's three volumes, Stent and Calendar, Taylor, Watson, and Zubay and Marmur.

Articles

Beerman, W., and U. Clever, "Chromosome Puffs." *Scientific American,* 210(4):50 (offprint #180), 1964.

Brawerman, G., "Eukaryotic Messenger RNA." *Ann. Rev. Biochem.,* 43:621, 1974.

Brown, D. D., "The Isolation of Genes." *Scientific American,* 229(2):201 (offprint #1278), 1973.

Chambon, P., "Eucaryotic Nuclear RNA Polymerases." *Ann. Rev. Biochem.,* 44:613, 1975.

Crick, F. H. C., "The Genetic Code." *Scientific American,* 207(4):66 (offprint #123), 1962.

Crick, F. H. C., "The Genetic Code III." *Scientific American,* 215(4):55 (offprint #1052), 1966.

Fraenkel-Conrat, H., "The Genetic Code of a Virus." *Scientific American,* 211(4):46 (offprint #193), 1964.

Hurwitz, J., and J. J. Furth, "Messenger RNA." *Scientific American,* 206(2):41 (offprint #119), 1962.

Miller, O. L. Jr., "The Visualization of Genes in Action." *Scientific American,* 228(3):34 (offprint #1267), 1973.

Nirenberg, M. W., "The Genetic Code II." *Scientific American,* 208(3):80 (offprint #153), 1963.

Paule, M. R., "Comparative Subunit Composition of the Eucaryotic Nuclear RNA Polymerases." *TIBS,* 6:128, 1981.

Perry, R., "Processing of RNA." *Ann. Rev. Biochem.,* 45:605, 1976.

Stein, G. S., J. S. Stein, and L. J. Kleinsmith, "Chromosomal Proteins and Gene Regulation." *Scientific American,* 232(2):46 (offprint #1315), 1975.

Watson, M. D., "Attenuation: Translational Control of Transcription Termination." *TIBS,* 6:180, 1981.

TRANSLATION OF GENETIC INFORMATION

23

INTRODUCTION

In the Central Dogma model, the translation of genetic information into protein structure (RNA → Protein) uses mRNA as a template in polypeptide synthesis to specify the incorporation of amino acids. Although the term *translation* was coined in the 1950s, it encompasses a biochemical phenomenon (protein synthesis) that has been of prime interest to biochemists since the early years of this century. Though Fischer established in 1902 that proteins are amino acid polymers, it was not until after the middle of the century that the biochemical details of the process that covalently links the protein building blocks began to be elucidated.

The first breakthrough was made in 1955 by Mahlon Hoagland when he identified the *activated form* of amino acids, suspected for many years to be required for protein synthesis. From studies on protein synthesis in rat liver, Hoagland proposed that the cytoplasmic enzymatic reaction for amino acid activation requires ATP and an amino acid and produces an *aminoacyl-adenylate* (activated amino acid) and inorganic pyrophosphate. This long-awaited revelation was followed in 1957 by a second finding, by Hoagland, Paul Zamecnik, and Mary Stephenson, that amino acid activation also uses a small RNA (tRNA) found in the soluble fraction of a cell (and originally called soluble RNA). This discovery of a role for a nucleic acid as a receptor molecule in amino acid activation (to accept the amino acid residue and form an aminoacyl-tRNA) added a surprising new dimension to protein synthesis and aided in revolutionizing views about their biosynthesis. The role of tRNA was the first cellular biochemical function identified for an RNA; thus,

$$\text{Amino acid} + \text{ATP} \rightleftharpoons \text{AMP}\sim\text{amino acid} + \text{PP}_i$$

$$\text{AMP}\sim\text{amino acid} + \text{tRNA} \rightleftharpoons \text{tRNA}\sim\text{amino acid} + \text{AMP}$$

as with DNA, the biological functioning of RNA was established belatedly in biochemical research.

The dominant role of ribonucleic acids in protein synthesis became apparent as the developing scenario for the process implicated (1) tRNAs as being essential in the process of amino acid activation, (2) rRNA (since synthesis occurs on ribosomes containing over 50 percent RNA), and (3), subsequently, mRNA as the coding agent of a gene. Crick, appreciating the significance of the role of tRNAs, postulated an *adaptor hypothesis,* which essentially proposed that, for protein synthesis, an amino acid needs to be adapted to have a specific surface capable of recognizing and combining with (by hydrogen bonding) an amino acid specifying template, which was later identified as mRNA. The adaptation involves a chemical modification of an amino acid by its covalent bonding to an adaptor molecule (a tRNA) whose structure recognizes an mRNA template and establishes an RNA-RNA interaction. The adaptor hypothesis expanded previous concepts about protein synthesis, which had centered on a straightforward linking of one amino acid to another, by focusing attention on the importance of RNAs in directing polypeptide synthesis.

This fresh biochemical approach to the study of protein synthesis blended in an exciting way with the molecular studies on transcription and the genetic code as they emerged. The outstanding contributions collectively made by these interrelated studies during the decade from 1955 to 1965 were instrumental in ushering in the era of molecular biology. With respect to protein synthesis, the complementary nature of the concomitant research on procaryotic, e.g., *Escherichia coli,* and eucaryotic, e.g., rat, rabbit, and yeast, systems did much to bolster faith in the underlying theme in molecular biology of the basic biochemical unity among living systems.

AMINO ACID ACTIVATION

The molecular adaptation of an amino acid for protein synthesis requires a specific activating enzyme, called an *aminoacyl-tRNA synthetase,* which catalyzes both the synthesis of an aminoacyl-adenylate and the transfer of its amino acid residue to the appropriate tRNA (Figure 23-1). The energy needed to link the carboxyl group of an amino acid to the α phosphate of an ATP is furnished by a pyrophosphate cleavage of ATP. Since the inorganic pyrophosphate produced in the reaction is subsequently hydrolyzed,

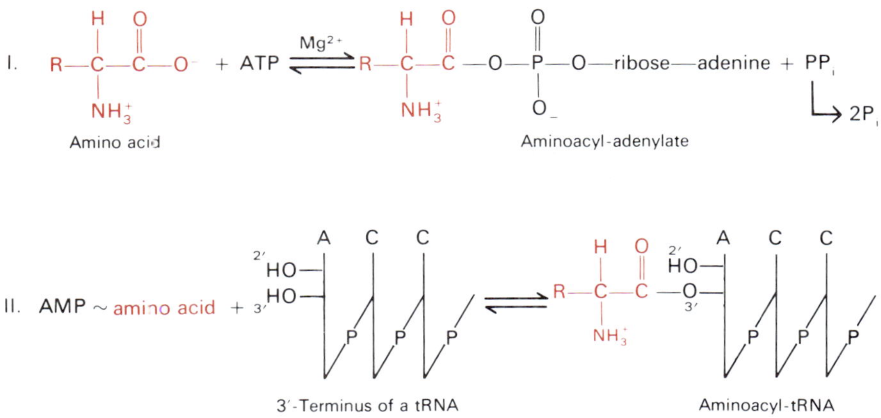

Figure 23-1
Aminoacyl-tRNA synthetase activation of an amino acid.

two high-energy phosphate bonds are consumed in this initial step in amino acid activation. Some of the energy released is conserved in the mixed anhydride bond $\left(-\overset{\overset{\textstyle O}{\|}}{C}-O-\overset{\overset{\textstyle O}{\|}}{P}-\right)$ of the aminoacyl-adenylate which, as an enzyme-bound intermediate, does not dissociate from the synthetase.

In the second step of the activation process, the amino acid of the adenylate derivative is transferred to a hydroxyl group of the 3′-terminal adenyl nucleotide of a tRNA. Although the *aminoacyl-tRNA* product is depicted as a 3′-derivative (Figure 23-1), it is uncertain whether the carboxyl group of an amino acid forms an ester bond with the 2′- or 3′-hydroxyl group. Published data indicate that the amino acid bonds first to the 2′-OH, then transfers to the 3′-OH. Acyl migration between the two hydroxyl groups occurs readily and rapidly, making it difficult to establish the chemical details of the reaction. The ester bond formed in the synthesis of an aminoacyl-tRNA has a free energy of hydrolysis comparable to that of the terminal phosphate of ATP. Hence, it is the hydrolytic cleavage of the PP_i produced in the first step that drives the overall activation process:

$$\text{Amino acid} + \text{ATP} + \text{tRNA} + H_2O \rightleftharpoons \text{aminoacyl-tRNA} + \text{AMP} + 2P_i$$

To ensure the critical fidelity necessary for the incorporation of the correct amino sequence into an mRNA-determined polypeptide structure, the activation process must be highly specific, and therefore every cell needs at least 20 different aminoacyl-tRNA synthetases and at least an equal number of tRNAs. With regard to tRNAs, studies reveal that it is not unusual for a particular cell

Table 23-1
Molecular information on some aminoacyl-tRNA synthetases

Amino acid specificity	Source	MW	Molecular structure
Isoleucine	*E. coli**	118,000	Single polypeptide
Proline	*E. coli*	94,000	Two identical protomers (α_2)
Methionine	*E. coli*	173,000	Four identical protomers (α_4)
Phenylalanine	*E. coli*	180,000	Four protomers ($\alpha_2\beta_2$)
Phenylalanine	Yeast	180,000	Four protomers ($\alpha_2\beta_2$)
Tryptophan	Pancreas (beef)	108,000	Two identical protomers (α_2)

* *Escherichia coli*

type to have more than one tRNA species for some amino acids. For example, as many as five tRNAs for Leu and Ser have been isolated from yeast, and 80 tRNA species have been identified in *E. coli.*

Each synthetase (Table 23-1) has a high degree of specificity for the amino acid it activates and for the tRNA that will accept the amino acid. The enzyme is even more selective in its attachment of the amino acid to the tRNA. Indeed, though isoleucyl synthetase can be induced to form an enzyme-bound AMP-valine, the enzyme cannot attach the valyl residue to isoleucyl-tRNA ($tRNA^{ile}$); instead, it promotes the hydrolysis of AMP-valine, i.e., the synthetase complex has a built-in mechanism for correcting errors in amino acid activation.

Over 70 tRNAs are now sequenced and, in view of the extensive degeneracy of the genetic code, many of the different codons are recognized by distinctive tRNAs. Although several tRNAs can be specific for a set of codons of a given amino acid, there does not appear to be a synthetase for each of the tRNAs, i.e., they can share the same activating enzyme.

MOLECULAR FEATURES OF tRNAs

As additional biochemical information on tRNAs was uncovered, it became increasingly apparent that these macromolecules are remarkably versatile in their biological functioning. For amino acid activation, a tRNA associates with a particular aminoacyl-tRNA synthetase (a specific RNA-protein interaction) and is specific for only one of the 20 aminoacyl-adenylates. For ribosomal polypeptide synthesis, the same biomolecule (now an aminoacyl-tRNA) must recognize the correct codon in an mRNA strand (codon-anticodon interaction) and its appropriate site (aminoacyl, or A, site) on a ribosome's surface. The molecular information that specifies these distinct functions is contained in a tRNA structure composed of about 75 to 90 ribonucleotides. In addition to their role in protein synthesis, some tRNAs are also implicated in other biological phenomena, e.g., in the regulation of genes of some biosynthetic pathways in procaryotes (p. 494) and in the reverse transcriptase reaction (p. 493), serving as the RNA primer for DNA synthesis. The attempts to explain the various structure-function relationships of these small RNAs make these biomolecules one of the most intriguing challenges in molecular biology.

Cloverleaf Model of tRNAs

Since Robert Holley (Nobel Prize, 1968) determined the first nucleotide sequence of a transfer RNA ($tRNA^{ala}$ of yeast, p. 236), over 70 other sequences from a variety of sources have been elucidated. A gratifying conclusion derived from these sequencing studies is that the cloverleaf structure (based on maximum intramolecular hydrogen bonding) originally conceived of by Holley is applicable to all tRNAs. These two-dimensional structures of tRNAs also have common molecular features (Figure 23-2), a phe-

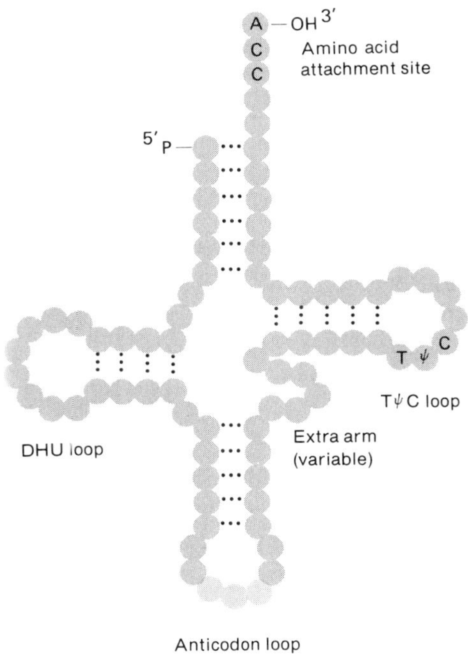

Figure 23-2
Some common features of the cloverleaf structures of tRNA molecules.

A

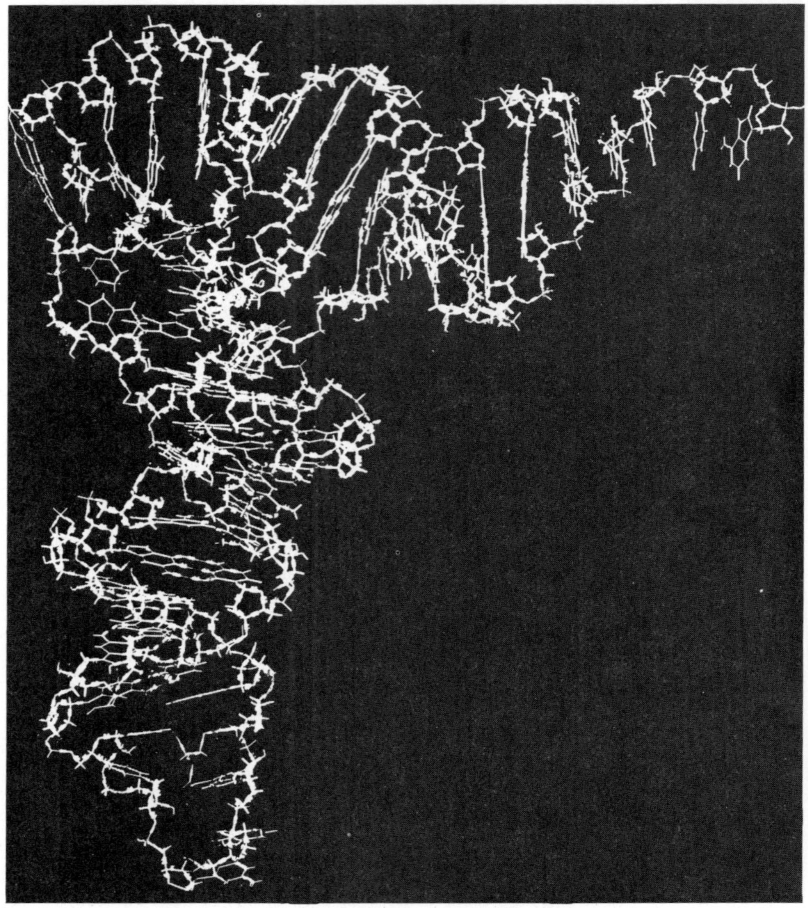

Figure 23-3
(A) Photograph of a model of the phenylalanyl tRNA of yeast (provided by Dr. Alexander Rich). (B) Drawing of the three-dimensional structure of yeast phenylalanyl tRNA. Purines are shown as rectangular slabs, pyrimidines as square slabs, and hydrogen bonds as lines between slabs. (Copyright 1982 by Irving Geis)

nomenon that is not unexpected since tRNAs, except for their individual specificities, have the same general function in protein synthesis.

All tRNAs have a significant percentage (7 to 15) of modified bases (p. 227); the roles of these unusual bases in the functioning of the molecules remain unclear. The modified bases of tRNAs prompted Crick to remark: "It almost appears as if tRNA were nature's attempt to make an RNA molecule play the role of a protein." What he meant by this statement made in the 1960s is that the unusual bases are probably needed to stabilize a nucleic acid molecule that has the intricate folding of the three-dimensional structure of a globular protein.

The 5′-termini of tRNAs are phosphorylated, and the 3′-terminal sequences (where the amino acid attaches) are always pCpCpA-

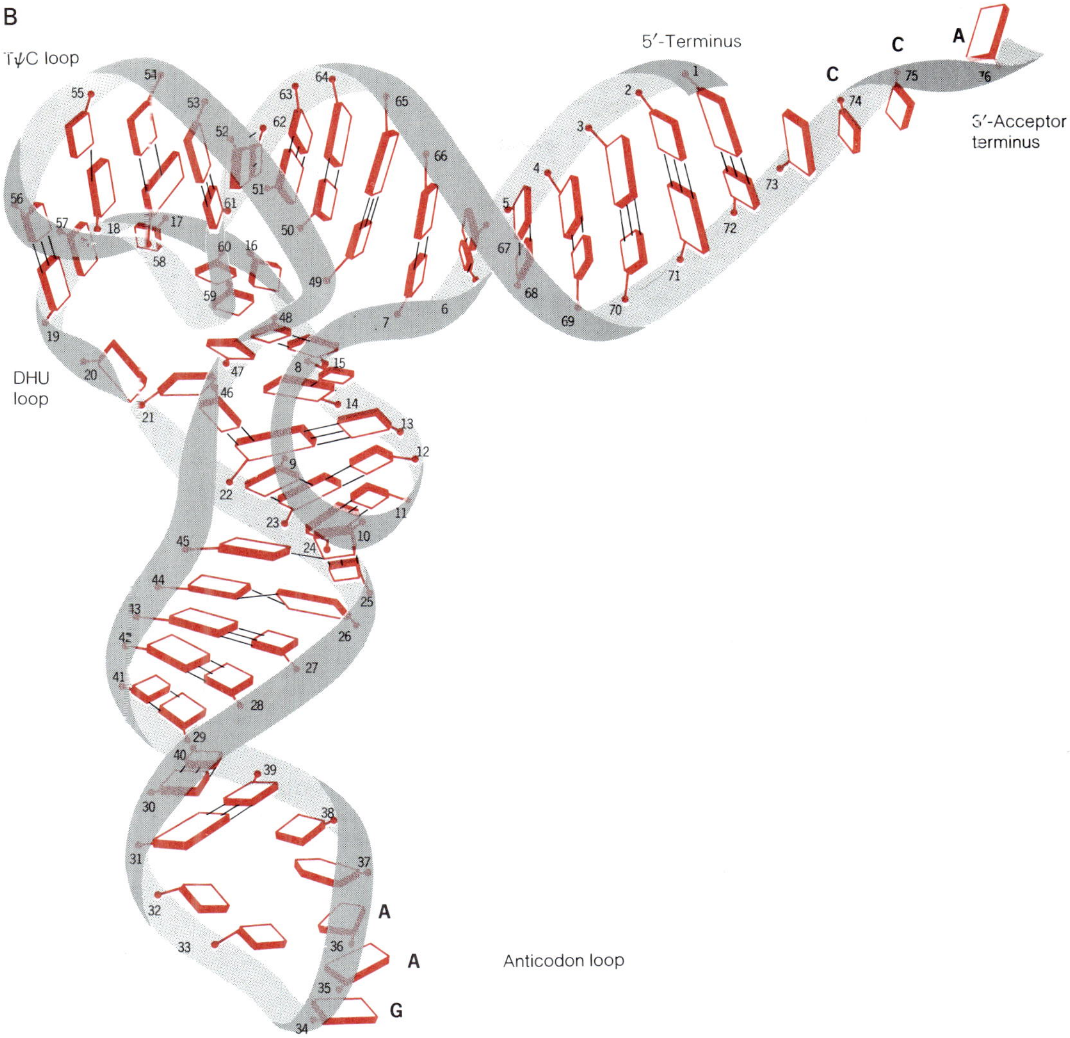

OH. In cloverleaf structures (Figure 23-2), about half of the nucleotides are base-paired to form double-helical sections and the other nucleotides constitute five characteristic single-stranded features of the molecules. One single-stranded region is the 3′-terminal sequence of four bases, and the other four (starting from the 3′-terminus) are the *TψC, extra, anticodon,* and *DHU loops.* The TψC loop, which may be involved in ribosomal binding, is named for the ribothymidine, pseudouridine, and cytidine sequence in its composition of seven nucleotides. The extra loop, not a distinct cloverleaf, has a variable number of bases. The anticodon loop, also composed of seven nucleotides, contains a sequence of three bases (anticodon) that recognizes a specific codon(s) in an mRNA being translated. The base sequence of an anticodon loop is usually

5′—pyrimidine—pyrimidine—anticodon—modified purine—variable base—3′

The DHU loop has eight to twelve bases, of which several are dihydrouracil (DHU); this loop is believed to be involved in the tRNA-activating enzyme interaction.

L-Shaped Conformation of $tRNA^{phe}$

A notable achievement by Alexander Rich and Sung-Hou Kim and colleagues is their elucidation of the three-dimensional structure of yeast phenylalanyl tRNA. The conformation of the molecule (Figure 23-3), determined at 3-Å resolution by x-ray crystallography, has an L-shaped appearance, with the 3′-terminus at one end and the anticodon loop at the other and a distance of 70 Å between the two ends. The TψC and DHU loops form the corner of the L-shaped molecule. Thus, the three cloverleafs postulated for the two-dimensional structure are present in the molecule's three-dimensional structure.

An interesting finding is that many of the hydrogen bonds that establish the L-shaped structure are not formed by the same pairing patterns found in the A·T and G·C base pairs of DNA (Figure 23-4). For example, in the pairing of 1-methyladenine 58 and thymine 54 in $tRNA^{phe}$, one hydrogen bond is between the amino group of C-6 of adenine and the keto group of C-2 of thymine (in DNA, it is the keto group of C-4 of the pyrimidine) and the other is between N-7 of adenine (N-1 in DNA) and N-3 of thymine. Too, in $tRNA^{phe}$, some bases hydrogen bond with two others; for example, guanine 22 base-pairs with cytosine 13 and 7-methylguanine 46 (Figure 23-5). In this triplet arrangement, the G·C pairing is like that in DNA; however, the G^{m7}·G base pair is distinctive. These novel types of base interactions will probably prove to be a general structural feature of other tRNAs.

WOBBLE HYPOTHESIS

In translation, interaction between an anticodon of tRNA and a codon of mRNA is accomplished by hydrogen bonding between antiparallel, complementary bases of the two triplet sequences (Figure 23-6), similar to that between the bases of the two strands of DNA. With the evidence of the degeneracy of the code (partic-

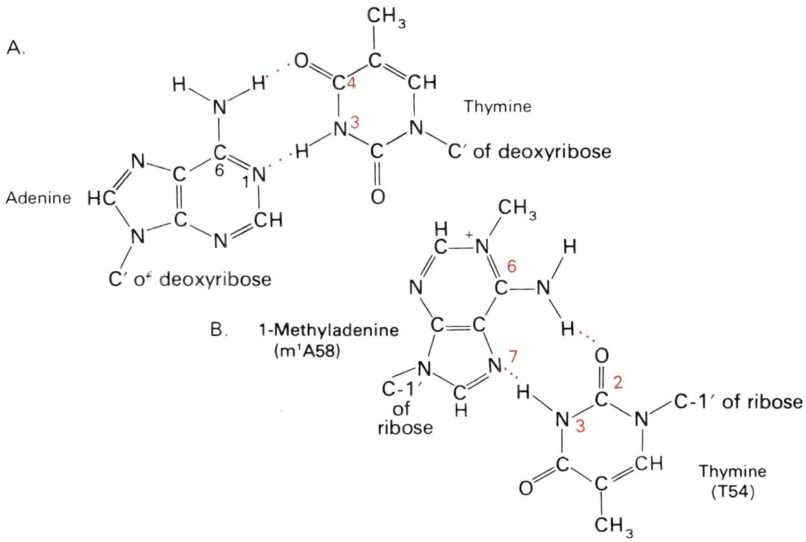

Figure 23-4
Base pairing between (A) an adenine and a thymine of DNA and (B) a 1-methyladenine and a thymine of phenylalanyl tRNA of yeast.

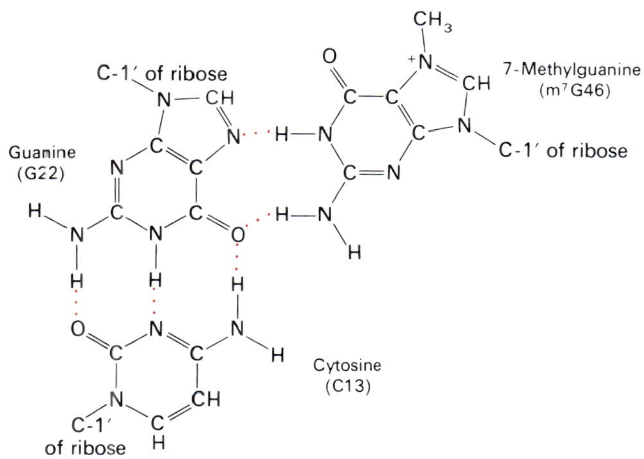

Figure 23-5
Base pairing between a cytosine (C13), guanine (G22), and 7-methylguanine (m^7G46) residue in phenylalanyl tRNA of yeast.

Figure 23-6
Recognition between a leucyl codon and its anticodon.

Table 23-2
Wobble hypothesis: allowable pairings between the first anticodon base and the third codon base

Anticodon of tRNA position 1 (5′-terminus)	Codon of mRNA position 3 (3′-terminus)
A	U
C	G
U	A or G
G	U or C
I	U, C, or A

ularly the variable position 3) and the identification of inosine (I) in the anticodon of $tRNA^{ala}$ (p. 243), it became suspected that the codon-anticodon base pairings are more flexible than those observed in DNA. Crick, by examining models for the various types of hydrogen bonding that can exist between two bases, came to the conclusion that the base pairs shown in Table 23-2 are sterically similar to the well-known Watson-Crick A·T and G·C bondings, as they occur in nucleic acid structures. Crick incorporated his findings into a *wobble hypothesis,* which assumes that, although the steric considerations for base pairing are stringent for

Three alanyl codons (mRNA)	5′—G—C—U—3′	5′—G—C—A—3′	5′—G—C—C—3′
Anticodon of yeast tRNA^{ala}	3′—C—G—I—5′	3′—C—G—I—5′	3′—C—G—I—5′

Figure 23-7
Recognition between the anticodon of an alanyl tRNA of yeast and three alanyl codons.

the first two bases of a codon (positions 1 and 2), there is some degree of pairing freedom for the third base (position 3). This proposal of steric freedom, defined as *wobble,* in the pairing of position 3 (3′-base) of a codon and position 1 (5′-base) of an anticodon proved to be correct. The tRNA for alanine, sequenced by Holley, has a 3′-C-G-I-5′ anticodon that binds to three of the four alanyl codons (Figure 23-7), as predicted by the wobble concept (Figure 23-8). In the case of tRNA^{phe}, which has the anticodon

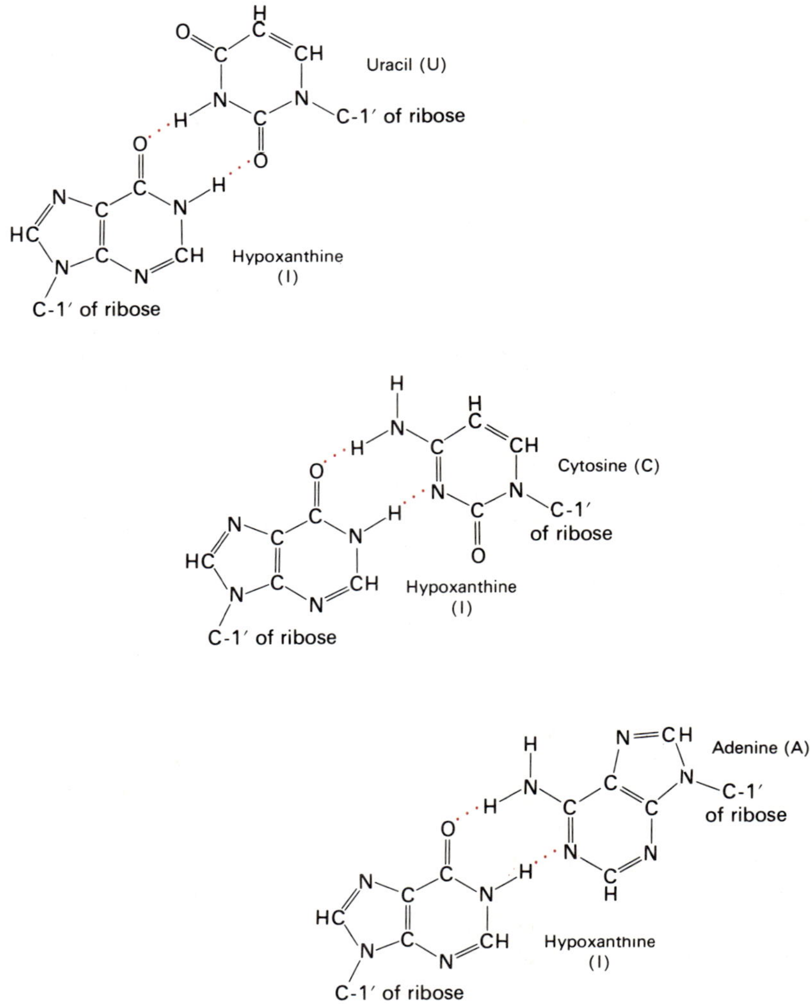

Figure 23-8
Base pairing of hypoxanthine with uracil, cytosine, or adenine.

Phenylalanyl codons (mRNA)	5′—U—U—U—3′	5′—U—U—C—3′
	: : :	: : :
Anticodon of $tRNA^{phe}$	3′—A—A—G—5′	3′—A—A—G—5′

Figure 23-9
Recognition of the anticodon of a phenylalanyl tRNA with the two codons of the amino acid.

3′-A-A-G-5′, the molecule recognizes both of the codons (5′-U-U-U-3′ and 5′-U-U-C-3′) of phenylalanine because of the guanine in position 1 of the anticodon (Figure 23-9). If the first position of an anticodon is known, it is possible to predict the number of codons a tRNA recognizes, i.e., three for I, two for U or G, and one for A or C. Because the wobble hypothesis is restricted to the base-pairing properties of position 3 of a codon, it is not applicable to degeneracy of the code involving positions 1 and 2. The wobble hypothesis did much to clarify the degeneracy of the code and the biological functioning of anticodons. The insights into the new types of *in vivo* hydrogen bonding and into the three-dimensional structure of $tRNA^{phe}$ have also considerably expanded the understanding of the molecular structure and function of nucleic acids.

POLYPEPTIDE SYNTHESIS IN *E. COLI*

Polypeptide synthesis occurs on ribosomes, where mRNAs and aminoacyl-tRNAs (bearing the amino acids to be incorporated) come together to effect the translation of genetic information. As illustrated in Figure 23-10, the 70 S ribosome of *E. coli* is composed of two subunits. The 50 S subunit has two types of RNA (23 S and 5 S) and 34 proteins in its structure, and the 30 S subunit has 16 S rRNA and 21 proteins. A eucaryotic ribosome has a molecular structure analogous to that of *E. coli,* e.g., a mammalian 80 S ribosome is composed of a 40 S and a 60 S subunit. A mammalian ribosome, however, contains more proteins than does an *E. coli* ribosome, i.e., about 30 to 40 proteins associated with the 40 S and 60 S subunits, respectively; it also has an additional 5.8 S RNA species in its larger (60 S) subunit.

After the broad aspects of protein synthesis were known, it has taken years of study with *E. coli* and eucaryotic systems to establish the series of molecular events that occur in the process. The production of polypeptides proceeds in four phases (activation, initiation, elongation, and termination), each of which requires a specific set of biomolecular compounds (Table 23-3). The following presentation on the initiation, elongation, and termination phases is based on the results obtained from the penetrating studies on protein synthesis in *E. coli,* a process similar to that occurring in

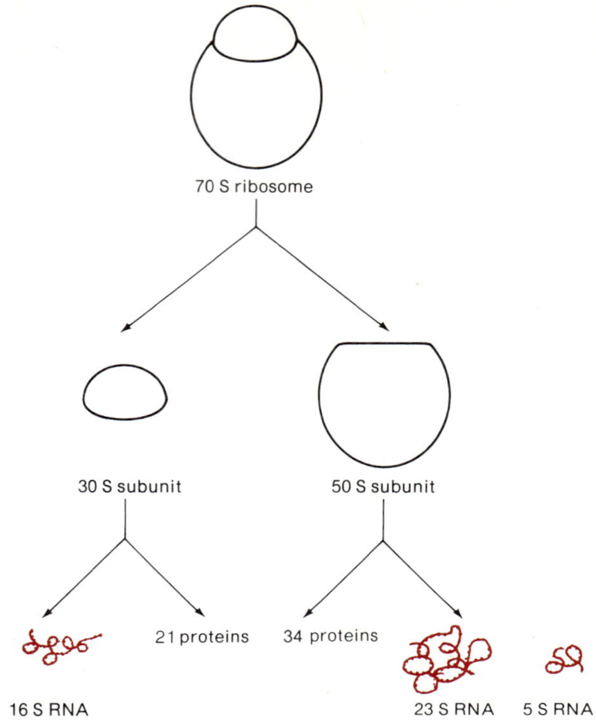

Figure 23-10
Structural components of a ribosome of *Escherichia coli*.

the cytoplasm of eucaryotes. It should be noted that the following discussion is descriptive in nature and does not attempt to encompass the complex molecular mechanisms involved in each phase of polypeptide synthesis.

Initiation of Polypeptide Synthesis

To commence polypeptide synthesis, an *initiation complex,* consisting of a 70 S ribosome, an mRNA, and an initiation tRNA carrying N-*formylmethionine* (p. 76) must first be produced. Thus, protein synthesis in *E. coli,* which proceeds from the N- to the C-terminus of the new polypeptide, requires a specific initiating amino acid with a blocked amino group.

N-*Formylmethionine*

In *E. coli* there are two distinct methionyl tRNAs, and both have a 3′-U-A-C-5′ anticodon that recognizes the unique 5′-A-U-G-3′ codon for methionine. When methionine is attached, one of the two molecular species of tRNA is formylated (Figure 23-11) by a

Table 23-3
Components required for each of the four phases in polypeptide synthesis (*Escherichia coli*)

Phase	Components
Activation of amino acids	Amino acids tRNAs Aminoacyl-tRNA synthetases ATP, Mg^{2+}
Initiation of synthesis	fMet-tRNA Initiation codon (AUG) of mRNA 30 S ribosomal subunit 50 S ribosomal subunit Initiation factors (IF1, IF2, and IF3) GTP, Mg^{2+}
Elongation of polypeptide chain	70 S ribosome Codons of mRNA Aminoacyl-tRNAs Elongation factors (EFT_u, EFT_s, and EFG) GTP, Mg^{2+}
Termination of synthesis	70 S ribosome Termination codons (UAA, UAG, and UGA) of mRNA Release factors (RF1 and RF2) GTP, Mg^{2+}

specific formylase that uses N^{10}-formyltetrahydrofolate (p. 175) as a one-carbon donor. It is in this manner that *E. coli* produces both an aminoacyl-$tRNA^{f-met}$ for the initiation step and an $tRNA^{met}$ for the subsequent incorporation of methionyl residues. Since many *E. coli* proteins have methionine as the amino-terminal amino acid, the formyl group is removed by an enzyme (deformylase) after syn-

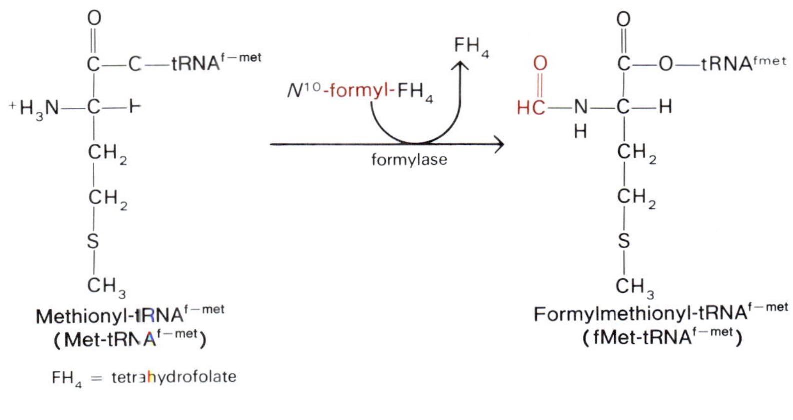

Figure 23-11
Formylation of methionyl-$tRNA^{f-met}$.

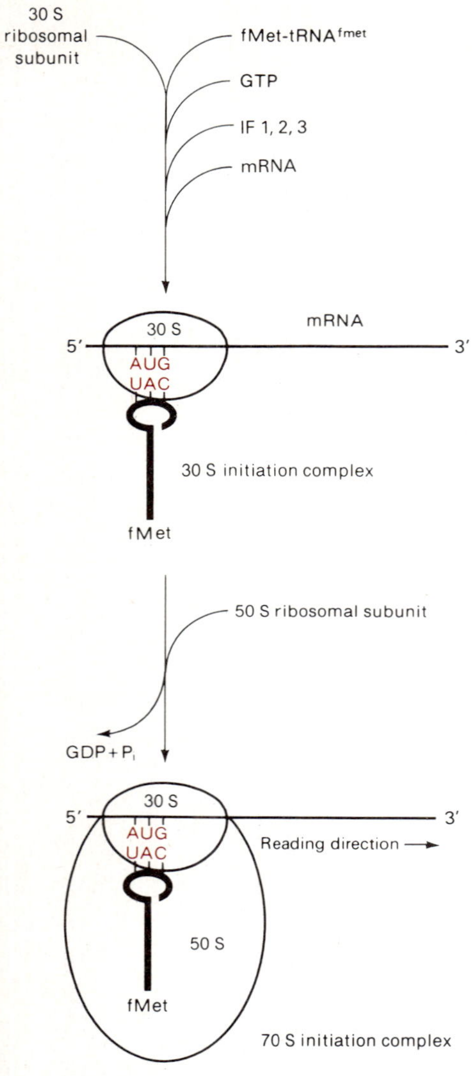

Figure 23-12
Simplified diagram of the initiation phase in polypeptide synthesis in *E. coli*.

thesis commences. Also, because other *E. coli* proteins do not have methionine at the amino termini, another enzyme(s) exists that cleaves the residue (and possibly others) from some of the newly synthesized polypeptides. Hence, minor processing of nascent polypeptides by selective enzymatic cleavage is a feature of protein synthesis, i.e., *post-translational modification.* Evidence establishes that methionine, not *N*-formylmethionine, is the amino acid that initiates protein synthesis in eucaryotes. Like *E. coli*, eucaryotes have two methionyl tRNAs; however, they lack the formylase. Interestingly, one of the two types of methionyl-eucaryotic $tRNA^{met}$ can be formylated by the *E. coli* formylase. It appears that, during the course of evolution, eucaryotes lost the ability to produce the enzyme, but one of the methionyl tRNAs retained the vestigal ability to be formylated.

Formation of an Initiation Complex

The assembly of an initiation complex is a process that requires three protein *initiation factors* (*IF1, IF2,* and *IF3*). The assemblage begins with the formation of a *30 S initiation complex* (Figure 23-12), which entails the association of an fMet-$tRNA^{f-met}$, an mRNA, and a 30 S ribosomal subunit. The participation of IF1, IF2, and IF3, as well as GTP, is required for the formation of the complex. As indicated, the interaction occurs at the 5′-terminal region of an mRNA; however, in polycistronic messages (p. 485), the complex can also assemble at an initiation site (AUG) in the interior of the mRNA.

The final step involves the joining of a 50 S ribosomal subunit to the 30 S initiation complex. In the production of this 70 S *initiation complex,* GTP is hydrolyzed to GDP and P_i, and the three initiation factors are released. The functional 70 S ribosome of a complex exists only during active protein synthesis because, after polypeptide production, it dissociates into 30 S and 50 S subunits, which are then independently reutilized to form new initiation complexes.

Postulated Role of 16 S rRNA

For the production of a 30 S initiation complex, the 3′-terminal region of the 16 S RNA component of a 30 S ribosomal subunit is believed to be the binding site for the mRNA. The pyrimidine-rich sequence at the 3′-end of 16 S rRNA (Figure 23-13) is postulated to hydrogen bond, in an antiparallel fashion, with a complementary sequence of purines on the 5′-side of the initiating AUG codon of a message. A minimum purine sequence consisting

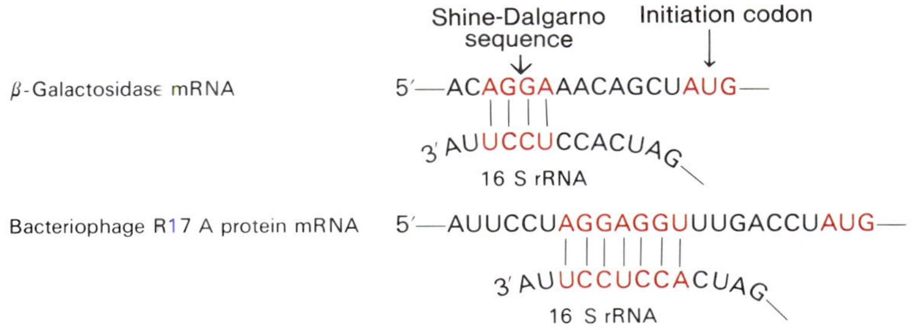

Figure 23-13
Proposed binding sequence of two mRNAs with the 3′-terminal sequence of 16 S rRNA.

of AGGA (about eight to 13 bases removed from the AUG codon) and called the Shine-Dalgarno sequence (p. 494) seems to be needed for the interaction of the mRNA with its *ribosomal binding site.* This particular pattern of mRNA-16 S rRNA binding probably aligns the initiating AUG codon of the message for proper binding with the anticodon in the 30 S-fMet-tRNA^{f-met} complex. In a polycistronic mRNA, containing two or more messages, there is a ribosomal binding site to the 5′-side of each message; hence, such mRNAs can accommodate as many initiation complexes as there are messages to be translated.

Elongation of a Polypeptide Chain

Once the 70 S initiation complex is formed, the incorporation of amino acids into polypeptide structure begins. This *elongation phase* of protein synthesis is mRNA directed, and the incorporation of each amino acid involves three sequential steps. An aminoacyl-tRNA first binds to the ribosome, after which *peptide bond formation* occurs; the last step, called *translocation*, prepares the protein-synthesizing complex for the next round of amino acid incorporation. A distinctive set of *elongation factors (EFs)* and the hydrolysis of 2 GTP are required for each cycle in the elongation process.

Binding of an Aminoacyl-tRNA

On a ribosome's surface, there are two binding sites for tRNAs; one is the *peptidyl (P) site,* specific for a peptidyl-tRNA bearing a peptide chain, and the other is the *aminoacyl (A) site,* which binds an aminoacyl-tRNA. In a 70 S initiation complex, fMet-tRNA^{f-met} occupies the P site (Figure 23-14), and the incoming tRNA binds to the A site. Which aminoacyl-tRNA attaches is determined by the codon positioned at the A site, e.g., the codon UUU specifies Phe-tRNA^{phe}. The binding of an aminoacyl-tRNA requires its asso-

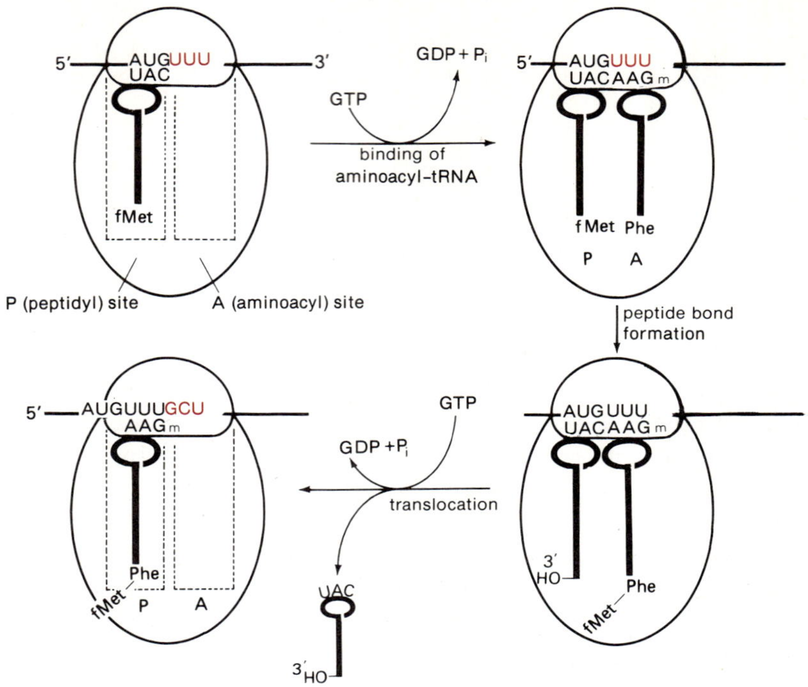

Figure 23-14
Schematic of the elongation phase of polypeptide synthesis.

ciation with an EFT_u-GTP complex (EFT_u is a temperature-unstable elongation factor) and promotes the hydrolysis of GTP to GDP and P_i. The EFT_u-GTP complex is regenerated by a two-step process (Figure 23-15). A second elongation factor, EFT_s (temperature-stable) first interacts with an EFT_u-GDP complex (released after a tRNA binds to a ribosome) to form EFT_u-EFT_s and then GTP replaces EFT_s to produce EFT_u-GTP complex.

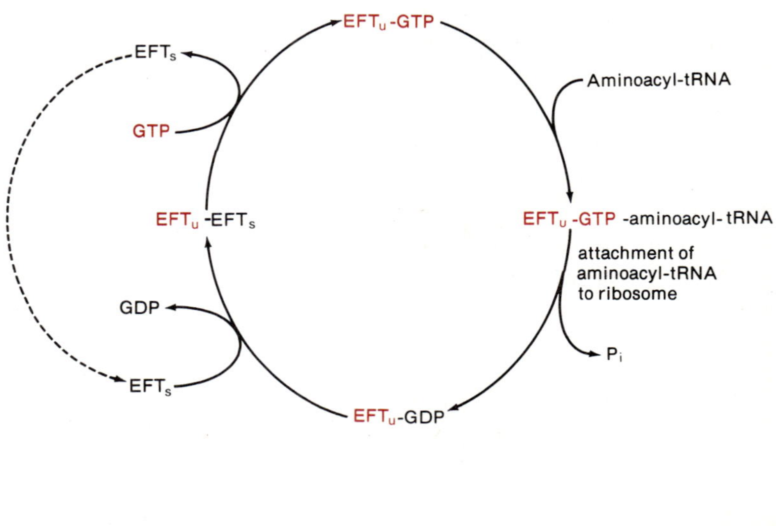

Figure 23-15
Diagram of the regeneration of EFT_u-GTP complex. (EFT_u and EFT_s represent temperature-unstable and temperature-stable elongation factors, respectively.)

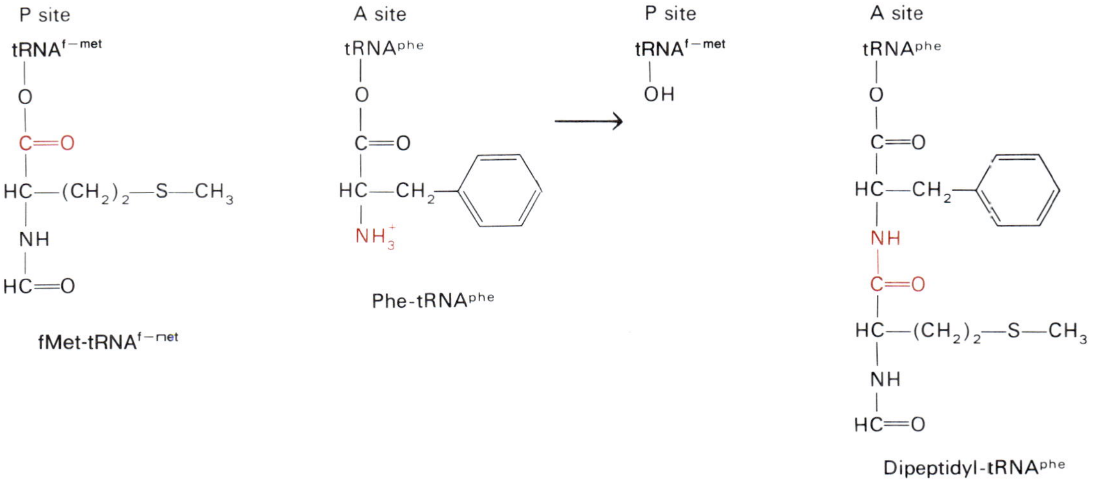

Figure 23-16
Schematic of peptide bond formation in polypeptide synthesis.

Formation of a Peptide Bond

When the P and A sites on a 70 S initiation complex are occupied (Figure 23-16), a peptide bond is formed between the two amino acid residues attached to the tRNAs. The enzyme that catalyzes the reaction is *peptidyl transferase,* a site on (i.e., group of proteins of) the 50 S subunit. For peptide bond formation (Figure 23-16), the *N*-formylmethionyl residue at the P site is transferred and bonded to the NH_3^+ group of the amino acid residue attached to aminoacyl-tRNA at the A site, producing a dipeptidyl-tRNA. In subsequent cycles of elongation, this reaction pattern is repeated each time, with the aminoacyl-tRNA at the A site accepting the growing polypeptide chain attached to the tRNA at the P site.

Translocation

At this point in an elongation cycle (Figure 23-14), a tRNA without its amino acid is attached at the P site and a peptidyl-tRNA at the A site. What follows next is the highly complex *translocation phase,* which results in (1) the tRNA with no amino acid being released, (2) the peptidyl-tRNA being relocated to the P site, and (3) the next *reading frame* (codon) assuming its position at the A site. Translocation uses a third elongation factor, *EFG,* or *translocase,* which functions in the movement of the peptidyl-tRNA to the P site. Transfer of a peptidyl-tRNA requires the prior formation of an EFG-GTP-ribosome complex and takes place with the hydrolysis of GTP to GDP and P_i. After translocation, the A site is vacant, and the next amino acid to be incorporated is determined by the new codon at the site, e.g., GCU to be recognized by Ala-$tRNA^{ala}$.

The elongation cycle is then ready to be repeated. It is in this manner that, codon after codon, the genetic instructions contained in an mRNA are translated in a 5′ → 3′ direction into a biologically functional polypeptide.

Termination of Polypeptide Synthesis

The signal to terminate protein synthesis once a nascent polypeptide is produced is furnished by a *termination codon* (*UAA, UAG,* or *UGA*) at the end of an RNA message. This *termination phase* uses one of two *release factors* (*RF1* or *RF2*), proteins that specifically recognize the three-base sequence of a termination codon. RF1 is specific for UAA or UAG and RF2 for UAA or UGA. Both release factors function in association with GTP, which is hydrolyzed in the process. It is thought that the RF-GTP complex induces the release of a newly synthesized polypeptide by triggering the hydrolytic cleavage of its carboxyl terminal amino acid from the final aminoacyl-tRNA still bound to the ribosome. Thus, the termination phase ends the NH_2 → COOH synthesis of a polypeptide.

Possible Role of GTP in Translation

Whereas ATP is selectively used for the activation of amino acids, GTP is the only nucleoside triphosphate utilized for ribosomal protein synthesis. In fact, GTP is needed for each of the three phases of polypeptide synthesis. It is doubtful that the hydrolysis of GTP that repeatedly occurs during polypeptide synthesis provides energy to drive reactions, since there is no evidence for the nucleotide's utilization to form covalent bonds. Its main role appears to be in the noncovalent binding of the various factors to a ribosome, e.g., EFT and RFs. One explanation is that GTP, in each of the translation phases, allows for the specific ribosomal binding of individual elongation factors and that its hydrolysis to GDP and P_i is required for the release of the bound factors, freeing them for reparticipation in protein synthesis.

EUCARYOTIC PROTEIN SYNTHESIS

Although, as previously indicated, eucaryotic polypeptide synthesis follows the same pattern as that just described for *E. coli*, there are distinct differences between the two systems. That methionine and not *N*-formylmethionine initiates eucaryotic protein synthesis is one difference that has already been noted (p. 516). Another is that for eucaryotic chain termination only one release factor (RF1), which recognizes all three termination codons, is required (as opposed to two factors for the *E. coli* system).

The most enigmatic difference to date, however, involves the number of initiation factors needed for formation of the eucaryotic initiation complex. Whereas the *E. coli* system requires three factors (IF1, IF2, and IF3, p. 516), the comparable eucaryotic process requires at the least the nine listed in Table 23-4. The list does not include factors that enhance the activity of eIF2 (Co-eIF2 factors) or those that enhance the dissociation of 80 S ribosomes into their subunits *(ribosome-dissociation factors)*. Note that two of the factors are oligomeric proteins. The current scheme (which is continuously subject to change as new factors are described) for the assemblage of the eucaryotic initiation complex is presented in Figure 23-17. The step-by-step process is similar to that described for formation of the *E. coli* complex (Figure 23-12), i.e., complexing of an initiating tRNA, the smaller ribosomal subunit, an mRNA, and, finally, the larger ribosomal subunit. With its requirement for six of the nine initiation factors, the addition of mRNA to the complex is the most complicated molecular step in the process. Note

Table 23-4
Known eucaryotic initiation factors isolated from rabbit reticulocytes

Factor	Molecular weight
eIF1	15,000
eIF2	≈150,000 (3 protomers)
eIF3	≈700,000 (9 protomers)
eIF4A	50,000
eIF4B	80,000
eIF4C	18,000
eIF4D	17,000
eIF4E	24,000
eIF5	≈150,000

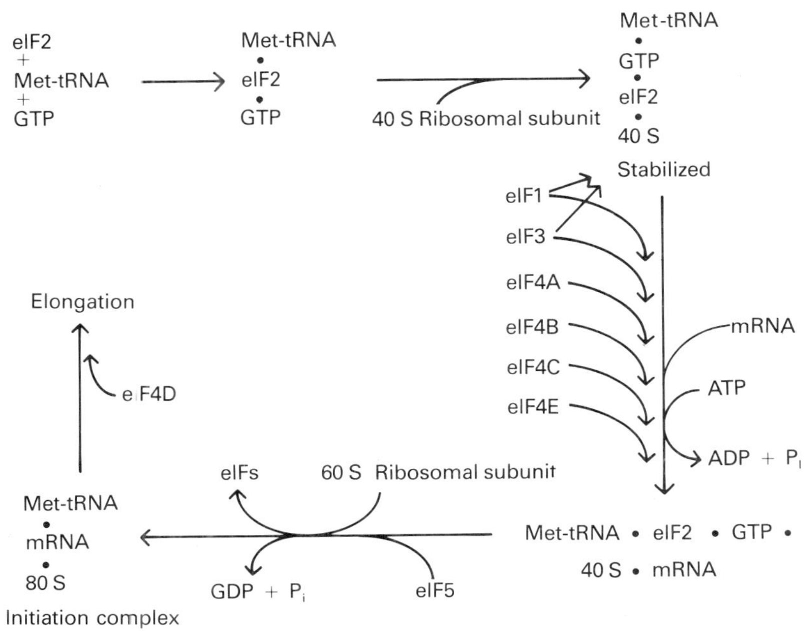

Figure 23-17
Proposed assembly of the initiation complex for eucaryotic polypeptide synthesis. (Adapted from R. Benne and J. W. B. Hershey, *J. Biol. Chem.*, 253:3078, 1978, and *Methods Enzymol*, LX:33, 1979)

in Figure 23-17 that ATP hydrolysis is required for mRNA addition. Thus, for the eucaryotic system, both ATP and GTP are necessary, whereas the comparable *E. coli* process requires only GTP. Because many novel molecular events remain to be clarified, eucaryotic protein synthesis continues to be an active area of research.

INTRACELLULAR TRANSPORT OF PROTEINS

Since many newly synthesized polypeptides are destined to be associated with particular cellular organelles (e.g., mitochondria or lysosomes) or with the exterior membrane or to be secreted (e.g., hormones), there is an obvious need for molecular mechanisms to guide such proteins to their appropriate intracellular locations. In recent years extensive research efforts have been made to elucidate the underlying biochemical processes involved in intracellular protein transport. That transported proteins reach their assigned cellular locations is of critical importance because mistakes in transport can severely affect cellular metabolism, and the cumulative damage that results can be fatal to an organism with the transport defect. For example, *I-cell disease* is a rare inborn error (Mendelian recessive) characterized by cellular export of at least eight enzymes which should be transported to lysosomes. Since the function of lysosomes is to degrade cellular debri (p. 59), lack of their normal component of degradation enzymes results in cells accumulating *inclusion* bodies, i.e., becoming I-cells. These inclusion bodies are bloated lysosomes filled with cellular debri, e.g., glycolipids, that cannot be degraded. The medical consequences of I-cell disease are exceedingly serious, and patients usually die in childhood.

Signal Hypothesis

The initial phase in the intracellular transport of eucaryotic proteins is the sequestering of nascent polypeptides in the lumen of the endoplasmic reticulum (ER). Since much of a cell's protein synthesis takes place on ribosomes associated with the ER membrane (RER, p. 57), it was suspected by scientists studying protein transport that polypeptide synthesis and transport into the ER are interrelated phenomena. Evidence to substantiate this idea was provided by a number of investigators, notably Günther Blobel and George Palade. It was from such studies that the *signal hypothesis*

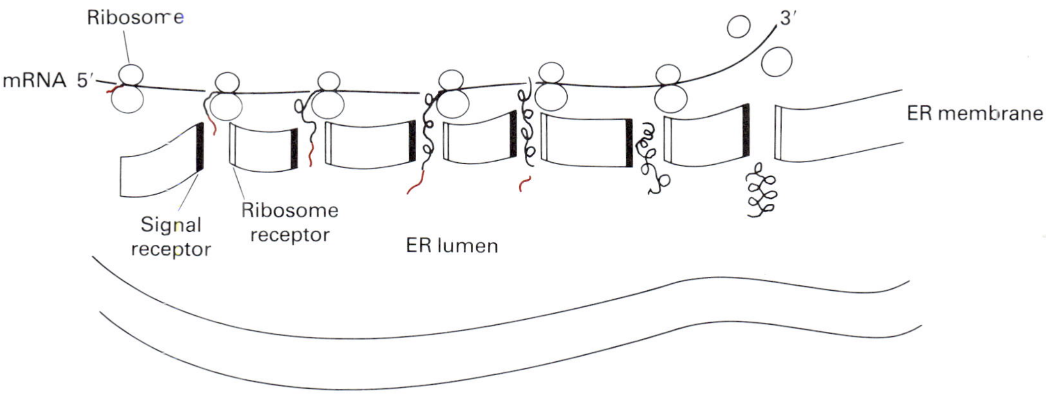

Figure 23-18
Schematic of the transfer of a newly synthesized polypeptide across the endoplasmic reticulum (ER) membrane, as proposed by the signal hypothesis.

was developed. As illustrated in Figure 23-18, the hypothesis provides a model to explain how new polypeptides scheduled for intracellular transport are routed into the ER lumen.

A polypeptide to be transferred across the ER membrane is synthesized with a *signal sequence* as its N-terminus. A signal sequence is composed of *15 to 30 amino acid residues*, many of which have *hydrophobic* side chains. These structural features are well illustrated by the signal sequence of insulin (a transported protein), which is:

$$^{+}H_3N\text{—Met—Ala—Leu—Trp—Met—Arg—Leu—Leu}$$
$$\text{—Pro—Leu—Leu—Ala—Leu—Leu—Ala—Leu}$$
$$\text{—Trp—Gly—Pro—Asp—Pro—Ala—Ala—Ala—}$$

Insulin is synthesized as preproinsulin (p. 543), and the *pre* prefix refers to the 24-residue signal sequence depicted above. Note that 21 of the residues have hydrophobic side chains (a higher percentage than is found in most signal sequences).

For transport of a polypeptide into the ER lumen, the signal sequence (first segment of the polypeptide produced) attaches to a specific site, called the *signal receptor*, located on the ER membrane. The hydrophobicity of the signal sequence is postulated to be the molecular key for the polypeptide's interaction with the membrane (which is also a hydrophobic structure). A second recognition site is the *ribosome receptor*, which serves to anchor the organelle to the membrane. As illustrated in Figure 23-18, the interaction between the signal sequence and the ER membrane is believed to open a pore, or channel, in the membrane through which the polypeptide, as its synthesis proceeds, is transported into the ER lumen. The signal sequences of most polypeptides are removed by a membrane enzyme *(signal peptidase)* during the transport process; the peptidase is located on the luminal side of the ER.

Thus, the molecular instructions for transport into the ER (in the form of a hydrophobic sequence) are furnished by the polypeptide. Although the general features of the model are accepted, it is expected that continued research will provide modifications about the molecular details of transport. To date, for example, there is little evidence for the membrane pores postulated; therefore, whether the membrane actually opens is a matter of debate. Also, a signal sequence need not be the N-terminus since, in the case of ovalbumin, the sequence is located in the interior of the polypeptidyl sequence.

Glycosylation and Protein Transport

Once in the ER lumen, the biochemical processes that prepare individual polypeptides for their cellular distribution are initiated. Whereas hydrophobicity provides the first molecular instructions for intracellular transport, it is *glycosylation* (addition of carbohydrates) that establishes the molecular patterns polypeptides need to continue their intracellular routing. The addition of oligosaccharide units, which convert polypeptides into glycoproteins, commences in the ER lumen and continues when they are transported from the ER to the Golgi apparatus (p. 59). Current evidence supports the contention that the particular oligosaccharide unit associated with a glycoprotein furnishes the molecular instructions for its cellular destination. Acquisition of oligosaccharide units by polypeptides has been compared to the assignment of zip codes to mailing addresses, with each type of oligosaccharide (of which there are many) representing a distinct zip code.

Glycosylation begins soon after a nascent polypeptide enters the ER lumen. Carbohydrates bind to either the amide group of an asparagine or the hydroxyl group of a serine or threonine (p. 219). Oligosaccharides attached to asparaginyl residues are referred to as *N-linked* and those to seryl or threonyl residues as *O-linked.* Of the two types of linkage, much more is currently known about synthesis of N-linked oligosaccharides than about that of O-linked. Implicit in the following presentation about N-linked glycosylation is that the molecular instructions dictating which oligosaccharide unit a polypeptide will attain reside in the sequence and structure of the protein, e.g., which Asn(s) will be glycosylated and what molecular species of oligosaccharide it will bear.

Core Glycosylation

In the ER lumen, an N-linked oligosaccharide is not added to a polypeptide by a series of one-carbohydrate additions but, instead, as an intact unit—called the *common oligosaccharide core* (Figure 23-19). However, construction of this core (containing nine mannosyl, three *N*-acetylglucosylamine, and three glucosyl residues) occurs by a sequential addition of single carbohydrates to *dolichol phosphate*, the *lipid carrier* of the oligosaccharide (Figure 23-19). (Note that the first carbohydrate, GlcNac, is added as a phosphorylated derivative.) Dolichol is an unusually long-chain lipid containing 17 or more isoprene units (p. 378). When the oligosaccharide unit is synthesized, the *activated oligosaccharide core* transfers the common core to a polypeptide.

After attachment, the oligosaccharide core does not remain intact since it is subjected to enzymatic trimming (removal of carbohydrates). Within three to five minutes after the core is added, for example, trimming of two of the three glucosyl residues occurs. Trimming commences in the ER and continues when the polypeptides are transferred to the Golgi apparatus. For some polypeptides, trimming produces the required oligosaccharide units; for others, trimming *and* subsequent addition of new carbohydrates are needed for these polypeptides to acquire their characteristic glycosylated patterns. It is in the Golgi apparatus that most of the final trimming and additions take place (called *terminal glycosylation*).

Fuc	fucose
Gal	galactose
Glc	glucose
GlcNac	*N*-acetylglucosamine
Man	mannose
NAN	*N*-acetylneuraminate (sialate)

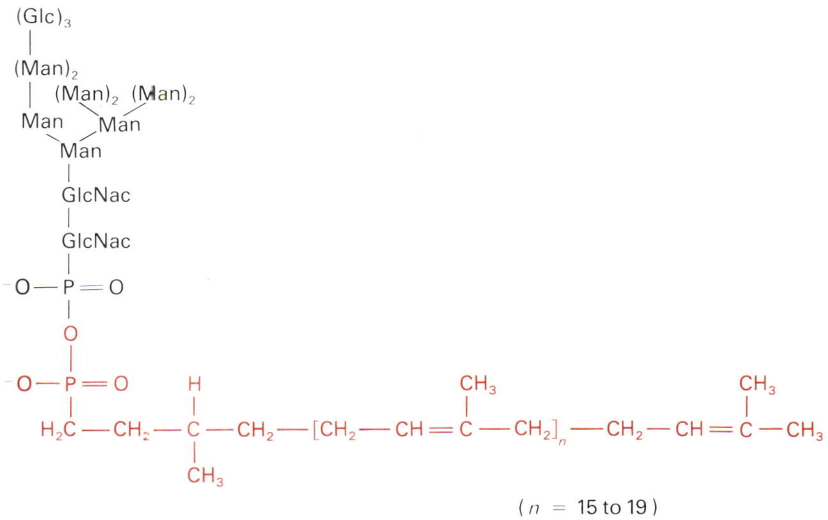

Figure 23-19
Activated oligosaccharide core. The dolichol phosphate (lipid carrier) moiety is shown in red.

Terminal Glycosylation

In the Golgi apparatus, glycosylation is completed and the glycoproteins are sorted and packaged for transport to their individual cellular locations. Little detailed information is currently available about the biochemical mechanisms involved in the "sorting and packaging"; however, with respect to N-linked oligosaccharides, a unified concept about the types of units attached is developing.

As a rule, N-linked oligosaccharides have the same *inner core,* which is the branched pentasaccharide (containing three mannose and two *N*-acetylglucosamine) depicted in Figure 23-20. Apparently, trimming of the common core can proceed to the level of the inner core. The N-linked oligosaccharides generally fall into one of the two following classifications. The *simple mannose-rich units* are those that possess the inner core (1) with short or long mannose oligosaccharides attached (chicken ovalbumin in Figure 23-20) or (2) with one or a few carbohydrates attached (human immunoglobulin M or IgM). Those oligosaccharides with *N*-acetyllactosamine (disaccharide unit of Gal and GlcNac) linked to the mannosyl residues of the inner core are called *complex* N-*acetyl-*

Figure 23-20
Four oligosaccharide units of glycoproteins.

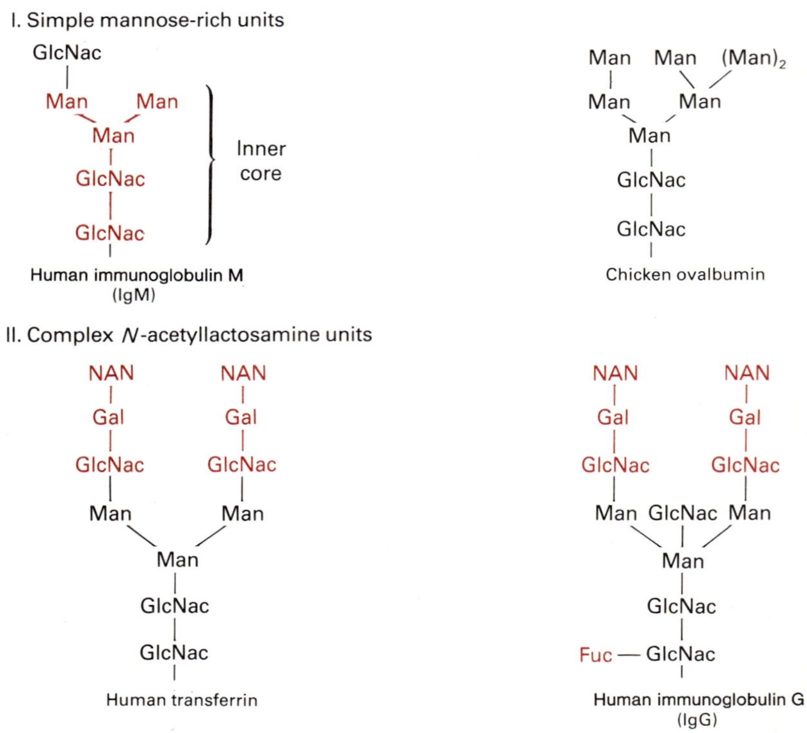

lactosamine units, since they generally have additional sialate (NAN) residues bonded to their galactosyl residues. The oligosaccharide units of human transferrin and immunoglobulin G (Figure 23-20) are examples of complex *N*-acetyllactosamine units which, unlike the simple units, have been found only in animals. Note that the two examples of complex units have three carbohydrates (galactose, sialate, and fucose) not found in the common core.

The four oligosaccharides illustrated in Figure 23-20 are but a few of the many types known. Since each unit contains specific information regarding protein transport, elucidation of the molecular functions of these oligosaccharides remains a scientific challenge. Recent studies on the defective enzymes associated with I-cell disease provide an excellent example of the medically important information currently being obtained about the roles of individual oligosaccharide units in glycoprotein transport. As stated previously (p. 522), an I-cell is defective because it exports lysosomal enzymes. These exported enzymes, it has been determined, lack *mannose 6-phosphate*, which is present in the oligosaccharide units of normal lysosomal enzymes. Thus, it appears that mannose 6-phosphate is a necessary signal for the transport of lysosomal enzymes from the Golgi apparatus to the lysosomes.

THE ϕX174 GENOME

The information presented in this chapter and the two previous ones concerning the flow of genetic information from DNA to protein summarizes, in part, a golden age in molecular biology that continues still. In the last two decades, whenever it seemed that the spectacular advances provided through molecular studies were beginning to subside, exciting new announcements expanded the frontiers of molecular research further, adding fresh insights into the molecular basis of life. In the 1960s, examples of such announcements included Jacob and Monod's model for gene regulation, Nirenberg's identification of the first codon, Sutherland's two-messenger hypothesis to explain a hormone's molecular action, and Temin and Baltimore's discovery of reverse transcriptase. Each of these contributions has had an electrifying effect on biological research and has introduced molecular concepts about significant biological phenomena.

The announcement in 1977 (*Nature*, 265:687) of the nucleotide sequence of the genome of the bacteriophage ϕX174 (Figure 23-21) by Frederick Sanger and collaborators, was another of

A

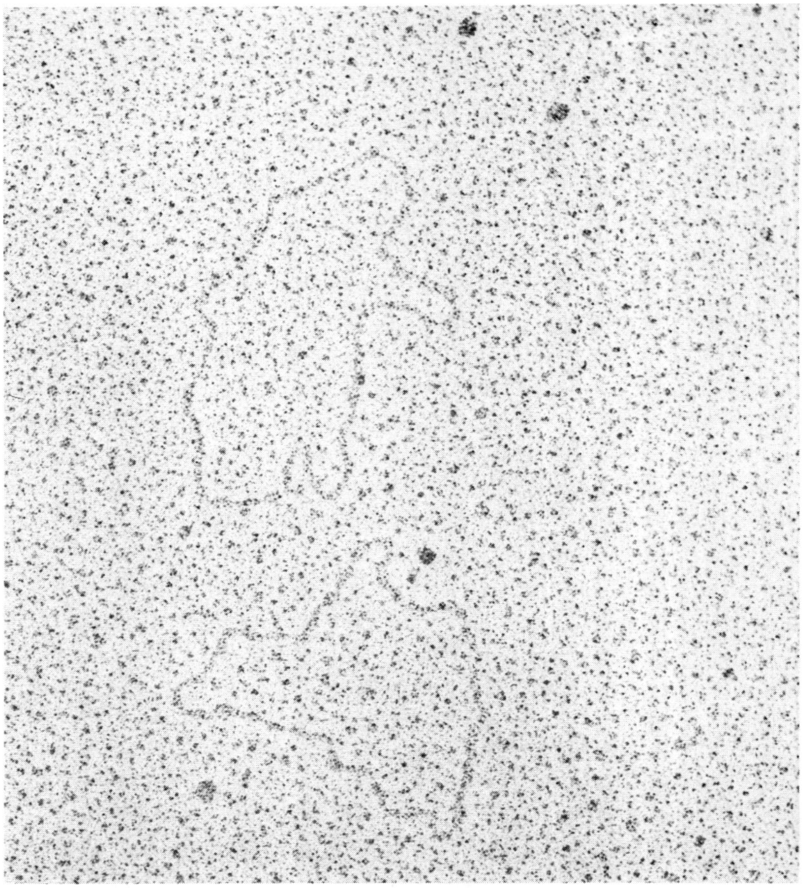

B

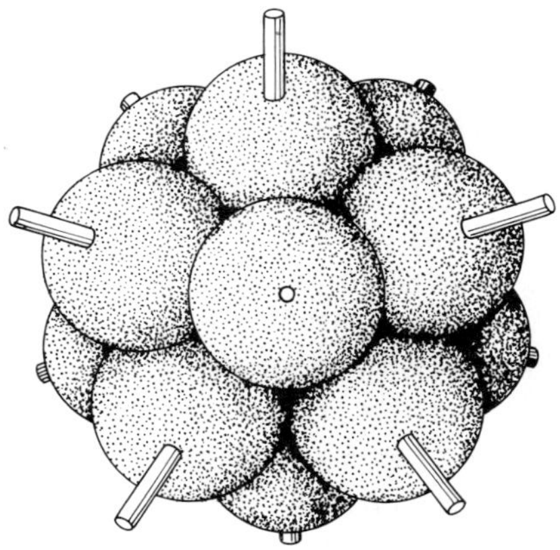

Figure 23-21
(A) Electron micrograph of the single-stranded genome of bacteriophage ϕX174, 100,000X magnification (courtesy, Dr. Alix R. Bassel); (B) drawing of the bacteriophage ϕX174.

these scientific milestones. Now that genetic sequences are obtainable, a new era in the molecular study of life has emerged. The particularly exciting aspect about the elucidation of the ϕX174 genome is that the protein sequence and genetic data that have accumulated over the past two decades on this single-stranded DNA phage could be assimilated and compared with the base sequence of its genome. Thus, all established concepts of genetic information storage, regulation of transcription, and the genetic code (principles that have guided molecular studies on transcription and translation) are now verifiable at the genic level.

Analysis of the ϕX174 Genome

The sequence of ϕX174 DNA, attained principally by the plus-and-minus method of Sanger (p. 456), was analyzed in conjunction with the accumulated evidence on the phage genes and their products to establish the genetic makeup of the DNA molecule. Knowledge of the genetic code was important in this comparative study. The sequences similar to known ribosomal binding and promoter regions, along with the initiation codon (ATG in DNA), helped to identify the beginnings of genetic messages, and the termination condons (TAA, TAG, and TGA in DNA) permitted recognition of the ends of these messages. The triplet reading frames that code for the known amino acid sequences of ϕX174 proteins were also inferable from the code and therefore were useful in establishing the location of genes in the DNA molecule. Collectively, these types of analyses permitted an intelligent interpretation of the coding capacity of the genome. Whereas many of the conclusions are based on collaborative evidence and are considered established, others will require additional confirmation. However, the overall evaluation is considered a generally accurate view of the biological functioning of the genome.

Coding Capacity of ϕX174

Table 23-5 lists the genes of ϕX174 and information about their size and the molecular weight of the proteins for which they code. Figure 23-22 is a companion illustration that schematically depicts the physical location of the genes in the 5,375-nucleotide sequence. (Because the following discussion is based on the 1977 report, it should be noted that the nucleotide content of the ϕ X174 genome has since been revised upward to 5,386 $\pm$ 1.)

Table 23-5
Proposed coding capacity of ϕX174

Gene	No. of nucleotides	MW of protein
A	1,536	56,000*
[A*]	?	35,000†
B	(360)‡	13,845*
C	?	7,000†
D	456	16,811§
E	(273)	9,940*
J	114	4,097§
F	1,275	46,400*
G	525	19,053§
H	984	35,800*
Noncoding and C	485	
Total	5,375	

*Calculated from nucleotide sequence.
†Determined by SDS-disc gel electrophoresis. (In this electrophoretic procedure, performed in the presence of sodium dodecyl sulfate, the molecular weight of a polypeptide determines its rate of migration in the gel.)
‡Parentheses indicate an overlapping sequence not included in the total number of nucleotides.
§Calculated from amino acid sequence.

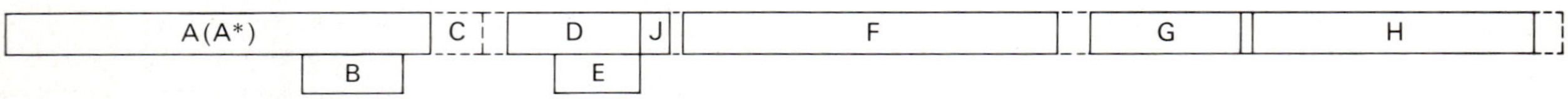

Figure 23-22
Proposed coding capacity of bacteriophage ϕX174. Genes that code for proteins are identified by letters. A* is an additional protein that is coded by gene A; nucleotide sequences between genes are indicated by dashed lines.

The A and B Genes

Analysis of the nucleotide sequence of the A gene (the largest) reveals an unexpected genetic feature of the ϕX174 genome. Although the 1,536 nucleotides of the gene code for the A protein (512 amino acids; MW ≈ 56,000), the code for the B protein (120 amino acids; MW ≈ 13,845) is also contained within the same sequence. Hence, *gene B is located within gene A.* The reading frame of the coding sequence for the B protein, which begins and ends within the A gene, is one nucleotide to the left of that of the A protein (Figure 23-23). The revelation of this economical packaging of genetic information is a novel observation, one that would not have been predicted by the previously existing views of the genetic code.

Figure 23-23
Hypothetical reading frames for the A and B genes of ϕX174.

Reading frame for B protein

—ABCDABCDAB—

Reading frame for A protein

viewed as three-letter words:

CANYOUTRYEELMANYOURATE

Gene A is also responsible for the production of a third protein (A*), which has a MW of about 35,000. Transcription of the mRNA coding for A* protein begins at an internal initiation site in the A gene, and the reading frame of the coding sequence is the same as that of the A protein. The biological significance of A* remains debatable even though it possesses the enzymatic activities associated with the A protein (which is now believed to participate in the termination of the replication and release of newly synthesized single-stranded genomes of ϕX174). Another genetic feature of the A gene is that the origin of replication of the genome is located within its nucleotide sequence.

The C, D, E, and J Genes

The C gene is situated between the A and D genes (Figure 23-22), but, on the basis of available data, the exact length of the nucleotide sequence of the gene remains uncertain. In Table 23-5, the C gene sequence is included in the column for the noncoding nucleotides. A possible initiation codon for the C protein coding sequence is one that overlaps the termination codon of the A protein sequence (Figure 23-24). Biochemical evidence suggests that the C protein is small with a MW of about 7,000 (Table 23-5).

The D gene is well defined, since the amino acid sequence of the D protein (needed for the assembly of the ϕX174 capsid, or coat) is known. The D protein has 151 amino acid residues and a MW of 16,811. The sequence indicated between the C and D genes (Figure 23-22) includes not only the C gene terminal region (of uncertain length) but also a sequence that precedes the initiation site of mRNA synthesis for the D gene. Analogous to the relationship between the A and C genes, the termination codon for the D gene overlaps with the initiation condon for the J gene (TAATG), which codes for a small basic protein whose sequence of 37 residues (MW $\approx$ 4,097) is known.

This definite placement of the D and J genes adjacent to each other seemingly conflicted with genetic data that placed the E gene between D and J genes. Based on a comparative analysis of the genome and biochemical data, it was concluded that the *E gene is contained within the D gene* (Figures 23-22 and 23-25). Thus, both the A and D genes contain another gene within their nucleotide sequences. The E gene, which codes for a protein of 91 amino acid residues (MW $\approx$ 9,940), has a reading frame for its coding sequence that is one nucleotide to the right of that of the D gene (Figure 23-25).

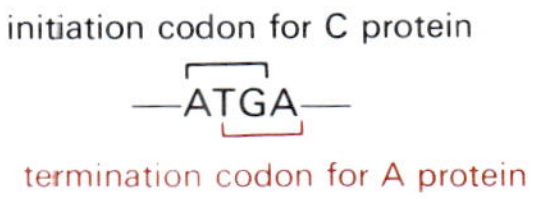

Figure 23-24
Proposed overlap of termination and initiation codons of the A and C genes, respectively.

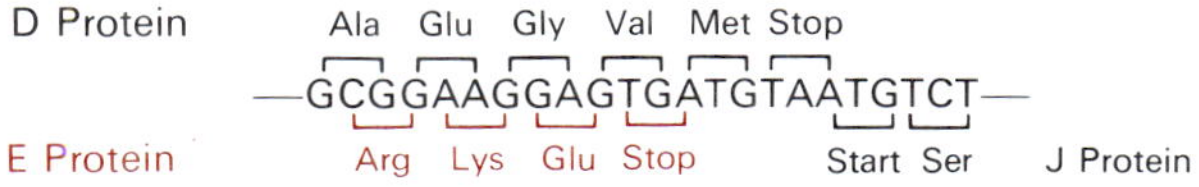

Figure 23-25
Nucleotide sequence depicting the termination of the D and E genes and initiation of the J gene of ϕX174.

The F, G, and H Genes

After the J gene, there is a sequence of 39 nucleotides (not translated) before the initiation of the F gene, which codes for the F protein of about 242 amino acids (MW ≈ 46,400). Since most of the amino acid sequence of the F protein (capsid component) is known, the nucleotide sequence of the F gene is well established. Between the F and G genes, there is a long, untranslated sequence of 111 nucleotides. The G gene codes for the G protein (whose amino acid sequence is also known), which has a MW of 19,053. The makeup of the genome is completed with the inclusion of the H gene, located between the G and A genes. This gene codes for the H protein which, based on nucleotide sequence data, contains 326 amino acid residues (MW ≈ 35,800). After termination of the H gene, there are 66 untranslated nucleotides before the initiation of the A gene.

Comments on the ϕX174 Genome

The announcement of the sequence of ϕX174 was a stellar culmination of years of study on the genetic and biological functioning of the bacteriophage. The comparative analyses allowed by knowledge of the sequence of the genome support the current concepts about the various types of molecular information contained in genetic material. The knowledge of initiation, termination, and sense codons and about promoter, leader, and ribosomal binding sequences, collectively contributed significantly in making genetic sense of the sequence of 5,375 nucleotides. A remarkable feature of the announcement was that it not only satisfactorily confirmed existing molecular interpretations of a gene and its control and expression, but also added new knowledge about molecular genetic information.

The sequence of the ϕX174 genome revealed for the first time how the coding efficiency of a limited number of nucleotides can be increased by having a DNA sequence (genes A and D) serve the dual purpose of coding for two functional proteins. Whether this coding phenomenon is limited to the small genomes of bacteriophages is unknown; however, there exists the intriguing possibility that this coding pattern may be of broad biological significance. The overlap of termination and initiation codons (also observed in the *trp* operon of *E. coli*) is another interesting coding interaction that may be shared by many genomes.

SUMMARY

Since the mid-1960s, the determination of the nucleotide sequences of over 70 transfer RNAs (tRNAs) has revealed several structural similarities among the molecules. All tRNAs have a phosphorylated 5′-terminus, a 3′-terminal CCA-OH sequence, and a significant percentage of modified bases. Their cloverleaf structures (based on maximal intramolecular hydrogen bonding) are also very similar; all have a single-stranded sequence of the 3′-terminus and four single-stranded loops, *viz.,* the TψC, extra, anticodon, and DHU loops. The elucidation of the three-dimensional structure of a tRNA, by x-ray diffraction studies, has confirmed the presence of the four loops in the molecular conformation of a tRNA. The conformation, which is L-shaped, also reveals some unusual base pairings and several bases that hydrogen bond not with one but with two other bases in the molecule.

The wobble hypothesis states that, with regard to the base pairing between a codon (mRNA) and an anticodon (tRNA), the pairing is stringent for the first two bases of the codon but not for the third (3′-terminus base). The hypothesis is derived from the knowledge that various types of hydrogen bonding, other than those between A·T(U) and G·C pairs can exist between two bases. The molecular data on known anticodon sequences and on the degeneracy of the genetic code substantiate the hypothesis.

There are four phases in polypeptide synthesis. The first step involves the activation of amino acids, which is catalyzed by aminoacyl-tRNA synthetases. These synthetases, which require Mg^{2+} as a cofactor, catalyze the formation of an aminoacyl-adenylate (AMP-amino acid), utilizing an amino acid and ATP as substrates. The amino acid moiety of the AMP-amino acid is then transferred to the 3′-terminus of a tRNA to produce an aminoacyl-tRNA. Each aminoacyl-tRNA synthetase is highly specific for the particular amino acid it activates and for the tRNA that accepts this amino acid. The activation step requires the expenditure of two high-energy phosphate bonds.

The second phase of polypeptide synthesis involves the formation of an initiation complex. In *E. coli*, the initiation aminoacyl-tRNA carries *N*-formylmethionine as its amino acid moiety (fMet-$tRNA^{f\text{-}met}$). The formation of the initiation complex commences with the association of fMet-$tRNA^{f\text{-}met}$, an mRNA, and a 30 S ribosomal subunit. The formation of the initiation complex is completed by the addition of a 50 S ribosomal subunit to produce the 70 S initiation complex. Three initiation factors (IF1, IF2, and IF3) and GTP are required for the formation of the initiation complex.

The elongation of the polypeptide chain constitutes the third phase of synthesis. This complex phase encompasses the following three sequential steps: (1) binding of an aminoacyl-tRNA to the A site on the ribosome, (2) peptide bond formation, and (3) translocation (which prepares the synthesizing complex for the incorporation of the next amino acid). In the formation of the first peptide bond, the *N*-formylmethioninyl residue, located at the P site on the ribosome, is transferred and bonded to the amino acid attached to the aminoacyl-tRNA at the A site. In each subsequent peptide bond formation, the growing polypeptide chain is transferred from the tRNA at the P site (placed there by translocation) to the aminoacyl-tRNA at the A site. Two elongation factors (EFT_u and EFT_s) and GTP are required for aminoacyl-tRNA binding to the ribosome; peptidyl transferase (of the 50 S ribosomal subunit) catalyzes peptide bond formation, and elongation factor G (EFG) and GTP are needed for translocation.

The final phase is the termination of polypeptide synthesis, which is signaled by a termination codon (UAA, UAG, or UGA) at the end of an RNA message. A release factor (RF1 or RF2) interacts with a termination codon to bring about the hydrolytic release of the nascent polypeptide chain. GTP is required for the termination process.

Although mammalian polypeptide synthesis is similar to that of *E. coli*, there are notable differences. Mammalian systems utilize methionine, and not *N*-formylmethionine, to initiate protein synthesis and require only one release factor instead of two. With respect to formation of the initiation complex, the mammalian process involves at least nine factors, whereas that of *E. coli* needs only three.

For intracellular transport, newly synthesized proteins are first sequestered in the lumen of the endoplasmic reticulum (ER). Transfer across the ER membrane relies on the hydrophobic signal sequence (the first 15 to 30 amino acid residues of the N-terminus) of a nascent polypeptide which, by interacting with a signal receptor located on the membrane, initiates transport across the membrane. The signal sequence is removed during the transport process. Once the polypeptide is in the ER lumen, glycosylation (addition of oligosaccharides) begins, and it is the distinctive oligosaccharide pattern associated with a new polypeptide that furnishes the molecular instructions for its cellular distribution. The first carbohydrate unit attached to an asparaginyl residue (N-linked oligosaccharide) of a glycoprotein is a common oligosaccharide core containing 15 carbohydrates; soon after its addition, this core undergoes trimming. The trimming and subsequent addition of new carbohydrates continue when the polypeptide is transferred from the ER to the Golgi apparatus. The two types of N-linked oligosaccharides are called the simple mannose-rich units and the complex N-acetyllactosamine units. A serious inborn error, I-cell disease, results from the defective transport of at least eight lysosomal enzymes which lack the mannose 6-phosphate present in the oligosaccharide unit of normal lysosomal enzymes.

The remarkable achievement of obtaining the 5,375-nucleotide sequence of the ϕX174 genome confirmed many of the current molecular hypotheses about a gene and the mechanisms that control its expression. Additionally, the findings provided new knowledge about the coding capacity of a genome. Two novel and important revelations are that a given nucleotide sequence can code for two different proteins, i.e., can contain two structural genes, and that the termination codon of one gene can overlap the initiation codon of another.

QUESTIONS

1. Which of the following statements are true about transfer RNAs?

(a) A majority of tRNAs, but not all, have modified bases.

(b) The 5′-terminus has a triphosphate group attached to it.
(c) In the anticodon, the base at position 3 is the wobble base.
(d) The anticodon is located in a single-stranded region of the cloverleaf structure.
(e) About 50 percent of the nucleotides are base paired in the cloverleaf structure.
(f) The sequence at the 3′-terminus, but not that at the 5′-terminus, is a single-stranded feature of the cloverleaf structure.
(g) Anticodons do not contain modified bases.
(h) All aminoacyl-tRNAs specifically bind to the A site on a ribosome during protein synthesis.
(i) There are a total of 20 tRNAs, one for each aminoacyl-tRNA synthetase.
(j) All tRNAs have the same CCA-OH sequence at the 3′-terminus.

2. Select the answers to the comments below about the process of protein synthesis in *E. coli* from the following list.

EFG	IF2
EFT_s	IF3
EFT_u	Peptidyl transferase
IF1	RF1

(a) Required for formation of an initiation complex
(b) Forms a complex with an aminoacyl-tRNA and GTP
(c) Required for the regeneration of an EFT_u-GTP complex
(d) Required for translocation
(e) Required for peptide bond formation
(f) Required for termination of synthesis
(g) Component of the 50 S ribosomal subunit

3. (a) In the red blood cells of rabbits, it requires about three minutes to synthesize the α-chain of hemoglobin (146 amino acid residues). How many amino acids are incorporated per second?
(b) How long would be required to synthesize a polypeptide containing 146 residues in an *E. coli* cell, which can incorporate 20 amino acids per second under optimal growth conditions?

4. (a) How many *nucleoside triphosphates,* starting with amino acid activation, are needed to synthesize a polypeptide containing 150 amino acid residues?
(b) If the polypeptide synthesized is one of six identical protomers of an enzyme, how many *nucleoside triphosphates* are needed for the synthesis of the enzyme?
(c) How many *high-energy bonds* are consumed in the synthesis of the hexameric enzyme?
(d) How many *nucleoside triphosphates* are needed to synthesize the RNA message that codes for the enzyme (assume the message commences with AUG)?

5. Below is a portion of a hypothetical strand of DNA that carries genetic information:

—T—A—C—T—T—A—C—C—C—A—C—A—C
—G—A—T—G—A—A—G—T—C—G—A—A—C
—T—T—A—C—T—

(a) What is the base sequence of the mRNA molecule coded for by this portion of the DNA sequence?
(b) With the use of Table 22-2, p. 480 determine the amino acid sequence of the peptide coded for by this portion of mRNA.
(c) Shift the codon-reading frame in the RNA message one base to the right. What amino acid sequence is now coded for by the same mRNA?

6. On p. 482 the statement was made that a minimum of 32 tRNAs are required for cytoplasmic protein synthesis (and also for that of *E. coli*). Given the genetic code (Table 22-2), the wobble hypothesis (p. 510), and known information about cytoplasmic protein synthesis, explain how was this number of tRNAs derived.

7. Which of the following statements are currently believed to be true about the intracellular transport of proteins?
(a) The signal sequence of a nascent polypeptide has no amino acids with hydrophilic side chains.
(b) The signal receptor anchors the ribosome to the ER membrane.
(c) In the lumen of the ER, the signal sequence is glycosylated.
(d) In the glycosylation of polypeptides, only the amino acid side chains with hydroxyl groups are glycosylated.
(e) Dolichol phosphate is the lipid carrier of the common oligosaccharide core that is added to all polypeptides in the ER lumen.
(f) All trimming of the common oligosaccharide core occurs in the Golgi apparatus.
(g) Simple mannose-rich units contain only mannose.
(h) Complex *N*-acetyllactosamine units have Gal and GlcNac residues.

SUGGESTED READINGS

Arnstein, H. R. V., ed., *Synthesis of Amino Acids and Proteins.* Vol. 7 in MTP International Review of Sciences, Biochemistry Section. Baltimore, Md.: University Park Press, 1975. This volume contains advanced review articles on protein synthesis.

Ingram, V. M., *Biosynthesis of Macromolecules*. Menlo Park, Calif.: W. A. Benjamin, 1972. A well-written introductory text on protein synthesis.

Moldave, K., and L. Grossman, eds., *Nucleic Acids and Protein Synthesis*. Parts G and H, *Methods Enzymol.* Vols. 59 and 60. New York: Academic Press, 1979. These two volumes contain a comprehensive collection of 130 articles about the biochemical methods used to study protein synthesis.

Nomura, M., A. Tissières, and P. Lengyel, eds., *Ribosomes*. Cold Spring Harbor, N.Y.: Cold Spring Harbor Laboratory, 1974. Advanced reviews on the assembly, structure, and function of ribosomes.

Schimmel, P., D. Söll, and J. Abelson, eds., *Transfer RNA: Structure, Properties, and Recognition*. Cold Spring Harbor, N.Y.: Cold Spring Harbor Laboratory, 1979. An excellent reference source for transfer RNAs.

Söll, D., J. Abelson, and P. Schimmel, eds., *Transfer RNA: Biological Aspects.* Cold Spring Harbor, N.Y.: Cold Spring Harbor Laboratory, 1980. The companion volume that completes the most authoritative review available on transfer RNAs.

Weissback, H., and S. Pestka, eds., *Molecular Mechanisms of Protein Biosynthesis*. New York: Academic Press, 1977. Another good book on the biochemistry of protein synthesis.

The following references listed at the end of Chapter 21 are also recommended: Lewin's Vols. 1 and 2 and Watson.

Articles

Allfrey, V. G., and A. E. Mirsky, "How Cells Make Molecules." *Scientific American*, 205(3):74 (offprint #92), 1961.

Bielka, H., "The Eukaryotic Ribosome." *TIBS*, 3:156, 1978.

Brimacombe, R., G. Stöffler, and H. G. Wittman, "Ribosome Structure." *Ann. Rev. Biochem.*, 47:217, 1978.

Järnefelt, J., J. Finne, T. Krusius, and H. Rauvala, "Protein-Bound Oligosaccharides of Cell Membranes." *TIBS*, 3:108, 1978.

Kornfield, R., and S. Kornfield, "Comparative Aspects of Glycoprotein Structure." *Ann. Rev. Biochem.*, 45:217, 1976.

Kurland, C. G., "Structure and Function of the Bacterial Ribosome." *Ann. Rev. Biochem.*, 46:173, 1977.

Lodish, H. F., "Translational Control of Protein Synthesis." *Ann. Rev. Biochem.*, 45:39, 1976.

Marshall, J. J., "Manipulation of the Properties of Enzymes by Covalent Attachment of Carbohydrate." *TIBS*, 3:79, 1978.

Mitra, S. K., "Recognition Between Codon and Anticodon." *TIBS*, 3:153, 1978.

Rich, A., "Transfer RNA: Structure and Biological Function." *TIBS*, 3:34, 1978.

Rich, A., and S. H. Kim, "The Three-Dimensional Structure of Transfer RNA." *Scientific American*, 238(1):52 (offprint #1377), 1978.

Schimmel, P. R., and D. Söll, "Aminoacyl-tRNA Synthetases: General Features and Recognition of Transfer RNAs." *Ann. Rev. Biochem.*, 48:601, 1979.

Sinsheimer, R. L., "Single-Stranded DNA." *Scientific American*, 207(1):109 (offprint #128), 1962.

Weissbach, H., and S. Ochoa, "Soluble Factors Required for Eucaryotic Protein Synthesis." *Ann. Rev. Biochem.*, 45:191, 1976.

Yanofsky, C., "Gene Structure and Protein Structure." *Scientific American*, 216(5):80 (offprint #1074), 1967.

HORMONES: INTERCELLULAR CHEMICAL MESSENGERS

24

INTRODUCTION

Interwoven in the previous information presented on cellular biochemistry is the recurring theme of the critical importance of organizational design and control of metabolic activities for maintaining the integrated coordination of life processes. One of the most productive research areas in molecular biology is the study of the basic biochemical mechanisms that cooperatively accomplish the remarkable feat of maintaining the ordered, yet responsive, metabolic environment which all forms of life require. Molecular studies on *Escherichia coli* (an organism with a comparatively low degree of cellular organization) have clearly established that bacteria, like higher forms of life, are endowed with complex systems of homeostatic controls (Chapter 22) which regulate the cell's biological activities. Within a single cell, whether procaryote or eucaryote, such controls are referred to as *intracellular homeostasis.* With the evolution of multicellular living systems, maintenance of the vital homeostatic state required the concomitant development of *extracellular homeostasis,* an additional level of coordination needed to regulate the metabolic functions of the various components of an organism. Through the eons of biological time, the biochemical mechanisms that originally allowed for molecular communication between the organelles and intracellular fluid of single-cell eucaryotes evolved into the present complex and sophisticated series of regulatory physiological processes possessed by many species, including humans. Because extracellular homeostasis had its origin in the ancient development of life, it can be viewed as an evolutionary extension of intracellular homeostasis. The evolutionary heritage of extracellular homeostasis is becoming increasingly appreciated, since it is now known that the metabolically important molecular expression of many of the extracellular homeostatic processes in vertebrates occurs in the intracellular environment of a cell or, more descriptively, in the basic *milieu de la vie* (environment of life).

In vertebrates, extracellular homeostasis is regulated by two complex systems, namely, the *endocrine system* and the *central nervous system.* While appearing to be functionally distinct, these two organ systems are in fact intimately related and, through the hypothalamus, share a common control center for much of the regulation of physiological functions. The endocrine system is composed of a group of organs that secrete specific chemical messengers called *hormones,* whose collective role is to regulate the internal biological functioning of an animal. The term *hormone* (from the Greek word *horman,* meaning *to set in motion*) was first used by Ernest Starling in 1905 to describe the physiological action of *secretin,* a chemical messenger which, when released from the duodenum, subsequently stimulates the secretion of a bicarbonate-rich fluid by the pancreas. This stimulatory role of secretin was intensively studied by Starling and his brother-in-law, William Bayliss. The novel finding made by these two eminent physiologists is that this particular stimulation of the pancreas is mediated by a *chemical messenger carried by the blood.* Until these studies, only neural stimulation of glandular secretions was known. From these studies, performed during the first decade of the century, Bayliss and Starling defined a hormone as a chemical transmitter that is produced and released by an organ, then transported by the blood to act on another organ. The term *organ,* as used in this statement, is defined as a body part specialized for a certain function. Although not specifically stated, the classification recognizes that a hormone may act on more than one responsive *(target)* tissue, e.g., growth hormone. This classical definition applies to most hormones but does not strictly encompass all; some *tissue hormones,* e.g., prostaglandins (p. 557), are thought to be produced by one cell type in a tissue and act on another cell type(s) within the same tissue.

Hormones are biologically very potent molecules and are therefore effective in minute quantities, *viz.,* at blood levels of 10^{-7} to 10^{-10} *M.* Because they occur in small amounts, their isolation poses a formidable task. It is not unusual for several tons of an endocrine tissue, e.g., hypothalamic, to be processed to isolate a milligram or two of a particular hormone. However, in spite of the inherent difficulties, most of the known mammalian hormones have been purified (Table 24-1), and the amino acid sequences of

Table 24-1
Some mammalian hormones

Hormone*	Source	Discovered	Purified
Polypeptide releasing factors			
CRF	Hypothalamus	1958	1959
TRF	Hypothalamus	1962	1962
FSHRF	Hypothalamus	1964	1966
LRF	Hypothalamus	1960	1961
PRF	Hypothalamus	1969	
GHRF	Hypothalamus	1959	1965
GIF	Hypothalamus	1968	1970
Polypeptide			
ACTH	Anterior pituitary	1924	1948
TSH	Anterior pituitary	1922	1948
FSH	Anterior pituitary	1926	1931
LH	Anterior pituitary	1926	1940
LTH	Anterior pituitary	1929	1933
GH	Anterior pituitary	1921	1948
Vasopressin	Posterior pituitary	1895	1954
Oxytocin	Posterior pituitary	1901	1954
Insulin	Pancreas-β cells	1889	1926
Glucagon	Pancreas-α cells	1930	1953
Secretin	Duodenum	1902	1961
Angiotensin	Liver/plasma	1939	1955
Steroids			
Aldosterone	Adrenal cortex	1934	1952
Glucocorticoids	Adrenal cortex	1935	1938
Testosterone	Testes	1889	1931
Estrogen	Ovary	1925	1929
Progesterone	Corpus luteum	1925	1934
Amine			
Acetylcholine	Autonomic nervous system	1921**	1868
Thyroxin	Thyroid	1895	1915
Triiodothyronine	Thyroid	1951	1952
Epinephrine	Adrenal medulla	1895	1897
Norepinephrine	Adrenal medulla	1948**	1904

*Abbreviations:
CRF—corticotropin releasing factor
TRF—thyrotropin releasing factor
FSHRF—FSH releasing factor
LRF—LH releasing factor
PRF—prolactin releasing factor
GHRF—growth hormone releasing factor
GIF—growth hormone release-inhibitory factor
ACTH—adrenal corticotropin
TSH—thyroid-stimulating hormone
FSH—follicle-stimulating hormone
LH—luteinizing hormone
LTH—prolactin
GH—growth hormone

**Recognized as a hormone

Source: R. H. Williams, ed., *Textbook on Endocrinology,* 5th ed. Philadelphia: W. B. Saunders, 1974.

many protein hormones have been determined. The structural features of several of the protein hormones are also known.

One active area in endocrinology is a fruitful blend of physiology and biochemistry (often referred to as *biochemical physiology*) that seeks a molecular understanding of the mechanism of action of the individual hormones produced by the endocrine system. Since the following discussion is primarily concerned with a few of the hormones and their molecular functions, it should be noted that this presentation is an incomplete, abbreviated view of endocrinology. Because the overall function of the endocrine system relies on the interrelated performances of its many components, any simplified and dissected discussion of endocrinology provides only a limited appreciation of a complex biological system whose complete understanding continues to challenge scientists.

GENERAL CLASSES OF HORMONES

Mammalian hormones are classified according to the following three chemical groupings: *polypeptides, steroids,* and *amines* (Table 24-1). The polypeptide hormones are a diverse group of regulators and include the release and release-inhibitory factors, produced by the hypothalamus, which stimulate and inhibit,

Figure 24-1
A releasing and a release-inhibitory factor.

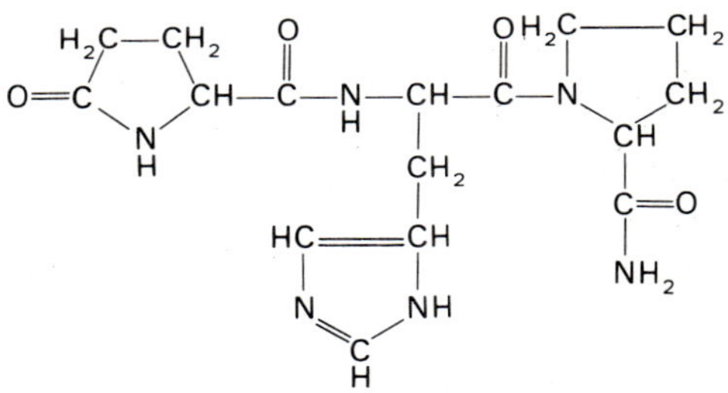

Thyrotropin releasing factor
(Pyroglutamylhistidylprolineamide)

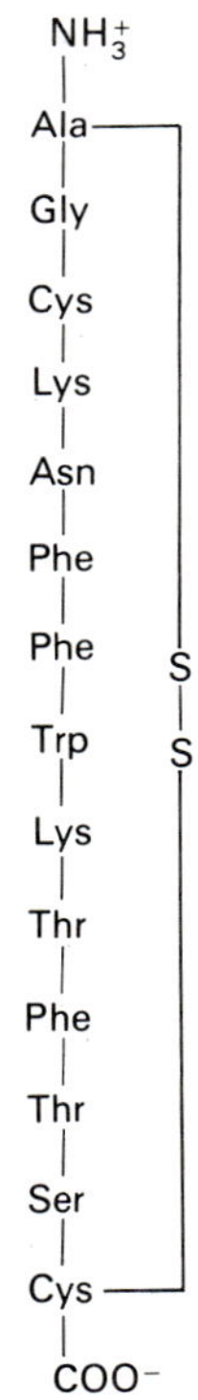

Growth hormone release-inhibitory factor

respectively, the release of polypeptidyl hormones from the anterior pituitary (Figure 24-8). The polypeptide hormones produced by the anterior pituitary are larger than the hypothalamic release or inhibitory factors. For example, the thyrotropin releasing factor (Figure 24-1) is a tripeptide, and thyrotropin is a polypeptide containing 220 residues (MW ≈ 28,3000). Similarly, the growth hormone release-inhibitory factor (Figure 24-1) and the growth hormone (MW ≈ 21,000) have 14 and 191 residues, respectively. As discussed in Chapter 19, the steroid hormones (Figure 24-2) are synthesized from cholesterol (p. 386) and include the female sex hormones *(estrogens),* male sex hormones *(androgens),* and adrenal cortical hormones. Although about 50 adrenal cortical hormones are produced by the adrenal cortex, only a few are believed to be secreted in measurable quantities. The amines are small molecules, several of which are derivatives of the amino acid tyrosine (Figure 24-3).

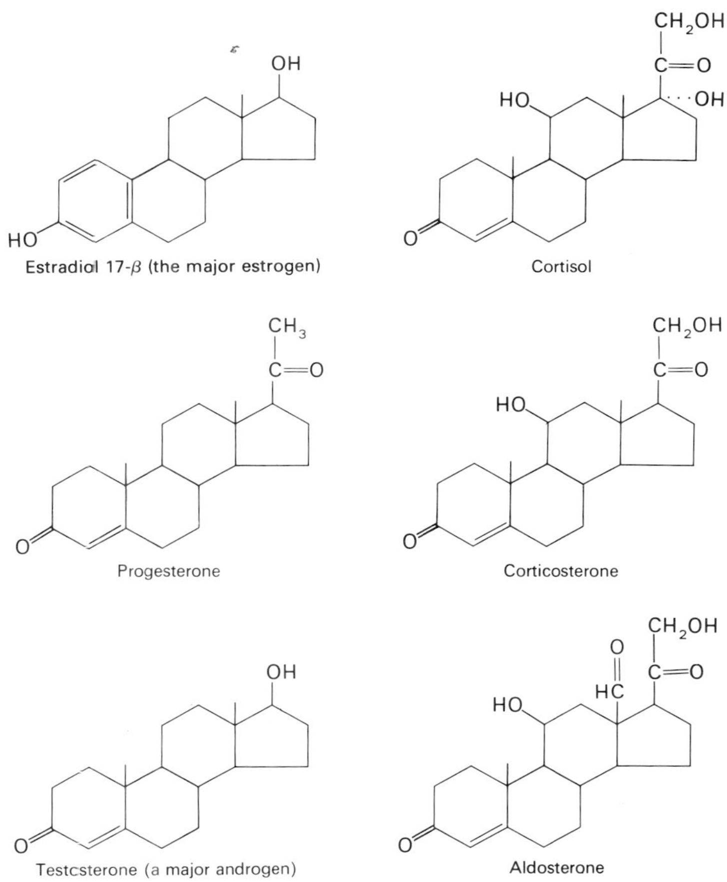

Figure 24-2
Some steroid hormones.

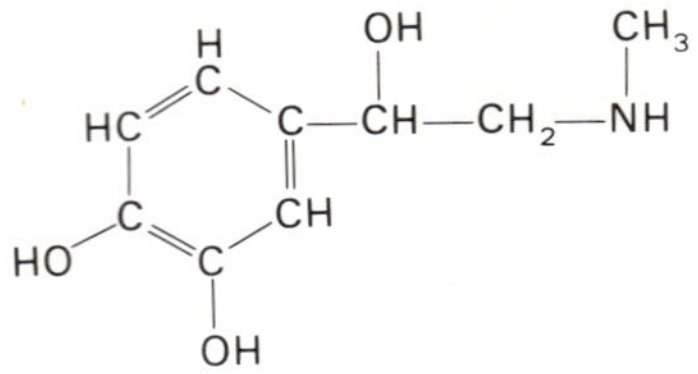

Epinephrine (adrenaline)

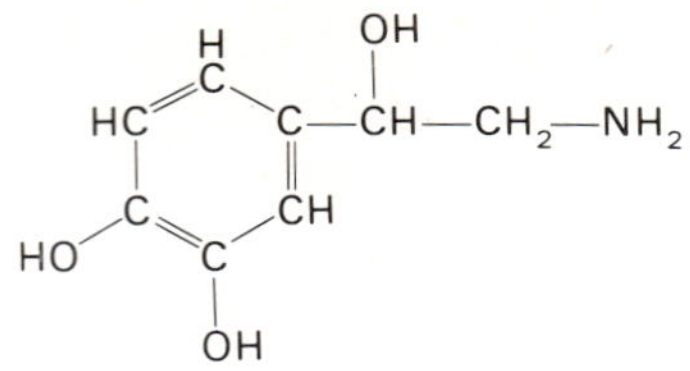

Norepinephrine (noradrenaline)

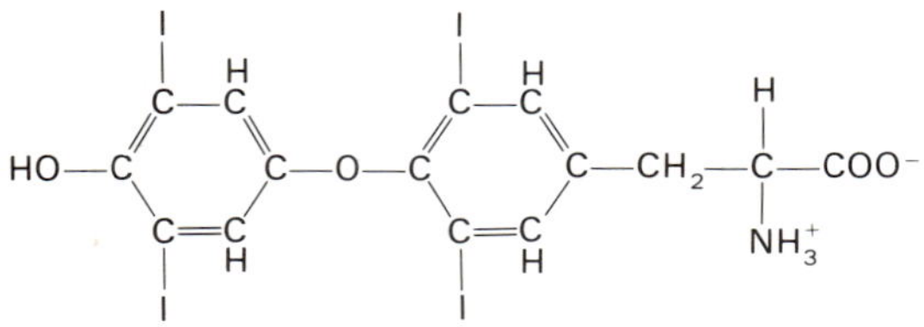

Thyroxine

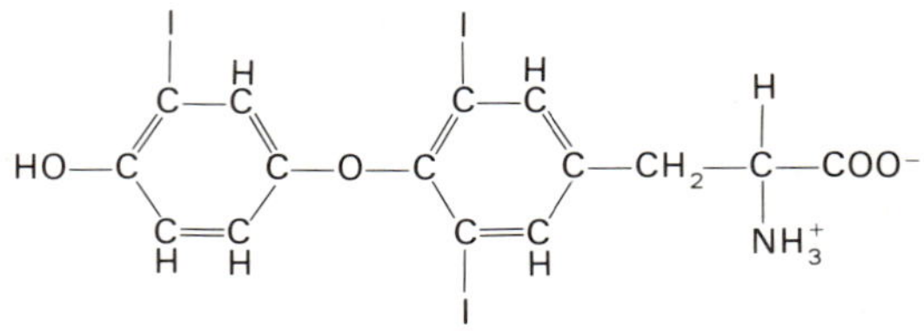

Triiodothyronine

Figure 24-3
Some amine hormones.

The identification of all mammalian hormones is still incomplete since active derivatives of known hormones, as well as new compounds with hormone-like activity, continue to be discovered. It is now realized, for example, that some steroid hormones, when transported into certain organs, undergo metabolic transformations which increase, decrease, or even qualitatively modify their known biological activities. Thus, a steroid hormone may be structurally altered to produce other steroid regulators having different biological activities.

Epinephrine and norepinephrine stimulate the heart and increase blood pressure, the flow of blood through coronary vessels, and the metabolic rate.

PROHORMONES (HORMONOGENS)

Thyroxine and triiodothyronine—thyroid hormones that are among the major regulators of metabolic rate.

Some polypeptide hormones, like certain enzymes, are synthesized in an inactive form, i.e., as *prohormones,* or *hormonogens.* Prohormones are examples of proteins with extracellular functions, such as the enzymes chymotrypsin and trypsin (p. 147), whose biological activities are dormant until activated by peptidases. Table 24-2 lists some of the known prohormones.

Table 24-2
Some prohormones

Prohormone	Source
Proinsulin	Pancreas β cells
Proparathyroid hormone	Parathyroid
Angiotensinogen	Liver
Progastrin	Stomach

Proinsulin

Studies in the early 1970s by Donald F. Steiner and colleagues on the synthesis of insulin in the pancreas revealed that two polypeptides are synthesized which react with antibodies prepared against pure insulin. The smaller of the two polypeptides proved to be insulin, and the larger, insulin-like molecule was eventually identi-

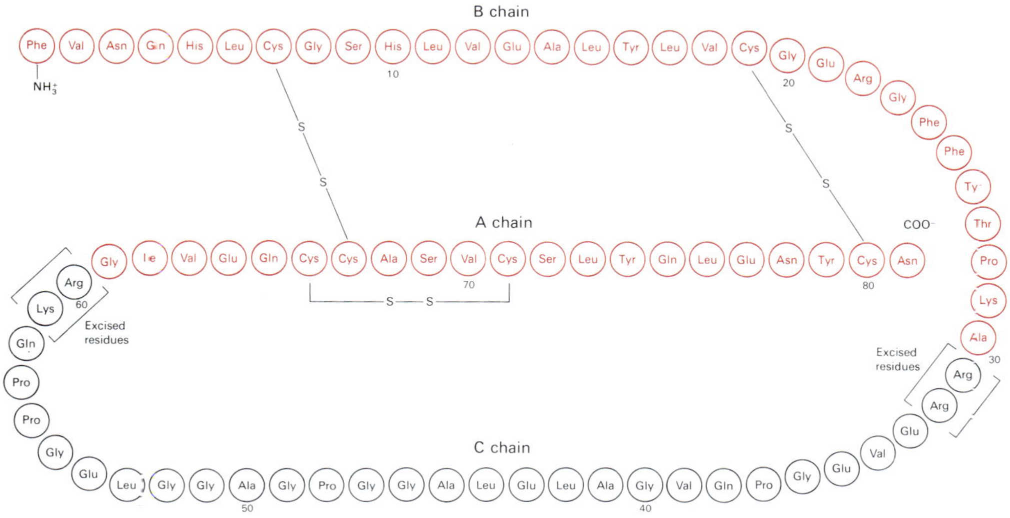

Figure 24-4
Amino acid sequence of proinsulin (bovine).

fied as the precursor form of insulin, which has only traces of insulin activity. The primary structure of proinsulin (Figure 24-4), which was established by enzymatic and chemical studies, reveals that the bovine prohormone has 30 more amino acid residues than the A and B chains of insulin, which together have a total of 51 residues. As seen in Figure 24-4, the carboxyl- and amino-terminal segments of proinsulin are the A and B chains of insulin, respectively, and both proinsulin and insulin have the same three disulfide bonds. The production of insulin from proinsulin involves two endopeptidase-catalyzed cleavages, each of which excises two basic amino acid residues. The products of these selective cleavages are insulin (A and B chains) and a C chain.

Recently it was reported that, in the β cells of the pancreas, the initial polypeptide synthesized on the ribosomes is *preproinsulin,* which is proinsulin with 24 additional amino acids on the amino terminus, i.e., $^{+}H_3N$-(prepeptide)-B chain-(C chain)-A chain-COO^{-}. As discussed previously (p. 523), the prepeptide is the signal sequence that guides the nascent polypeptide through the endoplasmic reticulum membrane. From the lumen of the endoplasmic reticulum, proinsulin (signal sequence is cleaved during membrane transport) is transported to the Golgi apparatus, where it is then processed into insulin and the C chain. Insulin (as well as the C chain) is stored in Golgi vesicles until the hormone, in response to endocrine signals, is released into the blood. In the case of insulin, both synthesis of the precursor molecule and its conversion into the active hormone occur in the pancreas.

$^{+}H_3N$—Asp—Arg—Val—Tyr—Ile—His—Pro—Phe—His—Leu—Leu—Val—Tyr—Ser—R

Angiotensinogen (inactive)

↓ renin (in plasma)

$^{+}H_3N$—Asp—Arg—Val—Tyr—Ile—His—Pro—Phe—His—Leu—COO^- + $^{+}H_3N$—Leu—Val—Tyr—Ser—R

Angiotensin I (relatively inactive)

↓ converting enzyme (in lungs)

$^{+}H_3N$—Asp—Arg—Val—Tyr—Ile—His—Pro—Phe—COO^- + $^{+}H_3N$—His—Leu—COO^-

Angiotensin II (active)

R is the remainder of the amino acid sequence.

Figure 24-5
Activation of angiotensinogen.

Aldosterone—most important known regulator of sodium, potassium, and water metabolism.

Angiotensinogen Activation

Antiotensinogen is a prohormone that is synthesized in the liver and then released into the blood. The activation of this prohormone involves a two-step mechanism to produce *angiotensin II,* which stimulates aldosterone secretion by the adrenal cortex. Angiotensin II is also the most potent vasoconstrictor known. The first step in the activation process (Figure 24-5) involves *renin,* a peptidase produced by the kidneys and released into the blood. Renin specifically cleaves a leucyl-leucyl peptide bond in angiotensinogen to produce *angiotensin I,* a relatively inactive decapeptide. In the second step, another peptidase, called *angiotensin converting enzyme,* removes the C-terminal dipeptide from angiotensin I to yield active angiotensin II; the converting enzyme is found predominantly in the lungs. This endocrinological interplay between three organs to produce a hormone is an example of one of several distinct physiological mechanisms that occur in extracellular homeostasis. It is such variations in individual hormonal systems that make any broad generalization concerning the functioning of the endocrine system difficult.

ENDOCRINE CONTROL SYSTEMS

Among the components of the endocrine system are such seemingly unrelated organs as the pituitary, thyroid, ovary, testes, pancreas, and liver; however, in this highly integrated network, no single hormonal system operates independently of the others. The removal or malfunction of one endocrine organ (gland) affects the functions of the others, e.g., hypo- and hyperthyroid conditions. Thus, in these interrelated patterns of intercellular communication, all individual hormonal systems must function properly to ensure normal extracellular homeostasis.

Simple Control Loops

The control systems that regulate the release of the hormones produced by the endocrine system include a variety of mechanisms of varying complexities. The simplest system for the control of hormonal release is depicted in Figure 24-6. As indicated, the hormone is released from the endocrine gland, and stimulation of the target organ subsequently results in a *negative feedback* on the hormone's release. The mechanism operates as a *simple closed-loop system* and is controlled by the alteration of a variable in the extracellular fluids. This type of regulation functions in the case of insulin, whose release is controlled by the level of glucose in the blood plasma. High glucose levels (after a meal) trigger the pancreatic (endocrine gland) release of insulin, which effects a rapid reduction in blood glucose levels. The mode of action of insulin is believed to be enhancement of the uptake of glucose into fat and muscle cells (target cells). The decrease in blood glucose, caused by the insulin, serves as a negative feedback signal to halt further release of the hormone. No hormonal system, though, operates in such a simple manner because external factors can also affect the system. For example, amino acids can stimulate the glucose-induced release of insulin, and glucagon (another pancreatic hormone) counterbalances the action of insulin by stimulating the breakdown of glycogen in the liver to replenish the glucose level in the blood. Glucoreceptors in the intestine, responding to ingested glucose, are also involved in the stimulation of insulin release. This control system between the intestine and the pancreas is interesting because it performs an anticipatory function, i.e., alerts the organism's metabolic system to an influx of dietary glucose. Thus, even the so-called simple control loops are subject to physiological complexities.

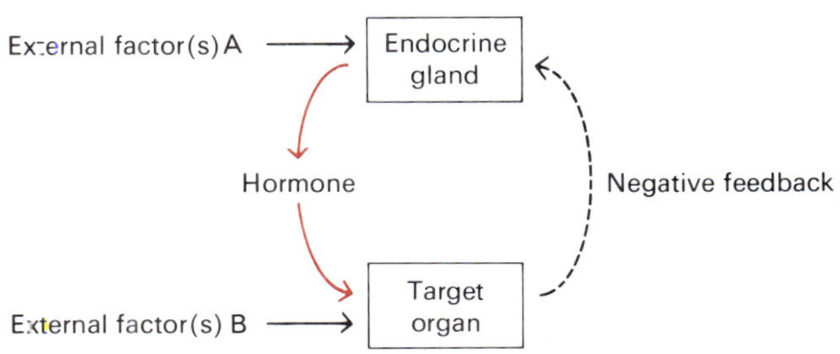

Figure 24-6
Simple endocrine control system.

Hypothalamic-Pituitary System

In contrast to the simple closed-loop system, the hypothalamic-anterior pituitary system constitutes a highly complex operation. This system includes a hierarchy of organization and function that involves several levels of hormone-target tissue interactions. The basic features of the hypothalamic-anterior pituitary system, which selectively controls the production of a variety of hormones, are depicted in Figure 24-7. The initial stimulus is furnished by the central nervous system, and the primary acceptor of this signal is the hypothalamus. The hypothalamus responds by secreting a hormone releasing factor which stimulates the anterior pituitary to release a hormone (hormone 1). Because of the close anatomical relationship between the hypothalamus and the pituitary in the brain, the two glands are often called the *hypothalamic-pituitary unit.* When a *trophic hormone,* e.g., thyrotropin, is secreted by the anterior pituitary and released into the blood, it *acts on a specific endocrine gland or glands.* In turn, the endocrine gland releases a

Figure 24-7
Hypothalamic–anterior pituitary control system.

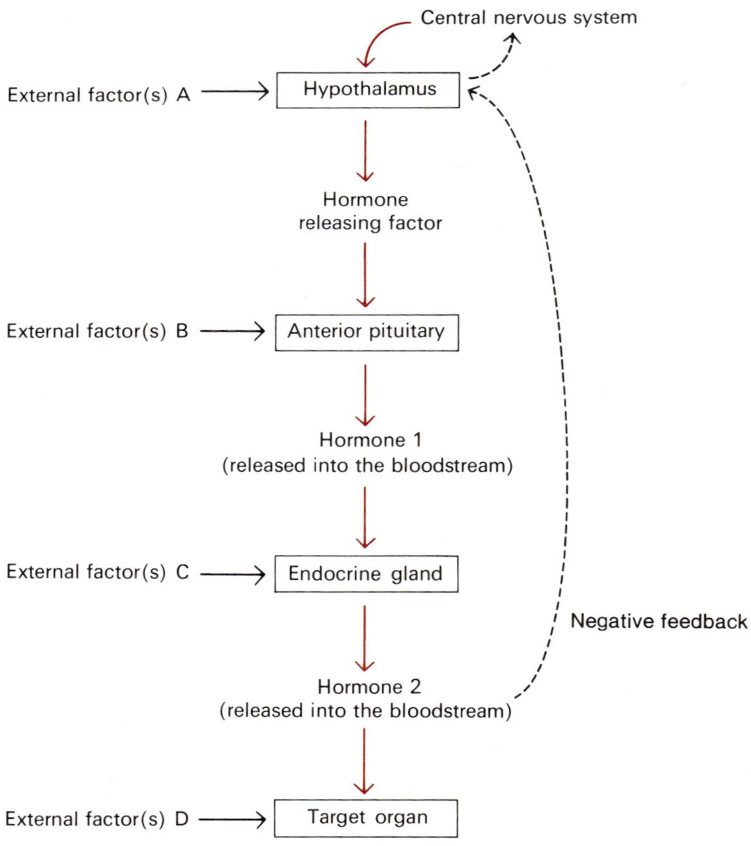

hormone (hormone 2) that is transported by the blood to a target organ. The negative feedback control of the system is generally (but not always) accomplished at the level of the hypothalamus by the final endocrine gland product (hormone 2). The anterior pituitary also secretes hormones that have no specific trophic action, e.g., growth hormone, since they act directly on a target organ. As noted in Figure 24-7, the system can be influenced by different molecular external factors at all levels of its operation.

A more detailed representation of the hypothalamic-pituitary endocrine system, emphasizing its overall physiological significance, is illustrated in Figure 24-8. The hypothalamus and pituitary

Figure 24-8
The hormonal network controlled by the hypothalamic anterior pituitary control system.

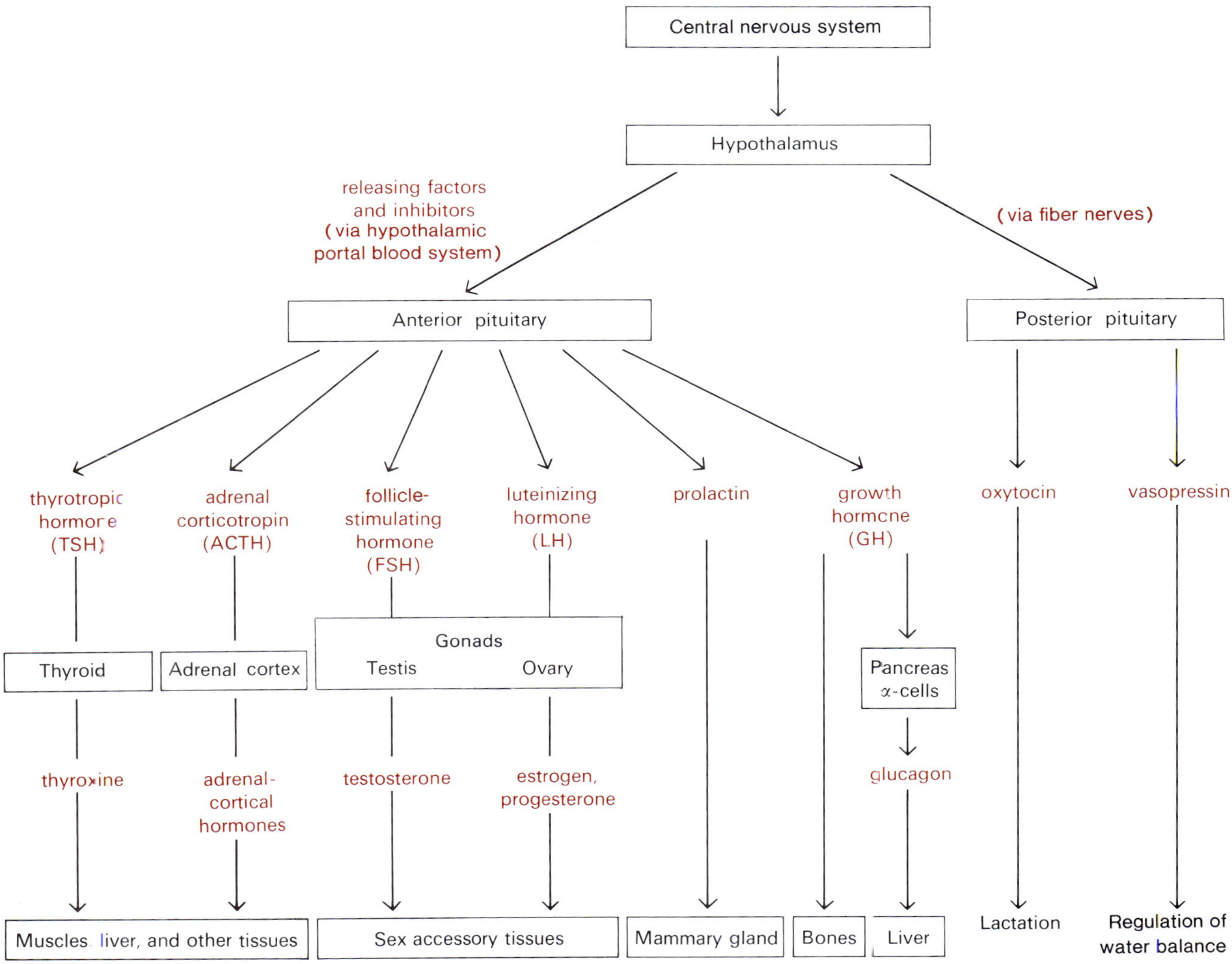

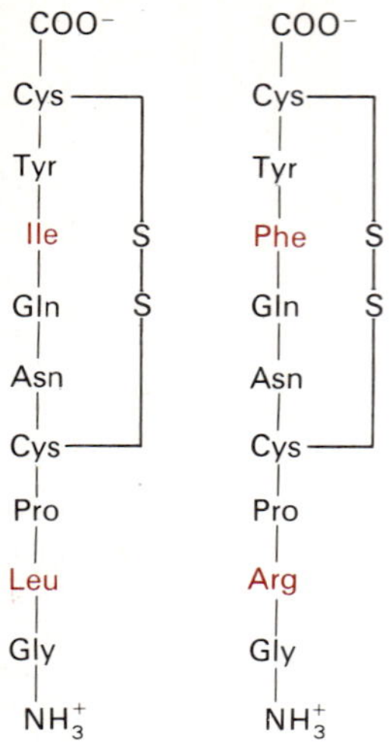

Oxytocin (bovine) Vasopressin (bovine)

Figure 24-9
Two hormones produced by the posterior pituitary.

control a large network of diverse hormonal functions, and, although the pituitary is known as the *master gland,* the title rightly belongs to the highest center of the system, the hypothalamus. Because the hypothalamic-pituitary system entails the interaction of the nervous system with the endocrine system, the study of its physiological functioning is called *neuroendocrinology.*

An endocrinological relationship between the hypothalamus and the posterior pituitary (Figure 24-8) not previously mentioned involves the release of *oxytocin* and *vasopressin.* These two hormones (Figure 24-9) are synthesized in the hypothalamus and then transferred via axons, bound to a protein (MW $\approx$ 30,000) called *neurophysin,* to the posterior pituitary, from which they are released into the blood. Vasopressin is an *antidiuretic hormone which regulates water balance,* and oxytocin *influences lactation.* Note that although each of the two hormones has a distinct physiological function, they both are nonapeptides that are identical except for two amino acid residues - a striking example of how seemingly minor chemical variations in structures can profoundly affect the biological activity of biomolecules.

MOLECULAR BASIS OF HORMONAL ACTION

One of the most welcomed scientific advances made in recent years is the long-awaited breakthrough toward understanding the mechanism of hormonal action at the molecular level. These pioneering investigations, which initiated a new era in hormone research, will undoubtedly rank among the most beneficial contributions made by science this century. Two different models have now been proposed to account for the molecular action of hormones; one applies principally to polypeptide and amine hormones and the other to steroid regulators. The two mechanisms may not ultimately explain all hormonal actions; however, they are now serving well as the Rosetta stone for molecular research on hormones.

The Rosetta stone, a piece of black basalt bearing a trilingual inscription in Demotic, Egyptian hieroglyphic, and Greek, was found near the Rosetta mouth of the Nile River in 1799. The stone provided M. Champollion with the first clue toward deciphering hieroglyphics.

TWO-MESSENGER HYPOTHESIS

It was the studies on the hormone epinephrine (Figure 24-3) by Earl Sutherland (Nobel Prize, 1971) and associates that provided the first well-defined molecular interpretation of hormonal action. Epinephrine, produced and released by the adrenal medulla, is the hormone that elicits the "fight or flight" response in vertebrate animals. Among the various metabolic events initiated by epinephrine is the breakdown of glycogen *(glycogenolysis)* in the skeletal muscles (to yield glucose 1-phosphate for intracellular energy production) and, to a lesser degree, in the liver (to provide a quick extracellular supply of blood glucose). With respect to glycogenolysis in the liver, Sutherland and other prominent biochemists discovered that phosphorylase, the enzyme that catalyzes the breakdown of glycogen (p. 269), exists intracellularly in active tetrameric (*phosphorylase* a) and inactive dimeric (*phosphorylase* b) forms. Interconversion between these alternate states of the enzyme is determined by two other enzymes: *phosphorylase kinase* phosphorylates specific seryl residues of phosphorylase *b* to produce the active form (phosphorylase *a*) and, conversely, *phosphorylase phosphatase* removes these same phosphate groups to yield the inactive form (phosphorylase *b*). Further investigations revealed that epinephrine stimulates the intracellular activity of phosphorylase by enhancing the conversion of phosphorylase *b* into *a*. Two additional findings by Sutherland, which proved to be of prime importance, were (1) epinephrine's effect on glycogen breakdown is mediated at the level of the cellular membrane and (2) the actual stimulatory agent within a liver cell is the nucleotide *3′-5′ cyclic adenosine monophosphate* (*cyclic AMP*, or *cAMP*), discovered by Sutherland in the late 1950s. As additional biochemical insights into epinephrine-induced glycogenolysis were gained, a pattern of sequential reactions emerged, and the studies culminated with Sutherland's proposal of the *two-messenger hypothesis.* The hypothesis explains, in molecular terms, epinephrine's role in both the stimulation of glycogen breakdown and the inhibition of glycogen synthesis in the same cell. These two biochemical effects, simultaneously exerted by epinephrine on carbohydrate metabolism, cooperatively assure the availability of liver glucose to supply the blood with sugar in times of stress.

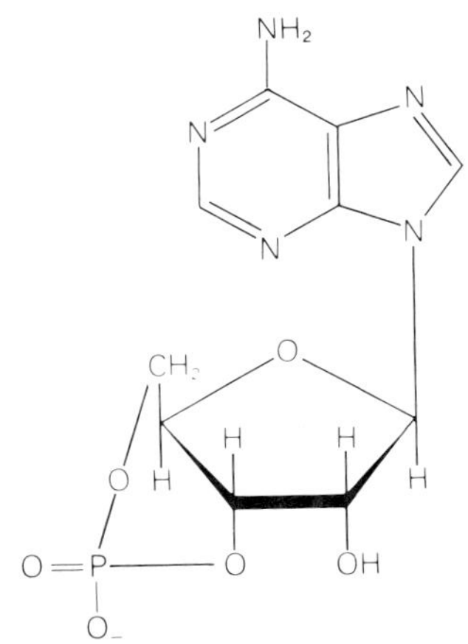
Cyclic AMP

Glycogenolysis

In the two-messenger hypothesis (Figure 24-10), epinephrine (hormone) is assigned the role of the *extracellular*, or *first, messenger* which, after its release into the blood, binds to *specific receptor sites* on the outer surface of the membrane of liver cells (a target organ). The fact that the blood concentration of released epinephrine is only about 10^{-9} *M* exemplifies the high degree of specific recognition (selective affinity) between hormones and their

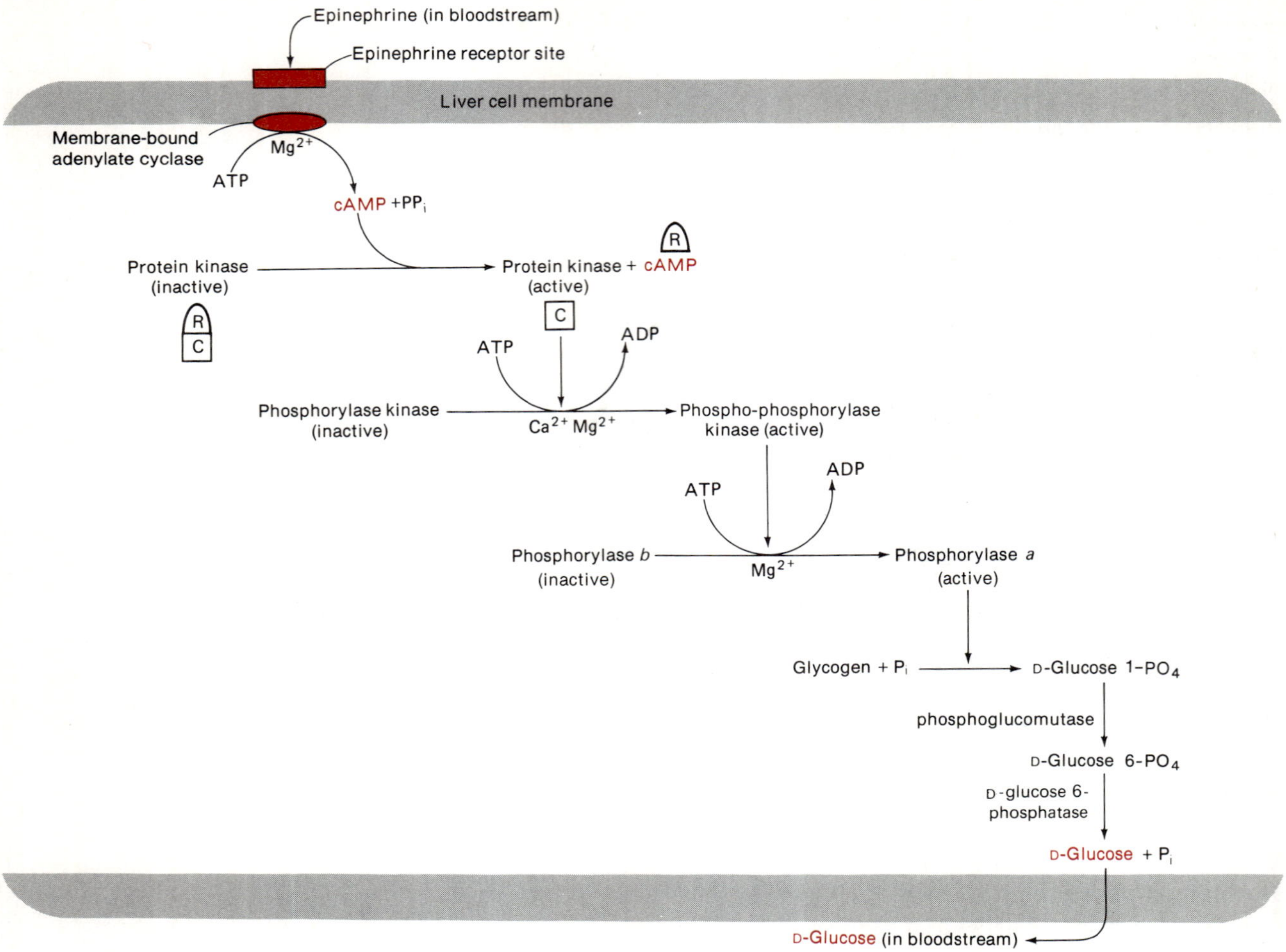

Figure 24-10
Schematic of the stimulation of glycogenolysis in the liver by epinephrine.

receptor sites. The binding of epinephrine to liver cells, it is postulated, induces specific conformational changes in membrane components which stimulate *adenylate cyclase* (an enzyme bound to the inner membrane) to synthesize cyclic AMP. The enzyme is specific for ATP and converts the nucleoside triphosphate into cyclic AMP and PP_i (Figure 24-10). The hormone-induced synthesis of cyclic AMP, the *intracellular,* or *second, messenger,* triggers the following cascade of cellular events, which terminates in an increased rate of glycogenolysis.

The first step in the sequence is the binding of cyclic AMP to the regulatory subunit (R) of *protein kinase,* which releases the active (catalytic) form of the enzyme (C) from association with its inhibitory regulator. Protein kinase (an enzyme that phosphorylates proteins) then catalyzes the activation of *phosphorylase kinase* in a reaction that uses ATP to phosphorylate the inactive form of the enzyme. (Note that calcium is required for the reaction; its role is discussed on p. 552.) With the activation of phosphorylase kinase, production of phosphorylase *a* proceeds by the phosphorylation of phosphorylase *b*, with ATP again serving as the phosphate donor. Increased phosphorylase activity in turn accelerates the breakdown of glycogen to glucose 1-phosphate, a product that is converted by isomerization and then by dephosphorylation into available free glucose. Hence, in epinephrine-induced glycogenolysis, (1) the released hormone delivers its physiological message to the liver cells by promoting the synthesis of cyclic AMP, which then (2) relays the message intracellularly by activating protein kinase, which initiates two phosphorylation reactions that result in enhanced glycogenolysis.

Inhibition of Glycogen Synthesis

The hormonal action of epinephrine to prevent the utilization of glucose to synthesize glycogen is illustrated in Figure 24-11. The primary cellular responses to epinephrine are the same as those described above, i.e., increased cyclic AMP synthesis and activation of protein kinase. However, in the case of glycogen synthesis, protein kinase inactivates *glycogen synthase* by phosphorylating

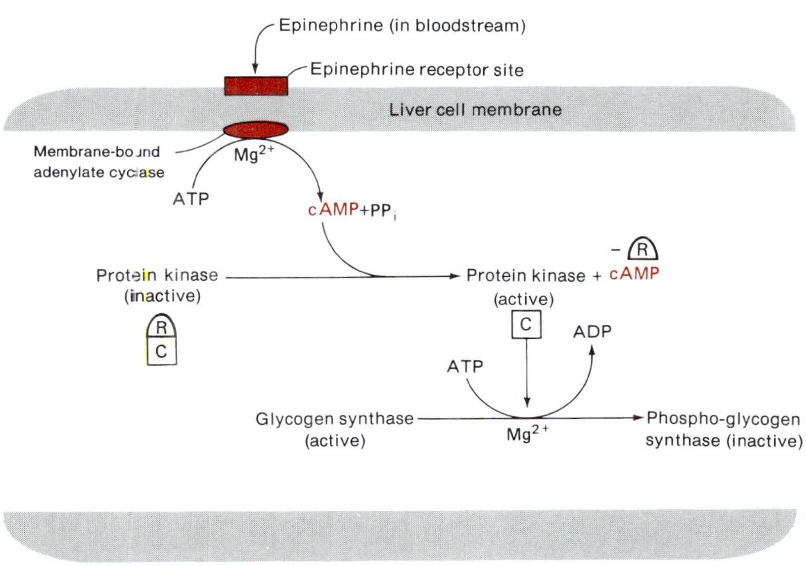

Figure 24-11
Schematic of inhibition of glycogen synthesis in the liver by epinephrine.

the active form of the enzyme. For the biochemical events that account for epinephrine's control of the synthesis and degradation of glycogen, it is apparent that each of the mechanisms relies on the phosphorylation of enzymes to accomplish its task. It is also apparent that protein kinases need not have a singular substrate; indeed, this class of enzyme is probably common to all hormone-cyclic AMP systems, with the substrates of each activated protein kinase being adapted to the particular hormonal response and to the specific cell type in which they occur. For example, epinephrine also stimulates the breakdown of triacylglycerols in fat cells to produce fatty acids (sources of metabolic energy), and the previously described synthesis of cyclic AMP and activation of protein kinase promote the intracellular stimulation of a lipase by phosphorylation to degrade the triacylglycerols.

Release from Epinephrine Activation

The cellular responses to epinephrine are alleviated when the adrenal medulla is no longer stimulated to release the hormone and, as a result, the blood level of the endocrine messenger declines. It is believed that, at low circulatory levels, epinephrine dissociates from its cellular receptor sites and the intracellular synthesis of cyclic AMP sharply decreases. The inactivation of cyclic AMP in cells previously stimulated occurs by a reaction, catalyzed by a phosphodiesterase, that cleaves the molecule's intraphosphodiester linkage to produce AMP. Intracellularly, the normal metabolic state is gradually restored by the inactivation or activation of the enzymes affected by the hormonal response: phosphatases restore phosphorylase to its inactive form and glycogen synthase to its active form.

ROLE OF CALCIUM IN PHOSPHORYLASE KINASE REGULATION

On p. 551, Ca^{2+} was noted as being associated with the phosphorylase kinase reaction because calcium affects glycogenolysis in a manner analogous to that described for cAMP-induced phosphorylation (Figure 24-10). This role of Ca^{2+} links muscle contraction to glycogenolysis, since muscular action increases the ion's intracellular level, which in turn activates phosphorylase kinase to stimulate glycogen breakdown. Thus, phosphorylase kinase is subject to activation by both phosphorylation and Ca^{2+}. Studies on the kinase of white anaerobic muscle provide the following proposal to explain regulation of the enzyme (and, hence, of glycogenolysis) under normal conditions by Ca^{2+} and in stressful situations by cAMP and Ca^{2+}. Note that although the proposal is based on extensive *in vitro* results, its *in vivo* validity remains to be established.

Molecular Structure of Phosphorylase Kinase

Phosphorylase kinase contains four molecular species of protomers, called α, β, γ, δ, and the active form of the enzyme is $(\alpha,\beta,\gamma,\delta)_4$, which has a molecular weight of approximately 1,300,000. The α and β protomers can be phosphorylated by protein kinase (Figure 24-10); γ is the catalytic protomer and δ the Ca^{2+}-binding protomer. The δ protomer, it was recently determined, is *calmodulin,* a Ca^{2+}-dependent regulatory protein broadly distributed throughout different tissues. Research on calmodulin (discovered and described in 1970) is currently of special research interest because it is the multifunctional regulatory protein involved in many Ca^{2+}-regulated cellular phenomena that have long been recognized, e.g., regulation of certain key enzymes in intermediary metabolism. That the common protein for mediating Ca^{2+} regulation can be part of the molecular structure of an enzyme is a new biochemical facet of calmodulin. (The δ protomers of phosphorylase kinase, it s estimated, account for one third to one half of the calmodulin content of rabbit skeletal muscle.)

Phosphorylase kinase Protomer	MW
α	145,000
β	128,000
γ	45,000
δ	17,000

Kinase Regulation by Ca^{2+}

When muscle contraction occurs under normal circumstances, released Ca^{2+} binds to the δ protomers (calmodulin) of dephosphorylated phosphorylase kinase, the predominant form of the enzyme in muscle cells not hormonally stimulated. There are four Ca^{2+}-binding sites per δ protomer—one low-affinity (dissociation constant $\approx 6 \times 10^{-7}$ *M*) and three high-affinity ($\approx 1.6 \times 10^{-8}$ *M*) sites. The binding of three or four Ca^{2+} (Ca^{2+}_{3-4}) to each of the δ protomers stimulates the enzyme's activity five- to tenfold. However, to obtain maximum activation by Ca^{2+}, another calcium-binding protein (*troponin C,* which shows many homologies to calmodulin) must interact with the dephosphorylated phosphorylase kinase complex. When Ca^{2+}-bearing troponin C associates with the complex, the enzyme's activity increases 75- to 250-fold. Thus, in the activation of the dephosphorylated enzyme by Ca^{2+}, the δ protomers can account for a five- to tenfold increase in enzyme activity, but the major effector is tropinin C, which can increase the activity much more.

Kinase Regulation by cAMP and Ca^{2+}

Under conditions of stress (flight or fight response), when regulation of phosphorylase kinase activity is hormonally controlled by cAMP, the activation pattern of Ca^{2+} is significantly altered. Once phosphorylated, the enzyme is much more sensitive to activation by the δ protomer. Hence, the binding of one or two calcium ions to each of the δ protomers (Ca^{2+}_{1-2}) is sufficient to bring about maximum activation of phosphorylase kinase (which is over fourfold that effected by the concertive action of the δ protomers and troponin C on the dephosphorylated enzyme). When phosphorylase kinase is phosphorylated, addition of troponin C does not increase activation; thus, with regard to the phosphorylated kinase, the δ protomer is the dominant Ca^{2+}-dependent regulator.

Comments on Double Second Messengers

The above discussion on phosphorylase kinase regulation illustrates how both cAMP and Ca^{2+} serve as second messengers to stimulate glycogen breakdown. This interrelated function of cAMP and Ca^{2+} in intracellular regulation is not unique. Indeed, a variety of cAMP-mediated responses are known to involve Ca^{2+}, e.g., steroid synthesis by ACTH, insulin release by glucose, and growth hormone release by its release factor (GHRF). The inclusion of an additional second messenger in the control of an intracellular metabolic function satisfies a criticism that the original epinephrine-cAMP model did not allow for a plasticity of response. With two regulators, a finer, more sophisticated intracellular control of hormonal response is attainable. Future research may reveal additional second messengers, e.g., phosphate ions, which would provide even more biochemical flexibility in a cell's response to a hormonal stimulus.

COMMENTS ON MOLECULAR INSIGHTS INTO HORMONAL ACTION

A pertinent feature of the cAMP-mediated hormonal action is the magnitude of amplification that the original endocrine signal, issued in the form of minute quantities of a circulating hormone, undergoes to create dominant physiological effects. The advantageous use of intracellular catalytic reactions to magnify the initial hormonal message is the key to the effectiveness of the regulatory functions controlled by the endocrine system. On a molecular basis, the epinephrine studies also reveal that extracellular homeostasis relies on the highly specific recognition between a hormone and its receptor sites (protein or lipoprotein), i.e., a structurally complementary relationship between two molecules. Hence, the physiological functioning of a hormone is determined by a specific interaction between a molecule in the blood and another molecule associated with a particular membrane. Whereas the terms *target organ* and *target tissue* have been used in this chapter, *target cell* is now more frequently used to describe more precisely the molecular aspects of hormonal function.

Finally, it should be noted that the role of cyclic AMP in regulatory processes had its origin early in biological history. Although discovered as a second messenger in an extracellular homeostasis process, cyclic AMP was found to be an important regulator of intracellular homeostasis in procaryotes. As discussed in Chapter 22, the various regulatory roles identified for the nucleotide in *E. coli* testify to its major metabolic importance in probably all forms of life. From an evolutionary viewpoint, cyclic AMP is a common link relating primitive intracellular homeostasis to complex extracellular homeostasis.

HORMONAL ACTION OF STEROIDS

A second molecular model of hormonal action is the one that has been proposed for steroid hormones (Figure 24-12). Whereas a peptide or amine hormone is postulated to interact with a cell by binding extracellularly to the plasma membrane, steroid hormones (fat-soluble biomolecules) can penetrate the membrane of their target cells and enter the cytosol. Thyroid (amine) hormones, it should be noted, are exceptions, since their mode of action is like that of the steroid hormones. The molecular events that transpire after the entry of a steroid hormone into a cell were first described in the early 1970s by Elwood V. Jensen and associates, based on their findings on the effect of estradiol (an estrogen) on mammalian uterine tissue. In the cytosol of uterine cells, estradiol binds noncovalently to a specific *estrogen-receptor protein* (MW $\approx$ 200,000), and the interaction initiates a conformational change in

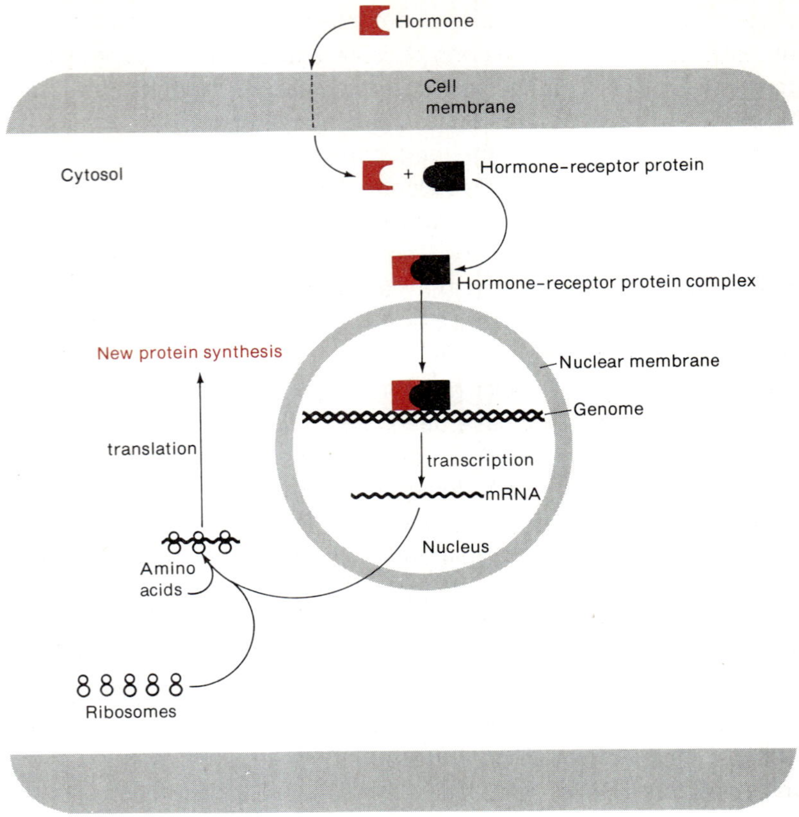

Figure 24-12
Schematic of the proposed molecular mechanism of steroid-hormone action.

the protein. The hormone-receptor protein complex then enters the nucleus of the cell, where it binds to the genome, i.e., to particular site(s) on chromosomes. When the estradiol-uterus and epinephrine-liver models are compared, three differences are noted between the general molecular actions of steroid and peptide hormones. Steroids (1) enter into target cells, (2) have specific cytosolic protein receptors, and (3) exert their hormonal influence at the chromosomal level.

A steroid hormonal system that strongly supports the estradiol-uterine model is the progesterone-oviduct system. Studies on progesterone in the laboratory of Bert W. O'Malley have revealed that, in chick oviduct cells, a progesterone-receptor protein complex is formed and that the complex, upon binding to the genome, initiates an increased production of specific oviduct proteins, e.g., ovalbumin and avidin. It has been calculated that, under progesterone influence, an oviduct cell is capable of synthesizing 1,000,000 ovalbumin molecules per minute. The progesterone studies are especially revealing because they relate the chromosomal binding of the hormone to gene expression. In terms of the Central Dogma that, in part, describes the flow of genetic information (DNA → RNA → Protein), chromatin-bound progesterone brings about an increased production of specific mRNAs, which results in the synthesis of particular oviduct proteins.

Although the above illustrations of steroid hormone action emphasize the turning on of genes (as depicted in Figure 24-14), the genetic control exerted by a steroid hormone may also bring about repression of gene expression. For example, estradiol 17-β represses synthesis of the follicle-stimulating hormone (FSH) by ovine pituitary, i.e., turns off the FSH gene.

As in the case of cAMP-mediated systems, it is anticipated that a detailed elucidation of the molecular action of steroid hormones will reveal a more complicated intracellular situation than the one depicted in Figure 24-12. Many steroids, and certain peptide hormones, elicit *pleiotropic responses* (multiple phenotypic expressions) from their target tissues, e.g., increases in general RNA and protein synthesis and changes in membrane structure and function. In view of the numerous changes induced by many steroids in their target cells, the proposed intracellular action (influencing genetic expression) of these hormones may be expanded in the future to include cytosolic effects, e.g., control of cytosolic protein synthesis. Nevertheless, the molecular model proposed for steroid hormone action, as well as that for peptide and amine hormones, constitutes an invaluable contribution toward a more comprehensive understanding of the molecular forces controlling the complex physiological processes in biological systems.

PROSTAGLANDINS

One class of metabolic regulators that deserves special attention is the *prostaglandins* (cyclic derivatives of 20-carbon fatty acids), particularly those of the essential fatty acid arachidonic acid (p. 207). Although prostaglandins (prostate gland was one of the original sources of these compounds) have been known since the 1930s, it has only been since the 1960s that the broad range of their potent biological activities has been recognized. As more prostaglandins and their derivatives (most of which are tissue hormones, p. 538) are identified and studied, it is apparent that they are produced in most body tissues and hence affect almost every organ system. Prostaglandins affect smooth muscle and blood

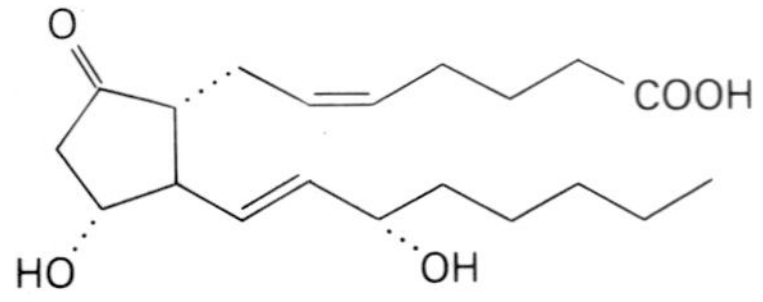

Prostaglandin E_2

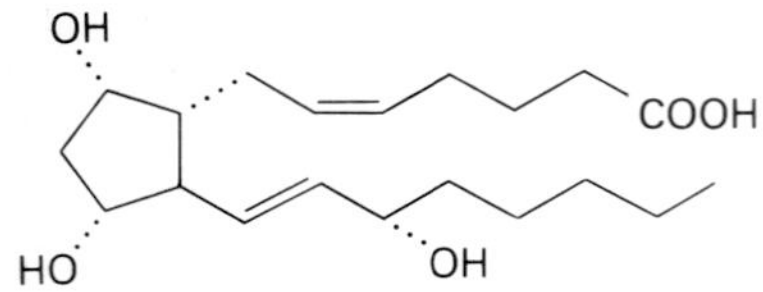

Prostaglandin $F_{2\alpha}$

pressure, and often the activities of individual prostaglandins oppose one another. For example, *prostaglandin E_2 (PGE_2)* dilates blood vessels and bronchi, and *prostaglandin* $F_{2\alpha}$ ($PGF_{2\alpha}$) constricts these smooth muscle tissues. Prostaglandins have a half-life of five minutes or less in blood (and about 30 seconds in saline) and are metabolized during passage through the lungs. The short half-life of these tissue hormones is thought to ensure their transient and limited response at the intermediate site of production. From a medical point of view, they are potentially the most revolutionary therapeutic substances yet discovered. Some of the immediate possible therapeutic applications of prostaglandins include their use in the treatment of high blood pressure, coronary thrombosis, bronchial asthma, anaphylactic shock, edema, gastric hyperacidity, peptic ulceration, nasal congestion, and infertility. Some prostaglandins are also effective labor inducers in pregnant women. Since 1964, when Bengt I. Samuelsson and associates discovered that arachidonic acid is the precursor of prostaglandin E_2, basic research on prostaglandins has continued to uncover new findings that are proving to be of invaluable importance to the medical sciences.

Figure 24-13 depicts the reactions involved in the biosynthesis of three types of prostaglandins from arachidonic acid. The initial, and key, reaction, which yields a *cyclic endoperoxide* intermediate, is catalyzed in a two-step process by the *endoperoxide synthetase complex.* In the first reaction, *fatty acid cyclo-oxygenase* catalyzes the addition of one molecule of oxygen to C-9 of arachidonate and of a second to C-15. The bond formation between C-8 and C-12 accompanying this oxygenation produces the five-membered endoperoxide ring structure characteristic of prostaglandins. In the second reaction, the first intermediate, *PGG_2,* is converted into *PGH_2* by a *hydroperoxidase.* Although all tissues producing prostaglandins have the synthetase complex, the biochemical fate of the PGH_2 synthesized is determined by tissue-specific enzymes. For example, in a tissue producing prostaglandin E_2, the enzyme *endoperoxide isomerase* is present and converts PGH_2 into PGE_2 (Figure 24-13).

The synthesis of PGE_2 from arachidonate was the first pathway elucidated for prostaglandin production. In 1975 Samuelsson and his associates announced the isolation of a new prostaglandin, *thromboxane A_2 (TxA_2),* from blood platelets. A majority of the PGH_2 produced in platelets is converted into TxA_2 by the enzyme *thromboxane synthetase* (Figure 24-13). The following year, John R. Vane and colleagues identified yet another type of prostaglan-

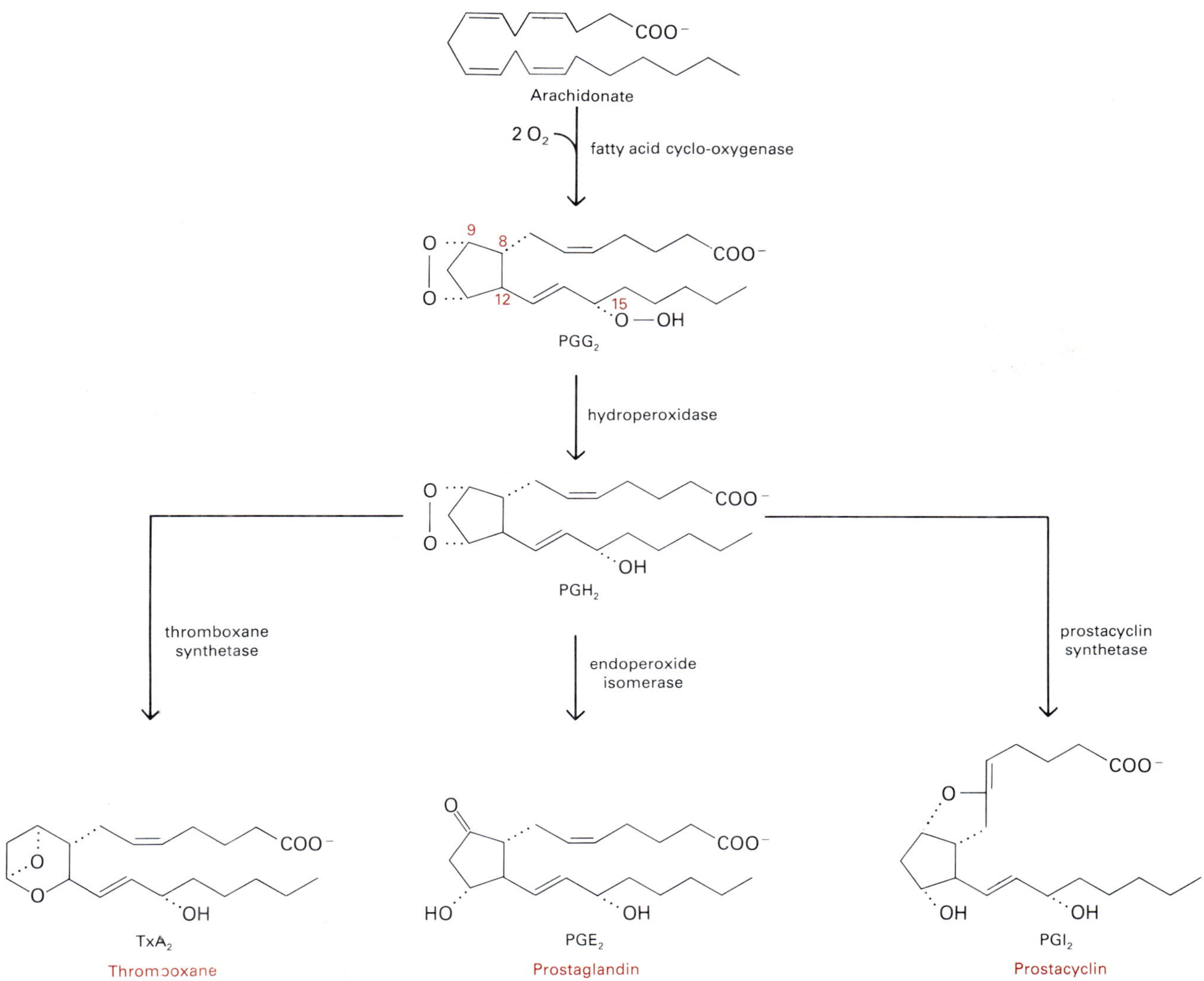

Figure 24-13
Production of PGE_2, PGI_2, and TxA_2 from arachidonic acid.

din, *prostacyclin I_2 (PGI_2)*, which is produced primarily in vascular tissues, e.g., blood vessels. Prostacyclin I_2 synthesis from PGH_2 is mediated by *prostacyclin synthetase* (Figure 24-13). Thus, tissues are differentially endowed with enzymes that transform endoperoxides into specific types of prostaglandins.

The striking symmetry and diversity displayed by the actions of known prostaglandins constitute a critical balance required for the physiological functions performed by these compounds. Thromboxane A_2 and prostacyclin I_2 are medically important examples of how such a balance operates *in vivo.* Thromboxane A_2 is a highly effective blood vessel constrictor (vasoconstrictor) and platelet aggregator; conversely, prostacyclin I_2 is a potent vasodilator and inhibitor of platelet aggregation. As little as 5 ng of TxA_2 per ml induces platelet aggregation, and 1 ng of PGI_2 per ml inhibits aggregation. Platelets are the blood cells that first appear at the site of an injury, where they mass (aggregate) to produce a temporary plug that serves as a base on which the stronger fibrin clot (p. 585) subsequently forms. The aggregating response initiated by TxA_2, therefore, is critical for this defensive role of platelets. However, for maintenance of normal blood flow, TxA_2-induced aggregation of platelets would quickly prove fatal. Thus, a vital opposing role of PGI_2 is to prevent platelets from aggregating on blood vessel walls (a site of PGI_2 production). Unlike other prostaglandins, PGI_2 is not metabolized during passage through the lungs. In fact, minute amounts of PGI_2 are produced by the lungs and released into the blood (which qualifies the prostacyclin as a true hormone), where it performs a second critical role by ensuring that platelets do not aggregate and block blood circulation. Thus, PGI_2 and TxA_2 are continuously engaged in a "tug of war" with respect to platelet aggregation.

The potential therapeutic value of thromboxanes and prostacyclins in the treatment of cardiovascular diseases is indicated in Figure 24-14. The medical interest in prostacyclins is particularly intense because of the multiple clinical uses that can be made of its biological actions. For example, strokes and certain heart

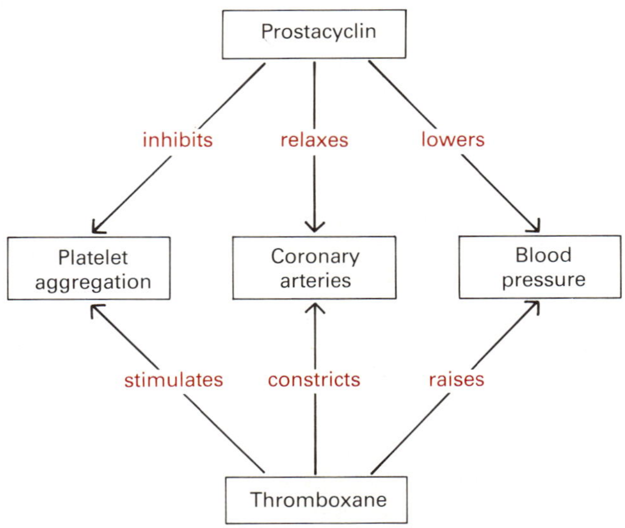

Figure 24-14
Opposing effects of prostacyclin and thromboxane on the cardiovascular system.

attacks result from small blood clots blocking normal blood flow to either the brain or the heart. The antiaggregation action of prostacyclins could be utilized to aid in preventing clots which in turn would prevent such attacks. Long-term studies are now under way to assess this proposal. To date, results from these studies (which involve over 1,000 heart attack victims) are providing evidence that prostacyclin treatment does significantly lower the risk of fatal attacks.

A recent addition to the growing family of prostaglandins of medical importance is *leukotriene C,* identified by Samuelsson and his associates. Leukotriene C is structurally an unusual molecule, containing the amino acid *cysteine* covalently bonded to a derivative of arachidonic acid. Since leukotriene C (found in leucocytes) is not synthesized via the endoperoxide synthetase complex, it is not considered by some to be a true prostaglandin. One prominent physiological effect of leukotriene C is the constriction of air passages in the bronchi during an asthma attack. Since the release of leukotriene C is believed to be triggered by a response to allergic stimuli, e.g., a drug or pollen, it has been postulated that inhibition of leukotriene C synthesis by a drug would provide a therapeutic means of combating asthma attacks and inflammatory allergic responses. A compound has been synthesized that may prove to be such a drug, since *in vitro* studies have established that it inhibits leukotriene C synthesis but not that of the prostaglandins.

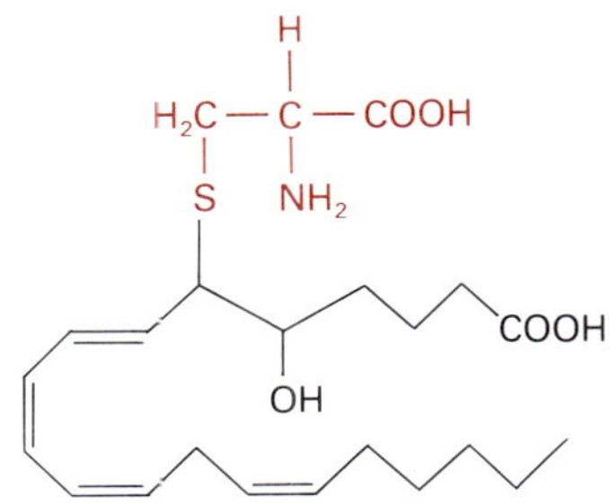

Leukotriene C

An additional finding of interest provided by the biochemical studies on prostaglandins is that *aspirin* inhibits prostaglandin synthesis by acetylating fatty acid cyclo-oxygenase. The identification of the *in vivo* action of aspirin solved one of the long-standing mysteries in pharmacology.

SUMMARY

Hormones are the biomolecules that regulate the metabolic functions of an organism and represent the evolutionary development of extracellular homeostasis. A hormone is defined traditionally as a chemical compound which, after being produced and released by one organ, is transported by the blood to act on another organ. Hormones, because they are biologically very potent compounds, are effective in exceedingly small quantities, i.e., 10^{-7} to 10^{-12} *M.* In vertebrates, extracellular homeostasis is mediated by the endocrine system (group of organs that produce and secrete hormones) and the central nervous system. There are three types of mammalian hormones: amines, e.g., epinephrine; polypeptides, e.g., insulin; and steroids, e.g., sex hormones. Many polypeptide hormones are synthesized as prohormones, or hormonogens, i.e., as inactive forms, which are then cleaved by specific peptidases to yield the active hormones. Insulin and angiotensin are examples of polypeptide hormones that are synthesized as prohormones. The release of hormones is controlled by a variety of mechanisms which vary in complexity. Whereas the release of some hormones, e.g., insulin, is controlled by relatively simple mechanisms, the release of others, e.g., anterior pituitary hormones, involves highly complex mechanisms.

In recent years, two models that account for the molecular basis of hormonal action have been proposed; one applies principally to amine and polypeptide hormones, the other to steroid hormones. The two-messenger hypothesis, proposed for amine and polypeptide hormones, was derived from studies on the liver's glycogenolytic response to epinephrine. Biochemical studies support the proposal that when epinephrine (first messenger) is released into the blood, the hormone interacts with specific receptor sites on the outer membranes of its target organs, e.g., liver cells. This hormone-receptor interaction triggers the intracellular synthesis of cyclic AMP (second messenger), which in turn initiates a series of biochemical reactions (predominantly the phosphorylation of enzymes) that results in the activation of phosphorylase (which degrades glycogen) and the inactivation of glycogen synthase. As a result, glycogenolysis is enhanced and the glucose 1-phosphate produced is converted into available free glucose for release into the blood. The hormonal action of many amines and polypeptides is now known to involve cyclic AMP as the second messenger.

In some two-messenger systems, Ca^{2+} is also used as a co-second messenger. For example, in muscle glycogenolysis, Ca^{2+} and cAMP are responsible for the activation of phosphorylase kinase. Under normal conditions (when muscle cells are not hormonally stimulated), Ca^{2+} alone stimulates glycogenolysis by binding to the δ protomer of phosphorylase kinase. The δ protomer is calmodulin, a multifunctional protein involved in Ca^{2+}-regulated cellular phenomena. Maximum stimulation of the kinase activity under normal conditions occurs when another Ca^{2+}-binding protein of muscle (troponin C), as well as Ca^{2+}, binds to the enzyme.

The mode of hormonal action of steroids differs from that of amine and polypeptides, since steroids enter their target cells, bind to specific cytosolic protein receptors, and exert hormonal influence when the steroid-receptor protein complex binds to sites on chromosomes. Thus, by the activation of genes, steroids directly influence the synthesis of proteins and other biochemical processes in their target cells.

A prostaglandin is an example of a tissue hormone, which is one that is produced by one cell type in a tissue and acts on another cell type within the same tissue. The many prostaglandins (cyclic derivatives of fatty acids) that have been described are currently of great medical interest since they are potentially the most important therapeutic agents to be discovered.

For example, thromboxane A_2 is a highly effective vasoconstrictor and blood platelet aggregator, and prostacyclin I_2 is a potent vasodilator and inhibitor of platelet aggregation. *In vivo,* thromboxane A_2 is critically important for blood clot formation and prostacyclin I_2 is equally as important because it prevents thromboxane A_2-induced platelet aggregation from interfering with normal blood flow. Thus, the counteracting activities of the two prostaglandins play important physiological roles. A recently described member of the prostaglandin family is leukotriene C, which constricts air passages in the bronchi during an asthma attack or allergic reaction. Drugs that specifically inhibit leukotriene C synthesis would be exceedingly useful in combating asthma and allergic responses.

QUESTIONS

1. Which of the following:

Aldosterone	Oxytocin
Epinephrine	Testosterone
Estradiol	Thyrotropin releasing factor
Growth hormone	Thyroxin
Insulin	Vasopressin

   (a) are amine hormones?
   (b) are polypeptide hormones?
   (c) are steroid hormones?

2. Select answers to the statements about epinephrine-induced glycogenolysis listed below from the following list:

Adenylate cyclase	Phosphodiesterase
cAMP	Phosphorylase
Epinephrine	Phosphorylase kinase
Glycogen synthase	Protein kinase
Phosphatase	

   (a) First messenger
   (b) Second messenger
   (c) Binds to receptor site on outer membrane of target cell
   (d) Catalyzes the synthesis of cAMP
   (e) Catalyzes the inactivation of cAMP
   (f) Catalyzes the conversion of phosphorylase *b* into *a*
   (g) Catalyzes the conversion of phosphorylase *a* into *b*
   (h) Catalyzes the activation of protein kinase
   (i) Catalyzes the activation of phosphorylase kinase
   (j) Catalyzes the inactivation of phosphorylase kinase
   (k) Catalyzes the activation of glycogen synthase
   (l) Catalyzes the inactivation of glycogen synthase
   (m) Is the active form of the enzyme that is phosphorylated
   (n) Is the inactive form of the enzyme that is phosphorylated

3. Fill in the blanks below with your choice from the following list:

α	δ
β	Troponin C
γ	

   In the structure of phosphorylase kinase, (a) ________ is the catalytic protomer, (b) ________ the Ca^{2+}-binding protomer, and (c) ________ and (d) ________ the protomers that are phosphorylated. The (e) ________ protomer is calmodulin. For activation of the dephosphorylated kinase (normal circumstances) in white muscle, the two regulatory proteins are (f) ________ and (g) ________, with (h) ________ providing maximum activation. For activation of the phosphorylated kinase (hormonally controlled), (i) ________ is the major regulatory protein, and addition of (j) ________ does not increase the activation.

4. Which of the following statements are true about the proposed hormonal action of a steroid?
   (a) Enters its target cell
   (b) Transported into its target cell by specific cytosolic protein receptor
   (c) Covalently bonds to its protein receptor
   (d) Initiates cAMP synthesis in its target cell
   (e) Is phosphorylated in its target cell
   (f) Exerts its hormonal influence by activating genes
   (g) Influences cellular protein synthesis
   (h) Binds to the nuclear membrane of its target cell

5. Norepinephrine and epinephrine (Figure 24-3) are catecholamines (dihydroxyphenylamines) that are generally synthesized from dietary tyrosine. For their synthesis, tyrosine is first converted into 3,4-dihydroxyphenylalanine (DOPA), which is then decarboxylated to produce dopamine. (Dopamine is a neurotransmitter in the central nervous system and a potent vasodilator.) Dopamine serves as the direct precursor of norepinephrine which in turn is methylated to produce epinephrine.
   (a) Write the four general reactions (drawing structures) for the synthesis of epinephrine from tyrosine.
   (b) Which reactions are catalyzed by hydroxylases?
   (c) Is dopamine a catecholamine?

6. Which of the following statements about prostaglandins are true?
   (a) Prostaglandins are cyclic derivatives of C_{18} fatty acids.

(b) Prostaglandins have short physiological half-lives.
(c) Arachidonate is the parent compound of prostaglandins.
(d) Thromboxane A_2 is produced from PGH_2 in blood platelets.
(e) Thromboxane A_2 is a vasoconstrictor.
(f) Prostacyclin I_2 is produced from PGH_2 in vascular tissues.
(g) Prostacyclin I_2 induces aggregation of platelets.
(h) Prostacyclin I_2 is a vasodilator.
(i) Prostacyclin I_2 is unusual because it contains cysteine in its structure.
(j) Leukotriene C is used to treat asthma attacks.

7. Given the following numbered hydrocarbon structure related to steroid hormones:

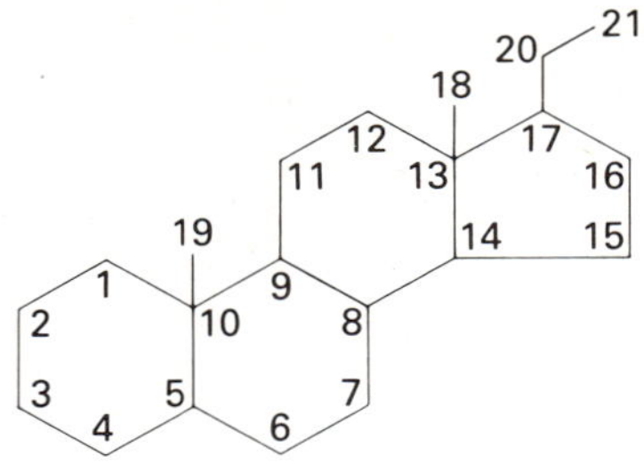

draw the structures of the steroids involved in the following postulated pathways for the synthesis of aldosterone.

11-Deoxycorticosterone ⟶ Corticosterone (Figure 24-2)
11-Deoxycorticosterone ⇣ 18-Hydroxy-11-deoxycorticosterone
Corticosterone ↓ 18-Hydroxycorticosterone
18-Hydroxy-11-deoxycorticosterone ⇣ 11-Deoxyaldosterone
18-Hydroxycorticosterone ↓ Aldosterone (Figure 24-2)
11-Deoxyaldosterone ⇢ Aldosterone (Figure 24-2)

SUGGESTED READINGS

Austin, C. R., and R. V. Short, *Mechanisms of Hormone Actions.* Vol. 7 in the *Reproduction in Mammals* series (paperback). New York: Cambridge University Press, 1980. A collection of authoritative essays describing current knowledge about hormone action.

Beers, R. F. Jr., and E. G. Bassett, eds., *Polypeptide Hormones* (12th Miles International Symposium). New York: Raven Press, 1980. A book that covers the rapid advances being made in the field of polypeptide hormones.

Bondy, P. K., and L. E. Rosenberg, eds., *Metabolic Control and Disease,* 8th ed. Philadelphia: W. B. Saunders, 1980. Article by J. D. Baxter and K. M. MacLeods on the molecular basis of hormone action is highly recommended.

Li, C. H., ed., *Hormonal Proteins and Peptides,* 7 vols. New York: Academic Press, 1973–1979. An excellent reference source for information on the polypeptide hormones.

Robinson, G. A., R. W. Butcher, and E. W. Sutherland, *Cyclic AMP.* New York: Academic Press, 1971. A comprehensive review written by the discoverer of cyclic AMP and two of his colleagues.

Schulster, D., S. Burstein, and B. A. Cooke, *Molecular Endocrinology of the Steroid Hormones.* New York: Wiley, 1976. Another recommended book on the biological functions of steroid hormones.

Villee, D., *Human Endocrinology: A Developmental Approach.* Philadelphia: W. B. Saunders, 1975. Endocrinology from fetus to adult.

Williams, R. H., ed., *Textbook of Endocrinology,* 5th ed. Philadelphia: W. B. Saunders, 1974. Thirty-eight authoritative contributions make this book an outstanding reference source.

Articles

Attardi, B., "Genetic Analysis of Steroid Hormone Action." *TIBS,* 1:241, 1976.

Bailey, J. M., "Prostacyclins, Thromboxanes and Cardiovascular Disease." *TIBS,* 4:68, 1979.

Bornstein, J., "Biological Actions of Synthetic Part Sequences of Human Growth Hormones." *TIBS,* 3:83, 1978.

Chan, S. J., and D. F. Steiner, "Preproinsulin, A New Precursor in Insulin Biosynthesis." *TIBS,* 2:254, 1977.

Czech, M. P., "Molecular Basis of Insulin Action." *Ann. Rev. Biochem.,* 46:359, 1977.

Davidson, E. H., "Hormones and Genes." *Scientific American,* 212(6):366 (offprint #1013), 1965.

Guillemin, R., and R. Burgus, "The Hormones of the Hypothalamus." *Scientific American,* 227(5):24 (offprint #1260), 1972.

Hems, D. A., "Short-Term Hormonal Control of Liver Glycogen Metabolism." *TIBS,* 2:241, 1977.

Klee, C. B., T. H. Crouch, and P. G. Richman, "Calmodulin." *Ann. Rev. Biochem.,* 49:489, 1980.

O'Malley, B. W., and W. T. Schrade, "The Receptors of Steroid Hormones." *Scientific American,* 234(2):32 (offprint #1334), 1976.

Pastan, I., "Cyclic AMP." *Scientific American,* 227(2):97 (offprint #1256), 1972.

Pike, J. E., "Prostaglandins." *Scientific American,* 225(5):84 (offprint #1235), 1971.

Samuelsson, B., M. Goldyne, E. Granström, M. Hamburg, S. Hammarström, and S. Malmsten, "Prostaglandins and Thromboxanes." *Ann. Rev. Biochem.,* 47:997, 1978.

Schally, A. V., D. H. Coy, and C. A. Meyers, "Hypothalamic Regulatory Hormones." *Ann. Rev. Biochem.,* 47:89, 1978.

Tager, H. S., and D. F. Steiner, "Peptide Hormones." *Ann. Rev. Biochem.,* 43:509, 1974.

BASIC ASPECTS OF HUMAN NUTRITION

25

INTRODUCTION

Previous chapters emphasized the biochemical mechanisms by which *phototrophs,* harnessing the radiant energy of the sun, and *chemoorganotrophs,* utilizing biomolecules as fuels, produce the chemical energy (ATP) needed for life processes. In the case of chemoorganotrophs, it is apparent that the sources of chemical energy utilized must be obtained from their surroundings. What have not been specifically noted, except for water and the B-vitamins, are the other substances (organic and/or inorganic) which all organisms must secure from their surroundings to carry out their biological functions, including the production of energy. The importance of some of these substances has already been indicated, e.g., iron (hemoglobin, myoglobin, and cytochromes), magnesium (cofactor for many enzymatic reactions), and phosphate (intracellular buffering system). Collectively, those substances that must be obtained by an organism to provide the raw materials for the structures and functions of life constitute the organism's *nutritional requirements.* The nutrients required by the many different life forms vary widely, with some needing remarkably few and others many.

As photosynthetic organisms, plants are relatively self-sufficient, requiring only sunlight, water, CO_2, and certain mineral ions, e.g., nitrates and phosphates. In the case of legumes (p. 396), the nitrate (nitrogen) requirement is satisfied by a symbiotic relationship between the plants and *Rhizobia,* bacteria which fix atmospheric nitrogen.

Some chemoorganotrophs have remarkably few nutritional requirements. For example, the procaryote *Escherichia coli* requires only an energy and carbon source, e.g., glucose, and some inorganic salts [$MgSO_4$, K_2HPO_4, and $Na(NH_4)HPO_4$] for growth. Another example is the eucaryote *Neurospora crassa* (common bread mold), which needs sucrose, several inorganic salts, and the B-vitamin biotin. Not all procaryotes or lower forms of eucaryotes, however, have such simple nutritional needs. The procaryotic lactobacilli, e.g., *Lactobacillus bulgaricus* (used to prepare yogurt), have complex nutritional requirements that include—in addition to a carbohydrate and inorganic salts—purines, pyrimidines, most of the B-vitamins, and many amino acids.

With the exception of mineral needs, the nutritional requirements of an organism generally reflect its biosynthetic capabilities. The amino acid requirement of an organism exemplifies this point. Because organisms require 20 amino acids for protein synthesis, they must acquire those they cannot synthesize. Some biological systems, e.g., plants and *E. coli,* have the biosynthetic capacities to produce all 20 amino acids and thus have no requirement for them. On the other hand, other organisms require some (or many) of the amino acids because they lack the enzymatic machinery (metabolic pathways) to synthesize them. Many lactobacillus strains require 16 to 18 of the 20 amino acids; these bacteria have an amino acid requirement that exceeds the nine needed by humans.

The topic of human nutrition is a matter of continuing importance, extending from an individual's self-concern to the feeding of the world's population. Of the problems facing humankind, food production and malnutrition rank among the most severe because humans, who are not as biosynthetically versatile as plants or *E. coli,* must rely on their diets for the approximately 50 nutrients they need to maintain normal life processes. The precise number of nutrients required is still uncertain because it is experimentally difficult to document a human need for some of the substances thought to be required. An example is the nutritional status of certain trace elements, e.g., tin and nickel, for which no human deficiencies have been observed. If these elements are essential, they are present in sufficient amounts in all diets.

The nutrients required by human beings are generally classified as *macro-* or *micronutrients.* The macronutrients are *carbohydrates, proteins,* and *lipids* (fats)—the three nutrients used by the body in large amounts (gram quantities) to supply energy, building materials, and necessary biomolecules, e.g., amino acids. The micronutrients are the *vitamins* and *minerals,* which are required in small amounts (mg or μg quantities). One of the two nutritional factors not included in the above categories is *water,* essential for all forms of life. Whereas it usually requires months to develop a serious vitamin deficiency, lack of water becomes a life-threatening situation within a few days. The other nutritional factor is *fiber,* composed mainly of indigestible polysaccharides, which has an important role in normal digestion. The classification of fiber *(roughage)* as a nutrient is still a matter of debate, since its main role appears to be mechanical and not chemical, i.e., it does not participate in the chemistry of life.

U.S. RECOMMENDED DAILY ALLOWANCES

In the United States, the Food and Nutrition Board (under the auspices of the National Academy of Sciences' National Research Council) maintains a Committee on Dietary Allowances, which continually monitors and reassesses all pertinent information about nutrition, e.g., latest research findings, nutrient content of foods, and citizens' eating habits. About every five years, the Committee publishes a revised guideline, called *U.S. Recommended Daily Allowances (U.S. RDA),* for every nutrient about which a valid nutritional evaluation can be made. The recommendations are made for 17 different population groups, classified on the basis of age, sex, height, and weight (and pregnancy). The current recommended allowances for young adults are listed in Table 25-1. Because recommended allowances are often mistaken for *minimal requirements* (amounts needed to avoid clinical symptoms of deficiency), it should be noted that the quantities listed in the U.S. RDA include a *large safety margin.* The recommendations are scientific estimates that provide for an excess of all nutrients for at least 95 per cent of the population and therefore represent quantities much larger than those experimentally determined as minimal requirements. For example, although the U.S. RDA for vitamin C is 60 mg, only 10 mg are required per day to avoid scurvy, the clinical manifestation of a deficiency. The U.S. safety factor for vitamin C is significantly larger than that recommended by the analogous British or Canadian RDA, which is 30 mg.

Note that the values for the fat-soluble vitamins listed in Table 25-1 are expressed as *equivalents,* rather than as specific amounts. Because these vitamins occur in different forms with varying biological potencies, each of their allowances is expressed

Table 25-1
U.S. recommended daily allowances (1980) for the average male and female between ages 19 and 22

Nutrient	Male	Female
Protein	56 g	44 g
Fat-soluble vitamins		
Vitamin A	1 mg RE*	0.8 mg RE*
Vitamin D	7.5 μg†	7.5 μg†
Vitamin E	10 mg α-TE‡	8 mg α-TE‡
Water-soluble vitamins		
Vitamin C	60 mg	60 mg
Niacin	19 mg	14 mg
Pantothenic acid (estimate)	10 mg	10 mg
Vitamin B_6	2.2 mg	2 mg
Riboflavin	1.7 mg	1.3 mg
Thiamin	1.5 mg	1.1 mg
Folic acid	0.4 mg	0.4 mg
Biotin (estimate)	0.3 mg	0.3 mg
Vitamin B_{12}	3 μg	3 μg
Minerals		
Calcium	800 mg	800 mg
Phosphorus	800 mg	800 mg
Magnesium	350 mg	300 mg
Zinc	15 mg	15 mg
Iron	10 mg	18 mg
Copper (estimate)	3 mg	3 mg
Iodine	150 μg	150 μg

*RE = retinol equivalent (1 retinol equivalent = 1 μg retinol or 6 μg β-carotene).
†As cholecalciferol.
‡α-TE = α-tocopherol equivalent (1 α-TE = 1 mg *d*-α-tocopherol).
Source: Food and Nutrition Board, National Academy of Sciences—National Research Council.

as equivalent to the amount established for the most common or active form of the vitamin. For vitamins A, D, and E, the equivalency standards are retinol (p. 577), cholecalciferol (p. 581), and *d*-α-tocopherol (p. 586), respectively. Use of equivalents to express the RDAs of fat-soluble vitamins was introduced in 1980. Prior tables stated the allowances as *international units (IU),* which refer to biological potencies (as opposed to specific quantities). In the 1974 tables, the allowances for vitamins A, D, and E are listed as 5,000, 400, and 30 IU, respectively. The vitamin D content of irradiated milk, for example, contains 400 IU of vitamin D_2 (ergocalciferol, p. 581). Because the use of international units had become so well established, it will undoubtedly take time for equivalency allowances to be universally adopted.

The following discussion of macro- and micronutrients required by human beings uses the U.S. RDA as a guide and is presented as a practical view of some aspects of nutrition—a subject that, in spite of its continuous popularity, is so often poorly understood.

MACRONUTRIENTS AS ENERGY PROVIDERS

A person's daily intake of calories is generally furnished by a combination of carbohydrates, proteins, and lipids (the three macronutrients). Carbohydrates are the predominant source of calories for most of the world's human population; for most Americans, protein and lipids provide the majority of calories. If alcoholic beverages are consumed regularly in appreciable amounts, then *ethanol* can be considered a macronutrient. A six-pack of 12-oz. cans of beer contains about 900 Calories and 1½ oz. of scotch about 100. (In nutrition, the kilocalorie measure used by biochemists is called a Calorie.) Thus, with reference to the caloric content of a six-pack, it contains 900 Cal (nutritional *large calories*), or 900 kcal (900,000 nutritional *small calories*).

As emphasized in Chapters 15 to 17, mitochondrial processes for aerobic respiration readily use pyruvate (carbohydrates), catabolic products of amino acids (proteins), and fatty acids (lipids) as fuels. It is the free energy (ΔG), expressed as calories per mole, released by the oxidation of the macronutrients that provides a person with the chemical energy to produce the ATP used for the biological functions necessary for normal life processes. The recommended daily caloric need varies with a number of factors, e.g., sex, age, weight, daily activity, and climate. The recommendations for the *U.S. RDA reference woman* (128 lbs.) and *man* (154 lbs.) are *2,000* and *2,700* Cal, respectively. An obese condition is generally attributed to a diet that provides a daily intake of calories that exceeds a person's caloric expenditure (output), with some of the excess energy taken in being stored in the form of fat (one pound of body fat is equivalent to 3,500 Cal, or kcal). In any effective weight-losing program, the underlying principle is to take in fewer calories than are expended on a daily basis, accomplished by dieting and/or a program of exercise.

The term *empty calories* refers to food items that provide a person with an energy-producing nutrient, usually sugar, and nothing or little else nutritionally, e.g., many types of soft drinks. One nutritional concern about such foods is that, when consumed daily as a significant portion of a diet, they often substitute for foods that furnish a larger number of needed nutrients for the same caloric intake, e.g., soft drinks substituting for milk or natural fruit juices.

A diet continuously high in empty calories will lead to dietary deficiencies and/or obesity.

DAILY REQUIREMENT OF MACRONUTRIENTS

Protein is the only macronutrient for which there is a recommended daily allowance (Table 25-1). The abundance of carbohydrates in diets precludes any nutritional concern about a deficiency of this nutrient. Also, because of the physiological efficiency of gluconeogenesis (p. 282), low levels of carbohydrates are tolerated, e.g., high-protein diets. With respect to lipids, only very small amounts of the three essential fatty acids, *linoleic, linolenic,* and *arachidonic,* are needed. As previously discussed (p. 377), both linolenic and arachidonic acids can be produced by the body if linoleic acid is available; thus, linoleic acid is the only truly essential fatty acid since it cannot be synthesized by humans. Because the requirements for these fatty acids are readily met in the diets of Americans (vegetable oils and fish being very good sources), there is no nutritional concern about these nutrients. ''Vitamin F,'' sold in health food stores, is not a vitamin but rather one or more of the essential fatty acids; as a dietary supplement, ''vitamin F'' merely adds unnecessary lipids to the diet.

Linoleic acid:

$$H_3C(CH_2)_4CH\overset{cis}{=}CHCH_2CH\overset{cis}{=}CH(CH_2)_7COOH$$

Linolenic acid:

$$H_3CCH_2(CH\overset{cis}{=}CHCH_2)_2CH\overset{cis}{=}CH(CH_2)_7COOH$$

Arachidonic acid:

$$H_3C(CH_2)_4(CH\overset{cis}{=}CHCH_2)_3CH\overset{cis}{=}CH(CH_2)_3COOH$$

AMINO ACID REQUIREMENTS

The dietary requirement for proteins is not based on the macronutrient's caloric content but on a need for the amino acids they contain. The nine amino acids required by humans are presented in Table 25-2, and, as noted, *adults require eight* and *growing children additionally need histidine.* The term *essential* is used to describe the nine amino acids needed, and the other eleven, which the body can make, are called *nonessential.* However, because all 20 amino acids are essential for protein synthesis, certain nutritionists prefer the terms *indispensable* (for essential) and *dispensable* (for nonessential), adjectives they believe better describe the amino acid needs as dietary.

Table 25-2
Amino acids required by humans

Amino acid	U.S. RDA for adult males (grams)
Histidine*	Unknown
Isoleucine	0.84
Leucine	1.12
Lysine	0.84
Methionine	0.70
Phenylalanine	1.12
Threonine	0.56
Tryptophan	0.21
Valine	0.96

*Required by growing children. Recent evidence suggests it may also be required by adults.

HIGHER- AND LOWER-QUALITY PROTEINS

The types of protein, as well as quantities, consumed determine how well the daily requirements of the nine essential amino acids are met. Proteins from animal sources (eggs, milk, meat, and fish) are called higher-quality proteins because they supply adequate amounts of all the essential amino acids. Plant proteins are called lower-quality proteins since they have a low content (limiting amount) of one or more of the essential amino acids. The three most common limiting amino acids are lysine, methionine, and tryptophan (Table 25-3). Although plant proteins have limiting amounts of some (but not all) essential amino acids, it is a mistake to assume that they are poor protein sources. Many populations, as well as vegetarians, have no problem obtaining adequate amounts of needed amino acids from vegetable and grain diets. With such diets, it is necessary to consume larger total quantities of protein than would be required from diets containing animal protein and to include diverse vegetable protein sources. In this manner, various plant proteins *complement* one another to provide the amino acid requirements.

A serious malnutrition problem, however, can arise when there is reliance on a diet that predominantly contains only one source of plant protein, e.g., corn or rice, because large quantities of food must be eaten to ensure an adequate daily intake of protein. The gravity of the nutritional problems increases when there is a reduced daily intake in such diets as a result of crop failure, drought, or overpopulation. A reduction in caloric intake in a diet that furnishes marginal amounts of protein creates *protein-calorie malnutrition* because some of the protein eaten must be used to supply the body with energy. On a worldwide basis, this type of ''protein crisis'' is considered the most critical nutritional problem. Serious efforts are being made to alleviate the protein malnutrition that plagues or threatens certain populations of the world. With the use of modern technology, vegetable proteins are being supplemented to overcome amino acid deficiencies, e.g., addition of methionine to soybean ''milk'' and lysine to wheat. Plant breeders who previously stressed *yield per acre* in their breeding programs

Table 25-3
Limiting amino acids in some plant proteins

Food	Amino acid(s)
Cereal grains and millets	Lysine and threonine
Sunflower seeds	Lysine
Rice and soybeans	Methionine
Peanuts	Methionine, lysine, and threonine
Legumes (peas and beans)	Methionine and tryptophan
Green leafy vegetables	Methionine

are now turning their attention to *nutrition per acre,* e.g., Opaque corn, which has a higher lysine content than regular corn.

Protein deficiency is not a general problem in societies that have a diet high in animal protein, such as in the United States. The Americans' love of meat provides high excesses of the essential amino acids in their diets. Probably few Americans realize that a one-quarter-pound hamburger patty or the meat patty and cheese slice of a cheeseburger (not including the protein in the bun) adequately supplies the daily allowance of the required amino acids.

MICRONUTRIENTS

The requirement for micronutrients was not recognized until 1912, when the British scientist Frederick G. Hopkins (Nobel Prize, 1929) showed that experimental animals became sick when fed only carbohydrates, proteins, and lipids (the three macronutrients) in purified form. He postulated that certain ''accessory factors'' found in natural foods, e.g., milk, are also required for proper nutrition. That was the same year that Casimur Funk (p. 163) identified thiamin, which cures beriberi (which means *I cannot,* symbolizing the incapacitated condition created by thiamin deficiency). In the following decades, Hopkins's proposed ''accessory factors'' were to be identified as two classes of micronutrients called the *water-soluble* and *fat-soluble vitamins.* The micronutrients also include the *macro-* and *microminerals.* Recognition that more than three nutrients are required by animals ushered in an era of study that resulted in a more complete and thorough understanding of nutritional requirements. The nutritional importance of micronutrients is evident in Table 25-1, in which 19 of the 20 RDAs listed are for micronutrients.

WATER-SOLUBLE VITAMINS

B-Vitamins

Nutritional information on the B-vitamins and their biochemical roles as coenzymes are presented in Chapter 10. With respect to the U.S. RDA for the B-vitamins, it should be noted that the quantities listed for biotin and pantothenic acid are *optional RDAs.* No human deficiencies have been described for these two B-vitamins because intestinal bacteria provide adequate amounts; thus, no true RDAs have been established. Listing of these two vitamins in the U.S. RDA was made in response to claims by food enthusiasts that the two nutrients are required in large amounts daily, i.e., could be present in inadequate amounts in average diets. The RDAs for biotin (0.3 mg) and pantothenic acid (10 mg) suggest that daily allowances, if known, would be mg quantities, as are those of the other B-vitamins.

Vitamin C (Ascorbic Acid)

No other vitamin, with the possible exception of vitamin E, is as generally misunderstood as is vitamin C. It is ironic that the oldest therapeutically used vitamin, furnished in the 1750s in the form of limes to British sailors (thus the epithet "limeys") to prevent scurvy, is still a subject of controversy. *Vitamin C* is *L-ascorbic acid* (Figure 25-1), a lactone of a sugar derivative of glucose. As a strong reducing agent, ascorbic acid readily loses two hydrogen atoms to become *L-dehydroascorbic acid* (Figure 25-1), which also has vitamin C activity; however, if the lactone ring is hydrolyzed to produce *L-diketogulonic acid,* vitamin activity is lost. Unlike the B-vitamins, ascorbic acid occurs in relatively large amounts in animal and plant tissues (1 mg ascorbic acid per 100 ml human blood plasma), and this is one reason why vitamin C, although water soluble, is not classified as a B-vitamin. Although vitamin C is known to have general metabolic functions, e.g., prevention of infant anemia, maintenance of capillary integrity, and formation of connective tissue, no specific biochemical function has been established for the nutrient. Most higher animals and all plants have no requirement for the vitamin because they synthesize it from glucose. The only vertebrates that are known to require vitamin C are monkeys, guinea pigs, red-vented bulbuls (birds), Indian fruit bats, certain fish, and humans.

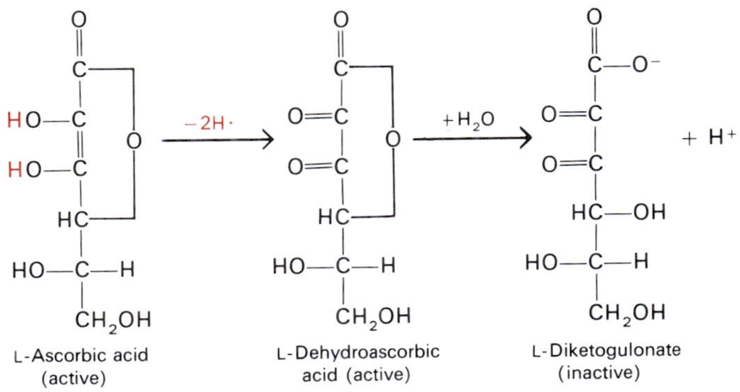

Figure 25-1
Conversion of ascorbic acid into diketogulonate.

As a dietary nutrient, the vitamin C allowance is readily obtainable from many foods, especially certain fruits (citrus and strawberries) and vegetables (cabbage, potatoes, tomatoes, and broccoli). One of the richest sources is peppers (green and red); in fact, ascorbic acid was first isolated from Hungarian sweet peppers by Albert Szent-Györgyi (Nobel Prize, 1937). A daily consumption of only 1 ounce of cantaloupe, 1 tablespoon of green peppers, or 2 ounces of tomato juice is sufficient to obtain the 10 mg of vitamin C needed to prevent scurvy. Vitamin C is sensitive to heat and air, and therefore cooking (especially overcooking of vegetables) destroys an appreciable amount of its activity. The vitamin's susceptibility to heat explains why uncooked foods, e.g., oranges and raw cabbage, are often recommended as dietary sources of the nutrient. Too, because acidic conditions protect the vitamin, citrus fruits and their juices, and tomato juice, are considered good, stable sources.

Currently, vitamin C is being touted as a panacea for a variety of human ills. For many years, daily ingestion of 1 g or larger quantities of the vitamin has been advocated by some (not by nutritionists) as a preventive measure against the common cold. In recent years, because Nobel laureate Linus Pauling has actively championed the role of vitamin C as a cold preventative, much public interest has been generated in the matter. The main argument used by vitamin C advocates is that, although 10 mg per day prevents scurvy (obvious clinical symptoms of a deficiency), a much larger intake is needed to maintain the good health necessary to avoid such illnesses as colds. *Megadoses* (gram quantities) of the vitamin are often recommended by these enthusiasts as a satisfactory daily intake. Unfortunately for these advocates, nutritional studies do not support their claims because 30 mg per day (British and Canadian RDA) provide the body's tissues with saturating levels of the vitamin. Excessive amounts of consumed vitamin C, being water soluble, are not maintained or stored in the body but are excreted.

Although large daily intakes of vitamin C have not proved toxic, the safety of prolonged use of large doses has yet to be established. An unexpected concern about vitamin C intake was reported recently. An individual who had ceased a prolonged daily intake above 250 mg developed scurvy (although a daily intake of at least 30 mg was being ingested). The report raises the possibility that the body's physiological mechanisms can adjust from regular to higher daily intakes of vitamin C but readjusts poorly to the reverse situation.

FAT-SOLUBLE VITAMINS

The four fat-soluble vitamins are A, D, K, and E, and they can also be regarded as lipids. Vitamins A, K, and E are *terpenoids,* and vitamin D is a *steroid.* All four are *isoprenoid compounds,* since they are synthesized from *isoprene units* (p. 379). Unlike the B-vitamins and vitamin C, fat-soluble vitamins can be stored in the body, e.g., an adult's liver can store enough vitamin A to last several months or longer. However, because fat-soluble vitamins are storable, excessive intakes can result in toxic conditions *(hypervitaminoses).*

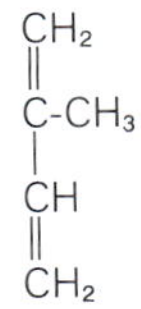

Isoprene

Note that in Table 25-1 there is no RDA for vitamin K because, as in the case of biotin and pantothenic acid, intestinal bacteria synthesize the body's requirement. No vitamin K deficiencies have been produced except by oral ingestion of medicines, e.g., sulfa drugs or antibiotics, that suppress intestinal synthesis. The same can be said for vitamin E, since an eight-year nutritional study failed to produce a deficiency in human volunteers; the RDA for vitamin E was determined from studies on experimental animals. Vitamin E is not a problem in human nutrition because it is ubiquitous in foods. Since humans produce vitamin D with the aid of sunlight, vitamin A is the only fat-soluble vitamin that might pose a dietary problem.

Vitamin A (Retinol)

General Information

Vitamin A_1, or *retinol,* (Figure 25-2) is the common form found in mammals and marine fish; another form, *vitamin A_2,* or *retinol$_2$,* occurs in freshwater fish. Plants do not have vitamin A but, instead, have *carotenoids* (α, β, and γ carotenes, p. 347), which serve as precursors, or *provitamins,* of vitamin A for animals consuming them. When ingested, the symmetrical β-carotene molecule (Figure 25-2) is enzymatically cleaved into two molecules of vitamin A_1 in the intestinal mucosa and liver. The yellow color of carotenoids characterizes certain foods (carrots, pumpkins, and cantaloupes) as rich sources of provitamin A; another visual indication of carotenes is the deep green of many vegetables (spinach, broccoli, and romaine lettuce).

Although *vitamin A* is known to function in *visual processes,* it probably has other metabolic roles, since all tissues are affected by a deficiency. One general role is believed to be in the transport of calcium ions across certain membranes. Young children are

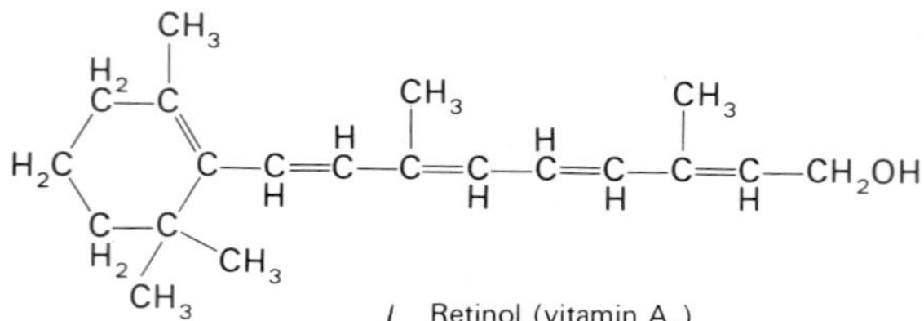

Retinol (vitamin A_1)

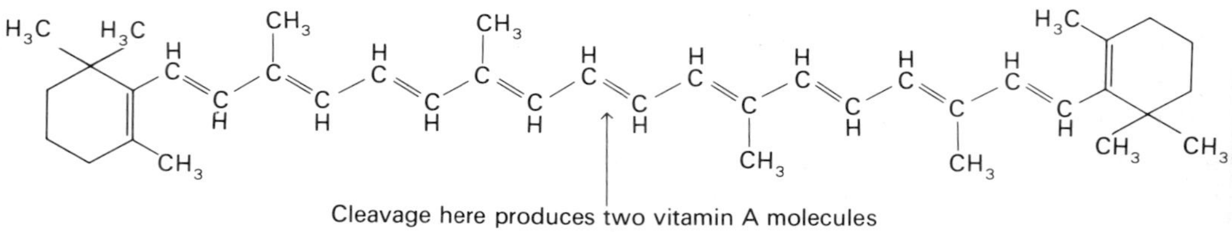

β-Carotene (provitamin A)

Figure 25-2
Vitamin A and provitamin A.

more susceptible to vitamin A deficiency than adults (who store the vitamin better) and can develop a condition called *xerophthalmia* ("bad eyes"). If not corrected, xerophthalmia results in blindness. In adults, vitamin A deficiency is first noted as night blindness. Lack of dietary vitamin A is one of the world's most serious nutritional problems, and it is estimated that at least one million cases of blindness occur every year in certain parts of the Middle East and Asia because of vitamin A deficiency.

About 1 mg of vitamin A is required daily by an adult. Continuous excessive intake of the vitamin is dangerous because it results in fragile bones and in abnormal fetal development. In the case of plant carotenoids, a dietary oversupply, e.g., eating too many carrots on a daily basis, results in *carotenosis*—a condition readily diagnosed by the yellowing of the skin.

Mode of Action in the Visual Process

The function of retinol (vitamin A_1) in the visual process was established by George Wald, who was awarded a Nobel prize in 1966 for his distinguished contribution to science. In the retina of the human eye, there are two types of photoreceptor cells—the *cone cells* and the *rod cells.* There are about three million cone cells and one billion rod cells in a human retina. Cone cells are adapted for high light intensities and are responsible for color vision; rod cells respond to low light intensities and do not sense color. Rod cells are involved in night vision, which is affected by vitamin A deficiency. The outer segment of a rod cell (Figure 25-3) is an elongated structure containing about 500 stacked, disk-like membranous vesicles (flattened sacs, each about 160 Å thick), which serve as light receptors. About 80 per cent of the protein in the membranes of these vesicles is *rhodopsin,* a conjugated protein that absorbs light (a photoreceptor molecule). Rhodopsin, which has a molecular weight of 25,000 to 35,000, is composed of the protein *opsin* and its prosthetic group, *11*-cis-*retinal* (Figure 25-4), which is covalently bonded to opsin by Schiff-base formation (p. 170) between its aldehyde group and the ϵ-amino group of a specific lysyl residue in the protein. 11-*cis*-Retinal is a derivative of vitamin A_1. In the production of 11-*cis*-retinal from vitamin A_1 (*all*-trans-*retinol*), the alcohol group of the vitamin is oxidized to an aldehyde in a reaction catalyzed by *retinol dehydrogenase* (Figure 25-4) to yield *all*-trans-*retinal,* which is then isomerized by *retinal isomerase* to produce the 11-*cis* derivative.

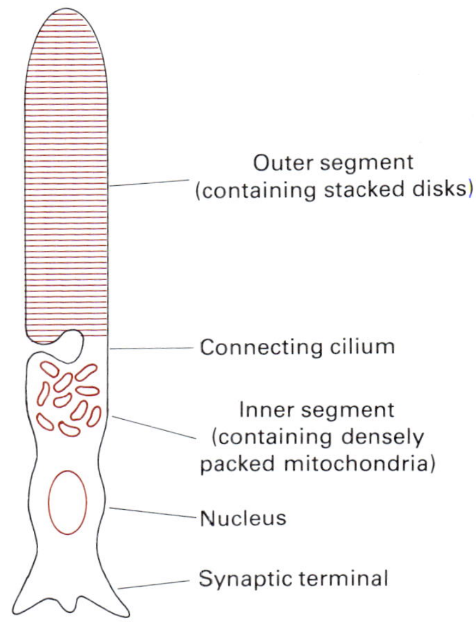

Figure 25-3
Schematic of a retinal rod cell.

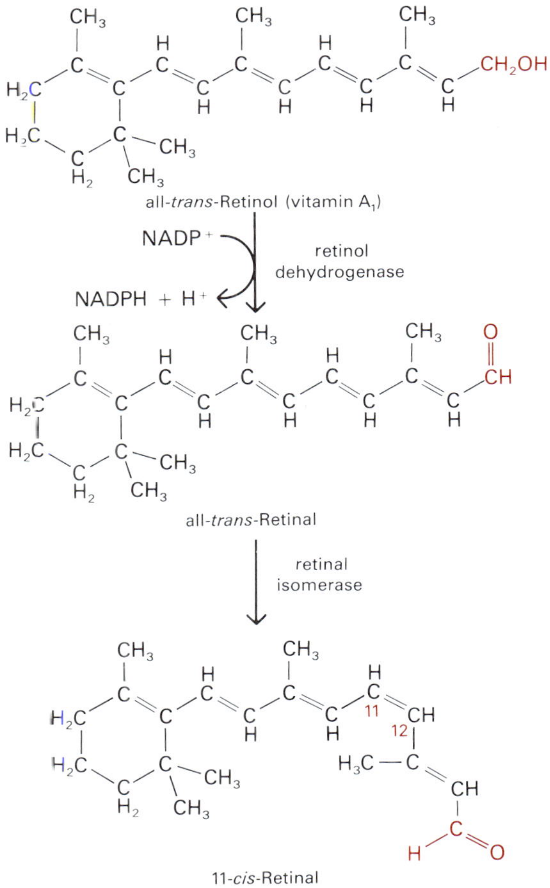

Figure 25-4
Conversion of all-*trans*-retinol into 11-*cis*-retinal.

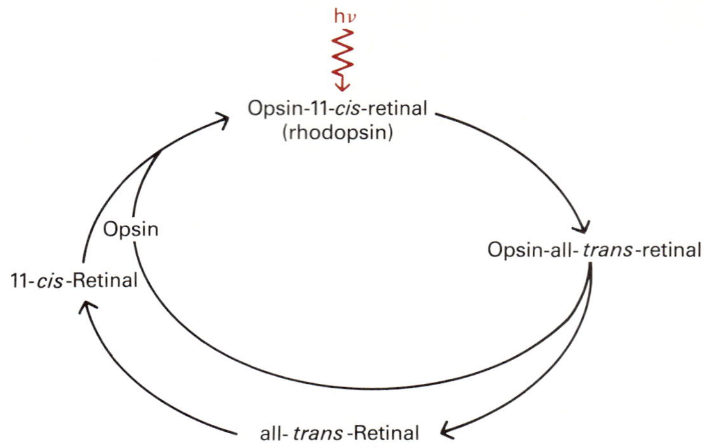

Figure 25-5
Visual cycle in rod cells.

Light absorption by rhodopsin results in a photochemical event that induces the isomerization of 11-*cis*-retinal to all-*trans*-retinal (Figure 25-5), producing *bleached rhodopsin.* This isomerization is the primary event in visual excitation. As indicated by the structures in Figure 25-4, the geometry of the retinal group is significantly altered by its transformation from the 11-*cis* to the all-*trans* isomer, and, as a consequence of this change in conformation, the Schiff-base linkage between opsin and retinal becomes destabilized. Within about a millisecond, a sequence of conformational changes occurs which terminates in the hydrolysis of the Schiff-base linkage to yield opsin and all-*trans*-retinal (Figure 25-5). It is during this series of changes that a nerve impulse is generated in the retina; this impulse initiates the molecular events that subsequently result in light being perceived by the brain.

To regenerate rhodopsin, all-*trans*-retinal is converted into 11-*cis*-retinal, which then recombines with opsin to complete the visual cycle (Figure 25-5). all-*trans*-Retinal may be reduced, by *retinal reductase,* to all-*trans*-retinol, which is then converted into 11-*cis*-retinal via the retinol dehydrogenase and retinal isomerase reactions (Figure 25-4). There is evidence, however, that all-*trans*-retinal may be directly isomerized to 11-*cis*-retinal by a photochemical event.

Vitamin D

General Information

For humans, the most important forms of *vitamin D,* which is critical for the *absorption and utilization of Ca^{2+} for bone formation,*

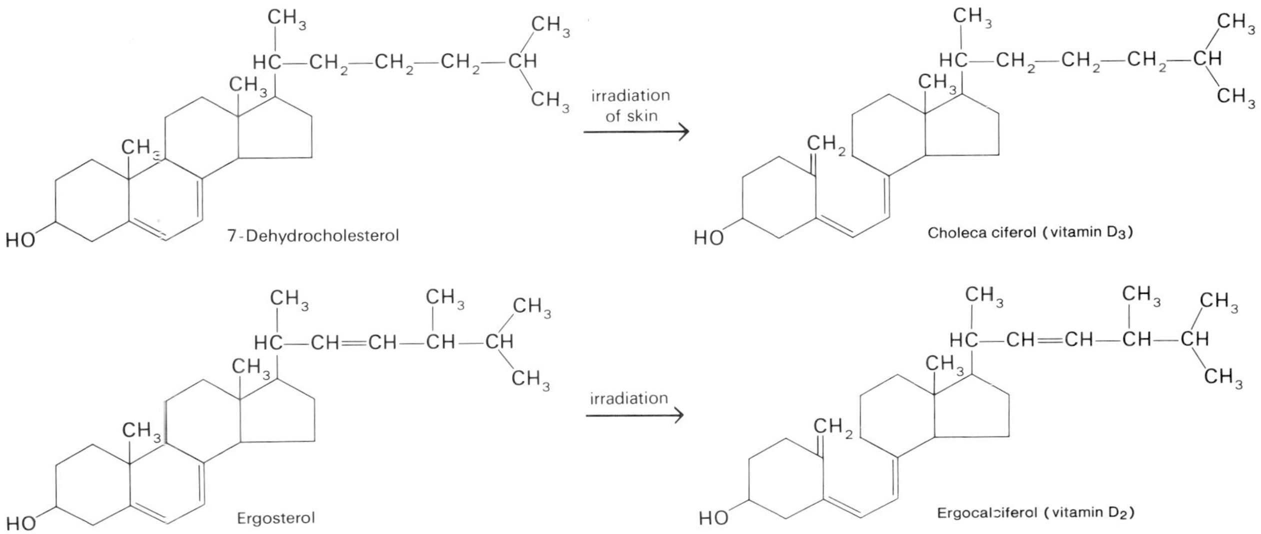

Figure 25-6
Production of cholecalciferol and ergocalciferol.

are *cholecalciferol (vitamin D_3)* and *ergocalciferol (vitamin D_2)*. Cholecalciferol is synthesized by irradiation (sunlight) of its natural precursor, 7-dehydrocholesterol (Figure 25-6), which is found in skin (p. 384). For adults, exposure to sunlight for 30 minutes a day is believed to satisfy the daily requirement (about 10 μg) of the vitamin. For children, vitamin D is vitally important to avoid malformation of bones during growth (rickets). Irradiated milk, which contains the RDA of vitamin D per quart, provides a readily available source. The vitamin D supplementation of milk (which otherwise has little or none of the vitamin) is accomplished by ultraviolet irradiation of added *ergosterol,* a yeast sterol, to produce ergocalciferol (Figure 25-6). Irradiated milk was developed as a safeguard against vitamin D deficiency in infants and children because few natural foods, excepting fish liver oil, contain the nutrient.

Excessive intake of vitamin D, like that of vitamin A, results in brittle bones, irritability, vomiting, and weight loss. British studies show that as little as six to ten times the RDA, if ingested regularly, can be toxic for small children, and twice those amounts are dangerous for adults. Because of the serious toxic effects of both vitamins A and D, self-prescription of these vitamins is precarious.

Mode of Action

In the 1970s, studies by Hector F. DeLuca, Anthony W. Norman, and Robert H. Wasserman and their associates successfully identified the biochemical processes by which cholecalciferol (vitamin

D_3) is converted into its biologically active form in the body and the mode of action of this active derivative. Cholecalciferol, following its ingestion or synthesis from 7-dehydrocholesterol in skin, is metabolized in the liver to *25-hydroxycholecalciferol* (Figure 25-7), which is then converted into *1,25-dihydroxycholecalciferol* [1,25$(OH)_2D_3$] in the kidneys. [Although the biologically active form of the vitamin is 1,25$(OH)_2D_3$, there have been some reports that other cholecalciferol derivatives, e.g., 24,25-dihydroxycholecalciferol, also have biological activity.] Cholecalciferol and its derivatives are *seco-steroids,* i.e., ring A is not rigidly fused to ring B (Figure 25-7); as steroids, they are structurally unique. Because of its conformational mobility, ring A of a seco-steroid exists in two equilibrated conformers, and this unusual structural feature of 1,25$(OH)_2D_3$ may prove to be important for the molecule's function.

Now that the *in vivo* mode of action of vitamin D is understood, it is proposed that 1,25$(OH)_2D_3$ is actually a hormone and not a vitamin. This proposal is based on the accumulated findings that the cholecalciferol derivative fits the traditional description of a hormone as a chemical messenger produced by an organ or gland and transported by the blood to its target tissue (p. 538). As depicted in Figure 25-8, 1,25$(OH)_2D_3$ is produced in the kidneys (organ) and transported by the blood to intestinal mucosa and bone (target tissues), where it functions in the processes for the absorption, reabsorption, and mobilization of calcium and phosphate ions. In its mode of action, 1,25$(OH)_2D_3$ behaves as a steroid hormone by entering its target cells and inducing the synthesis of specific proteins (p. 555). Studies reveal that, in intestinal mucosa, 1,25$(OH)_2D_3$ stimulates production of *calcium-binding protein* (CaBP), which is responsible for the increase in calcium absorption.

As indicated in Figure 25-8, production of 1,25$(OH)_2D_3$ is under the control of parathyroid hormone (PTH). The hormone modulates synthesis of *25-hydroxycholecalciferol 1-hydroxylase* (1-hydroxylase), the enzyme that catalyzes the reaction producing 1,25$(OH)_2D_3$. Production of PTH, in turn, is controlled principally by the circulating levels of serum calcium. There is an inverse relationship between the level of serum calcium and PTH production, e.g., high serum Ca^{2+} results in low PTH production. High levels of 1,25$(OH)_2D_3$ also affect the production of 1-hydroxylase by exerting a feedback control on PTH synthesis (Figure 25-8). The following examples illustrate the physiological interplay involving PTH, 1,25$(OH)_2D_3$, and serum Ca^{2+} that influences production of

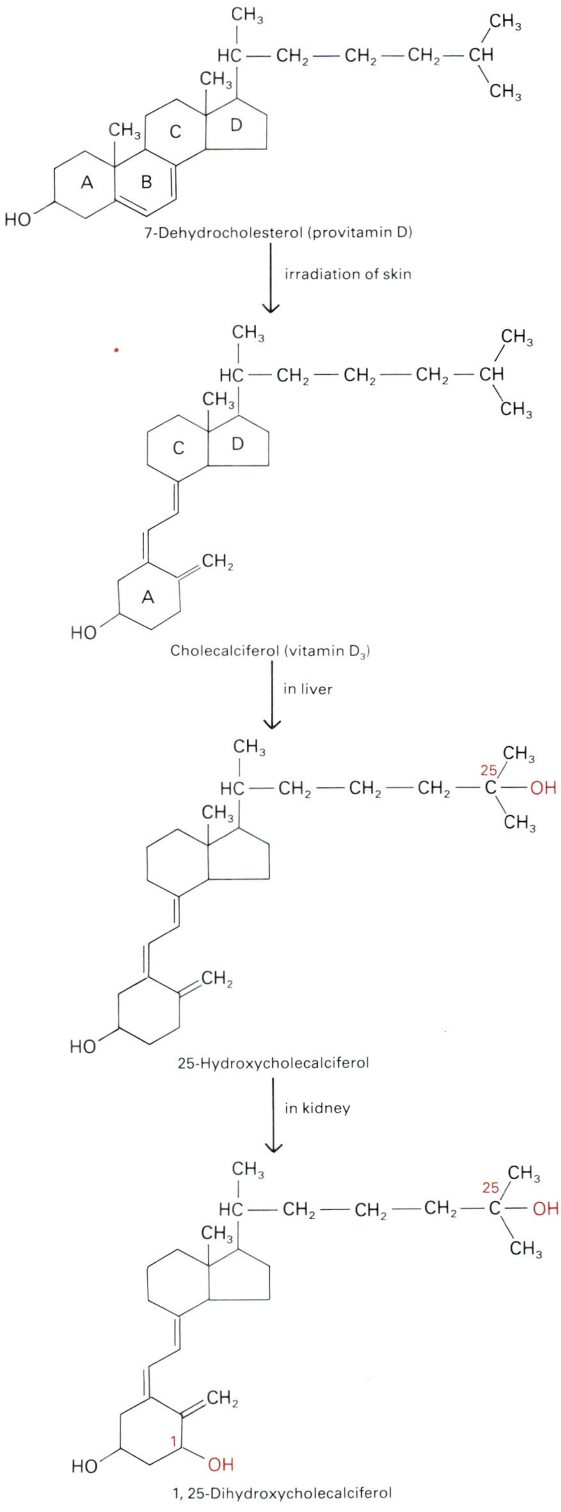

Figure 25-7
Conversion of 7-dehydrocholesterol into 1,25-dihydroxycholecalciferol.

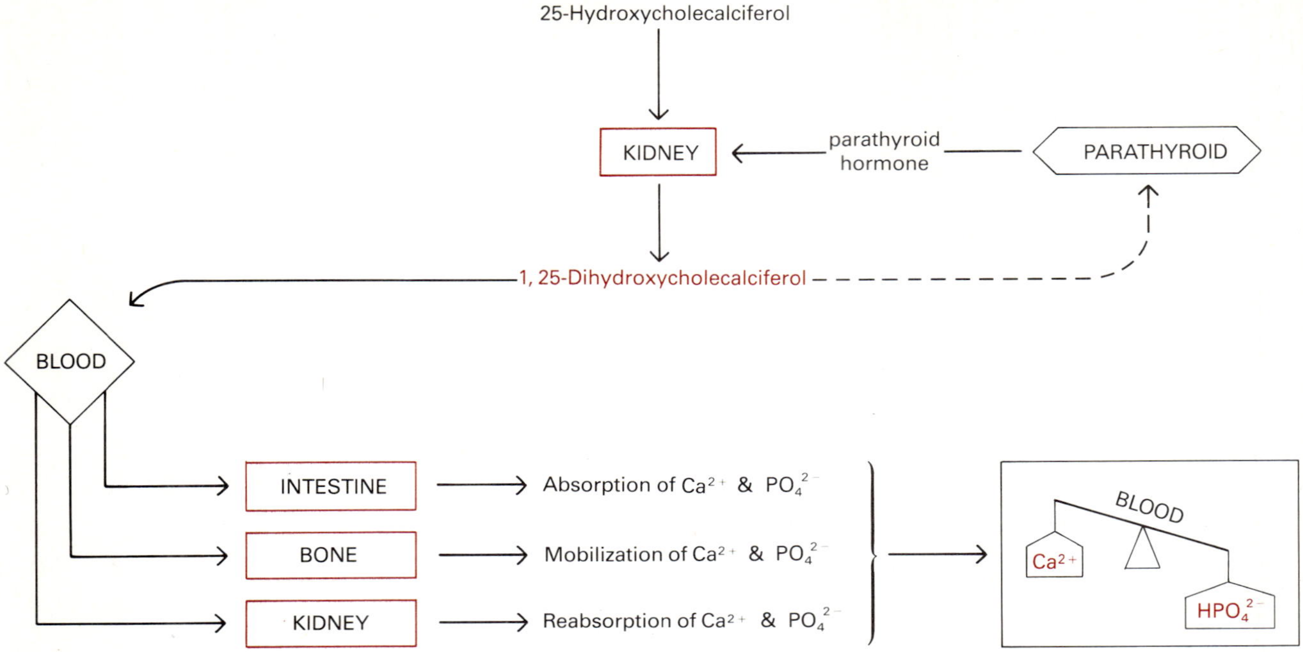

Figure 25-8
Schematic of the proposed hormonal function of vitamin D. (Adapted from A. W. Norman, J. W. Coburn, and K. Schaefer, *Contr. Nephrol.*, 14:190, 1978)

the 1-hydroxylase: cholecalciferol deficiency → $1{,}25(OH)_2D_3$ deficiency → decreased Ca^{2+} absorption → low serum Ca^{2+} → increased PTH production → increased 1-hydroxylase synthesis. Conversely the course of events would be: high level of $1{,}25(OH)_2D_3$ → increased Ca^{2+} absorption → high serum Ca^{2+} → decreased PTH production → decreased 1-hydroxylase synthesis. In the first example, low serum Ca^{2+} stimulates PTH production and, in the second example, the high serum level of both calcium and $1{,}25(OH)_2D_3$ acts on the parathyroid to decrease synthesis.

Vitamin K

General Information

Dam was on a lecture tour of the United States and Canada when the Nazis occupied his homeland, Denmark, in 1940. He spent the war years at the University of Rochester, and it was during this period that he received his Nobel prize.

Vitamin K was named for the Danish word *koagulation* by its discoverer, Henrik Dam (Nobel Prize, 1943) and, as the name implies, is a nutritional factor needed for *normal blood clotting.* Two active forms, *vitamin K_2* and *vitamin K_3* (menadione), are known (Figure 25-9). Vitamin K_2 with its long, isoprenoid side chain is believed to be the biologically active form of the vitamin. Menadione is a synthetic product. As stated previously, intestinal flora provide humans with their normal vitamin K requirements. Because vitamin

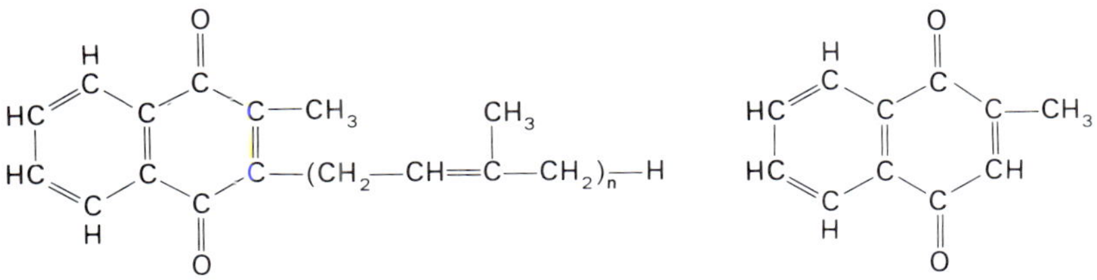

Figure 25-9
Two forms of vitamin K.

K is produced by many bacteria and most plants (neither of which has blood), it most probably has other biological functions. For example, there is evidence that vitamin K may serve as an electron carrier in some specialized routes of electron transport since, like CoQ (p. 320), it is a quinone that can be reversibly oxidized and reduced.

Mode of Action

It was only recently that the biochemical function of vitamin K in the process of blood clotting was elucidated. In the latter stages of the cascade of reactions involved in blood clotting, the vitamin (along with calcium and a phospholipid surface) is required for the conversion of *prothrombin* into *thrombin.* Prothrombin, a glycoprotein, is a zymogen (p. 147), and its cleavage by the protease *Factor X_a* (activated Factor X) produces thrombin, another protease. In the next reaction, *fibrinogen* is cleaved by thrombin to yield *fibrin,* which subsequently produces a *fibrin clot* (blood clot). Vitamin K (in its reduced form) is required for the chemical modification of prothrombin that precedes its cleavage by Factor X_a. This modification entails the γ-carboxylation of ten specific glutamyl residues located in the N-terminal region of prothrombin (Figure 25-10). The molecular details of this enzymatic reaction and of the role of vitamin K remain obscure. The carboxylations, which produce dicarboxylic glutamyl residues, increase the binding of Ca^{2+} to prothrombin (Figure 25-10) which, in turn, is required for the interaction of prothrombin (substrate) and Factor X_a

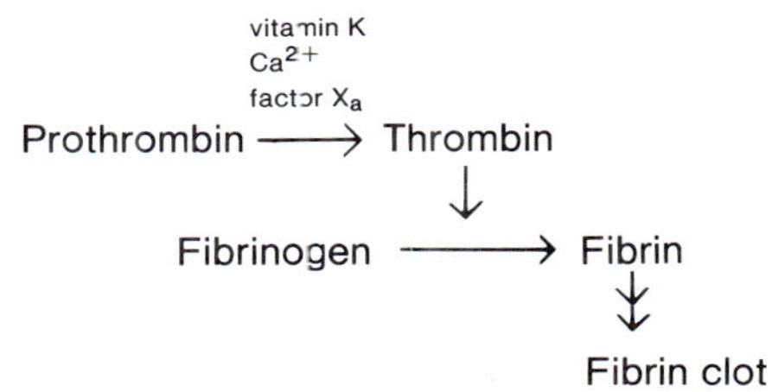

Figure 25-10
Vitamin-K-dependent γ-carboxylation of a glutamyl residue in prothrombin and subsequent binding of a calcium ion.

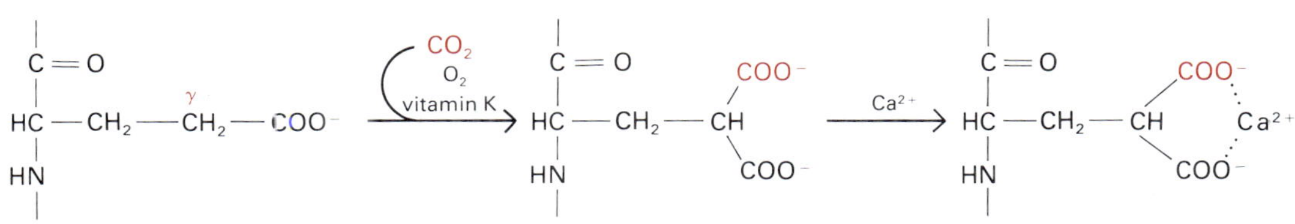

(enzyme). Thus, in this manner, vitamin K and calcium participate in the preparation of prothrombin for its conversion into thrombin. Although the biochemical function of vitamin K was first described in studies on thrombin production, it is now known that γ-carboxylation reactions, mediated by vitamin K, also occur in some of the earlier reactions of the blood clotting process, as well as in bone metabolism.

Vitamin K Antagonists

Warfarin's name is derived from the initials of the Wisconsin Alumni Research Foundation, which sponsored the research on the compound.

Two antagonists of vitamin K are *dicumarol* (dicoumarol), first isolated from moldy clover hay, and *warfarin,* the synthetic analogue of vitamin K (Figure 25-11). Both antagonists prevent blood clotting. Dicumarol is often given to patients who have suffered heart attacks caused by blood clots, as a preventive measure against further clotting in the blood vessels. Warfarin is a rodenticide which, when ingested by a rodent over a period of time, causes its death by producing internal hemorrhaging.

Vitamin E (Tocopherol)

Vitamin E was first isolated in 1922 as a factor from vegetable oils that cured infertility in male and female rats grown on a diet of milk. Because of its identification with reproduction, the vitamin was named tocopherol, from the Greek work *tokos,* meaning *to*

Figure 25-11
Two vitamin K antagonists.

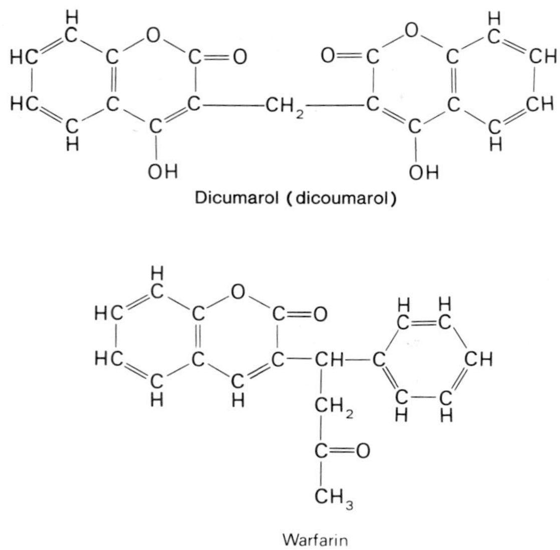

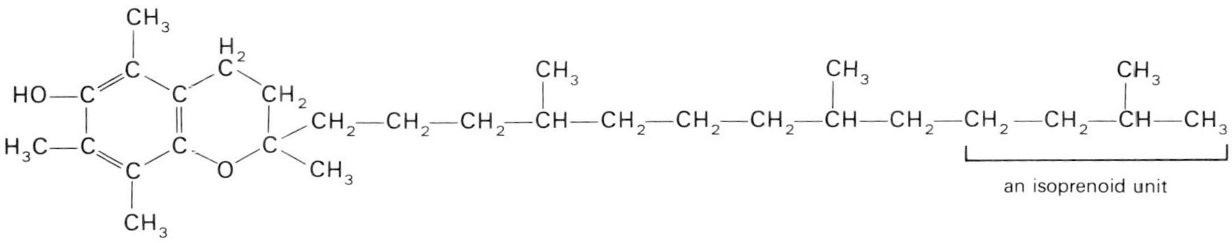

Figure 25-12
Vitamin E (*d*-α-tocopherol).

bear young. At least eight tocopherols have been identified in plants; *d*-α-tocopherol (Figure 25-12) is the most abundant and active of the known forms. Vitamin E, since its discovery, has been one of the most enigmatic nutrients with regard to human nutrition. Although deficiencies in experimental animals result in marked clinical manifestations, e.g., infertility in rats, heart damage in calves, and retarded growth in rabbits, no human deficiency has ever been observed. Tocopherol's biological function(s) in humans remains uncertain; however, because it is an antioxidant, the vitamin is now believed to protect the highly unsaturated fatty acids in cell membranes from oxidation. Because of this protective property, tocopherols are often added to commercial oily foods to prevent their oxidation (it is oxidation that makes such foods turn rancid).

No nutritionist worries about vitamin E deficiency because all diets provide adequate amounts, with vegetable oils being excellent sources. However, like vitamin C, vitamin E enjoys a mystique that, for some food enthusiasts, makes the nutrient the answer to many human medical problems. Many of the clinical deficiencies noted in experimental animals have been transposed to the human condition to exploit vitamin E's efficacy, e.g., increases sexual potency and prevents miscarriages (infertility in rats) and cures and prevents heart disease (heart damage in calves). Even though a great variety of health-giving qualities have been attributed to the vitamin, nutritional scientists would be pleased to establish even *one* unequivocal biological function for the nutrient. The current lack of a well-defined biological role(s) for vitamins C and E, such as are known for the other water- and fat-soluble vitamins, appears to be an invitation for some nonscientific speculators to propose and propagate their ideas (often profitably) about the functions of these two nutrients. Because vitamin E, unlike vitamin C, is fat soluble, there is always the potential danger of hypervitaminosis.

MINERAL REQUIREMENTS

The other broad class of micronutrients is the *macro-* and *micro-minerals* (Table 25-4). The term *mineral* was used in the nineteenth century to describe the inorganic nutrient (originally thought to be a single requirement) in the ash remaining after combustion of animal and plant materials. Today, a *nutritional mineral* is generally defined as an *inorganic substance (atomic element or ion) that is essential in small quantities for life processes.* Although a variety of minerals are required by animals, some are needed in such small amounts that it is proving difficult to establish their requirement in humans. Cobalt is an example of an essential nutritional factor required in very minute amounts. Americans obtain cobalt (in the form of vitamin B_{12}) from animal products. Because plants and cereals lack vitamin B_{12}, vegetarians face a possibility of developing a deficiency (called ''vegetarian back''); however, by drinking one or two glasses of milk daily, enough vitamin B_{12} is consumed to prevent a deficiency.

Macrominerals

The *macrominerals,* those elements required in the largest amounts, are *calcium, phosphorus, magnesium, sodium, potassium,* and *chloride.* In the U.S. RDA listings (Table 25-1), there are no recommended allowances for sodium, potassium, or chloride because these three minerals are so abundant in foods, as NaCl and KCl, that diets furnish excessive amounts. The requirements for phosphorus (800 mg) and magnesium (300 to 350 mg) are also easily met with a normal diet. Deficiencies of phosphorus are essentially unknown, and the average American diet has about 210 mg of magnesium per 1,000 calories consumed.

Calcium, because it is needed for *formation and maintenance of bones and teeth,* is the macromineral that receives the most atten-

Table 25-4
Minerals in human nutrition

Macrominerals	Calcium, chloride, magnesium, phosphorus, potassium, and sodium
Microminerals (trace elements)	
Requirement established:	Chromium, copper, fluoride, iodine, iron, manganese, molybdenum, selenium, and zinc
Requirement postulated:	Nickel, silicon, tin, and vanadium

tion. The U.S. RDA for calcium (800 mg) is about twice that recommended by the United Nations and Canada. Some American nutritionists think that the U.S. requirement should be made more realistic by lowering the RDA, since calcium is not obtainable in quantity from most foods other than milk and cheese. In fact, milk and milk products are listed as one of the basic four food groups (p. 591) because of their calcium content. The concern of nutritionists is that, since the U.S. RDA for calcium is not currently being met by many Americans (who otherwise eat diets that meet RDA standards), worry and alarm may arise over imagined deficiencies. It should be stressed, however, that appropriate calcium intake is an important nutritional consideration for growing children, pregnant women, nursing mothers, and also older people, who often do not eat large quantities of food.

A metabolic role of bones that is not often recognized and that is of nutritional importance is their capacity to store minerals such as calcium, phosphorus, and magnesium. If dietary sources of these three macrominerals are temporarily low, the body is able to draw upon reserves present in bone material. These reserves, however, are not limitless and need to be replaced to avoid defective bone metabolism.

Microminerals (Trace Elements)

The *microminerals* are *trace elements* which, like vitamins, are required by animals in amounts that range from milligram quantities to a very small fraction of a microgram. Among the trace elements, only some are known to be required by humans—those established either by nutritional criteria or by biochemical evidence. The latter group, not identified by deficiencies, includes *chromium* (required for glucose metabolism), *manganese* (important for bone formation), *selenium* (part of glutathione peroxidase structure), and *molybdenum* (enzymatic cofactor). Other trace elements are required by experimental animals and are suspected of being human nutrients but, as yet, have not been proved essential. The microminerals of uncertain nutritional status include *nickel, tin, vanadium,* and *silicon.*

The five nutritionally established microminerals are *iron, zinc, copper, iodine,* and *fluoride;* a U.S. RDA is set for all except fluoride (Table 25-1). Most natural supplies of drinking water furnish the fluoride needed daily, and those that do not can be *fluoridated,* often to an extent that does not exceed natural fluoridation. Fluoride is important in *preventing tooth decay in children* because, when fluoride is available for formation of teeth, calcium is deposited and retained better in the relatively soft teeth of a child. Fluoride toothpastes and/or fluoride treatments at a dentist's office are additional means of supplying the nutrient to children.

The other halogen nutrient, *iodine (iodide),* still remains a nutritional concern because deficiencies of this nutrient are not yet eradicated in the United States. A dietary deficiency of iodine, required by the thyroid gland, results in *endemic goiter* (swelling of the neck because of an enlarged thyroid gland). It is ironic that although iodized salt, which provides good protection against iodine deficiency, has been available for years, goiters are still a medical problem in some parts of the country. Unfortunately, many food processors do not use iodized salt and often schools and food chains do not buy the supplemented salt. It has been calculated that for a penny or less per year per person, goiters could be permanently eradicated in the United States.

There are no immediate concerns about copper and zinc in the American diet because animal proteins are good sources of zinc and certain vegetables (potatoes, dried peas, and beans) and grains contain copper. Some evidence has appeared recently, though, that the RDA of 3 mg for copper may be marginal for humans and that some diets may not be meeting the requirement.

Iron, important for oxygen transport (hemoglobin) and other critical biological processes (e.g., electron transport), is a nutrient that currently concerns nutritionists. Although a diet may have an adequate iron content, a deficiency can occur because of the poor intestinal absorption of the mineral (about 10 per cent). This absorption factor in iron utilization is receiving serious attention because recent studies reveal that women of menstrual age and children (but not adult males or children under six) have an iron shortage (not a deficiency). The RDA for women is set at 18 mg, taking into account that adolescent to middle-aged women need more iron than adult males (10 mg). The Food and Nutrition Board now recommends that women take an iron supplement. However, because the dietary intake of iron by women with shortages ranges from 8 to 12 mg daily, large amounts of supplementary iron are not necessary and, in fact, not recommended because of the constipation problem they create.

CURRENT NUTRITIONAL STATUS OF AMERICAN DIETS

How well do the eating habits of Americans fulfill their need for the approximately 50 nutrients? The concensus among nutritionists is that the daily intake of *calcium, riboflavin, vitamin A, vitamin* B_6, and *folic acid* may be marginal for many Americans. These suspicions are not surprising since, in recent years, Americans have tended to drink less milk (very good source of calcium and riboflavin) and to eat fewer foods that are good dietary sources of many vitamins. With regard to B-vitamins, it is difficult to obtain all the daily requirements from a limited number of dietary items because, although found in many foods, they are seldom present in large amounts. Whereas nutrients like calcium and vitamin A can be obtained easily from milk and carrots, respectively (i.e., a single dietary source), B-vitamin requirements most often are collectively met by a combination of foods.

Any nutritional concern about possible marginal intakes of the five nutrients mentioned above can be corrected by including more milk and dark green, leafy vegetables (good sources of vitamin A, vitamin B_6, folic acid, and riboflavin) in a diet. The dark green color of a vegetable often indicates a good source of vitamin A; for example, one cup each of romaine (dark green) and iceberg (light green) lettuce have 0.21 and 0.036 mg RE of vitamin A, respectively. One stalk of broccoli and one cup of boiled spinach, both dark green vegetables, have 0.90 and 2.92 mg RE of vitamin A, respectively. The same amounts of broccoli and spinach have 0.36 and 0.25 mg of riboflavin, respectively (a cup of green beans and a good-sized baked potato have 0.11 and 0.07 mg, respectively). As for folic acid (first isolated from spinach leaves), leafy vegetables are a primary dietary source.

FOUR BASIC FOOD GROUPS AND BALANCED DIETS

Based on the nutrient content of individual foods and the nutritional requirements of humans, *four basic food groups* (Table 25-

Table 25-5
Basic four food groups

1. Milk group	Two or more glasses of milk or servings of cheese, ice cream, or other milk products daily
2. Meat group	Two or more daily servings of meat, fish, poultry, and/or eggs—with dry beans, peas, or nuts as alternates
3. Vegetables and fruits	Four or more daily servings of vegetables from a group comprising dark green or yellow vegetables, citrus fruit, tomatoes, etc.
4. Bread and cereals	Four or more daily servings of either enriched or whole grain products

5) are now recommended for daily inclusion in a diet. The groupings, called the *Basic Four,* are designed as a guide to what foods and in what quantities each should be eaten daily to obtain a balanced nutrition. The caloric content of the group requirements, based on standard servings, is estimated to be from 1,250 to 1,750 Cal and, therefore, is not expected to furnish all dietary calories or to account for a person's total daily intake of food. Part of the bread-and-cereal requirement, for example, may be acquired in the form of sweet rolls, cookies, or pies, each having a high sugar content.

There has been some criticism that the Basic Four is too abbreviated for a nutritional plan, and that eating items of the four groupings in proper quantities daily is no guarantee of proper nutrition. This criticism stems from the fact that it is possible to select narrowly from each of the groupings and not obtain a balanced diet. It is assumed, however, that day-to-day dietary selections vary enough to provide a good representation of each of the four groups over a period of time. The Basic Four, in spite of criticisms by some professionals, is a valuable general guide to those who have little or no knowledge of nutrition.

A standard nutritional maxim is that *no normal person who eats a balanced diet has a need for vitamin supplements,* in contrast to claims by some about the extra need of a particular vitamin or group of vitamins. One grave concern to nutritionists, however, is faddish interests in one particular nutrient (e.g., vitamin E or "muscle-building protein") because overemphasis of a single (or group of) nutrient(s) de-emphasizes the essentiality of a wide variety of other dietary factors. For proper nutrition, adequate amounts of *all* nutrients are needed; excesses of some do not substitute for the need of others. Contrary to some claims, there is no one "perfect food" or "perfect nutrient." Another standard nutritional truth is that *the greater the variety of food items in a diet, the greater are the chances of obtaining a balanced diet.* Nutritionists would also warn a person who is trying to lose weight by lowering caloric intake to be certain that the restricted diet is not also lowering the intake of essential nutrients. In low-calorie diets, vitamin and mineral supplements are often recommended as a guarantee that these requirements are being met. Although debates about nutrition and proper diets will undoubtedly continue between nutritionists and food enthusiasts, an individual who has a good basic understanding of nutrition and common sense should feel confident about his or her diet and should not be confused by the many claims constantly being made.

SUMMARY

The nutrient requirements of humans are classified as macronutrients and micronutrients. Macronutrients are biomolecules required in gram quantities, whereas micronutrients are biomolecules or minerals needed in milli- or microgram quantities. The macronutrients, which serve as the primary source of calories, include carbohydrates, proteins, and lipids. Proteins also provide the nine essential amino acids (histidine, isoleucine, leucine, lysine, methionine, phenylalanine, threonine, tryptophan, and valine). Lipids serve as the source of the three essential fatty acids (linoleic, linolenic, and arachidonic acids).

Micronutrients include the vitamins and minerals. With regard to the vitamins, there are two categories—water soluble and fat soluble. The water-soluble vitamins include the B-complex, which includes biotin, thiamin (B_1), pyridoxine (B_6), riboflavin (B_2), niacinamide, pantothenic acid, folic acid, and vitamin B_{12}. Another required water-soluble vitamin, not included in the B-complex, is vitamin C. The fat-soluble vitamins are A, D, K, and E. The mineral requirements are classified as macro- and microminerals. The macrominerals include the following six elements, which are required in quantities that range from hundreds of milligrams to grams: calcium, phosphorous, magnesium, sodium, potassium, and chloride. The micronutrients, or trace elements, required in milli- or microgram quantities, include those that are established as nutrients and those that are postulated to be required (based on evidence from studies on experimental animals). The established microminerals include chromium, copper, fluoride, iodine, iron, manganese, molybdenum, selenium, and zinc; those believed to be needed are nickel, silicon, tin, and vanadium.

In the United States, the Committee of Dietary Allowances continuously assesses all available scientific evidence about nutrition and publishes a revised guideline, called the U.S. Recommended Daily Allowance (U.S. RDA) about every five years. The RDAs for the 20 nutrients listed in the table are scientific estimates of daily requirements that include a large safety margin, i.e., they are not minimal requirements. The nutrients included in the U.S. RDA include protein, the nine water-soluble vitamins, three of the fat-soluble vitamins (A, D, and E), and seven of the minerals (calcium, phosphorus, magnesium, iron, zinc, copper and iodine).

Retinol (vitamin A_1) functions in the visual process as 11-*cis*-retinal, the prosthetic group of rhodopsin (major membrane protein of rod cells). In the primary event of visual excitation, light absorbed by rhodopsin isomerizes 11-*cis*-retinal to all-*trans*-retinal. This isomerization results in a series of conformational changes that terminate in the hydrolysis of rhodopsin to yield the protein opsin and all-*trans*-retinal. It is during this series of changes that a nerve impulse is generated in the retina and then transmitted to the brain.

The mode of action of vitamin K in blood clotting is to serve as a cofactor in reactions that bring about the γ-carboxylation of glutamyl residues of certain proteins involved in the process. In the case of prothrombin, its conversion into thrombin requires that ten glutamyl residues in the N-terminal region of the protein be γ-carboxylated (vitamin-K-dependent reaction) to produce dicarboxylic glutamyl residues, which bind Ca^{2+}. This binding of Ca^{2+} is a prerequisite for the interaction of prothrombin (substrate) with Factor X_a (a protease) to produce thrombin.

Cholecalciferol (vitamin D_3) is synthesized in the skin from 7-dehydrocholesterol and is then converted into 25-hydroxycholecalciferol in the liver. The active form of the vitamin is 1,25-dihydroxycholecalciferol, which is produced from 25-hydroxycholecalciferol in the kidney. 1,25-Dihydroxycholecalciferol functions in the processes for the absorption, reabsorption, and mobilization of calcium and phosphate ions. It is proposed that vitamin D is a hormone and not a vitamin since its active form, 1,25-dihydroxycholecalciferol, is produced in one organ (kidney) and transported by the blood to its sites of action (intestinal mucosa and bone).

QUESTIONS

In lieu of the usual end-of-the-chapter questions, the following quiz to assess general nutritional knowledge is substituted. The questions are a sampling of those that have been asked in surveys designed to evaluate nutritional concepts. Although there may be exceptions to some of the right answers listed in the back of the book, there is an established scientific basis for each answer.

True or false:

1. The chemicals added to our manufactured food take away much of its value for health.
2. Food grown with chemical fertilizers is just as healthy as food grown with natural fertilizers.
3. Chemical sprays that farmers use make our food a danger to health, even if the sprays are used carefully.
4. There is no difference in food value between food grown in poor worn-out soil and food grown in rich soil.
5. Synthetically made vitamins are just as nutritionally good as natural vitamins.
6. Much of our food has been so processed and refined that it has lost its nutritive value.
7. If people feel tired and rundown, they probably need more vitamins and minerals.
8. Older people need about the same amounts of vitamins as young people.
9. Healthy, active children benefit by taking a concentrated sweet, such as candy, for energy.
10. Nondairy cream substitutes are high in polyunsaturated fats.
11. Corn oil margarine has more polyunsaturated fatty acids than one made of predominantly hydrogenated vegetable oils.
12. Strawberries, cantaloupes, and green peppers are good sources of vitamin C.
13. Gelatin desserts are sources of good-quality protein.
14. One slice of bread contains more calories than a two-ounce serving of roast beef.
15. Toasted bread has fewer calories than untoasted bread.
16. Potato peel contains more vitamins than does the white flesh inside.
17. Potatoes are a good source of protein.
18. Eating late-night meals and snacks is more likely to cause weight gain than eating the same amounts of food earlier in the day.
19. Adults do not need the calcium in dairy products because their bones and teeth are already formed.
20. Physically active people require much more protein than do moderately active or inactive people.
21. Certain combinations of foods, such as fruits and vegetables, cancel each other's nutrient values when digested at the same time.

22. To achieve a balanced diet, you must eat food from each of the basic food groups at every meal.
23. As a source of nutrients, lettuce is a poor buy.
24. Excessive intakes of vitamin A or D are poisonous.
25. The liquid protein used for diet purposes is a good source of the essential amino acids.

SUGGESTED READINGS

Deluca, H. F., ed., *The Fat-Soluble Vitamins.* New York: Plenum Publishing, 1978. An authoritative book about this class of vitamins.

DeLuca, H. F., *Vitamin D: Metabolism and Functions.* New York: Springer-Verlag, 1976. Written by the scientist who has contributed greatly to our current knowledge about the vitamin.

Deutsch, R. M., *Realities of Nutrition.* Palo Alto, Calif.: Bull Publishing, 1976. An excellent text on nutrition for the layperson interested in the subject.

Fleck, H., *Introduction to Nutrition,* 3rd ed. New York: Macmillan, 1976. A good basic textbook on the subject.

Lowenberg, M. E., E. N. Todhunter, E. D. Wilson, J. R. Savage, and J. L. Lubawski, *Food and People,* 3rd ed. New York: Wiley, 1979. Interesting reading about how food is used by people and about the political and economic aspects of food and its production.

Pryde, L. T., *Chemistry of Pesticides, Foods and Drugs.* Menlo Park, Calif.: Cummings Publishing, 1973. An introductory textbook summarizing information about foods, additives, use of pesticides, and other topics of recent interest.

Robinson, C. H., and E. S. Weigley, *Fundamentals of Normal Nutrition,* 3rd ed. New York: Macmillan, 1978. Another good basic textbook.

Runyan, T. J., *Nutrition for Today.* New York: Harper and Row, 1976. A broad coverage of nutrition, with topics ranging from the nutritional status of an individual to that of the world.

Williams, S. R., *Essentials of Nutrition and Diet Therapy,* 2nd ed. St. Louis: C. V. Mosby, 1978. A recommended paperback on nutrition.

Winick, M., *Nutrition in Health and Disease.* New York: Wiley, 1980. A nutrition textbook designed for premedical and other health profession students.

Articles

Boerma, A. H., "A World Agricultural Plan." *Scientific American,* 223(2):54 (offprint #1186), 1970.

Brown, L. R., "Human Food Production as a Process in the Biosphere." *Scientific American,* 223(3):160 (offprint #1196), 1970.

Dowling, J. E., "Night Blindness." *Scientific American,* 225(2):34 (offprint #1053), 1971.

Felig, P., "Amino Acid Metabolism in Man." *Ann. Rev. Biochem.,* 44:933, 1975.

Fernstrom, J. D., and R. J. Wurtman, "Nutrition and the Brain." *Scientific American,* 230(2):84 (offprint # 1291), 1974.

Harpstead, D. D., "High-Lysine Corn." *Scientific American,* 225(2):34 (offprint #1229), 1971.

Hauschka, P. V., J. B. Lian, and P. M. Gallop, "Vitamin K and Mineralization." *TIBS,* 3:75, 1978.

Loomis, W. F., "Rickets." *Scientific American,* 223(6):76 (offprint #1207), 1970.

Nishikimi, M., and S. Udenfriend, "Scurvy as an Inborn Error of Ascorbic Acid Biosynthesis." *TIBS,* 2:111, 1977.

Norman, A. W., and H. L. Henry, "Vitamin D to 1,25-Dihydroxycholecalciferol: Evolution of a Steroid Hormone." *TIBS,* 4:14, 1979.

Pirie, N. W., "Orthodox and Unorthodox Methods of Meeting World Food Problems." *Scientific American,* 216(2):27 (offprint #1068), 1967.

Scrimshaw, N., "Food." *Scientific American,* 209(3):72 (offprint #1153), 1963.

Stenflo, J., "Vitamin-K-Dependent Carboxylation of Blood Coagulation Proteins." *TIBS,* 1:256, 1976.

Stenflo, J., and J. W. Suttie, "Vitamin-K-Dependent Formation of γ-Carboxyglutamic Acid." *Ann. Rev. Biochem.,* 46:157, 1977.

Young, V. R., and N. S. Scrimshaw, "The Physiology of Starvation." *Scientific American,* 225(4):14 (offprint # 1232), 1971.

RECOMBINANT DNA RESEARCH

26

INTRODUCTION

When biological research of the twentieth century is assessed by future science historians, there is little doubt that the development of recombinant DNA research will rank among the most influential, as well as among the most controversial, of scientific contributions. The historical accounts of the birth of this area of research should be especially interesting in view of the wide attention it received soon after it was announced in 1972. In science it was readily considered by many to be "the technique that would revolutionize biology." Additionally, an equally impressive claim was soon forthcoming by some that the industrial applications of recombinant DNA research would generate its own "economic boom in the 1980s." The worlds of science and finance were not the only segments of society conscious of the new technology. On the negative side, serious concerns were voiced about the potential hazards the novel research might bring about—concerns that eventually earned the attention of the U.S. Congress. Thus, unlike the unheralded introductions of most scientific techniques, that of recombinant DNA research received such immediate and broad attention that it became a *cause célèbre.* The 1970s may well be remembered as the decade of the recombinant DNA debate.

When the topic of recombinant DNA research was introduced on p. 453, the discussion centered on the construction of a recombinant DNA molecule and its establishment in an *E. coli* cell. What were not discussed were the research potentials that were quickly perceived and that served as the seeds for the debate mentioned above. A few of these potentials, as envisioned in the early 1970s, are recalled below.

From a positive viewpoint, the new research offered scientists the long-awaited means by which to study the structure and function of eucaryotic genes, which previously were not amenable to molecular analysis. It would now be possible, for example, to isolate, amplify, and analyze chromosomal fragments containing particular genes, i.e., those responsible for rRNAs, antibodies, or hemoglobins (p. 608). Another possible beneficial contribution would be the utilization of bacteria containing recombinant DNA molecules as "microfactories" for the production of hormones or rare blood factors (a possibility that rapidly became a reality, p. 613).

Another positive use envisioned for the research was the construction of bacterial strains that could clean up oil spills by degrading the organic components of crude oil. Also proposed was the incorporation of nitrogen-fixing (*nif*) genes into the genomes of nonlegumes, e.g., corn and wheat, so that these valuable food sources could fix their own nitrogen and not need fertilizers. These two possibilities received much attention from both the press and scientists.

However, as scenarios were being proposed emphasizing the beneficial potential of the research, others were being brought forth about its hazardous potentials. For example, because *E. coli* is a normal inhabitant of the lower intestine of humans and many animals, accidental release of *E. coli* harboring potentially hazardous recombinant DNA molecules into or beyond a laboratory environment could prove to be dangerous to humans and/or animals. This concern was expressed in scenarios that envisioned recombinant DNA molecules possessing oncogenic viral DNA or genes that produce bacterial toxins, e.g., diphtheria or tetanus toxins. (It should be noted that subsequent research established that the *E. coli* strains used in the research do not colonize in the human gut; indeed, the strains mandated for the research are so enfeebled by mutations that they cannot survive outside their immediate laboratory environment.) It was also imagined that bacterial strains could be constructed which would harbor not only genes to produce toxins but also genes to make the bacteria resistant to antibiotics, i.e., an ideal agent for bacteriological warfare.

Thus, because recombinant DNA research was a new, powerful technology of unprecedented capabilities, the hypothetical pro-and-con proposals about its future uses soon became the focal point for the debate. The research became a much-touted example of scientific knowledge that could be used for good or harmful purposes—another notable example being nuclear power. Because of the emerging Jekyll-Hyde image of recombinant DNA research, the National Academy of Sciences undertook, in 1974, an analysis and assessment of the overall implications of the new research. One result of the evaluation was the adoption by the National Institutes of Health of a set of guidelines pertaining to the physical and biological containment measures that must be taken by all federally funded scientists conducting recombinant DNA research.

The biohazard symbol, which is displayed to identify equipment, containers, rooms, and experimental animals used in recombinant DNA research.

Although concerns about the potential dangers persisted until about 1977, most of the initial fears about the research had essentially subsided by the end of the decade. Several factors accounted for this dissipation of concern. For example, as more recombinant DNA research was undertaken by scientists, it became increasingly apparent that the technique, as utilized under the guidelines, does not pose any hazardous problems or create dangerous situations. (The continued relaxing of the original guidelines reflects the comparable relaxing of the original concerns.) Also, in view of the dangers envisioned by critics, authorized attempts were made to test the validity of some of the most dire predictions. For example, DNA of the polyoma virus (a rodent oncogenic virus) was inserted into two different vectors, a plasmid and a bacteriophage (p. 455), and the two different types of recombinant DNA molecules were independently established in *E. coli.* The bacteria harboring the molecules were then injected into and also fed to mice (which are very susceptible to the virus). The results of these studies revealed that, contrary to prediction, the mice did not develop polyoma-induced tumors when the viral DNA was introduced as an insert of a recombinant DNA molecule. (Feeding and injecting mice with the recombinant DNA molecule itself also did not produce tumors.) Thus, the negative results obtained when such potentially dangerous conditions were deliberately set up aided greatly in allaying concerns about the research.

In the 1980s, wide interest in recombinant DNA research continues. However, in this decade, the major emphasis is on the scientific and economic rewards this research and related biotechnology have already produced or continue to promise. Hence, within the first ten years of its existence, recombinant DNA research underwent a thorough scrutiny by both the public and science that transformed its image. Originally viewed by critics as a potential threat to society, the research is now generally spoken of as an exceedingly potent and valuable advance in modern science. The remainder of the chapter presents a few of the reasons for the interest and acclaim recombinant DNA research is currently receiving.

CONSTRUCTION OF RECOMBINANT DNA MOLECULES

Figure 21-17 on p. 454 depicts a version of one of the original protocols for the construction of a recombinant DNA molecule with eucaryotic genomic DNA as the insert. Since the early 1970s other procedures have been developed and, because of the enormous scientific interest in this area of research, new methods for constructing recombinant DNA molecules continue to be reported, e.g., use of single-stranded viruses as vectors (p. 620). The two methods described in this section are ones that rely not on the insertion of eucaryotic DNA *per se* but, rather, on the incorporation of the nucleotide sequence containing the coding information for the protein to be produced. The particular value of these two procedures will become apparent in the section on intervening sequences.

Cloning of Complementary (Copy) DNA (cDNA)

One protocol developed for the construction of a recombinant DNA molecule is illustrated in Figure 26-1. The experimental strategy makes use of the messenger RNA (mRNA) of the gene to be cloned. The mRNA with its polyA tail (p. 497) is used as a substrate for *reverse transcriptase* (p. 438). As noted, the enzyme, after synthesizing a strand of DNA complementary to the mRNA, called *complementary (copy) DNA (cDNA),* continues synthesis by using the cDNA as a template. The resultant product, therefore, is a double-stranded cDNA molecule *(ds cDNA)* containing the coding information of the gene that produced the original mRNA template. Note that the ds cDNA molecule is synthesized with a single-stranded (ss) loop in its structure. To remove this loop, the ds cDNA is treated with S1 nuclease (which degrades ss DNA), creating a 3′- and a 5′-terminus. The ds cDNA is now ready to be prepared for its insertion into the vector.

The plasmid vector (p. 453) into which the cDNA will be inserted is the commonly used *E. coli* plasmid pBR322. This plasmid is a small, circular double-stranded DNA molecule (4,362 base pairs and MW of 2.6×10^6) bearing two genes for antibiotic resistance, ampicillin (*amp*^{r}) and tetracycline (*tet*^{r}). Since the plas-

Figure 26-1
A strategy for cloning cDNA. (Adapted from the protocol used to close mouse seminal vesicle genes activated by testosterone; courtesy of Dr. Steven Harris)

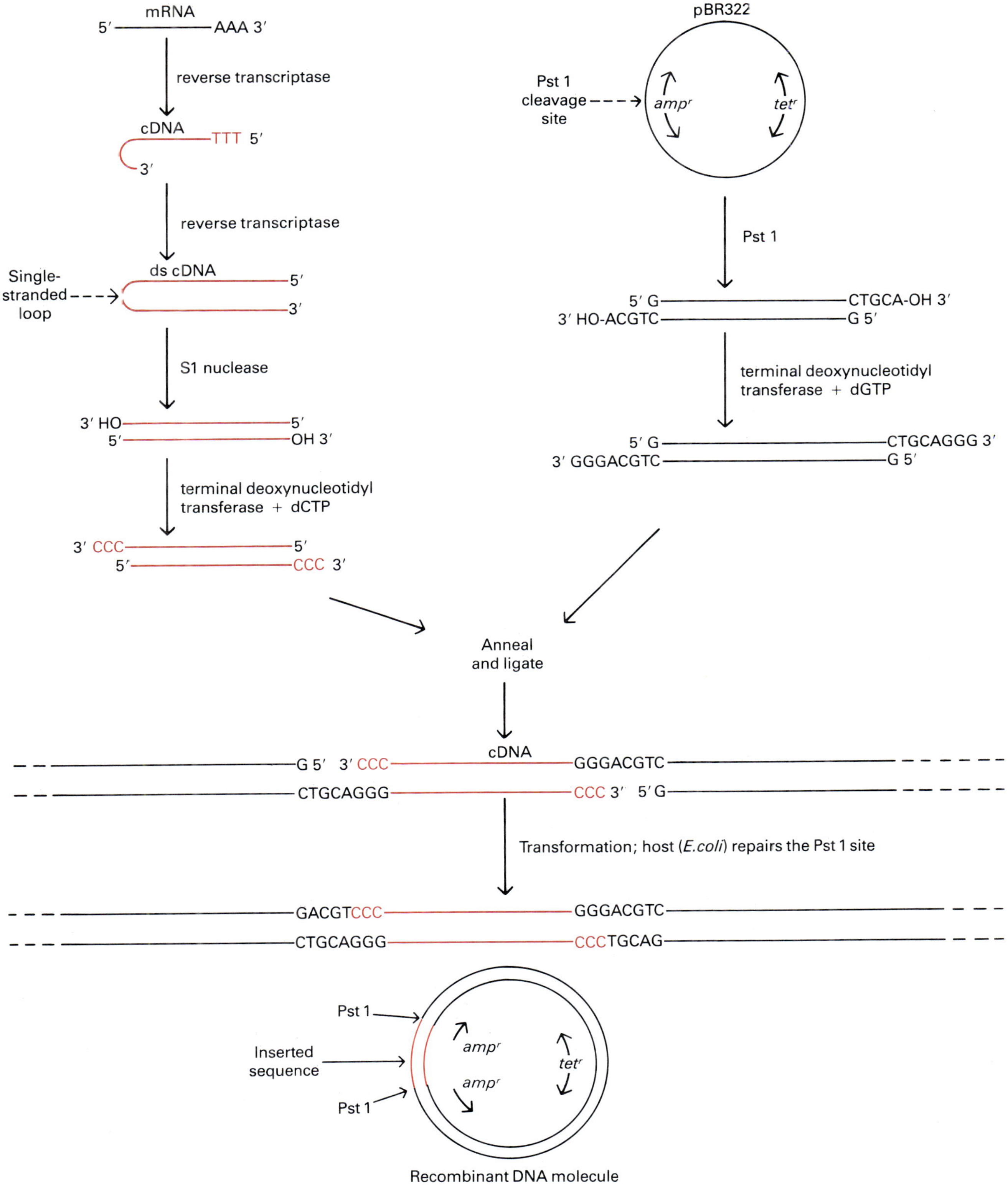

mRNA
5′ AAA 3′
reverse transcriptase
cDNA
TTT 5′
3′
reverse transcriptase
ds cDNA
5′
3′
Single-stranded loop
S1 nuclease
3′ HO 5′
5′ OH 3′
terminal deoxynucleotidyl transferase + dCTP
3′ CCC 5′
5′ CCC 3′
pBR322
Pst 1 cleavage site
amp^r
tet^r
Pst 1
5′ G CTGCA-OH 3′
3′ HO-ACGTC G 5′
terminal deoxynucleotidyl transferase + dGTP
5′ G CTGCAGGG 3′
3′ GGGACGTC G 5′
Anneal and ligate
cDNA
G 5′ 3′ CCC GGGACGTC
CTGCAGGG CCC 3′ 5′ G
Transformation; host (E.coli) repairs the Pst 1 site
GACGTCCC GGGACGTC
CTGCAGGG CCCTGCAG
Pst 1
Inserted sequence
Pst 1
amp^r
amp^r
tet^r
Recombinant DNA molecule

mid has a single cleavage site for the restriction endonuclease Pst 1, cleavage by the enzyme yields a linear double-stranded DNA molecule (Figure 26-1). The endonuclease recognizes the following palindromic sequence and cleaves at the sites indicated:

$$5'\text{—C—T—G—C—A}\overset{\downarrow}{\text{—}}\text{G—}3'$$
$$3'\text{—G}\underset{\uparrow}{\text{—}}\text{A—C—G—T—C—}5'$$

Note in Figure 26-1 that cleavage occurs within the amp^r gene, i.e., the antibiotic-resistance gene is split.

At this stage of the protocol, annealing of the cDNA and cleaved pBR322 cannot proceed because the ds cDNA lacks the single-stranded 3′-termini required for base pairing with the pBR322 molecule. To remedy the situation, advantage is taken of *terminal deoxynucleotidyl transferase* (p. 462), which adds deoxyribonucleotides onto the 3′-termini of DNA molecules. Thus, the enzyme is used to add a short sequence of deoxycytidine nucleotides (dCMPs) to the 3′-ends of the cDNA; complementary sequences of deoxyguanosine nucleotides (dGMPs) are also added to the 3′-ends of the linear pBR322 molecule (Figure 26-1). The two DNA molecules, now possessing 3′-termini complementary to each other, are annealed and ligated (p. 447). As depicted, the two single-stranded 3′-termini produced by Pst 1 cleavage of pBR322 remain unpaired in the recombinant molecule, i.e., the molecule has two single-stranded gaps. These gaps can be repaired by *in vitro* techniques or, more simply, by the repair system of the host *E. coli* after the recombinant DNA molecule has been taken up (p. 455).

Cleavage within the amp^r gene of pBR322 by Pst 1 is a significant feature of the experimental strategy because the antibiotic-resistance genes of the plasmid are used as *genetic markers* in screening for *E. coli* cells possessing the recombinant DNA molecule. Cells lacking the plasmid are sensitive to both ampicillin and tetracycline, i.e., do not harbor the amp^r and tet^r genes. However, those containing pBR322 with the ds cDNA insert are ampicillin sensitive (because of the split amp^r gene) and tetracycline resistant (intact tet^r gene). Thus, in the design of construction strategies, use is made of antibiotic-resistance genes (which are carried by many plasmids) to facilitate screening for cells bearing the recombinant DNA molecule.

Cloning of Chemically Synthesized Genes

Another procedure for the production of recombinant DNA molecules (1) makes use of the advances that have been made in the

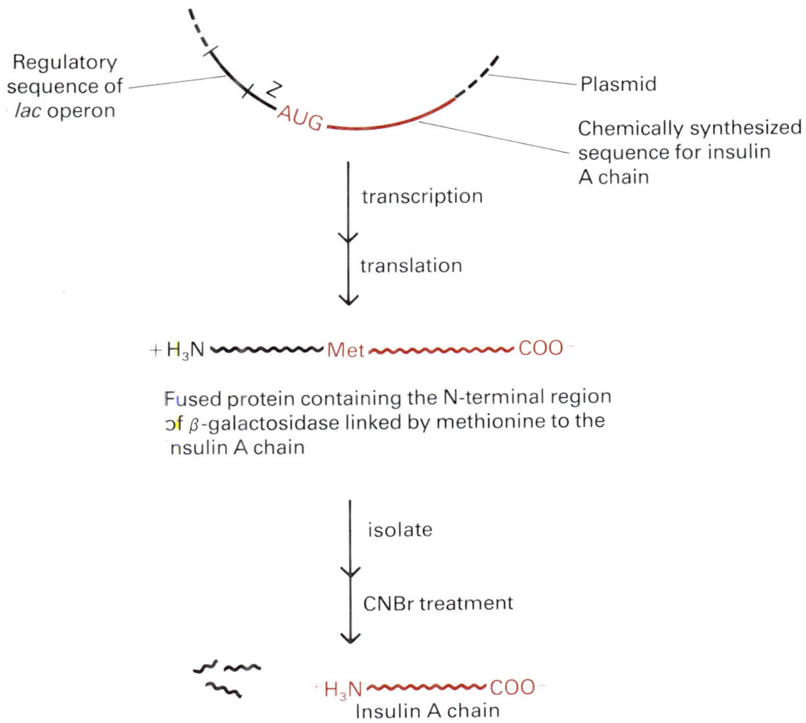

Figure 26-2
Schematic of insulin A chain production by a recombinant DNA molecule with an insert containing a portion of the *lac* operon joined to a synthetic deoxynucleotide sequence.

chemical synthesis of polydeoxynucleotides and (2) takes advantage of the molecular knowledge gained about bacterial operons. An outline of the protocol is illustrated in Figure 26-2. The recombinant DNA molecule depicted is one constructed for production of the A chain of human insulin (p. 604). Since the coding sequence of the chain is derivable from its known 21-amino-acid sequence, chemists are able to synthesize a polydeoxynucleotide containing the coding information for the chain. For reasons which will soon become apparent, the methionyl codon (AUG) is incorporated as the codon preceding those for the A chain. The synthetic sequence is then joined to a portion of the *lac* operon (p. 485) of *E. coli* containing the regulatory region and a segment of the first structural gene, Z, which specifies β-galactosidase. Thus, the insert to be incorporated into a plasmid contains a chemically synthesized coding sequence whose *in vivo* expression will be under the control of the *lac* regulatory system. (A similar insert utilizing the analogous portion of the *trp* operon, p. 494, has also been constructed.) After the recombinant DNA molecule containing the insert is constructed and established in *E. coli* (p. 455), the *fused* (or *chimeric*) *protein* produced is composed of the N-terminal sequence of β-galactosidase linked to the A chain of insulin by a methionyl residue.

That transcription of the A chain sequence is controlled as part of the *lac* operon is a clever genetic manipulation that capitalizes on the inducibility of the operon. By growing the *E. coli* harboring the recombinant DNA molecules under conditions that induce the *lac* operon, synthesis of the fused protein is greatly enhanced.

As indicated in Figure 26-2, the fused protein is isolated from *E. coli* and then cleaved *in vitro* with *cyanogen bromide* (p. 96). The cleavage step reveals the rationale for including the Met codon in the synthesized sequence. Since cyanogen bromide cleaves on the carboxyl side of methionyl residues, the treatment releases an intact A chain (which has no methionine) from the fused protein. The described procedure, therefore, is restricted to the cloning of coding sequences lacking methionine. Although the application of fused-protein production is generally limited, it is nevertheless of importance for the commercial manufacture of several small polypeptide hormones, e.g., somatostatin I and II and thymosin alpha 1 (p. 613), which have no methionine.

The B chain of insulin, whose 30-amino-acid sequence also has no methionyl residues, is produced by the same procedure. After the two independently produced chains are isolated following their release by cyanogen bromide, they are mixed and chemically combined to form the three disulfide bonds that complete the production of the insulin molecule (Figure 26-3). (It should be noted that, because of random oxidation of the sulfhydryl groups, not all the A and B chains combine to produce biologically active insulin.) Insulin production by this procedure has been successfully accom-

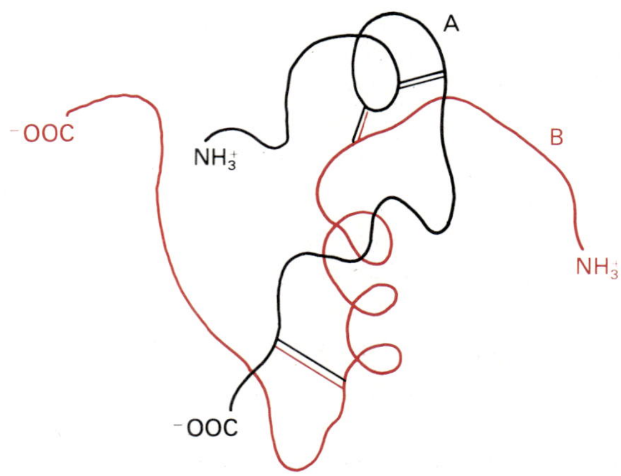

Figure 26-3
Schematic of the three-dimensional structure of human insulin. The A and B chains are depicted in black and red, respectively. The two black-red lines represent disulfide linkages between the chains, and the double black line represents the intrachain disulfide linkage of chain A.

plished, and, by 1981, clinical trials on the hormone were under way.

The protocol synthesizing fused proteins was initially developed to counteract concerns (described previously) that *E. coli* serving as "microfactories" to produce hormones could prove dangerous to humans and animals if the bacteria were accidentally released from the laboratory environment. As is apparent, independent production of the A and B chains of insulin, each as a fused protein, avoids such criticism since the hormone itself is produced under *in vitro,* and not *in vivo,* conditions. (As will be noted in the subsequent section on intervening sequences, these initial fears proved to be unwarranted.) A more pragmatic reason for the synthesis of fused proteins is that small polypeptides produced from recombinant DNA molecules are generally degraded rapidly by *E. coli* peptidases. Thus, production of small polypeptides, e.g., the small polypeptide hormones mentioned above, as fused proteins assures that adequate amounts of the intact molecules can be ultimately obtained.

EUCARYOTIC GENES

Once recombinant DNA research was under way, it rapidly began to fulfill the earlier hopes and proclamations that the new technology would furnish a wealth of important biological data. An outstanding example of the impact the research has already had on biological thinking concerns the novel and impressive information it has provided about the structure and expression of eucaryotic genes.

Intervening Sequences

A startling revelation that recombinant DNA research brought forth about the structure of most (but not all) eucaryotic genes is that their nucleotide sequences are fragmented into discrete coding regions (called *exons*) which are separated from one another by noncoding *intervening sequences (introns).* These features of a discontinuous eucaryotic gene (also referred to as a *split gene*) are schematically illustrated in Figure 26-4. Hence, the widely held assumption that the coding information of a eucaryotic gene, like that of a procaryotic gene (p. 479), is contained in a continuous sequence did not prove to be true. The discovery of this common, fundamental difference between procaryotic and eucaryotic gene

The term *exon* is derived from *expressed.*

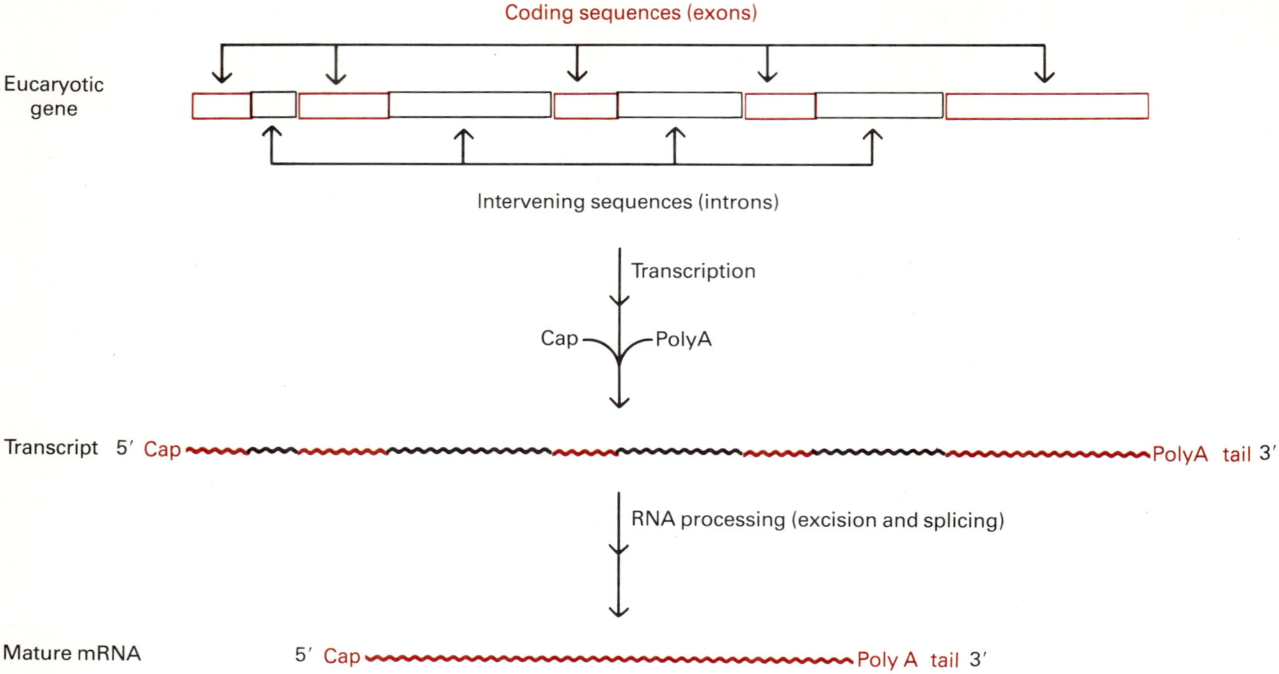

Figure 26-4
Schematic of the transcription of a fragmented eucaryotic gene and the subsequent processing of the transcript to produce a mature mRNA.

structure radically altered many preconceived ideas about the latter (which, for the most part, relied on procaryotic analogies).

How the dispersed coding information of a fragmented gene becomes the continuous coding sequence of an mRNA is another new and intriguing finding provided by recombinant DNA research. As indicated in Figure 26-4, when such a gene is transcribed, both the coding and noncoding sequences are contained in the original transcript, i.e., the entire gene is transcribed. This transcript is called *heteronuclear RNA (hn RNA).* As a eucaryotic transcript, it then undergoes ``capping'' (p. 497) and addition of a polyA tail (p. 497). What follows next is a series of events, called *RNA processing,* during which the intervening sequences are excised (removed) from the transcript and the coding sequences are spliced (joined) to produce the *mature mRNA.* Processing, like transcription, takes place in the nucleus, and only the mature mRNA is transported into the cytoplasm, where its continuous coding sequence is translated into a polypeptide structure. What biochemical and molecular events transpire during RNA processing remain a challenging enigma, one that is currently being actively investigated.

Eucaryotic Genes as Inserts

The discovery of intervening sequences occurred during the recombinant DNA debate and, for promoters of the research, provided a welcomed relief. As stated in the chapter's introduction, there was a concern that recombinant DNA molecules harboring eucaryotic genes specifying proteins, e.g., hormones, could pose problems for humans exposed to bacteria harboring such genes. When it was recognized that most eucaryotic transcripts for polypeptide synthesis require processing, it was also realized that it is highly improbable that procaryotes possess the eucaryotic enzymatic machinery for processing. Thus, the potential dangers voiced about *E. coli* harboring eucaryotic genes ceased to be a valid concern because without proper processing, procaryotic synthesis of eucaryotic proteins would be defective.

Although intervening sequences provided a finding that placated critics of recombinant DNA research, it also created a problem (faulty synthesis) for the scientists and commercial interests intending to use the technique to produce eucaryotic proteins. Solutions to the problem, however, were soon forthcoming, e.g., the two methods for constructing recombinant DNA molecules described earlier in the chapter. Inserts of cDNA prepared from mature mRNAs or of artificial genes that transcribe directly into mRNAs conveniently bypass the problem of intervening sequences in native DNA.

R-Loop Structures and Intervening Sequences

With regard to these new insights into eucaryotic gene structure, what recombinant DNA data provided evidence for intervening sequences? The key observations were obtained by analysis of the annealing (hydrogen bonding) patterns of mRNAs to the DNA segments possessing the genes responsible for the messages. Since mRNA is the complementary copy of the genetic information contained in one strand of a double-stranded gene (called the *sense strand*), its base sequence can hydrogen bond with the sense strand. When an RNA bonds with double-stranded DNA, an *RNA-DNA heteroduplex* is preferentially formed, and the portion of the DNA strand displaced by the RNA appears in electron micrographs as a single-stranded loop. The heteroduplex and the single-stranded loop associated with it are referred to as an *R-loop structure.* If the gene has a continuous coding sequence, then the R-

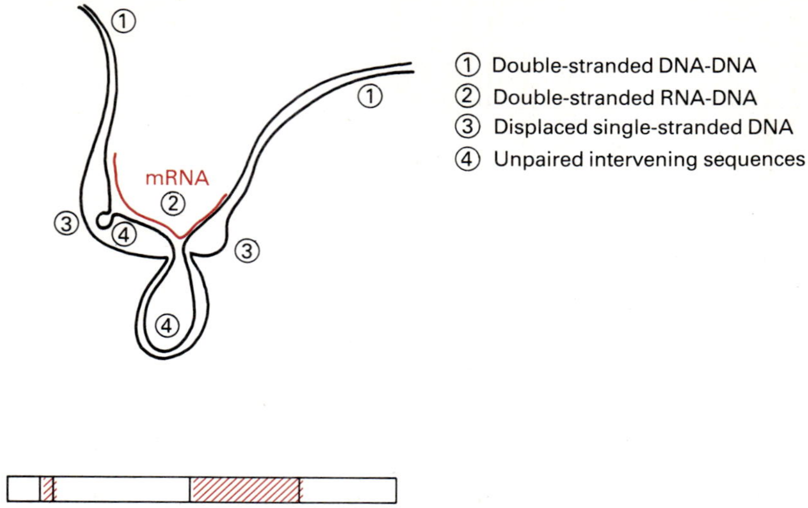

β-Globin gene with its two intervening sequences (red diagonal lines)

Figure 26-5
Schematic of the RNA-DNA heteroduplex obtained by Leder and his associates when they annealed the cloned *β*-globin gene of mouse with *β*-globin mRNA, and of the proposed structure of the *β*-globin gene.

loop structure has (as determined by electron micrographs) a single displaced DNA loop (equivalent in size to that of the mRNA displacing it). When cloned eucaryotic genes were isolated and tested with their respective mRNAs, an unexpected type of R-loop structure was obtained. Figure 26-5 is a schematic of an electron micrograph of the annealing pattern obtained by Philip Leder and his associates when they hybridized the cloned *β*-globin gene of mouse with the *β*-globin mRNA (containing the coding information for the *β* chain of hemoglobin). Studies on the globin genes by these investigators were among the pioneering efforts to use recombinant DNA research to elucidate eucaryotic genetic structures.

As illustrated in Figure 26-5, the two displaced single-stranded DNA loops identify what region of the double-stranded DNA annealed to the mRNA. What is unusual about the pattern is that the region of RNA-DNA hybridization contains *two* displaced single-stranded loops separated by double-stranded DNA. Note also that the single-stranded DNA sequence annealed to the mRNA in the left loop of the illustration also possesses a small loop. From this annealing pattern, it was concluded that the *β*-globin gene contains its genetic information not in a continuous coding sequence but, rather, in *three* distinct sequences (exons) which are separated by *two* noncoding intervening sequences (introns). As illustrated in Figure 26-6, the first intervening sequence (small loop

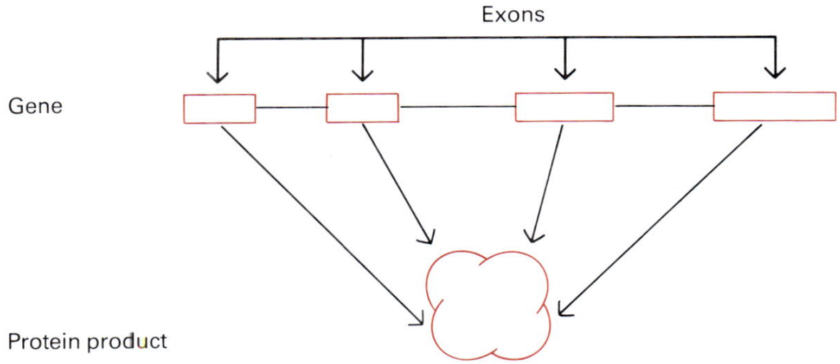

Figure 26-6
Schematic of the proposed exon-protein domain interrelationship.

in annealed DNA strand) is located near the beginning of the gene, and there is a much larger intervening sequence (large double-stranded loop separating the two displaced loops) in the interior. (The electron micrograph data, it should be noted, do not provide the information about the orientation of the intervening sequences within the gene; other studies established these conclusions.)

The evidence for intervening sequences is strengthened by the finding that when the original transcript (hn RNA) of the β-globin is annealed to genomic DNA, a *single* displaced loop is observed in the RNA-DNA heteroduplex. This result provides additional evidence that the entire gene (exons and introns) is transcribed and that hn RNA is processed to produce the mature mRNA.

As more cloned genes were subjected to RNA-DNA heteroduplex analysis, it soon became apparent that intervening sequences were a common feature of almost all eucaryotic genes studied, with histone genes being a notable exception. From the studies, it also became evident that gene fragmentation patterns vary. For example, whereas the β-globin and other globin genes have three coding regions, ovalbumin and conalbumin have eight and 17, respectively. To date, the most fragmented gene reported is that of α-collagen, which has 52 exons. As indicated in Figure 26-6, the size of individual introns also varies, and these units can range in size from 10 to 10,000 nucleotides. Variability is also noted in the ratio of the total nucleotide content of exons to that of introns; some genes are mainly composed of exons and others of introns. For example, although the message for conalbumin contains about 2,500 nucleotides, the conalbumin gene (17 exons and 16 introns) has about 10,300 base pairs ($\approx$ 10.3 kb); hence, about 75 per cent of the gene consists of noncoding nucleotides.

Biological Significance of Fragmented Genes

Now that introns and exons have been discovered and described, it is logical that the question about their biological significance be asked. Although the answer(s) to the question is not yet available, there are nevertheless an abundance of proposals. Currently, the ideas put forth by Walter Gilbert (Nobel Prize, 1980) are generating the most attention. The essence of Gilbert's proposal is that the fragmented gene arrangements identified in eucaryotes are beneficial for evolutionary change. A *coding region,* or *exon,* he believes, codes for a specific *functional domain,* e.g., a catalytic or critical structural sequence, of the protein specified by the gene. Figure 26-6 illustrates the concept. As indicated, each of the four domains of the protein product is coded for by an exon of the gene. The proposed evolutionary advantage of packaging units of genetic information into discrete coding regions separated by noncoding sequences is that exons can be excised *en toto* from one location of the genome and inserted into another. The evolutionary shuffling of sets of coding information, it is believed, can lead to new assemblages of exons that create new functional genes. The Gilbert proposal is distinctly different from a widely accepted concept (supported by biochemical and genetic data) that most new genes arise as a result of duplication of an existing gene. The exon-domain hypothesis is currently being tested by the comparing of known exons with the structures and functions of the proteins they code for, e.g., hemoglobins and antibodies. Some other thoughts about the biological significance of fragmented genes include their roles in the regulation and/or developmental expression of genes.

Regulation of Eucaryotic Gene Expression

In addition to the illuminating information recombinant DNA research has already furnished about eucaryotic gene structure, it is also beginning to yield clues about the molecular elements involved in the regulation of their expression. Although the understanding of eucaryotic gene expression is far from complete, some pieces of the scientific puzzle are being assembled, and these data are summarized in Figure 26-7.

Genes Transcribed by Pol II

The diagram in part A of Figure 26-7 depicts some of the available information about the regulation of genes transcribed by *pol II,* the eucaryotic RNA polymerase that transcribes *unique genes* (those

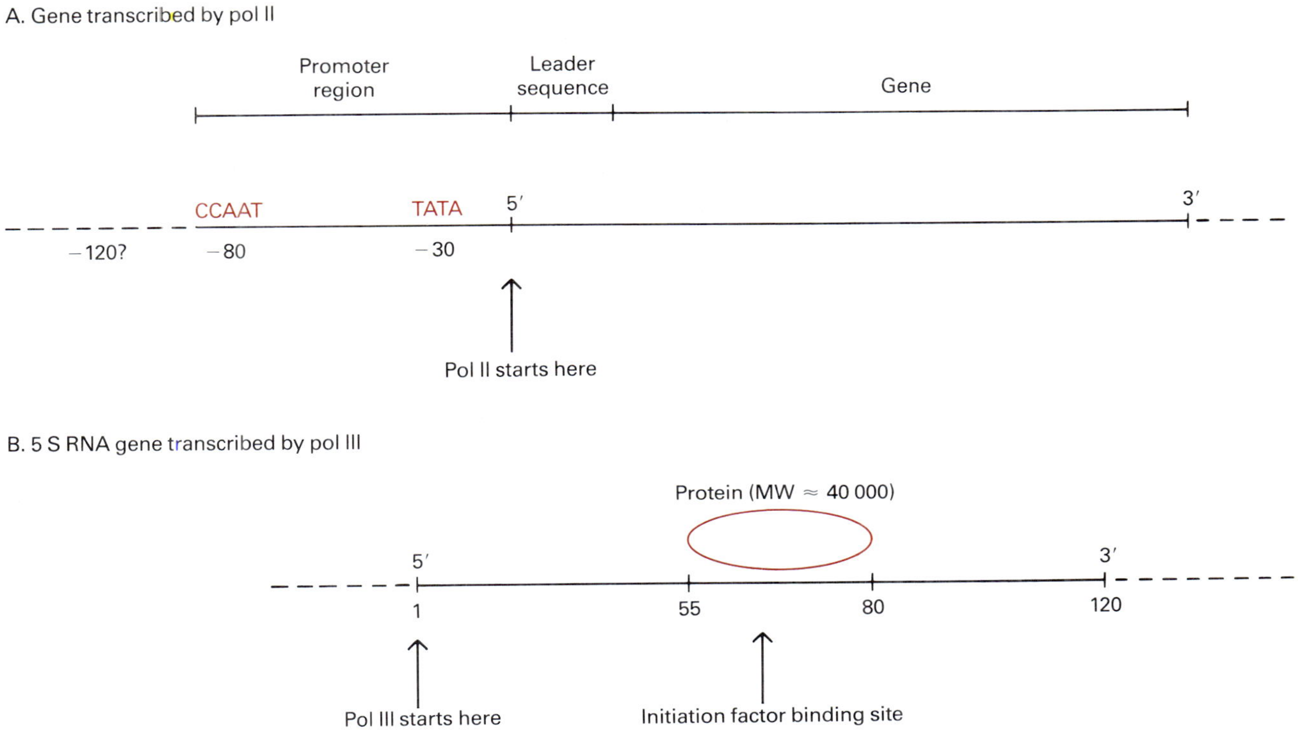

Figure 26-7
Schematic of the proposed regulation of the expression of two types of eucaryotic genes.

developmentally regulated). The elements regulating their expression, like those of bacterial genes (p. 485), are located in the nucleotide sequence flanking the 5′-terminus of the gene, i.e., are *upstream.* Another similarity with procaryotic regulatory systems is that at about position −30 (30 base pairs upstream from the initiation site for transcription) there is a *TATA sequence* (called the *Goldberg-Hogness box*). This particular sequence is known to be part of the promoter regions for bacterial genes and essential for their expression (p. 477). Since a TATA box has been identified in about the same position for nearly 60 eucaryotic genes, it is believed to serve as a eucaryotic promoter also. The importance of the sequence for transcription was demonstrated by Pierre Chambon and his associates, who showed that, by converting the TATA sequence of the conalbumin gene into TAGA (by mutation), *in vitro* transcription of the gene is essentially abolished. That other control sequences are involved was derived from *in vivo* data, such as those of Steven McKnight and associates, who determined that the region around the CCAAT sequence at position −80 is necessary for efficient transcription of a gene. Their studies with a viral thymidine kinase gene showed that deletions in this region severely affect accurate transcription of the gene when it is introduced into frog oocytes. Deletions of nucleotide sequences at about position −120 also affect transcription; in studies on a histone gene, deletions in this region significantly reduce transcription. There is some evidence, as well as the distinct possibility, that other upstream sequences are also involved in this type of eucaryotic gene expression.

To summarize the above data, it is now apparent that transcription of eucaryotic genes is a multistep process. It appears that the TATA box is needed to direct pol II to fix transcription about 30 bp downstream and that the more remote upstream sequences, e.g., −80 and −120 positions, are critical for efficient and effective *in vivo* transcription.

5 S RNA Gene Transcribed by Pol III

Donald Brown and his colleagues, who have extensively studied a gene transcribed by *pol III* (RNA polymerase III), have discovered molecular details about its expression (Figure 26-7) that differ from those described above for pol II genes. The gene (of which there are 40,000 copies in a frog oocyte) specifies the 5 S RNA associated with ribosomes (p. 513). The 5 S RNA gene is one of the relatively few genes transcribed by pol III; others include the tRNA genes and some genes of unknown function. The gene producing the ribosomal RNA is only 120 nucleotides in length. To determine which sequences control expression of this small gene, Brown's group isolated the 5 S RNA gene with its flanking sequences and systematically removed sequences from the 5′- and 3′-termini. These shortened DNAs were then cloned, isolated, and tested *in vitro* to determine if they produced transcripts. Their findings were surprising because they discovered that removal of the first 55 nucleotides from the 5′-terminus or about 40 from the 3′ terminus of the gene is required before transcription is inactivated. These results mean that the nucleotides from about positions 55 to 80 are needed for initiation of transcription (and not the sequence flanking the 5′-terminus, as is required for pol II transcription). However, the flanking region of the 5 S RNA gene does influence the exact starting site for transcription.

Further studies revealed that for transcription to initiate at the 5′-terminus, a cellular protein (MW $\approx$ 40,000) must bind to this central region of the gene. Hence, it appears that the control elements located outside the pol II genes are situated within the 5 S

RNA gene. This interaction of an initiation protein with an interior sequence may be a unique feature of 5 S RNA transcription because other pol III genes do not appear to need the protein for transcription.

The role of the protein is not limited to its binding to the gene; it also binds to the 5 S RNA product in a 1:1 ratio. It is believed that the RNA-protein complex is the storage form of the ribosomal RNA, which is synthesized in large quantities in oocytes prior to production of the other ribosomal RNAs. Thus, while waiting for the subsequent developmental production of these RNAs and other ribosomal components (so that ribosomes can be assembled), 5 S RNA is stably stored as an RNA-protein complex.

COMMERCIAL USE OF THE RESEARCH

The adaptation of recombinant DNA research to commercial ventures has been as impressive as its rewarding application to basic biological research. Some of the scenarios originally developed for industrial uses of the technique have already been realized, especially those envisioning the production of medically important proteins. The successful insertion of various genes (natural and synthetic) into recombinant DNA molecules and their satisfactory expression by host cells are evidenced by the fact that, by 1981, approval had been granted by NIH for the large-scale production (up to 2,000 liters) of the following human hormones: insulin, somatostatin, growth hormone, and thymosin alpha 1. Large-scale production of human leukocyte and fibroblast interferon is also under way. *Interferon* (a protein) is an *antiviral agent* whose production is induced in a cell following the cell's infection by a virus. Besides interferon's important potential in combating viral infections, claims have also been made for its usefulness in cancer therapy. Clinical testing is currently under way to evaluate the efficacy of interferon as an anticancer agent.

Since commercial and economic interests in the technique are inseparable, recombinant DNA research is now regarded as a multibillion dollar growth industry, and a number of "gene-splicing" companies have been established. Many of these companies are financially backed by the large pharmaceutical, chemical, and/or oil companies. In this section, three illustrations of how recombinant DNA research is or can be applied to a broad spectrum of beneficial uses are presented.

Single-Cell Protein

Single-cell protein (SCP) is the general term applied to *protein derived from bacteria or other single-cell organisms.* Currently, there is broad commercial interest in capitalizing on the ability of many microorganisms to utilize simple organic compounds or biological by-products and waste as nutrients for growth. Large-scale production of SCP by the growing of microorganisms on what are now considered unconventional sources of nutrients, e.g., cellulose or stockyard manure, is an innovative example of "killing two birds with one stone," i.e., utilizing waste materials to produce protein that in turn can be sold as animal feed supplement.

In England, a commercial attempt is being made to produce SCP by growing the bacterium *Methylophilus methylotrophus* on methanol, which can be used as its carbon and energy source. (The magnitude of the economic venture is evidenced by the fact that the world's largest fermenter was built for the project.) One initial drawback to the economic feasibility of this process, however, concerned the inefficient mode of nitrogen assimilation used by *M. methylotrophus.* As depicted in part A of Figure 26-8, the organism (as do many other species of bacteria) incorporates inorganic nitrogen (NH_4^+) via the *glutamine synthetase* reaction (p. 310), which converts glutamate into glutamine. Note the ATP requirement for the reaction. For glutamate synthesis, the nitrogen, which is incorporated into the amide group of glutamine, is then transferred to α-ketoglutarate in a reaction catalyzed by *glutamate synthase.* When this two-step mechanism for nitrogen assimilation is compared to the *glutamate dehydrogenase* reaction (part B of Figure 26-8), it is apparent that the latter reaction accomplishes the same metabolic function in a more direct and therefore more energy-efficient manner. Because of the ATP required for its nitrogen assimilation, *M. methylotrophus* needs more methanol (primary energy source for cellular ATP synthesis) for growth than if the glutamate dehydrogenase reaction were available to the organism. From a commercial viewpoint the additional methanol required because of inefficient nitrogen assimilation prevents the SCP produced from having a market advantage over agriculturally produced soybean protein (a competitor as an animal feed supplement). Researchers working on the SCP project solved the nitrogen assimilation problem, however, and the solution, described below, is a pertinent example of an industrial application of recombinant DNA research.

The basic aim of the researchers was to alter the way *M. meth-*

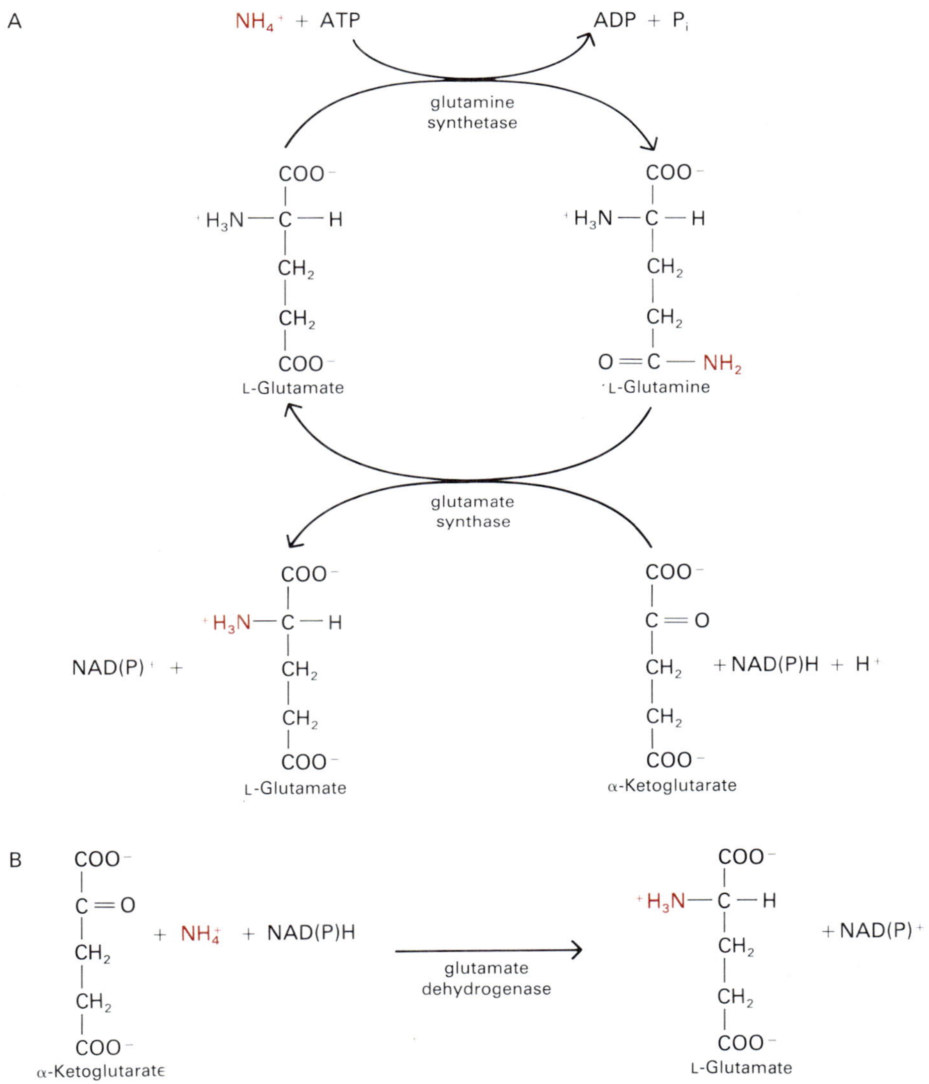

Figure 26-8
Pathways of ammonia assimilation by bacteria.

ylotrophus incorporates inorganic nitrogen, and achieving this goal required substitution of the glutamate dehydrogenase reaction for the naturally occurring mode of assimilation. To inactivate the organism's nitrogen assimilating system, the gene responsible for glutamate synthase (Figure 26-8) was mutated. Since this enzyme catalyzes the organism's major reaction for producing glutamate, the mutation is lethal. To bypass the problem of lethality, the scientists isolated a *temperature-sensitive mutant,* i.e., the mutated enzyme does not become inactive unless the growth temperature is raised to 37° C or higher. Thus, the organism functions "normally" at temperatures below 37° C, but lacks glutamate synthase activity if the temperature is raised to 37° C or higher.

The second phase of the research project involved isolation of the glutamate dehydrogenase gene from *E. coli* and its insertion into a plasmid. The recombinant DNA molecule produced was then introduced into the mutated *M. methylotrophus.* The end result is that nitrogen assimilation (at 37° C or higher) by the mutant harboring the recombinant DNA molecule occurs principally by the more efficient glutamate dehydrogenase reaction. As a consequence, the carbon efficiency of the newly constructed bacterium is 3 to 5 per cent better than that of the naturally occurring organism. Although the percentage of improvement appears small, it is considered to be significant enough to make the SCP process economically feasible. The genetic manipulations performed to make *M. methylotrophus* a more effective organism for industrial use are sterling examples of genetic engineering and of how such molecular techniques are being used for commercial purposes.

Ti Plasmid of *Agrobacterium tumefaciens*

With the introduction of recombinant DNA research, its potential use in agriculture for crop improvement was instantly recognized, e.g., incorporation of nitrogen-fixing genes into commercial nonlegumes (p. 598). As attractive as such ideas were, it was also recognized that the establishing of new genes in plant (eucaryotic) genomes via bacterial recombinant DNA molecules was not a feasible undertaking. Recently, however, studies by plant pathologists on crown gall (an economically important plant disease) provided findings that immediately revived interest in plant genetic engineering.

Crown gall disease is caused by the bacterium *Agrobacterium tumefaciens* which, following its infection of a plant at a wound site, induces a plant tumor. (Crown gall tumors have long been viewed as a model for plant cancer.) In studies on the pathogenicity of *A. tumefaciens,* scientists discovered that the genes responsible for tumor induction are carried on a plasmid, called *Ti* for *tumor-inducing,* harbored by the bacterium. An equally intriguing finding was that, following infection, the Ti plasmid becomes stably incorporated into the genome of the infected plant cells. Indeed, once the plasmid is incorporated, the presence of *A. tumefaciens* is no longer needed for tumor formation or maintenance, since the plasmid genes responsible for the disease are now part of the cells' genome.

It was readily realized that discovery of the Ti plasmid and of its ability to incorporate into plant genomes provided plant scientists with a *natural vector* for incorporating foreign genes into plants. Because of the Ti plasmid, plant molecular biology is currently one of the most active areas of recombinant DNA research.

Figure 26-9 illustrates a hypothetical protocol for incorporating

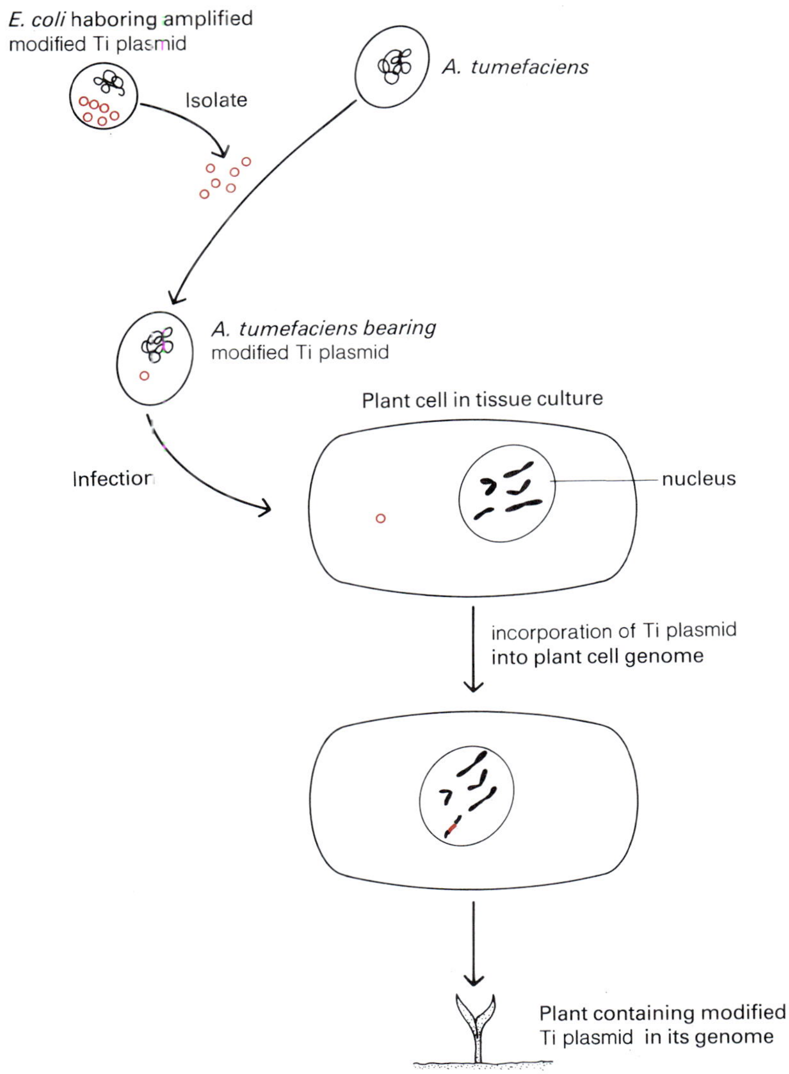

Figure 26-9
Hypothetical protocol for the incorporation of a modified Ti plasmid into a plant's genome.

a genetically manipulated Ti plasmid into a plant genome. (Although the protocol as such has yet to be used, individual steps described have been successfully accomplished with the native or a modified plasmid.) As indicated, a modified Ti plasmid with a gene insert is constructed. The gene could be, for example, one that bestows resistance to a pathogenic plant fungus. Of critical (and obvious) importance is that the plasmid's tumor-inducing genes be mutated to abolish its ability to form plant tumors. To produce the desired Ti recombinant DNA molecule, advantage is taken of the well-established *E. coli* system, i.e., the Ti plasmid is introduced into *E. coli,* which then serves as its host while the molecule is being genetically modified. Once the modified Ti plasmid is constructed and amplified, it is isolated from *E. coli* and reintroduced into *A. tumefaciens.*

Plant cells growing in tissue culture are next infected with *A. tumefaciens* carrying the modified Ti plasmid. Like natural Ti plasmids, they incorporate into the genomes of plant cells. As the plant cells grow and form masses (called *callouses*), shoots appear. These shoots can be grown into normal, healthy plants bearing the incorporated recombinant molecule in their genomes and, more important, having pollen and seed that also contain the modified plasmid.

Although this protocol or a variation of it may be developed, there are yet many problems remaining to be solved or circumvented. For one, since *A. tumefaciens* infects only dicotyledons (plant species whose seeds germinate to form two leaves), its host range does not include many important crops, e.g., cereal grains (which are monocotyledons). Also, although foreign genes have been established in plant genomes, their expression in plants appears to be low; thus, much remains to be learned about the molecular signals plants use for gene expression. Additionally, the tissue culture technique used in the protocol has yet to be established for many of the plant species being considered for genetic improvement.

Finally, with regard to the incorporation of nitrogen-fixing genes (of which there are approximately 17), most plant scientists believe that nitrogen-fixing nonlegumes of commercial importance would produce inferior crops. (Recall the significant percentage of a plant's ATP synthesis required to fix nitrogen, p. 397.) Since nitrogen fixation by *Rhizobia* in legumes has a strict requirement for anaerobic conditions, it is further realized that more than the incorporation of the nitrogen-fixing genes will be required to mimic the naturally evolved process.

Hoof-and-Mouth Disease Vaccine

One of the most promising beneficial applications of recombinant DNA is that of vaccine production. Current methods for preparing vaccines generally entail the use of pathogens, i.e., bacteria or viruses, that are either dead or attenuated (alive but lacking virulent properties). When injected into humans or animals, a vaccine elicits antibody production against specific antigens characteristic of the pathogen used to prepare the vaccine. Thus, as a result, humans or animals become immune to the disease since they have acquired the ability to produce the antibodies that protect them from the live, virulent form of the pathogen. Because the antigens are proteins associated with a particular pathogen, e.g., cellular membrane or viral capsid proteins, these proteins could be produced by recombinant DNA methods for use as vaccines and substituted for those prepared with the pathogens. The gene responsible for the antigenic protein can be isolated, inserted into a vector, and established in *E. coli,* which can then produce the protein for use as a vaccine. From the viewpoint of health safety, the recombinant DNA approach to vaccine production avoids the possible problem that not all pathogens in a vaccine are dead or attenuated. There are, unfortunately, tragic examples in medical records of vaccines that, instead of providing immunity, produced the disease because they contained contaminating live virulent pathogens.

Commercial production of vaccines, like that of hormones, is another once-hypothetical scheme that has become a reality. The first vaccine to be produced is one against hoof-and-mouth disease which, worldwide, is one of the most devastating diseases of livestock. The economic and nutritional potential of this vaccine, with regard to future world production of meat, cannot be overestimated.

Hoof-and-mouth disease, caused by a virus, is highly contagious and incurable. Cattle, pigs, goats, and sheep are susceptible to the disease, which causes blistering of the feet, tongue, and mouth of an afflicted animal and results in lameness and weight loss. To contain the disease, it is necessary to kill and bury infected animals; often an entire herd is destroyed even if only one animal is infected. Some years ago in England, for example, it was necessary to kill more than 280,000 animals to contain a major outbreak.

The key to the success of this venture resides in knowing that VP-3 (viral protein 3), one of the four proteins of the viral coat, induces antibody production that confers immunity to the disease. Hence, by incorporation of the VP-3 gene into a recombinant DNA molecule, the protein is now being produced (by *E. coli* harboring the molecule) for use as a vaccine. Although there are six other major strains of the hoof-and-mouth disease virus and although viruses often undergo mutations that alter their antigenic properties, recombinant DNA technology offers a promising means of effectively containing, if not eliminating, one of the ancient scourges of livestock production.

TWO NEWER HOSTS

Whereas the above discussions emphasized applications of the technology, it should also be noted that the basic protocols continue to evolve. *Escherichia coli* continues to serve well as a host for much of the laboratory research but, for large-scale fermentation processes, microorganisms already used for such purposes are being considered as substitutes for *E. coli.* Two host strains receiving much attention are a soil bacterium, *Bacillus subtilis,* and a yeast, *Saccharomyces cereviciae.* Yeast, of course, has a long, established industrial history, and *B. subtilis* has been adapted in recent years by the Japanese for fermentation processes. Because *B. subtilis* is a soil bacterium, there can be no criticism about its being a danger to humans and animals or about the possible production of endotoxins (characteristic of many *E. coli* strains). Endotoxins, if present as contaminants in hormone preparations or vaccines, could be pyrogenic (induce fever). Thus, in time, it is anticipated that those microorganisms best suited for fermentation processes will replace *E. coli* as the industrial host for recombinant DNA technology.

PHAGE M13 AS A VECTOR

New vectors are also being introduced. The use of the single-stranded DNA bacteriophage M13 (p. 443) as a vector is becoming commonly adopted. It is advantageous in DNA sequencing studies to use M13 because an investigator then deals only with single-stranded DNA, i.e., does not need to work with double-stranded DNA, which requires strand separation to obtain the sequences to be analyzed (p. 462). Also, because the progeny of the modified phage is secreted into the medium by the host cell, purification of the molecules (from the medium) is an easier task than isolation of recombinant DNA molecules from extracts of lysed cells.

CONCLUDING COMMENTS

Information presented in this chapter offers but a glimpse of the world of recombinant DNA research, which is rapidly establishing itself as a successful and influential force in science and industry. The topic was selected for the book's closing chapter because the recounting of its short history (which was often stormy but always promising) also serves to relate a story about the genesis of a new and exciting era in biological research. The beneficial applications already found for the newly developed technique continue to illustrate what the original proponents envisioned when they claimed that the novel science would revolutionize biology.

SUMMARY

When recombinant DNA research was introduced in the early 1970s, a debate soon developed about the potential benefits and hazards the new technique might generate. The issue received much attention from the press, the public, and scientists. By the end of the decade, however, many of the original fears had subsided and, although great interest in the research persists, it is the beneficial rewards to science and society that are being emphasized.

Two methods for the construction of recombinant DNA molecules involve the insertion of a double-stranded copy DNA (cDNA) or of a chemically synthesized gene into a plasmid. For the preparation of cDNA, the mRNA of the gene to be cloned is used as a substrate for reverse transcriptase. For the chemical synthesis of genetic DNA, knowledge of a polypeptide's amino acid sequence and the genetic code allows chemists to synthesize double-stranded DNA containing the genetic information for the sequence. The chemical method is proving useful for the production of certain polypeptide hormones for medical use, e.g., human insulin and somatostatin I.

From recombinant DNA research has come the revelation that, for most eucaryotic genes, their nucleotide sequences are fragmented into discrete coding regions (called exons) which are separated from one another by noncoding intervening sequences (called introns). The number (and size) of exons and introns vary considerably in different genes; for example, the β-globin gene has three exons and two introns and the conalbumin gene has 17 exons and 16 introns. The biological significance of fragmented genes remains unknown. However, one hypothesis proposes that exons code for domains of proteins, e.g., a catalytic sequence, and that new functional genes can be created by the evolutionary shuffling of exons. Data on eucaryotic gene regulation provide evidence that genes transcribed by RNA polymerase II are regulated by the nucleotide sequence flanking the 5′-terminus of the gene, i.e., at the TATA box (30 bp upstream), in the CCAAT region (80 bp upstream), and at a position about 120 bp upstream. In the case of the small 5 S RNA (ribosomal) gene, transcribed by RNA polymerase III, a cellular protein must bind to the interior of the gene in order for transcription to occur. This protein also binds to 5 S RNA, the transcriptional product, and the complex is believed to be the storage form of the ribosomal RNA.

Commercial applications of recombinant DNA research include the production of hormones, e.g., insulin and growth hormone, and interferon (an antiviral agent). In England, the use of the bacterium *Methylophilus methylotrophus* to produce single-cell protein (SCP) from methanol has profited greatly by an alteration of the organism's mode of nitrogen assimilation, which is energy inefficient. The natural mode was abolished by mutation, and the glutamate dehydrogenase gene of *E. coli* was introduced into *M. methylotrophus* via a recombinant DNA molecule. This genetic engineering of the bacterium makes its nitrogen assimilation efficient. In agricultural research, the Ti (tumor-inducing) plasmid of *Agrobacterium tumefaciens,* which incorporates into a plant cell's genome, is currently being exploited as a natural vector for the future incorporation of foreign genes into plants. Of great significance to the worldwide production of livestock is the production of a vaccine for hoof-and-mouth disease, caused by a virus. The vaccine is produced by *E. coli* harboring a recombinant DNA molecule containing the gene for viral protein 3—a viral coat protein that induces antibody production which confers immunity to the disease.

Two new hosts being developed for commercial use are *Bacillus subtilis* and *Saccharomyces cereviciae.* The single-stranded DNA bacteriophage M13 is a new vector that is now frequently being used in recombinant DNA research.

QUESTIONS

1. Select your answers to the statements and questions below from the following list:

   ss animal DNA
   ss bacterial DNA
   ss plant DNA
   ss plasmid DNA
   ss synthetic DNA
   ss viral DNA
   ds animal chromosomal DNA
   ds bacterial chromosomal DNA
   ds plant chromosomal DNA
   ds plasmid DNA
   ds synthetic DNA
   ds viral DNA

   In the example given about the construction of recombinant DNA molecules using cDNA as inserts, the insert was (a) ________ and the vector was (b) ________.

   In the example given about the cloning of the A chain of human insulin, the insert was (c) ________ and (d) ________ and the vector (e) ________.

   In crown gall disease, the Ti genes, which are (f) ________, incorporate into (g) ________.

   (h) If pBR322 is used as a vector, which of the above DNAs could serve as inserts?
   (i) If phage M13 is used as a vector, which of the above DNAs could serve as inserts?

2. Which of the following statements are true about the structure and expression of eucaryotic genes?
   (a) All eucaryotic genes are discontinuous.
   (b) The coding sequences of a gene are called exons.
   (c) The noncoding sequences of a gene are called intervening sequences.
   (d) A gene always has an equal number of exons and introns.
   (e) To date, genes have been found to contain between two and ten exons.
   (f) An exon sequence is always longer than that of an intron.
   (g) Both the exon and intron regions of a gene are transcribed.
   (h) A mature mRNA contains only exon transcripts in its sequence.
   (i) One hypothesis is that introns code for protein domains.

3. Which of the following statements refer to the expression of a gene transcribed by RNA polymerase II (pol II) or of the 5 S RNA gene transcribed by RNA polymerase III (pol III)?
   (a) Regulatory elements are located upstream from the gene.
   (b) Promoter region is believed to be the Goldberg–Hogness box.
   (c) For initiation of transcription, a protein must bind to the central region of the gene.

(d) TATA box is believed to be needed to fix the starting site of transcription.
(e) Initiation protein binds to the transcription product.
(f) The 5′-flanking region of the gene is not needed for initiation of transcription.

4. Give an example of a chimeric DNA, chimeric RNA, and chimeric protein.
5. Why is it advantageous for a recombinant DNA molecule to bear an antibiotic-resistance gene, e.g., *tet*^{r}, when it is being maintained and propagated in *E. coli* (or in another host)?

SUGGESTED READINGS

Broda, P., *Plasmids.* San Francisco: W. H. Freeman, 1979. A book devoted to a discussion of the circular DNA molecules commonly used as vectors in recombinant DNA research.

Chakrabarty, A. M., *Genetic Engineering.* Boca Raton, Fla.: CRC Press, 1978. Emphasizes procedures utilized in genetic engineering research.

Conger, B. V., ed., *Cloning Agricultural Plants via In Vitro Techniques.* Boca Raton, Fla.: CRC press, 1981. Reference guide on the principles and practices of cloning agricultural plants.

Scott, W. A., and R. Werner, *Molecular Cloning of Recombinant DNA.* New York: Academic Press, 1977. A good introduction to recombinant DNA research.

Setlow, J. K., and A. Hollaender, eds., *Genetic Engineering: Principles and Methods.* New York: Plenum Publishing, 1979. A collection of authoritative papers on recombinant DNA research.

Wu, R., ed., *Recombinant DNA. Methods Enzymol,* Vol. 68. New York: Academic Press, 1980. A compilation of recombinant DNA techniques, ranging from the isolation of the DNA to be cloned to the expression of cloned genes.

Articles

Abelson, J., "RNA Processing and the Intervening Sequence Problem." *Ann. Rev. Biochem.,* 48:1035, 1979.

Clowes, R. C., "The Molecules of Infectious Drug Resistance." *Scientific American,* 228(4):18 (offprint # 1269), 1973.

Ehrlich, S. D., "*Bacillus subtilis* and the Cloning of DNA." *TIBS,* 3:184, 1978.

Flavell, R. A., D. M. Glover, and A. J. Jeffreys, "Discontinuous Genes." *TIBS,* 3:241, 1978.

Freifelder, D., ed., *Recombinant DNA.* San Francisco: W. H. Freeman. 1978. Readings from *Scientific American.*

Gilbert, W., and L. Villa-Komaroff, "Useful Proteins from Recombinant Bacteria." *Scientific American,* 242(4):74 (offprint #1466), 1980.

Gordon, J., and M. A. Minks, "The Interferon Renaissance: Molecular Aspects of Induction and Action." *Microbiol. Rev.,* 45:244, 1981.

Grobstein, C., "The Recombinant DNA Debate." *Scientific American,* 237(1):22, 1977.

"Industrial Microbiology." *Scientific American,* 245(3)—September 1981 issue:
1. A. L. Demain and N. A. Solomon, "Industrial Microbiology."
2. H. J. Phaff, "Industrial Microorganisms."
3. D. A. Hopwood, "The Genetic Programming of Industrial Microorganisms."
4. A. H. Rose, "The Microbiological Production of Food and Drink."
5. Y. Aharonowitz and G. Cohen, "The Microbiological Production of Pharmaceuticals."
6. D. E. Eveleigh, "The Microbiological Production of Industrial Chemicals."
7. E. L. Gaden Jr., "Production Methods in Industrial Microbiology."
8. W. J. Brill, "Agricultural Microbiology."

Kourilsky, P., and P. Chambon, "The Ovalbumin Gene: An Amazing Gene in Eight Pieces." *TIBS,* 3:244, 1978.

Sinsheimer, R. L., "Recombinant DNA." *Ann. Rev. Biochem.,* 46:415, 1977.

ANSWERS

Chapter 3

1. (a) antilog (−7.4) = antilog (−8 + 0.6) = $10^{-8} \times 3.99 \approx 4 \times 10^{-8}$ *M*; (b) 2.5×10^{-7} *M*; (c) 3.2×10^{-2} *M*; (d) 5.0×10^{-5} *M*; (e) 5×10^{-3} *M*; (f) 1.6×10^{-3} *M*

2. (a) 4.0; (b) ≈ 7.0 (10^{-7} *M* [H^+] of H_2O > 10^{-8} *M* [H^+] of HCl); (c) $-\log(2 \times 10^{-3}) = -\log 2 - \log 10^{-3} = -0.30 - (-3) = 2.7$; (d) $[H^+]^2 = 2(1 \times 10^{-3}$ *M*$) = 2 \times 10^{-3}$ *M* = 2.7; (e) 2.2

3. (a) 7.4; (b) 2.5; (c) 1.4

4. (a) 1.58; (b) 0.39; (c) 0.06 *M* HPO_4^{2-} and 0.04 *M* $H_2PO_4^-$; (d) 0.028 *M* HPO_4^{2-} and 0.072 *M* $H_2PO_4^-$; (e) pH 7.4

5. (a) 0.2 liter × 0.1 mole/liter = 0.02 mole of H_3CCOOH (HA) in 250 ml
   0.05 liter × 0.1 mole/liter = 0.005 mole of NaOH, producing 0.005 mole of Na acetate (A^-) in 250 ml
   pH = 4.76 + log [0.005]/[0.015]
   pH = 4.76 + log 1/3
   pH = 4.76 − log 3
   pH = 4.76 − 0.48 = 4.28 ≈ 4.3
   (b) 4.76

6. (a) lactic acid = 0.21 *M*, Na lactate = 0.29 *M*
   (b) formic acid = 0.18 *M* and Na formate = 0.32 *M*

Chapter 4

1. all of them

2. (a) alanine—1, threonine—2, and serine—1
   (b) and (c)

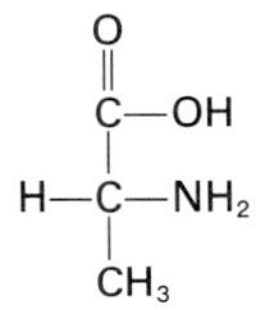

D-Alanine

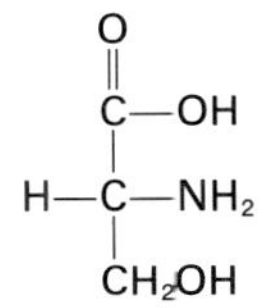

D-Serine

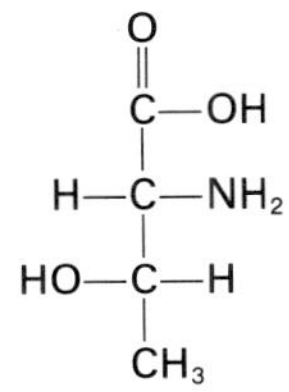

D-Threonine

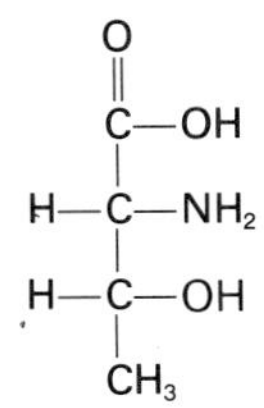

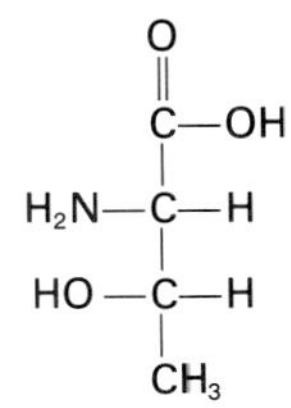

3. d, e, f, and i

4. electrostatic attraction: c and f
   electrostatic repulsion: b and h
   hydrogen bonding: a, e, g, and i
   hydrophobic interaction: d and j

Chapter 5

1. Cytoplasm is the non-nuclear portion of a cell's contents; cytosol is the soluble (structureless) fraction of the cytoplasm.

2. b, c, d, g, and h

3. b, c, f, and l

4. a, b, and i

5. a and i

6. c and e
7. f and h

Chapter 6

1. (a)

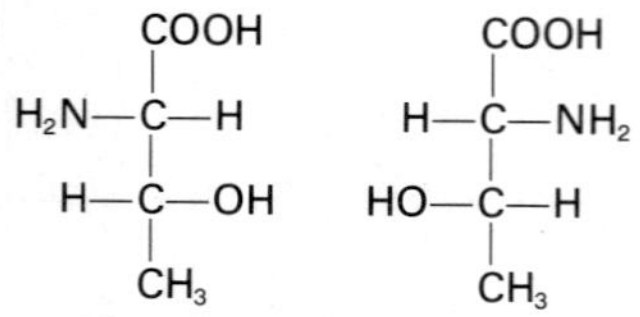

L-Threonine D-Threonine

(b) L- and D-threonine
(c) L-*allo*- and D-*allo*-threonine

2. (a) Glycine:

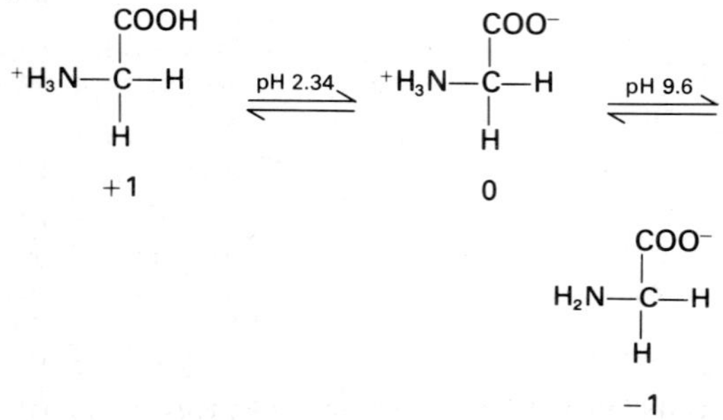

(b) Aspartic acid:

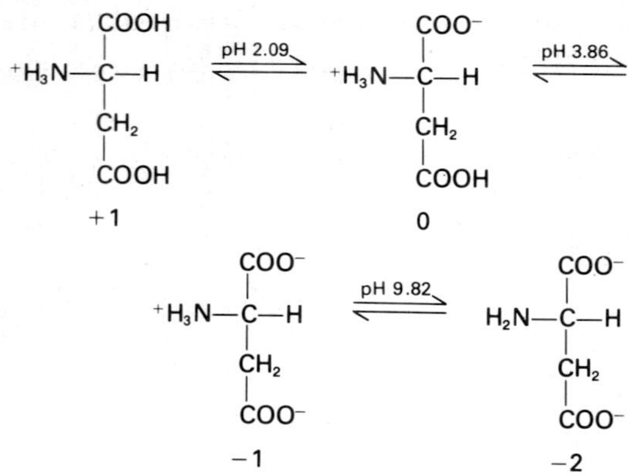

(c) Lysine:

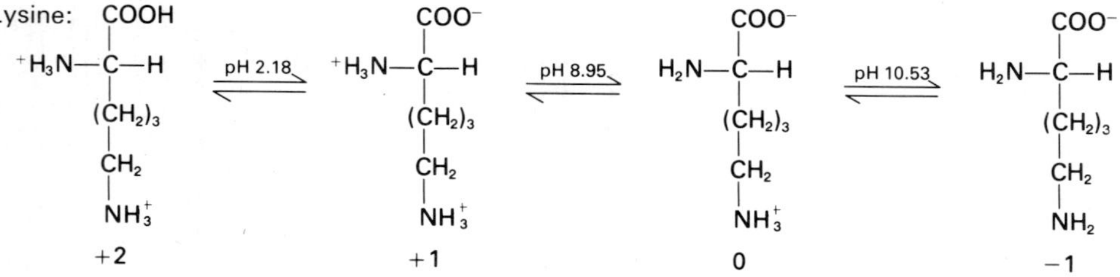

(d) pH 1: Gly = +1; Asp = +1; Lys = +2
(e) pH 7: Gly = 0; Asp = −1; Lys = +1
(f) Gly = 5.97; Asp = 2.97; Lys = 9.74

3. proline
4. (a) Asp and Glu; (b) His; (c) none
5. electrostatic repulsion
6. (a) +4; (b) 0; (c) 0; (d) 2; (e) 1; (f) 4; (g) 3; (h) no
7. (a) 2–12 and 17–24; (b) 1; (c) 13–16; (d) electrostatic attraction; (e) hydrophobic interaction; (f) disulfide bond formation; (g) electrostatic repulsion; (h) hydrogen bonding; (i) hydrophobic interaction; (j) hydrogen bonding
8. Top row, left to right: (a) Asp or Glu and N-terminal residue; (b) Tyr and Asp or Glu; (e) two Cys; (c) two Phe Bottom row, left to right: (c) two Ile; (c) two Ala; (d) two Ser; (a) Arg and C-terminal residue

Chapter 7

1. (a)

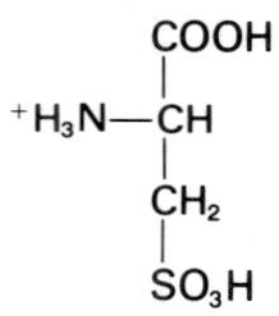

Cysteic acid

(b)

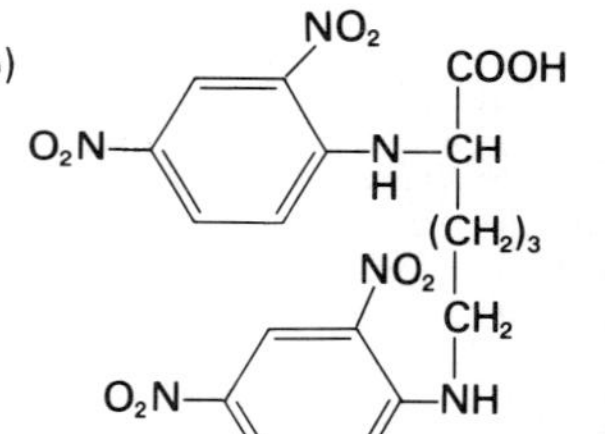

α,ε-(DNP)$_2$-lysine

(c)

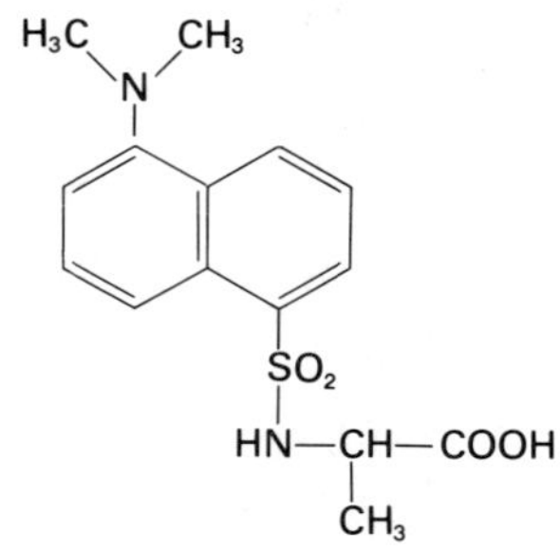

Dansyl alanine

(d)

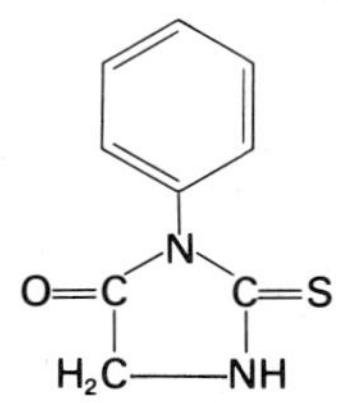

PTH-glycine

2. CO_2 release
3. (a) (b) (c)

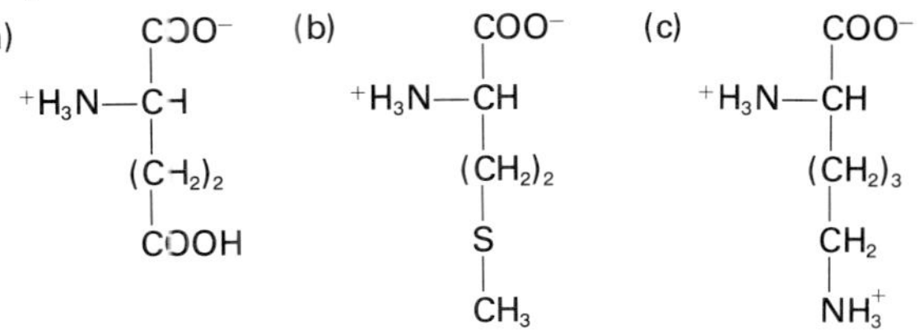

Glutamate Methionine Lysine

4. Homoserine has one more carbon ($-CH_2-$) than serine.

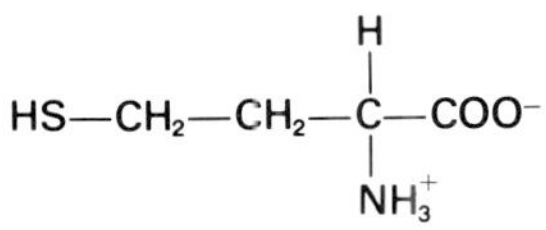

5. ⊕ b, c, a
6. (a) N-terminal Gly (Lys); (b) none; (c) C-terminal Arg; (d) N-terminal Gly (Lys)
7. (a) 4; (b) no (2 N-terminal Gly); (c) yes: Gly-Ala-Lys
8. (a) 3; (b) yes (only 1 N-terminal Gly); (c) yes: Leu-Arg
9. $^{+}H_3N$-Ala-Thr-Arg-Val-Val-Met-Leu-Phe-COO^{-}
10. H_2N-Gly-Ser-Arg-Pro-Leu-Asn-Glx-Lys-Tyr-Met-COOH

Chapter 8

1. myoglobin: d; hemoglobin: c, g, and i; both: a, b, f, and h; neither: e and j
2. e
3. c, f, and g
4. increased: b, d, and f; decreased: a, c, and e
5. no significant effect
6. HbA
7. yes (β chain not required)
8. Although deoxygenated HbA accommodates the R group of β-6-Val of HbS, its lack of this Val residue prevents formation of long filaments of deoxygenated Hb.
9. (a) 50 per cent; (b) 0 per cent

Chapter 9

1. b and d
2. b and d
3. (a) 1, 3, and 5; (b) 1 through 6
4. $v = V_{max}/(1 + K_M/[S])$ (a) 50%; (b) 33.3%; (c) 66.7%; (d) 9.1%; (e) 90.9%
5. (a) $K_M = 7.5 \times 10^{-3}$ M and $V_{max} = 7.9 \times 10^{3}$ μmoles min^{-1} mg^{-1}
(b) $K_M = 7.5 \times 10^{-3}$ M and $V_{max} = 6.4 \times 10^{3}$ μmoles min^{-1} mg^{-1}
(c) noncompetitive
6. isomerase: i and k; lyase: b, f, and h; hydrolase: a, d, and j; transferase: c and l
7. partially digests (hydrolyzes) the protein in foods

Chapter 10

1. (a) riboflavin; (b) none; (c) thiamin; (d) folic acid; (e) vitamin B_{12}; (f) riboflavin; (g) nicotinamide; (h) thiamin; (i) vitamin B_{12}; (j) biotin and pantothenic acid; (k) nicotinamide; (l) biotin; (m) folic acid
2. (a) NAD^+, $NADP^+$, FAD, and CoA; (b) TPP, CoA, and lipoamide; (c) TPP and pyridoxal phosphate; (d) NAD^+, $NADP^+$, FAD, and FMN
3. (a) FAD and FMN; (b) NAD^+ and $NADP^+$; (c) pyridoxal phosphate; (d) biocytin; (e) pyridoxal phosphate and FMN; (f) TPP; (g) CoA; (h) biocytin and FH_4
4. $H_3CCH_2OH + NAD^+ \rightarrow H_3CCHO + NADH + H$
$^{-}OOCCH_2CH_2COO^{-} + FAD \rightarrow {}^{-}OOCC{=}CCOO^{-} + FADH_2$
5. (a) transamination, decarboxylation, and racemization (b) pyruvate carboxylase
6.

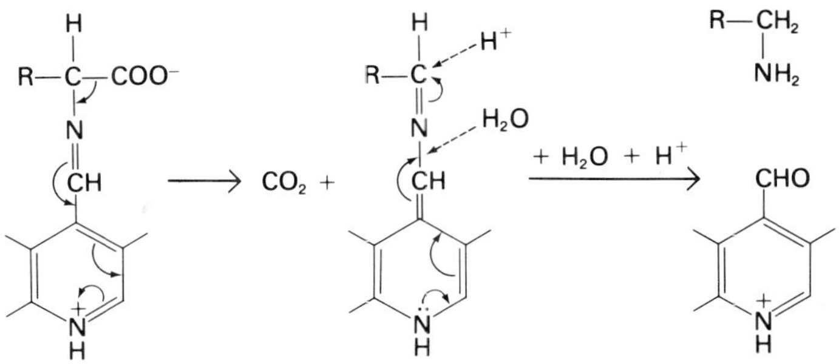

Chapter 11

1. (a) aldohexoses; (b) 4; (c) 1, 3, and 5; (d) 3; (e) 1 and 4; (f) 6

(g)

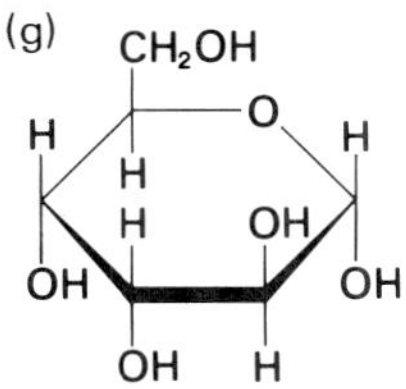

2. (a) 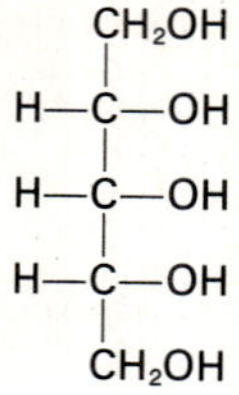 (b) 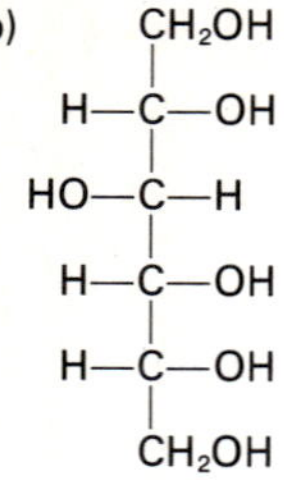 (c) 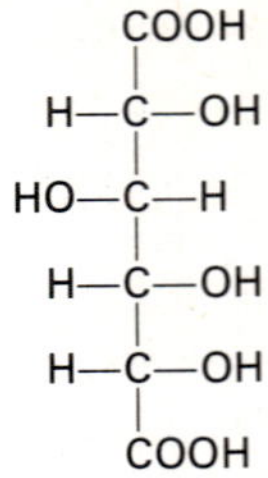

3. (a) D-glucose; (b) C-1 and C-6; (c) β; (d) α

4.

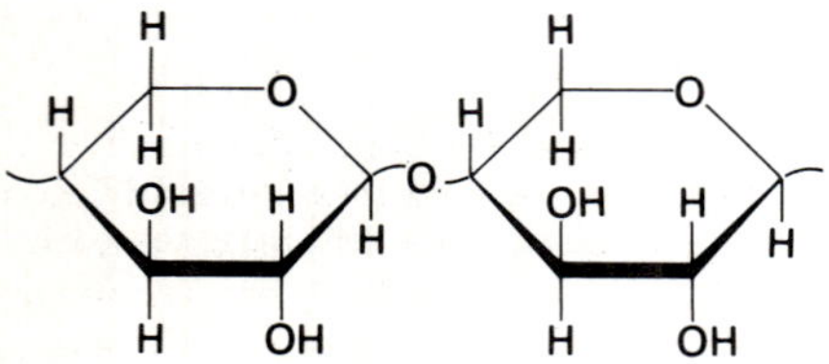

5.

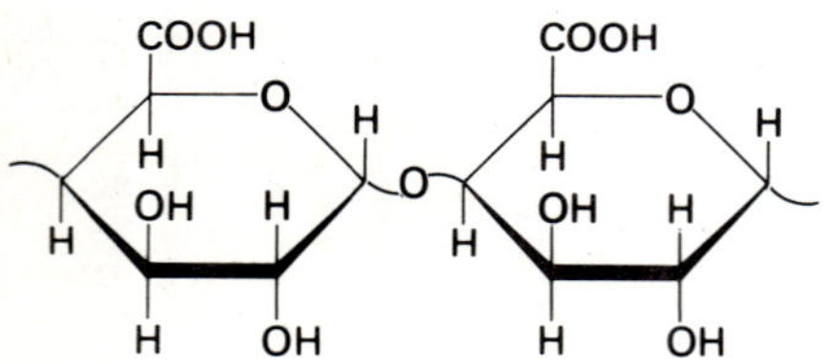

6. (a) Cornstarch $\xrightarrow[\text{(hydrolysis)}]{H_2O}$ D-glucose $\xrightarrow{\text{glucose isomerase}}$ D-fructose

(b) Sawdust → cellulose $\xrightarrow{\text{cellulase}}$ D-glucose
Sawdust → hemicellulose $\xrightarrow{\text{hemicellulase}}$ D-xylose

(c) catalyzes the partial digestion of starch in foods
(d) lactase (Lactose in milk ⟶ D-galactose and D-glucose)

7.

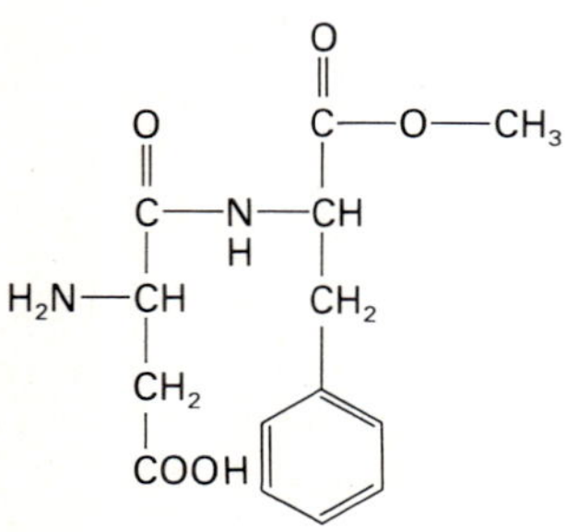

Chapter 12

1. b and d
2. a and e
3. f and g
4. b, c, d, and e
5. (a) β-(1-4); (b) β-(1-4); (c) C-2 of NAN and C-3 of Gal; (d) mannose: C-4 through C-9 and pyruvate: C-1 through C-3
6. a, b, c, and d
7. b, c, and e
8. d and f

Chapter 13

1. (a) 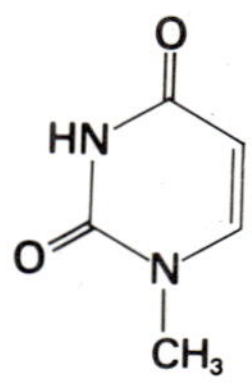 (b) 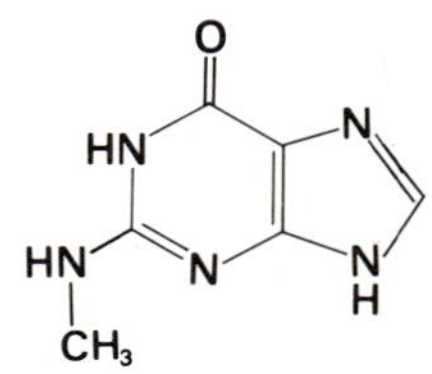 (c) 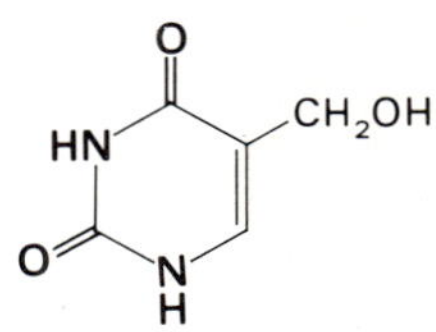 (d) 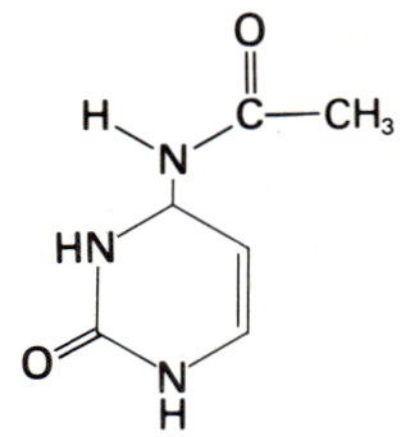

2. b, c, and d

3.

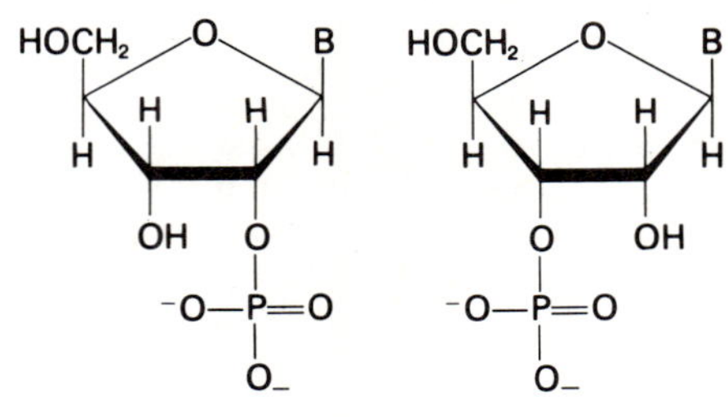

4. (a) tRNA; (b) rRNA; (c) all three; (d) mRNA; (e) tRNA; (f) rRNA; (g) tRNA

5. (a) a mixture of 5′-monophosphonucleosides and a nucleoside (5′-terminal nucleotide)
(b) a mixture of 3′-monophosphonucleosides and a nucleoside (3′-terminal nucleoside)

6. (a) no hydrolysis
(b) 5′-A-Up-3′, Cp, 2(A-Up), A-A-Up, Up, and A-Gp
7. More energy is needed to disrupt the three hydrogen bonds of a G·C base pair than the two of an A·T base pair.
8. f
9. all true except e

Chapter 14

1. b, d, e, f, i, j, k, and o
2. yes; (a) 14.6 kcal mol^{-1}; (b) 14.5 kcal mol^{-1}
3. (a) $[A]_{eq} = 0.01 \times 0.2\ M = 0.002\ M$
$[B]_{eq}$ and $[C]_{eq} = 0.99 \times 0.2\ M = 0.198\ M$
$K_{eq} = [B][C]/[A] = [0.198][0.198]/[0.002] = 0.0392/0.002 = 19.6$
(b) $\Delta G^{\circ\prime} = -1{,}363 \log K_{eq} = -1{,}363 \log 19.6 = -(1{,}363)(1.29) = -1{,}760$ cal mol^{-1}
(c) $+1{,}760$ cal mol^{-1}
4. (a) Creatine + $P_i \rightleftharpoons$ phosphocreatine + H_2O $\quad \Delta G^{\circ\prime} = +10.3$ kcal mol^{-1}
ATP + $H_2O \rightleftharpoons$ ADP + P_i $\quad \Delta G^{\circ\prime} = -7.3$ kcal mol^{-1}
Creatine + ATP $\rightleftharpoons$ Phosphocreatine + ADP $\quad \Delta G^{\circ\prime} = +3.0$ kcal mol^{-1}
(b) $+0.4$ kcal mol^{-1}
(c) -7.5 kcal mol^{-1}
5. $\Delta G = \Delta G^{\circ\prime} + 1{,}363 \log [ADP][P_i]/[ATP] = -7{,}300$ cal mol^{-1} $+ 1{,}363 \log (1.4 \times 10^{-4})(1 \times 10^{-3})/(1.85 \times 10^{-3})$ $\Delta G = -7{,}300 + 1{,}363 \log 7.57 \times 10^{-5} = -7{,}300 + 1{,}363\,(-4.121)$
$\Delta G = -7{,}300 + (-5{,}617) = -12{,}917$ cal mol^{-1}, or -12.9 kcal mol^{-1}
6.

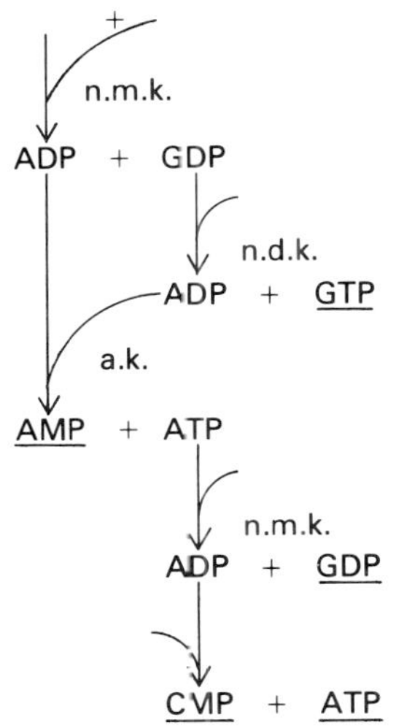

Chapter 15

1. NADH + $H^+ \rightarrow NAD^+$: a and d; $NAD^+ \rightarrow$ NADH + H^+: f; ATP → ADP + P_i: c; ATP → ADP: b and g; ADP → ATP: e and h
2. (a) hexokinase and 6-phosphofructokinase
(b) D-glyceraldehyde 3-phosphate dehydrogenase
(c) L-lactate dehydrogenase
(d) phosphoglycerate kinase and pyruvate kinase
(e) D-glyceraldehyde 3-phosphate dehydrogenase
(f) enolase
3. (a) TPP; (b) biotin; (c) TPP; (d) D-glucose 1,6-bis-PO_4; (e) GTP; (f) ATP; (g) acetyl-CoA
4. (a) hexokinase, 6-phosphofructokinase, and pyruvate kinase; (b) pyruvate dehydrogenase, glyceraldehyde 3-phosphate dehydrogenase, and lactate dehydrogenase; (c) pyruvate kinase, hexokinase, phosphoglucomutase, 6-phosphofructokinase, and phosphoglycerate kinase; (d) phosphoglucomutase, pyruvate decarboxylase, alcohol dehydrogenase, and lactate dehydrogenase
5. (a) no; (b) yes
6. (a) and (b): lower-than-normal affinity for oxygen
7. G—S—S—G + NADH + $H^+ \rightarrow$ 2G—SH + NAD^+
Oxidized glutathione → Reduced glutathione
8. (1) transketolase; (2) transaldolase; (3) transketolase

Chapter 16

1. (a) pyruvate dehydrogenase, isocitrate dehydrogenase, and α-ketoglutarate dehydrogenase; (b) pyruvate dehydrogenase, isocitrate dehydrogenase, α-ketoglutarate dehydrogenase, and malate dehydrogenase; (c) succinate dehydrogenase; (d) aconitase; (e) fumarase; (f) succinyl-CoA synthetase; (g) pyruvate dehydrogenase and α-ketoglutarate dehydrogenase; (h) citrate synthase
2. a, b, d, and h
3. thiamin, niacinamide, riboflavin, and pantothenic acid
4. (a) -18.6 kcal mol^{-1}; (b) -10.6 kcal mol^{-1}
5. -8.1 kcal mol^{-1}
6. Ile, Leu, Phe, and Tyr
7. (a) C-2 and C-3 of oxaloacetate; (b) all four carbon atoms of oxaloacetate
8. T = c, e, i, and j; G = f and h; B = a, b, and g; N = d
9. (a) succinate dehydrogenase; (b) catalytic oxidation of pyruvate by certain di- and tricarboxylic acids (those referred to in parts c and d); (c) and (d) succinate; (e) yes—especially the fumarate result, which supported the postulate that fumarate is converted into succinate

by an oxidative pathway not requiring succinate dehydrogenase; (f) in a cycle not inhibited, oxaloacetate is regenerated at the end of each cycle; in an inhibited cycle, oxaloacetate cannot be regenerated (1 oxaloacetate + 1 ''pyruvate'' → 1 citrate →→ 1 succinate)

Chapter 17

1. (a) NADH dehydrogenase and CoQ; (b) cyts. aa_3, b, c, and c_1; (c) cyt aa_3; (d) NADH dehydrogenase; (e) cyts. aa_3, b, c, and c_1; (f) cyt aa_3; (g) NADH dehydrogenase; (h) cyts. c and c_1; (i) cyt. b; (j) cyt aa_3
2. b, c, d, and e
3. (a) 3ATP; (b) 1 ATP; (c) 30.7%; (d) -20.3 kcal mol^{-1}; (e) 85.3%
4. (a) -6 kcal mol^{-1}; (b) no
5. (a) -6 kcal mol^{-1}; (b) 0.46 kcal mol^{-1}
6. (a) 262.8 kcal; (b) 5.3%
7. (a) 30 acetyl-CoA, 27 NADH, and 27 $FADH_2$; (b) 90 NADH, 30 $FADH_2$, and 30 GTP; (c) 465 ATP; (d) 465 + 30 (GTP) − 6 (activation of three fatty acids) = 489; (e) 3,570 kcal

Chapter 18

1. (a) P_{700}, chlorophyll a, and ferredoxin; (b) P_{680}, Mn^{2+}, chlorophylls a and b, and H_2O; (c) cytochrome f, plastoquinone, plastocyanin, and P_{700}; (d) ferredoxin; cytochrome f; (e) plastocyanin
2. (a) -10.4 kcal mol^{-1}; (b) yes; (c) 4
3. (a) -12.7 kcal mol^{-1}; (b) yes
4. -4.6 kcal mol^{-1}
5. 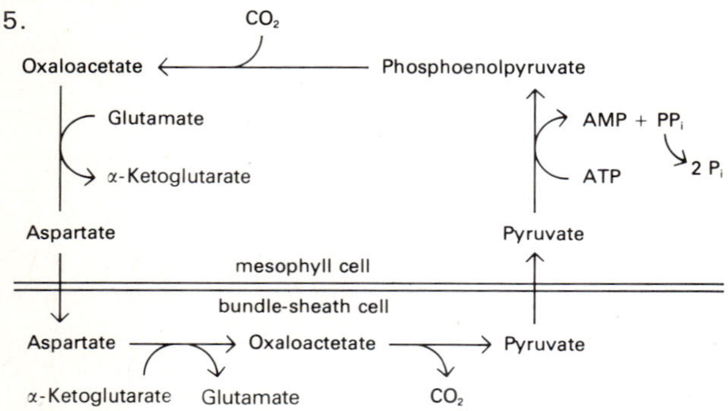

6. $C_6H_{12}O_6 + 6\ O_2 \rightarrow 6\ CO_2 + 6\ H_2O$
   $\Delta G^{\circ\prime} = -686$ kcal mol^{-1}
   therefore, $6\ CO_2 + 6\ H_2O \rightarrow C_6H_{12}O_6 + 6\ O_2$
$\Delta G^{\circ\prime} = +686$ kcal mol^{-1}
$\Delta G^{\circ\prime}$ of 1 CO_2 → 1 (CH_2O) = +686/6 = +114 kcal mol^{-1}

7. 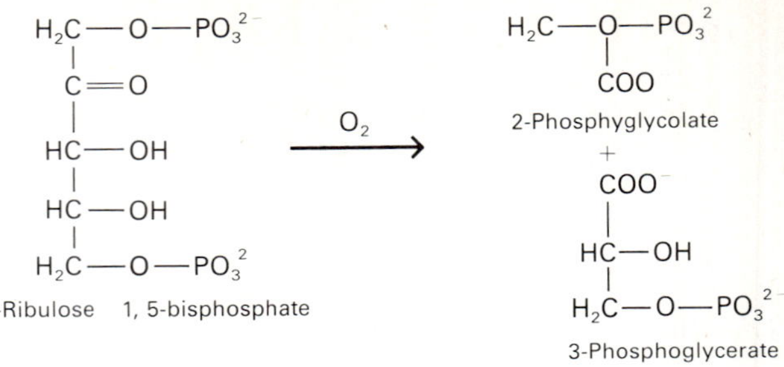

Chapter 19

1. (a) ATP-citrate lyase; (b) acetyl-CoA carboxylase; (c) acyl carrier protein; (d) ACP-acyltransferase; (e) ACP-malonyltransferase; (f) β-keto-ACP synthase; (g) β-keto-ACP reductase; (h) β-hydroxy-ACP dehydratase; (i) β-enoyl-ACP reductase; (j) β-enoyl-ACP reductase and β-keto-ACP reductase; (k) ACP-malonyl transferase, ACP-acyltransferase, β-enoyl-ACP reductase, β-hydroxy-ACP dehydratase, β-keto-ACP reductase, and β-keto-ACP synthase
2. (a) mevalonate; (b) acetate; (c) and (d) 3,3-dimethylallyl pyrophosphate and 3-isopentenyl pyrophosphate; (e) geranyl pyrophosphate; (f) 3-isopentenyl pyrophosphate; (g) farnesyl pyrophosphate; (h) squalene; (i) lanosterol
3. 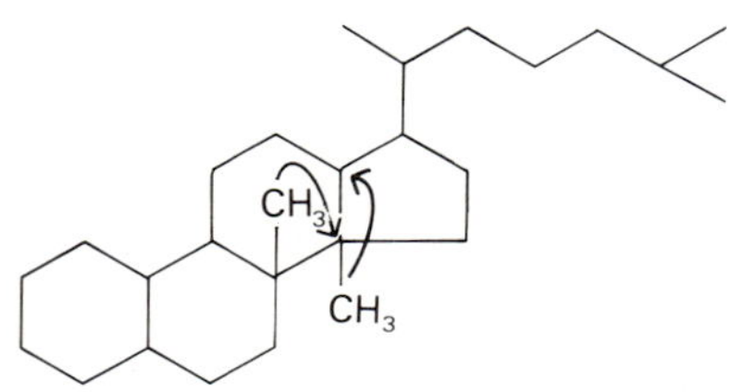
4. a, c, e, and h
5. (a) 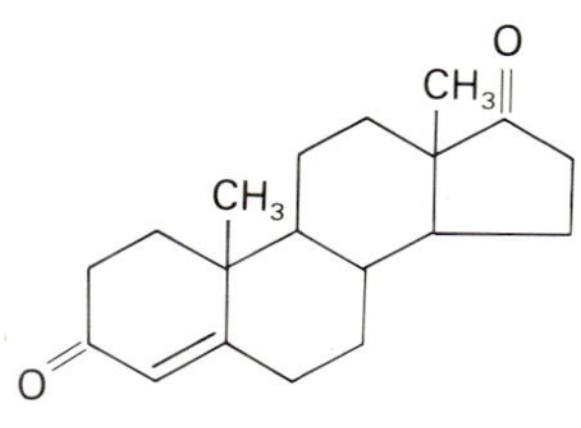

(b)

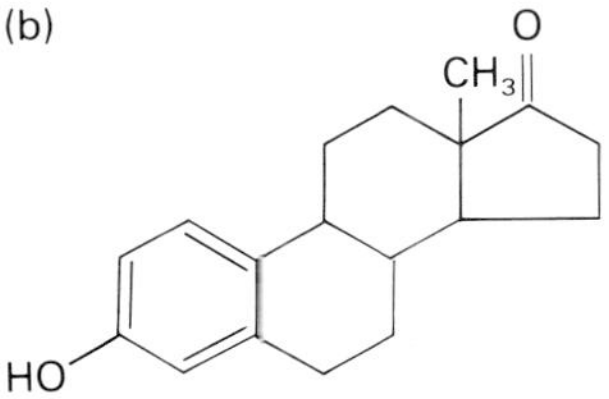

(c)

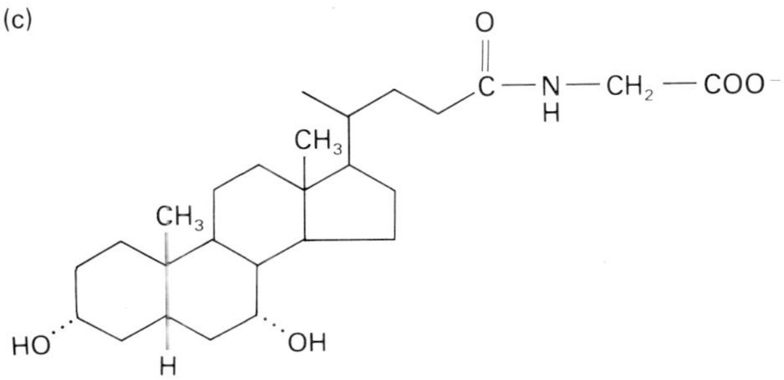

(d)

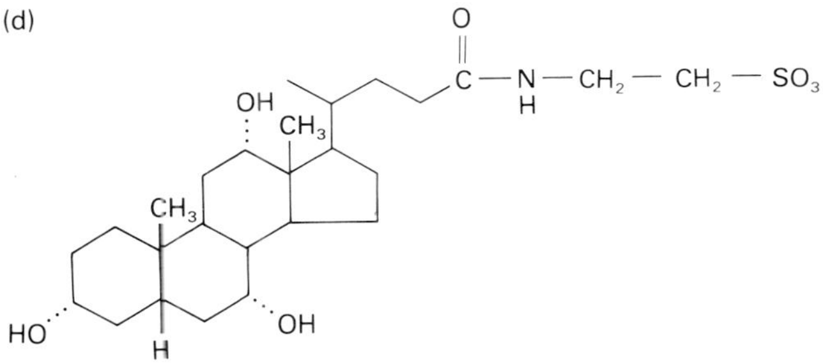

6. (a) 2; (b) 1, 2, and 3; (c) none

Chapter 20

1. a, d, e, h, i, j, k, and l
2. (a) 3; (b) 1; (c) 1; (d) 1; (e) 2; (f) 3
3. (a) ornithine transcarbamoylase; (b) argininosuccinate synthetase; (c) ornithine transcarbamoylase; (d) arginase; (e) argininosuccinase; (f) arginase; (g) argininosuccinate synthetase; (h) citrulline; (i) ornithine; (j) arginine
4. (a) 5; (b) 2
5. (a) Asp; (b) Gln; (c) Gly; (d) Gln; (e) N^{10}-formyl-FH_4; (f) CO_2; (g) N^{10}-formyl-FH_4; (h) ATP; (i) GTP; (j) ATP
6. $2\,O_2^- + 2\,H^+ \xrightarrow{\text{dismutase}} O_2 + H_2O_2$
   $H_2O_2 \xrightarrow{\text{catalase}} H_2O + \frac{1}{2}\,O_2$
7. 7
8. α-carbon
9. (a) carbamoyl phosphate synthetase; (b) aspartate transcarbamoylase; (c) orotidine phosphate decarboxylase; (d) aspartate transcarbamoylase; (e) dihydroorotate dehydrogenase; (f) orotate phosphoribosyl transferase; (g) aspartate transcarbamoylase, carbamoyl phosphate synthetase, and dihydroorotase; (h) and (i) orotate phosphoribosyl transferase and orotidine phosphate decarboxylase

Chapter 21

1. a, b, d, e, and f
2. two molecules of double-helical DNA, with one molecule having two old (parent) strands and the other two new (daughter) strands
3. (a)

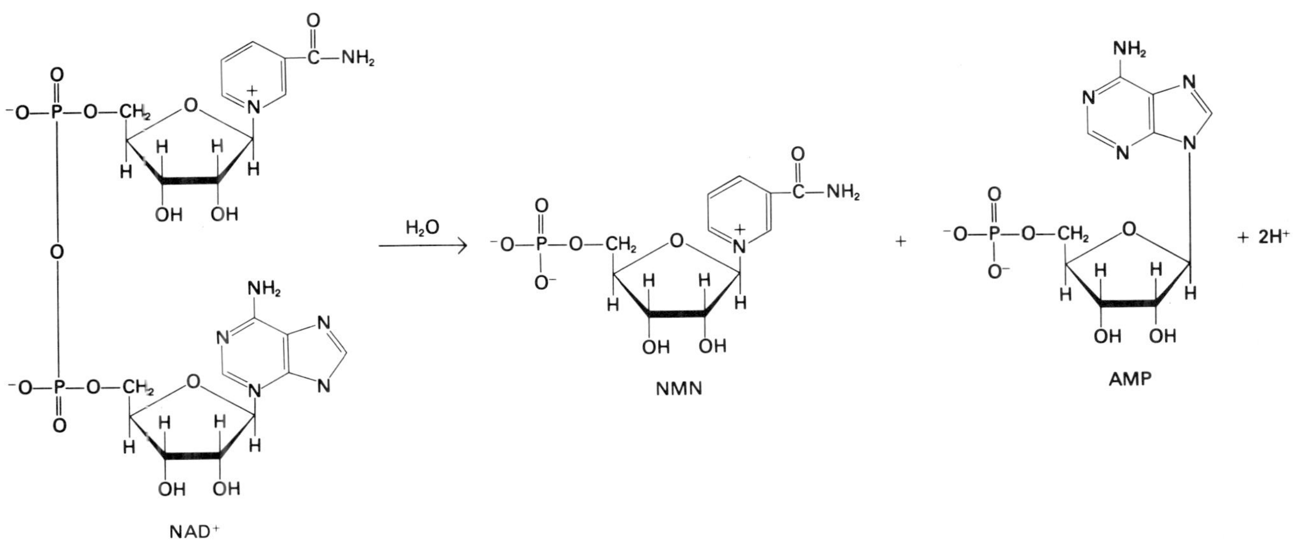

(b) phosphoanhydride bond

4. 5′—A—G—A—T—A—T—C—T—3′
3′—T—C—T—A—T—A—G—A—5′

5. (a) an RNA polymerase, DNA polymerase I and III, and DNA ligase; (b) DNA polymerase I and DNA ligase

6. 5′—G—G—A—C—T—A—T—C—G—T—?—3′

7. (a) ddATP = 1,8 and 10 nucleotide(s); (b) ddCTP = 3 and 7; (c) ddTTP = 2 and 6

8. (a)

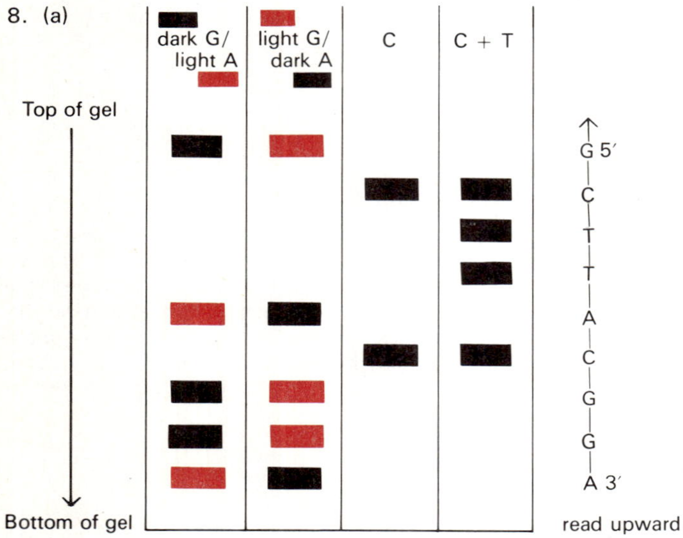

(b)

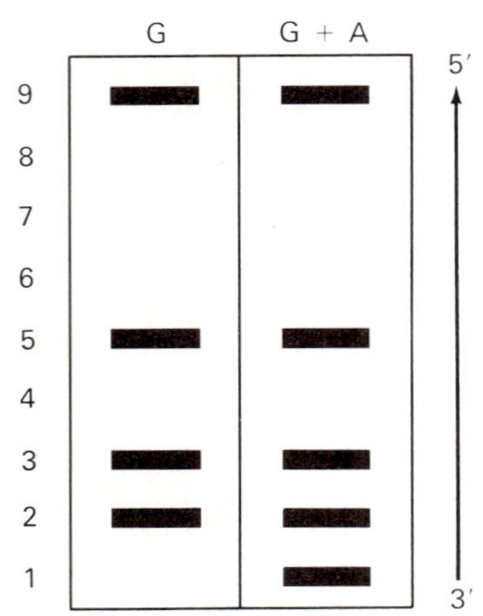

Chapter 22

1. b, c, g, h, and i

2. A doublet code would have 16 possible codons (4^2), and a code relying on a sequence of four nucleotide bases would have 256 possible codons (4^4). Since the code needs to account for 20 amino acids, a doublet code would provide too few codons and one of four bases too many.

3. (a) and (b) The *lac* operon would probably be constitutive, i.e., enzymes would be produced regardless of growth conditions, since a mutation in the I gene would result in an altered (ineffective) repressor molecule and a mutation in the operator region would decrease its affinity for the repressor molecule. (c) and (d) Very little or no expression of the operon would occur since the cAMP-CAP complex is needed for the initiation of transcription.

4. (a) B; (b) P; (c) B; (d) E; (e) E; (f) B

5. (a) effector; (b) sensor protein; (c) sensor gene; (d) set of integrator genes; (e) activator message set; (f) activator protein; (g) sensor gene; (h) receptor sequence; (i) receptor sequence

6. (a) Ser, Gly, and Gln; (b) UGA; (c) Met (*N*-formylmethionine—see Chapter 23), Asn, and Met; (d) repressor molecule (I gene) and β-galactosidase (Z gene)

7. M = b, f, and g; C = a, e, and j; B = c and h; N = d and i

8. The R gene for the *trp* operon specifies a repressor molecule, which by itself does not interact with the *trp* operator region (allowing for the constitutive transcription and translation of the *trp* genes). When intracellular levels of tryptophan (end product) become high, the amino acid (EP) interacts with the *trp* repressor molecule (R). The complex (R-EP), which has a high affinity for *trp* O, binds to the operator region and prevents transcription. When intracellular levels of tryptophan decrease, dissociation of the amino acid from the repressor occurs (R-EP → R + EP) and the operon is once again transcribed.

Chapter 23

1. d, e, f, and j

2. (a) IF1, IF2, and IF3; (b) EFT_u; (c) EFT_s; (e) EFG; (e) peptidyl transferase; (f) RF1; (g) peptidyl transferase

3. (a) about 0.8; (b) 7.3 seconds

4. (a) 450; (b) 2,700; (c) 3,600; (d) 453

5. (a) -A-U-G-A-A-U-G-G-G-U-G-U-G-C-U-A-C-U-U-C-A-G-C-U-U-G-A-A-U-G-A
(b) H_3N-Met (fMet)-Asn-Gly-Cys-Ala-Thr-Ser-Ala-COO^-
(c) ^+H_3N-Met(fMet)-Gly-Val-Leu-Leu-Gln-Leu-Glu-COO^-

6.

Sets of 4 codons:	8 × 2 tRNAs	16
Set of 3 codons (Ile):	1 × 1 tRNA	1
Sets of 2 codons:	12 × 1 tRNAs	12
One codon (Met and Trp):	2 × 1 tRNA	2
tRNA for initiating Met:	1 tRNA	1
		32

7. h

Chapter 24

1. (a) epinephrine and thyroxine; (b) growth hormone, insulin, oxytocin, thyrotropin releasing factor, and vasopressin; (c) aldosterone, estradiol, and testosterone

2. (a) epinephrine; (b) cAMP; (c) epinephrine; (d) adenylate cyclase; (e) phosphodiesterase; (f) phosphorylase kinase; (g) phosphatase; (h) cAMP; (i) protein kinase; (j) phosphatase; (k) phosphatase; (l) protein kinase; (m) phosphorylase kinase and phosphorylase; (n) glycogen synthase

3. (a) γ; (b) δ; (c) α; (d) β; (e) δ; (f) δ; (g) tropinin C; (h) tropinin C; (i) δ; (j) tropinin C

4. a, f, and g

5. (a)

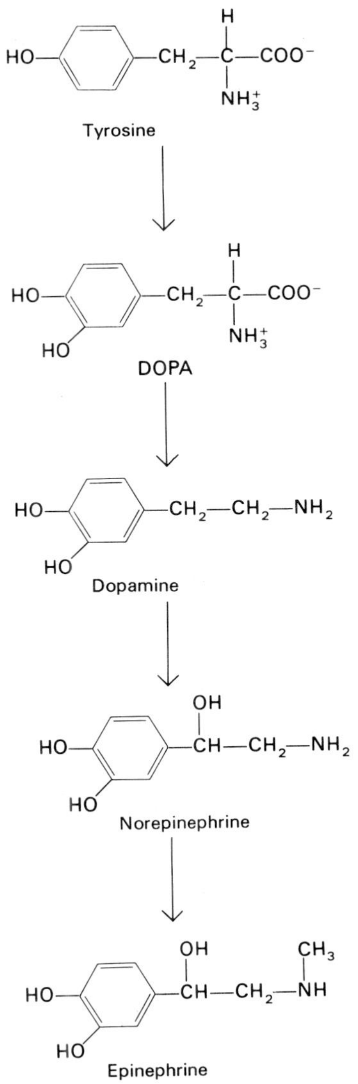

(b) synthesis of DOPA and of norepinephrine
(c) yes

6. b, c, d, e, f, and h

7.

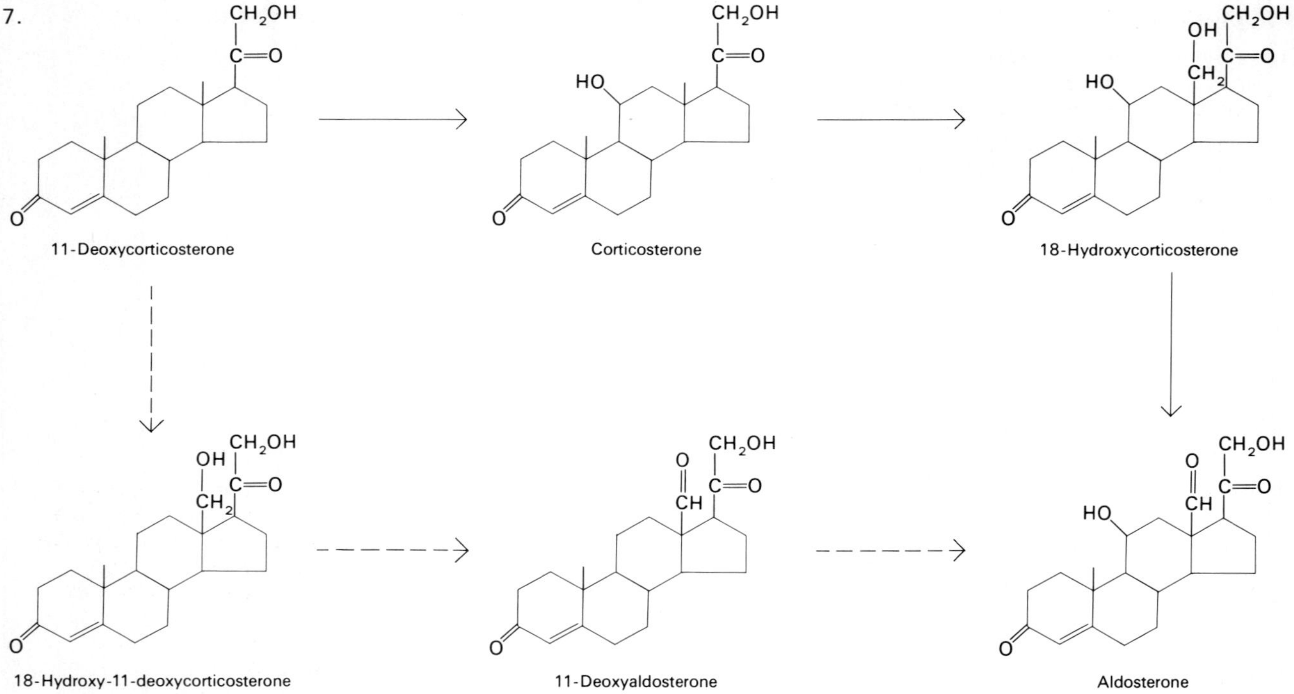

Chapter 25

Questions 2, 5, 8, 11, 12, 17, 23, and 24 are true.

Chapter 26

1. (a) ds synthetic DNA; (b) ds plasmid DNA; (c) ds synthetic DNA; (d) ds bacterial chromosomal DNA; (e) ds plasmid DNA; (f) ds plasmid DNA; (g) ds plant chromosomal DNA; (h) all ds DNA listed; (i) all ss DNA listed
2. b, c, g, and i
3. (a) pol II; (b) pol II; (c) pol III; (d) pol II; (e) pol III; (f) pol III
4. chimeric DNA: recombinant DNA molecule; chimeric RNA: mRNA containing the transcript for the N-terminal region of β-galactosidase (gene Z of the *lac* operon) and for the A chain of human insulin; chimeric protein: protein produced from the preceding chimeric mRNA
5. Addition of the appropriate antibiotic to the growth medium confers a selective advantage to those cells harboring the recombinant DNA molecule since sensitive cells, i.e., those that have lost the molecule, are killed. Under these conditions the recombinant DNA molecule becomes indispensible for a cell's viability.

INDEX

Page numbers in italics refer to the structure of a compound or to the principal presentation.